Reprint Publishing

For People Who Go For Originals.

www.reprintpublishing.com

COMPARATIVE MORPHOLOGY AND BIOLOGY

OF THE

FUNGI

MYCETOZOA AND BACTERIA

DE BARY

𝔏𝔬𝔫𝔡𝔬𝔫

HENRY FROWDE

OXFORD UNIVERSITY PRESS WAREHOUSE

AMEN CORNER, E.C.

COMPARATIVE MORPHOLOGY AND BIOLOGY

OF THE

FUNGI

MYCETOZOA AND BACTERIA

BY

A. DE BARY

PROFESSOR IN THE UNIVERSITY OF STRASSBURG

THE AUTHORISED ENGLISH TRANSLATION BY

HENRY E. F. GARNSEY, M.A.

Fellow of Magdalen College, Oxford

REVISED BY

ISAAC BAYLEY BALFOUR, M.A., M.D., F.R.S.

Fellow of Magdalen College, and
Sherardian Professor of Botany in the University of Oxford

WITH 198 WOODCUTS

Oxford

AT THE CLARENDON PRESS

1887

PREFACE TO THE ENGLISH EDITION.

I DO not deem it necessary to say anything in explanation of the reasons which have led to the preparation of this translation of Professor de Bary's book on the morphology and biology of Fungi, Mycetozoa and Bacteria. It brings within reach of all English-speaking students the most thorough and comprehensive treatise upon these groups which has appeared in any language; and the picture that is presented of the state of our knowledge of the subject at this time, along with the suggestions and indications of the lines upon which further investigation is especially wanted, will, it is to be hoped, not only instruct readers, but also stimulate them to research.

To render adequately some of the precise terminology has been a serious difficulty in the translation. The terms which have been adopted are consistently used, and the occasional notes, along with the 'Explanation of Terms' which I have added, should prevent all misconception of their signification. It must be remembered that the definitions and synonymy given at the end of the book have reference to terms only as they are used in the text; they are not exhaustive. The extension of the original meaning of Berkeley's term 'sporophore' and its use as the equivalent of the German 'Fruchtträger' is a prominent innovation to which attention may be directed in this place..

Several friends have been so kind as to give me their opinion and criticism upon questions of terminology, and I have specially to acknowledge suggestions from Professor Bower, Mr. A. W. Bennett, Dr. S. H. Vines, and Professor Marshall Ward; for help in some difficulties I have to thank the author himself.

I. B. B.

OXFORD: *March* 1, 1887.

PREFACE.

———◆◆———

I PUBLISHED a work in the year 1866 entitled Morphology and Physiology of the Fungi, Lichens and Myxomycetes as the second volume of Hofmeister's Handbook of Physiological Botany. This work was intended to give a systematic and critical account of the state of our knowledge of the portions of natural history indicated by the title at that time. It had its mistakes and its deficiencies; the index too was omitted, but for this I was not responsible. At the same time it was not without its value; it paved the way for further advances and was favourably received.

Some years ago I was urged by many persons to prepare a new edition of my work. Other occupations and duties long prevented me from setting about this task and interrupted and delayed it repeatedly after it had been begun; and when I addressed myself more closely to the work some four years ago, it soon became apparent that a new edition in the strict sense of the word would not satisfy modern requirements. Hence the progress of the work resulted in the production of a new book, which can only be partially regarded as a new edition of the earlier one, though this for brevity's sake is always cited in the present work as the first edition.

The reasons for the change had their origin in the considerable additions to the material to be discussed. Eighteen years ago it was comparatively easy to give a detailed description of the state of our knowledge of the morphology and physiology of the Fungi within a moderate compass. Since that time the amount of matter has increased greatly, and with it the number of questions and controverted points which have to be considered; an account which is not to be confined within the narrow limits of a text-book readily assumes large proportions and renders division of labour desirable.

The physiology of the Fungi has received more comprehensive treatment than the morphology, partly in general treatises on vegetable physiology, those especially of Sachs and Pfeffer, and partly in the extensive modern literature of the chemistry of fermentation. There is no recent detailed critical survey of their morphology; in giving such a survey in the following pages, with brief allusions only to strictly physiological topics, I believe that I shall best meet both present requirements and the wishes of most of my readers.

No comprehensive account of the morphology of any portion of the vegetable kingdom, and least of all of the Fungi, can be satisfactory without constant reference to the phenomena known as biological, that is to their habits of life and adaptations.

These must therefore be handled at some length, and in their turn again lead up to the consideration of questions of physiology.

I ought perhaps to have carried the change still farther, and have omitted a variety of anatomical and histological details, the introduction of which was suitable and necessary eighteen years ago, but which in the present state of botanical science may be regarded as superfluous or at least not indispensable. Still they can do no harm, and may possibly or even probably be of service. I have therefore taken some matter of this kind from my former work, and added it to the main text of the present publication in the smaller type employed for the description of some other details.

One change which I have ventured to introduce may and perhaps will be objected to. It is, that I have omitted the section which treated of the origin of the Fungi, Myxomycetes and Bacteria, and that I set out in all cases from the assumption that these plants are like all other plants the product of germs, each of which is derived from parents of the same species and owes its existence in every species to processes of development in the parents or to organs belonging to them. It is known that other views have prevailed with regard to the origin of the plants described in this work, and are still entertained by a few persons. It may be observed in passing that the early botanists, of whom mention is made in Ehrenberg's Epistola de Mycetogenesi, considered the Fungi to be merely lusus naturae and no plants at all. There are some who still think that Fungi and Bacteria are certainly plants, but that they are or may be produced by spontaneous or heteromorphous generation (abiogenesis, heterogeny), that is from inorganic matter showing only chemical predisposition to organisation, like crystals in the mother-lye, or else from commencements which are organised but which proceed from organisms that are not themselves either Fungi or Bacteria. The former of these two views requires no further consideration in this place. The other will be discussed in Chapter V, p. 270, in the special case of Sprouting Fungi and Yeast-fungi; it assumes in general terms, that constituent portions of living cells belonging to higher organisms, 'vesicles, granules,' the microsomata of modern terminology, can continue an independent life after the death of the living body of which they formed a part, and develope under favourable conditions into Fungi and Bacteria. These forms may then develope their specific germs, and a progeny from these germs specifically resembling the parents. H. Karsten and his adherents represent views of this kind, and A. Wigand has supported them at the present day. Their most logical development is to be found in A. Béchamp's theory of the microzymes. These are very minute bodies, 'granulations moléculaires,' which are contained in the substance (protoplasm) of animals and plants of the most different kinds and grades of organisation, and not only develope independently after the death of the parent-organism, but enjoy an almost unlimited duration of vitality, since they may lie during entire geologic periods in such a rock as chalk, and yet retain the power of development. These microzymes give rise in a suitable medium to Bacteria, Sprouting Fungi, and similar forms, and since the localities in which they originate are of very frequent occurrence they are to be found everywhere. Béchamp published his theory in the Reports of the Academy of Paris twenty years ago; he reproduced it in the Transactions of the Medical Congress at London in 1881, and in a large new work, Les Microzymas dans leurs rapports avec l'hétérogénie, l'histio-génie, la physiologie et la pathologie, Paris, 1883.

Theories of the kind here described, and others more or less like them, are constantly recurring from time to time on the subject of the origin of the Fungi and Bacteria. They appeared in earlier times with still greater breadth of application. Fifty years ago it was believed that not only minute organisms but that Fungi of the size of the Uredineae were produced from the altered substance of other organisms, in the case of the Uredineae from phanerogamous plants; two hundred years ago maggots were supposed to be bred from putrid flesh. It is easy to understand how such ideas of spontaneous generation should have been prevalent in ancient times. Even their repeated recurrence in modern times and with our modern know-ledge is also capable of explanation. It must be assumed that organisms did once come into being of themselves without parents, being produced from organisable but not yet organised matter. It must moreover be allowed, that this may still happen at any moment and perhaps does actually happen; its impossibility cannot be proved. To produce actual proof of an original formation of a living being is a matter of the highest interest, and has as powerful attraction for the biological investigator, as the prospect of producing the homunculus in the phial for the alchemist. But the experience of centuries has shown that whenever the homunculus really appeared in the flask, it proved to be a small imp which had been secretly introduced into it from without; and speaking seriously, the result was always of this kind. In every single instance exact investigation has shown that the organisms which were supposed to have had no parents proceeded from germs produced from parents of the same species as themselves; it has also shown how they were formed and whence they came. Those who maintained that direct proof had been given of generation without parents have been driven back step by step into narrower territory, and upon minuter and at last upon the minutest objects, from simple inorganic matter to the organised mini-mum, the 'atome structuré vivant'; in other words they were reduced to seek their proof where it is still most difficult to say whether it is to be found or not. This is what has happened in all researches into the origin of the Fungi, as soon as each individual case was rigidly examined. We have had ocular demonstration of the fact since the year 1860, through the labours especially of Pasteur and his school. That there is no generation without parents is therefore a maxim of experience; it is in distinct accord with the present state of our knowledge, after making allowance for all conceivable possibilities, and we must set out from this principle in a book which is concerned with real knowledge.

There is not much to be said by way of preface with regard to the plan of this work. I have endeavoured to make my remarks intelligible even to those who are only beginning the study of the Fungi, but I have assumed that my readers are masters of such a previous general knowledge of botanical science as is to be obtained by a course of study in a University, or by the use of good text-books. The reader is here referred to such works, especially to Sachs' Text-book and Lectures, and Goebel's Outlines, and also to Prantl's and Luerssen's smaller compendia, and among works not in German to Van Tieghem's Traité de Botanique.

A few descriptions only of individual Fungi will be found scattered through the volume; others must be sought in our at present imperfect floras, in Saccardo's

Sylloge Fungorum for example, in Winter's Die Pilze Deutschlands, P. A. Karsten's Mycologia Fennica, and Fries' more important publications, which latter must always be indispensable; books also of descriptive lichenology must be consulted. The new edition of Leunis' Synopsis by A. B. Frank, and in its special line Frank's Pflanzenpathologie, are works to be recommended to the student. I must not be supposed to express an unfavourable opinion of other works because I do not mention them; my only purpose is to point out some books which will be of service to those who require direction in their reading.

I have endeavoured to make myself as well acquainted as was possible with the special literature of my subject and to draw it into my service. I have made some use even of publications which appeared at the close of the year 1883 and the beginning of 1884 while my book was being printed. That some particulars have been omitted altogether, or been forgotten for the moment in the preparation of some of the sections, is what I should have expected beforehand, even if I had not subsequently noticed their omission. I must plead in excuse the extraordinarily large number of mycological communications of every possible size which have been published during the last twenty years.

The number of publications on the Fungi is so large that it was impossible to quote them all, and to have done so would have transgressed the limits of practical utility. The more important references given in the text will in every case show where further details are to be obtained, especially with the assistance of Hoffmann's Mycologische Berichte from 1865 to 1872, and Just's Jahresbericht, which has appeared regularly since 1873. Special notices of the literature of the subject will be found in the separate sections of the work.

The illustrations are for the most part the same as those in the earlier work. A certain number are new and drawn by myself; a smaller portion have been taken from other authors whose names are in every case given in the explanations appended to the figures. I express my warmest thanks for the permission to use these illustrations, and no less warm thanks to all those who have lightened my labours by their communications and by assistance of other kinds.

A. DE BARY.

STRASSBURG : *June* 30, 1884.

TABLE OF CONTENTS.

———•———

FIRST PART. FUNGI.

DIVISION I. GENERAL MORPHOLOGY.

CHAPTER I. HISTOLOGICAL CHARACTERISTICS.

SECTION   PAGE

I. General construction. Hyphae. Growth-forms: Filamentous Fungi, Fungus-bodies, Sprouting Fungi (Chytridieae, Laboulbenieae) . . 1

II. Structure of Fungus-cells: Protoplasm, Cell-nuclei, Cell-contents . . 5

III. Cell-membrane: Structure, Material composition, Excretion of calcium oxalate 8

Remarks. Literature 11

CHAPTER II. DIFFERENTIATION OF THE THALLUS.

1. General Survey.

IV. Mycelium and sporophore 17

2. Mycelium.

V. Mycelia. Filamentous mycelium. Haustoria 18

VI. Mycelial membranes 21

VII. Mycelial strands (Rhizomorphae, Agaricus melleus) 22

VIII. Sclerotia: Structure, origin, germination 30

Details. Historical remarks 40

IX. Sclerotioid bodies: Transitory resting states: Xyloma . . . 42

Literature 43

3. Sporophore.

X. General characteristics 45

XI. Simple sporophores 46

XII. Compound sporophores. General differentiation. Course of growth . . 48

XIII. Structure of mature compound sporophore 57

CHAPTER III. SPORES OF FUNGI.

XIV. Introduction. General characteristics and distinctions 59

I. Development and Scattering of Spores.

XV. General phenomena of development 6c

XVI. Intercalary and acrogenous spore-formation. Basidia. Sterigmata . . 61

SECTION		PAGE
XVII.	Dissemination of acrogenously formed spores. Abscision. Abjection. Disappearance of the sporiferous structure	68
XVIII.	Endogenous spore-formation:	
	(a) Sporangia of Phycomycetes	73
XIX.	(b) Asci	76
XX.	Dissemination of endogenously formed spores :	
	(a) Phycomycetes	81
XXI.	(b) Ascogenetic spores. Ejection. Mechanical arrangement for ejection of spores. Simultaneous and successive ejection. Mechanism of simultaneous ejection	84
XXII.	Puffing in Discomycetes	89
XXIII.	Peculiar mode of ejection in Pyrenomycetes	91
XXIV.	Force of ejection	91
XXV.	Reported peculiarities in Lichen-fungi	93
XXVI.	Successive ejection from asci	93
XXVII.	Solution and swelling of asci	96
XXVIII.	Maturation of ascospores after ejection	97
XXIX.	Combinations of different kinds of spore-development ; sporidesms .	98

II. Structure of ripe Spores.

XXX.	Spore-membrane. Exosporium, episporium, endosporium. Germ-pores. Gelatinous envelopes and appendages.—Protoplasm, nucleus, content.—Swarm-spores	99

III. Germination of Spores.

XXXI.	Tube-germination and sprout-germination.—Germ-tube. Primordium of the mycelium. Promycelium and sporidia	109
	Historical remarks on Chapter III	115

DIVISION II. COURSE OF DEVELOPMENT OF FUNGI.

CHAPTER IV. INTRODUCTION.

XXXII.	General course of development in Algae, Mosses, Ferns, and Phanerogams. Homologies and affinities. Form-genera and form-species in Fungi. Tulasne's pleomorphy. Gradual recognition of the course of development and of the homologies in Fungi. Main or Ascomycetous and other series	119
XXXIII.	Closer consideration of the course of development of the higher plants. Archicarp; fructification.—Spore, sporocarp, sporophyte.—Sexual and asexual segment of the development. Homology of a member of the development independent of its sexual function. Interruption and restoration of the homologies.—Alternation of generations.—Homology interrupted and not restored . . .	121
XXXIV.	Agreement of the course of development of the Fungi with that of plants which are not Fungi. Meaning of pleomorphy. Misunderstandings and the way to remove them	126

SECTION		PAGE
XXXV.	Terminology. Spores. Gonidia, &c.	128
XXXVI.	Review of the chief groups of Fungi	132

CHAPTER V. COMPARATIVE REVIEW OF THE SEVERAL GROUPS.

Peronosporeae.

| XXXVII. | . . . . | 132 |

Ancylisteae.

| XXXVIII. | . . . . | 139 |

Monoblepharis.

| XXXIX. | . . . . | 140 |

Saprolegnieae.

| XL. | . . . . | 141 |

Mucorini.

XLI.	General course of development	145
XLII.	Zygospores. (*a*) Mucoreae and Chaetocladieae. (*b*) Piptocephalideae— Azygospores	147
XLIII.	Typical gonidiophores. Mucoreae.—Chaetocladieae.—Piptocephalis and Syncephalis	151
XLIV.	Accessory gonidia. Acrogonidia (chlamydospores, stylospores). Choanephora.—Gemmae.—Serial gemmae, sprout-gemmae . .	154
	Doubtful Mucorini. Historical remarks. Literature . . .	156

Entomophthoreae.

| XLV. | . . . . | 158 |

Chytridieae.

XLVI.	Common characteristics of the present group of Chytridieae. Sporangia. Resting spores	160
XLVII.	Rhizidieae.—Polyphagus. Less well known forms . . .	162
XLVIII.	Cladochytrieae	165
XLIX.	Olpidieae	166
L.	Synchytrieae	167
LI.	Comparative review	169
LII.	Doubtful Chytridieae. Tetrachytrium. Hapalocystis . . .	170
	Literature	171

Protomyces and Ustilagineae.

LIII.	Protomyces	171
LIV.	Ustilagineae. Conformation. Compound sporophore . .	172
LV.	Development of resting spores	174
LVI.	Structure and germination of resting spores	176
LVII.	Gonidia of Tuburcinia and Entyloma	179
LVIII.	Course of development and homologies	180

Ascomycetes.

GENERAL CHARACTERS. SPOROCARPS.

SECTION		PAGE
LIX.	Structure of sporocarp. Apothecium, perithecium, cleistocarp . .	185
LX.	Apothecia	187
LXI.	Perithecia	190
LXII.	Cleistocarpous forms.—Elaphomyces.—Tuberaceae.—Onygena.—Myriangium	193

ORIGIN OF SPOROCARP.

LXIII.	Review of the main facts. 1. Eremascus.—2. Genera which have at first a distinct archicarp with no distinct envelope.—3. Polystigma with archicarp in the primordium of the sporocarp.—4. Xylaria and allies with a temporary Woronin's hypha.—Genera without a distinct archicarp	197
LXIV.	Separate descriptions.	
	1. Erysipheae	201
	2. Eurotium	203
	3. Penicillium	204
	4. Gymnoascus. Ctenomyces	206
	5. Ascobolus	206
	6. Pyronema	208
	7. Sordaria. Melanospora	210
	8. Collemaceae	211
	9. Forms with archicarp imperfectly examined . . .	214
	10. Polystigma	215
	11. Xylarieae	216
	12. Sclerotinieae	218
	13. Pleospora herbarum	220
	14. Claviceps, Epichloe, Cordyceps, Nectria and others . .	220
	15. Ascodesmis	221
	16. Sphyridium, Baeomyces, Cladonia and others . . .	221
	Early investigations into the development of the sporocarp in Lichens .	222

THE COURSE OF DEVELOPMENT IN ASCOMYCETES.

LXV.	Statement of facts. Species without gonidia; others with a regularly intercalated formation of gonidia. Different forms of gonidia in the same species. Microgonidia and megalogonidia. Pycnidia, pycnospores, stylospores.—Examples of species which have been fully examined	223
LXVI.	Homologies of the members of the development in Ascomycetes. Controversy regarding the sexual organs	231

DETERMINATION OF IMPERFECTLY KNOWN ASCOMYCETES.

LXVII.	The points on which the question turns	238
LXVIII.	Archicarps, sporocarps	239
LXIX.	Spermogonia and spermatia	240
LXX.	Doubtful spermatia	242
LXXI.	Gonidia, gonidiophores, pycnidia	244
LXXII.	Combinations of different forms. Examples of the kind . . .	248

SECTION   PAGE.

LXXIII.   Occurrence of known or supposed members of the development of some species outside the normal connection. Tendency of these to constantly similar production. Hence a possible reduction or splitting of a species 251

LXXIV.   Determination of organs of propagation which have been considered to be rudimentary 255

Literature of sections LIX—LXXIV 261

DOUBTFUL ASCOMYCETES.

LXXV.   Introduction. Helicosporangium. Papulaspora. Laboulbenieae . 263

LXXVI.   Exoascus 265

LXXVII.   Saccharomyces 267

LXXVIII.   Affinity between Exoascus and Saccharomyces. Possible relations of the latter group to the Ascomycetes 269

Confusion between the 'Yeast-plants' 270

Literature 272

Uredineae.

LXXIX.   Aecidia-forming and Tremelloid Uredineae. Sporocarps (aecidia) and spermogonia of the aecidia-forming species. Course of development of Endophyllum 274

LXXX.   Aecidia-forming Uredineae with gonidia: teleutopores, uredo . . 279

LXXXI.   Uredineae with imperfectly known course of development . . 282

LXXXII.   Tremelloid Uredineae 283

LXXXIII.   Relationship between Uredineae and Ascomycetes 285

Literature 285

Basidiomycetes.

LXXXIV.   Introduction 286

HYMENOMYCETES.

LXXXV.   Conformation of unveiled sporophore 287

LXXXVI.   Veiled sporophore. Velum. Annulus. Volva 289

LXXXVII.   Structure of full-grown sporophore 297

LXXXVIII.   Hymenium. Cystidia. Basidia 300

GASTROMYCETES.

LXXXIX.   Comparative account of the differentiation of the sporophore . . 308

XC.   Special morphology, history of development and anatomy of the same 312

COURSE OF DEVELOPMENT AND AFFINITIES OF BASIDIOMYCETES.

XCI.   Course of development of perfectly known form. Exobasidium, Tremellineae, Typhula, Coprinus, Agaricus melleus, Crucibulum and Cyathus, Sphaerobolus 328

XCII.   Gonidia of Basidiomycetes which have been thoroughly examined . 331

XCIII.   Imperfectly known and doubtful gonidia 333

XCIV.   Homologies and affinities of Basidiomycetes 337

Literature 341

DIVISION III. MODE OF LIFE OF FUNGI.

CHAPTER VI. Phenomena of Germination.

1. Capacity of germination and power of resistance in Spores.

SECTION   PAGE

XCV.   Duration of capacity of germination in spores. Resting state. Power of resistance to mechanical injuries, withdrawal of water, extreme temperatures 343

2. External conditions of germination.

XCVI. 349

CHAPTER VII. Phenomena of Vegetation.

1. General conditions and phenomena.

XCVII.   Conditions affecting growth. Temperature 352
XCVIII.   Nutrient substances. Other chemical constituents of the substratum. Effects of Fungi on the substratum. Fermentations, oxidations. Ferment-excretion 353

2. Nutritive adaptation.

XCIX.   Distinction of 1. pure saprophytes, 2. facultative parasites, 3. obligate parasites either *a*, strictly obligate, or *b*, facultative saprophytes . 356

3. Saprophytes.

C. 357

4. Parasites.

CI.   Adaptation between parasite and host. Predisposition of host. Endophytic and epiphytic parasites 358
CII.   Attack of parasite on host 360
CIII.   Growth of parasite after it has seized on the host and reactions of the host on the parasite. Destroying and transforming parasites . . 366

Parasites on Animals.

CIV.   Facultatively parasitic Aspergilli and Mucoreae; obligately parasitic Entomophthoreae, Laboulbenieae, Cordyceps, Botrytis Bassii . . 369
CV.   Imperfectly known parasites on animals : Saprolegnieae . . . 375
CVI.   Fungi of skin-diseases 376
CVII.   Actinomyces. 'Chionyphe Carteri' 377

Parasites on Plants.

a. *Facultative parasites.*

CVIII.   Fungi of rotting fruit. Sclerotinieae. Pythieae. Nectriae. Hartig's wood-destroying Hymenomycetes 379

b. *Obligate parasites.*

SECTION   PAGE

CIX. Facultatively saprophytic : Peronosporeae, Mucorini, Ustilagineae, Exo-basidium, Myxomycetes ; strictly obligate : Peronosporeae, Erysi-pheae, Uredineae, &c. 385

CX. Autoecism and metoecism 386

CXI. Growth and extension of parasites in substance of larger plants . . 389

CXII. Behaviour of these parasites to separate tissues and to the parts of the cells of the host 392

CXIII. Reactions of the plants attacked 394

Lichen-forming Fungi.

CXIV. Formation of the Lichen-thallus by the growing together of certain Algae and Ascomycetes and a few Hymenomycetes which attack them. Enu-meration of Lichen-forming algal forms as at present known . . 395

CXV. First beginning of the Lichen-thallus 398

CXVI. Conformation and structure of the Lichen-thallus. Fruticose, foliaceous and crustaceous forms. Distinctions in anatomical structure : 1. Heteromerous thallus. 2. Graphideae and similar forms. 3. Granular crustaceous thallus of Thelidium and others. 4. Coenogonium-form. 5. Collemaceæ or Gelatinous Lichens. 6. Hymenomycetous Lichens 401

CXVII. Soredia 415

Pseudo-lichens.—Historical remarks. Literature 416

SECOND PART. MYCETOZOA.

CHAPTER VIII. MORPHOLOGY AND COURSE OF DEVELOPMENT.

Myxomycetes.

CXVIII. Spores. Germination. Swarm-cells 421

CXIX. Plasmodia 423

CXX. Transitory resting states. Cysts. Sclerotia 427

CXXI. Development of sporophores and sporangia 429

CXXII. Structure of mature sporophores and sporangia ; sporophores of Cera-tieae ; simple sporangia ; aethalia 434

Acrasieae.

CXXIII. 441

Affinities of the Mycetozoa.

CXXIV. 442

Doubtful Mycetozoa.

CXXV. Bursulla. Vampyrellae. Nuclearia. Plasmodiophora . . . 446

CHAPTER IX. Mode of life of Mycetozoa.

SECTION   PAGE
CXXVI.   Conditions of germination   448
CXXVII.   Phenomena and conditions of vegetation. Causes of movement of
            plasmodia. Enclosing of solid bodies   449
CXXVIII. Process of nutrition   451
            Literature   453

THIRD PART. BACTERIA OR SCHIZOMYCETES.

CHAPTER X. Morphology of Bacteria.

CXXIX.   Structure of cells. Cell-aggregates and growth-forms . . .   454
CXXX.   Course of development. Endosporous and arthrosporous forms. De-
            velopment of endospores. Special description of some Bacilli . 459
CXXXI.   Development of arthrosporous forms   468
CXXXII.   Controversy respecting species in Bacteria   472
CXXXIII. Systematic position of Bacteria   474

CHAPTER XI. Mode of life of Bacteria.

CXXXIV.   Capacity and conditions of germination in spores   476
CXXXV.   General conditions and phenomena of vegetation.—Temperature.
            Nutrient substances. Oxygen. Aerobia, and anaerobia. Effect
            of substances not serving as nutrients. Oxygen and nutrient sub-
            stances as inciters of movement   478
CXXXVI.   Special vital adaptations. Saprophytes. Parasites. Parasites on
            plants. Parasites on animals. Exciting causes of disease. Life-
            history of Bacillus Anthracis as an example of facultative parasites.
            Doubtful obligate parasites ; Spirochaete Obermeyeri ; Nosema
            Bombycis.—General remarks on Bacteria which excite disease . 481
            Literature   489

EXPLANATION OF TERMS   491
INDEX   503

ERRATA.

Page 1, line 15, *for* rudimentary branch *read* branch-primordium.
  ,,   9, line 8 from bottom, *for* Muscaria *read* muscaria.
  ,,   ,, line 3 from bottom, *for* erinaceus *read* Erinaceus.
  ,,   18, line 2, *for* or *read* and.
  ,,   29, lines 1 and 8, *for* fructification *read* sporophore.
  ,,   34, line 5 from bottom, *for* commencemet *read* commencement.
  ,,   50, line 5 from bottom, *for* Coprineae *read* Coprini.
  ,,   51, line 4, *after* Clavarieae *insert* Calocera.
  ,,   ,, line 15 from bottom, *for* Coprineae *read* Coprini.
  ,,   57, line 16 from bottom, *for* Helvetia *read* Helvella.
  ,,   64, line 22, *after* sterigma *read* , as in Coprinus.
  ,,   76, line 12, *for* orientated *read* originated.
  ,,   99, line 5, *for* sporiderms *read* sporidesms.
  ,, 130, line 4 from bottom, *for* rudimentary *read* commencing.
  ,, 162, line 5 from bottom, *for* tubular-rhizoid processes, *read* tubular rhizoid-
            processes.
  ,, 177, line 4 from bottom, *for* sometines *read* sometimes.
  ,, 181, line 7, *for* Urocystis *read* Ustilago.
  ,, 192, line 6 from bottom, *for* Verrucariae *read* Verrucaria.
  ,, 244, last line, *for* polytrichum *read* polytricha.
  ,, 268, line 3, *for* Mycodermain, *read* Mycoderma in.
  ,, 277, line 7 from bottom, *for* sempervivi *read* Sempervivi.
  ,, 286, last line, *for* 53 *read* 48.
  ,, 306, line 14, *for* violacea *read* foliacea.
  ,, 330, line 12, *for* 25° C. *read* 25° C.
  ,, 349, line 9, *for* winter, of *read* winter of.
  ,, 352, note 2, *for* 294 *read* 262.
  ,, 368, lines 25 and 26, *for* fir *read* silver-fir.
  ,, 376, line 17, *for* **Mentagrophytes**, Rob.; *read* **Mentagrophytes**, Rob.);
  ,,   ,, line 18, *for* pityriasis versicolor [1]). *read* pityriasis versicolor [1].
  ,, 391, line 2 from bottom, *for* bypodites *read* bypodytes.
  ,, 399, line 8 from bottom, *for* peridermis *read* periderm.
  ,, 406, line 16 from bottom, *for* malacca *read* malacea.
  ,, 427, line 18, *for* Stemonites *read* Stemonitis.
  ,, 428, line 5 from bottom, *for* homogenous *read* homogeneous.
  ,, 434, lines 18 and 19, *for* serpula *read* Serpula.
  ,, 441, line 17 from bottom, *for* serpula *read* Serpula.

FIRST PART.

FUNGI.

———◦⊶◦———

DIVISION I. GENERAL MORPHOLOGY.

CHAPTER I. HISTOLOGICAL CHARACTERISTICS.

SECTION I.   The **thallus**, which in most Fungi is the whole body of the plant not serving directly as an organ of reproduction, begins as a tubular germ-cell (*germ-tube*, Keimschlauch) which, by continued growth progressing in an apical direction accompanied by repeated formation of lateral branches, developes into a branched body of cylindric thread-like form, the *hypha*.   Both growth and branching follow the laws which prevail generally in the vegetable kingdom.   The branching is usually monopodial, in a few cases only it is dichotomous, as in Botryosporium, species of Peronospora, and some Mucorini.

In some groups, especially the Saprolegnieae, Peronosporeae and the Zygo-mycetes, the thallus or hypha of the Fungus, like the thallus of the Siphoneae, is an unsegmented branched tubular cell up to the time when organs of reproduction are formed.   But in the great majority of cases it becomes a branched row of cells owing to the incessant formation of transverse walls concurrently with the growth of the hypha at its apex.   The segmentation either takes place only in the apical cell for the time being and at the place of insertion of the rudimentary branch, so that each branch is made up of an apical cell and segment-cells of the first order only, as in Penicillium[1] and Botrytis cinerea, or new intercalary partition-walls are formed in the segment-cells of the first order.

In the more simple Fungi the branched hypha alone constitutes the thallus; such forms are termed Hyphomycetes, *Filamentous Fungi* (Fadenpilze), or Haplomycetes[2]   The body of the largest Fungi, the Mushrooms and Lichens of ordinary parlance, are also composed of hyphae, but their ramifications meet and cohere to form larger aggregates.   Such a body, which appears as if formed by the union of Filamentous Fungi, may be termed a *compound Fungus-body* (zusam-

---

[1] Löw in Pringsheim's Jahrb. VII. p. 473.—Brefeld, Schimmelpilze, II. p. 27.
[2] From the Greek word ἁπλοῦς, meaning simple.

mengesetzer Pilzkörper) or simply a Fungus-body to distinguish it from that of the simple Filamentous Fungus. Both are *growth-forms* (Wuchsformen) comparable with those growth-forms in the higher plants which are known as the tree, shrub, and herb. Many species appear only in the filamentous form, as succeeding chapters will show; others assume both forms according to their stage of development and external conditions; all have the filamentous form in their earliest stage. It is obvious that intermediate forms will be found between these two chief ones.

It has been assumed above that a hypha or a Filamentous Fungus is the product of a single germ-cell, and this is often actually the case. It has been repeatedly shown that even a compound Fungus-body may be composed of the ramifications of a single filament proceeding from a single germ-cell. But this is not always the case, or at least cannot always be proved, owing to the frequent coalescence of similar hyphal branches with one another (Fig. 1), which takes place in the

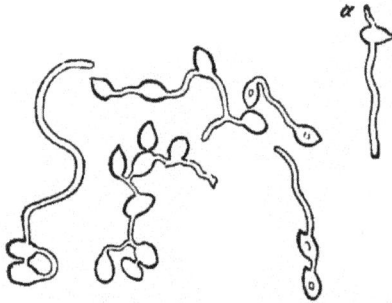

FIG. 1. Germinating gonidia of *Nectria (Spicaria) Solani,* Reinke; *a* developing into an isolated hypha, in the rest the hyphae have coalesced. Magn. 390 times.

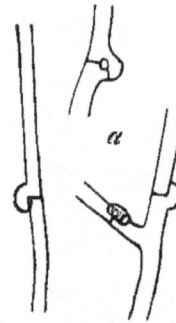

FIG. 2. Clamp-connections of the mycelium of *Hypochnus centrifugus,* Tul. Magn. 390 times.

following manner:—the lateral wall or the extremity of a branch or of a segment-cell of the branch places itself on another branch or cell, and the membranes of both disappear at the point of contact, so that the cavities and protoplasmic contents of the two cells become united into one. Coalescence in this way may take place between the branches of the same hypha, and also between such as are growing together but were originally distinct, being the product of distinct germ-cells. The forms which result from such coalescence are very various; H-shaped cross links or bridges, loops of various form and number, even network of many narrow meshes are found. One curious form must be mentioned here, the *clamp-connections* (Schnallenverbindungen), first observed by Hoffmann (Fig. 2). They occur only on hyphae with transverse segmentation, and chiefly in the Basidiomycetes (many Agaricineae, species of Polyporus, Typhula, Hypochnus, Cyathus, Hymenogaster, &c.). A clamp of this kind when fully formed is usually a nearly semicircular protuberance like a short branch which springs from one cell close to a transverse wall, and is closely applied to the lateral wall of the adjoining cell in such a way that the transverse wall cuts the middle of the plane of contact at a right angle. Sometimes the protuberance does not lie close on the lateral wall at all points, but forms an eye-hole. Brefeld observed the origin of these formations in Coprinus, and found that the protuberance extends itself from the one of the two adjoining cells to the other, and then coalescence takes place, so that the two cells enter into open communication with one another through the protuberance, and finally the protuberance is separated from the cell on which it first arose by a new wall which usually coincides with the plane of the lateral wall. The protuberance generally continues in open communication with the second cell, with which it coalesced; but here too subsequent separation is sometimes, though seldom, effected by the formation of a wall, and it is only to this exceptional case that Hoffmann's original term 'clamp-cells' is properly suited. In Coprinus according to Brefeld the first cell, the cell which puts out the protuberance, is always the one on the apical side of the transverse wall. In this case therefore the whole structure is formed almost exactly in the reverse way to that which is suggested by its appearance when fully formed; whether this is so in all other cases has yet to be ascertained.

The growth of the compound Fungus-body, so far as it depends on the formation of new cells and not on the expansion of the old, is due simply to the growth in length of the united hyphae and to the formation of new branches on them; these branches are formed partly on the surface of the body, partly in its interior, where they thrust themselves in between the branches previously formed.

In the fully developed state of these compound forms it is in most cases easy to see the fine fibrillation due to their construction out of hyphae; the course of single hyphae and their ramifications may often be followed with the aid of the microscope for considerable distances, whether they lie parallel to one another or whether they cross one another repeatedly and are intertwined.

In other cases the entire thallus or separate parts of it appear to have an entirely different composition. Here the tissue when fully formed consists of isodiametric roundish or polyhedral cells, and especially in thin sections no longer appears to be an arrangement of hyphae, but resembles the ordinary parenchyma of the higher plants. Examples of this tissue are to be found in the pileus of Russula and of Lactarius, in the rind of the peridium in many of the Lycoperdaceae, in many sclerotia, in the stipe of the Phalloideae, in many Lichens, and in some other cases. But if we examine this tissue more closely and follow the history of its development, we see plainly that it is really formed from and consists of hyphae, and that it owes its parenchymatous structure simply to the firm union of the hyphae, and to the form, expansion, and displacement of their cells. The parenchyma of the higher plants is formed by cell-division, the partition-walls as they successively arise being placed in turns in one of three or two directions in space. From this difference in origin the Fungus-tissue of which we have been speaking is distinguished by the name of *pseudo-parenchyma*. Successive cell-divisions in two or three directions occur only exceptionally in the formation of the thallus of the Fungi, as in pycnidia and perithecia (on this see Division II).

The *union* (Verbindung) of the hyphae to form the compound Fungus-body is for the most part brought about by their *interweaving* (Verflechtung) one with another, and the direction and closeness of the weft vary according to the species. The hyphae of the flocky felt-like tissue of Polyporus fomentarius[1], of Daedalea, of the stipe and pileus of the Amaniteae, of the medullary layers of many Lichens, &c. are loosely woven, leaving broad interstices usually filled with air; those of the firm tissue, often

---

[1] Amadou of commerce.

hard as horn or wood, in the dark rind of the dry Pyrenomycetes, of the Tuberaceae, of many sclerotia, &c., leave scarcely any intercellular spaces. Every intermediate stage is to be found between the loose accidental intertwining of the socially growing Hyphomycetes and the firm structure of the Fungi which have a definite form, and states the farthest apart from one another sometimes occur in the same species.

When the hyphae run parallel to one another, as in the stipe of Agaricus Mycena, Coprinus and other species, their connection may be brought about by *cementation* (Verklebung) or *concrescence* (Verwachsung) of the membranes, and in the same way they often gain much firmness where they are interwoven with one another. In hard tissues, as the rind of many non-fleshy Fungi, and in the masses of gelatinous tissue described on page 9, the outer surfaces of the hyphae are often inseparably grown together, or are cemented together by a narrow slip of a firm homogeneous substance; in fleshy Fungi the hyphae are often united by an intervening substance which softens in water and allows them to be artificially separated. This cementing substance may be called *intercellular substance*. Whether it is to be regarded as a part of the cell-membranes themselves or as an entirely distinct body from them, is a question which in the case of the Fungi has not yet been specially examined; there is therefore the less ground for assuming other laws than those which prevail in the histology of other plants. Lastly, the coalescence above mentioned of branches originally distinct also adds to the firmness and solidity of the compound Fungus-body; its occurrence is shown in fleshy and gelatinous species by the frequency of H-shaped connections, though no special researches have been made into their mode of formation.

Of the exceptional cases referred to above in which the Fungus-thallus is not formed of hyphae, the first to be noticed is that of the forms recently termed by Nägeli *Sprouting Fungi* (Sprosspilze). This name, like that of Filamentous Fungi, indicates a growth-form; and this is the only form in some Fungi, as in the species of the genus Saccharomyces known as Yeast-fungi, or it is peculiar to particular states of other species which otherwise appear as filamentous or compound forms. In the latter cases it may be a matter of doubt, for reasons to be hereafter discussed, whether the sprout is to be regarded as a vegetative or as a reproductive organ.

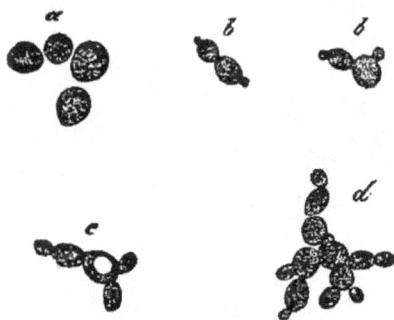

FIG. 3. *Saccharomyces Cerevisiae; a* cells before sprouting, *b—d* cells sprouting in a fermenting saccharine solution ; stages of development in the order of the letters. Magn. 390 times.

The characteristic features of the Sprouting Fungi are as follows (Fig. 3). A cell grows to a certain size and shape, the latter being usually spherical or somewhat ovoid, and puts out an excrescence or sprout which remains connected with it by a narrow base; this new formation is of the same nature as the parent-cell and is separated from it by a transverse wall either before or after it has reached its proper size. The process of sprouting may be repeated in the daughter-cell and in every succeeding generation, the number of which is unlimited in presence of sufficient nutriment. The number of sprouts that can be produced by each active cell and the spots at which they appear are not certainly determined, though with regard to the latter it is possible to give certain rules in individual cases; as a rule the sprouts may arise simultaneously at several spots in the mother-cell, or one after another from the same spot. If all the generations of sprouts continue united to one another, it is obvious that their combination simply forms an irregularly branched hypha, which is distinguished from ordinary segmented hyphae only by the constriction at the narrow point of insertion of the sprout-cells. This is in fact the condition in which the cells remain as long as they are not disturbed; but eventually the full-grown sprouts separate easily from one another, and a slight movement leaves only isolated sprouts or small aggregates of them.

The second exception is found in those simplest forms of the Chytridieae, in which the entire thallus consists of a single round cell which ultimately produces spores; the less simple species of this group are allied by their structure to the unicellular Filamentous Fungi described in sections XLVI-L.

The Laboulbenieae may perhaps be reckoned as a third exception, but their character is still very imperfectly known (see Division II).

The view given above of the **structure and growth of the thallus** of the Fungi has been already distinctly indicated in Ehrenberg's Epistola de Mycetogenesi (Nov. Act. Acad. Nat. Cur. X); it is also clearly expressed in Vittadini's Monogr. Lycoperdineorum (Mém. Acad. Turin. Ser. II, V, p. 146, 1841); and the views of later writers (Montagne, Esquisse organographique, &c. sur les champignons, Paris, 1841, German edition, Prague, 1844, and Schleiden, Grundz. 3rd edition, II, p. 34) are in accordance with it. Schleiden, and Unger after him (Anat. u. Physiol. d. Pflanzen, p. 149), call the weft of distinct hyphae felted tissue, 'tela contexta.' Thoroughly to establish and work out this view required fresh anatomical investigations, and to these Bonorden and Schacht have given the chief impulse and contributed the first more important materials. See Bonorden, Allgem. Mycologie, Stuttg. 1851, and Schacht, Die Pflanzenzelle, p. 134.

On the **clamp-connections** see Hoffmann in Bot. Ztg. 1856, p. 156; Tulasne, Carpol. I, 115; Bail in Hedwigia, I, 96, 98, &c.; Brefeld, Unters. über Schimmelpilze, III, especially p. 17; Eidam in Cohn's Beitr. z. Biol. Bd. II, 229.

The distinction of **pseudo-parenchyma** or apparent parenchyma was introduced by myself in the first edition of this work. The term may be retained as it has become familiar; but it should be remembered, that it ought only to be used to indicate the appearance of the close short-celled tissue of the Fungus with reference to the ordinary conformation of the parenchyma of the higher plants. If the latter tissue is characterised not, as is most usual, by the form of its cells, but by their structure and by the functions indicated by their structure, then the pseudo-parenchyma is neither more nor less to be compared with it than any other aggregate of cells in the Fungus which serves for metabolism.

The term **Sprouting Fungus** may be used as a general expression for the growth-form which it designates. As the Saccharomycete with this form of growth which has chiefly come under consideration is yeast, the term Yeast-fungus has usually been employed instead of Sprouting Fungus; but this leads to confusion, and the word Yeast-fungus should not be used for the growth-form, but should be confined to the special cases to which it is suited.

The **Schizomycetes** will be described in the third part of this work.

Section II. The cells of the Fungi agree in all important points with the cells of other plants as regards structure, growth, and mode of division.

The **protoplasm**, which in most cases fills uninterruptedly the interior of the young cell, encloses in the full-grown cell of the Fungi, as of other plants, one or several sap-cavities (*vacuoles*). Comparatively large vacuoles are often separated from one another by thin films ·of protoplasm, which in elongated cylindrical cells are not unfrequently placed across the cell, like transverse septa of cell-membrane, and this position has before now caused them to be mistaken for transverse cell-membranes[1]. The greater part of the 86–94 per cent. of water, which Schlossberger and Döpping found in fleshy mushrooms, is to be placed to the account of the watery cell-sap.

Errera discovered a remarkably large amount of *glycogen* in the cells of many Fungi. This substance permeates the protoplasm, renders it unusually refringent, and may be recognised under the microscope by this quality and by the characteristic reddish-brown colour which it assumes in the presence of iodine. It occurs especially in the asci of the Discomycetes and Truffles, but Errera found it also in the vegetative cells of some of these Fungi, of some of the Mucorini, and of certain Hymenomycetes, &c.

Nuclei are found in many cells connected with reproductive processes in the Fungi, in asci for example and basidia, and their relations to the formation of daughter-cells are in some cases at least clearly understood; but there is some uncertainty with regard to the nuclei in the vegetative cells of the thallus, owing to their minuteness. On the one hand the presence of nuclei in the vegetative cells is probable where they have not been directly observed, because the reproductive cells which have nuclei are formed directly from vegetative cells and are distinguished from them only by their greater size, which may be the only reason why their nuclei are clearly seen, and because the nuclein which is characteristic of cell-nuclei has been shown by macrochemical methods to be present in cells, in which the presence of a morphological nucleus is not or has not been certainly ascertained. In conformity with this, Strasburger with the help of colouring reagents detected nuclei in the cells of the thallus and in the spores of the Saprolegnieae, and Schmitz had previously asserted their existence in a number of other Fungi, as for instance in Oidium lactis, in the Peronosporeae, Mucorini, and Saccharomyces; to these may be added Penicillium glaucum (Strasburger) and especially the gonidial state of Peziza Fuckeliana (Botrytis cinerea). But on the other hand the objects under consideration, except in the Saprolegnieae, are of such minute size, that the satisfactory discrimination of true nuclei from other small bodies contained in the protoplasm, and like them perhaps rendered more distinct by colouring reagents, is extremely difficult, and can only be obtained after renewed investigations. The accounts in our possession make it distinctly probable that the protoplasm of the elongated vegetative cells of the Fungi which have been examined contains several or even many small nuclei, the division and multiplication of which is not in direct morphological connection with the vegetative cell-division. Only the short vegetative cells of Saccharomyces according to Schmitz are uninucleate. The reproductive cells to which we have referred above have one or more nuclei according to the species; the connection of the nuclei with the formation of daughter-cells, as far as it is known, will be described below along with the phenomena of reproduction.

The protoplasm of the cells of the Fungi contain no *chlorophyll* or analogous

---

[1] See Reisseck in Botan. Zeitg. 1853, p. 337.

colouring matter or *amylum-grains*, nor, as far as is known, any vehicles for colouring matters nor their homologous *plastids* (A. Meyer's trophoplasts).

It would appear that the formation of fatty matters takes the place very generally of the amylogenesis which holds in plants containing chlorophyll; these matters always form a percentage of the dry substance of vegetating Fungi, and may amount with diminution of the proteid substances to 50 per cent. of the material stored up in the resting-states, to 35 per cent. in the fatty sclerotia of plants like Claviceps, and to 50 per cent. in the Moulds (? Penicillium) in the resting or involution-stage, that is, after the close of vegetation. During the time of active vegetation the fatty substances are disseminated in the form of minute drops in the protoplasm of the cells of Fungi, as they are in the cells of other plants, and help to make it look granular or turbid; in the resting states (periods of involution), in which reserve-material is stored up, they may collect into large strongly refringent drops which occupy the largest part of the cell-cavity. Examples of the latter case are the sclerotia of Claviceps, the thallus of Sphaeria Stigma, Fr., S. discreta, Schw., S. eutypa, Fr., Vermicularia minor, old moulds, many spores, &c. &c.

In many cases the collections of fatty matter are colourless or only faintly coloured, but sometimes they are very highly coloured, if after the analogy of cases which have been carefully and chemically examined we may venture to apply the term fatty substance to bodies, of which we only know with certainty that they agree with fatty aggregates in outward appearance and in the ordinary microchemical reactions. If the bodies in question are really to be regarded as chemically definite fats, it still remains to be decided whether the colours belong to the fats themselves, or are derived from distinct colouring matters which would in that case be attached to the aggregates of fatty matters as their vehicles. With this reservation and pending the requisite strict chemical examination, we may designate as coloured aggregates of fatty substances those microchemically fat-like bodies which produce the characteristic colouring from yellow to brick-red in so many Fungi—Uredineae, Tremellineae, Stereum hirsutum, Sphaerobolus, Pilobolus, many Pezizas as P. aurantia, P. fulgens[1], and various other kinds. They are found thinly disseminated in the protoplasm of actively vegetating and growing cells, imparting to it a uniform colouring; after the death of the cells they often run together into larger drops; in older cells also they sometimes assume this form spontaneously. In the Uredineae, and according to Coemans in species of Pilobolus also, the red colouring matter shows a characteristic reaction, becoming intensely blue when treated with sulphuric acid, then quickly passing into a dirty green and then gradually losing all colour, a reaction which is seen in the similar red colouring matter of many parts of plants which do not belong to the Fungi, and in the red pigment-spots (eye-spots) of some of the lower forms of animal life. This reaction is not found in the other Fungi mentioned above. These facts sufficiently point to a different material composition of the bodies in question in different cases: some of them were spectroscopically examined by Sorby.

Van Tieghem discovered *crystalloids* of albuminoid substance (*mucorin*) in the gonidiophores and zygosporophores of most of the Mucorini. J. Klein found

---

[1] P. fulgens, Fr., was named P. cyanoderma in the first edition of this book.

them previously in Pilobolus. They have the form of octahedra or truncated triangular plates, are extruded from the protoplasm as it is preparing to form spores or sporangia and do not pass into these with the protoplasm; they subsequently disappear gradually in the decaying sporophore.

SECTION III. The **cell-membrane** in the Fungi remains thin and delicate to the last in most of the quickly-growing short-lived species; in others, especially in the long-lived solid Mushrooms and Lichens, it is thickened to a variable but often considerable extent, and is in that case stratified like other membranes. Formation of pits has been only rarely observed; fibriform, both spiral and annular, thickenings have been seen only in the capillitium of Batarrea (see Division II).

In their consistence and very limited power of swelling, which has not however been accurately determined, the membranes of many Fungi are very similar to the non-gelatinous cellulose-membranes of the higher plants. The elementary composition of the cellulose has also been ascertained by macrochemical analysis after proper purification in a number of cases, in Polyporus igniarius, P. fomentarius, P. officinalis, Agaricus campestris, Daedalea quercina, and Amanita muscaria. But the colourless, non-gelatinous and apparently pure cell-membranes of Fungi of every age are generally distinguished from the typical cellulose-membranes of the higher plants by being insoluble in ammoniacal solution of cupric hydrate, and by the absence of their characteristic reactions with iodine; they are not coloured blue by iodine and sulphuric acid or by Schulze's solution, or only after special and prolonged preparation, in the course of which they often display strong resistance to acids. We may therefore properly distinguish their substance by the special name of *Fungus-cellulose*. It remains still undecided whether their peculiar qualities are due to the presence of foreign deposits in their substance or to some other causes.

Membranes however are not wanting among the Fungi which display the typical blue reactions with iodine; such are all the membranes in the Saprolegnieae, in Protomyces macrosporus, in the thallus of the Peronosporeae, in the young state of some species of Mucor (M. Mucedo and M. fusiger), and in the cells of the resting perithecium of Penicillium glaucum (Brefeld). Clavaria juncea sometimes, but not always, shows the violet coloration with iodine and sulphuric acid; and this is the case also with the sterile forms known as Anthina pallida, A. purpurea, and A. flammea, which probably belong to Clavaria or its allies. Other Clavarieae show only the Fungus-cellulose. H. Hoffmann's observations on Amanita phalloides and Agaricus metatus may also be considered in this connection.

The non-gelatinous membranes of the Fungi, which are always colourless when young, often become coloured as they grow older, especially in long-lived forms, assuming usually various shades of brown from the lighter to the very darkest brown, more rarely some other colour, as the rosy red of the thallus of the mould, Dactylium macrosporum Fr., the blue of the surface of the thallus of Peziza fulgens, the green of Peziza aeruginosa, and Phycomyces nitens; the varied coloration of the membranes of the spores may also be mentioned here. The colours of the Lichen-fungi will be noticed further on in Division III. Apart from the Lichens, the colouring matter penetrates uniformly through the whole of the membrane or through certain lamellae of it.

The coloration of the membrane is accompanied with increased firmness and in most cases with exceptional power of resisting the action of concentrated sulphuric acid, phenomena which taken together recall the similar behaviour of the sclerosed, lignified, and suberised membranes of the higher plants. With the *coloration* therefore we may also speak of the *sclerosis* of the membranes. We learn also from other sources that the colouring at least is due to the interposition of substances which can be withdrawn by solvents from the membrane which then remains behind colourless, as we can withdraw the colouring deposits from the sclerosed membranes of Pteridophytes, or lignin and suberin from lignified and suberised cell-walls. We cannot at the present day speak of lignification in the strict sense of the word in connection with the membranes of the Fungi, since they do not show Wiesner's reactions when treated with anilin compounds and with phloroglucin. Phenomena approaching at least to suberisation in the strict sense of the word appear to occur sometimes, according to C. Richter's observations on Daedalea quercina. The greater part of the Fungi have not been subjected to any close examination on these points; the purely empirical expressions, coloration and sclerosis, may therefore serve for the present as general designations of the phenomena.

But there is another kind of membrane in the Fungi which is distinguished from those hitherto described by its *gelatinous* or even *mucilaginous* nature. The membrane in the dry state is hard and cartilaginous and swells by absorption of water to several times its former volume in the dry state; its consistence therefore in the moistened vegetating state is that of a tough or soft jelly. The outer layers of many filamentous mycelia have this gelatinous constitution, which is very conspicuously seen when the plants are cultivated in a fluid. The hyphae, when examined by transmitted light, appear to have a delicately thin membrane which seems to be surrounded by a hyaline fluid; further examination discloses either a distinct gelatinous sheath round each hypha, or a diffuse gelatinous mass in which the branched hyphae are all imbedded. Zopf observed this in Fumago for example. The phenomena present themselves in a very beautiful form in the Sclerotinieae when cultivated in saccharine solutions. The membranes of Saccharomyces Cerevisiae must be of this nature according to the observations of Nägeli and Löw.

Soft gelatinous membranes also characterise in many cases large masses of tissue of definite shape, which appear at first sight to be slimy mucilaginous masses and may be designated *gelatinous tissue* (Gallertgewebe) or *gelatinous felt* (Gallertfilz). Beautiful examples of this formation may be seen among the larger mushrooms in the gelatinous bodies of the Tremellineae, in the gelatinous layers of the peridium of the Gasteromycetes, as Geaster hygrometricus, Melanogaster, Hysterangium, the Phalloideae, Mitremyces, &c. (see Division II), in Bulgaria and Cyttaria, in the greasy slimy superficial layers of the pileus in such Hymenomycetes as Amanita Muscaria, Agaricus Mycena of the section Glutinipedes, Boletus luteus, &c., and in the young mycelial strands of Agaricus melleus (section VII). Membranes of the viscid gelatinous type are found in the elements of most Lichen-fungi, in those of the sclerotia of Sclerotinia and Typhula gyrans (section VIII), of the thallus of Hydnum erinaceus, of the massive sclerotium-like thallus of Polystigma and of the mycelium of Hysterium macrosporum (Hartig). In the three last-named cases and in many Lichen-fungi (see also Division III) the gelatinous membrane-lamellae are coloured blue directly by a watery solution of iodine. It is a matter of course that there should be intermediate forms between the highly gelatinous and the non-gelatinous membranes, such as are found for instance among the sclerotia. To the above-mentioned examples drawn from the vegetative parts of the thallus must be added the organs of reproduction, spores and the parts which immediately produce them,—a not less rich and varied contingent, of which more will be said in the chapters which deal with these organs.

We know very little of the chemical composition of the gelatinous membranes of the Fungi. From the few tolerably precise investigations and from analyses which have been made it seems probable that they are for the most part composed of one or more carbo-hydrates or mixtures of carbo-hydrates nearly allied to cellulose, but with great capacity for swelling. The membranes of the Lichen-fungi (Cetraria, Ramalina, Usnea, and Cladonia) are changed by boiling in water into a homogeneous jelly known as *lichenin*, the dry substance of which is isomeric with cellulose. According to Nägeli and Löw the membranes of yeast-cells (Saccharomyces Cerevisiae), after boiling repeatedly in water, pass partly into a mucilage which they term 'yeast-mucilage'; the analysis of its dry substance gave a formula very near $3(C_6 H_{10} O_5) + H_2 O$. When the membranes of the Lichen-fungi, especially Cetraria islandica, and of the asci of many of them are coloured blue by the direct action of iodine, the reaction is due to the carbo-hydrate which is mixed with the lichenin (itself not turning blue with iodine) and which can be extracted from it; its formula is also $C_6 H_{10} O_5$, and it was named by Dragendorff *lichen-starch*[1]. Most of the gelatinous membranes, like the yeast-mucilage, do not take the blue colour; they require further examination.

The gelatinous membranes also appear to be in many cases the seats of colouring matters, for instance of the scarlet-red of the surface of the pileus of Amanita muscaria, of the yellow of Boletus luteus, and of others, so that we might conclude that the characteristic colours of the Fungi, with the exception of the reddish-yellow mentioned above, were in almost all cases confined to the membranes. But to microscopic examination in the cases named and in some others the colour appears so pale and so uniformly distributed over the whole cell, that it is difficult to decide with any certainty whether it belongs to the membranes or to the contents, or whether it is distributed uniformly through them both.

A review of the anatomy of the membrane leads naturally to the mention of certain bodies, which are separated out from the cells and are imbedded in or more usually deposited on the membranes, or are interposed in the interstices of the hyphal weft; these are resinous excretions, lichen-acids, and especially calcium-oxalate. The lichen-acids will be noticed again in Division III.

Resinous excretions, the histogenetic relationships of which need not be discussed in this place, are known in great abundance as a coating of the hyphae which compose the sporophores of Polyporus officinalis, the mushroom of the Larch, and form sometimes 79 per cent. of the mass of this plant. Bauke[2] found the hyphae of a Diplodia furnished with a brown 'resin-like' covering. Zopf gives a similar account of species of this genus at page 48 of his work on Chaetomium which will be cited

---

[1] Berg, Zur Kenntn. d. Cetraria islandica (Diss. Dorpat. 1872).—Nägeli u. Schwendener, Das Mikroskop, Aufl. 2, 1877, p. 518.

[2] Pycniden, p. 35.

later. Both the old mycelium and the walls of the perithecia of Eurotium are marked by a similar reddish yellow or golden yellow covering.

Calcium oxalate is a substance so generally found in the Fungi that it is quite unnecessary to enumerate instances of its occurrence. I have noticed its absence in the Peronosporeae, in many Hyphomycetes, in species of Bovista and Lycoperdon, and in some Lichens which will be mentioned in Division III. The abundance with which it occurs on or between the cells of the plant varies according to the species, the individual, and the age; it is often more easy to find in young specimens than in older ones. It not unfrequently appears in the form of regular quadrate octohedra, but more commonly in that of slender needles, or irregularly shaped nodules, or angular granules (Figs. 4 and 5). These occur also on reproductive cells, as in the Mucorineae. When they appear on or in the surface of the plant, they often give it

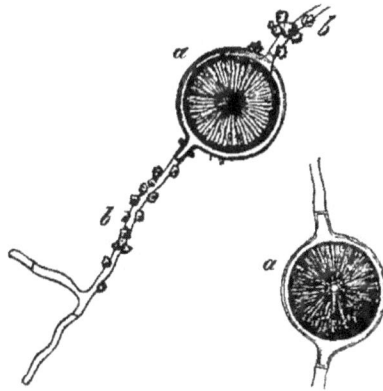

FIG. 4. Hyphae from the surface of a mycelial strand of *Phallus caninus*; *a* bladder-like cells filled with a crystalline sphere of calcium oxalate, *b* small irregular aggregates of the same salt on the outer surface of the hyphae. Magn. 390 times.

FIG. 5. Extremity of a hypha of the mycelium of *Agaricus campestris*, covered with small acicular crystals of calcium oxalate. Magn. about 390 times.

a chalky white appearance; this we see in many mycelial strands of Agaricus campestris, in the Phalloideae, in the thallus of Corticium calcareum and Psoroma lentigerum. The occurrence of the calcium oxalate inside the cells, though it has been observed several times, must be regarded as very exceptional. Small rod-like crystals are occasionally found in the vesicular cells of the stipe and pileus of Russula adusta. On the narrow cylindrical hyphae of the mycelium of Phallus caninus solitary large spherical or flask-shaped vesicular cells are found, which are almost filled by a large glistening sphere of calcium oxalate with a radiating crystalline structure (Fig. 4).

Structure of the membrane. I wrote at some length in the first edition of this work on the subject of the structure of the membranes of the vegetative cells of the Fungi, because it was important at that time to prove its conformity with like parts in other plants, in opposition to statements, especially of Schacht, founded on the minuteness of the objects in question, and assuming a much greater general simplicity in them. It will be well to repeat here the matter which was then produced, with some abbreviations and additions, notwithstanding the fact that it is now twenty years old, and that modern optical resources have made us acquainted with many further details in the objects observed; many fresh examples also might be adduced, but they are not required,

The young membranes of many woody and leathery Mushrooms, especially the

Gastromycetes and Hymenomycetes (Polyporus, Thelephora, &c.), is often comparatively thick, and in an older state is not unfrequently much thickened, even to the
obliteration of the lumen. The cells for example of the pileus of Polyporus fomentarius, of Crucibulum vulgare[1] and many other species, have in some parts the appearance
of solid cylinders, in others have a distinct cavity. The thickened membranes are
either firm and brittle or flexible, or gelatinous and soft. Where the thickening is
slight, as on the lateral walls of many Filamentous Fungi (Dematieae, Botrytis cinerea,
Peronospora), the membrane is usually homogeneous and not stratified, and even
the transverse walls are generally undivisible or with ·difficulty divisible into two
lamellae. But strongly thickened walls often show very distinct stratification without
as well as with the aid of reagents which cause swelling of their substance, such
as solution of potash or Schulze's solution or sulphuric acid. Good examples are the
thallus and gonidiophores of Cystopus, and the cells of the firm rind of the mycelial
strands of Agaricus melleus; to these may be added the thickened membranes which
sometimes occur in Pilobolus in consequence of retarded growth (Coemans). The
membranes of many dry resting Fungus-tissues (Polyporus zonatus, P." versicolor,
Daedalea, Trametes Pini, Lenzites betulina, the stout hyphae of Thelephora hirsuta,
the threads of the capillitium of Bovista plumbea, Geaster, Tulostoma and many
others) often show at least two distinct layers, an outer and firmer one which is
frequently of a bright colour, and an inner softer and more transparent layer.
Further stratification cannot usually be detected in these cases even with the use
of artificial means such as boiling in potash, though they may be seen sometimes in
an older pileus of Polyporus officinalis. Here may be seen, when the plant is examined
in water, an outer thin and apparently firm layer, and an inner thicker and evidently
soft layer; the outer layer is not sensibly altered when warmed in a solution of potash,
but the inner swells strongly, so as to protrude like a drop on the surface of fracture
beyond the outer layer, and at the same time often separates into numerous delicate
lamellae.

Very beautiful stratification is also shown in the cells of many Fungi in which the
membrane is gelatinous and is capable of swelling strongly in water. In Geaster
hygrometricus the inner layer of the outer peridium, which bursts in a stellate manner,
consists of straight cell-rows of equal length closely packed together and standing
parallel to one another and perpendicularly to the outer layer; they have a thick
membrane which is hard and cartilaginous in the dry state, but swells in water to
a tough gelatinous consistence and shows in a transverse section three to five lamellae
with different refringent power. The outermost lamellae of the cells in adjoining
rows are pressed close upon one another, and the bounding lines form a clearly
defined network on the transverse section. This structure is often obliterated in
old specimens.

An exactly similar stratification to that which has been described in Geaster is
found in the tissue of Hysterangium clathroides[2], which when dry is cartilaginous
but swells and becomes gelatinous in water, and also in the inner substance of many
sclerotia, as in the Sclerotinieae and in Typhula gyrans.

The lower part, the stipe, of the branching body of Calocera viscosa consists of
rows of cells all running nearly parallel to the longitudinal axis of the Fungus. Thin
transverse sections through the stipe give therefore circular or polygonal sections
of the individual cells. The outermost of the three concentric layers of tissue which
compose the stipe is in the fresh state of a viscid gelatinous consistence, and is
formed of slender rows of thick-walled cells which appear at first sight to be imbedded
in a soft homogeneous jelly. But if thin cross sections of the dried stipe are allowed
to swell slowly in water, it becomes apparent in this case also that the jelly is formed
of as many gelatinous layers of membrane in close contact with one another at all

---

[1] Sachs in Bot. Ztg. 1855.   [2] See Tulasne, Fungi hypogaei.

points as there are rows of cells. If the sections are kept for a long time in water, the delicate bounding lines of the lamellae disappear and the lamellae themselves coalesce into a homogeneous mass.

The above cases establish the occurrence of lamellae of different thickness and capacity for swelling in thickened cell-membranes; but it also follows from the facts which have been given, that the apparently homogeneous substance between the cells of these Fungi, like the pseudo-intercellular substance in many Fucoideae, Florideae, and others, is not to be regarded as a secreted homogeneous substance distinct from the cell-membrane, but originates in the close contact and partial coalescence of the outer gelatinous thickening-layers of all the hyphae.

The tissues of many Fungi (Melanogaster, Tremella, Exidia, Guepinia, Dacryomyces, Bulgaria, Thelephora mesenterica, Mitremyces, Cyttaria, Panus stypticus), the peridia of the Phalloideae, young Nidularieae, the surface of many Hymenomycetes, as Agaricus Mycena sect. Glutinipedes, Fr., Amanita muscaria, Boletus luteus, and many others, are of gelatinous constitution, and agree in structure with those of Calocera, Hysterangium and other forms described above; but the interstitial gelatinous substance appears in most cases to be really a homogeneous mass, and has not yet been separated into portions belonging to the individual cells. This may perhaps yet be done in many of these forms; at the same time it would appear from the published observations on Calocera and from the close affinity and agreement in structure between Calocera, Guepinia, and Tremella, and between Hysterangium and Phallus, &c., that we are justified in considering the homogeneous gelatinous substance of all the Fungi mentioned above as simply a product of the coalescence of soft gelatinous thickening-layers of the cell-membranes. H. Hoffmann seems to take this view[1], as he speaks of the gelatinous substance in the outer portions of the pileus of the fleshy Hymenomycetes as a product of the deliquescence of the membrane.

The threads of the capillitium in all species, as it would seem, of Lycoperdon (L. pusillum, L. Bovista, L. giganteum; see Division II) are delicately pitted. The thick transverse walls of Dactylium macrosporum, Fr. which are formed of two semi-lenticular lamellae have the large pit in their centre, just in the same way as it occurs in the transverse walls of filiform Florideae like Callithamnion. I have never seen similar pits in other Filamentous Fungi; their transverse walls are usually delicate, and in some cases, as in Botrytis cinerea, they appear to be thinner in the centre than at the margin.

Fungus-cellulose. In my first edition I gave the name of Fungus-cellulose to the substance of the greater part of the non-gelatinous membranes of the Fungi for the reasons given above. C. Richter has recently arrived at the conclusion that there is no special modification of cellulose requiring to be distinguished by such a name, and that the membranes supposed to contain it are composed of ordinary cellulose with foreign, possibly albuminoid admixtures. He shows that the membranes of Fungi like Agaricus campestris, Claviceps, Polyporus spec., Daedalea quercina, and Cladonia, which do not show the characters of ordinary cellulose even when treated in the customary manner with boiling solution of potash, Schulze's solution, or chromic acid, if subjected to longer maceration in a 7–8 per cent. potash solution do give the ordinary reactions of cellulose, turning blue with iodine and sulphuric acid and with Schulze's solution, and being soluble in ammoniacal solution of cupric hydrate. The maceration must continue for at least 2–3 weeks, sometimes, as in Daedalea, for as many months. These observations are a welcome confirmation of the near affinity of the substance of the membranes of the Fungi to ordinary cellulose which was indicated by macrochemical analysis; but they merely prove that the membrane of these Fungi is altered by maceration with potash in the way described. Whether

---

[1] Icon. analyt. fungorum, pp. 12, 25.

this alteration consists in the removal of some substance which was present from the first must remain uncertain; such an explanation has not been proved and others are at least possible. Without going further into the question here, we may merely recall the fact, that ordinary cellulose is coloured blue by iodine when certain reagents have produced certain changes in it; but zinc chloride, for instance, does not in this process remove some admixture which is present in the cellulose and prevents it from turning blue. Old threads of linen and cotton which have been repeatedly washed become blue at once in a dilute solution of iodine; the changes in their original condition which are thus indicated cannot consist in the simple removal of any substance. From such considerations it appears to me that the cause of the peculiar character of the Fungus-cellulose is not yet ascertained, and the harmless special name for it seems still to be desirable.

Coloration. The colouring-matters which are peculiar to the Fungi, i. e. which are produced in their metabolism, are chiefly if not solely the yellow and the reddish yellow which are partly attached to the fatty or fat-like contents of the cells, and partly disseminated through the membranes. It is not therefore too much to say, that all tints peculiar to the Fungi, which do not belong to the first category, proceed from the specific colour of the membranes.

An exception to this rule, which may perhaps be regarded as only apparent, is said to occur in some normally colourless moulds and parasitic forms, which growing in water on a substratum containing soluble red and violet colouring matters take up these unaltered in such a manner that even their cell-contents are correspondingly coloured. Fresenius[1] makes a similar statement with respect to species of moulds growing among red-coloured Micrococcus prodigiosus, Cohn; I found the same thing in Eurotium and in species of Mucor growing on red fruits and in Phytophthora infestans on red and blue potatoes. But I am now doubtful, firstly whether the colouring of the cell contents is in the protoplasm or in the cell-sap or in both, and secondly whether it is present in the living Fungus, or appears only in those of its cells which have been killed in making the preparation and have then taken up the colouring matter into their protoplasm.

One thing remains to be noticed here which has never yet been explained, the colouring of Peziza aeruginosa, P. (Chlorosplenium aeruginosum of Tulasne)[2]. This Fungus is found on wood with the green rot so common in forests, the colouring matter of which has been frequently examined since Vauquelin's time and most recently by Prillieux. The green colouring matter of this wood is usually contained in its cell-walls, but sometimes forms according to Prillieux in amorphous masses in the cavities of the wood elements. This is in many cases all that is to be seen; over wide spaces on and in the wood there is no trace of a coloured or uncoloured Fungus to be seen; (Gümbel, Fordos, and myself). If a Peziza occurs on and in wood of this kind, almost all parts of it are coloured green, and the colour is in the membranes and perhaps also in the interior of the cells of the Fungus, and often in such quantity that the Fungus is more deeply coloured than the wood itself. Single fructifications of the Peziza however rising from the wood are sometimes uncoloured, a pure white, in their upper parts which are farthest from the surface of the wood. These facts taken together led to the view that the green colouring matter is a product of the decomposition of the wood without the co-operation of the Peziza, which takes it up unaltered when it settles in the wood. The fact that Peziza aeruginosa only grows, as far as is known, on this green-rotting wood, and on no other substance, is not in itself a valid objection to this view. But there are on the other hand so many established instances of specific decompositions effected by certain Fungi, that the repeated confirmation of the above-mentioned fact and the absence of other species of Fungus were constantly suggesting the idea that the

---

[1] Beitr. 80.   [2] Carpol. III, p. 188.

green colour of the decaying wood must be a consequence, and the colouring matter a product of the Peziza which grows in and upon it. The question is still undecided ; but I have myself recently observed the important fact, that green-rotted wood is sometimes found in which microscopic examination can discover no evident coloration of the wood elements, but shows the presence inside them of numerous intensely green hyphae which most certainly belong to Peziza aeruginosa. All our observations show that the Fungus wherever it occurs always contains the green colouring matter ; it occurs only in green-rotting wood, and the view that the wood owes its colour to the Fungus must be allowed to be probable. The fact that wood is found with this particular form of decay but unquestionably free from the Fungus appears to be quite irreconcilable with this view ; but the objection disappears if we suppose with Cornu that the hyphae of the Peziza which vegetate in the wood are short-lived, and convey all their colouring matter to the wood when they die. It ought not to be difficult to settle this question by artificial cultivation.

One striking case of coloration may be added here, though strictly speaking it does not belong to our present subject. The tissue of the pileus of certain Boleti, especially Boletus luridus which in the uninjured state is yellow, assumes a blue colour as soon as it comes into contact with the outer air. Schönbein has carefully examined this phenomenon, and finds that it is a substance capable of being extracted from the Fungus by alcohol and probably of a resinous character which turns blue in the air. The blue colour appears in the alcoholic solution under the same conditions as it does in a solution of guaiac-resin, and since it has been proved that the colour is produced in the latter by combination with ozonised oxygen, Schönbein assumes a similar cause of the blue colour in the Fungus. The alcoholic extract from the Boletus does not by itself become blue when exposed to the air ; there must therefore be another substance contained in the Fungus, which ozonises the oxygen of the atmosphere, and then effects a combination with the resin, giving off the oxygen to it in the state of ozone. Phenomena of a similar kind observed in other cases confirm this conjecture. Thus both the tincture of guaiac and the alcoholic extract of Boletus turn blue at once, if they are allowed to fall in drops on the fresh tissue of some of the Agarici which do not themselves turn blue, especially Agaricus sanguineus. The watery juice of Agaricus sanguineus squeezed out from the plant and filtered produces the blue colour at once in both tinctures. From these facts it may be concluded that a number of fleshy Fungi contain a substance soluble in water, which absorbs oxygen and gives it up to other bodies in the state of ozone. The Boleti which turn blue contain this substance with another resinous substance, which like guaiac-resin is turned blue by ozone.

Literature of sections II and III :—

1. Cell-structure of the Fungi ; structure and chemical composition of the cell-walls.

SCHACHT, Die Pflanzenzelle, p. 136 ;—Id. Lehrbuch d. Anat. d. Pfl.

COEMANS, Monogr. du genre Pilobolus in Mém. des savants étrang. Acad. Bruxelles, XXX.

CASPARY, Monatsber. d. Berliner Acad. Mai, 1855.

H. HOFFMANN in Bot. Ztg. 1856, p. 158.

H. v. MOHL in Bot. Ztg. 1854, p. 771.

DE BARY, Unters. über d. Brandpilze ;—Id. Ueber Anthina in Hedwigia, I. 36 ;— Id. in Bot. Ztg. 1854, p. 466.

BRACCONOT in Ann. de Chimie, XII. 172.

PAYEN, Mémoire sur le développement des végétaux in Mémoires présentés à l'Acad. des sc. de France, IX (1846), p. 21.

MULDER, Physiol. Chemie, Braunschw. 1844-51, pp. 202, 203, where Fromberg's results are also given.

SCHLOSSBERGER, Ueber d. Natur d. Hefe (Ann. d. Chem. u. Pharm. 51, p. 206).

SCHLOSSBERGER u. DÖPPING, Beitr. ∠. Kenntn. d. Schwämme (Ann. d. Chem. u. Pharm. 52, p. 116.

A. KAISER, Chem. Unters. d. Agaricus muscarius L. (Inaugural-Diss., Göttingen, 1862).

BURGERSTEIN in Sitzungsber. d. Wiener Acad. 70.

C. RICHTER in Sitzgsber. d. Wien. Acad. 83, p. 494.

NÄGELI u. LÖW, Ueber d. chem. Zusammensetzung d. Hefe (Sitzungsber. d. Bayr. Acad. zu München, Mai 4, 1878).

NÄGELI u. SCHWENDENER, Das Mikroskop, Aufl. 2, p. 518.

FÜISTING in Bot. Ztg. 1868, p. 660.

See also the literature of the Lichens in Division III.

2. Cell-nucleus, cell-division.

SCHIMTZ, F., Ueber d. Zellkerne d. Thallophyten (Sitzgsber. d. Niederrh. Ges. Aug. 4, 1879).

ZACHARIAS, Ueber d. Beziehung d. Nucleus, &c. (Bot. Ztg. 1881, p. 169), where other literature is noticed.

STRASBURGER, Zellbildung u. Zelltheilung, Aufl. 3, 1880, p. 221, Taf. XIV.

3. Glycogen.

ERRERA, L., L'épiplasme des Ascomycètes et le glycogène des végétaux (Thèse, Bruxelles, 1882);—Id. Sur le glycogène des Mucorinées (Bull. de l'Acad. de Bruxelles, Nov. 1882).

4. Cell-contents, fat, crystalloids.

NÄGELI, Ueber d. Fettbildung bei d. niederen Pilzen (Sitzgsber. d. Bayr. Acad. München, 1879, p. 287).

ROSTAFINSKI in Bot. Ztg. 1881, p. 461.

SORBY, On comparative vegetable Chromatology (Proc. Roy. Soc. London, XXI, p. 442). See also Just's Jahresber. 1873.

VAN TIEGHEM, Nouvelles recherches sur les Mucorinées (Ann. d. sc. nat., Sér. 6, I, p. 24). See also the literature at the end of sect. XLIV.

5. Resinous secretions.

HARZ in Bull. Soc. Imp. de Moscou, 1868.

6. Green rot in wood.

VAUQUELIN in Ann. du Mus. d'hist. nat. VII, p. 167 (1866).

GÜMBEL in Flora, 1858, p. 113.

BLEY in Arch. d. Pharmac. 1858.

FORDOS in Comptes rend. Acad. d. sc. Paris, 87, p. 50 (1863).

ROMMIER in Comptes rend. Acad. d. sc. Paris, 66, p. 108 (1868).

PRILLIEUX in Bull. Soc. Bot. de France, 1877, p. 167.

CORNU in Bull. Soc. Bot. de France, 1877, p. 174.

7. Boleti which turn blue.

SCHÖNBEIN in Verhandl. d. naturf. Ges. Basel, 3 (1856), p. 339;—Id. in Abhandl. d. K. Bayer. Acad. VII (1855), and in Bot. Ztg. 1856, p. 819;—Id. in Bull. de l'Acad. de Belgique, Sér. 2, VIII, pp. 365, 372, and in Comptes rend. Jul. 16, 1860.

It does not fall within the scope of this work to go into the details of chemical analysis; the reader is referred for these to—

HUSEMANN und HILGER, Die Pflanzenreiche, Aufl. 2. See also the first edition of A. and Th. Husemann.

FLÜCKIGER, Pharmacognosie d. Pflanzenreichs, Aufl. 2, Berlin, 1883. The work contains exact accounts of Claviceps, Polyporus officinalis, Cetraria, &c.

G. DRAGENDORFF, Die qualitative u. quantitative Analyse von Pflanzen u. Pflanzentheilen, Göttingen, 1882.

J. KÖNIG, Chemische Zusammensetzung d. menschlichen Nahrungs- u. Genussmittel, Berlin, 1878 (Edible mushrooms).

CHAPTER II. DIFFERENTIATION OF THE THALLUS.

1. GENERAL SURVEY.

SECTION IV. The thallus of the greater part of the Fungi which are composed of hyphae is differentiated into two chief parts, a vegetative part known by the name of *mycelium* since the time of Trattinick[1], and the *sporophore*[2] (Fruchtträger, receptaculum of Leveillé, encarpium of Trattinick), which bears and produces the organs of reproduction and springs often in great numbers from the mycelium. It need scarcely be said that there are many gradations in the sharpness of this differentiation. The views and the terminology drawn from the species in which the differentiation is sharply defined have often been transferred to those in which it is less distinct. In simple filamentous forms, as Protomyces for instance and Entyloma, in which the reproductive cells are formed directly as segments of hyphae which are not further differentiated, we speak of these cells being formed directly on the mycelium. In many cases this distinction between mycelium and sporophore may be said to be only arbitrary.

Owing to the peculiar mode of life of the Lichen-fungi, the differentiation in many of them is to some extent different from that of the rest of the Fungi, and the traditional terminology therefore, which will be considered in Division III, is also different.

The mycelium is that part of the thallus which spreads in or on the substratum, derives nourishment from it and attaches the Fungus to it. In accordance with these functions it resembles the root-bearing rhizomes of the higher plants, and still more the rhizoids of Mosses in various points of form and growth. The sporophores may be compared to the flowering or fruit-bearing shoots of higher plants in respect of their function to which their form corresponds, and which consists essentially in the formation of organs of reproduction.

---

[1] Fungi austriaci, 1805.
[2] See note at beginning of section X regarding the use of the term sporophore.

2. THE MYCELIUM.

SECTION V.   The mycelia in their original form are always free hyphae; they either retain this character during their whole life, or the hyphae as they grow become at most loosely interwoven with one another without forming bodies with a definite shape and outline, *filamentous* or *floccose mycelia*; or the hyphae form by their union elongated branching *strands* (fibrous or fibrillose mycelia), or *membranous expansions*, or tuber-like bodies, *sclerotia*.

The **filamentous mycelia** are much the most common, and in the majority of Fungi they are the only known form.   Their character has been already described in speaking in the first chapter of the hyphae of the Fungi generally.   The branching of the mycelial filaments in all cases that have been observed with certainty is monopodial.   The phenomena of coalescence of hyphal cells that were originally free, and of clamp-connections which were described above, appear as a rule in their most striking form in filamentous mycelia.

Differences in the structure of mycelial filaments must necessarily depend chiefly on the presence or absence of a regular system of transverse walls, and this, as has been already intimated, varies in the different groups. (See also Chapter V.) Every species in each of the two chief categories thus obtained exhibits as a rule its own peculiar phenomena of growth and differentiation, provided the normal conditions of growth remain unchanged, and by these phenomena the several species and groups of species can be distinguished from one another.   These differences relate to the average size and special form of the cells, the divergence of the branches, the phenomena of coalescence and the like.   Owing to the diminutive size of the objects, they are usually very inconspicuous even under the most favourable conditions of growth, and to ascertain them with certainty requires careful observation.   They are liable also to so many changes from external causes that the determination of a mycelium, which under favourable conditions of development has well-marked characters, without its sporophore may be attended with considerable difficulty in practice, if it has to be observed under less favourable circumstances.   Much advance has been made in this point of late years through the careful examination of individual species, so that we may expect that the morphological characters of the mycelia of many species and groups of species will in time be clearly determined.

Some filamentous mycelia, belonging to species from very distinct groups, are distinguished by having special organs of attachment and suction, known as *haustoria*; these are peculiar branches which attach the mycelium firmly to the substratum, and in most cases also evidently serve to take up nutriment from it.   Such organs are found in many, but by no means in all, parasitic species living on plants and belonging to very different groups, as the Peronosporeae, Piptocephalis, the Uredineae, and Erysipheae.

The mycelial filaments of these Fungi spread themselves on or among the cells of the host; the haustoria are formed on them as special lateral branches which force their way into the interior of the cells; they vary in form according to the species and are more or less, often extremely, unlike the extracellular hyphae.   Organs of attachment, which at least very nearly resemble the haustoria of these parasites, are found in a few other non-parasitic mycelia; they will be noticed again subsequently.

Careful investigations into the formation of the mycelia of distinct species of non-parasitic Fungi are to be found especially in Brefeld's Untersuchungen über Schimmelpilze.

The formation of **clamp-connections** described above may be taken as an example of a peculiarity which is characteristic of the larger groups. It occurs, as far as we at present know, almost exclusively in the Basidiomycetes and chiefly in the Agaricineae; it is found in the Tuberaceae, but apparently in no other Ascomycetes. Its occurrence in Peziza Sclerotiorum, as stated in my first edition, seems not to be confirmed in more

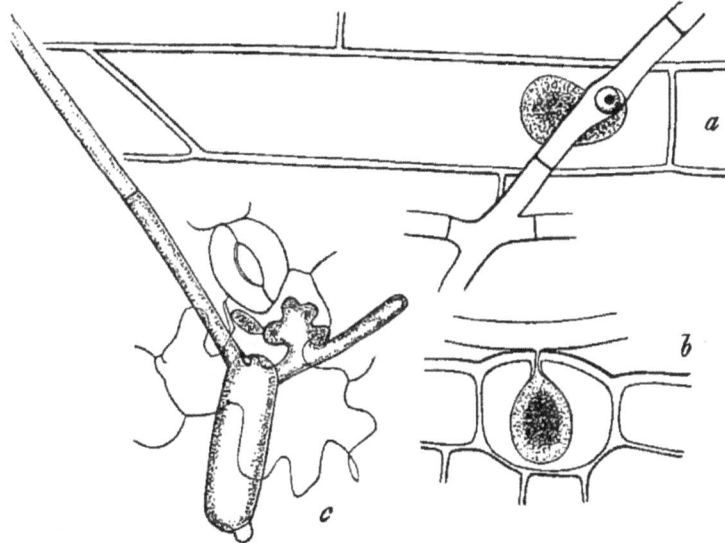

FIG. 6. *a* and *b Podosphaera Castagnei*, Lév. *a* epidermal cells of *Melampyrum sylvaticum*; a branched mycelial hypha is creeping over the surface and has sent a haustorium into one of the cells (surface view). *b* vertical section through epidermal cells with mycelial hypha and a haustorium which has penetrated into a cell. *c* a spore (gonidium) of *Erysiphe Umbelliferarum* putting forth germ-tubes on the epidermis of *Anthriscus sylvestris*. The smaller germ-tube on the right is sending a haustorium from the lobed attachment-disk into an epidermal cell. *a* and *b* magn. 600, *c* 375 times.

recent times. It is at present uncertain whether it is a feature of all the Basidiomycetes or only of all the Agaricineae, and the more so as according to Brefeld it is frequent in one species of the genus Coprinus, but comparatively rare in all the rest.

A greater number of distinctly marked characters have been observed in the mycelia of parasitic Fungi, especially the Erysipheae, Peronosporeae, Uredineae, and Ustilagineae than in other forms, and they have been observed for a longer time. Such characters occur chiefly in the formation of the **haustoria** of many species and groups of species in those divisions; the following are examples of them.

The mycelial filaments of the **Erysipheae** (Figs. 6, 7) are furnished with transverse walls, and their numerous but distant branches spread themselves over the epidermis of phanerogamous plants, being generally closely applied to it, but at the same time easily separable from it. At certain circumscribed spots, however, they are firmly attached to the substratum, and in these spots they are provided with a haustorium

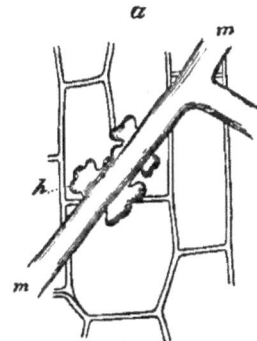

FIG. 7. *Erysiphe (Oidium) Tuckeri*. Mycelial hypha with lobed attachment-disk on the surface of a grape. After v. Mohl (Bot. Ztg. 1853, Tab. XI). Magn. 570 times.

which, springing as a branch from a cell of the mycelium in the form of a very delicate tube, pierces the outer wall of the nearest cell of the epidermis and enters its cavity; there it enlarges into an ellipsoid or somewhat elongated persisting vesicle filled with protoplasm, which in Erysiphe graminis is branched in a peculiar manner. The mycelial filament according to the species is either not altered at all at the point of origin of a haustorium, or is at most only slightly enlarged there ; or it has a flat nearly semicircular protuberance the height of which is at most equal to its own diameter ; or it has a protuberance in the form of a bluntly lobed disk scarcely exceeding the breadth of the filament, which is pressed closely down on the epidermis, and appears on both sides or only on one side beyond the flank of the filament. These lobed attachment-disks were first discovered by Zanardini in·Erysiphe Tuckeri.

The thick mycelial tubes of the **Peronosporeae**, which are usually without transverse walls and spread among the cells inside living plants, often clinging close to the outer surface of their walls, send haustoria into the cells, which have very different forms in the different species. In Cystopus (Fig. 8 *A*), Peronospora nivea, P. pygmaea, P. densa and others, they are like those of the Erysipheae but much smaller, and they usually or perhaps in all cases only make a deep indentation in the walls of the cells ; in P. parasitica they are lobately branched tubes, and their vesicular club-shaped branches often quite fill the cells of the host ; in most of the pleuroblastic Peronosporeae (Fig. 8 *B*) they are slender filiform lateral branches of the intercellular filaments with many curved and winding ramifications in the interior of the cells. In Phytophthora infestans, which inhabits the potato, branches of the mycelium which in this case scarcely deserve a separate name force their way at various spots, especially in the sprouting tubers, into the cells of the host.

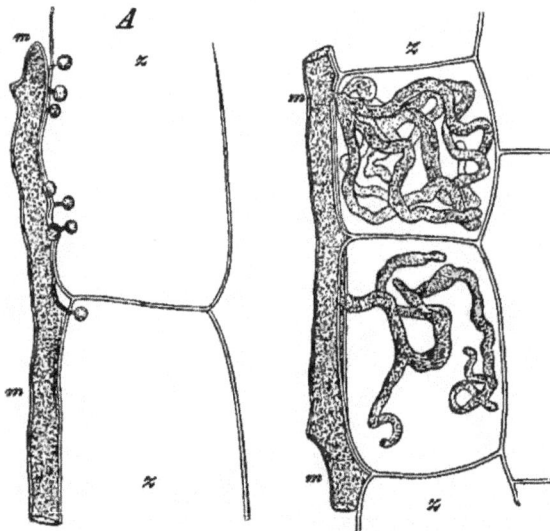

FIG. 8. Mycelial tubes *m* creeping about in the intercellular spaces with their haustoria penetrating into the cells *z—z; A* of *Cystopus candidus,* from the pith of *Lepidium sativum, B* of *Peronospora calotheca* from the pith of *Asperula odorata.* Magn. 390 times.

The intercellular mycelium of the **Uredineae** has a variety of haustoria formed like those of the Peronosporeae, especially the pleuroblastic species, and with them should be mentioned also the winding intracellular mycelial branches of the Ustilagineae.

The haustoria of **Piptocephalis, Syncephalis,** and **Mortierella** are very different from those which we have hitherto been considering. Piptocephalis Freseniana is parasitic on the larger Mucorini, and its mycelium, like that of its hosts, consists of tubes without transverse walls. If a growing filament of the mycelium of the parasite comes in contact with a mucor-tube, either with its apex or with its lateral wall, it spreads out slightly at the point of contact and thus attaches itself firmly as with a cupping-glass to the Mucor. A tuft of filiform radiating branching processes of such extreme delicacy that nothing is known of their minute structure now shoot forth from the middle of the surface of attachment into the cell of the host. The length of these suction-filaments is about equal to the diameter of the mucor-tube (see section XLIII). Van Tieghem and Le Monnier describe similar arrangements in Mortierella and Syncephalis, only in these genera the tubes which enter the cells of the host are not so different from those of the rest of the mycelium.

An allied case, though with important points of difference, is that of the clusters of haustoria in **Chaetocladium Jonesii,** a form which is usually parasitic on species of Mucor like the genera just described, and resembles them in structure. The tubes of this Fungus, both of its mycelium which spreads in the substratum and of the part of its thallus which rises above it, become firmly attached at the point of contact to the mucor-tubes which they encounter, and enter into open communication with them at this point by the dissolution of the cell-membranes and complete coalescence of the protoplasm of both plants. At these points of union they now put out small vesicular projections, which in strong specimens appear in numbers close together and form clusters which may reach the size of a pin's head. It is obvious that these vesicles do not, like the haustoria in the previous cases, serve as organs of attachment and nutrition, for organs of the kind are rendered unnecessary by the union of the parasite and the host. They are evidently storehouses of food-material, and the fertile branches of the thallus spring chiefly from them. But in relation to the morphological points at present under consideration they are in their nature essentially branches of the mycelium, which however stand in the closest and most exclusive relation to the physiological function above mentioned.

Organs of attachment of an unusual kind resembling haustoria are peculiar to the species of **Sclerotinia** which have been examined, S. tuberosa, S. Sclerotiorum, S. ciborioides, S. Fuckeliana, and also to the gonidial state of this species known as Botrytis cinerea. Under conditions to be described in the sequel the mycelium of these plants, often when still quite young, forms short branches on which arise tufts of secondary branches, which becoming closely clustered together are divided by numerous transverse walls into short segments with membranes that become dark brown with time. The clusters may be of the size of a pin's head, and have then been mistaken for sclerotia, with which however they have no connection. They are formed when the mycelium under conditions of plenteous nourishment is growing on a solid substratum, such as a plate of glass, which it cannot penetrate, and they apply themselves closely to the substratum. On substances into which the plant penetrates, such as the parts of plants which are suited to it, the tufts are not formed at all or are only feebly developed, in which case their branches soon pass into the substance of the host and grow there into slender branches of the mycelium. Brefeld gives figures of these formations in his Schimmelpilze[1].

SECTION VI. The mycelial hyphae of many Fungi, when the conditions are favourable, become interwoven with one another and form **membranous layers** which may be of considerable extent and thickness.

This is the case with such Hyphomycetes as Aspergillus niger, A. clavatus, and Penicillium glaucum, which in their simpler condition have a filamentous and floccose mycelium, if they grow on the surface of a moist nutritive substratum. They sometimes form large expansions on the surface of fluids, and may be lifted off them like a cloth. The free surface of the mycelium is in these cases usually clothed with the filiform sporophores.

A second series of examples is supplied by many, perhaps by the larger part, of the solid and especially of the woody and wood-inhabiting Hymenomycetes, the mycelia of which form very thick membranes or crusts, sometimes of considerable breadth and some millimetres in thickness, on the free surface of the substratum or in clefts inside carious stems of trees. Sporophores spring on the one side directly from the membranes, and on the other single filaments or bundles of filaments branch off from them and penetrate into the substratum. Other instances occur here and there in other groups, and are mentioned in special publications[2].

Apart from the exceptional case of Agaricus melleus which will be described below, the only general remark of importance upon the structure of these mycelial

---

[1] Schimmelpilze, IV, t. IX.
[2] See the literature cited at the end of the chapter.

membranes, which our present knowledge enables us to add to what has been already said, is that they only occur in Fungi with septate hyphae; the structure of the mycelium varies of course in particular points in each species.

Special generic and specific names have in former times been repeatedly given to mycelial membranes which are only known in the sterile state. Persoon's genus Mycoderma[1] may be composed to a great extent of forms of this kind which belong to the Hyphomycetes or to the Ascomycetes. Racodium cellare of Persoon[2] which forms the well-known olive-brown coating on old casks in cellars is, as far as we know, a mycelium formed of loosely interwoven filaments, the origin and reproductive organs of which are still quite unknown.

The mycelial membranes named by Tode and Persoon Athelia and Xylostroma are of a firmer kind. The Athelieae are the sterile states of the Thelephoreae (Thelephora, Hypochnus), in part perhaps their undeveloped sporophores; the Xylostromeae, which occur as broad flat formations of a woody or leathery texture in the decaying stems of trees, are the like states of firm wood-destroying Hymenomycetes, such as Polyporus abietinus, Thelephora hirsuta, Th. crocea, Schrad., Th. setigera, Fr., Th. suaveolens, Trametes Pini, Daedalea quercina, and other species of these and allied genera.

SECTION VII. The hyphae of the mycelia of many Fungi unite together into **strands**, which in their form, branching, and mode of spreading in the substratum look to the unassisted eye more or less like the roots of higher plants. Even some species of Hyphomycetes, those for instance known as Acrostalagmus, show a tendency to this kind of formation. But it is most frequent among the Fungi which have compound sporophores, such as the Phalloideae, many Lycoperdaceae, the Hymenogastreae, Nidularieae, and Sphaerobolus, in many of the Agaricineae, as A. campestris, A. praecox, A. dryophilus, A. aeruginosus, A. metatus, A. androsaceus, A. Rotula, A. platyphyllus and A. melleus, and amongst Ascomycetes, such as Elaphomyces, some species of Genea, Peziza Rapulum, Bull., and P. fulgens; the endophytic mycelium of Polystigma stellare, Lk., may be added to the list. It is evident from these examples that the formation of strands is not necessarily found in all the species that belong to the cycles of affinity indicated by the above names; on the contrary it may be wanting in one of two nearly allied species and be found in the other.

The strands, as has been said, spread themselves out in and on the substratum, growing at the apex and putting out similar branches, the arrangement of which scarcely follows any exact rule even in the same species. In each case the strands may either be in part free and tapering, or they may in part unite to form a coarser or finer net-work, or they may in part lose themselves in a loose filamentous web, or a single strand or several combined may expand into membranes, which spread over the substratum or spin themselves round bodies contained in it. Fresh strands may then take their rise from these expansions. This variation of form is essentially dependent on the character of the environment and its influence on the nutrition of the Fungus, as is well shown in the case of Agaricus melleus to be hereafter described.

In most cases which have been examined the strands are composed of uniform

---

[1] Mycol. Europ. p. 96.   [2] Syn. Fungor. 701.

hyphae with transverse septation, which vary according to the species. They generally run parallel to the longitudinal axis of the strands and are straight or undulating, and are either grown together by their lateral walls, as in Polystigma stellare, Agaricus Rotula and A. metatus, &c., or they are loosely woven together, as in Elaphomyces, the Nidularieae, Scleroderma, and the Hymenogastreae.

The structure of such Phalloideae as have been examined, of the Lycoperdaceae and of some Agarici, is somewhat more complicated. The strands of **Phallus impudicus** creep in the ground and may be several feet in length and 2 mm. in thickness. A transverse section through the stronger branches shows a thin, firm, white, outer layer or rind enclosing a thick cylinder of a brownish colour and gelatinous appearance (the medulla). The central and larger portion of the medullary substance consists of a felt of tough gelatinous character, in which the hyphae run longitudinally and are slightly sinuous and of unequal thickness. The outer portion of the medullary substance is exclusively formed of thicker hyphae. The rind is composed of a few layers of broad thin-walled hyphae wound firmly round the medullary cylinder in narrow spiral coils. It is easy to see that these hyphae spring as branches from the peripheral elements of the medullary tissue, then curve outwards and join the tissue of the rind. They form on their surface short distant branchlets which make the strands appear as if clothed with short hairs. The entire surface of the strands is covered with calcium-oxalate.

The strands of **Agaricus platyphyllus**[1] are very like the above in thickness, appearance, and structure, only the hyphae all run in the longitudinal direction and their walls are on the whole of firmer texture.

The strands of **Phallus caninus** resemble likewise those of Ph. impudicus, but are smaller in every respect. Here too all the hyphae run parallel to one another in the stouter parts which may be 1 mm. in thickness, and the white rind is distinguished from the yellowish gelatinous medullary substance which contains no air by more loosely interwoven hyphae, by air-filled interstices, and by the copious deposit of calcium-oxalate on the hyphae and in the vesicular cells described on page 11. **Clathrus** shows similar characters as far as my observation has gone. The rind and the medulla are often less distinctly separated from one another in the more slender branches of higher orders, but the former is always distinguished by its covering of calcium-oxalate. The strands of the **Agarici** (Agaricus campestris, A. aeruginosus, and A. praecox) and those of the **Lycoperdaceae** have the appearance of the slenderer branches of Phallus caninus and the same structure in all important points. The presence of the calcium-oxalate varies according to the genera and species, as was stated on page 11.

The formation of strands reaches its highest development, as far as is at present known, in the mycelium of **Agaricus melleus**. An excellent description of the structure and growth of this plant by Jos. Schmitz was published, with some additions by myself, in the first edition of this work; its life-history was elucidated by R. Hartig, and our knowledge of it was subsequently completed by Brefeld's cultures. There is the more reason for giving an account of the results of these investigations in this place because Agaricus melleus is the only one of the forms which we are at present considering in which the course of development has been followed from beginning to

---

[1] See Fries, Icones sel. Hymenomycetum, I, t. 61.

end. Agaricus melleus is chiefly a parasite on living European Abietineae (see Division III). It makes its way into the roots or the base of the stem beneath the ground, and the mycelium spreads in the cambium zone and in the young bast, forming

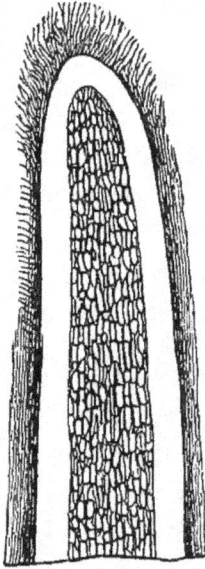

FIG. 9. *Agaricus melleus.* Median longitudinal section through the growing apex of a subterranean mycelial strand, seen by transmitted light. Magn. 40 times.

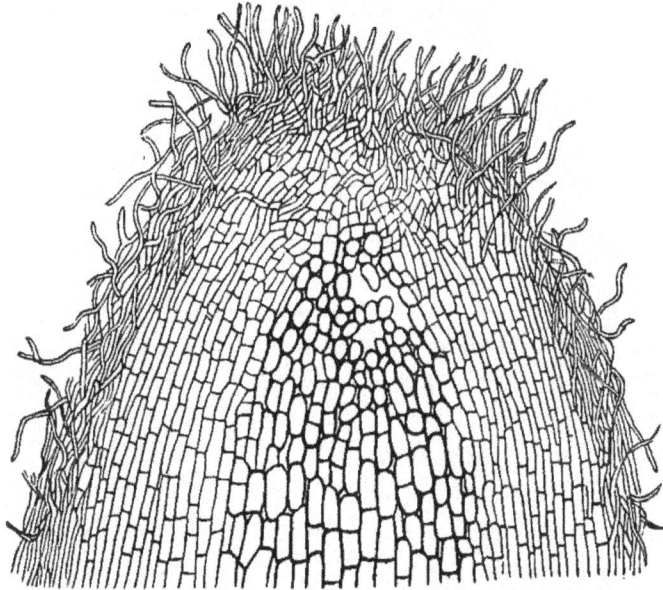

FIG. 10. *Agaricus melleus.* Thin median longitudinal section through the extremity of the growing apex of a subterranean mycelial strand. Magn. 250 times, but the drawing completed under higher magnifying power.

compressed or membrane-like expanded networks of strands at the cost of the sap-containing layers of tissue, and also sends out a large number of single hyphae from these strands into the rind and wood, and especially into the medullary rays, where they spread widely. From these *intra-matrical*, especially *subcortical*, parts other strands may proceed which develope as *extramatrical* strands usually in the soil, and are therefore *subterranean*, and branch and spread the Fungus over wide distances from one tree to another. These strands become more than 3 mm. thick and are round on the transverse section; they can also develope into enormous masses in moist rotting timber.

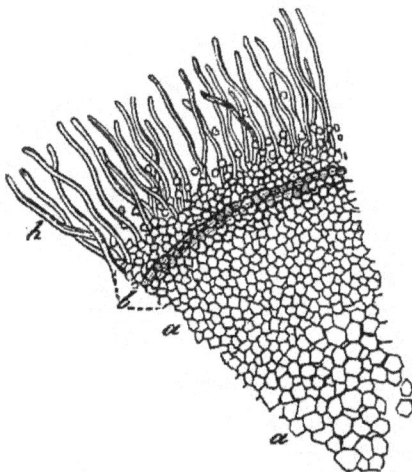

FIG. 11. *Agaricus melleus.* Transverse section through a young branch of a subterranean mycelial strand in about the lower half of Fig. 9. *a* the axile large-celled tissue passing towards the outside into the later-formed rind. The outer limit of the rind is at *b*; outside *b* is the covering of gelatinous felt with numerous spreading hair-like branches *h*. Magn. 190 times.

The cylindrical subterranean strands consist when fully formed of a dark-brown, brittle, usually smooth peripheral tissue or rind enclosing a white finely-felted medulla. The rind, which in stout specimens has the thickness of paper, is formed in its outer portion of about twelve or more layers of cell-rows (hyphae) running down the length of the strand, and connected with one another laterally without interspaces.

The cells of the hyphae are 2–4 times longer than broad, and have a firm brown membrane and a polygonal transverse section in conformity with the absence of intercellular spaces; the cells of the outer layers are narrower than those of the inner and have much thicker walls. The stratification of their membranes becomes more conspicuous when they are treated with potash.

The medulla consists chiefly of slender tough longitudinal hyphae about 1.5 mm. in thickness, which form acute angles with one another as they interweave, and have air in their interstices. Their membranes are comparatively firm; septation and branching are rarely seen in full-grown specimens. The slender medullary hyphae are in connection with the innermost layers of the rind; longitudinal sections show them arising as numerous branches from the cells of these layers, and making their way between them or running directly from them in oblique or transverse course to the medullary tissue. The longitudinal arrangement of the layers bordering on the medulla is thus rendered irregular to a degree which varies in each specimen and is sometimes considerable.

I have observed the development of the subterranean strands on adventitious branches, which it is not difficult to obtain from old specimens if cultivated in a damp chamber. The apex of such a branch, as it rapidly elongates (Figs. 9 and 10), is conical in form and colourless for a distance of some millimetres. It consists of a weft of delicate hyphae rich in protoplasm, the terminal branches of which form at the apex a loosely tangled tuft rendered slimy by the gelatinous swelling of the membranes. The apical tissue is continued downwards in the periphery of the branch into the gelatinous felt which covers it and which will be described presently, and in the middle into a short-celled irregular tissue of interwoven hyphae without interspaces, which forms the real conical growing point of the body of the strand. Active meristematic cell-multiplication, which cannot be followed in detail on account of the close interweaving of the hyphae, takes place in the uppermost region; close beneath this, where the strand begins to grow broader, there is partly elongation and extension of the elements of the tissue, partly formation of new elements. The former affects first and chiefly the axile portion of the strand, which occupies a third part of the total thickness; its cells subsequently undergo a few divisions close beneath the growing point and expand rapidly to a breadth of about 12–20 μ and 2–8 times that length; they continue thin-walled, are filled chiefly with hyaline cell-sap and are arranged in straight longitudinal rows. They diminish gradually in breadth as they approach the peripheral tissue (compare Figs. 10 and 11). An evident elongation of the cells takes place in the peripheral tissue, which serves to show more clearly their arrangement in longitudinal rows, but the increase in breadth is only small. As the circumference of the strand increases with every successive transverse section from the apex of the cone to the fully formed cylinder, and the hyphae are in close contact with one another without interstices, there must necessarily be an interpolation of new hyphal branches between those already formed.

The development of the definitive structure of the strand begins with the passage into the ultimate cylindrical form. The increase in the breadth of the large axile cells ceases near the apex, where the peripheral layers of the circumference increase considerably; the consequence is that the axile cells are torn from one another especially laterally, and intercellular spaces are formed between them, which serve from the first to conduct air (Fig. 10). The spaces widen most in the centre of the axile strand; in the simplest case a single large axile cavity is formed, and narrow air-spaces join on to it on the side of the periphery; in other cases single rows of cells remain in the centre of the cavity separated for the most part from the adjacent tissue and therefore quickly drying up; here too therefore there is really an air-filled axile cavity; its diameter varies much, but it is always at least half as large as that of the strand, and in strong specimens may reach a much larger relative size.

The wall which encloses the cavity consists in its immediate vicinity of the original large axile cells; these form about six irregular layers round the cavity, the cells of the outer layers becoming gradually narrower, as was stated above, and it is these layers which give rise to the medulla of the fully formed strand in the way which will be described presently. Outside of the zone which produces the medulla are the numerous layers of the close compact tissue, which ultimately forms the rind of the strand. This tissue does not however extend to the surface; this is occupied by a supplementary stratum of about six layers of hyphae with narrow lumina and thick gelatinous walls, which have coalesced into homogeneous mucilage, the gelatinous felt mentioned above. The hyphae of this tissue run for the most part longitudinally, and join with the hyphae of the apical tuft. They also give off spreading branches from the surface. From these must be distinguished other spreading branches, also provided with gelatinous walls, which spring from the hyphae of the rind ,beneath the gelatinous felt, and pass transversely through it towards the outside. Their number and distinction vary in individual specimens, and according to Hartig they are of special importance when the Fungus finds opportunity for adopting a parasitic life. A sharply defined boundary line between the innermost elements of the gelatinous felted layer and the outermost of the later rind cannot be drawn in the earlier stages of development.

The assumption by the tissue of its ultimate form begins with the thickening and turning brown of the walls of the hyphae. It advances on the transverse section from without inwards, and its first beginnings may be followed upwards to the base of the young apical cone. As the coloration advances the gelatinous felt which covers the rind dries up and usually no trace of it remains in older strands. At the same time the formation of the ultimate medullary hyphae begins inside; these arise, as is shown in Fig. 12, as slender lateral branches from the cells of the zone which produces the medulla, and from the innermost layers of the rind which are not sharply distinguished from it; these branches elongate and ramify and enter the axile cavity, and becoming woven together there fill it up in the manner which has been already described. As the zone which produces the medulla always consists of several layers of cells, the hyphae which proceed from its outer layers into the cavity must force their way between the inner layers, which may become much displaced and squeezed together, and this gives rise to the irregularly constructed boundary zone between the medulla and the rind which was mentioned above. A subterranean strand may form branches of the same kind in varying number and with no regular arrangement. At the point where a branch subsequently appears, a new formation in the form of a thick cushion of pseudo-parenchyma is developed within the inner layers of the rind by shoots from its cells, and the growing point of the strand emerges in a few days from the cushion, and breaking through the rind of the parent-strand grows in the manner which has already been described. Its final medullary hyphae become continuous with those of the medulla of the parent-strand. Schmitz first observed that, at least when old strands are cultivated in a damp chamber, the place of every future branch is indicated some days before its emergence by the appearance of a floccose tuft of hyphae $\frac{1}{2}$—1 mm. in size, arising partly beneath, partly also according to Hartig out of the surface of the parent-strand, which decays and disappears as the branch is formed.

The stronger subcortical strands and the membrane-like expansions in the living tree are said by Hartig to be similar to the subterranean ones just described in structure and development, except in respect of certain differences arising from difference of form, the somewhat smaller masses of tissue, and the fainter tinge of brown on the outer layers of the rind or the entire absence of that colour. Very delicate mycelial membranes and slender tufts of ramifying branches, which frequently arise on the edge of the larger mycelial body, have a more simple structure and consist only of hyphae of the rind. There is one important peculiarity in all these strands and expansions, that the numerous hyphae which stand out like hairs from the surface force their way into the tissue of the rind and wood, and spread and ramify there, and are organs by which the Fungus takes up its food. They often form bladder-like swellings in the tracheides of the pine-wood which they decompose, reminding one of the inner layers of the rind of the strand, and their number in the tracheides may be so large as to fill them with a tissue of bladder-like cells[1].

Brefeld has completed our knowledge of the life-history of the mycelium of Agaricus

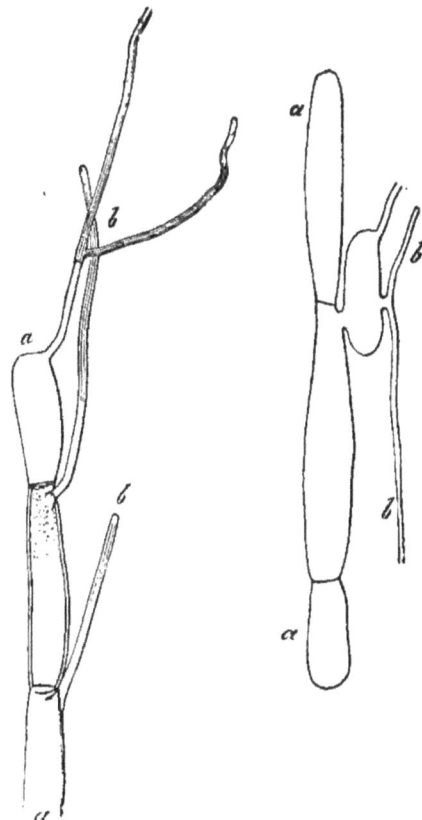

FIG. 12. *Agaricus melleus.* Subterranean mycelial strand. Isolated pieces of the large-celled originally axile tissue *a—a*, from which the definitive medullary hyphae *b* grow as branches. Magn. 390 times.

melleus by growing it from spores in an artificial nutrient solution (decoction of plums). A delicate branched radiating primary mycelial hypha was developed in about eight days from the germ-tube which issued from the spore cultivated on a microscopic slide. Thick tuft-like branchlets from single branches of the hyphae or from several adjacent ones then appeared in the centre of the circular expansion which was some millimetres in size; these tufts raised themselves erect and became united together into clews as large, according to the figures, as a good-sized pin's head, after the manner of the sclerotia to be described below in

---

[1] R. Hartig, Die Zersetzungserscheinungen d. Holzes, p. 59.

section VIII. The clews assumed a pseudo-parenchymatous structure owing to the swelling of the cells of the hyphae, and the greater part of their surface acquired a brown colour. Then one or several growing points appeared on most of the clews, always at isolated uncoloured spots in the part which did not project above the nutrient solution, and from these points mycelial strands were developed of the subterranean form just described. The primary mycelial hypha ceases to grow when the formation of strands commences; the strands also cease to grow as soon as the supply of nutriment is exhausted. When cultivated on bread or with a larger supply of the nutrient solution they developed vigorously and branched copiously, and showed all the important points of the subcortical form described above. They remained uncoloured beneath the substratum, and cessation of growth in length was followed by specially copious development of gelatinous hyphae spreading from the surface, and forming on the top of the fluid thick membrane-like patches of wefted covering with a vesicular pseudo-parenchymatous structure, and with the cell-walls coloured brown where they were in contact with the air. After a winter rest of several months' duration a large number of strands of the subterranean form were again produced from the cultivated specimens, being fed by them, and they were seen to make their way into the roots of living pines, where their further subcortical development was also observed.

The development of the sporophores, which will be described in Division II, begins according to Hartig on strands of both kinds in the same manner as the formation already described of similar branches on the strands.

Further details and variations, the great abundance of which is not to be wondered at, considering the great variety of form and adaptation displayed by the strands of **Agaricus melleus**, are to be found in the works of Hartig and Brefeld which are cited further on. I have endeavoured to correct my own former statements, and some also of those writers themselves, from these researches and some more recent ones of my own. Some statements have not been satisfactorily explained even by these investigations; among them a former remark of mine in the first edition of this book, that old and strong specimens of the subterranean form 'have often an uneven and wrinkled rind, in which through subsequent luxuriance of growth the number of the cell-layers is considerably increased and their arrangement is disturbed. I often but not always found inside these specimens a brown zone concentric with the rind from which it is divided by a narrow layer of ordinary medullary tissue, and enclosing a strand of the latter tissue. This zone consists of hyphae with brown membranes very tightly interwoven with one another, but in other respects resembling the ordinary elements of the medulla, into which it passes without a break. Eschweiler's account of the structure of the Rhizomorphae is founded on the examination of such specimens.'

Future investigations will perhaps clear up these less important points. Greater interest attaches to the question, whether the first development of the mycelium observed by Brefeld, and especially the primary formation of the subterranean strand, is an invariable occurrence in Agaricus melleus, or whether perhaps the subcortical formations do not proceed directly from the mycelial hyphae, when the spores germinate on a substratum which renders parasitic growth possible, that is upon the living root of a conifer.

The history of our knowledge of the mycelium of Agaricus melleus is somewhat remarkable. Before R. Hartig discovered that the strands belonged to this Hymenomycete, they were supposed to represent a distinct species of Fungus which was named by Roth Rhizomorpha fragilis, or the two forms, the subterranean and the subcortical, were made two distinct species, Rhizomorpha subterranea and Rh.

subcorticalis, Persoon. The attempts to find the fructification of these Fungi led to the most divergent views ; but there is no need happily to repeat here and criticise the complete enumeration and examination of them which was given in the first edition of this work. Some writers, as P. de Candolle, Eschweiler, Acharius, and more recently Fuckel, endeavoured to prove the Rhizomorphae to be true Pyrenomycetes, and assigned them perithecia, some of which, according to Tulasne, were in fact merely galls, while others belonged to real Pyrenomycetes, which had grown on or close to the strands of the Agaric. Otth regarded as their fructification a species of Stilbum or Graphium which is sometimes found on old strands in the form of small black bodies of the thickness of a bristle and 3-4 mm. in length and giving off spores by abjunction, a view which was supported by the resemblance of their structure to that of the rind of the strands, and which after all may in a limited sense still be correct. The question can only be decided by the history of the development of the Stilbum ; but this is not known, and the species may for the present be considered with greater probability to be a parasite on the strands.

Other writers, as Palisot de Beauvois, and in more recent times Caspary and Tulasne, looked upon the Hymenomycetes, especially the woody Polyporeae, as the sporophores of the Rhizomorphae, partly because the two were so closely associated in their growth, and partly because these observers confounded the strand-like or membranous mycelia of the former plants with the strands of Agaricus melleus, the characteristic structure of which they did not properly distinguish. Hence Caspary, for instance, brings the Rhizomorphae themselves into genetic connection with the sporophores of quite different species of Polyporus, Trametes Pini, and Agaricus ostreatus.

The name Rhizomorpha we now know to be superfluous ; it may and should be dispensed with, as I have myself done above. The same may be said of the name Xylostroma mentioned in the preceding pages, as also of the names Himantia, Ozonium, Hypha, Hyphasma, Fibrillaria, Ceratonema, all of Persoon, Byssus, Dill., Dematium, Lk. (in part), Corallofungus, Vaill. They were applied, as is well known, since the time of Palisot de Beauvois to sterile mycelial strands which sometimes attained a great size in damp woods, cellars, and mines, but their connection with distinct forms of sporophore, owing to the slight attention which was formerly paid to the study of mycelia, was never actually decided.

The forms, which Fries[1] regarded as a distinct genus **Anthina**, may also be mentioned in connection with sterile mycelial strands of doubtful affinity. The Anthinae, of which I am here speaking, and from which I exclude the section Pterula, Fr. because these appear to be fertile, are cylindrical or ribbon-shaped bodies an inch high on the average and about 1 mm. in thickness, which grow erect from a floccose mycelium largely developed in decaying wood and leaves, and branch in their upper part dichotomously or in a palmate manner. They are either of a bright red colour (A. flammea, A. purpurea) or pale brown (A. pallida). They consist of a strand of parallel hyphae firmly united together by a homogeneous connecting substance, and are formed by the union of the hyphae which spread abundantly through the substratum. The bundle is divided at the upper end, or its hyphae separate from one another and spread on all sides and form the bifurcating or palmate extremities. Specimens are not unfrequently found with the upper end of the plant bent down towards the ground, and there separated into a floccose mycelium or even into net-like anastomoses. I have myself never seen a sporophore in these forms, though Fries says of A. flammea, 'affusa aqua secedunt sporidia.' The small cells laterally attached to the hyphae, which I have occasionally found in A. pallida, and which I formerly spoke of as spores, I am now inclined to regard as very doubtful structures.

---

[1] Pl. homon. 169.

SECTION VIII. The name **sclerotium** has been given to certain thick tuber-like bodies formed on the primary filamentous mycelium which proceeds from the germinating spore; these, which are storehouses of reserve-material, become detached from the mycelium when their development is complete, usually remain dormant for a considerable time, and ultimately expend their reserve-material in the production of shoots which develope into sporophores.

The sclerotia are generally exposed on the surface of the substratum, or they are formed on the walls of broad fissures in it or sometimes even in the close tissues of phanerogamous plants.

Their form and average size vary much according to the species, the latter being dependent also on the quantity and quality of the food supplied to them. The sclerotia for example of Typhula variabilis are small spheres usually of the size of a mustard-seed, those of Sclerotinia Sclerotiorum differ extremely in shape and may be smaller than a pea or as large as a hazel-nut, or form shapeless cakes sometimes an inch in breadth; the sclerotia of species of Claviceps are horn-shaped blunt triangular bodies which may be more than an inch long and some millimetres in thickness, or scarcely 1 cm. in length and 1 mm. in breadth, according to the species and the nutrition.

The structure of these bodies in their mature resting state is, in some points of chief importance, the same in all the species. They consist chiefly of a uniform compact tissue, the *medulla*, which, with a single exception noticed below in paragraph *d*, is surrounded by an outer layer of peculiar structure forming the *rind* or outer coat. Both parts contain comparatively little water. The medulla is a close weft of hyphae or a pseudo-parenchyma, the elements of which are pale-coloured or colourless and contain a large quantity of reserve food-material; this in some cases takes the form of a great thickening along with gelatinisation of the membranes of the cells, the lumina of which are narrow and contain little solid matter, as in species of Sclerotinia and in Typhula gyrans, &c., or their cell-walls continue thin and the food-material is in the form of large accumulations of fatty matters, as in Claviceps, or of fine-grained protoplasmic substances, as in Coprinus stercorarius and others. Exact investigation of the reserve material has been made only in the case of the sclerotia of Claviceps[1]. The rind is composed of one or more layers of cells which have their membranes wholly or partially sclerosed and dark-coloured, and are poor in solid contents.

Within the limits of this general structure, which is common to these bodies, there are special structural arrangements which vary much in the different species. Outward likeness is not always accompanied by agreement in their internal structure, which may also be like or very unlike in nearly related species.

The following details, most of which appeared in the first edition of this work, will illustrate these points.

a. The sclerotia of the **Sclerotinia-Pezizeae** (Peziza tuberosa, P. Sclerotiorum, P. Fuckeliana, P. Candolleana, P. ciborioides, P. baccarum, &c.) have a thin, black, smooth or rough rind, and a medulla which in the dry state is of a white or whitish colour. The latter is a firm gelatinous tissue of cartilaginous texture without any

---

[1] Flückiger, Pharmacognosie d. Pflanzenreichs ; see before on page 17.

air-conducting passages, as in P. Fuckeliana, or with comparatively few of them. Its hyphae are cylindrical and septate, and interwoven with one another in every direction ; hence in thin sections of the sclerotia their lumina appear in all possible forms according as the section passes through them transversely, obliquely, or longitudinally (Figs. 13, 14). The cells in the moist state contain little else than a watery fluid ; in the dry state they contain air. Towards the rind the hyphae are divided into short cells, and in sections therefore most of the cells have a circular outline.

The rind consists of isodiametric roundish-cornered cells which have firm dark-brown membranes and adhere closely to one another. In small forms (Fig. 13) it is composed of one or two layers of cells, in larger (Peziza tuberosa, P. Sclerotiorum, Fig. 14) of three or four or more layers, and then the cells are usually arranged in irregularly radiating rows perpendicular to the surface. It can be easily shown in most cases that the elements of the rind are those segments of the medullary hyphae which lie nearest to the surface of the sclerotium.

The breadth of the hyphae varies in different species and sometimes in different individuals.

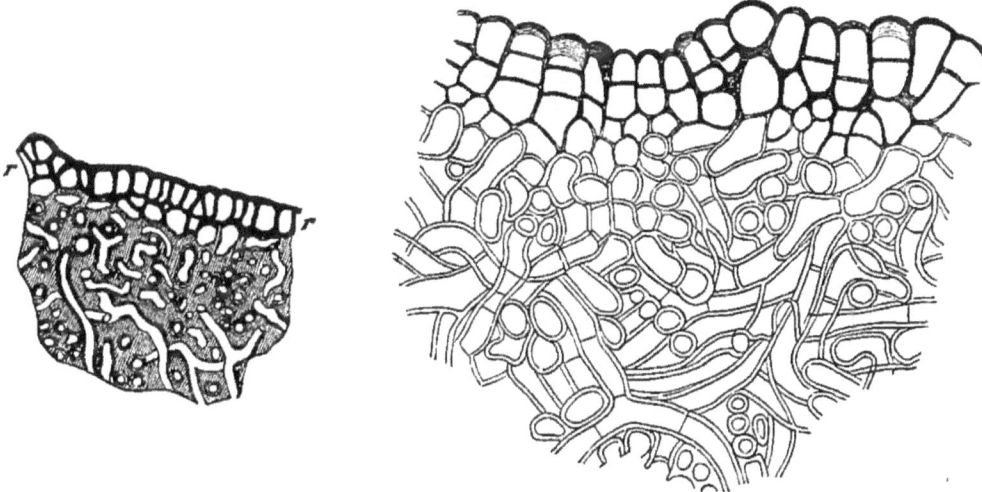

FIG. 13. Piece of a thin transverse section through a sclerotium of *Sclerotinia Fuckeliana* ; *r* the rind. Magn. 390 times.

FIG. 14. Thin section through a mature sclerotium of *Sclerotinia Sclerotiorum*, Libert, showing the rind and adjoining medullary tissue. Magn. 375 times.

Many of the forms which belong to this group occur on the surface of the part of the plant on which they grow, others inside them in their decomposing substance. The former (Peziza tuberosa, and P. Sclerotiorum frequently) show the structure, which has been described, quite perfectly. Some of the latter, as P. Sclerotiorum, often enclose isolated dead cells or larger portions of the tissue of the part of the plants, which they inhabit, in their own substance, as Corda pointed out. The foreign bodies thus enclosed are irregularly and inconstantly distributed through the medulla, and are sometimes surrounded by a layer of dark-brown cells of the rind.

The smaller sclerotia of this type, which are found growing on decaying leaves (Peziza Candolleana, Lev., P. Fuckeliana), regularly take possession of the substance of the leaf at the points where they are developed. They are weal-like swellings on the leaf, formed of the tissue-elements of the sclerotium, among which the dead elements of the leaf are interposed, though more or less displaced and separated from one another. The way in which the sclerotium takes possession of the tissue of the leaf is different in different species. The sclerotium of P. Fuckeliana for example (Fig. 19) inhabits only the parenchyma and epidermis of the leaf of the grape-vine, but sometimes it grows even over the hairs on the leaf and so appears to be spikey; it often appears along the veins of the leaf, but always outside the wood-bundles. I found the sclerotia of P. Candolleana on oak-leaves, but there also only in the parenchyma. But the sclerotium of a small Peziza which inhabits the leaves of Prunus insinuates itself among all the elements of the veins of the leaf.

b. The structure of the sclerotia of several of the Hymenomycetes, especially **Agaricus cirrhatus**, P. (?), **A. tuberosus**, Bull., and **Hypochnus centrifugus**, Tul., differs little from that of the first type. The chief difference is that the cell-walls in the rind are not a dark but a yellow brown; the surface of the rind is in most cases tolerably smooth, but in Hypochnus centrifugus it is uneven or felted over with the remains of the hyphae which surround the sclerotium in its younger state. The hyphae of the medullary tissue and their membranes are of varying thickness according to the species; they are in most cases chiefly filled with a watery fluid or with air; in Hypochnus centrifugus they contain drops of oil. I have never found tissue-elements of the host enclosed in the medulla even of those of the above mentioned sclerotia, which had developed in the interior of decomposing parts of plants (Mushrooms).

c. A somewhat different structure from the above is seen in a sclerotium in Rabenhorst's Herb. mycol. Nr. 1791, incorrectly named Sclerotium stercorarium, and of doubtful origin. Its white medullary tissue consists of thin-walled cylindrical hyphae which contain a watery fluid, and are usually rather loosely interwoven, the interstices being filled with air. Towards the surface the medulla passes gradually into an outer covering of many layers of narrower hyphae, which mostly run parallel to the periphery and form a tissue without interstices. The inner layers of this tissue are colourless, towards the outside the membranes become gradually yellow-brown, and those of the outermost layers are so considerably thickened that the lumina are much reduced in size. The whole sclerotium is thus surrounded by a firm uneven rind composed of several layers.

d. The light-yellow **Sclerotium muscorum**, which also belongs to some Agaric, consists of a web of broad thin-walled hyphae with narrow interstices containing air. The hyphae, which are not arranged in any order, are composed partly of elongated cylindrical and partly of short vesicular cells. The latter contain a clouded homogeneous yellowish protoplasm, or a watery fluid in which drops of yellow oil are suspended. The surface of the sclerotium appears to the naked eye of a darker colour than the centre, but under the microscope the structure is seen to be the same throughout, and the medulla and rind are not clearly distinguished. Single surface-cells project here and there as cylindrical papillae.

e. The snow-white medulla of the sclerotium of **Coprinus stercorarius**, Fr. has a similar structure to that in Sclerotium muscorum. It is a pseudo-parenchyma composed of broad irregularly roundish or elongate-ovoid cells and single cylindrical hyphae; all the cells are very thin-walled and filled with a colourless, uniformly and finely granular, somewhat strongly refractive protoplasmic substance, which issues forth from injured cells and spreads through water and makes it turbid. These cells form a close tissue which is hard in the dry state, and has more or less narrow interstices filled with air. The cells of the medulla become suddenly smaller towards the circumference. The surface of the sclerotium is formed of a firm apparently black rind which is wrinkled in the dry state. Where this rind borders on the medulla it shows four or five irregular layers of small cells of the shape and size of the outermost cells of the medulla, but with brown membranes and apparently always clear watery contents. This layer is surrounded by the more superficial rind consisting of three or more layers of large cells usually of irregular roundish outline, which at the periphery have some resemblance to the largest of the cells of the medulla, and contain a watery fluid or air within a slightly thickened wall of a dark violet-black colour. Many of the superficial cells of the rind project irregularly above the rest and some are prolonged into short irregular hairs or papillae, while in others the membrane is irregularly torn on the outer side of the cell and the exterior surface is thus rendered rough and uneven.

f. The sclerotia of some **Typhulae**, T. phacorrhiza. T. gyrans, T. Euphorbiae, Fuckel, T. graminum, Karst., &c. have the gelatinous medulla with cartilaginous consistence of the type *a*, with slight differences as regards the thickness and firmness of their membranes in the several species. The hyphae contain a clear watery fluid which sometimes has granules sparingly distributed through it; in T. graminum only they are densely filled with homogeneous turbid protoplasm. The rind in these species is a single layer of cells of uniform height connected by their sides without interstices, which are evidently the peripheral segments of the medullary hyphae, unlike them as they may be in structure. The cells are tabular or shortly prismatic in shape, their lateral walls often curved and sinuous; the inner and lateral walls are slightly, the outer walls very strongly thickened in the manner of the outer wall of the epidermal cells in vascular plants, and have their outer surface smooth (Fig. 15 *c*) or warted (Fig. 15 *a*, *b*). The rind is thus remarkably like the firm epidermis without stomata of many vascular plants.

g. In the sclerotia of **Typhula variabilis**, Riess, and **Peziza Curreyana** the structure of the rind is essentially the same as in the last type, but the white or in P. curreyana the rose-red medullary tissue is a weft of cylindrical hyphae with

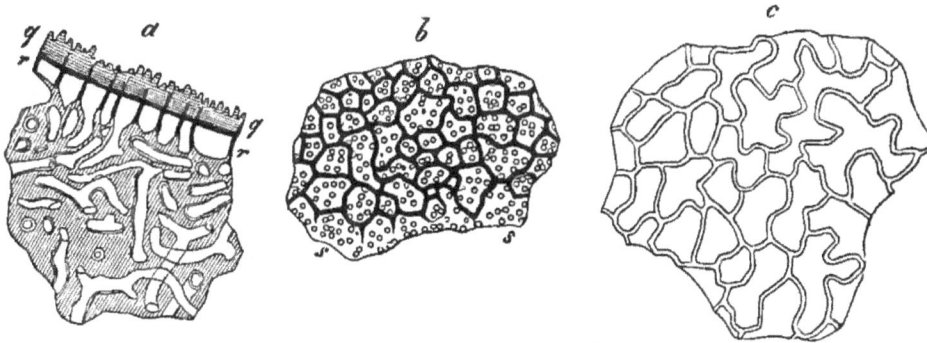

FIG. 15. *a* and *b* sclerotium of *Typhula phacorrhiza*. *a* piece of a thin transverse section; *r—r* rind-cells, *q—q* outer layers of the same. *b* piece of the rind flattened out, seen from the outside; at *s* the outside of outer layers only is seen without the lateral walls of the cells. *c* cortical layer of the sclerotium of *Typhula gyrans* flattened out and seen from the outside. Magn. 390 times.

air-spaces. The hyphae are thin-walled in Typhula with dense, granular contents; in P. curreyana the membrane is thickened and stratified, and the medullary tissue is more compact towards the periphery as is the case in type *a*.

h. The sclerotia of species of **Claviceps**, the blunt trilateral horn-shaped bodies which develope in the flowers of the Gramineae and Cyperaceae at the expense of the ovary and are known by the name of *ergot*, consist in the mature state chiefly of a dirty-white medullary tissue surrounded by a violet-brown rind. The medulla has the character of a pseudo-parenchyma formed of cylindric prismatic cells, which are on an average from one to four times as long as broad. The cells are arranged in straight or sinuous longitudinal rows, and the history of their development shows that they possess the characters of the fungal hyphae. This may also be clearly seen even in ripe sclerotia, in the interior of which clefts and fissures are often found clothed or loosely filled with a thin felt; sections show that the felt is composed of interwoven hyphae, which spring as branches from the rows of cells of the compact tissue and have the same characteristics, except that their weft is looser. The medulla has usually shorter and broader cells towards the periphery of the sclerotium than in its centre. The cells are everywhere provided with tolerably thick colourless membranes, and are usually firmly grown together on every side without interspaces. They contain large colourless drops of oil.

[4]   D

The medulla is surrounded at first by an inner stratum of the rind which is at every point firmly connected with it, and is composed of one or two layers of cells with contents showing no oil, and with membranes that are strongly thickened often more on the outside than on the inside and of a dark violet-brown colour. This inner stratum of the rind is enclosed by an outer portion formed of a few or even as many as twenty layers of longitudinally arranged or irregular branched rows of cells; the cells are narrow and their membranes are of a pale violet-brown. This is the thin pale-violet coating, often with longitudinal stripes or interruptions, which clothes the surface of the fresh sclerotium and may be easily broken or rubbed off from the firm inner rind.

All sclerotia, it would appear, develope as secondary formations on a primary sporogenous filamentous mycelium. They arise from a single branch of a mycelial filament which has quickly produced a tuft of many branchlets; this is the case in Coprinus stercorarius, Typhula variabilis, and T. gyrans. In others, as in the forms of Sclerotinia, several adjacent branches of the primary mycelium take part in the formation from the first. In both cases the young sclerotium soon rises above the substratum as a small tuft of loosely tangled hyphal branches in all essential points similar to the primary. Then in Sclerotinia Sclerotiorum the filaments of the tuft grow vigorously and branch and coalesce repeatedly by means of H-shaped cohesions, and thus the tuft itself developes into a dense white ball of the size of the sclerotium; till this size is reached, the structure of the hyphae remains as it was originally, the new branches are often slenderer than the primary hyphae and their character is uniform in all parts of the ball. The interstices in the tissue contain air, the surface is rendered finely hairy by the presence of slender spreading branchlets of the hyphae, and the whole body is soft and can be easily compressed into an extremely small compass. But from this time by new formations in its interior, and afterwards by expansion of the cells already formed, the tissue constantly increases in size and firmness. Lastly the thickening of the membranes commences, which is characteristic of the species, and this is accompanied with a partial disappearance of the air-spaces and the differentiation into medullary and cortical layers. This process of development begins in the interior of the tissue and advances rapidly towards the circumference. The outermost layer of the white ball takes no part in it, but remains for a time as a white felted covering on the rind which is distinct from it, and ultimately shrinks to nothing and disappears. The ripe sclerotium becomes detached from its felted environment as a body of sharply defined form and outline.

Sclerotinia Fuckeliana exhibits the same phenomena when cultivated upon a microscopic slide; but, as might be expected, these are modified by the nature of its environment when the sclerotium is developed spontaneously inside the tissue of a phanerogamous plant. Peziza ciborioides also behaves in a similar manner, but shows some specific variations in its development.

The tuft of hyphae which is the commencemet of a sclerotium of Typhula variabilis rises above the substratum in which the primary mycelium has spread its ramifications[1]. The branches of the tuft become woven together into a smooth white spherical body of small size, which is attached to the substratum by a short and slender stalk. The sphere enlarges rapidly by the formation of new cells and

---

[1] The substratum in the natural way of growth is formed of decaying leaves in winter and spring; Brefeld employed nutrient solutions for the purpose of artificial culture.

branches in all its parts. It is entirely formed at first of uniform thin-walled much branched hyphae, which are rich in protoplasm and closely woven together, but not without air-spaces. Those parts only of the hyphae which form the outer moist surface of the sphere have no interstices between them ; if we examine a sclerotium in a very early stage of its development we see that the surface is formed of a layer of short cells of uniform height, which are segments of hyphal branches running in numbers through the periphery. These cells are at first thin-walled and filled with protoplasm, like the other portions of the hyphae, and their walls are colourless. Growth by formation of new hyphal branches continues for some time longer in the centre of the sphere, and thus the medulla enlarges its circumference considerably, while its hyphae grow to twice their original size ; but the intertwining of the hyphae remains as it was before. From an early stage in the development no new cells are introduced between the previously existing cells of the superficial layer ; but these cells stretch in every direction, and sufficiently strongly in that of the surface of the sphere to remain united together into a layer without interstices. Their radial or lateral walls assume in this way the undulated inflated outline mentioned above in paragraph *f,* while their outer membrane becomes thickened to form the covering described in the same place, and takes the permanent yellow colour or passes through yellow and brownish yellow to a dark brown ; the protoplasm disappears. This formation of rind is continued also at the point of insertion of the stalk over a layer of cells which lies in the direction of the surface of the sphere, and the sphere is thus divided off from the stalk and is ultimately detached from it, while the stalk dries up. The development of Typhula gyrans follows a similar course. All that is known of the development of other sclerotia, excepting that of Claviceps, agrees with the processes above described, though the final differentiation is accompanied by certain variations in detail, as will be inferred from the statements in paragraphs *a* to *g* ; in this matter Brefeld's careful description of Coprinus stercorarius should be consulted.

The sclerotia, of the development of which we have been speaking, are not all formed on morphologically definite spots of the primary mycelium, and their number varies according to the state of its nutrition. When several begin to be formed near each other, they may unite as they grow into one body ; this leads to the formation of the irregular cakes and crusts mentioned above, especially in Peziza Sclerotiorum, though it is observed in other species also, as in Coprinus stercorarius, and in a less degree in Typhula gyrans. Much water is expelled from all the above sclerotia when the differentiation and final development commence in them, and appears on their surface in large clear drops. The whole process of development may under favourable conditions be accomplished from beginning to end in a few days.

The sclerotia of Claviceps (ergot), concerning which Tulasne's labours have given us more exact information, show several variations of detail in their development arising from the peculiar parasitism of the Fungus (Figs. 16, 17). The primary mycelium occupies at first the base of the young ovary in the flower of the Gramineae and Cyperaceae. In ordinary cases, to which we will at present confine our attention, it spreads rapidly through the entire ovary, with the exception of its apex and some-times also of the inner layers of its wall ; the ovary is thus changed into a white Fungus-body of nearly its own shape, with a surface marked with deep narrow curved furrows; and as gonidia are formed on the surface in a way that will be after-wards described (see Division II) the body may be termed a *gonidiophore* (Fig. 16). The hyphae of the Fungus-body must necessarily make their way for some distance from the ovarian base into the floral pedicel, for it is difficult to conceive of any other mode of supplying food to the Fungus; but we have no exact information on this point. When the gonidiophore is fully formed, the beginning of a sclerotium makes its appearance in the torus at its base and on the mycelium, which is supposed to spread through it, in the form of a small somewhat elongated fungal body enclosed in the white tissue and distinguished by its greater density (Fig. 16 *b, s*). It is formed at first of slender delicate separable hyphae which are continuous with those in the vicinity, but are somewhat firmer and more closely compacted. Its

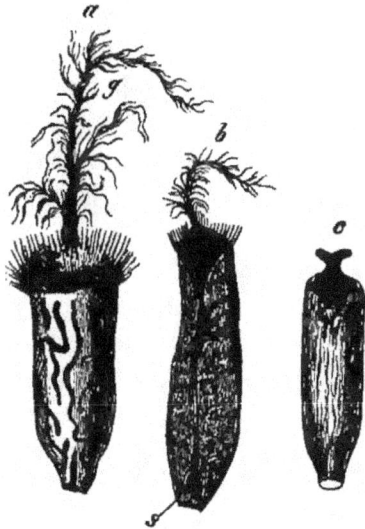

FIG. 16. *Claviceps purpurea*, Tul. *a* young ovary of *Secale cereale* penetrated and covered with the gonidio-phore, seen from without; the hairs of the ovary and the remains of the style *g* project at the apex from the fungal investment. *b* longitudinal section through a similar stage in the development from *Secale*; *s* commencement of the sclerotium. *c* similarly young state of the Fungus on the pistil of *Glyceria fluitans*, the Fungus projecting beyond the apex of the ovary. After Tulasne. Slightly magnified.

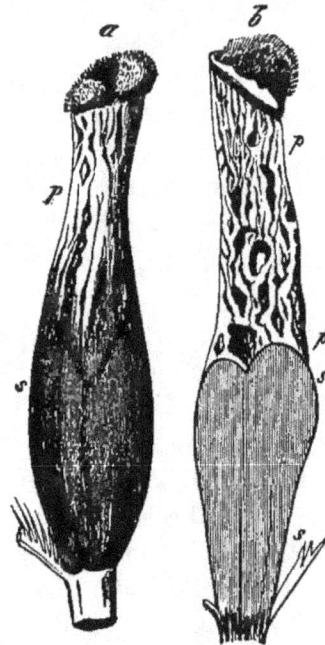

FIG. 17. *Claviceps purpurea*, Tul., on *Secale cereale*. *a* seen from without. *b* median longitudinal section. The sclerotium *s* rests on the torus and carries up the dry-ing gonidiophore *p* on its apex. After Tulasne. Slightly magnified.

surface soon acquires a violet hue, the superficial cells beginning to assume the character of the future rind. It now increases in thickness and elongates into the well-known horn-shaped body, which is attached at its base to the torus and projects above from between the paleae. The course of its development still requires more exact investigation. Its growth in the longitudinal direction is no doubt maintained by continued addition at the base. The increase in thickness of each transverse section above the base must in a great measure be due to the expansion of cells already formed, since these are more than four times broader in the fully developed parts than in the younger. The gonidiophore ceases to grow as soon as the sclerotium begins to be formed, and being detached from the torus as the sclerotium enlarges it is carried up like a cap on its apex, and there shrivels up and sooner or later falls off (Fig. 17).

The development of these sclerotia is slow; it required for instance about four weeks, according to Tulasne's observations, in the months of July and August in the flowers of Brachypodium sylvaticum. In this case, when the sclerotia began to develope, drops of a saccharine fluid were observed to appear; but it is uncertain to what extent they were connected with the formation of the sclerotia or with that of the gonidia.

It sometimes, but rarely, happens that the Fungus is developed beneath the point of attachment of the ovary; when this happens the ovary preserves its proper form and is carried up between the paleae on the apex of the sclerotium, and there withers away before its time.

The sclerotia when fully formed and matured pass into a *resting state,* the duration of which varies in species and in individuals with external and internal causes, as is the case with seeds, tubers, and rhizomes. It depends on the habits of life of the species in the natural condition whether the period of rest is confined for example to winter, as is the case with Claviceps, Peziza Curreyana, and P. Duriaei, or to summer, as in Typhula gyrans, T. variabilis, and T. phacorrhiza, or is not constantly connected with the time of year. In the former case the time of rest may only be slightly shortened in some species by changing the external conditions, as the example of Claviceps especially shows.

Sclerotia if kept dry will retain their power of development for a long time unimpaired; those of Peziza Sclerotiorum for more than a year, according to Brefeld for several years, those of Claviceps for about a year; after about that time those of the latter species and of Peziza Fuckeliana also usually lose their vitality.

The external conditions for the further development of the sclerotia are the ordinary general conditions for germination, sufficient supply of water and oxygen and a suitable temperature. The usual procedure under such conditions is as follows. The sclerotium first absorbs water and swells, and then in a longer or shorter time, often not till after some months, it sends out shoots at the cost of the food-material stored up within it, and these develope directly into the sporophores characteristic of the species. The sporophores in the species which form sclerotia, with the exception which will be named further on, are compound structures. They accordingly make their appearance as bundles of hyphae springing as branches from the elements of the sclerotium, and in two ways according to the species.

In one case (Figs. 18, 19) represented by Claviceps, Sclerotinia Fuckeliana and S. Sclerotiorum, Typhula gyrans and T. phacorrhiza, the bundle of hyphae arises at a certain spot in the medullary tissue and from branches of it which originate beneath the rind; the latter has no share in the new formations but is pierced through by the advancing bundle of hyphae. A more minute description of the special circumstances observed in Sclerotinia will be given in Division II.

In the second case the bundle of hyphae is formed by the ramification of outgrowths from the cells of the rind, as Brefeld rightly states in the case of Coprinus stercorarius; whether it is always a single cell that produces these shoots must for the present remain undecided. Agaricus cirrhatus also shows this formation.

Typhula variabilis is in some respects intermediate between the two cases; here according to Brefeld the initial bundle of hyphae appears on the surface of the rind, neither springing apparently from a cell of the rind, nor causing an evident broad fissure in the rind, and formed therefore in all probability from a single branch proceeding from a peripheral medullary hypha and piercing through the rind.

The exceptional case mentioned above, in which the product of the sclerotium is not a compound structure, is the formation of simple filamentous gonidiophores, known by the name of Botrytis cinerea, from the sclerotia of Peziza Fuckeliana. In most of the cases which I have myself examined a bundle of hyphae shoots out from the subcortical medullary region, and where it has broken through the rind the hyphae spread in different directions, and each developes into a gonidiophore. But it sometimes happens that the cells of the rind develope directly into gonidiophores.

In none of the sclerotia that have been examined is the origin of the shoots connected with a definite predestined morphological spot. Any fragment of the larger sclerotia, if not too small, can under ordinary circumstances produce them, as Tulasne showed in the case of Claviceps, and Brefeld especially in that of Coprinus stercorarius.

The number also of the shoots that may proceed from a sclerotium is not definite in any species; and some species can produce an almost unlimited number

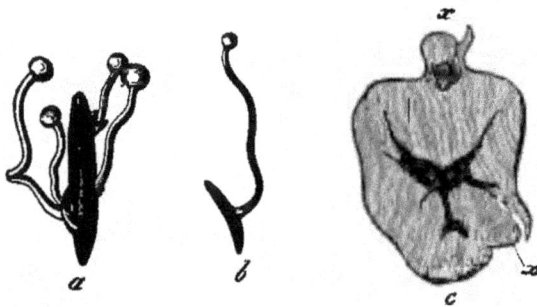

FIG. 18. a and c Claviceps purpurea. b C. microcephala, T. a and b sclerotia with mature sporophores. c transverse section through a sclerotium with the young sporophores emerging from the interior. After Tulasne, a and b nat. size, c slightly magnified.

FIG. 19. Sclerotinia Fuckeliana, a very small specimen. s transverse section through a sclerotium, from which a sporophore cut through lengthwise has proceeded. The dark spots in the sclerotium are the dead cells of the vine-leaf which it has occupied; the spots and dots at p are calcium oxalate aggregations. Magn. 20 times.

of these primordia (Anlagen) of sporophores on their sclerotia, others cannot do this. Vigorous specimens of Coprinus stercorarius, according to Brefeld, may produce hundreds of primordia, of which however few are ever perfected, and if those already formed are intentionally and repeatedly destroyed hundreds of fresh primordia as repeatedly make their appearance. Other species are less productive; Sclerotinia Sclerotiorum seldom has two dozen sporophores even on strong plants; species with small sclerotia have usually one only or very few.

The size of the individual sclerotia on one and the same species, other conditions being the same, generally causes a difference in the number of the sporophores which commence and complete their development, and in the vigour of growth of the latter. Larger sclerotia are on the whole more productive than the smaller. Claviceps purpurea produces 20–30 sporophores from such large sclerotia as are formed upon the ears of Secale cereale, but only one or a few weakly ones from the small sclerotia upon the spikelets of Bromus, Lolium, and Anthoxanthum. Similar differences arising from the size of the sclerotia are observed also in Sclerotinia Sclerotiorum and in Coprinus. The relation between size and productiveness is the same in fragments of a sclerotium as in their corresponding sclerotia. It is natural to assume, without closer inquiry into the metabolism, that the cause of these phenomena lies in the difference in quantity of the reserve-material at the disposition of the plant according to the size of the sclerotia or their fragments, and that the not infrequent irregularities and apparent exceptions to the rule are due, other things being equal, to differences in quantity or quality in the reserve-material, which may occur also, be it remembered, where the size of the sclerotia or of its fragments is the same.

The formation of the primordia and the further development of the sporophores is accompanied by the solution, transformation, and consumption of the food-material stored up in the sclerotia. The process begins at the point of origin of a primordium and spreads by degrees through the medullary tissue. In Claviceps, according to Tulasne, the oil disappears and its place is taken by watery fluid, the cell-membranes become thinner and ultimately very delicate, and the cells separate readily from one another. In the sclerotia of Sclerotinia Fuckeliana, S. Sclerotiorum, S. tuberosa, Typhula gyrans, &c. which are gelatinous with a cartilaginous consistence, the gelatinous thickening-layers of the hyphae become softer and pale and by degrees scarcely recognisable, so that the innermost layer only of the membrane can still be clearly seen as a delicate pellicle. The former firm union of the hyphae naturally comes to an end at the same time; and a mass of granular matter which turns yellow with iodine collects in the cavities of the cells, and diminishes again in quantity as the sporophores increase in number and size. In Coprinus stercorarius, according to Brefeld, the granular protoplasm of the cells is replaced by a watery fluid, and the membranes become pale and undistinguishable. Ultimately in all these cases the medullary tissue almost entirely disappears. The rind at first takes no perceptible part in these changes; it remains behind after the disappearance of the medulla as a soft sac which collapses and decays.

These processes take a longer or a shorter time in different cases. In Brefeld's culture of Coprinus stercorarius they were over in 7–10 days. In most species they take much longer time. Sclerotinia Sclerotiorum, for instance, may put out new sporophores one after another during some months from one sclerotium, and develope them slowly before the supply of food is exhausted. I have found sclerotia of Agaricus cirrhatus (see on page 37), which had developed one or more sporophores and fully matured them, not sensibly different in consistence and structure from others which had not yet produced any; they might therefore repeat the production of successive sporophores and perhaps during a considerable time; but this point remains to be determined.

Some sclerotia, as those of the Sclerotinieae, of Coprinus stercorarius and Claviceps, are able in the mature state and as long as they retain their vitality to form new rind over wounds, such as cut surfaces which reach to the medullary tissue, provided they are exposed to the air but are protected from desiccation. The new rind resembles the old ordinary tissue in all essential points. It is formed by the medullary hyphae exposed by the wound sending out branches, which become woven together into a delicate felt and cover the surface of the wound. The inner layers of this covering which are next the uninjured medullary tissue then develope into a new rind, while the outer ones dry up and disappear. If such wounded places are kept in a nutrient solution, the branches put out by the medullary hyphae on the exposed points may, in the Sclerotinieae at least, develope into vegetating mycelial hyphae instead of forming new rind. The sclerotia of the Typhulae in which the tissues are very distinctly differentiated appear not to be capable of these acts of regeneration.

The following remarks are intended to illustrate and complete what has been stated above.

The production of the primordia of sporophores from the cells of the rind of Coprinus stercorarius is given on the authority of Brefeld. It is not strange in itself, even in presence of the facts illustrated in Figs. 18 and 19, that the superficial cells of the sclerotium should remain capable of further development and of branching, and that the ordinary distinct division of labour between the protecting rind and the medulla should in some cases not be observed. There is therefore no antecedent difficulty in admitting a third mode of production such as Brefeld gives in the case of Sclerotinia Sclerotiorum, in which the medullary cells and cells of the rind both participate. But I have not admitted this case into my account because the facts will not bear this interpretation. Young shoots always spring in this species from the medulla in the peculiar manner which will be described in Division II, and burst through the rind to reach the surface. In somewhat older specimens, such as those which are very beautifully and correctly portrayed in Brefeld's Table viii. Fig. 9, the true state of the case is obscured by the circumstance that the superficial cells of the sporophore from the point of emergence are very like those of the rind of the sclerotium in shape and in their dark colour, so that the new cells appear to be directly continued into the superficial cells. Thin sections even in more advanced states of growth under sufficient magnifying power show that the case is as I have stated it, and exhibit clearly the arrangement of the black superficial cells, which are the extremities of hyphal branches proceeding from branches of the emerging tuft of hyphae and passing to the surface in diverging curved directions.

That the shoots from the sclerotia in the cases described above should always have been termed primordia of sporophores requires no special explanation, even in the case of Sclerotinia Sclerotiorum where they may under unfavourable circumstances develope in the ground into long branched strands. Even the normal sporophores of this Peziza may be branched, and the branches of the strands may under favourable conditions of development revert to the normal sporiferous state. Brefeld indeed saw them on several occasions produce a filamentous mycelium which subsequently formed sclerotia; but these are monstrous developments such as occur also elsewhere, the exceptional cases that confirm the rule.

It is true that phenomena have been reported in connection with the formation of shoots from sclerotia, which vary from the descriptions in the text; but more searching investigation is needed in all these cases. Thus Tulasne saw sclerotia of Hypochnus centrifugus, which had been placed in damp sand in the end of April, produce in August and September a filamentous mycelium like a spider's web, which subsequently developed the ordinary sporophores of Hypochnus. As regards the connection of the mycelial hyphae with the sclerotium it is merely stated that they spread in every direction from its surface. The consistence and structure of the sclerotia remained unchanged after the production of shoots; hence Tulasne rightly considers our knowledge of them as not yet complete.

Tulasne has already pointed out that Léveillé's older statement, that a floccose mycelium is first produced from the sclerotia of Agaricus grossus, A. stercorarius, A. racemosus, and A. tuberosus, and afterwards sporophores from the mycelium, is founded on a mistake.

Another exceptional occurrence, demanding more critical investigation, is described by Micheli[1] as taking place in Peziza Tuba, Batsch, a species which seems scarcely to have been examined since his time. The sclerotium as it lies in the ground puts

---

[1] Nova plantarum genera (1729), p. 205, 'Fungoides, No. 5.'

forth a number of erect sporophores in the spring, and forms a new sclerotium destined to produce sporophores in the succeeding year. The exhausted sclerotium of the previous year is usually still in existence when the new one is formed, so that the underground portion of the Fungus consists of three small tubers of unequal size.

Historical remarks. Although it has long been known that the sporophores of certain Fungi, species of Typhula and Agaricus and some others, are developed from small tuber-like bodies, our more exact knowledge of the nature of the sclerotia is derived from an excellent publication of Léveillé which only appeared in 1843, and even this work attracted little notice till Tulasne again called attention to the subject and threw new light upon it by his work on Claviceps in 1853. Up to that time the greater part of the sclerotia were considered to be independent representatives of distinct species, and the name Sclerotium was introduced by Tode[1] to designate the genus formed by the supposed species, each with its own specific name.

Some fifty species of Sclerotium were described by Fries in his Systema mycologicum and his Elenchus; the number was subsequently increased to eighty and additions to it are still made by writers, who prefer the hasty publication of imperfect observations to more prolonged investigation.

It appears, as has been shown above, that we are at the present time acquainted with the development and especially with the sporophores of a considerable number of sclerotia. Others are less perfectly known, in some only the mature sclerotium has been seen. Undescribed sclerotia are still not unfrequently found in examining Fungi. Appended is a list of the species of Fungi which are at present known or supposed to form sclerotia, together with the old specific names of the sclerotia wherever they have been ascertained.

1. **Peziza** tuberosa.—P. Tuba, Batsch (Micheli,'l. c.), P. Sclerotiorum, Lib. Sclerotium compactum, S. varium), P. Candolleana, Lev. (Sclerotium Pustula), P. Fuckeliana (Sclerotium echinatum, Fuckel); the two last named Pezizas are in all probability identical, and to them belong the gonidiophores known as Botrytis cinerea, P. (B. erythropus, Lev.), and the 'Sclerotium durum' from which these spring. The little Peziza mentioned in par. *a*, p. 32, as growing on the veins of the leaves of Prunus is very near to these species; its sclerotia found on the same leaves were incorrectly named Sclerotium areolatum, Fr. in my first edition.

Peziza ciborioides, Fr. (Hoffmann).—P. baccarum (Schröter).

P. Curreyana, Berk. (Sclerotium roseum, Kneiff).

P. Durieana, Tul. (Sclerotium sulcatum, Desm.).

The above Pezizas with some others have been made a separate genus **Rutstroemia**, by Karsten (Mycol. fennica) and **Sclerotinia** by Fuckel (Symbol. mycolog.).

Peziza ripensis, Hansen.

2. **Claviceps** purpurea, Tul., C. microcephala, Tul., C. nigricans, Tul. (Sclerotium Clavus, DC.).—C. pusilla, Cesati.

Hypomyces armeniacus, Tul.

Vermicularia minor, Fr., also **Xylaria** bulbosa, P. (see Tul. Carpol.).

3. **Typhula** lactea, Tul.—T. Todei, Fr.—T. caespitosa, Ces.—T. Euphorbiae, Fuckel (Sclerotium Cyparissiae, DC.?), T. graminum, Karst. (Sclerotium fulvum, Fr.), T. variabilis, Riess (Sclerotium Semen, Tode if the cortex is dark-brown, Sclerotium vulgatum, Fr. if it is yellow).—T. erythropus (Sclerotium crustuliforme, Dsm.).—T. phacorrhiza (Sclerotium scutellatum, A. S.).—T. gyrans (Sclerotium complanatum, Tode). I give the names in the two last species on the authority of Fries, Hymenomycetes Europaei, 1874. Léveillé had given the name of Clavaria juncea to the sporophores growing out of Sclerotium complanatum, and in my first edition I gave

---

[1] Fungi Mecklenburgenses selecti, p. 2.

the names Clavaria complanata and C. scutellata after the sclerotia to the two species which are scarcely distinguishable except by the sclerotia.

Pistillaria micans (Sclerotium laetum, Ehr.).—P. hederaecola, Ces.

Clavaria minor, Lév. (which also belongs to Typhula).

4. **Hypochnus** centrifugus, Tul.

5. **Coprinus** stercorarius, Fr. (Sclerotium stercorarium), C. niveus, Fr. (Hansen.)

Agaricus racemosus, P. (Sclerotium lacunosum).—A. tuberosus, Bull. (Sclerotium cornutum).—A. cirrhatus, P. (?) is the name which I have given above to the small white Agaric which grows from Sclerotium fungorum. Other Agarics are also said to form sclerotia: A. tuber regium, Fr., A. arvalis (Sclerotium vaporarium).—A. grossus, Lév., A. fusipes, Bull., A. volvaceus (from Sclerotium mycetospora, Nees in Nov. Acta Nat. Cur. XVI, 1), &c. Sclerotium pubescens, P., Sclerotium truncorum, Fr. were supposed to be connected with such Agarics, on which point see Léveillé and Tulasne. The statements and determinations are many of them doubtful, and more accurate investigations are required.

6. **Tulostoma** pedunculatum, Tul. (Schröter).

There are a large number of tuber-like compound Fungus-bodies the real character of which is still doubtful; our ignorance of their structure or development makes it impossible to decide whether they are sclerotia or some other formation. Among these are Pietra fungaja of South Italy, which is formed of the mycelium of Polyporus tuberaster, Jacq. rolled up into solid masses with bits of soil, stones, and the like; and the tuberous fungoid bodies named Mylitta, Sclerotium stipitatum, Berk., Sclerotium Cocos, Schweinitz, which grow beneath the surface of the ground to the size of a fist or a head and are known only in the sterile state, with some others. Swellings in the substance of phanerogamous plants such as the tubercles on the roots of the Leguminosae, which were once mistaken for sclerotia, require no further notice here.

SECTION IX. Besides the sclerotia which have been described above with well-marked characters morphological and biological there is a motley assemblage of compound Fungus-bodies, which approach the sclerotia in their biological character, but cannot be classed with them from a strict morphological point of view. Such bodies may be termed **sclerotioid,** or, for brevity's sake, simply sclerotia, if we do not thereby infer their identity with true sclerotia. The biological agreement between these bodies and sclerotia consists in similarity of structure, in their being storehouses of reserve-material and in their normally passing through a period of rest, after which they proceed to a further development. Morphologically they are

1. **Transitory resting stages of mycelia,** which under favourable circumstances again develope into filamentous mycelia. Such are the small fatty tubers which are the resting stage of the mycelium of Hartig's Rosellinia quercina, and perhaps formations like Sclerotium Cocos and others mentioned above as of doubtful character.

2. **Perithecia,** which when developed enter upon a long period of rest, and assume at the same time the form and structure of a sclerotium; these do not ultimately produce sporophores, but develope in their interior the asci, which are the characteristic organs of reproduction in perithecia. Of this kind are some species of Pleospora and Penicillium, which will be fully described in Division II. The 'sclerotia' of the Aspergilli of Wilhelm are certainly homologous with the perithecia of Penicillium and are also biologically analogous with them.

3. The bodies, which may still retain the old name of **xyloma,** and which differ for the most part from sclerotia only in their less definite shape and outline, and in

not sending out sporophores as branches, but in producing sporogenous receptacles in their interior. Well-known examples are the Ascomycetes of the genera Rhytisma, Polystigma, Phyllachora and many of their allies which live on leaves. They develope in the substance of the leaves, which they attack during the summer, a thallus very closely resembling in many respects the sclerotia of Sclerotinia Fuckeliana, and primordia of sporogenous receptacles are formed at the expense of the reserve-material in the interior of the thallus after it has passed the winter on a dead leaf; examples of these will be described in greater detail in Division II.

Literature of mycelia.

Filamentous mycelia, haustoria, &c. See the lists of authors under the several genera and groups in Division II.

Erisyphe:—

v. MOHL in Bot. Ztg. 1853, p. 585.
DE BARY, Beitr. zur Morph. u. Phys. d. Pilze III.

Mycelial strands. Xylostroma, &c. :—

ROSSMANN, Beitr. z. Kenntn. d. Phallus impudicus (Bot. Ztg. 1853, Nr. 11).
TULASNE, Fungi hypogaei, p. 2 (Fungor. Carpol. I, pp. 99, 120, &c.).
H. HOFFMANN in Bot. Ztg. 1856, p. 155.
PALISOT DE BEAUVOIS in Ann. du Mus. d'hist. nat. VIII (1806), p. 334.
DUTROCHET in Nouv. Ann. du Mus. d'hist. nat. III (1834), p. 59.
TURPIN in Mém. de l'Acad. d. Sc. XIV (1838).
(Turpin and Dutrochet both describe the development of Cantharellus Crucibulum, Fr. from a reticulately branching mycelium.)
FRIES, Plantae homonemeae, p. 213.
LÉVEILLÉ in Ann. d. sc. nat. sér. 2, XX, p. 247.
DE BARY, Ueber Anthina (Hedwigia I, p. 35);—Id. Beitr. z. Morph. u. Phys. d. Pilze I (1864).
JUNGHUHN in Linnaea, 1830, p. 388.
R. HARTIG, Wichtige Krank. d. Waldbäume, Berlin, 1874, p. 46 ;—Id. Die Zersetzungserscheinungen d. Holzes, Berlin, 1878.

Rhizomorphae. (See also Streinz, Nomenclator. Bail, Tulasne, II, cc., Palisot de Beauvois, l. c.) :—

EHRENBERG, De Mycetogenesi, l. c. p. 169.
ESCHWEILER, Commentatio de generis Rhizomorphae fructificatione, Elberfeld, 1822.
NEES v. ESENBECK in Nov. Act. Ac. Nat. Curios. XI, 654 ; XII, 875.
J. SCHMITZ, Ueber d. Bau &c. d. Rhizomorpha fragilis, Roth. (Linnaea, 1843, p. 478, tt. 16, 17).
TULASNE, Fungi hypogaei, p. 187 ;—Id. Fungor. Carpol. I.
BAIL in Hedwigia I, 111, and Ueber Rhizomorpha und Hypoxylon in Nov. Act. Ac. Nat. Cur. Bd. 28 (1861).
LASCH, Bemerk. über Rhizomorpha (Hedwigia I, 113).
OTTH, Ueber d. Fructification d. Rhizomorpha (Mittheilungen d. Naturf. Ges. Bern, 1856).
v. CESATI, in Rabenh. Herb. Mycol. Nr. 1931.
CASPARY, Bemerk. über Rhizomorphen (Bot. Ztg. 1856, p. 897).
FUCKEL in Bot. Ztg. 1870, p. 107.
R. HARTIG, Wichtige Krankh. d. Waldbäume, 1874.
BREFELD, Unters. über Schimmelpilze III, 136.

Sclerotia :—

DE CANDOLLE in Mém. Mus. d'hist. nat. II, 420.

LÉVEILLÉ, Mém. sur le genre Sclerotium (Ann. sc. nat. sér. 2, XX).

CORDA, Icon. fung. III.

TULASNE, Mém. sur l'Ergot des Glumacées (Ann. sc. nat. sér. 3, XX, 1853).

KUHN, Krankh. d. Culturgewächse, p. 113, t. V.

    (A list of the many treatises on the sclerotium of Claviceps = Secale cornutum
    or ergot, will be found in Tulasne, l. c. and in Handbooks of Pharmacognosy ;
    earlier ones in Wiggers, Dissert. in Secale cornutum, Göttingen, 1831.)

TULASNE in Ann. d. sc. nat. sér. 4, XIII (1860), p. 12 ;—Id. Select. Fungor. Carp.
Cap. VIII.

BERKELEY, Crypt. Bot. p. 256.

BAIL, Sclerotium und Typhula (Hedwigia I, 93).

v. CESATI, Note sur la véritable nat. des Sclerotium (Bot. Ztg. 1853, p. 73).

COEMANS, Rech. sur la genèse et les métamorphoses de la Peziza Sclerotiorum (Lib.
Bull. Acad. Belg. sér. 2, IX, Nr. 1).

WESTENDORP in Lib. Bull. Acad. Belg. VII, p. 80.

MÜNTER in Lib. Bull. Acad. Belg. XI, Nr. 2.

FUCKEL in Bot. Ztg. 1861 ;—Id. Enumeratio Fungor. Nassoviae 1 (1861), p. 100.
(Typhula, lapsu calami Claviceps Euphorbiae.)

KUHN in Mittheil. d. Landw. Instituts, Halle I, 1838.

HOFFMANN, Icon. anal. Fungor. Heft 3.

E. REHM, Peziza ciborioides (Diss. Gottingen, 1872).

W. TICHOMIROFF, Peziza Kauffmanniana (Bull. Soc. Nat. de Moscou, 1868, p. 294,
Tables 1–4).

ERIKSON, in Kongl. Landsbr. Akad. Handl. Tidskr. 1879 (Typhula graminum), 1880
(Peziza ciborioides).

BREFELD, Bot. Unters. ü. Schimmelpilze III (Coprinus), IV (Peziza, Typhula).

SCHRÖTER, Weisse Heidelbeeren (Peziza baccarum) (Hedwigia, 1879).

EIDAM, Botrytis-Sclerotien (Ber. d. Schles. Ges. Nov. 1877).

CATTANEO, Sulla Sclerotium Oryzae (Arch. del Laborat. Crittog. di Pavia, 1877).

SCHRÖTER in Cohn's Beitr. Bd. II (Tulostoma).

E. C. HANSEN, Fungi fimicoli Danici (Vedensk. Meddelelser af naturhist. Forening.
Kjobnhavn, 1876).

GASPARRINI, Ricerche sulla natura della pietra fungaja e sul fungo che vi soprannasce, Napoli, 1841.

    On the same subject: Treviranus in Vers. d. Naturforsch. in Bremen ;—Id. in
    Flora, 1845, 17.—Berkeley, Crypt. Bot. 288.—Tulasne, *l. c.*

Mylitta and allied forms :—

OKEN in Isis, 1825.

LÉVEILLÉ in Ann. d. sc. nat. sér. 2, XX, p. 247.

TULASNE, Fungi hypog. 197 ;—Id. in Sel. Fung. Carp. *l. c.*

CORDA, Icon. fung. VI, t. IX, 39.

BERKELEV, Crypt. Bot. 254, 288 in Gardener's Chron. 1848, p. 829.

BERKELEY, CURREY, HANBURY in Proceedings of Linn. Soc. London, III (1858),
102, and Trans. Linn. Soc. Lond. XXIII, 91, t. IX, X.

H. GORE, Tuckahoe or Indian Bread, in Annual Report of Board of Regents of
Smithsonian Instit. for 1881, Washington, 1883, p. 687.

Rosellinia quercina :—

R. HARTIG in Unters. a. d. Forstbot. Instit. z. München I (1880).

3. SPOROPHORES.

SECTION X. The sporophores[1] (Fruchtträger) are branches of the thallus of peculiar form which spring from the mycelium and produce and bear the organs of reproduction. The term organs of reproduction designates the germs of new individuals; by individual we understand the *bion* as Häckel uses the word, and the mother-cells which immediately produce them. These organs are distinguished according to the particular case by different names, spores, gonidia, basidia, asci, &c., and the structures that bear them may have corresponding names, as gonidiophores, &c. As a Fungus may have more than one kind of organ of reproduction, as will be shown in the sequel, more than one of the special forms just named may appear on the same species.

It has been already intimated that the sporophores have been compared with the flowers or inflorescences of phanerogams, because their development usually closes, as in the phanerogams, with the formation of a number of organs of reproduction, by which their growth is limited either absolutely or for a time. This limited growth is accompanied by a form and structure more sharply and characteristically differentiated than that of most mycelia, and in many cases by a comparatively large development. The sporophores therefore are not only the most characteristically constructed part of the plant, but also the most striking on account of their form and size; hence they used often to be taken for the whole plant, and are also at the present time the chief subject of description in botanical works.

It follows from what has now been said, that we have to distinguish in the sporophore generally between the points of origin of the organs of reproduction themselves and the other parts of the structure which may serve them as supports or envelopes or the like, and which in each case bear some conventional name. These parts are almost always raised above the substratum, and are firmly attached to it and fed from it through the mycelium. The mycelium sometimes, not always, has filiform or hair-like organs of attachment, *rhizoids*, in the shape of branches of the hyphae which spring from the base of the sporophore and complete the provision for its secure attachment and for the supply of nourishment, at least of that part of it which it obtains from water. They have received the name of *secondary mycelium* from their resemblance to the primary mycelium. It has not been ascertained in any instance, whether under favourable conditions they are capable of assuming the normal characters of a mycelium and producing sporophores; in many Fungi, for example,

---

[1] [Compounds of the Greek word καρπός and such terms as 'fruit' and 'fructification' are better used only for structures which have some direct connection with the sexual act; hence the term 'carpophore,' the more exact translation of the German Fruchtträger, as well as the terms 'fruit' and 'fructification,' which under this limitation would not convey the sense of Fruchtträger in this place, have been avoided and the more comprehensive expression 'sporophore' is introduced here as its rendering. Berkeley's specific use of 'sporophore' for what Léveillé had termed 'basidium' is no obstacle to the use of the word in the sense here indicated, as the term 'basidium' is now a generally accepted one. The more recent application of the word 'sporophore' to designate that stage in the life-history of a plant which is the product of an ovum and which as a whole or in part is concerned with the formation of spores need be no objection to its use here; the terms 'sporocarp' and 'sporophyte' sufficiently denote different cases of that spore-producing stage (see section XXXIII).]

species of Coprinus, Claviceps, Mucor stolonifer, and Syncephalis, this power does not exist.

Sporophores may be divided, according to their structure, into two groups : *simple* or *filamentous sporophores* (Fruchthyphen, Fruchtfäden), consisting of a single hypha or of a branch of a hypha, and *compound sporophores* (Fruchtkörper) in the sense assigned to the expression 'compound' in section I.

1. SIMPLE SPOROPHORES.

SECTION XI.   Simple sporophores are branches of mycelial hyphae which usually rise erect from them, and are themselves branched in a variety of modes characteristic of the different species.   When the hypha or its branch has grown to a length which has a fixed average in each species, seldom, as for example in the larger Mucorineae and Saprolegnieae, exceeding a few millimetres, the organs of reproduction, spore-mother-cells, spores, are produced at their extremities in forms which also vary in the different species and groups of species.   With the formation of these organs the growth of the sporophore ceases in most cases at once and for ever, as for instance in the sporangiophores of Mucor and in the gonidiophores of Peronospora (section XXXVII).

In other cases, such as that of the successive abjunction of spores which will be described in section XVI, the growth, that is the increase in size of the sporophore, comes virtually to an end with the commencement of abjunction ; but abjunction may continue at the same spots for a considerable length of time if sufficient nourishment is supplied.   The gonidiophores of Penicillium, Eurotium, and Aspergillus are examples of this kind (see section XVI).

In a third series of cases the first terminal formation of spores takes place at the extremity of the sporophore after its apical growth has ceased, and when this formation is completed a fresh growth in length of the sporophore begins at or close beneath it, and is soon stopped by a new formation of spores similar to the first one ; and on one hypha or branch the same process may be again and again repeated.

The second case is described, as was said above, in greater detail in section XVI. The first and the third may be illustrated by some examples in the present place, though their consideration will also be resumed in later sections.

The thick tubular aseptate simple sporophores of the different species of **Saprolegnia** abjoint their extremity, which is club-shaped and filled with protoplasm, by a transverse septum to form a spore-mother-cell (sporangium), in which numerous spores are formed by division of its protoplasm (section XVIII).   The spores escape when ripe by an opening at the apex of the sporangium, which elsewhere remains entire. This is all that happens in weakly specimens, which therefore represent our first case. In strong specimens on the other hand which have been duly fed, the transverse septum beneath the empty sporangium becomes convex outwards and projects into the sporangium, and assumes the character of a new tubular point, which grows on into the empty sporangium and often through the opening at its apex into the free space beyond, and finally transforms its terminal portion into a new sporangium. This process may be repeated several times on the same sporangiophore, so that several successive sporangia are nested within one another.

The allied genus **Achlya** differs partly from Saprolegnia in developing one or two opposite lateral branches close beneath the empty sporangium, which themselves put out branches bearing new sporangia. There is here therefore a cymose branching of the sporangiophores.

The gonidiophores in **Peronospora**, which are also without transverse septation, are repeatedly forked or monopodially paniculate. The branches are all at first narrowly conical, and when their longitudinal growth is completed their terminal portions swell into an ovoid form, as is seen in Fig. 20 *a*, and are abjointed to form gonidia, and with this the de-velopment of a gonidiophore of Peronospora comes to an end. But in the nearly allied genus **Phytophthora** after the abjunction of each goni-dium the narrow end of the branch which bore it swells slightly immediately beneath it, and elongating at the same time pushes the gonidium so much to one side that it pre-sently forms a right angle with the pedicel. Then in P. infestans the gonidiophore swells at the point of attach-ment of the gonidium into a small narrowly flask-shaped

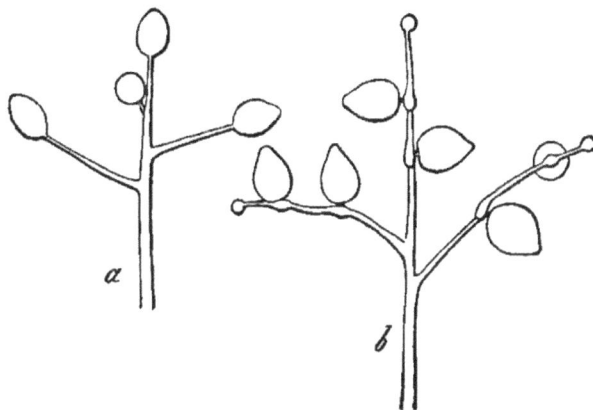

FIG. 20. *Phytophthora infestans*, extremity of two simple sporophores. *a* forma-tion of the first gonidia on the tip of each branch. *b* two ripe gonidia on each branch, with the beginning of the formation of a third. Magn. about 200 times.

vesicle, and its upper end elongates at the same time and again assumes the character of a gonidia-forming point. After a time a gonidium is formed on it in the manner described above, and the process is repeated usually three or four times on the same gonidiophore, or as many as twelve or fourteen times in luxuriant plants. Older simple gonodiophores therefore, when examined dry, are seen to bear a number of lateral nearly equidistant gonidia forming a right-angle with the gonidiophore, and each standing on a flask-shaped swelling (Fig. 20 *b*). As the ripe gonidia fall off instantly in water, preparations treated with water have the older branches of the gonidiophore swollen at inter-vals into the shape of a flask with a single unripe gonidium at most at its apex.

The simple sporophores of **Haplotrichum, Gonatobotrys,** and **Arthrobotrys** (Fig. 21) are short erect rows of cylindrical cells usually simple, but sometimes with single branches. The apex of the uppermost cell swells up considerably in Haplo-trichum, slightly in the other species, and puts out numerous crowded protuberances, which together form a small spherical head and develope into gonidia. This is the whole of the development in Haplotrichum. But in the other two forms the apex of the gonidiophore begins to lengthen again after the first head is matured and grows through it, and thus the head becomes a whorl surrounding the flanks of the gonidiophore; the growing end

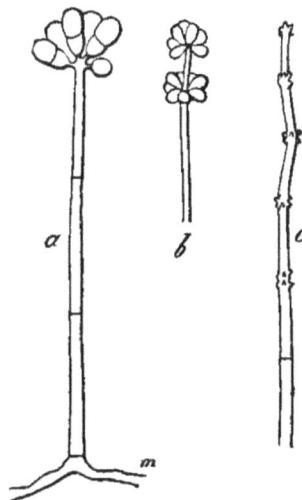

FIG. 21. *Arthrobotrys oligospora*, Fres. *a* sim-ple sporophore springing from the mycelial hypha *m* with the first head of gonidia. *b* second head above the first. *c* old sporophore with the trace of five successive heads. After Fresenius (Beitr). *a* and *c* magn. about 200 times, *b* less highly mag-nified.

of the gonidiophore attains to about the length of one of its lower segments, be-comes septate above the first head and then forms a new head at its extremity like the first. This proceeding may be repeated several times, till the gonidiophore is at length occupied by several whorls of spores at a cell-length's distance from one another, or shows their points of attachment when the spores themselves have dropped[1].

The gonidiophores of Sclerotinia Fuckeliana, known by the name of **Botrytis cinerea**, send out several lateral branches in a paniculate manner beneath their apex, the lower of which are themselves branched. The somewhat enlarged and rounded ends of the primary hyphae and of its branches abjoint many spores simultaneously on their surface. As these ripen the sporiferous terminal cells of the hypha as well the entire lateral branches die, dry up, and almost disappear, while the spores are clustered together without arrangement. But a new growth begins in the cell beneath the terminal cell ; it either simply elongates, and then forms a new sporiferous structure, or it sends out one or more strong lateral branches which behave in the same way as the primary hypha. Formation of sporiferous structures and prolification may take place repeatedly on the same hypha; traces of the branches that have been cast off are seen in the circular scars which project a little towards the outside[2].

2. Compound Sporophores.

SECTION XII. The chief forms of the compound sporophores of the Fungi are well-known to every one; the stalked umbrella-like and the sessile flabelliform or horse-shoe-shaped *pileus* of the Hymenomycetes (Champignon, Mushroom, Amadou-fungus), the club-shaped or shrubby Clavarieae, the *peridia* of the Bovistae and Truffles, the *cups* of Peziza, and lastly the simpler forms which issue as flat or pulvinate bodies from the surface of dead or living plants and are comprised under the terms *layer*, *stromata*, or *receptacula*.

Some of the last more simple forms may be regarded as transitions to simple sporophores, being indeed aggregates of these and exhibiting a more or less compact and characteristic union, which may however vary in one and the same species. Such are the *gonidial layers* of Cystopus and Hypochnus centrifugus, Tul. The gonidiophores of Penicillium are sometimes single hyphae, sometimes are united together into tufts to which Link gave the name of Coremium; and the same is the case with the gonidiophores of the insect-killing Sphaeriaceae, in which the club-shaped branching tufts, which are often of considerable size, are known by the name of Isaria[3].

But by far the largest number of compound sporophores, and it is with these that we are chiefly concerned here, show much more constant and more distinct differentiations. Amid the great diversity of individual peculiarities one character may be regarded as almost universal, namely, that a compound sporophore produces its characteristic organs of reproduction (spore-mother-cells) in large quantities, and that they are grouped in a definite manner and at definite spots upon it. The spore-mother-cells form there continuous strata or aggregates of some other shape, either by themselves or accompanied by accessory organs usually termed *paraphyses*. These aggregates are conveniently included under the general name of

---

[1] Fresenius, Beitr. t. III, V.—Corda, Prachtflora.—Coemans, Spicilège, No. 8.—Woronin, Beitr. III, t. VI.

[2] Fresenius, Beitr. t. II.

[3] Bot. Ztg. 1867, 1.

hymenium or *sporogenous layer*, and are thus distinguished from the rest of the sporophore. At the same time descriptive mycology, following convenience and tradition, is in the habit of employing special appellations for the hymenia of the several orders and reserving the word hymenium for the Hymenomycetes.

The structure of the hymenia will be described in later chapters. Many points also in the structures and especially in the development of the sporophores, in the narrower sense of the word, must be reserved for future consideration, partly because a comparative examination of their first inception presupposes a previous discussion of sexual relationships, partly because we have frequently to deal with facts which are characteristic of single divisions, and which it will be more convenient to discuss when we are engaged with these in Division II. Meanwhile it may be well to notice here a few phenomena of very general occurrence.

It is only in certain cases that a compound sporophore begins as a terminal or intercalary portion of a single hypha, which portion then developes into its ultimate form by successive cell-divisions in every direction and by further differentiations and growth in definite directions, somewhat after the manner of the anther of a Phanerogam, if such a comparison is admissible. Some *pycnidia* among the Pyrenomycetes (see Division II) show this exceptional behaviour.

The general rule here, as in the development of sclerotia and mycelial strands, is that the sporophore proceeds from the union of branches of the hyphae, and grows by the elongation and branching of these according to a general plan and in directions determined by the species, and that new hyphal branches are introduced between those previously formed in agreement with the original design. This earliest stage, which may be called the meristematic stage, and in which new segments and new hyphal branches are added, is succeeded in every section of the sporophore by a stage of increase of volume of the existing elements and of their permanent differentiation, the amount of which is very different in different cases, and reaches its highest point in the Gastromycetes and especially in the Phalloideae.

The hyphal branches which form the compound sporophore originate in some cases in a single branch of the mycelium, which may have the morphological significance of an archicarp or homologue of a female sexual organ with its immediate supporting structure, as in Eurotium, or have no sexual relationship, as was pointed out above in the sclerotia of Coprinus and Typhula variabilis, and in the sporophores of some species of Coprinus which were shown by Brefeld especially to be produced without the intervention of sclerotia.

In the majority of better-known cases the formation of the compound sporophores begins with the union of two or several or many hyphal branches of different origin. This is the case with some of the sporocarps of the Ascomycetes which will be described at length in Division II, with the very simple hymeniophore of Exoascus Pruni, with most of the compound sporophores mentioned above as growing from sclerotia (the various species of Peziza, Claviceps, Typhula gyrans, &c.), and the compound sporophores of Agaricus melleus which have their origin, according to Hartig[1], in the mycelial strands in the same way as the ordinary mycelial branches. Most of the Hymenomycetes which are not fleshy might be added to the list, inasmuch as their compound

---

[1] l. c. above, p. 28.

sporophores, as far as they have been observed[1], always begin their development as comparatively large compact tufts of hyphae springing from the mycelium, and we may even venture to assume that the great majority of compound sporophores take their origin, as here described, from many hyphae. At the same time it must be acknowledged that it has been found possible to follow them back to their very first beginning with perfect certainty only in the few cases which have been noticed above.

Many inconspicuous compound sporophores, such as the gonidiophores of Uredineae and the stromata of many small Pyrenomycetes, remain as it were in the stage of the tufts of hyphae and pass into their ultimate form without further remarkable phenomena of growth. But where a larger structure is produced, the course of development, amid great variety of detail, discloses two chief types, which closely resemble the two types of growth above described, for the mycelial strands on the one hand and for the sclerotia on the other. In the one type, as in the formation of sclerotia, the growth is nearly *uniform for a long time in all parts* of the structure; then comes the second chief stage, the ultimate development by internal differentiation. The compound sporophores of the Gastromycetes show this mode of proceeding in the most marked manner. In the other type the general course is *progressive*[2], just as it is in the mycelial strands or in the single hyphae, advancing in the direction of fixed spots in the surface, which are themselves pioneers in the advancing growth and maintain it by formation of new cells; as any section becomes removed from these spots growth in it ceases, and its component elements assume their definitive character. According to the form of the whole structure and of the superficial portion of it in which progressive growth occurs, this growth may be said to advance towards the apex, to be *apical* (*acropetal*), or to be *marginal*, and the peripheral progressively growing spots by another usage may be termed growing points or margins; or growth is progressive towards the whole of the free surface of the structure which bears the hymenium, as in the horse-shoe-shaped pilei of Polypori which are several years old, and in other Hymenomycetes also with various modifications and limitations.

Growth thus on the whole progressive does not preclude intercalary areas of new formation and extension from making their appearance between portions in which these processes had ceased; but the actual occurrence of these areas has never been distinctly proved in any of the cases which belong strictly to this type.

On the other hand, the combination of the two types of growth has been ascertained in a considerable number of species. We find, for instance, internal differentiation and subsequent progressive growth in the compound sporophore which is the chief product of Amanita. The young stipe of the Coprineae[3] has a transverse intercalary zone in the part just below the apex, in which there is continued formation of new cells by (meristematic) division; the velum also has intercalary growth, but all the growth of the pileus and the final elongation of the stipe is progressive. In the Xylarieae, Cordyceps, &c. the growth of the club-shaped sporophore is progressive

---

[1] Hartig, Zersetzungserscheinungen d. Holzes, p. 21 (Polyporus annosus), p. 32 (Trametes Pini), p. 41 (Polyporus fulvus), p. 50 (P. mollis), p. 98 (Hydnum diversidens), &c.

[2] Goebel in Arbeiten d. Bot. Instit. zu Würzburg, II, 354.

[3] Brefeld, Schimmelpilze, III.

(acropetal), and the perithecia are subsequently formed in them by internal differentiation; other instances might be given.

Some compound sporophores with acropetal growth are normally and often uniformly branched, as in many Clavarieae, the Xylarieae, and Thamnomyces. The ramifications appear always to arise as bifurcations of the growing points, but this has not yet been sufficiently investigated. Inquiry is also needed in the cases of supposed monopodial and always slightly irregular branching seen in Agaricus racemosus, Isaria brachiata, and some others. Peziza Sclerotiorum often has irregulary disposed exogenous branches on the stalk-like portion of a compound sporophore, especially on the part near the ground. We are not concerned here with branching which is purely adventitious, artificially excited, or monstrous.

The duration of growth under normal and favourable conditions differs much in different species. In the small delicate fleshy Coprini the whole process, from the commencement to full ripeness and decay, may be completed, according to Brefeld, in 8–10 days, while the solid woody Polyporeae maintain a progressive growth for years; Trametes Pini, for example, according to Hartig for 50–60 years. There are all possible intermediate cases between the two extremes. Long-lived species necessarily advance and remain stationary periodically with the change of the seasons.

Apart from this latter influence, the course of growth under uniformly favourable conditions is marked by the same order in the Fungi as in the higher plants. First there is the laying down of new parts by formation through division of meristem of new tissue-elements accompanied by a small increase in volume; then the differentiation of the tissues takes place, and lastly the final elongation and increase of volume. The first two operations are performed slowly and steadily at least relatively to the third, the one passes gradually into the other; and the transition to the third stage is also gradual in the forms which are not fleshy, such as the Xylarieae and the leathery and woody Hymenomycetes. But in the forms with fleshy succulent substance, especially the Hymenomycetes and Phalloideae, the transition to the third stage is often abrupt, and this stage itself is traversed with relative and absolute rapidity. Of the 8–10 days occupied by the growth of the small Coprineae mentioned above the last tenth at most was devoted to the final elongation and expansion, the former stages requiring 7–9 of the 8–10 days. In the case of many other fleshy Hymenomycetes, such as Amanita, the time necessary for the first two stages is much longer than this, and Schäffer is not much beyond the mark in estimating it at a year in Phallus impudicus [1]; but more exact determinations are desirable. But the final elongation and expansion are accomplished in all these cases, under favourable circumstances, in a few days at most. The proverbial rapidity with which the succulent Fungi shoot up from the ground, as is the case with the green vegetation in spring, is chiefly due to the abruptness of the final extension of the structure which has long been in existence and slowly and gradually developed.

It has in no case been shown with perfect certainty that the peripheral extremities of the hyphae, which take the lead in progressive growth, remain the same during the whole process, and that the compound sporophore therefore is built up of the united branches of a constant number of monopodially branched primary

---

[1] Der Gichtschwamm mit dem grünschleimigen Hute, Regensburg, 1760, p. 7.

hyphae destined from the first for the particular structure. In small compound sporophores with an apex which continues narrow, being composed of only a few hyphae, like those of Typhula to be noticed presently, it may, though it need not necessarily, be assumed that such is the case. In the much more frequent cases in which the advancing apex or margin becomes constantly broader with uniform thickness and separation of its elements, new hyphal branches must be introduced one after another between the original ones, or take the place of them in sympodial succession. In long-lived species with periodical cessation of growth in the cold or dry season most of the extremities of the branches must die away, and be replaced when growth begins again by branches of deeper origin which thrust themselves between them. The position of the dead extremities or the portions of new growth limited by them may then be seen as zones in the older structure.

The consideration of special phenomena of development is deferred to a later page for the reasons assigned above. But it will be well to give a more detailed description in this place of some examples at least of progressively growing compound sporophores, because this mode of growth is very general both in the different divisions of the Fungi and in sporophores of very different morphological value. The examples are for the most part those of the first edition of this book.

1. The stalk-like compound sporophores of **Typhulae** which form sclerotia, Typhula variabilis especially, begin on the sclerotium as the bundle of firmly united parallel hyphae with their extremities curving dome-like towards each other, which was noticed above on page 38. The compound sporophore increases in length. The united extremities of the hyphae in the dome-shaped apex continue all the while very delicate and full of protoplasm, and comparatively small-celled. As the segments of the hyphae are further removed from the apex of the growing sporophore they increase steadily over a certain distance in length, breadth, and thickness, and the whole structure increases in compactness and firmness; no further augmentation takes place at its base. From these facts it appears that the growth in length of the sporophore, so far as it depends on formation of new cells, takes place at and close beneath the apex by the apical growth of the united hyphae; this is therefore the growing point. Then the cells produced at the growing point elongate in the order in which they are formed and assume their ultimate form. As the elongation commences the primordia of scattered unicellular hairs make their appearance as branches on the superficial hyphae of the lower sterile portion, and on the upper part the dense weft of the hymenial layer. At length the activity of the growing point ceases and with it the growth of the whole sporophore. In the interior of the parts more remote from the growing point there appears to be no further formation, or at any rate no considerable formation of new cells, either by division of previously formed cells or by addition of new hyphal branches introduced between those already in existence[1].

2. The compound sporophores of **Sclerotinia Sclerotiorum** (Fig. 22), the early stages and special structure of which will be again described in Division II, burst forth as cylindrical bodies from the sclerotium, grow in this form to a length of 10 mm. more or less, and then increase in breadth at the apex in such a manner as to pass through the shape of a club into that of a stalked funnel-shaped cup, which may finally have its margin turned outwards. The young cylinder consists chiefly of a bundle of nearly parallel hyphae; the slender delicate-walled extremities of the hyphae

---

[1] See also Brefeld, Schimmelpilze, III.—Reinke u. Berthold, Die Zersetzung d. Kartoffel durch Pilze, p. 58.

are in the apex of the structure and form its growing point, in which growth in length continues, while it dies out in the parts below it as these become successively further removed from it and the cells of the hyphae have grown longer and thicker. The parallel arrangement of the hyphae is not everywhere maintained; firstly, a number of short hyphal branches bend round obliquely upwards and outwards close below the apex, and terminate in the free lateral surface, where they form a cortical layer to the compound sporophore; and secondly, in the first stages of the development, when the sporophore is scarcely visible to the naked eye, the hyphae surrounding the longitudinal axis are somewhat more loosely united together than in the periphery and have their extremities slightly curved towards the axis. In this state the primordial sporophore grows to a certain length by continued apical growth in a straight line. Then the growth in length of the axile hyphae is retarded while that of the peripheral hyphae advances rapidly. In this way a narrow canal, which can only be seen under the microscope, is formed in the apex of the cylindrical sporophore with its upper margin bounded by the slightly incurved extremities of the hyphae. And while growth constantly proceeds in the direction of this margin, successive formations of new and similar elements take place in it, and behind it intercalary additions (which however gradually cease to appear), together with the first formation and

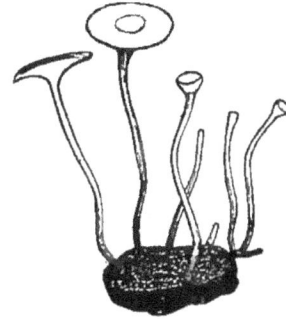

FIG. 22. *Peziza (Sclerotinia*, Fuckel) *Sclerotiorum.* Lib. Sclerotium with emerging sporophores of different ages. Nat. size.

development of the hymenium and other parts: thus the cylindrical body gradually assumes the form of the stalked funnel. The final growth in thickness of the stipe, which is however only slight, occurs chiefly in the periphery, and as the axile hyphae share but little in it, a narrow canal is formed which traverses its length.

In other species of stalked **Peziza,** P. nivea for instance, I have not myself observed the first stages in the formation of the cups; they are not however difficult of observation, and several accounts have been given of the way in which they grow for a time by formation of new elements in their originally involute margin, and at length assume their final form by an expansion of their tissue-elements advancing in the direction of the margin.

3. The sporophore of **Stereum hirsutum** (Fig. 23), which is described as a pileus divided in half without a stalk and laterally attached, is generally an irregularly roundish flat disk, the larger part of which stands out at right angles from the substratum, while the other and often very small part is firmly fastened to it; if the substratum

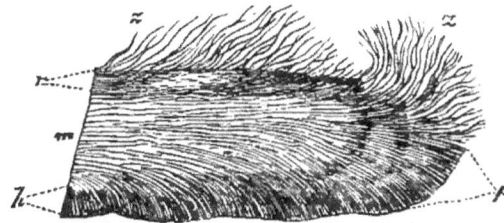

FIG. 23. *Stereum hirsutum,* Fr. Vertical radial section through the margin of a fresh pileus slightly magnified and giving a partly diagrammatic representation of the course of the hyphae; *p* the advancing margin with two zones behind it, *h* the hymenial layer, *m* medulla, *r* rind, *z* covering of hairs.

is vertical, the projecting part of the Fungus has a horizontal direction, its upper surface is thickly covered with hairs, and on its under surface is the hymenium. We need not here take notice of other and more irregular forms which are of frequent occurrence.

The sporophores first appear in the form of semicircular gray tufts of hyphae 1-2 mm. in breadth. These are formed on stout mycelial filaments which spread in large numbers through the dead wood inhabited by the Fungus. The tufts are formed of numerous hyphae, which spread from a central point with tolerable regularity like the spokes of a wheel. They are closely interwoven at the centre, but separated towards the circumference by constantly increasing interspaces, and thus the surface is covered with spreading hairs. These appear under the microscope to be colourless or of a uniform brownish colour, while the hyphae of the central tissue is coloured by granules of a reddish-yellow pigment. As the development proceeds the lower half of the hemispherical body, lower that is in relation to the substratum which is supposed to be vertical, takes a reddish-yellow tint and its surface becomes smooth and velvety. Thin radial sections following the direction of the hyphae show that, as far as the last-mentioned character extends, numerous hyphae, most of which contain reddish-yellow pigment-granules, have grown from the central weft to the surface and have thrust themselves in large numbers everywhere between the previously formed hairs and enclosed them. The upper half of the hemispherical compound sporophore retains its original character. Now begins a vigorous longitudinal and apical growth of the hyphae which run into the margin of the reddish-yellow under surface of the young pileus, while those which terminate in its middle portions elongate but little or not at all. Hence the upper surface becomes concave and the horizontal portion of the pileus raises itself from the substratum, while the growth of the hyphae advances at its margin. Sections show that the margin consists of a massive and compact layer of truncated rather thick hyphal extremities, which incline slightly towards the under surface and usually contain reddish-yellow pigment-granules. These extremities join on to the perfect hyphae of the pileus close to the point of origin of the latter ; these perfect hyphae being distinguished from them by their pellucid contents, but not by greater thickness, and running in radiating lines parallel to the surface of the pileus. The differentiation of the tissue of the pileus begins close behind the advancing margin, and results in a lower colourless medullary stratum, and an upper thin rind-stratum distinguished by membranes of a clear brown colour. Numerous hairs begin to be developed on the upper surface nearer the margin, and the hymenium on the lower surface. The former are simple stout hyphal branches which either spread or are curved backwards; the outermost of them project beyond and mostly cover the growing margin. Numerous branches run obliquely and with a slight curve close behind the margin towards the hymenial surface. The nearer the base of the pileus, the more numerous are the hymenial elements which are introduced between those previously formed, and the more decidedly do they assume the vertical position as regards the surface of the original constituents of the hymenium. The portion of the pileus which is attached to the substratum shows essentially the same mode of growth as that which projects from it, only the hairs on its outer surface penetrate as rhizoids into the substratum. Measurements by J. Schmitz and microscopic examination show that the enlargement of the pileus takes place only next the margin[1].

4. The unveiled umbrella-shaped pileus with central stalk of the **Agarici** (Fig. 24) appears at first on the mycelium as a small cylindrical, ovoid, or even spherical body, pointed at the upper end and consisting throughout of very delicate firmly united hyphae running longitudinally. At a very early stage, when the entire structure in the specimens which I have examined is ½–2 mm. in length, the extremities of the hyphae at the upper end spread in every direction as they grow and at the same time branch copiously. This gives rise to a small hemispherical head separated from the lower portion by a shallow annular furrow, the primordium of the pileus (Fig. 24 *a*). A vigorous growth then commences in the extremities of the hyphae which form its margin, and they constantly elongate, but at the same time retain their original thickness, and continue as closely woven together as at first ; there must therefore be a constant introduction of new branches between the earlier ones in the direction of the surface of the pileus. The hyphae which run towards

---

[1] See also R. Hartig, Zersetzungserscheinungen d. Holzes, p. 130, t. XVIII.

the apex of the pileus soon cease to lengthen; they become the tissue of the middle of the pileus, while as the margin advances the hyphae which run into it send out numerous straight or curved branches upwards and outwards, which in their turn soon cease to elongate and form the general tissue of the pileus (Fig. 24 *b*). Closely crowded branches from the under surface of the layer which runs into the margin grow at the same time and in the same centrifugal succession from a curved base perpendicular to the under surface of the pileus; these are the beginnings of the tissue that bears the hymenium and of the hymenium itself. They are at first of uniform length, and the surface of the hymenium is smooth at first, as Hoffmann rightly affirms in opposition to a former incorrect statement of mine, though it only continues so for a short time. The elongation of the hymenial hyphae which grow vertically downwards takes place in alternating radial bands in varying degree. In some it continues longer, and they project beyond the smooth surface as the trama of the lamellae, on which the hymenial elements arise in the position already described, advancing from the base toward the free edge. The hyphal extremities cease to elongate at an earlier period in the intervals between the lamellae, and become directly elements of the hymenium.

During this growth by terminal and marginal formation of new constituents, the parts at a distance from the growing point or margin enlarge by expansion of their cells, and the tissue which is at first uniform is differentiated at the same time into the several layers of the mature sporophore. It is readily observed that this process of expansion also advances in the stipe from below upwards, and in the pileus from the centre to the margin. To this expansion of the originally very small elements to several times their former size is due in great part,

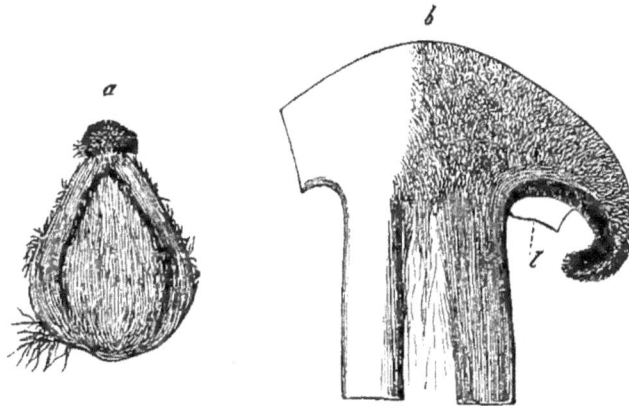

FIG. 24. *Agaricus (Collybia) dryophilus*, Bull. Radial longitudinal section showing the course of the hyphae. *a* a quite young and entire specimen 1·3 mm. in height; first beginnings of the pileus. *b* older specimen with the pileus 2·5 mm. in breadth; *l* piece of a lamella. Slightly magnified.

especially in rapidly growing fleshy sporophores, that enlargement which may be seen with the naked eye. In Agaricus (Mycena) vulgaris for instance I succeeded in determining, by measurement of the cells and counting their number on the transverse section, that the increase in length and breadth of the stipe, which becomes on an average 50–60 mm. long, from the time when its length was about 3 mm. and its cells could be exactly measured, must be almost exclusively due to extension of the cells. I obtained the same result in the case of Nyctalis parasitica; the conclusion was similar in the case of Agaricus (Collybia) dryophilus, Bull., but less precise on account of the very unequal length of cells placed at the same height. Exact measurements can scarcely be made in the pileus owing to the curvatures and want of uniformity in the cells, but here too it is evident that there is an expansion of the tissue-elements, which often exerts considerable force and advances towards the margin. It appears to me to be doubtful whether there is also any formation of new cells by transverse division of the primary cells of the hyphae and by production of new branches in parts removed from the margin. It does not take place in either of the two cases just mentioned, but they are too isolated to allow of our drawing a general conclusion from them. Branches often occur on hyphae which are much expanded or are in the act of expanding, which are little if at all thicker than the primary hyphae and are rich in protoplasm, and therefore look as though they were of recent formation. Whether they are really new branches, or only branches formed at an early period but not sharing in the expansion, must remain undecided. In the hymenium certainly new elements are introduced between the earlier ones usually for some time after its first formation.

A distinct *epinasty* prevails at first in the general growth of the pileus; the parts belonging to the upper side grow more vigorously than those of the lower, and according to the position and breadth of the annular zone which is most strongly epinastic, either the margin of the pileus is rolled inwards or the whole of the under or hymenial surface approaches or even touches the stipe; or both effects are produced, and this is most frequently the case. Subsequently, when growth is coming to an end, the epinasty changes to *hyponasty*, the under side grows more strongly than the upper, and the entire pileus expands with more or less rapidity in each separate case from its original bell-like or conical form into that of an umbrella, while the incurved margin may even become curved upwards and outwards.

We cannot enter here into the variations which occur in the development and form of the pileus in the several groups and species, and indeed we possess a very limited number of exact observations on them. The account which has just been given is founded on my own examination of Agaricus (Mycena) vulgaris, Pers., A. (Collybia) dryophilus, Bull., and Nyctalis parasitica, Fr., on the study of the history of development of Agaricus (Clytocybe) cyathiformis and Cantharellus infundibuli-formis in conjunction with Woronin, and on the works of H. Hoffmann. Hoffmann indeed makes the hyphae of the middle of the pileus in the section Mycena not run radially towards the surface and there terminate, but parallel to the surface of the pileus (for so I understand his expression 'horizontal'); and in this small point our otherwise conformable accounts differ. It is possible that different species vary in this respect. The course of the hyphae can be readily seen to be as I have stated it in Agaricus vulgaris, when the pileus is still young, but not in the older states; then the whole of the superficial tissue of the pileus assumes the character of a tough gelatinous felt, which may be removed as a coherent membrane from the pileus, and in which the hyphae have no particular arrangement.

In most of the sporophores of which we are here speaking, especially those of a fleshy consistence, growth proceeds without interruption and soon reaches its ter-mination; it may go on more slowly or be arrested for a time, if the conditions are unfavourable, and afterwards recommence; more serious injuries, especially persistent drought and cold, stop it altogether. The power of withstanding such unfavourable influences varies much in the different species. On the other hand, as has been already stated, the pileus in many leathery and woody forms, such as the Xylarieae and especially the Hymenomycetes, has the power of recommencing suspended growth with the return of favourable conditions. During each stationary period in the Hymenomycetes the hyphal extremities in the margin and upper surface of the sporophore, which for the most part die off, assume in many cases another colour and usually a darker one than that of the rest of the tissue, which is seen therefore in sections to be divided by dark lines into the zones already mentioned (Fig. 23). The tissue of the sterile surface also has often a different colour at the beginning from that which it has at the end of a period of growth; and at the commencement of the period of growth it often swells suddenly into a cushion, which runs quite round the margin of the pileus and flattens out again towards the margin with the continuation of growth. The periods of rest and growth are thus here as elsewhere indicated on the sterile surface of the Fungus, as in some other plants, by zones concentric with the margin of the pileus, and usually answer exactly to the interior zones but are sometimes less distinctly marked. It is scarcely necessary to mention examples of these 'pilei zonati,' since they occur in many of the most common and best-known species, such as Stereum hirsutum, Polyporus zonatus, P. igniarius, P. fomentarius, P. Lenzites and their allies. The hymenial side in most of these Fungi increases in circumference only as the margin of the pileus continually advances, but no increase in thickness is thereby brought about.

On the other hand there is an addition to the free hymenial surface in every period of growth in many species of Polyporus, especially in the Fomentarii of Fries, in Polyporus fomentarius for example, P. igniarius and P. Ribis, and in Trametes Pini, Corticium quercinum and allied forms. Sections therefore through older specimens show zones on the hymenial side as well as in the rest of the tissue of the pileus ; each zone answers to a zone in the substance of the pileus and is its continuation, the youngest hymenial zone being continued into the outermost marginal zone of the pileus.

Persoon[1] and Fries[2] call the zones of these Polypori annual zones. They may no doubt be correctly compared in certain points to the annual rings of Dicotyledons, but it has never been distinctly proved that only one new zone is formed each year in these plants. There is no doubt that many zones are formed in the course of a year in most of the other zoned mushrooms. J. Schmitz has shown this in detail for Stereum hirsutum, and there are a certain number of many-zoned pilei in the Hymenomycetes which only last one year.

SECTION XIII. The **structure of the mature compound sporophores** either continues to be evidently hyphal, or it becomes pseudo-parenchymatous in the sense of the word explained in section IV. When a compound sporophore is much differentiated both kinds of structure may occur in different regions of it and in different strata.

In the first case the course of the ramification of the hyphae may be quite irregular and they may interlace in every direction, as in most sclerotia; this happens in sporophores in which the growth is not peripherally progressive, and in the small structures mentioned on page 51 in which this mode of growth is at least only feebly developed, as in the Uredineae and in endophytic Ascomycetes (Rhytisma, Polystigma, Epichloe). The pileus of Morchella and Helvetia may be mentioned in this connection. But where the growth is distinctly progressive, whether apical or marginal or towards the surface, the great mass of the hyphae usually follow a course corresponding to these directions, and in large compound sporophores the surface of sections or broken pieces may often appear fibrillate even to the naked eye. The course of the hyphae in these cases either corresponds exactly with the form and direction of growth of the parts, as for example in Stereum hirsutum (Fig. 23) and in the stipes of the smaller Agarics, or the hyphae are sinuous and interwoven with a larger or smaller number of ramifications spreading in the most different directions, and forming what at first sight appears to be an entanglement of threads, such as we see for example in the tissue of the pileus of many Agarics. But sections made in the right directions will usually show that here the primary hyphae follow a course which answers to the general rule. There are however some exceptions ; a striking instance of the latter kind is found in Polyporus annosus[3], which has persistent progressive growth of the margin of the pileus and of the hymenial projections, yet, except in the outermost margin of the latter, the course of the slender interwoven

---

[1] Essbar. Schwämme, p. 17.

[2] Epicris. p. 463.

[3] R. Hartig, Zersetzungserscheinungen d. Holzes, p. 21.

hyphae and of their ramifications appears to be entirely without arrangement. R. Hartig's description of Polyporus fulvus should also be consulted [1]

Compound sporophores with distinct pseudo-parenchymatous structure could only be illustrated by a number of individual cases all differing in many points from one another, but such an enumeration would be out of place here.

One feature common to most, if not all, compound sporophores of all types of structure is the more or less distinct separation of a peripheral layer from the inner tissue. Compound sporophores with much internal differentiation, among the Gastromycetes especially, exhibit many peculiarities in connection with this point, which will be noticed again in later sections in describing individual cases. The separation in compound sporophores with progressive growth, and also in some small ones with only slight differentiation, usually consists in the fact that an inner less compact and firm mass, which may be termed the *medulla* or *medullary mass*, is surrounded in the parts which do not directly bear the organs of reproduction, as in the sclerotia, by a peripheral *rind* or *cortical layer*, in special cases termed also the *pellicula* or *cutis*, which is the outer boundary of the whole structure. When the compound sporophore forms organs of reproduction directly on its surface, the hymenial layer takes the place of the cortical. Both the medulla and the rind may be separated again into subordinate layers.

The rind is distinguished from the medulla either by the structure, size, and firmness of union alone of its elements, their arrangement (the fibrillation) being similar in both, or their arrangement also is different.

In the first case the rind is usually of a firmer texture than the medulla owing to the less breadth and closer union of its elements. This is its character in very many fleshy or cartilaginous Mushrooms, such as the larger Clavarieae, Calocera, many Agaricineae and Pezizeae, and in the stroma of Rhytisma. The cells of the rind have also not unfrequently coloured sclerosed walls, which are wanting in those of the medulla, as for instance in Peziza hemisphaerica, Rhytisma, Stereum hirsutum, &c. In other forms the rind is distinguished from the medulla by gelatinous cell-walls, as in the pileus and stipe of Agaricus (Mycena) vulgaris, in the pileus of Russula integra, in Panus stypticus, and many other Agaricineae, the outer covering of which is a tough gelatinous felt, while the interior tissue is not gelatinous. A different arrangement of the elements of the rind from that of the medulla occurs frequently in compound sporophores with hyphal structure; the hyphae of the medulla follow in their course the form of the sporophore, but numerous curved branches with their convexity towards the apex pass off from its hyphae in the direction of the surface, where they terminate in copious ramifications and close union with one another. The extremities themselves either form a tangled weft, as for instance in Auricularia mesenterica and species of Polyporus, or else they are placed perpendicularly to the surface, so that the rind appears to be formed of palisade-like cells or cell-rows, as in Peziza Sclerotiorum, in the large-celled tissue of the surface of the stipe of Helvella crispa and H. elastica, in the outer and inner surface of the hollow stipe of H. esculenta and Guepinia contorta [2], in the smooth surface of the pileus of Polyporus lucidus (Fig. 25), and in that of P. fomentarius.

---

[1] R. Hartig, Zersetzungserscheinungen d. Holzes, p. 40.
[2] Dacryomyces contortus, Rabenh. Herb. Mycol. Nr. 1984.

A very large majority of Fungi have spreading *hairs* on their surface, which arise as branches from the hyphae of the compound sporophores and show this origin even where its final structure is pseudo-parenchymatous. Some of them come from the hyphae of the surface itself, some originate at a greater or less depth beneath it and pass obliquely to the outside through the layers of tissue that cover their point of origin. They are simple cells or cell-rows and branched or unbranched, and scarcely yield to the hair-formations of the higher plants in variety of form, direction, size, colour, structure, and thickening of their membranes; the most varied series of these formations is to be found in Peziza and the allied Ascomycetes, and in Erysiphe and Chaetomium.

In many cases the hairs are closely combined in tufts, which appear to the naked eye according to the species as bristles, scales, or warts, as for instance on the surface of the pileus of Polyporus hirsutus and P. hispidus, of Hydnum auriscalpium, Tremellodon gelatinosus, &c., or as cylindrical tufts expanding into the shape of a funnel at their extremity, such as are found on the sterile surface of the pileus of Fistulina hepatica, and from their shape were once described as rudiments of the tubuli of the hymenial surface [1] If the superficial felting of hair is very thick, it may be a question whether it should not be considered to be a cortical layer, and the determination must rest on what is suitable in each particular case.

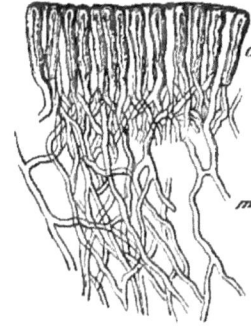

FIG. 25. *Polyporus lucidus.* Thin longitudinal section through the surface of the pileus; *c* rind, *m* medullary hyphae. Magn. 190 times.

Where the compound sporophore is very close to the substratum, single hairs or tufs of hairs often assume the character of *rhizoids*.

Here would naturally be the place to speak of the Lichen-thallus and especially of its fruticose and foliose heteromerous forms; but it will be more convenient to reserve this part of the subject to Division III.

CHAPTER III. SPORES OF FUNGI.

SECTION XIV. The propagation of the Fungi, in the widest sense of that word which implies the production of new bions through a mother-individual, is generally effected by the abjunction, and in most cases by the complete separation, of cells from the maternal structure, which then develope into daughter-bions if the necessary conditions are present. The single cell thus abjointed from the mother and capable of this development into one or more than one bion we term here a *spore*; empirically we fix that moment and condition of its development, in which abjunction from the mother as its nutrient source is effected, as the moment and condition of its *ripeness* or *maturity*; the commencement of the further development of the ripe spore is its *germination*.

---

[1] Fries, Syst. Mycol. I, 396.

It is a universal histological law that every spore is the daughter of one or some-times more than one mother-cell, which is consequently called the *spore-mother-cell.*

There are many and great differences between spores and their mother-cells in respect of their special qualities, structure, and mode of formation, and in respect of their position in the life-history of the species and the homologies which result from it, and several different kinds of spores may be formed in the course of development of a single species. Hence there are several categories, kinds, or forms of spores and spore-mother cells, that may be distinguished according to these different points of view. In their terminology these distinctions are expressed sometimes by adjectives, sometimes by compounds of the word spore, as *swarm-spores, ascospores,* &c., some-times also by special words, *gonidium, ascus, basidium,* and others. Each of these terms signifies a spore or its mother-cell in the general meaning of the word above indicated, with a definite specific reference. A fuller explanation of the terms and an account of the reasons for their adoption are reserved for Chapter IV.

The distinction between spores and their mother-cells on the one hand and vegetative cells on the other is naturally drawn first of all from cases in which the differences are most distinctly marked, and these constitute the large majority. It is to be expected that in a large and much graduated series of forms some would occur in which those differences would be less marked, sometimes indeed be almost obliterated. As examples of this may be mentioned the vegetative form described on page 4 as Sprouting Fungus, in which each sprout may be quite rightly termed a spore in the above acceptation of the word, and in the *gemmae* formed by the abjunction of cells with the power of germination from the vegetative hyphae in the Mucorini, Tremellineae, and Ascomycetes, which will be subsequently described. No confusion would be caused by a consistent use of the word spore in these cases. Whether it would be convenient that other terms should be introduced in its place is a matter to be determined by judicious agreement in each case.

Many of the peculiar characters of spores and many of the phenomena attending their formation and ripening recur in the most different groups of the Fungi, and would necessarily be included in a general survey. Others again are confined to individual groups and can only be fully discussed with them. At the same time we shall, I believe, get a clearer view of the whole subject if the second series of characters is considered along with the first, and those points only are reserved for special description in Division II which are quite irreconcilable with such general treatment ; among these are especially questions concerning homologies and sexual relations which are still in many respects obscure and debatable and must be considered in each individual case.

1. DEVELOPMENT AND SCATTERING OF SPORES.

Section XV. According to our present views on the origin of cells, every cell is the daughter of a mother-cell, and except in the cases of conjugation and rejuvenescence is formed by a process of division which takes place in the mother-cell [1]. Either all

---

[1] I pass over for the present the conjugation and coalescence of cells for reasons of convenience which have been partly intimated above; this subject will be considered in Division II. Reju-venescence I exclude because it is not the formation of a new cell, but only the transformation of a previously existing cell.

the constituents of the mother-cell enclosed within its cell-wall take part in this process, and the whole of the mother-cell is parcelled out into daughter-cells, or one portion of protoplasm only including the nucleus is separated from the other parts and applied to the formation of daughter-cells, while the remainder is not used for this purpose, but is subsequently turned to account in some other way. The former process bears the traditional name of *cell-division*; the latter, from an original misapprehension, is termed *free cell-formation*. The expressions *total* and *partial division* would better describe the phenomena as they are understood at the present day.

Both modes of cell-formation occur in the formation of the reproductive cells which we are considering; in the majority of cases the division is total. Asci afford beautiful examples of the so-called free cell-formation. In some cases, as in the formation of the spores of Mucor and the Saprolegnieae, it may be a question to which of the two types they belong. Excepting such doubtful cases, all cases of total division are cases of bipartition with formation of firm partition-walls. Of the details of the process of division we know no more than we know in the case of the division of vegetative hyphae; we can only say that a partition-wall makes its appearance.

The following are the chief modes of formation of spore-mother-cells and spores, distinguished from one another partly by the differences which have just been indicated and partly by peculiarities of form which present themselves in the process of division.

SECTION XVI. 1. **Intercalary formation** (Intercalare Bildung). A delimitation (Abgrenzung) takes place of portions in the growing hyphae, and the cells constituting them become distinguished by their form and structure, acquire the characters of spores or spore mother-cells, and at length become free by the gradual decomposition or swelling of the parts that support or connect them. Their position is not uniform in the cases that are known to us.

Normal formations of this kind are the resting spores of Protomyces, Cladochytrium, the spores of Entyloma, and the not infrequent cases of the formation of gemmae already briefly mentioned [1].

2. **Acrogenous abjunction** [2] (Acrogene Abgliederung). Delimitation of the extremities of hyphal branches with limited growth takes place by means of transverse septa for the purpose of forming reproductive cells, which are therefore placed, at least during their formation, at the apex of a stalk or sporiferous structure named since Léveillé *basidium* or *sterigma* (*ascus suffultorius* of Corda). In more highly differentiated species the basidium is the peculiarly formed terminal cell of a hyphal branch, and the spores are frequently the extremities of slender stalk-like ramifications of this cell. In this case the actual stalks of the spores are termed sterigmata in a narrower

---

[1] [See the sections in which the Mucorineae and Tremellineae are described.

[2] The distinction between ‘Abgliederung’ and ‘Abschnürung’ (see section XVII) finds no expression in English botanical terminology, whilst the idea implied in both has been rendered by the term ‘abstriction.’ If this term could be restricted to what is designated by ‘Abschnürung,’ it might be retained as satisfactory; but unfortunately it has come to be used so generally in the sense of ‘Abgliederung,’ that to avoid confusion it is altogether discarded in the text and the term ‘abjunction’ (with the verb to ‘abjoint’) is introduced as rendering ‘Abgliederung,’ and ‘abscision’ (with the verb to ‘abscise’) is used to express ‘Abschnürung.’]

sense of the word and the cell from which they spring is the basidium. In species of more simple character both expressions are used according to convenience. Among the manifold variations in individual cases which must be left for special description there are at the same time a number of generally recurring phenomena according to the mode of abjunction, the numerical relations, and the ultimate shedding of the abjointed portions.

As regards the *form* which may be exhibited by the phenomenon, the cross septum may appear beneath the apex of the sporiferous cell, the apex itself being usually expanded: the portion thus delimited is the spore; the breadth of its base is about equal to that of the sporiferous cell. The simplest examples are most uredospores (Fig. 26) and the teleutospores of Uromyces. A second case is that in which branches grow at certain points from the sporiferous cell and these are either abjointed at the point of insertion, which usually becomes much constricted after the manner of the

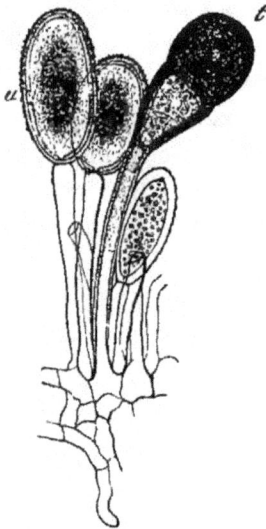

FIG. 26. *Puccinia graminis.* Small piece of a hymenium; *u* uredospores with four germ-pores in their equator, *t* a pair of teleutospores, the upper with a germ-pore in its apex. Magn. 390 times.

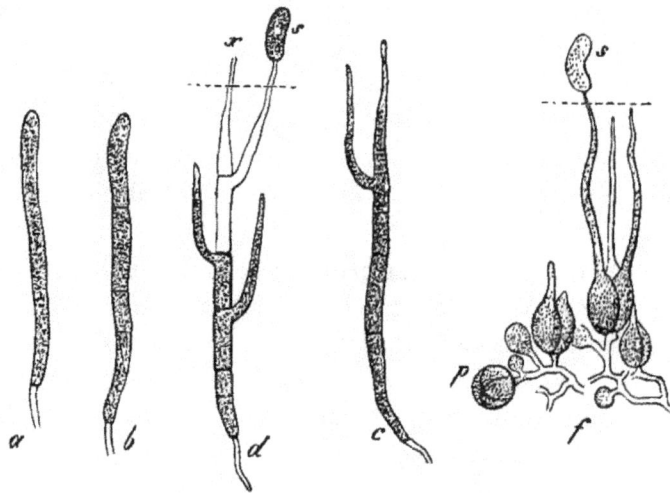

FIG. 27. *a—d Auricularia Auricula Judae.* Development of basidia and spores; successive stages of the development according to the letters. *a* cylindrical terminal cell of a hypha from which several basidia are formed by transverse division *b*; each of the basidia puts out a long narrowly conical sterigma (*c, d*) from its upper extremity, and the swollen apex of the sterigma is abjointed as a spore *s*; *x* a sterigma from which the spore has fallen. *f* the development of the basidia of *Exidia spiculosa*, Sommerf.; four basidia are formed from the cell *p* by divisions crossing one another in the cell; the other parts of the figure show younger and later states; *s* a spore. The dotted lines indicate the surface of the hymenium. *f* after Tulasne highly magnified. *a—d* magn. 390 times.

sprouts in the species of Sprouting Fungi (p. 4), or they elongate into slender stalks, sterigmata in the narrower sense mentioned above, and their swollen extremity forms by abjunction a spore. Examples, to be again noticed, are to be found in the Basidiomycetes, in Eurotium, Penicillium, Haplotrichum, Peziza Fuckeliana, &c. Intermediate cases occur, as might be expected, between the extremes and require no further description.

A sporiferous cell or basidium may produce only one reproductive cell by acrogenous abjunction, or several, even many, may be formed. The first is the case in most species of the former of the two categories just mentioned, for instance in the uredospores of Puccinia, Uromyces, and others. The basidia of Entomophthora are examples of the second case, and those of most species of Tremella, Exidia, and Auricularia Auricula Judae with long sterigma-shoots, the swollen apex of which becomes a spore by abjunction (Fig. 27).

The acrogenous abjunction of the greater number of propagative cells is either *simultaneous* or *successive*. It is simultaneous when a number of shoots make their appearance at the same time at the apex of the basidium, grow with the same rapidity, and experience abjunction at the same time, either at their point of insertion or beneath their apex which is borne by the stalk (sterigma). The protoplasm of the basidium is used up in the process and is not again renewed. Simultaneous abjunction is especially characteristic of the basidia in the hymenia of most Basidiomycetes (Hymenomycetes, Gastromycetes (Figs. 28, 29), Calocera, Dacryomyces). The basidia in these are generally club-shaped terminal cells of hyphal branches, and spores are abjointed at their broad apices, usually as the extremities of long sterigmata, more rarely as sessile sprouts of elongated or rounded shape, as in Geaster hygrome-

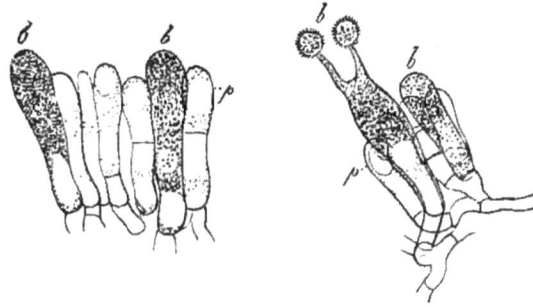

FIG. 28. *Octaviania carnea*, Corda. Thin sections through the hymenium. *b, b* basidia, ooe of them with two spores in the act of formation, *p* paraphyses. Magn. 390 times.

tricus, Scleroderma, Polysaccum, and Phallus. The number of spores is in the large majority of cases 4 to each basidium, in some cases 2, as in Calocera, Dacryomyces, some species of Hymenogaster and Octaviania, in a few cases 6–9, as in the Phalloideae, Geaster, and Rhizopogon. Slight variations from these regular numbers are not unfrequently found, especially in the species which do not form four spores on each basidium; in the Hymenogastreae, as in Hymenogaster Klotzschii, basidia are found which form only one spore.

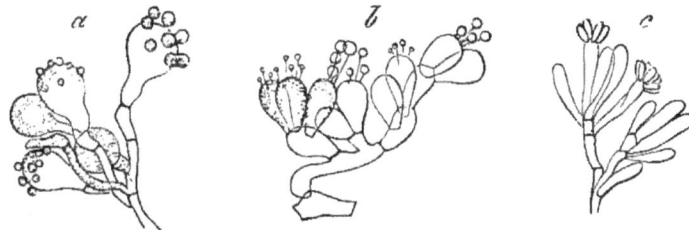

FIG. 29. Basidia of Gastromycetes oo their basidiophores. *a* basidia of *Geaster hygrometricus* with eigbt sessile spores. *b* four-spored basidia of *Lycoperdon pyriforme*. *c* four- to eight-spored basidia of *Phallus caninus*. Magn. 390 times.

Outside the group of Hymenomycetes basidia which produce many spores simultaneously occur in a great variety of forms on many gonidiophores, as in Peziza Fuckeliana, Botryosporium, Haplotrichum, and Gonatobotrys. The number of spores abjointed is normally higher in such cases than in the Basidiomycetes; they are usually placed close together on short stalks, so that we may speak of spore-heads formed simultaneously. The typically unicellular branches of Peronospora which abjoint gonidia may be included with the above, especially if we take into consideration the form distinguished by Cornu under the name of Basidiophora.

The simultaneously plurisporous basidia of the Basidiomycetes are usually more or less broadly club-shaped before the formation of spores, as has been already said ; in Calocera and Dacryomyces they are cylindrical thin-walled cells rich in finely-granular protoplasm, which either entirely fills their inner space or is interrupted by vacuoles. It may be assumed that there is a nucleus always present, though in the smaller forms it has been looked for in vain up to the present time. Where it has been observed, as in Dacryomyces, Calocera, Corticium calceum, and especially in the basidia of Corticium amorphum (Fig. 30) which become ¼ mm. in length, it is a spherical weakly refringent body (perhaps the nucleolus), lying in about the centre of the cell. It is not to be seen in the early states of the basidium, and it disappears when the formation of sterigmata commences. More exact investigation into its behaviour in spore-formation has yet to be made. When the basidium has reached its full size, the sterigmata make their appearance on its rounded apex as narrow subulate sprouts, and when they have arrived at a certain length their extremity, which up to this time is finely pointed, swells into a vesicle which gradually acquires the form, size, and structure of the mature spore. As the spore advances to maturity the protoplasm of the basidium passes into the swollen extremities,

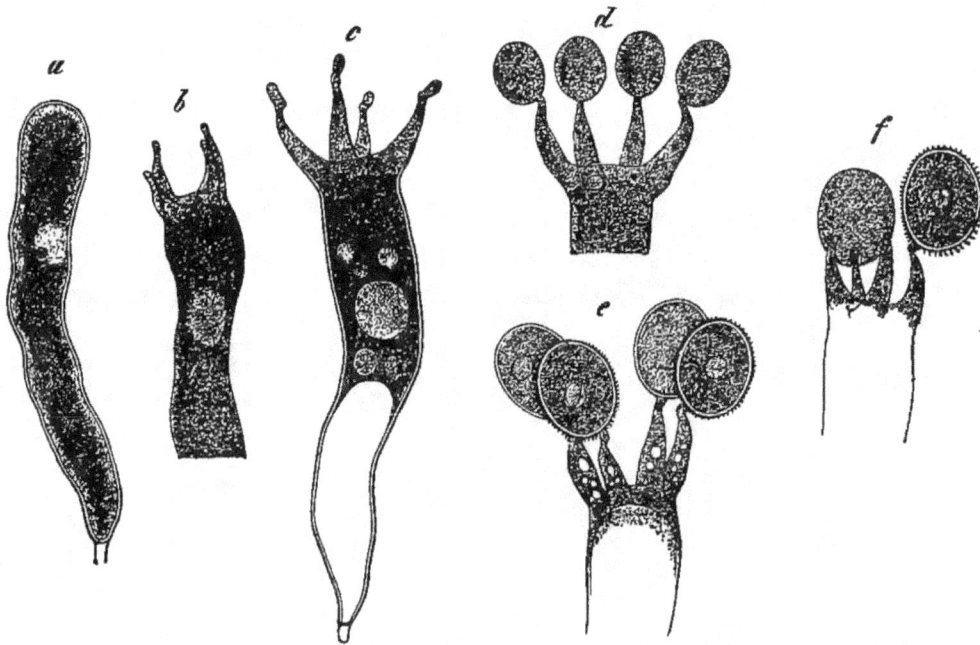

FIG. 30. *Corticium amorphum*, Fr. Development of the spores, the successive stages being in the order of the letters. *a* a nearly mature basidium with cell-nucleus. *f* basidium with two ripe spores, two others having already dropped off. Magn. 390 times.

and at length, when the spores are nearly matured, the delimitation of them by cross septa takes place ; the basidium has by this time given up the largest part of its protoplasm, but retains a thin parietal layer and is still turgescent. A clear central spherical portion may be distinguished in the young fresh spores of many species ; it remains to be determined whether this is a nucleus and has proceeded from the nucleus of the basidium.

The point of abjunction of the spore is either exactly at the apex of the sterigma or a little below the apex at a bend turned outwards (Fig. 30), so that the spore when detached takes with it the apical portion of the sterigma as a short stalk, or in a few cases, as in Bovista and some species of Lycoperdon, abjunction occurs at the point of insertion of the sterigma, which is consequently attached as a long stalk to the spore when the latter becomes free. In the cases mentioned above, where there are no sterigmata, abjunction as far as is known takes place in the way described.

Arthrobotrys (Fig. 21) will serve as an example of another form belonging to this category.

In basidia exhibiting successive abjunction of many propagative cells the process of abjunction is repeated several times on the same basidium. There are three very distinct *sub-forms of successive abjunction*, each of which has some special peculiarities connected with it. These sub-forms may be distinguished as the *sympodial* and the *serial* or *concatenate* (Reihen-weise, Kettenweise), the latter being again divided into the *simple* and the *branched*.

In the *successive sympodial* form (Figs. 31, 32) a single acrogenous spore is first of all abjointed from the extremity of the basidium or sterigma, which is always finely pointed. Then a new protuberance sprouts forth close to the point of insertion of this first spore, pushes it on one side and occupies the extremity of the sporophore, and is there abjointed as a new spore. A like proceeding may be repeated many times; the last spore formed is always at the apex, and its older sisters are ranged on the same plane or one after another below the apex. In very extreme cases the

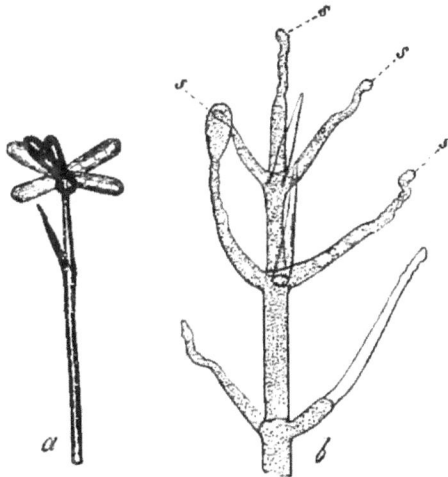

FIG. 31. *Dactylium macrosporum*, Fr. Extremities of sporiferous hyphae. *a* in a dry state with a head of spores above. *b* in water with the primordia of the youngest spores *s* at the extremities of the branches, the small unevennesses beneath being the points of attachment of the older spores which have become detached in the water. Magn. 300 times.

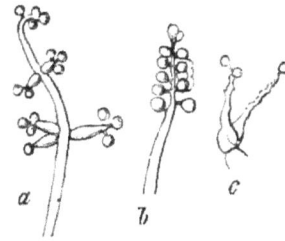

FIG. 32. *Botrytis Bassii*, Bals. *a* end of a young sporiferous hypha; short lateral branchlets have successively abjointed each 1-4 round spores. *b* end of an old branch which is producing spores by abjunction and is thickly covered with spores, the youngest of which are terminal. *c* two sporogenous branches, from which all the spores have fallen with the exception of the youngest and uppermost. *a* magn. 390, *b* about 700 times. See Bot. Ztg. 1867.

spores are very soon entirely detached and either fall off or remain adhering only to the one last formed, as happens in the formation of the gonidia of Epichloe typhina, in Claviceps and in the forms known by the name of Acrostalagmus of Corda. In other cases, as Botrytis Bassii and the small gonidia of species of Hypomyces and Hypocrea[1], each spore arises at least far enough above its predecessor for the points of insertion to occupy some space, and the spores therefore remain united into *sympodial successive heads*. If the spores are detached, their insertions form slight projections or even short stalks at the extremity of the sporophore (Fig. 32). If the sporophore were to elongate between each pair of spores a sympodial arrangment would be produced, like that of Phytophthora in Fig. 20.

---

[1] Verticillium agaricinum and its allies, Trichoderma viride, &c. See Tulasne, Carpol. III.

In the *simple successive serial* or *concatenate* forms the abjunction is repeated beneath the insertion of each propagative cell in the same direction and in the same form as in the case of the first cell. If the line of abjunction in that case was broad and transverse, the extremity of the sporophore beneath the youngest spore elongates to a definite extent and abjunction again takes place by formation of a new transverse wall; if the first sprout which becomes a spore has a constricted insertion, then after each abjunction a similar sprout with constricted base is developed beneath it from the persistent end of the sporophore and in turn undergoes abjunction. In this way a chain of similar segments is produced, in which the cells are younger in proportion as they are nearer to the extremity of the sporophore from which they are

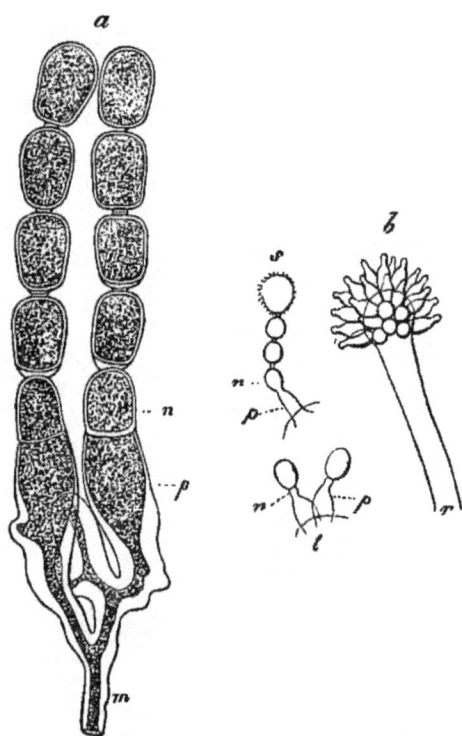

formed. The number of cells in a chain may be considerable, 20–30 or more. Examples are to be found in the gonidia of most of the Erysipheae, Cystopus, Penicillium, Cordyceps, and the Aecidieae, in the uredospores of Coleosporium, Chrysomyxa, and many others (Fig. 33).

Branched serial or *concatenate* forms arise when one or more outgrowths standing side by side on the apex of a filiform sporophore are abjointed, and then by repeated abjunctions produce a structure not unlike one of the Sprouting Fungi (Fig. 3). The first sprout-cell puts out one new protuberance from the apex which is remote from the sporophore, and this new cell and each succeeding one can do the same; a row of cells is thus formed in which the members are successively younger as the apex is approached. Each of them can then form one or more lateral sprouts below its apex which adjoins the cell next above it, and these new cells and their progeny are similar to the first (Fig. 34). According as the lateral sprouts on cells of successive orders are placed

FIG. 33. *a Cystopus Portulacae*; *m* a mycelial branch bearing two basidia which are producing gonidia by abjunction; the figure is further explained in the text on page 69. *b Eurotium Aspergillus glaucus*; *r* end of a sporophore covered with radiating sterigmata, on which the formation of spores is just beginning. *s* and *t* isolated portions showing single sterigmata *pp* with their spores; *n* youngest spore of a chain. *a* magn. 390, the rest 300 times.

singly or in a whorl of two or more members, chains are produced in which the branches vary in number and form. The cells of all orders are so many reproductive cells which are similar to one another in all important points and become ultimately detached from one another. Examples of this kind occur in the forms named by Fresenius[1] and Riess[2] Periconia, in which sprout-chains are collected together into a compact head at the extremity of a filiform sporophore, and in the small gonidial forms of Pleospora and species of Fumago and its allies in the Sphaeriaceae[3],

---

[1] Beiträge.   [2] Bot. Ztg. 1853.

[3] Tulasne, Carpol. II.

of which those named Cladosporium herbarum, Dematium herbarum[1] and Alternaria are the best known; to these may be added the delicate heads of Myriocephalum botryosporum[2] and many others.

Connected with these three kinds of acrogenous abjunction of spores is one which is less distinctly marked and which may be termed the mode of **cross-septation** (Querzergliederung). In this the terminal portion of a hypha or hyphal branch grows first of all to a certain length, and then ceases to elongate but is divided by cross septa into a number of spore-cells.

This mode of formation is seen most distinctly in the sporogenous branches of strong specimens of Oidium lactis[3] which rise into the air above the substratum. These branches have a cylindrical form and are many times longer than broad. When they have ceased to elongate they quickly divide by formation of cross septa into numerous cylindrical spores which are from one to two times longer than broad. In small specimens this cross-septation may extend over the whole plant, occurring even in the branches of the mycelium in the substratum. The formation of cross septa appears to commence in stronger individuals at the free apical extremity and to advance basipetally; but this point is quite as uncertain as the question, whether the sporogenous branch consists at first of a number of longer cells which are afterwards divided into the short members by repeated intercalary bipartition, or whether the latter are the first divisions formed either simultaneously or successively and in basipetal order in the branch which was up to the time of their formation unicellular; on these points further examination of the branches of this Oidium is to be desired. Of similar character are the gonidial mother-cells of Syncephalis and Piptocephalis[4], which spring simultaneously from the apex of the capitate extremities (basidia) of the sporophores and form a small clustered head. They have the form of elongated cylinders with rounded apices, and are divided after they have ceased to grow in length into several short cylindrical spores by transverse septa formed either simultaneously or successively and in basipetal order, but always very rapidly.

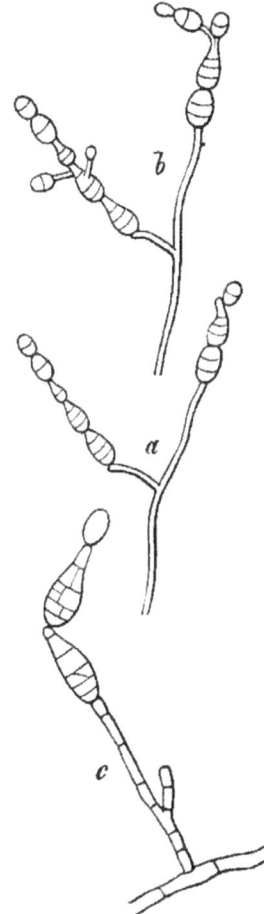

FIG. 34. Species of *Alternaria. a* and *b* extremities of a sporiferous hypha growing obliquely into the air from a specimen grown on a microscopic slide, *a* on the 4th Aug. at midday, *b* some 23 hours later; the two rows of spores which are still simple in *a* are branched in *b. c* a young sporophore on a mycelial filament submerged in water. The membranes of the pointed ovoid spores are yellowish brown where they show partition-walls in their interior and are colourless only at the upper pointed extremities. This is the case also with the youngest and still small spores, the sporophores and the mycelium. *a* and *b* magn. about 145, *c* 225 times.

---

[1] Löw in Pringsheim's Jahrb. VI, p. 494. Penicillium cladosporioides, P. viride, P. chlorinum, all of Fresenius (Beitr.), and P. olivaceum, Corda, are evidently the same form. Even if Tulasne's view that the plant belongs to Pleospora herbarum is not confirmed, its connection with one of the allied Sphaeriaceae is more than probable. [2] Fresenius, Beitr. t. V.

[3] Fresenius, Beitr.—Brefeld, Ueber Gährung in Thiel's Landw. Jahrb. V, 1876, t. II.

[4] De Bary und Woronin, Beitr. II. Brefeld, Schimmelpilze, I. Van Tieghem et Le Monnier in Ann. d. Sc. nat. Sér. 5, XVII, p. 370. See also below, sect. XLIII, Fig. 74.

The spore-formation in Ustilago and Geminella, which will be further considered in section LV, appears from Winter's observations on Geminella and Ustilago Ischaemi [1] to be nearly allied to the cases just described. More certainly is this the case with many acrogenously formed so-called *septate spores*, as those of Puccinia (Fig. 26 *t*) and Phragmidium, and many forms of Hyphomycetes, the systematic position of which has not yet been exactly determined, Trichothecium, Arthrobotrys Fusiporium, &c.; also forms which we now know as gonidiophores of the Pyrenomycetes, such as Fries' groups of the Dematieae and Sporidesmieae, Helminthosporium, for example, Cladosporium, Alternaria, Sporidesmium, Phragmotrichum, Polydesmus, Melanconium, Stilbospora, Coryneum, Exosporium, and very many other forms. See Figs. 21 and 34, and section XXIX.

Section XVII. The **inception** (Anlegung) **of acrogenously produced spores** takes place in every instance according to one or other of the processes which have now been described. In some also ripeness, that is the capacity for further normal development, and the size, form, and structure which indicate this capacity, is reached, as has been repeatedly stated above, when the delimitation is completed. This is the case, for example, in Corticium amorphum (Fig. 30) and in many, perhaps in all the Basidiomycetes, and to a certain degree in Cystopus Portulacae (Fig. 33); in the case of many other small spores attached by a very narrow stalk it is not possible to speak with certainty on this point, because the minuteness of the point of insertion renders it impossible to determine the exact moment when delimitation by the cross septum is effected. On the other hand, many cases are known in which after acrogenous delimitation the cross septum undergoes a considerable amount of growth before it is mature, and it obtains the necessary food for this purpose from the sporophore; this is the case, for example, in all the species of the Uredineae mentioned in the preceding sections, and in Eurotium and Penicillium, &c. In a rapidly growing successive chain in these species the majority of the younger members are still immature, and the nutrient material, so far as it comes from the sporophore, must pass by the younger cells to reach the older more distal ones.

Many acrogenous spores are *persistent* on the sporophore after they are mature, and are carried away from the place where they were formed only by accidental external mechanical agencies, as the teleutospores of Uromyces, Puccinia, and Phragmidium, the large gonidia of Hypomyces and many other septate forms above mentioned.

But the larger part of these spores are *detached* from the sporophore as soon as they are mature by the aid of internal causes, which during the process of ripening bring about certain changes in the original condition and thus render the ultimate separation possible. The three chief modes known to us in which this purpose is effected are the *disappearance of the sporiferous structure* (Schwinden der Träger), *abscision* [2] (Abschnürung), and *abjection* [3] (Abschleuderung).

The **disappearance of the sporiferous structure** is most common among the Gastromycetes, in which, when the spores are ripe, not only the basidia, but usually

---

[1] Flora, 1876, Nr. 10, 11.   [2] See note on page 61.
[3] See note on page 84.

also the rest of the hymenial tissue, becomes entirely dissolved by processes of decomposition not accurately known, and the spores are thus set at liberty. They lie at first in the place where they were formed; their subsequent fortunes are described in Division II. The history of the basidia, which make their appearance as branches of the simple sporophores and form gonidia in Peziza Fuckeliana ('Botrytis cinerea'), is essentially the same. They disappear entirely after the gonidia are ripe, and the latter cling in loose heaps to the place of their formation.

The process of **abscision** is the most common of the three and appears with the greatest variety of forms. Generally a transverse zone between the adjacent cells disappears or grows soft, and their separation is thus effected or made easy. The transverse zone which disappears is either a middle lamella of the cross septum between the two cells or it is a small stalk-cell, which is cut off from the young spore by a cross septum and then disappears, as in the uredo-chains of Coleosporium and Chrysomyxa and all the Aecidieae. The changes observed in the zone of separation are in one series of cases simply that it becomes gradually smaller and especially narrower and at length entirely disappears; in other cases it swells up into a jelly and becomes disorganised. The product of the swelling may in the latter case be persistent, and is then usually increased to a considerable extent by the gelatinisation of the lateral walls of the spores, which are therefore ultimately glued together by a gelatinous mucilaginous gummy substance; in other cases the products of disorganisation at length entirely disappear, and complete isolation of the spore is effected. It is natural to suppose that this process described as disappearance consists in a transformation into soluble compounds

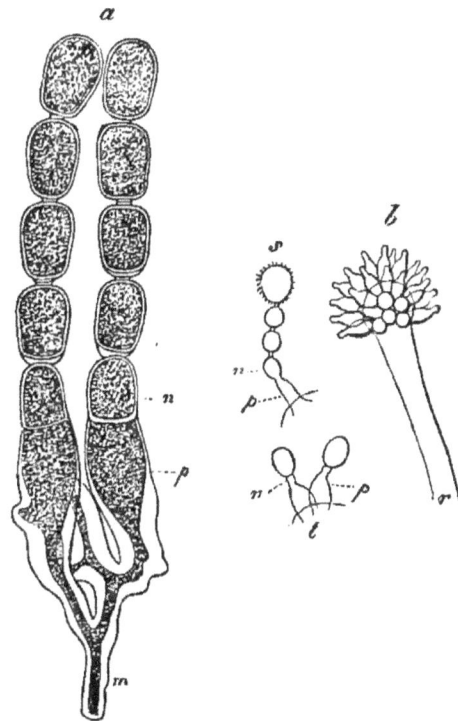

FIG. 35. *a Cystopus Portulacae*; *m* mycelial branch bearing two basidia which are producing gonidia by abjunction; the figure is explained in the text. *b Eurotium Aspergillus glaucus*; *r* extremity of a sporophore covered with radiating sterigmata, on which the formation of spores is just beginning. *s* and *t* isolated portions showing single sterigmata *pp* with their spores; *n* youngest spore of a chain. *a* magn. 390, the rest 300 times.

and a simultaneous osmotic absorption of these into the adjacent cells, especially in the many cases in which the spore about to be removed by abscision continues to grow while the disappearance is proceeding, and would seem therefore to be receiving more food. In some cases one might also suppose a process of combustion. Precise statements on these points are not possible in the present state of our knowledge.

These phenomena are well exemplified in the simple successive gonidial rows in the genus **Cystopus**, especially in C. cubicus and C. Portulacae (Fig. 35 *a*), which latter plant is more particularly referred to in this place. Delimitation of the rounded apical portion of the basidium (*p*) is effected by a broad cross septum to form a gonidium (*n*).

The septum appears first as an annular ridge on the inner side of the lateral wall of the basidium, and grows slowly into a plate of considerable thickness which is convex on the side towards the basidium and correspondingly concave on the other, and shows the bluish lustre of gelatinous membranes under the microscope. When it is fully formed the apex of the basidium elongates to form a new gonidium. The new portion thus formed is close under the transverse septum. In correspondence with the subsequently rounded form of its apex it is from the first somewhat narrower than the septum, and by its elongation it separates the septum at its margin from the lateral wall of the basidium and carries it upwards with the gonidium to which it belongs. Each gonidium accordingly has its slightly convex surface resting at first on its younger sister with the margin free, but attached to its apex by the broad middle portion. The gelatinous cross septum, to which the whole of the surface of attachment belongs, is continuous above with the lateral wall of the gonidium; and while this becomes slightly thickened as it developes, a membrane, which is not at first clearly defined, is formed on the inner surface of the septum and is also continued into the lateral wall which it resembles in

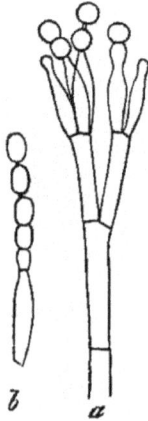

appearance; this is the persistent basal wall of the gonidium. At the same time the original gelatinous transverse septum begins to disappear from its margin inwards as if it melted away. There is now in all beyond the third and fourth youngest gonidia of a row only a quite narrow intermediate piece in the middle connecting each with its younger sister. This piece is of about the same height as the original septum, but the bluish glistening substance in it dwindles from below upwards into a small plate which becomes continually thinner and remains attached to the wall of the gonidium to which it belongs. As this process goes on the intermediate piece becomes pale and very slightly refringent, and after persisting for some time in this state at length disappears. There is no reason for regarding this delicate intermediate piece as a part of an outside membrane which covers the whole gonidial chain like a sheath, as I formerly did. The single gonidia show no important changes after they are detached from the gonidiophore beyond the thickening of their membrane already mentioned, which cannot be further pursued here.

Most of the propagative cells formed by acrogenous abjunction which have now been mentioned are detached by abscision in the manner described above, and, like the cells of Sprouting Fungi, they must become detached by disappearance of an original intermediate lamella just as is observed in Cystopus. Careful examination shows indications of this in almost all cases, but the details of the proceeding are often difficult to follow on account of the too small size of the parts, yet it can be seen very distinctly, in spite of their small size, in the successive gonidia-rows in Eurotium and Penicillium, notwithstanding their minuteness (Figs. 35 *b* and 36). Some further details may be seen in Zalewski's treatise, mentioned at the end of this chapter.

In a certain number of forms the separation is effected by the formation of a gelatinous or gum-like deliquescent substance both on the surface of separation and on the rest of the circumference of the spore. It must be supposed that this substance also is the product of an outer lamella of the spore-membrane which was not originally gelatinous; but the minuteness of the objects prevents this from being certainly ascertained. With the ordinary amount of moisture necessary for the growth of Fungi the deliquescent substance absorbs so much water that the spore when abjointed is easily detached; a drop of water washes it away at once, if kept dry it will continue to adhere to the gonidiophore. Spores abjointed close together in large numbers cohere through the coalescence of their gelatinous envelopes and form masses which break up again in water. In the case of spores successively abjointed on the free apex of one or several closely adjacent sterigmata, if the development takes place without interruption in a damp atmosphere, the gelatinous substance deliquesces and forms a spherical drop, in which the spores lie embedded as in a vesicle. And all this occurs alike in those cases where the successive abscision affects spores arranged in rows (gonidia of Nectria Solani[1]) and in those where they are in heads (Acrostalagmus cinnabarinus, gonidia of Claviceps and Epichloe). Where abscision of large numbers of spores takes place inside narrow receptacles provided with narrow orifices, their release from the receptacle is effected by the formation of a gelatinous or gummy substance which swells in water and emerges with the spores from the receptacle. Examples are to be seen in numerous gonidia-receptacles in the Pyrenomycetes. See Division II.

A description of the development of the spore-chains in the aecidium of **Chrysomyxa Rhododendri**[2] will show the mode in which the spores are shed in the Uredineae by the solution and disappearance of a stalk-cell or intermediate cell beneath each spore. The spores in each chain are formed by successive abjunction at the upper extremity of a short club-shaped basidium, from which at first an almost cylindrical spore-mother-cell is abjointed by a plane transverse septum. This cell, which is about one and a half times longer than broad, subsequently changes its shape; one side bulges considerably, the other only slightly, and the whole cell thus becomes irregularly barrel-shaped. It is then divided into two unequal daughter-cells by a plane partition-wall which runs from the angle formed by the basal cross septum and the more prominent side obliquely towards the flatter side, cutting off the lower third part of it; the lower of the two daughter-cells is a small wedge-shaped stalk-cell or intermediate cell, the upper is larger and develops into a spore. The spore is at first of a some-what complex and irregular form, as is sufficiently apparent from what has been said above and from the Fig. 37. It increases considerably in size, assuming in so doing a nearly spherical or ellipsoid figure, and becomes invested with a new membrane of considerable thickness, into the

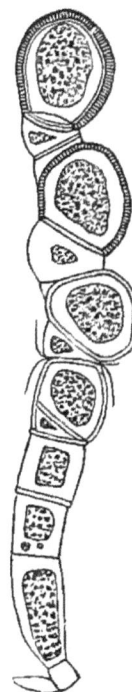

FIG. 37. *Chrysomyxa Rhododendri.* Basidium with a chain of spores from an aecidium; explanation in the text. Magn. 600 times.

structure of which we must not at present enter. The stalk-cell grows at the same time in height and breadth, remaining much lower on the side where its wedge-form thinned out originally than on the opposite and now convex side, and has an elliptic transverse section. Ultimately the stalk-cell disappears, its membrane and the outer primary lamellae of the membrane of the mother-cell and of the transverse septum swell up, become gelatinous, and finally vanish entirely with the cell-contents, and the spores are now isolated. The division into stalk-cell and spore is usually found in the third youngest spore-mother-cell on a basidium, more rarely in the fourth youngest. The gelatinous dissolution of the stalk-cell is usually far advanced in the

---

[1] De Bary, Kartoffelkrankheit, p. 41. Reinke und Berthold, Die Zersetzung d. Kartoffel durch Pilze, p. 39.

[2] Bot. Ztg. 1879, p. 803.

one belonging to the sixth youngest spore in the chain. Phenomena essentially the same occur in other species of the Uredineae, but with considerable variations in form in the different species[1].

Where filiform sporophores rise free into the air, a further mechanical arrangement is found which greatly assists the shedding and scattering of the abscised spores. It may be readily observed in the Hyphomycetes, in Peronospora, for example, Phytophthora infestans, and in the gonidiophores of Peziza Fuckeliana, &c. The hyphae of these Fungi are cylindrical in the moist and turgescent state, but collapse when dry and especially when the spores are ripe into a flat ribbon-like form[2], and the drier they are the more strongly do they become twisted round their own longitudinal axis. They are so highly hygroscopic that the slightest change in the humidity of the surrounding air, such for instance as may be caused by the breath of the observer, at once produces changes in their turgescence and torsion; the latter give a twirling motion to the extremity of the gonidiophore and the ripe spores are thereby thrown off in every direction.

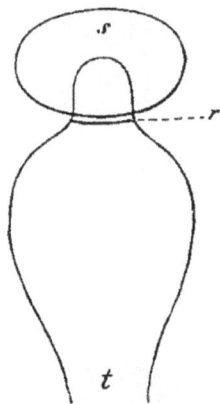

FIG. 38. *Pilobolus Cedipus,* Mont. Diagrammatical representation of a longitudinal section; *t* upper end of the sporiferous cell, *s* the cell which will be abjected; the transverse wall which forms its boundary below thrusts its convexity into the interior of the cell; *r* annular suture of dehiscence.

Abjection of acrogenous propagative cells is effected by a mechanism which we shall have to speak of again in section XXI. The cell which is to be abjected, whether spore or spore-mother-cell (for brevity we shall call it spore), is abjointed singly by a cross septum at the apex of a tubular and often comparatively large sporiferous cell, a basidium or a sterigma, which retains its parietal protoplasm still intact after the abjunction of the spore and is still turgescent in consequence of a continued supply of water in increasing quantity. Its membrane is highly extensible and elastic, and continues to stretch as the tension increases with the increased amount of water absorbed. But its cohesion is less over an annular area immediately beneath the cross septum than in any other part of the circumference, and if the tension reaches a certain point, it overcomes the resistance of the less coherent annular zone, the *suture of dehiscence*; the wall opens by a circular fissure, the pressure of turgescence is instantly relieved and the elastic wall contracts, especially in the direction of the transverse diameter, and this causes a large part of the fluid contents to be squirted out at the same moment with force through the fissure, and as it strikes full on the transverse septum, the spore that rests upon the septum is abjected with it. The basidium thus emptied collapses and perishes.

The process of abjection may be observed most completely in the acrogenously abjointed spore-mother-cells of Pilobolus crystallinus and its nearest allies, of which we shall speak again in later sections (Fig. 38). It occurs also, as Brefeld[3] has

[1] Bot. Ztg. l. c. p. 786. De Bary, Brandpilze, p. 59. Reess, Rostpilzformen d. Coniferen, Halle, 1869. R. Hartig, Wichtige Krankh. d. Waldbäume, t. IV, V.

[2] Fresenius, Beitr. t. II.

[3] Bot. Ztg. 1870, p. 161. Abhandl. d. Naturf. Ges. zu Halle, Bd. XII, 1. 1871.

shown, in the basidia of Empusa and species of Entomophthora. The ripe spores are thrown to a distance of 2–3 cm. and adhere by the remains of the ejected protoplasm to the bodies against which they strike. The ripe spores of Coprinus, especially C. stercorarius, are abjected from the basidia by the same mechanism, as Brefeld informs us[1]. They are attached, as Fig. 30 shows in the case of other Hymenomycetes, to the extremities of very slender sterigmata which spring four together from the apex of a basidium. The four spores of each basidium are abjected at the same moment, and a small drop of fluid which issues from the sterigma shows that it is open at the apex, while a small quantity is also seen to be attached to each spore as it drops. The similarity in the basidia and in the mode of formation of the spores in all the Hymenomycetes and other facts also make it probable that the process of abjection is widely spread, perhaps occurs universally, in this group of Fungi; but more extended investigations are still needed to clear up this point.

The following are some of the other facts just referred to. It has long been known that the hymenium of a Hymenomycete when turned upwards becomes gradually sprinkled over with free spores, and if it is turned downwards, spores fall from it in large quantities. Some of them fall in an exactly vertical direction, as appears from the fact that the spores which fall on a piece of paper placed under the free hymenium of an Agaric are arranged in radial lines answering exactly to the radial arrangement of the lamellae on the pileus.

These phenomena are in themselves quite compatible with simple abscision as described in the preceding pages, but they do not exclude the supposition that the spores were abjected with some small exertion of force, as Brefeld has also pointed out[2]. On the other hand a dispersion of the spores is observed in these Fungi in other directions than that of the vertical. The statement of Bulliard[3] has recently been confirmed by Hoffmann and de Seynes, that many spores fall from the hymenium of an Agaric when turned downwards far beyond the line which corresponds to the margin of the pileus. Hoffmann saw white clouds of spores rise like smoke from Polyporus destructor when there was a slight movement in the air, 'but when the air in the closed chamber was perfectly still no spores reached a glass plate hung at a distance of only three quarters of an inch above the plant, while those which fell on a glass plate two inches and a half beneath the Fungus covered nearly uniformly up to the margin a space of more than six times the circumference of the Fungus.' Other Hymenomycetes gave similar results. These observations point to abjection of the spores, but do not absolutely prove it, because the facts described might be due to movements of torsion in the sterigmata, such as were noticed above on page 72.

Lastly, it may be observed that the abjection of the spores in Leitgeb's Completoria[4] may also have been caused by the mechanism which has already been described; Leitgeb's own explanation I have not been able to understand.

SECTION XVIII. 3. **Endogenous spore-formation** (Endogene Sporenbildung). Many spores are produced inside mother-cells, the wall of which remains intact till the spores are ripe and forms a spore-receptacle or *sporangium*.

The sporangia are mostly acrogenous cells which are either persistent on their sporangiophore, or are removed from it by abscision, as in Cystopus and other

---

[1] Schimmelpilze, III, p. 65.
[2] Ib. p. 132.
[3] Champignons de France, I, p. 51.
[4] Sitzungsber. d. Wiener Acad. Bd. 84, July, 1881.

Peronosporeae; more rarely they are intercalary. Their spores are produced by division without formation of parting walls and conform to two chief types: 1. A parietal layer of protoplasm at least is not included in the division and is left behind in the sporangium, as is the case in asci. 2. No parietal layer remains behind, as in the sporangia of the Phycomycetes, which show much variation in details.

a. In the **sporangia of the Phycomycetes** the whole of the protoplasm, whether parietal and enclosing a vacuole or filling the lumen of the cell, is divided to form the spores. The number of spores directly produced by the division is not fixed in any species except perhaps in Tetrachytrium[1], and is often very large, as in Mucor, Pilobolus and large Saprolegnieae. The division usually appears to be simultaneous; but Büsgen observed under very favourable circumstances in Leptomitus lacteus and Mucor that the protoplasm was divided by very rapid bipartitions into successively smaller portions up to the final formation of spores. When the division is complete the future limiting surfaces of the spores are at first indicated by granular plates, but these are at once replaced by homogeneous delicate and narrow partition-layers, which often however become broader; these layers proceed probably from the blending of the grains and usually continue of a soft gelatinous consistence, being directly transformed into plates of cellulose only perhaps in species of Dictyuchus. In the Mucorini and in Dictyuchus clavatus the spores which lie between such dividing layers become invested at once with a firm cellulose membrane; in other species a distinct membrane does not appear before the spore leaves the sporangium. The early stages of the division in Aphanomyces show exceptional deviations from the ordinary type. No growth of the spore when it has once been separated off takes place inside the sporangium in any of the above cases.

The process of division may be observed in its greatest completeness in the sporangia of the larger **Saprolegnieae** which live in the water; in Saprolegnia, for example, Achlya and Leptomitus lacteus. The sporangium is a large club-shaped cell delimited by a transverse wall from the unicellular tubular sporangiophore. It is densely filled with a coarsely granular protoplasm, or may have a large axile vacuole. Shortly before the division the protoplasm becomes everywhere uniformly and finely granular and has small inconstant vacuoles at wide distances from each other. It is then suddenly divided by granular plates, which look like rows of granules when seen in profile, into numerous polyhedral or polygonal portions, the future spores, which in Leptomitus, as was said above, are formed by rapid successive bipartitions. The partition soon becomes more pronounced, the partition-streaks which were before granular now become homogeneous, and no longer appear as fine clear lines but grow broader as the spores are rounded off. With this the separation of the spores is complete in the cases which we are considering: the substance of the partition-plate which is derived apparently from the granules that were previously present continues homogeneous, soft and capable of swelling. Colouring reagents, as Fr. Schmitz[2] first discovered, show the presence of a number of nuclei in the sporangium as soon as it is delimited, and a division of them afterwards. Each spore obtains a nucleus, which has been directly observed to proceed in Leptomitus from the division of the original nuclei. These processes which lead at once to the formation of spores are however preceded by other separations in which the behaviour of the nuclei has not been clearly ascertained. The coarsely granular protoplasm

---

[1] Sorokin in Bot. Ztg. See also below, section LII.
[2] Sitzsber. d. Niederrhein. Ges. Aug. 4, 1879.

of the sporangium is first divided into portions resembling the future spores in number, position and size, and the division is effected by partition-plates which are at first granular in character and afterwards become broad hyaline stripes. These subsequently disappear, and the whole of the protoplasm assumes the uniformly finely granular appearance described above, and at once proceeds to the final division. The same or very similar phenomena, and among them the preliminary transitory separation, have been observed in Dictyuchus monosporus, but there in place of the final partition-plates firm cellulose membranes are formed, from which the spores subsequently escape. In Dictyuchus clavatus each spore is invested with a membrane of cellulose, but is separated from its neighbours by a thin layer of a hyaline substance, which is soft and gelatinous in water and must be formed from the partition-plates; it is still a question whether the membranes of the spores are obtained by differentiation from the plates, or are a later product from the spores themselves.

In the **Mucorini** with endogenous spore-formation which do not grow in water (Mucor, Pilobolus, &c.) the processes of division cannot be followed throughout under the microscope; but whatever can be learnt about them from dead material agrees so closely with the final stages of division in the Saprolegnieae, and especially with those of Dictyuchus clavatus, that we are justified in assuming that the process of division is quite similar. At first the spores are delimited as polyhedric bodies by very narrow partition-plates: by and bye each is rounded off and invested with its own cellulose membrane, as in D. clavatus, and separated from the others by a layer of gelatinous substance that swells in water. In some species of Mucor (M. plasmaticus of Van Tieghem) this intermediate substance is largely developed[1], occupying in the intact sporangium even more space than the spores themselves, and is finely granular. It may be doubted whether the entire mass is formed in such cases from the partition-plates; it is possible that it is exuded from the spore-forming protoplasm before the division, or comes also in part from the membrane of the sporangium (see section XX). Our present investigations still leave the point unsettled. Preliminary separations have not been observed with certainty in Mucor.

The description given above does not hold good of all the Saprolegnieae which have been examined, and is true of Phytophthora only among the allied Peronosporeae. The only point in which all agree is in the ultimate appearance of the hyaline partition-plates with capacity for swelling, and this remark applies also to the Chytridieae; some details with respect to the latter group which belong to this place will be given in section XLVI.

Aphanomyces deviates from the Saprolegnieae to a greater extent than any other of the allied forms. The spores are cylindrical, three times as long as broad with rounded ends, and lie one behind the other in a single row in the slender cylindrical filiform sporangia. At the commencement of their formation the granular parietal protoplasm, which is from the first uniformly distributed but always remains parietal, aggregates into dense transverse girdles, which are three or four times as long as broad and are separated from one another by shorter hyaline transverse zones, in which only a very thin almost entirely homogeneous parietal layer of protoplasm remains attached to the membrane. When the granules originally distributed in coarse irregular streaks have become uniformly distributed in the thick girdles, annular constriction appears in the parietal layer in the middle of each hyaline transverse zone, and advances in the centripetal direction till the zone separates into halves which become absorbed in the adjacent thick zones. These at the same time have become spores, and are separated from one another by hyaline interstices filled probably with a substance which is less dense and more capable of swelling. The behaviour of the nuclei in this proceeding has not been investigated. Further details will be found in Büsgen's treatise cited at the end of this chapter.

---

[1] Brefeld, Schimmelpilze, I, 16.

SECTION XIX. *b.* The asci (thecae) are in almost all cases the solitary terminations of hyphal branches; several or many of these asci, in most cases a very large number of them, arranged nearly parallel to one another and with hairs (paraphyses, section XII) between them are grouped together to form hymenia, which in the Discomycetes are open superficial layers on the sporophores, but in the Pyrenomycetes are in the interior of receptacles (perithecia) which are either closed or provided with a narrow orifice. See sections LIX–LXII, where some account is given of the exceptions to the rule thus briefly stated.

The formation of the asci is not essentially different from that of other terminal cells of hyphal branches. They are generally club-shaped, more rarely broadly ellipsoid or are stalked spheres as in Tuber, Elaphomyces, Erysiphe, Eurotium, and others; and when once orientated they grow on usually without interruption till they attain their definitive shape and size and then in most cases begin immediately to form spores. It is only in some species of Erysiphe that the formation of spores is preceded by a longer period of rest; it is indeed possible that young immature asci may pass through the winter period of rest in the case of species which, like Rhytisma and its allies, form their spores in spring, but this has never been directly observed.

In the very large majority of cases eight primordia of spores are formed simultaneously in each ascus; the facts connected with this proceeding were carefully investigated by myself in 1863[1] in certain species of Peziza, Helvella, and Morchella, by Strasburger[2] recently in Anaptychia ciliaris, and by Fr. Schmitz[3] in species of the same genera and of Ascobolus, Chaetomium and Exoascus, with the following results.

In a number of Pezizas (P. confluens, P., Fig. 39, P. pitya, P.) the young ascus is filled with finely granular protoplasm containing vacuoles; a nucleus may be seen in the centre of the protoplasm, as soon as the tube has reached about a third of its ultimate length, in the form of a clear spherical body, in the centre of which is another smaller and strongly refringent body. It has yet to be learnt whether the whole body should be called the nucleus and the inner and smaller body the nucleolus, or whether the latter alone is the true cell-nucleus.

As the ascus elongates the protoplasm moves into its upper extremity, and the lower portion, which may constitute three-fourths of the whole length, now contains only a more watery fluid and a thinner parietal layer of protoplasm. When the ascus has reached its full length, the commencement of spore-formation is indicated by the appearance of two smaller nuclei in the place of the original nucleus. In a later stage four nuclei are seen and then eight; the nuclei are always of similar structure but their size diminishes as their number increases. Their arrangement and Strasburger's observations on Anaptychia leave no doubt that they are produced by successive bipartitions from the primary nucleus. The eight nuclei of the last order are about equidistant from one another, and are each ultimately surrounded by a sphere of protoplasm, which is distinguished from the rest of the protoplasm by its greater transparency and has a very delicate line of delimitation. These portions of protoplasm

---

[1] Die Fruchtentwickelung d. Ascomyceten, p. 34.

[2] Bot. Ztg. 1872, p. 272. Zellbildung u. Zelltheilung, Aufl. 3, p. 49.

[3] See above on page 16.

are the commencements of spores ; they are formed simultaneously and soon become invested with firm membranes, and grow as they lie arranged in a longitudinal row inside the ascus to about double their original size. The protoplasm which surrounds them at first disappears rapidly in Peziza pitya as they increase in size, and like the protoplasm contained in the spores is always coloured yellow by iodine in this species. The protoplasm of the ascus before the spores are formed, and that within the spores at all times, shows the same reaction with iodine in Peziza confluens. But after the orientation of the spores the protoplasm of the ascus shows the characters of a substance, for which I formerly proposed the name of *epiplasm*, and which is distinguished from ordinary protoplasm by being more highly refringent, by its peculiar

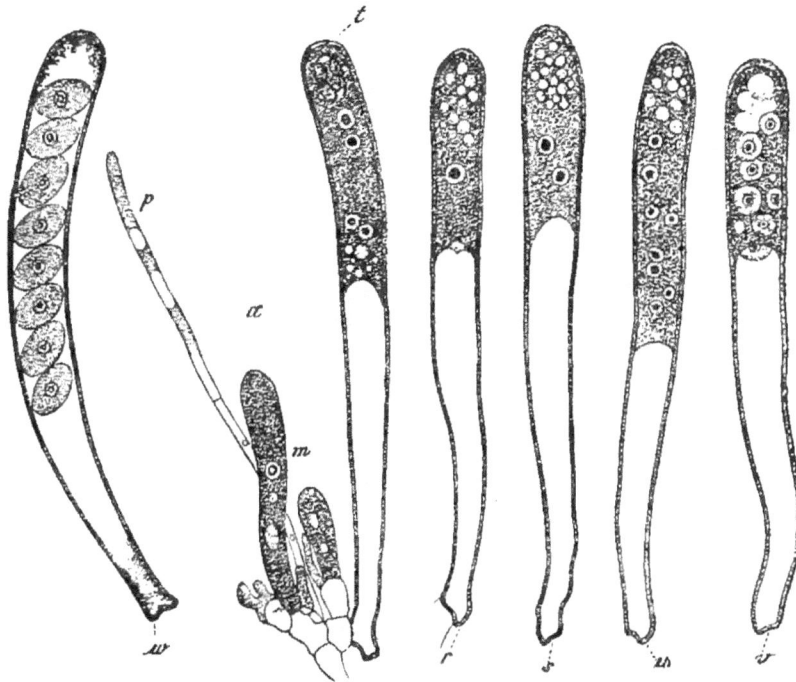

FIG. 39. *Peziza (Pyronema) confluens*, P. *a* a small portion of the hymenium ; *p* a paraphysis, which is only attached to the hyphal branches from which the three asci spring without originating in them. *r—w* full-grown asci, the successive stages of the development in the order of the letters ; in *r—u* the nuclei are multiplying, in *v* the spores are being formed, in *w* they are mature. *m* young asci. Magn. 390 times.

homogeneous and glistening appearance, and especially by the reddish brown or violet brown colour which it assumes when treated with very dilute solution of iodine. Errera [1] has recently shown that this reaction with iodine is due to the circumstance that the epiplasm contains a relatively large quantity of glycogen permeating a protoplasmic or albuminoid vehicle ; the term *glycogen-mass*, or shortly *glycogen*, may therefore be substituted for that of epiplasm.

In some other species with large asci (Peziza convexula, P. Acetabulum [2], and P. melaena, Helvella esculenta, H. elastica, and Morchella esculenta) the contents of the ascus which are at first uniform are differentiated before the spores are formed into protoplasm and glycogen-mass. The former aggregates in Peziza convexula into a

---

[1] See above on page 6.

[2] The species named Peziza sulcata ? in my work on the Ascomycetes belongs to P. Acetabulum.

transverse zone lying in the middle of the ascus, or in most other species into a mass which fills the upper third or fourth portion of the ascus; the remaining and especially the lowermost portion contains only the glycogen-mass which is marked by many vacuoles of varying size and arrangement. Sometimes, as in Morchella esculenta and Peziza Acetabulum, the uppermost portion of the ascus above the protoplasm is occupied by a layer of glycogen, the protoplasm filling a cavity with a sharply defined outline in the glycogen-mass. The nucleus always lies in the protoplasm and in or near its centre, at which point the spores are formed in these as in other species, the mode of formation being essentially the same as that described above. The young spore-primordia are in contact with one another in Peziza convexula and Morchella esculenta. Only the first stage in the division and the ultimate eight nuclei, round which the formation of spores takes place at once, have been directly observed in most of the above species; the other stages have been seen only in Peziza convexula. But the accounts which we possess and observations on the formation and division of nuclei in other plants compel us to assume, that the process in the eight-spored ascus are essentially the same in all cases and that the successive stages in the division of the nucleus have simply been overlooked, owing partly to the rapidity with which the operation is effected and partly to difficulties of observation of other kinds.

Numerous independent observations on a considerable number of **Discomycetes** with eight spores formed simultaneously in one ascus have established the presence of the primary nucleus before the formation of spores, the appearance of the young spores in the manner just described, and the occurrence or non-occurrence of differentiation of the glycogen-mass and protoplasm according to the species. There is therefore no reason to doubt, that the course of development above described prevails very generally in the group which contains the genera Peziza, Phacidium, Leotia, Ascobolus and Geoglossum. It is often difficult to follow it throughout in large asci, like those of Leotia lubrica, Geoglossum hirsutum, Helvella, &c., because the protoplasm of the young tube and of the spores is rendered opaque by the number of drops of oil. In very many other cases the minuteness of the asci and spores either prevents a complete observation or renders it difficult; but even in these cases a little attention will enable us to see the primary nucleus, the simultaneous appearance of the eight spores as portions of protoplasm with a fine boundary-line, and sometimes (Sclerotinia sp.) a nucleus in each of them. In the small asci of e.g. Peziza tuberosa, P. Sclerotiorum, P. calycina and Phacidium Pinastri, and also in some larger ones, as those of Lecidella enteroleuca, Pertusaria lejoplaca, Lecanora pallida and Sphaerophoron coralloides, the primary nucleus appears as a strongly refringent roundish body, which is either homogeneous or more pellucid and as if hollowed out in the centre; the clear, transparent, spherical space is not or not always (Peziza Fuckeliana) to be seen through the periphery.

It is more difficult to observe the formation of the spores in the asci of the **Pyrenomycetes** containing eight spores formed simultaneously, than in the Discomycetes, on account partly of the minuteness and delicacy of the organs, partly of the presence of oil globules which are present usually in large numbers in the protoplasm. Yet careful observation shows that the spores are formed in the way described above. A nucleus has rarely been seen in them (Sordaria fimiseda, Fig. 52). An oil globule was often taken for a nucleus by older writers. The primary nucleus on the contrary may be distinctly seen in many species before the spores are formed; it has the characteristics of the nucleus of Peziza calycina and P. tuberosa which have just been described, and always lies in the same position a little above the middle of the ascus, for example in the nuclei of Xylaria polymorpha, Nectria, Sphaeria obducens, Curcurbitaria Laburni, Pleospora herbarum, Sordaria fimiseda, De Not., and some others. The contents of the asci in most of the Pyrenomycetes which have been examined show only the yellow coloration of protoplasm with iodine; but the glycogen reaction is beautifully shown in Sphaeria obducens during or even before the formation of the spores, and in Pleospora herbarum, Sordaria fimiseda, and Sphaeria Scirpi after their formation. All these facts tend to show that the developement of the eight-spored asci in the Pyrenomycetes is essentially the same as in the Discomycetes, and that further observations will confirm this view.

The eight-spored asci of **Podosphaera Castagnei** have a large nucleus in the young state; this subsequently disappears, and the spores which are formed simultaneously have very distinct central nuclei and are imbedded in a glistening glycogen-mass.

Nuclei were found also by Fr. Schmitz in the asci and spores of **Exoascus Pruni**; in other respects the development of the spore is entirely the same in this species as in the Discomycetes (see also section LXXVI).

The **number of primordia of spores** laid down in the typically 8-spored asci is very constant; exceptions are comparatively rare, such as that of 9 spores in Cryptospora, Tul. and in Exoascus, and 13 developed normally in a single ascus of Peziza melaena. It more frequently happens, especially in the Pyrenomycetes and Lichen-fungi and according to Boudier in Ascobolus also, that some of the 8 spore-primordia remain undeveloped; most of the cases in which less than 8 spores have been found in species in which that is the typical number, may probably be thus explained. The abortion of individual spores is almost always an accidental phenomenon; it occurs regularly, according to Tulasne's account[1], only in Collema cheileum, where the mature ascus always (?) contains aborted as well as perfect spores, the aborted ones adhering irregularly to one another or to the perfect spores.

A larger or smaller number than 8 is the typical number of spores in the asci of some Ascomycetes; 1–2 for example in Umbilicaria and Megalospora, Mass.; 2 in Erysiphe guttata, Pertusaria sp. and Endocarpon pusillum; 4 in Erysiphe sp. and in Aglaospora profusa; 16 in Ascobolus sexdecimsporus, Crouan[2], Hypocrea rufa, P., H. gelatinosa, Tode, H. citrina, Tode, H. lenta, Tode, &c.[3]; 40, 50 and more in Diatrype quercina and D. verrucaeformis, Calosphaeria verrucosa, Tul., Tympanis conspersa, Fr., and T. saligna, Tode; the genera Bactrospora, Acarospora, and Sarcogyne of Massalongo have over 100 spores, most species of the genus Sordaria[4] have 8-spored asci, but in some the asci have 4, or 16–64, or even 128 spores. In some species again the number varies; the asci of Dothidea Sambuci, Fr. produce 2–4 spores, those of Erysiphe sp. and Pertusaria sp. 4–6, Sordaria fimiseda 4–8, S. pleiospora 16–64, and others might be mentioned; in Tuber the number of spores varies as much as from 1–6, and in Elaphomyces from 1–8. The history of the development of these asci has not been so accurately studied, if we except the asci of the two last genera, as that of the typically 8-spored asci; still all that is known of it and of the spores themselves, especially their simultaneous appearance, agrees with the account which has been given of the 8-spored genera. The genera nearest allied to these often have asci with 8 spores, Erysiphe for example, Diatrype, Aglaospora and Calosphaeria; and asci with 4 as well as 8 spores occur in some species of Sordaria and in Valsa ambiens, V. salicina and V. nivea, some in the same, some in separate perithecia. Hence it may very well be presumed that the formation

---

[1] Mém. sur les Lichens. See the literature cited at close of section LXXIV.
[2] Crouan in Ann. d. sc. nat. sér. 4 (1858).
[3] See Currey in Trans. Linn. Soc. London, XXII.
[4] G. Winter, Die deutschen Sordarien, Halle, 1873.

of spores in these instances differs from that in the 8–spored asci in no other respect than in the number of nuclear divisions and spore-primordia. Whether regular abortion of a certain number of original spore-primordia occurs in individual cases, where the number of perfect spores is small, is still uncertain. The formation of the spores too in Tuber and doubtless also in the rest of the Tuberaceae and in Elaphomyces differs much less from that in asci with 8 spores than appeared from my former observations, which were conducted with imperfect means. There also we find simultaneous orientation of spores and nuclei. The inequality in the number of the spores is due partly to the inequality in the number of the original spore-primordia or divisions of the nucleus, partly to the frequent want of uniformity in later developement and to the partial disappearance of spores after they are once formed.

In the full-grown stalked spherical ascus of **Tuber** (T. aestivum, T. melanosporum, T. brumale (Fig 40) and their allies) the protoplasm which is at first irregularly granular and interspersed with vacuoles becomes differentiated into a dense parietal strongly refringent layer of glycogen, which turns a brownish red with iodine, and an excentric spherical cavity filled with finely granular weakly refringent protoplasm which becomes

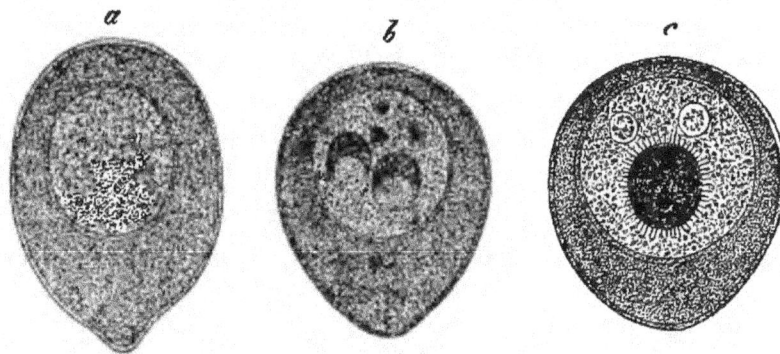

FIG. 40. *Tuber brumale*, Vitt. Full-grown asci isolated in water. *a* protoplasmic cavity separated from the layer of glycogen. *b* six young spores visible in the protoplasm. *c* shows one spore half matured and two which have remained quite small in the same position. Magn. 390 times.

yellow with iodine. The limiting layer of the glycogen-mass is very compact where it borders on the cavity, and its double contour is often so sharply defined that older writers supposed it to be the membrane of a special cell. The spores are formed in the protoplasm. Observations made by Errera have shown that there is one nucleus in the protoplasm, visible even in younger asci, which by successive divisions gives rise usually to 4-6 nuclei ; then as many spore-primordia appear simultaneously and in close proximity to one another round these nuclei in the form of small and very delicate cells. As the cells now begin to grow they move further apart and usually develope unequally, some outstripping the others, while some remain stationary at an early stage of development and at length disappear. Hence the frequent occurrence of quite delicate spore-primordia with others that are far advanced, which once led me to suppose that they were formed successively, and hence also the unequal number of ripe spores, varying from 1-4-6 in an ascus. The old drawings reproduced in Fig. 40 will sufficiently illustrate the subject for the present.

The asci of **Elaphomyces granulatus** are of similar form to those of Tuber, and contain before the orientiation of the spores a very transparent protoplasm forming a thin parietal layer round one or more vacuoles and turning yellow with iodine ; no glycogen-mass appears in them. I found in the lower third of a half-developed ascus where the great increase in breadth begins, a small but distinct nucleus with the structure of that in Peziza confluens; I could not see it in the ascus when fully formed, but the young spore-primordia on the other hand have a distinct nucleus. The spores lie close together and form a small group of usually six small round delicate cells, which occupy the apex or a part of one side of the ascus; they are all alike when quite young and were probably therefore formed simultaneously, but they develope very unequally; the mature asci contain from one to eight, usually six spores.

c. The **formation of spores in the sporangia of Protomyces macrosporus** (Fig. 41), if the expression is allowable in this case, takes place after they are laid on or in water. Before the water makes its way into them they have experienced complex changes, which cannot be further described in this place, and have assumed the form of spherical vesicles (Fig. 41 *b*) the walls of which are lined with a layer of dense granular protoplasm (*c*) enclosing a large central cavity filled with water. No nuclei have been seen in them. The layer of protoplasm now breaks up simultaneously all round the cell usually into hundreds of 'spores' (*d*), which when the separation is complete are small polygonal finely granular bodies parted by narrow hyaline streaks,

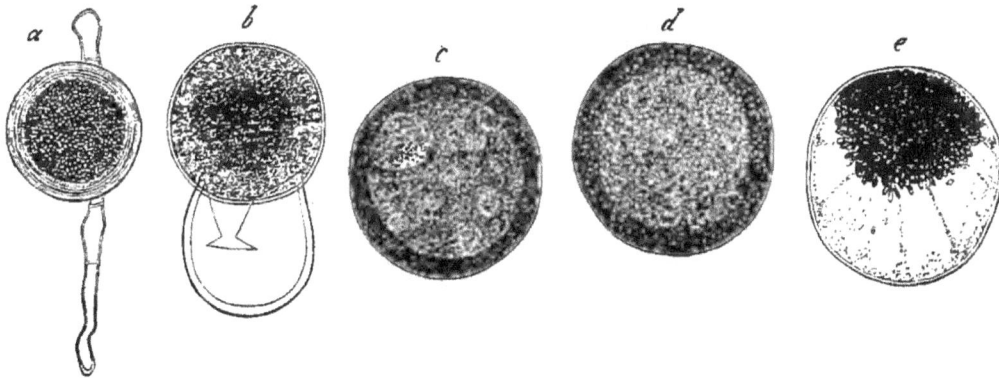

FIG. 41. *Protomyces macrosporus,* Unger. *a* mature resting-spore in the dormant state (see section LIII) with the remains of the hypha on which it was formed. *b* further development when cultivated in water; the protoplasm enclosed in an inner layer of the membrane (inner cell) swells up and escapes from the ruptured outer layers of the membrane. *c—e* development of the spores in the inner cell (sporangium) which has escaped from the outer cell. In *c* the protoplasm is parietal. In *d* the protoplasm is divided into spores. In *e* the spores form a cluster and are separated from the layer of protoplasm which still lines the wall. Magn. 390 times.

and presently assume the form of small cylindrical rods about 2.2 μ in length. The differentiation in the protoplasm described in my work quoted in section LIII as preceding the partition requires fresh examination. A granular parietal layer of protoplasm which permanently clothes the membrane and a small portion of hyaline substance between the spores, possibly of the nature of protoplasm, is not employed in the formation of the spores. The latter substance becomes visible, when the spores have taken the form of rods and have collected into a ball (*e*) on one side of the wall of the sporangium, as a series of radiating threads running from the ball of spores to the wall-utricle; but by degrees it disappears entirely and a watery fluid takes its place.

SECTION XX. **The spores which are produced endogenously are usually set free from their mother-cells** in some determinate manner as soon as they are ripe and fully grown. In a few cases, as in Elaphomyces, Eurotium and perhaps in Penicillium, they escape from the mother-cell before they have acquired the size and structure which usually precede germination, and they subsequently attain to these at the expense of dissimilar cells which had surrounded the sporangium. In extremely rare

[4]   G

and exceptional cases the release of the ripe spores is left to chance, there being no special arrangement made for it, and the spores may even germinate inside the mother-cell, the germ-tubes piercing or bursting through its wall, as may be seen in the sporangioles of Thamnidium and its allies.

The arrangements for the escape of the spores vary in different species.

a. The aquatic swarm-spores of the **Saprolegnieae** (with one partial exception to be noticed hereafter), of the **Peronosporeae** and **Chytridieae** make their exit through a narrow orifice, formed usually at the apex of the wall of the mother-cell by the sudden swelling and disappearance of a circumscribed portion of the wall of the mature sporangium. The spot is marked out in many species by gelatinous thickening of the membrane before it begins to swell. This is nowhere more conspicuous than in the sporangia of Phytophthora, in some species of Peronospora, and in some of the Chytridieae which have gelatinously thickened terminal papillae; in other cases, as Saprolegnia, the thickening has not been observed. While the place of exit swells, the entire contents of the sporangium, the mass of spores and the surrounding matter, absorb water and also swell[1]; and as the lateral walls of the sporangium are but slightly extensible, the spores which lie beneath the place of exit are first squeezed

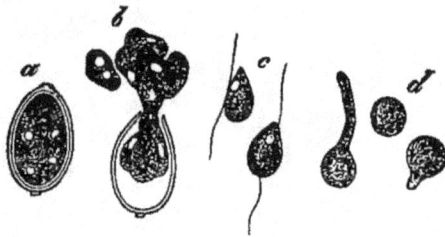

FIG. 42. *Phytophthora infestans,* Mont. *a* sporangium lying in water after the division is complete. *b* escape of the 10 (swarm-) spores from the sporangium. *c* spores in motion. *d* the same come to rest and beginning to germinate. Magn. 390 times.

out through it and the others follow. The proceeding may vary in individual cases, and it remains for investigation to determine to what extent the spores themselves, the intermediate parting substance (see p. 74) and perhaps also an inner layer of the wall of the sporangium, participate in the first general swelling caused by the absorption of water. In the cases which have been more carefully examined (Achlya, Saprolegnia and Phytophthora, Fig. 42) it can be seen directly that it is the hyaline substance surrounding the spores inside the firm wall which swells the most. It is also observed in most cases that a hyaline layer on the inner surface of the firm wall first comes into prominence, and increases in breadth and pushes the mass of spores towards the middle of the sporangium and the place of exit. The spores, even where as in some cases they show independent movements before they are set at liberty, are now virtually passive, and in Achlya especially they are evidently squeezed together as they escape from the sporangium by the limpid mass which surrounds them. It is therefore in the swelling of this mass that the expelling force resides; but it is still uncertain whether the mass consists entirely of the original soft partition-layers which must in that case suffer partial dislocation when the spores are discharged, or whether an innermost layer of the wall of the sporangium swells and some product from the spores themselves is also added.

The phenomena connected with this swelling at the place of exit occur only at a given moment after the formation of the spores is completed, and in water moreover which is perfectly pure and contains free oxygen. That the point of exit, which

---

[1] Walz, Bot. Ztg. 1870, p. 689.

has in many cases been formed some time before and gelatinously thickened, does not swell in the water before this moment must be due to its not having yet undergone the change which takes place in it after the spores are matured, supposing both phenomena to have a common cause or the cause of the change to lie in the ripe spores. In the latter which is the more probable case we are almost compelled to suppose that some secretion must proceed from the spores, which acts as a ferment altering and dissolving definite portions of the wall which had been previously prepared. The same view will apply with slight modification to the substance inside the persistent firm wall of the sporangium which is capable of swelling and takes an active part in the expulsion of the spores, and to the discharge of the zoospores of many of the Algae.

On the special features in the formation of the zoospores of Pythium, the formation of small heads in Achlya, Aphanomyces and Achlyogeton, and the coating of the spores in these genera and in Dictyuchus, which cannot be further described in this place, see section XL and the special literature there cited.

b. The upper and largest portion of the outer wall of the spherical sporangia of **Mucor** (including Thamnidium, Rhizopus, Absidia, Phycomyces, &c) and **Mortierella** is changed when the spores are ripe into a substance which dissolves in water, and in most of the mucor-forms is incrusted with a thin spiky coating of calcium oxalate. The presence of the smallest quantity of water causes the wall and the substance between the spores which is present in greater or less abundance to dissolve and liberate the spores (see p. 75). The lower portion of the outer wall which surrounds the point of insertion does not participate in these changes and remains after the dissolution of the rest of the sporangium as a ring or collar round its insertion; the basal wall is also persistent and forms in Mucor the strongly convex or even vesicular structure known as the columella.

In the allied genus **Pilobolus** the sporangium has at first the shape and structure and even the oxalate incrustation of that of Mucor. The upper and larger portion of its outer wall is very firm and of a bluish black colour; a comparatively narrow annular zone round the point of insertion is more delicate and colourless. The mass of spores contained in the sporangium is at first surrounded, especially at the point of insertion of the sporangium, by a colourless gelatinous layer lying between the spores and the wall, and endowed with great capacity for swelling in water. This layer appears to be developed to a greater or less extent according to species, but whether it is originally a part of the wall of the sporangium or formed like the spores from the contents of the sporangium is at present uncertain. If water reaches the thin basal zone of the outer wall it penetrates through it and causes the gelatinous layer which lines it to swell up at once, and the wall of the sporangium is consequently ruptured round the point of insertion and carried upwards by the continuously swelling substance. It is not known whether the membrane is still intact when the water makes its way through it, or whether fissures for the admission of the water are previously formed in it as the result of changes of form and varying moisture after maturity. In species like Pilobolus anomalus, Ces. (Pilaira, v. Tiegh.) with very long filiform sporangiophores, nothing further happens beyond the gradual solution of the gelatinous layer and the breaking up of the mass of spores; but in most species (P. crystallinus, P. oedipus, &c.) the ripe sporangium is abjected from its sporangiophore and adheres by means of the gelatinous layer to foreign bodies, while the spores swell to their full extent and are disseminated. The sporangiophore in these species [1] is a cell some millimetres in length, cylindrical in its middle portion, but inflated in its lower part and in its upper part especially just beneath the sporangium. It becomes more and more turgescent after the spores have matured and causes abjection of the sporangium by means of the mechanism described on page 72. The separation takes place in the line of an annular fissure, which is close beneath the insertion of the outer wall of the sporangium and is seen before the sporangium is flung off as a fine sharply marked line on the wall (Fig. 38). The delicate wall of the lower portion of the sporangium is ruptured at the moment of abjection, being struck by the ejected fluid, and thus the swelling of the gelatinous layer investing the spores is secured.

The sporangium is sometimes abjected with considerable force. The sporangia of Pilobolus oedipus, in which species, according to Coemans and Brefeld, the greatest amount of force is exerted, are thrown, as we learn from the former authority, to a height of more than 1.05 M. The process, as Coemans has also proved, is greatly dependent on the amount of light. Under favourable circumstances the development of the sporangiophore begins at midday or in the afternoon ; it is completed and the sporangia and spores are also formed during the night, and the sporangium is thrown off during the following morning at an earlier or later hour according to the greater or less amount of light. Exclusion of light does not entirely prevent abjection but may delay it 12–15 hours. P. oedipus shows this sensitiveness to light and the normal periodicity to a less extent than P. crystallinus. We must not enter further in this place into the connection between these phenomena and the very strong positive heliotropism of the sporangiophore.

The increasing turgescence of the sporangiferous cell before abjection, assuming that the superficial extent and elasticity of the membrane remain the same, may be caused either by increasing osmotic absorption of water on the part of the sporangiferous cell itself, or by the forcing of water from the mycelium into the passive sporangiferous cell, or by the combined operation of both these agencies. In my first edition I assumed that the latter of the two was the only operative cause, because a drop of water which increases in size is often seen to issue after abjection from the open sporangiferous cell before it finally collapses. More exact measurements are required for the confirmation of this view.

Section XXI. **The spores produced in asci and those of Protomyces macrosporus are set free** in one of two ways according to the species ; either by *ejection* [2] (Ausschleuderung, Ejaculation) or by *solution* or *gelatinous swelling* (Auflösung, gallertige Verquellung) of the asci.

The first process, the **process of ejection**, is found only in the case of spores which normally attain their full development inside the ascus. As they advance towards this state, the protoplasm around them and the glycogen-mass subsequently formed constantly diminish in quantity, being doubtless used to a great extent as material for the formation of the spores. We are not at present in possession of more

---

[1] Coemans in Mém. conc. de l'Acad. royale de Belgique, XXX.—J. Klein in Pringsheim's Jahrb. VIII, p. 305.—Brefeld, Schimmelpilze, I and IV.—Van Tieghem, Mucorinées. See also the literature cited in sections XLI–XLIV.

[2] 'Abjection' and 'ejection' have been adopted as renderings of 'Abschleuderung' and 'Ausschleuderung,' the throwing off or throwing out with force of spores from the sporogenous structure.

exact knowledge on this point. When the spores are ripe a considerable quantity of protoplasm richly interspersed with vacuoles still remains in some species, as Sphaeria Lemaneae[1], S. Scirpi and Sordaria fimiseda; in most cases, however, the residue is but scanty, but in all without exception the inner surface of the membrane is covered by an unbroken though often very thin layer of protoplasm. The chief portion of the contents of the ascus surrounding the spores consists of an apparently watery fluid.

The membrane, which when young is always delicate and not stratified, increases in thickness as the ascus matures, but often shows no signs of being divided into layers even in such large asci as those of Morchella esculenta, Peziza Acetabulum, P. pitya, P. melaena, and Ascobolus furfuraceus; in some species, especially in Lichen-fungi, it is distinctly stratified, and in a number of cases, which will be noticed more fully below, it has peculiar local thickenings at the apical extremity. It shows the reaction of Fungus-cellulose in most Fungi; yet a dilute solution of iodine produces a blue colour in not a few cases, either over the whole of the ascus, as in most Lichens and in Peziza convexula, P. cupularis and others[2], according to Coemans in some species of Ascobolus also, or only at the apex of the ascus, as in some instances which will be considered at greater length in the sequel.

The ejection of the spores from the asci is either *simultaneous* or *successive.*

The **simultaneous ejection of spores** is much the most common, occurring in nearly all the Discomycetes, in the Erysipheae, in some Sphaeriaceae and in the sporangia of Protomyces. Certain special modifications are said to occur in Lichen-fungi and will be noticed again further on; but except in these cases the ejection of the spores is due to the same mechanical arrangement as that which causes the abjection of the spores and sporangia of Empusa or Pilobolus. It has been carefully observed (with the exception of the case of Protomyces which must be at present disregarded) in club-shaped or ovoid asci which are broader towards their free extremity and contain four, eight, sixteen, or more rarely a larger number of spores. After the spores are matured the ascus with its parietal layer of protoplasm enclosing a constantly augmenting quantity of watery fluid expands considerably and becomes more turgid. The expansion may amount to five-fourths or four-thirds or even to twice or several times the original diameter of the ascus, i.e. the diameter at the time of the ripening of the spores, and takes place in the direction of the length as well as the breadth, affecting especially the upper and apical portion of the ascus. That the membrane of the ascus is almost entirely passive in this extension and continues to be perfectly elastic may be proved at any time by cutting it through or by extracting the water.

When the ascus begins to expand the spores move into its apical region, where they are closely packed together in the watery fluid and in the simplest and most common case are arranged in a single longitudinal row, the uppermost member of which is close beneath the apex; it is more unusual for them to form two or more irregular rows, as in Ascobolus and its allies. In some cases gelatinous appendages which will be described by-and-bye, serve apparently to keep the spores in their relative positions in this arrangement or at least to assist in doing so[3]. According to

---

[1] Woronin, Beitr. III.

[2] See also Nylander in Flora, 1865, p. 467.

[3] Zopf in Sitzgsber. d. Berliner naturf. Freunde, Feb. 17, 1880. Zopf's last work on this subject (Zeitschr. f. Naturaw. 56, Halle, 1884) could not be consulted.

Zopf the uppermost spore in some Sordarieae is even attached to an inwardly directed process from the membrane at the apex of the ascus. No such arrangements have been observed in the majority of cases, and the apical position of the spores is sufficiently explained by the consideration that the one-sided expansion of the apical region must produce currents in the contained fluid in the direction of the apex, which must push the spores suspended in it towards this expanding apex.

When the wall has reached a fixed maximum of extension, it suddenly gives way at a point of least cohesion near the apex, which is the *point of dehiscence*; at the same moment the elastic lateral wall contracts to the size spoken of above as the original size, and the apical portion of the fluid contents together with the group of spores is driven out through the fissure. Then the open ascus collapses and perishes.

The arrangement of the spores in the apical portion of the ascus before their ejection, when there are no special arrangements for attaching and securing them, is evidently the result of the conditions of space and form. In many Discomycetes for instance the spores are ellipsoidal or elongated, and their length greater than the breadth of the ascus; they lie therefore parallel in the ascus, in a single longitudinal row close behind one another, each placed obliquely and touching the wall of the ascus with both ends, the uppermost one having its upper extremity close to the apex (Figs. 39 *w* and 43). If the breadth of the ascus is much greater than the diameter of the spores the arrangement is more irregular; thus there is an irregular longitudinal row in Ascobolus pulcherrimus[1], two such rows in many Ascoboli[2] (Fig. 45), and an irregular ball crowded up into the apex of the ascus in the eight-spored asci of Exoascus Pruni[3] and in the many-spored asci of Ryparobius[4]. But the longitudinal arrangement is maintained in the comparatively very broad asci of Sordaria (Fig. 44), where it may be due to the attachment of the spores to one another.

The form of the fissure varies with the species, and it cannot always be distinguished with certainty.

A longtitudinal rent simple or lobed, passing over the apex and leaving a broad hole when the ascus is emptied, forms the opening in the asci of Exoascus Pruni, Peziza cupularis and Erysiphe[5], and according to Boudier[6] of Geoglossum, Helotium, Leotia, and Bulgaria sarcoides.

In many Pezizas, as P. convexula, P. confluens, P. granulata, P. abietina, P. vesiculosa, P. melaena, all the Ascoboli and Helvella crispa, the fissure is annular and runs close beneath the blunt summit of the wall of the ascus, which is therefore cut off like a lid, and when the spores are ejected is lifted off either all the way round or only on one side where the uppermost spore touches the wall of the ascus; the latter is the case, for instance, in Peziza vesiculosa and P. granulata. In larger Ascoboli the edge of the lid may be seen before the ejection of the spores as a distinctly marked transverse line[7]. In some forms, as P. abietina and P. vesiculosa,

---

[1] Woronin, Beitr. II, t. III.

[2] Boudier in Ann. d. sc. nat. sér. 5, X.

[3] De Bary, Beitr. I, t. III.

[4] Boudier, l. c.

[5] R. Wolff, Erysiphe; see the literature cited in section LXXIV.

[6] Loc. cit. p. 202.   [7] See Boudier's figures, l. c.

it is the apical and most extensible portion of the wall and chiefly the area forming the lid in that portion which is most distinctly coloured blue with iodine. In the Sordarieae also I frequently saw the ascus open by a comparatively tall lid.

There is a third series of cases in which the spores are ejected through an apical perfectly circular hole which before ejection of the spores is a circumscribed thinner or less compact portion of the wall of the ascus. In Rhytisma acerinum this hole is replaced by a minute mucro forming the uppermost extremity of the apex of the still closed ascus. In Peziza Sclerotiorum (Fig. 43), P. tuberosa, and their allies, the wall of the ripe but not turgescent ascus is more than twice as thick at the slightly convex apex before the spores are discharged than it is on the sides; it is also formed of two layers and is traversed in the middle by a longitudinal streak which is less strongly refringent and looks like a stopper inserted in the ascus. The apex of the turgescent ascus, ready for the ejection of its spores, is considerably broader and strongly convex outwards, with its wall not thicker than the lateral walls and with none of the internal structure just described. The spores are discharged through the stopper, and after the discharge there is an open passage in its place round which the form and structure of the non-turgescent state are once more restored. In these cases again the apical portion of the wall, which is most capable of stretching and is thickened when it is not in a state of tension, is that which turns blue with dilute solution of iodine, and the stopper which indicates the point of dehiscence is most intensely coloured.

The following remarks will further illustrate the above short account of the mechanism for the ejection of the spores.

a. The expansion of the ascus by increase in the amount of its fluid contents has been directly observed. That this is merely a passive stretching of the cell wall, and not a phenomenon of growth with permanent results, is shown by facts which are easily observed, namely, that the ascus contracts to its previous dimensions after discharging its spores, or if an artificial opening is made in its wall while its membrane

FIG. 43. *Peziza (Sclerotinia) Sclerotiorum.* Asci observed as they lay isolated in water. *a* a mature ascus before ejection of the spores. *b* the same after ejection. *c* another specimen in the same stage of development as *a*, cut through transversely. Magn. about 400 times.

at the same time increases in thickness, as is shown most clearly in the case of the strong local thickenings which have been described as occurring in Peziza Sclerotiorum. This species shows with peculiar distinctness that it is the apical region of the ascus which stretches most; but in all other cases attentive comparison will show that it is the apical region, or pretty well the apical half, which is most altered in form and size while the lower half is little or not at all affected. The directions of greatest extensibility and the shapes produced by them vary much in different species, as appears from a comparison of Figs. 43, 44 and 45. The enormous increase in volume of the asci of Sordaria may perhaps suggest actual growth, especially as they are comparatively rich in protoplasm after the spores are matured; the point requires further investigation, but it should be noticed that the contraction after the ejection of the spores is in this case also very considerable.

That it is the increase in the amount of fluid content which causes the expansion of the ascus is shown by the fact, that the expansion diminishes with a diminution in the amount of fluid in the cell and disappears, either suddenly and entirely, if the wall of the ascus opens spontaneously or is pierced artificially and the fluid escapes, or gradually through the slow operation of alcohol, glycerine, or saline solutions which withdraw water from the uninjured ascus. On the other hand the expansion of the asci (and ejection of the spores) is promoted by placing uninjured asci in water.

b. The great elasticity of the wall of the ascus is sufficiently shown by the facts above enumerated.

c. The spores are in many cases retained according to Zopf in the expanding apex of the ascus by a special apparatus of attachment[1]. In Sordaria Brefeldii a hollow cylindrical thick-walled process of the membrane, which turns blue with iodine, reaches from the apex into the lumen of the ascus. The spores, like those shown in Fig. 52, are provided with terminal appendages which connect them together in a row; the distal appendage of the uppermost spore attaches the entire row to the process from the wall of the ascus, ‘sometimes by thrusting itself into its cavity which it fills up, sometimes by closely grasping it. And this apparatus is further completed by another arrangement; the membrane of the ascus over a subterminal zone is capable of great swelling and can lay firm hold on the appendage borne by the chain of spores, as a hand grasps the throat.’ Similar apparatus may perhaps frequently be in use especially in the Pyrenomycetes, as is indicated by the structural features in the apices of asci which will be discussed in section XXVI. Our present knowledge does not allow us to speak with certainty on this point. In many cases, especially in the Discomycetes, there is no such apparatus present, the spores being suspended in the fluid of the ascus. The spores must have nearly the same specific gravity as the fluid; if not, they would change their position as the ascus changes its inclination, which they do not do. Most, if not all, spores produced in asci sink in pure water; the fluid contents of the ascus must therefore be of greater specific gravity than pure water, since it holds in suspension bodies of greater specific gravity than water. If increase in the amount of the fluid contents causes the apical portion of the ascus to stretch more than the other parts, currents must be set up in the fluid in the direction of the apex and continue as long as the expansion continues, and push the spores therefore permanently towards the apex. The arrangement of the spores may then be affected by special directions in the currents which we cannot at present determine, as well as by the conditions of space noticed above.

d. The ascus lined with a layer of protoplasm and preparing to eject its spores is in the condition of a cell in a state of constantly increasing turgescence, the characteristics of which may here be presumed to be known[2]. It is natural therefore to suppose that the increase in the amount of fluid contents is caused by absorption of water by endosmosis, and that this absorption is due to the operation of osmotically active substances, dissolved in the cell-contents, which cannot pass through the layer of protoplasm. All the facts that have been observed agree with this supposition, and especially the circumstance that volume and turgescence can be alternately diminished and restored in individual asci by careful removal of water by means of a saccharine solution or of glycerine, and by its reintroduction. The opposite view expressed in the first edition of this work was founded on the fact that the protoplasmic utricle in the asci which were examined was either injured or killed in the process of withdrawing the water, and it has been shown that isolated asci are very liable to suffer in this way. The presence of the substances which are active agents in inducing endosmosis is evidently coincident with the disappearance of the

---

[1] As cited on p. 85.

[2] Pfeffer, Pflanzenphysiologie, I, p. 50.—De Vries, Mechan. Ursachen d. Zellstreckung, Leipzig, 1877.

residue of protoplasm or glycogen-mass as the spores mature. I was unable to determine their character more nearly, and can now only state that neither sugar nor any acid reaction was ascertainable in the fluid contents of the asci of Peziza granulata, P. Sclerotiorum, and Ascobolus furfuraceus.

e. It is evident from what has now been said that, other conditions being the same, the ejection of the spores must be hastened by a lateral pressure operating on the ascus from without. This may readily be shown by experiment on isolated asci placed in water beneath a cover-glass. In the living Fungi the asci stand very many together in the hymenium usually with paraphyses between them, and there the lateral pressure on the asci increases in part with the advancing growth, as new asci are introduced between the previous ones, and in part with the addition of water; the hymenia in the Discomycetes which have paraphyses swell considerably in the direction of their surface and in greater proportion than the tissue of their sporophores.

f. All that has been said of the club-shaped asci may be applied in its main points to the spherical sporangia of Protomyces macrosporus, which are formed free in the water. The place of greatest extensibility, towards which the numerous 'spores' move, is in accordance with the shape of the sporangia a broad thin section or pit in the wall, in the middle of which the fissure ultimately appears as a gaping slit.

Section XXII. It has been already said that the **asci of the Discomycetes** of which we now proceed to speak, are arranged in superficial hymenia nearly vertically to the surface and between numerous paraphyses of uniform height, the extremities of which indicate the middle level of the hymenial surface. The asci of a hymenium are not developed simultaneously; during a period of time which varies in different species new asci grow up one after another from beneath between the paraphyses, while the older ones are ripening. When the asci approach maturity and begin to enlarge each one elongates so much that its apex projects above the surface of the hymenium, while its basal portion continues attached to the original place of insertion. After ejection of the spores the ascus shrinks and the apex usually returns to below the level of the hymenial surface. Where there are no paraphyses, as in Exoascus, the same phenomena are observed with the modifications which that difference naturally entails.

In the hymenia of Peziza, Helvella, Morchella, Bulgaria, Exoascus, and the majority of the Discomycetes when they are ripening, the individual asci are constantly discharging their spores in succession. If the Fungus is placed in a closed and damp chamber and a glass plate is set in front of the hymenium, spores are soon found lying usually eight together in a minute drop of fluid, and gradually the plate becomes thickly strewed with them. But besides this gradual emptying of the asci many of the Discomycetes have the peculiar habit of 'puffing' (Stäuben), that is, of suddenly discharging a whole cloud of spores, if they are shaken, or if the chamber in which they have been kept is opened. The phenomenon is of course produced by the simultaneous emptying of a number of asci. The Fungi on which my experiments were chiefly made—Peziza Acetabulum, P. Sclerotiorum, and Helvella crispa—do not puff when they are cultivated in a very damp and still atmosphere enclosed by a bell-glass; under these conditions only the continued gradual discharge of the spores takes place. As long as the Fungus remains shut up in the damp atmosphere no amount of shaking will cause it to puff, whether it is kept in the dark or in the light of day, or is suddenly brought from the dark into diffuse or direct sunlight; but it puffs as soon as it is removed from the damp chamber into a dry atmosphere. If the hymenium is only moderately damp, so that the tips of the ripe projecting asci look like a slight rime or a fine down on it, the puffing commences in a few seconds after the bell-glass or other covering is removed. If it has been kept very wet, the hymenium is covered with a thin layer of water and glistens more or less and is of a darker colour than in the moderately moist condition. In such a hymenium the puffing does not take place till the layer of water is evaporated and the slight rime-like appearance is observed; the puffing is accelerated by whatever accelerates the evaporation.

From these facts it appears that sudden loss of water is the proximate cause of the puffing. Since puffing occurs instantaneously in hymenia that are not wet, the withdrawal of water as soon as dry air comes in contact with the Fungus cannot produce it by causing a shrinking and contraction of the entire hymenium and a consequent increase of the pressure on the asci from without. All this could not possibly be brought about to any important extent in one or a few seconds of time, and some simple experiments and measurements are sufficient to convince us that the pressure which operates on the asci from without under long-continued desiccation is not at first increased, and eventually decreases to a considerable extent, but that it increases in proportion as the hymenium absorbs water.

The loss of water can only therefore cause the puffing by altering the state of tension in each ascus, either by lessening the expansion of the lateral walls and so increasing the pressure of the fluid contents on the place of dehiscence, or by lessening the power of the place of dehiscence to resist the pressure which remains unaltered. The correctness of this explanation is confirmed by the observation, that ejection takes place when ripe isolated asci lying in a little water are suddenly exposed to the operation of reagents like alcohol and glycerine which withdraw their water.

The above remarks leave little room for doubt that motion and shaking affect the puffing only by hastening the evaporation of the water. A hymenium which has just sent forth a cloud of spores can be induced to repeat the operation several times, if the plant is moved rapidly to and fro, and the less perfectly ripe asci are made to dehisce. But then, and in many cases after the first puffing, a rest of at least some hours is necessary, that a sufficient number of new asci may come to maturity to allow the puffing to be observed.

The phenomenon of puffing is absent from some Discomycetes; I have never been able to excite it in Peziza pitya, Morchella esculenta, or Exoascus Pruni; it is readily produced in the majority of species. I have observed it in Peziza melaena, P. tuberosa, P. aurantia, P. cupularis, P. badia, P. confluens, and Rhytisma acerinum, in addition to the species which have been already named. Many other observations have been recorded since the time of Micheli.

In Ascobolus and the genera which have been recently separated off from it ejection is never successive but always simultaneous from all the asci that are at any time ripe in the hymenium, and here too we have the phenomenon of puffing. The mechanism of the discharge and the conditions for the puffing are the same as those which have been described in the case of the other Discomycetes; but they are also dependent on the illumination to an extent which requires to be more closely examined.

SECTION XXIII. The process of ejection in the **Pyrenomycetes which discharge their spores simultaneously** was first correctly described in Sordaria by Zopf[1]. Numerous asci placed upright side by side in a thick tuft fill the swollen enlarged basal portion of a flask-shaped receptacle, the *perithecium*, which is continued upwards into a more or less elongated *neck*. In large forms, as S. fimiseda, the neck is more than a millimetre in length but much shorter in the smaller species, and is traversed longitudinally by a very narrow *canal*, not so broad as an ascus, which enlarges into a conical form at its inner end above the group of asci and is open to the air above at the outer end. Till the spores begin to ripen the asci are between narrowly cylindrical and club-shaped, and of the same height as the basal ventral portion of the perithecium. Then they begin to elongate one after another while they grow much broader at the apex. The only direction in which they can elongate is that of the canal of the neck. When the apex of the first ascus reaches the inner end of the canal, it enters it and swelling there to a broadly club-shaped form, and causing a corresponding enlargement of the canal, it continues to lengthen, till its apex is on a level with the outer mouth of the canal or a little above it; then its ejection takes place. Then the next ascus enters into the now empty canal, and so on one after another. The lower extremity of the ascus continues attached to its original point of insertion at the base of the perithecium until ejection. The elongation is therefore very considerable; in the case depicted in Fig. 44, for instance, it is more than six times the length attained by the ascus at the time the spores are ripe, and is at least three times that length beneath the widened upper part. The lower portion seems to become narrowed at the same time under the pressure of the neighbouring asci which are beginning to swell, but it is difficult to be quite certain on this point on account of the strong lateral pressing together of the parts.

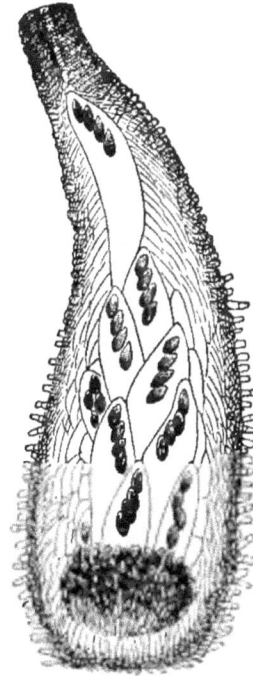

FIG. 44. *Sordaria minuta*, Fuckel (?). Form with 4-spored asci; small perithecium grown on a microscopic slide and observed in the living state lying in the culture-fluid, in optical longitudinal section; at the base of the perithecium is a dense group of asci, most of them with ripe spores; above this group are other mature asci in various stages of elongation preparatory to ejection, the uppermost having almost reached the opening of the neck. Magn. about 100 times.

The rapidity with which the elongations are accomplished is comparatively small. In a small specimen observed in water the movement of the apex about a spore's length ($= 17\ \mu$) occupied some 15 minutes, and the passage through the whole neck about 8 hours. In the specimen of Sordaria minuta (?) given in Fig. 44 the motion was quicker, a spore's length of 10 μ requiring some five minutes. How far light, heat and other external causes accelerate or retard the movement, and what are the specific differences which certainly exist, are points which have yet to be investigated.

SECTION XXIV. **The force with which the spores are ejected** does not appear to be great. In Bulgaria inquinans and Protomyces macrosporus they are sent straight

---

[1] As cited on p. 85.

upwards to a height of 1–2 cm., in Exoascus Pruni of 1 cm.; in the strongly puffing Fungi, such as Peziza vesiculosa, P. Acetabulum, Helvella crispa, and Ascobolus furfuraceus, they are thrown to a distance of more than 7 cm., in Sordaria fimiseda, according to Woronin, they travel 15 cm., in the smaller species of this genus about 2 cm., in Rhytisma acerinum only a few millimetres. The movements in the act of puffing in large hymenia were said by Desmazieres to produce an audible sound, but this has been doubted by recent observers; I have myself however heard a very perceptible hissing noise produced by strong specimens of Peziza Acetabulum and Helvella crispa.

The peculiar features in the old genus **Ascobolus** (including Saccobolus and others), which led to many false and even strange notions, are connected with the large size of the asci, the great prominence above the surface of the hymenium at the period of maturity, and the regular periodicity in their ripening and in puffing[1]. Coemans has given us a full account of how a number of asci ripen and eject their spores daily for several days together, when the hymenium has reached a certain point of development. The asci in consequence of their expansion begin to appear above the surface of the hymenium towards evening and continue to do so till the succeeding afternoon; between 1 and 3 o'clock the tension reaches its highest point, and the slighest shock causes ejection which is simultaneous in all the projecting asci. It is difficult to determine whether ejection takes place when everything around is perfectly still. The stillness is in fact always broken by a number of younger asci beginning to expand every afternoon in preparation for ejection on the following day. It is natural to suppose that there must be a direct relation between this regular daily periodicity and the light-period, and Coemans found that ejection was delayed 4–5 hours in the Ascoboli, when cultivated in darkness. Boudier and Zopf observed that the asci are to a high degree positively heliotropic when they are in process of expansion; their curvatures towards the source of light may extend through nearly 90°, but these curvatures almost, if not quite, entirely disappear after ejection or if the expansion is artificially stopped. The connection between all these points requires more exact investigation.

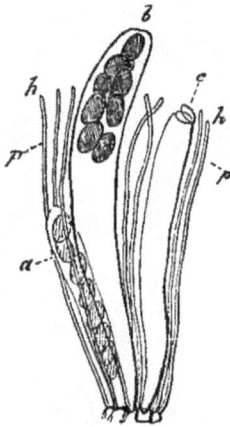

FIG. 45. *Ascobolus furfuraceus*, P. Portion of a section through the hymenium; *h—h* the upper surface of the hymenium shown by the extremities of the paraphyses *p*, *a* young ascus, *b* nearly ripe ascus projecting above *h—h*, *c* a similar one which discharged its spores during the observation and contracted with an open lid at the apex. Magn. 195 times.

When the asci are ready to eject their spores they are very much extended and their broad club-shaped apex rises considerably above the surface of the hymenium; this led to the erroneous idea which was reproduced by Boudier, that the asci became detached from their point of insertion and wandered up between the paraphyses; they really remain firmly attached, as in all the rest of the Discomycetes (Fig. 45). The projecting asci moreover are distinctly visible to the naked eye in the larger species as dark points, by reason of the dark violet-coloured spores in their apices. These points disappear at the moment of dusting, because the spores fly off and the empty tubes are drawn back beneath the surface of the hymenium. Older observers were led by these appearances to the mistaken notion that the entire asci were ejected from the hymenium, and hence the name Ascobolus.

---

[1] Crouan in Ann. d. sc. nat. sér. 4, VII (1857), p. 175.—Coemans. Spicilège, I (Bull. soc. bot. Belg. I, 1).—Boudier in Ann. d. sc. nat. sér. 5, X, p. 191

SECTION XXV. In most **Lichen-fungi** with open hymenia the mechanism for the ejection of the spores is similar to that which has now been described, though it differs from it in particular points which appear to me to require further investigation. The structure of the hymenia is essentially the same as in the Discomycetes; there is, according to Tulasne, the same turgescence of the mature ascus in both, and the same simultaneous ejection through one or more longitudinal fissures in its apex[1] The asci are emptied one after another as they ripen, and the spores, according to Tulasne, are flung outwards to a distance of about one centimetre; the sudden discharge of many asci at once has not been observed. The differences alluded to above are, that the apices of the asci do not project above the surface of the hymenium but continue on a level with it or a little beneath it, and that the pressure on the asci from without appears to be a chief cause of the bursting of the asci and of the ejection of the spores. The ejection of the spores is in fact due to the action of water, which causes a considerable swelling throughout the gelatinous hymenium in the direction of its surface, and consequently a lateral pressure on its turgescent asci. The pressure is moreover increased by the resistance which is offered to the superficial enlargement of the hymenium by the thallus which bears it, and which has less capacity for swelling by absorption of water, or by special thallus-margins or excipula circumscribing the hymenium, which, as Tulasne has shown, bend in such a manner when they absorb water, that they directly oppose the enlargement of the surface of the hymenium. Ejection of spores from an ascus withdrawn by isolation from the influence of these pressures, such as easily occurs in other Discomycetes, has never been observed, so far as I know, in any Lichen-fungus.

With regard to the Lichen-fungi which have perithecia, we only know that ejection from their asci also takes place[2], but the mechanism has not been properly investigated.

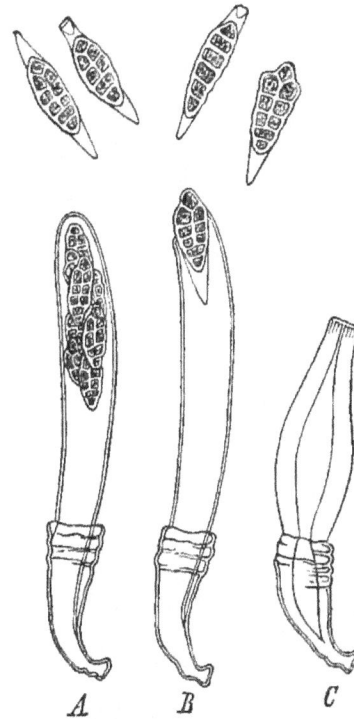

FIG. 46. *Sphaeria Scirpi. A* the ascus after elongation with the ruptured outer membrane at the base and the spores not yet ejected. *B* the last spore of an ascus sticking in the fissure waiting ejection; four already ejected are immediately above it. *C* ascus emptied of its spores. From Pfeffer's Physiology, after Pringsheim.

SECTION XXVI. **Successive ejection.** An isolated mature ascus of Sphaeria Scirpi is a broad short club-shaped body, as Pringsheim first showed[3], almost entirely filled by its eight large spores, which are crowded together in two irregular rows. It has an apparently homogeneous moderately thick wall with a double contour and lined with a layer of protoplasm. Ejection takes place under water. Before it

---

[1] Tulasne, Mém. sur les Lichens (Ann. d. sc. nat. sér. 3, XVII).

[2] Tulasne, l. c.—Stahl, Beitr. z. Entwickelungsgesch. d. Flechten, II, 1877.

[3] Jahrb. I, 189.

begins an extremely thin lamella of the wall, not previously distinguishable, suddenly splits off at the apex of the ascus, and the inner lamella issuing through it lengthens in a few seconds into a tube nearly three times as long and as broad as or broader than the original ascus; the lower part of the tube continues to be attached to the torn outer lamella (Fig. 46 *A*). The wall of the elongated tube is of about the same thickness as that of the original ascus-wall. The eight spores follow *pari passu* the apex of the elongating tube, keeping as nearly as possible their original grouping, the uppermost one being close to the apex. 'Soon the uppermost spore is seen to have moved into an aperture which has formed in the terminal point of the tube and then to be ejected through it with great force. As soon as this has happened, the tube shortens by about half the length of a spore, so that the second spore now touches the point of the tube and is pressed into the aperture, which it stops up. Then the tube elongates again to its original length and the second spore is then ejected with the same force as the first.' The entry of all the remaining spores into the opening and their successive ejection is accomplished in the same manner. Finally the empty tube, which is open at the apex, contracts rapidly to about $\frac{1}{3}$ of its length while the membrane continues to swell considerably, and eventually becomes disorganised without undergoing further changes. The whole proceeding occupies only a space of a few minutes. It is evident that we have in this case a modification of the discharging mechanism, the chief point in which must be that the width of the apical aperture is too small for a simultaneous discharge of the spores; but the question requires further investigation.

Successive ejection is at present certainly known only in a few Pyrenomycetes. That it should occur also in the open hymenia of Discomycetes is not probable; a brief statement of Crouan[1] about Vibrissea can scarcely be used here, and moreover needs confirmation. In the Pyrenomycetes with which we are concerned here the asci, like those represented in Fig. 44, are packed close together at the bottom of perithecia with a narrow canal-like orifice, and when they elongate they thrust themselves one after another through the canal without becoming detached from their insertion till their upper extremities are free from the perithecium, and then ejection takes place. Woronin was the first who described these proceedings correctly in Sphaeria Lemaneae[2]. His account holds good also of Sphaeria Scirpi, Phyllachora Ulmi, and Cordyceps militaris. The tops of the asci may be seen emerging one after another from the ripe perithecia of the latter species in a damp atmosphere, and ejecting their spores. Each rises about six times its own diameter above the mouth of the perithecium; in a few minutes the slender filiform spores fly one after another with the speed of an arrow from the tip of the ascus; each of these successive ejections is followed by a slight but permanent shortening of the ascus, which reaches the level of the orifice of the perithecium as the last spore is ejected.

The ejection of spores from the perithecia of Claviceps, which was observed by Tulasne[3] and may be seen with the naked eye as a puffing out of fine minute glistening needles, is undoubtedly produced in the same way as in Cordyceps, and my explanation of it at page 145 of my first edition was in the main incorrect. It would appear from the accounts which we possess that the same or similar processes are

---

[1] Ann. d. sc. nat. sér. 4, VII, p. 176.   [2] Beitr. III, p. 5.   [3] Carpol. I, p. 42.

common in the Pyrenomycetes, but further investigation of individual cases is desirable.

The last remark specially applies to a considerable number of Pyrenomycetes, in which the asci have the same structure as in Sphaeria Scirpi and S. Lemaneae and elongate in the same way if they are placed singly in water when they are mature ; among these are Sphaeria inquinans and S. obducens, Schm., Cucurbitaria Laburni, and some species of Pleospora[1]. See Fig. 47.

In all these species the membrane of the ascus consists of a thin outer layer with little power of swelling, and an inner soft gelatinous layer which swells to an unusual extent in water. If a ripe ascus is placed in water, the inner layer swells and breaks through the outer layer and protrudes in the manner described in the case of Sphaeria Scirpi. When the ascus is intact the inner layer is thin as compared with the cavity of the cell and appears to be tightly pressed between the unyielding outer layer and the protoplasmic utricle, which is tensely filled with fluid contents. As soon as the pressure upon it is relieved by the bursting of the ascus, it swells to such an extent in the direction of the longitudinal axis that the lumen is contracted into a narrow canal and the contents, whether spores or protoplasm, are driven out through the fissure. This happens in fully developed asci which are nearly mature, and in young half-grown asci ; in both the membrane in the uninjured state is thin as compared with the wide lumen.

In these species the ejection of the spores is rare, though it has been observed ; the asci when placed in water swell up at once, even after elongation, into a clouded gelatinous mass. This may be chiefly due to the fact that the asci of these land species which swell so readily have always been examined under very injurious conditions, in sections, for instance, or as crushed specimens suddenly placed in water, and not in their normal state —a treatment which the asci of Sphaeria Scirpi and S. Lemaneae which grow in water would be better fitted to endure. But the same remark applies also to other asci which

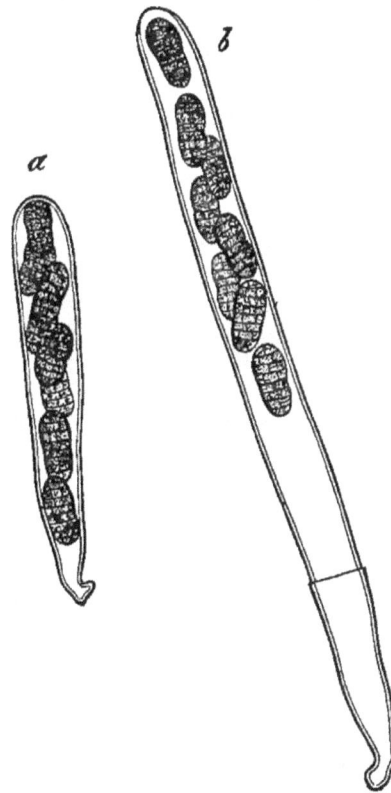

FIG. 47. *Pleospora herbarum*, Tul. (large form). *a* ripe ascus fresh from the perithecium with compound pluri-cellular spores. *b* the same after being placed in water, the inner membrane being extended and the outer ruptured. In this specimen the spores were ejected in the same way as in *Sphaeria Scirpi*, but they usually remain in the ascus in this species. Magn. 195 times.

do not suddenly elongate when isolated in water, as appears from the case of Cordyceps ; these and many others when isolated and immersed in water show more or less rapid gelatinous swelling of the walls of the asci.

Some of the asci of which we are speaking have characteristic thickenings on the apical portion of their walls ; in Cordyceps, Claviceps, and Epichloe typhina the apex is thickened and becomes a nearly cylindrical body, almost as long as the breadth of the ascus, pierced longitudinally by a very narrow canal, and set like a lid or cork on the thin lateral wall. If we recall to mind the thickened apex

---

[1] Tulasne, Carpol. l. c. and II, t. XXVIII, &c.—Currey in Microscop. Journal, Vol. IV, p. 198. —Sollmann in Bot. Ztg. 1863.

as it was described in Peziza Sclerotiorum, &c. (Fig. 43), which is extended by stretching into a thin membrane, it becomes a question whether the thickenings in the cases we are considering are not extended in the same way into thin membranes with the expansion of the ascus, and are to be considered therefore as reserve-pieces of membrane destined to be extended and to assist in the ejection of the spores, and comparable with the ring of cellulose in the vegetative cells of Œdogonium ; the matter at least deserves inquiry.

Thickenings of the apex such as those that have been described occur also in the asci of many Pyrenomycetes, in which ejection has never been observed. With these may be specially mentioned the conical projection in species of Rosellinia which has been recently examined by Crié[1], but is better understood and described by de Seynes[2]. In dried specimens of Rosellinia Aquila the cone is a cylindrically ovoid body projecting from the apex into the interior of the ascus, longer than the breadth of the apex of the ascus which it almost but not quite fills, and traversed by a narrow longitudinal canal ; in other words it is like a very thick annular ridge projecting from the inner surface of the wall of the apex : it is coloured dark blue with iodine, as has been often described. If the view expressed in the case of Cordyceps is correct, it is a question in the last-mentioned case also whether the thickenings at the apex are not reserve-pieces to assist in the ejection of the spores and destined to expansion. On the other hand, from Zopf's account of Sordaria Brefeldii (page 88) we might ask whether they possibly serve as means of fixing the spores in the apex of the ascus. All this requires investigation, in which each species must be separately examined, since ejection is by no means found in all the Pyrenomycetes.

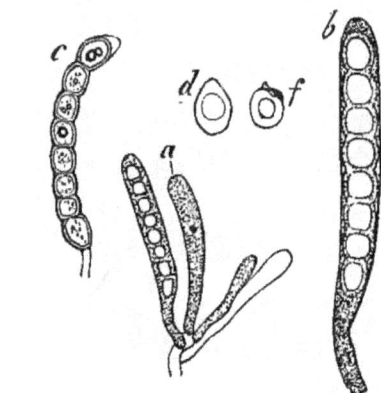

FIG. 48. *Sphaerophoron coralloides*, P. *a* young asci. *b* one of them more highly magnified. *c* a nearly ripe ascus. *d* outline of an isolated ripe spore. *f* outline of a similar spore from which all but a small portion of the dark violet episporium has been detached. *b* magn. about 700, all the rest about 390 times.

SECTION XXVII. **Liberation of the spores by solution or gelatinous swelling of the wall of the ascus** occurs, but not frequently, in free open hymenia. It appears, however, in the latter of the two forms to be characteristic of the Discomycete Roeslaria hypogaea[3]. It occurs in the first form, or with a disappearance of the ascus that cannot be more exactly defined, in Sphaerophoron (Fig. 48), Acroscyphus, and the Calycieae, to which the genera Lichina and Paulia which have perithecia are nearly allied, as has been shown by Montagne[4], Fresenius[5], and Tulasne[6]. The young spores in an early stage are almost as broad as the narrow and delicate asci, and are arranged in a single or in places in a double and uninterrupted row in the upper parts of the ascus, from the wall of which they are separated by only a thin layer of protoplasm (or glycogen-mass) *a, b*. They now enlarge more rapidly than the

---

[1] Comptes rendus, 88 (1879), pp. 759, 985.

[2] Comptes rendus, 88 (1879), pp. 823, 1043.—R. Hartig in Unters.—d. forstbot. Inst. z. München I, p. 20, t. II.

[3] Von Thümen, Pilze d. Weinstocks, p. 210.

[4] Ann. d. sc. nat. sér. 2, XV, 1841.

[5] Fresenius in Flora, 1848, p. 753.

[6] Mém. p. 77. See also Strasburger, Zellbildung u. Zelltheilung, 3rd ed. p. 54.

ascus-wall which surrounds them, and the protoplasm disappearing they at length entirely fill the cavity of the ascus; the wall of the ascus then forms a delicate septum between each pair of spores, and not unfrequently appears constricted between each pair *c*, but the septa are at length broken through in Sphaerophoron and most of the Calycieae, and the spores are thereby set free from one another and are collected together as a loose dust on the surface of the hymenium. In Lichina and Paulia the spores remain firmly united together.

The phenomenon which we are considering is very common in asci which ripen in closed receptacles. Chaetomium (Zopf) and Melanospora parasitica[1] may be mentioned first as trustworthy examples among the Pyrenomycetes. In these species the wall of the ascus swells when the spores are ripe into a copious jelly, which increases in volume by absorption of water to such an extent that it issues out of the orifice of the perithecium and brings out with it the spores which are imbedded in it. The tissue surrounding the asci also swells up and adds to the quantity of jelly employed in the way described. The mass of spores thus forced out of the perithecium collects at its orifice in the form of drop-like aggregations or twisted tendril-like filaments like a tough mass squeezed through a narrow tube. This mode of emptying of asci and perithecia is very common in Pyrenomycetes with thin-walled asci, occurring probably in species of Nectria[2], in Hypoxylon concentricum, Nummularia, Stictosphaeria, Eutypa, Quaternaria, and many other Xylarieae and Valseae described by Tulasne[3]. But more exact observation of individual cases is still required with reference to the considerations mentioned in this section. Perithecia without an orifice like those of Chaetomium fimeti and Cephalotheca tabulata[4] burst by the swelling of the gelatinous substance in a fixed manner.

The asci of Eurotium, Penicillium, Anixia truncigena[5], Onygena, Elaphomyces, and the Tuberaceae, which are also developed in closed receptacles but with no natural aperture, and opening only in eonsequence of decay or some accidental lesion, are dissolved or decomposed sometimes after temporary gelatinous swelling; they disappear entirely and allow the spores to enter free and unconnected into the cavity of the receptacle.

Section XXVIII. **The liberation of the spores takes place in almost all cases as soon as they have reached maturity,** that is as soon as they are fully developed, and this as a rule precedes the commencement of germination. In the few exceptions to the rule, which are rendered more remarkable by its generality and apparent necessity, the spores when they are liberated from the ascus increase considerably in size either at the expense of their environment or of the reserve material which they themselves contain, till they have reached the degree of development which strictly corresponds to the state of maturity. Elaphomyces is specially noteworthy in this respect[6], in which the spores, after they are set free from the perishing asci, grow to

---

[1] Kihlmann, Zur Entw. d. Ascomyceten (Act. soc. Fennicae), XIII, 1883.

[2] Janowitsch in Bot. Ztg. 1865.

[3] Carpol. II.

[4] Zopf in Sitzber. d. naturf. Freunde, see p. 85, and for Chaetomium, Nov. Act. Acad. Leopold Bd. 42, Nr. 5 (1881).

[5] H. Hoffmann, Icon. Analyt. III, p. 70.

[6] See De Bary, Fruchtentw. d. Ascomyceten, p. 33.

about double their former size inside the general receptacle without any essential change in their structure. A similar though less striking development is seen in Eurotium, Sphaerophoron, and the Calycieae (Acolium ocellatum).

Section XXIX. **Combinations of the different modes of formation and shedding of propagative cells** which have now been described bring about the appearance of the bodies which are often known as *septate spores*, but which it would be better to term *compound spores*, sporae compositae. In their case a spore-mother-cell or spore-initial-cell is developed either acrogenously or endogenously and then usually in an ascus, and developes by means of one or several successive bipartitions with firm partition-walls into a pluricellular body, in which each cell is an independent spore with power of germination. Such a two- or more-celled body formed of spores may remain persistent on its sporophore, or may be separated from it in one of the ways which have been described (see Fig. 34), or may be set free from a receptacle, while its members remain firmly attached to one another; and this is the case in the great majority of instances, and those the most typical. In these respects, therefore, its behaviour is the same as that of many simple spore-cells, which it resembles also in shape and size. Moreover it happens very frequently indeed in closely related species, and sometimes even in the same individual, that cells of exact morphological equivalence remain at one time undivided and form a simple spore, at another time become by division pluricellular compound spores. The teleutospores of Uromyces and Puccinia, the gonidia of Gonatobotrys which are one-celled bodies, and of Arthrobotrys which are two-celled (Fig. 21), are among the many examples of this kind, and especially in countless Ascomycetes with typical 8-spored asci the spore-primordia develope at one time into eight simple spores (Figs. 39, 43, 45), at another into pluricellular bodies, which escape in this form from the asci (Figs. 46, 47).

These facts have given rise to the phraseology which speaks on the one hand of simple unicellular spores, and on the other of septate or pluricellular spores (multiloculares, cellulosae of Corda, semen multiplex of Tulasne). If the expression spore is not to have a different meaning in different cases, but the same meaning in all cases, and this is what ought to be, it is obvious that it can only be applied to the single cell capable of germination; such expressions as pluricellular spores are therefore sheer nonsense, yet to get rid of them entirely would require the establishment of a perfectly new terminology, and to attempt this would be for many reasons a hopeless task. The evil may be to some extent lessened by the use of the expression compound spore and its correlative terms. The number of members (merispores) in a compound spore is different in different cases, two in Puccinia, Arthrobotrys, and Anaptychia ciliaris, three in Triphragmium, four in Phragmidium, Pleospora, and Sphaeria Scirpi, &c., and they are arranged in one row or in more.

I have several times called attention in other places[1] to the subject of terminology which has just been mentioned, and, as might be expected, without result owing to the overpowering influence of habit and want of thought in descriptive terminology. The whole matter may be made perfectly clear in form and expression if we start from the conception of a spore which is always presupposed in this work, and say that spores are produced from their primordia (primordial cells or initial cells) and these

---

[1] Brandpilze (1853); Flora, 1862, p. 63. See also my 1st ed., p. 123.

again from their mother-cells, such as asci and the like. Initial cells of the same morphological value may in case A become spores directly, in case B may become by division aggregates of spores, which were named above compound spores, but which it would be better perhaps to designate by· the term previously suggested of *spore-groups* or *sporiderms*. The descriptions would thus be rendered short and clear, and would stand the scrutiny of the attentive and thoughtful student. The result would justify the experiment.

That compound spores are produced by ordinary bipartition with septation of their initial cells, as represented in Fig. 49, has been recently confirmed by Strasburger[1]. I do not here go further into the subject of Körber's sporohlasts and the correlative terms[2]. The long filiform initial cells of species of Cordyceps[3] divide into a very large number of short cylindrical simple spores, but it is still a question whether the division is simultaneous or by successive bipartitions. In these Fungi moreover the compound

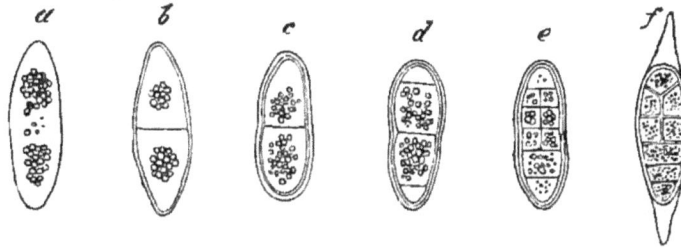

FIG. 49. *Sphaeria Scirpi*, D C. *a—e* successive stages of development of the spores in the order of the letters; all drawn from specimens inside recently isolated and uninjured asci. *f* ripe compound spores discharged from the ascus. After Pringsheim. *a—e* magn. 390, *f* 350 times.

spores, eight of which are formed in each ascus, break up of their own accord when they escape into their numerous component parts, each of which can germinate by itself; thus they are disuniting spore-groups, and their disunion may take place inside the ascus. We should be obliged to say in this case, according to the usual terminology, that each spore breaks up into many spores capable of germination. In nearly related genera, such as Claviceps, the filiform spores remain undivided till after germination, and put out germ-tubes. The spores also of Cenangium fuliginosum, Fr. behave in the same way as those of Cordyceps, while the spores of other species of Cenangium do not disunite[4]. De Notaris[5] gives an account of the disunion of the spores of Sporormia fimetaria, Not.

II. STRUCTURE OF THE RIPE SPORES.

Section XXX. In considering the structure of the ripe spore we must distinguish between *motionless spores* and those endowed with power of motion or *swarm-spores*.

The **motionless spores** are much the most common, and everything that was said in previous sections, with the exception of sections XVIII. *a* and XX. *a*, about the development and discharge of spores, referred only to them. In the mature state they are cells of extreme variety of form in the different species, most frequently round or longer than broad, in Claviceps and some others elongated cylindrical tubes.

---

[1] Zellbild. n. Zelltheil. ed. 3, p. 51.

[2] Syst. Lichen. German. Einleitung.

[3] Tulasne, Carpol. III; Bot. Ztg. 1867, 1.

[4] See Tulasne in Ann. d. sc. nat. XX, sér. 3, p. 135.

[5] Microm. Ital. Dec. V in Mem. R. Acad. d. Torino.

When ripe, and usually some time before they are ripe, they have a firm cell-membrane which may in very many cases be divided into two layers, an outer layer, the *episporium* or *exosporium*, and an inner, the *endosporium*, each of which may be itself stratified. In delicate or small spores the separation into two layers is very difficult to recognise, or cannot be recognised, before germination, and in some cases it never becomes apparent (Exoascus); in the latter case the cell-wall, often wrongly termed episporium, is a simple colourless or a coloured membrane.

In the numerous cases in which the two layers are to be distinctly seen, the outer one is usually a firm membrane, of various colours and various shades of colour, rarely quite colourless, and usually gives its colour to the whole spore. The surface is either quite smooth, as in most of the teleutospores of Puccinia and in many Pezizeae, or is more commonly furnished with thickenings projecting outwardly in the shape of warts, spikes, wrinkles, or a net-work of ridges, which vary in thickness and height in the different species, from the very delicate punctiform elevations of the gonidia of Puccinia coronata, Eurotium, &c. and reticulations of Peziza aurantia and Puccinia reticulata to the extremely thick warts of Genea, the spikes of Tuber melanosporum, Octaviania and Triphragmium echinatum, and the anastomosing ridges of Tuber aestivum. The outer layer is in these cases either homogeneous or stratified. In acrogenously formed spores a thin outermost lamella is often distinguishable from the rest, and is shown by the history of development to be the original delicate membrane of the spore-primordium, which developed with the spore, while the other lamellae are formed on its inner surface. The prominences on the surface of the spore often belong, as in uredospores and Corticium amorphum, exclusively to this outer envelope, which may be called the *primary lamella*. The compound or septate spores also are generally enclosed in the wall of the mother-cell, which developes with them as a close-fitting sac (Fig. 51 *t*). The episporium of some spores shows a differentiation perpendicular to the surface into portions of unequal density, a striation or areolation, along with or instead of the stratification. Fischer von Waldheim[1] has observed this in the Ustilagineae. The episporium of the aecidia of Phelonites strobilina, Peridermium Pini, Caeoma pinitorquum, Chrysomyxa and other Uredineae is particularly beautiful, appearing as if composed of small prismatic rods of denser substance perpendicular to the surface of the spore and connected together by narrower bands of a less dense and more transparent substance (Fig. 50). The convex extremities of the rods project outwards like warts. The structure is best seen when the episporium is made to swell by the application of sulphuric acid[2].

The endosporium is usually colourless or at least much paler than the coloured episporium, and is smooth and homogeneous or stratified; it is generally distinguished from the episporium by greater softness and delicacy, but by no means always by less thickness.

Some spores have a number of pores or pits disposed in regular order on their surface; the number of these is usually definite, but may vary within narrow limits in the same species. Many of these pores serve as places of exit for the tubular outgrowths from the spore at the time of germination, and may therefore be termed *germ-pores*;

---

[1] See sections LVI and LVII, and the literature cited there.
[2] See Reess, Die Rostpilzformen d. Coniferen; also Bot. Ztg. 1879, p. 803.

others perform no such function and are therefore only simple pores or pits. The position of the pores on the membrane is different in different species. The spores, for instance, of Sordaria fimiseda, de Not. have at their apex a germ-pore which is closed only by the outermost lamella of the membrane (Fig. 52). The germ-pores of the uredospores which I have examined, those of Puccinia for example and Uromyces, are sharply defined round holes in the endosporium, and are closed by the episporium on the outside; those of the teleutospores of the same genera appear to be pits in the episporium, which do not however extend into its outermost lamellae and seem to be closed on the inner side by the unpierced endosporium. Some, perhaps many, uredospores (those of Puccinia graminis for instance) show a pit at their point of attachment which is of no service for germination (Fig. 51). The spores of some Basidiomycetes, as Hymenogaster Klotzschii, are furnished with a pit at the same

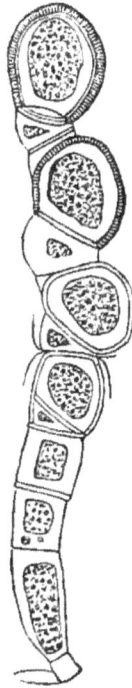

FIG. 50. *Chrysomyxa Rhododendri.* Basidium from an aecidium bearing a chain of spores. For the explanation see the accompanying text and on page 71. Magn. 600 times.

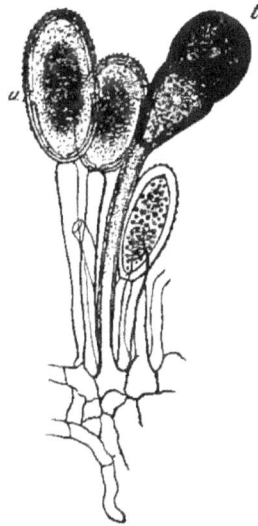

FIG. 51. *Puccinia graminis.* Small portion of a hymenium; *u* uredospores with four germ-pores in their aequator, *t* a pair of teleuto-spores, the upper one with a germ-pore in its apex. Magn. 390 times.

place which seems to belong to the episporium. The majority of the spores of the Basidiomycetes have no pit at this spot, but are prolonged into a small stalk which Corda[1] mistook for a pit. This stalk, as far as I could ascertain in the large-spored Corticium amorphum (see Fig. 30), is mainly a continuation or protuberance of the endosporium, over which the episporium is not continued or is present only as a very thin membrane; the stalk itself shows a narrow lumen, or its membrane is thickened to such an extent that the cavity has disappeared. There is no germ-pore in any of these spores at the point of insertion. The spore of Coprinus has one germ-pore, according to Brefeld[2], in the apex.

---

[1] Anleitg. p. XXXII. [2] Schimmelpilze, III.

The episporium of some Ustilagineae, especially Ustilago receptaculorum, shows also a broad pellucid mark covering $\frac{1}{3}$ or $\frac{1}{4}$ of the circumference of the spore, and corresponding to a thinner spot which passes gradually into the thicker and darker coloured wall around it.

Lastly, we may mention the delicate striae on the ovoid spores of Ascobolus furfuraceus and its allies. These, according to Janczewski, are thin places in the form of longitudinal slits which do not penetrate through the whole of the substance of the violet-coloured episporium, and often form acute-angled anastomoses. The colourless endosporium is smooth, homogeneous, and with its surface unbroken [1].

The various lamellae and strata of the spore-membrane, of which we have hitherto been speaking, are developed, as far as our present observations go, in the same mode and succession as those of vegetable cell-membranes generally; there is no reason therefore for entering further in this place into the details of their development than has been done in previous sections.

Besides the membranes, which have now been described, the spores of many Fungi have on their surface envelopes or appendages formed of a colourless transparent gelatinous substance, which swells greatly and usually dissolves and disappears under the influence of water, and shrivels up when treated with reagents which withdraw water. They may be termed *gelatinous envelopes* or *gelatinous appendages.* They are found alike in spores formed in asci and in acrogenous spores, in simple and in compound spores.

The ascospores of many of the Sphaeriaceae, as Massaria [2], some species of Sphaeria [3], Xylaria pedunculata first mentioned by Berkeley [4], and Fuckel's Hypocopreae and Coprolepeae, are surrounded with a gelatinous coating of varying depth and with very faint circumscription. A similar structure is found in Rhytisma Andromedae, Hysterium nervisequum, and other Hysterineae. The compound spore of Sphaeria Scirpi is enclosed in a delicate transparent sac, which fits close to its sides but is prolonged at either end into a long conical appendage [5] (Figs. 46, 49). The spores of many other Sphaerieae, species for example of Valsa and Melanconis [6], have a subulate appendage or a semicircular knob at their extremities. A semilenticular gelatinous appendage is found on the episporium on one side of the round spores of Peziza melaena and the ovoid spores of Ascobolus furfuraceus, P. and several of its nearest relatives [7], which swells into a hemispherical or spherical shape when the spores are discharged in water; Peziza convexula and Ascobolus immersus, P. [8] have the entire episporium encircled by a broad gelatinous coat. The common envelope, which encloses as in a sac the 8 or 16 spores in the ascus of the forms of Ascoboli distinguished as Saccoboli, must also be mentioned in this place.

Among acrogenously produced spores those of the plant known as Myxocyclus

---

[1] See Boudier, l. c. on page 92.—Janczewski in Bot. Ztg. 1871, p. 768.
[2] Fresenius, Beitr., and Tulasne, Carpol.
[3] Sollmann in Bot. Ztg. 1862, 1863.
[4] Mag. of Zool. and Bot. II, p. 224 (1838); see also Tulasne, Carpol. II.
[5] Pringsheim's Jahrb. I. t. 24.
[6] Tulasne, Carp. II.—Fresenius, Beitr. t. VII, 22, 21.
[7] Boudier in Ann. d. sc. nat. sér. 5, X.
[8] Coemans, l. c. on page 92.

confluens, Reess[1], have a broad gelatinous envelope round their large compound spore-mass. The spore-heads in Acrostalagmus and the heads of Myriocephalum botryosporum composed of crowded branching spore-chains are enclosed in a gelatinous envelope, and many other instances of the kind might be mentioned. To this place also belongs the apparently homogeneous often very soluble gelatinous substance which covers the gonidial layers in a great number of Ascomycetes, and in which their spores are imbedded. See above on page 70.

The morphological significance of the different appendages has still to be determined more exactly by the history of development. The gelatinous coatings and large gelatinous envelopes of entire hymenia must in the case of acrogenously produced spores be the gelatinous outer membranes of spores or spore-mother-cells, or the product of their coalescence. In the case of many spores produced in asci it is *à priori* probable that the appendages and gelatinous envelopes are also partial thickenings of the outermost lamellae of the membranes, or are due to the gelatinous character of the whole of the outer layer. This would be in accordance with Sollman's statements[2], but the confusion that reigns in them with regard to the most elementary principles of histology makes it impossible to trust them. The sac which encloses the spores of Sphaeria Scirpi is certainly the primary outermost lamella of the membrane which at first fits everywhere closely to the spore, and becomes expanded into the conical appendages at the extremities of the spore as it ripens. Zopf[3] maintains that such gelatinous appendages, especially in the Sordarieae, are portions of protoplasm which have not been used in the formation of spores, that is, speaking plainly, are the direct product of the protoplasm in the ascus which was not devoted to the spore-primordia. This view is not altogether new, for Kützing[4] before ascribed a similar origin to the entire episporium of Tuber; but it deserves consideration both as regards the gelatinous appendages and as regards the episporium as a whole, especially considering the analogy of the development of the oospores of Peronospora (Chapter V); it has not, however, been distinctly proved.

Appendages of a different origin from these gelatinous ones are peculiar to the ascospores of Sordaria fimiseda, S. coprophila, and others, and may even occur along with them[5]. The spores of the first of these two species are in their early stages small delicate ovoid cells prolonged below into a cylindrical stalk and rich in protoplasm. As all its parts increase steadily in size, soft gelatinous thickening of its membrane with fine longitudinal striation makes its appearance at both extremities of the spore, projecting outwards in the form of a sharply conical usually hooked process, and growing in size as the spore grows. When the spore has attained its full development the greater part of the protoplasm moves from its lower cylindrical into the upper ovoid portion, which is then divided off from the former by a transverse wall, while its membrane becomes thickened and stratified, and gradually acquires a dark violet colour; the cylindrical portion continues attached as a hyaline stalk to the dark spore (Fig. 52). As appendages of this kind arising from sterile sister or neighbour cells may be

---

[1] See Tulasne, Carp.—Fresenius, Beitr.

[2] Bot. Ztg. 1862 and 1863.

[3] Sitzungsber. d. Naturf. Freunde, Berlin, Feb. 17, 1880.

[4] Philosoph. Bot. p. 236.

[5] Woronin, Beitr. III.—Winter, Die deutschen Sordarien, Halle, 1873.

mentioned also the envelope-cells on the spore-clusters of Urocystis, which will be described in Chapter V, but not the germ-tubes which appear inside the ascus in Sphaeria praecox, and which were described by Tulasne [1] as filiform appendages.

We have already spoken, in concurrence with Zopf's views, of the physiological import of the gelatinous appendages as organs which may serve to attach the ascospores to one another and to the apex of the expanding ascus. This is evident in the case of the Sordarieae, in which the spores, each with its dark episporium, lie in a row in the ascus one behind the other and in contact with one another, and the conical gelatinous processes on each spore are firmly attached to those of its next

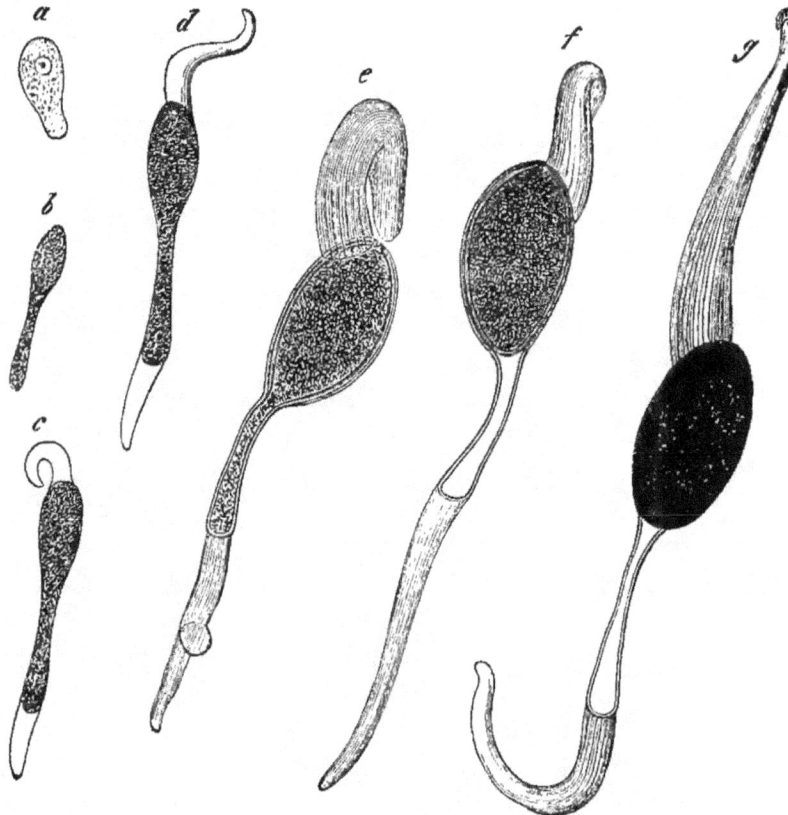

FIG. 52. *Sordaria fimiseda*, de Not. Development of the spores, the successive stages in the order of the letters. *a—f* from recently isolated uninjured asci; *f* a fully developed spore but with the yellowish brown membrane still transparent and the germ-pore visible above. *g* a ripe spore escaped from the ascus, with its membrane of a violet black colour. Magn. 390 times.

neighbour or are twisted round them. How far the function of these appendages is the same in other cases is a question which requires further careful investigation.

Of the *chemical nature* of the spore-membrane only some isolated facts are known, and the subject is still in need of more thorough examination. Hoffmann [2] has collected together a variety of details on this subject.

Most spore-membranes, according to the concurrent testimony of all observers, are distinguished by their great power of resisting decomposition and the influence of strong reagents, more especially concentrated mineral acids. Many are only slightly

----

[1] Carp. I p. 80.
[2] Pringsheim's Jahrb. II, p. 308.

affected even by concentrated sulphuric acid, the less so as a rule in proportion to the depth of their colour, and, as might therefore be expected, the episporium usually less than the endosporium. Others swell more or less strongly in sulphuric acid or wholly disappear. In very many cases the application of concentrated sulphuric acid is an excellent method of showing the minuter structural characters of the episporium, partly because it is rendered more transparent, partly because the other parts are destroyed or stand out from the episporium, which either bursts of itself or is readily made to burst.

Coloured episporia are usually more or less destroyed by boiling in potash; the reticulately thickened episporium of Tuber aestivum is entirely destroyed by this means according to Schacht[1], as is also the finely warted outermost lamella of the spores of many Uredineae[2]. These membranes therefore resemble to some extent the cuticle of the higher plants in their behaviour with reagents; but it has still to be determined whether they agree with it in other respects also, and are therefore of the nature of cork.

With a few exceptions, to be mentioned presently, spore-membranes are coloured yellow with iodine and sulphuric acid after maceration with potash or Schulze's solution, or, if not, are not coloured blue, resembling in this respect the majority of fungal hyphae.

These remarks apply to the single, simple spore in the strict use of the term, and to the compound spore also.

The gelatinous envelopes and appendages and the other gelatinous lamellae of which we were just speaking are affected by reagents in the same way as similar bodies in other organs and classes of plants. They are as a rule very perishable and soon disappear if the spores are sown in or on water.

The coloured episporium of the Ascoboli carefully studied by Boudier and Janczewski (l. c. on page 102) has peculiar characteristics which we must not attempt to describe in this place.

The entire membrane of the acrogenously produced spores of Peronospora behaves in exactly the same way as the cellulose of the higher plants towards iodine and sulphuric acid. Dilute solution of iodine is the only reagent which colours the whole of the spore-wall of Currey's Amylocarpus and the gelatinous envelopes of the spores of Xylaria pedunculata[3] an intense blue. The outermost lamella of the episporium of the spores of Corticium amorphum, Fr. (Fig. 30) with slender spike-like warts is coloured a beautiful bright blue with watery solution of iodine, and dark blue with iodine and sulphuric acid; the spike-like processes share in the coloration, but the inner thicker lamella of the episporium and the endosporium remain uncoloured.

The protoplasm of the spores of the Fungi is either dense and apparently homogeneous, or it has a greater or less number of granules or drops of oil disseminated through it, and in most cases appears to be colourless when a single spore is examined under the microscope: in a few cases only it is coloured by embedded pigments.

---

[1] Anat. u. Phys. II, 193. [2] De Bary, Brandpilze.
[3] Tulasne, Carp. I, II.

The fatty matter which it often contains appears in many cases in the form of spherical drops; such a drop, often surrounded by smaller ones, occupies the centre of the spore in Peziza Acetabulum, Helvella elastica, and other species. In many other cases smaller drops (f oil are distribu ed without arrangement in the protoplasm, or are collected at fixed spots in tolerably constant numbers. The best known and most remarkable examples of this kind are found in the ellipsoid spores of Peziza vesiculosa, P. Sclerotiorum, Helvella esculenta (Fig. 58) and their allies, which have one or more rarely two drops of oil in their foci. In P. tuberosa and P. hemisphaerica on applying iodine I saw spherical or irregularly shaped bodies which were not previously visible make their appearance at those points, and acquire the reddish brown colour of glycogen while the rest of the cell-contents became yellow.

A large proportion of the smaller granules, which are present often in considerable quantities in the protoplasm, may also consist of emulsionised fatty matter [1]. The reddish yellow pigment of the spores of the Uredineae and of Pilobolus may also be mentioned again in this place in connection with the fatty substances. See above, page 7.

If a nucleus is distinguishable in the young spores it can often be still seen in the same spores when they have reached maturity; but this is not always the case even where the protoplasm is not clouded by granules or large drops of oil.

It has already been said that a round pellucid body is to be seen in the centre of the protoplasm of some acrogenously produced spores, among the Hymenomycetes, for instance, and in the teleutospores of the Uredineae, the real nature of which is still undetermined, it being uncertain whether it is to be regarded as a nucleus or as a vacuole.

The 'nuclei' of older authors (before the year 1863) were for the most part drops of oil, the real nature of which can easily be determined by reagents. Corda and Tulasne, on the other hand, call the entire protoplasm of the spore the nucleus, which may be quite right in itself, but which is not compatible with the cell-terminology here set forth.

The protoplasm of the spore in the young state is rich in water, and when dry absorbs water rapidly from its environment. A spore lying in water appears under the microscope to be filled with it to turgescence. As it loses water it contracts, and if the wall is thin the membrane either sinks in irregularly or forms definite folds; round or ovoid spores take therefore the shape of a concavo-convex lens, the edges of which are often bent over towards each other, and the spore has thus the form of a boat. Thick-walled spores do not change their form in drying, or change it but little. In many cases an air-bubble is formed inside the protoplasm as it parts with water, as in Peziza abietina and P. melaena, in species of Sordaria, in Melanospora parasitica, &c.; so also in the gonidia of Cystopus (Hoffmann) and in the resting spores of Protomyces macrosporus. This is due to the fact that air, that is to say, some gas, is dissolved in the contents of the fresh turgescent spore, and is set free as soon as the quantity of water is brought down to a certain limit. The same result is produced if the spore in the water is exposed to the influence of reagents like alcohol, glycerine, or sulphuric acid which have the power of extracting fluids; the air-bubble disappears when water replaces these reagents.

---

[1] Hoffmann in Pringsheim's Jahrb. II, p. 308.

The spores of many Phycomycetes have the characteristics of autonomous motile cells and are therefore named *swarm-spores*, or *zoospores*, as having motion like animals. They are always formed endogenously by simultaneous division (section XVIII), and are liberated from the sporangium by the process of swelling described in section XX. Their origin and their development, at least up to the period of germination, take place only under water; the species which produce them are inhabitants of the water, or at least their sporangia find their way into water for the purpose of forming the spores. The swarm-spores of the Fungi are usually roundish or ovoid protoplasmic bodies without a firm cellulose-membrane. They generally contain one or more vacuoles arranged in a definite manner which varies according to the species; a nucleus has been found in recent times wherever it has been searched for and the size of the subjects permitted. A flagellum or one or two slender cilia with the power of lively motion spring from a definite spot in their surface as processes of the peripheral layer of protoplasm, and by their means the spore shows a movement of rotation in the water round its own axis and usually also a rapid movement of translation in space in all directions. A third movement of undulation and amoeboid change of shape alternates with the other two in most Chytridieae and in Monoblepharis. The mechanism of all these movements is not better known in the present instance than it is in other swarm-cells; they commence in some species (Saprolegnia, Pythium, the Chytridieae) inside the sporangium shortly before the liberation of the spores, and the cilia consequently are by that time already formed; in other cases, as Achlya and Cystopus, the cilia and the movements make their appearance after the spores have entered the water. The motion under favourable circumstances only lasts a short time in the swarm-spores of the Fungi, sometimes only one or a few minutes; the spores then come to rest, the cilia are drawn in or disappear, a delicate cellulose-membrane is formed, and either germination ensues or other changes take place in special cases which will be described further on. The direction of locomotion is not affected in most of the cases of which we are now speaking by rays of light falling on the spores on one side, but this is the case in certain Chytridieae; the spores of Polyphagus and Chytridium vorax[1] are phototactic, and the fact is all the more interesting, because these Fungi are parasitic on swarm-cells which are also phototactic and contain chlorophyll, namely Euglena and Haematococcus, and owing to the peculiarity just mentioned they are in a position to follow the movements of their hosts and to overtake them.

It is not desirable to discuss the little known mechanism of the movement of swarm-spores in this place, for that would necessarily involve the consideration of the same phenomena in the Algae, where they are more open to observation.

The zoospores of Fungi (Fig. 53) are usually ovoid or roundish lenticular with a thick blunt margin, often pointed at the extremity which is in front when the spore is in motion, and with one surface convex and the other slightly concave, so that when seen in profile they have the shape of a bean. A roundish pellucid spot, a vacuole in the granular protoplasm immediately beneath the surface, lies in the median line of the concave side a little nearer the anterior than the posterior extremity, and a long cilium springs from its anterior as well as from its posterior margin the anterior cilium being directed forwards when the zoospore is in motion, the

[1] Nowakowski and Strasburger. See Strasburger, Wirkung d. Lichtes und d. Wärme auf Schwärmsporen, Jena, 1878.

posterior being directed backwards and dragged behind. In Phytophthora infestans, according to older observations which perhaps need revision, both cilia spring from a point on the posterior margin of the pellucid spot.

These 'beanshaped' zoospores are peculiar to the Peronosporeae and Saprolegnieae. Similar but not the same forms are found in certain Chytridieae (see section XLVI). In the genera **Achlya, Achlyogeton,** and **Aphanomyces** the spores are discharged from an orifice in the sporangium without cilia and autonomous movement, and collect before the orifice in close lateral union with one another, forming the surface of a hollow sphere and a small head. As they pass from the sporangium on to the surface of the sphere, each of them becomes itself spherical and forms a thin firm cellulose-membrane. After a rest of some hours the protoplasm escapes from this cell-wall and then assumes the characters of the bean-shaped swarm-spore. It is only after the escape of the protoplasm that the cilia slowly develope at the spots which have been described, and as they develope the movement sets in, beginning as a slight swaying motion and passing by degrees into quicker rotation and finally into rapid movement of translation. In the sporangia of most species of the genus **Dictyuchus**, as described in section XVIII, the spores are separated by firmly united cellulose-septa; after some hours' rest the protoplasm issues from the cellulose which

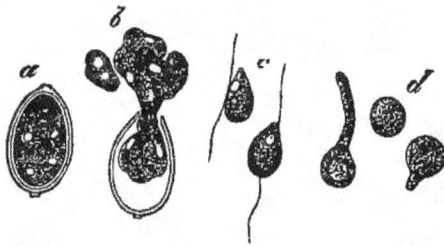

FIG. 53. *Phytophthora infestans,* Mont. *a* a sporangium lying in water after the division is complete. *b* escape of the 10 swarm-spores from the sporangium. *c* spores in the motile state. *d* spores at rest and beginning to germinate. Magn. 390 times.

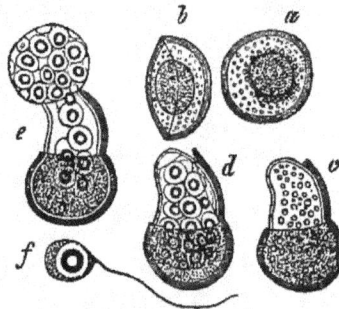

FIG. 54. *Cladochytrium Iridis.* *a* resting-spore with a brown membrane seen from the broad side. *b* the same rotated through 90°, in the centre is a large fatty spherical body. *c—e* successive stages of germination of a single specimen; the brown outer coat opens by a lid and the inner cell developes into a tubular receptacle of swarm-spores. *d* formation of the spores completed. *e* escape of the spores. *f* a single swarm-spore. *a—e* magn. 375 times. *f* 600 times.

surrounds it and then becomes a swarm-spore as in Achlya. But the spores are not discharged through a single orifice in the sporangium, but each pierces through the nearest spot of the lateral wall of the sporangium, and even also through the empty membranes of the adjacent sister-spores.

In the genus **Saprolegnia** a peculiar mode of proceeding is the rule. The spores are discharged from the orifice of the sporangium as ovoid motile bodies with the pointed extremity anterior in the swarming. This extremity is hyaline and two cilia project from its extreme point. The broad posterior portion is formed of granular protoplasm in which three small hyaline vacuoles lie immediately beneath the surface at laterally equidistant spots in the same transverse section. The spore comes to rest after a brief period of movement which does not last usually longer than a few minutes, assumes a spherical form and becomes invested with a thin cellulose-membrane; but some time after, some hours or even days, it emerges again from the state of rest, the protoplasm escapes as in Achlya out of the membrane and is transformed into a bean-shaped swarm-spore. Individual exceptions to this dimorphism of single spores occur in all species, in so far as a spore may omit the second swarming-period and pass directly from the first state of rest to germination.

The majority of the **Chytridieae** have small round swarm-spores which are capable of motion as they leave the sporangium (Fig. 54). Their protoplasm, which is otherwise tolerably homogeneous and transparent, contains one, or in exceptional cases two or more, round and comparatively large drops of oil which are colourless or coloured in shades of yellow and red according to the species, and are excentrically situated; a nucleus has been shown to be present in some cases[1], and it is more than probable that it is present in all. A single very long cilium arises from one point of the surface. The sudden curvature of the cilium occasions a backward and somewhat hopping movement of the spore often alternating with longer periods of rest, especially towards the end of the stage of movement; this stage either passes directly into that of rest or, as most frequently happens, into a stage exhibiting a creeping amoeboid motion, in which the cilium disappears or is dragged behind.

According to Cornu the zoospores of Monoblepharis behave in a similar manner to those of the Chytridieae. Further details and figures illustrative of these points will be found below in the sections of Chapter V which deal with the Peronosporeae, Saprolegnieae, and Chytridieae.

III. GERMINATION OF SPORES.

SECTION XXXI. Spores begin to **germinate** under certain necessary conditions, which will be considered further on, one only being mentioned here in passing, viz. a supply of water.

Since Prevost published his Mémoir on Caries and Ehrenberg his Epistola de Mycetogenesi the germination of a great number of representatives of most of the divisions of the Fungi has been observed and described. If the attempts made to procure germination have not hitherto in certain cases been successful, this is partly no doubt because special conditions are requisite for the purpose and these conditions have not yet been ascertained; with the study and determination of these the number of failures is constantly diminishing. On the other hand, beside the spores which have the power of germination, there are other cells in many species which resemble the spores in origin and structure, but which so persistently withstand all attempts to make them germinate that they must be considered to be incapable of germination. These organs are only mentioned here in passing; their further significance will be specially noticed below in section LXX.

The morphological process in germination consists generally in the fact that phenomena of development are exhibited in the spore, which are specifically distinct from those which lead to maturity.

These phenomena may vary, either in different kinds of spores, or in the same spore according to the external conditions; for instance the recently matured acrogenously formed spore (gonidium) of Phytophthora becomes the mother-cell of swarm-spores in pure water containing much free oxygen; on the contrary in nutrient solutions it usually puts out germ-tubes.

The changes in form which take place in germination group themselves naturally under two heads. First, the germinating spore becomes the mother-cell of new spores with or without important change of form, as in Protomyces, Phytophthora, and Cladochytrium (Figs. 41, 42, 54). We may in this case speak according to circumstances of spore-like spore-mother-cells or sporangia, instead of spores. This would depend on the object and requirements of each occasion, without prejudice to

---

[1] Strasburger, l. c. page 107.

the terminology here adopted. Secondly, the spore grows out into one or more tubular processes with the characteristics of hyphae, more rarely with those of the Sprouting Fungi. The two kinds are naturally connected together by intermediate forms, and an instance of this has been already in effect given in Fig. 54. Other instances and some partial exceptions in the simplest of the Chytridieae will be described in different places in Chapter V.

The modes of formation of sporangia in germination have been already considered in the foregoing sections; here, therefore, we have only the other

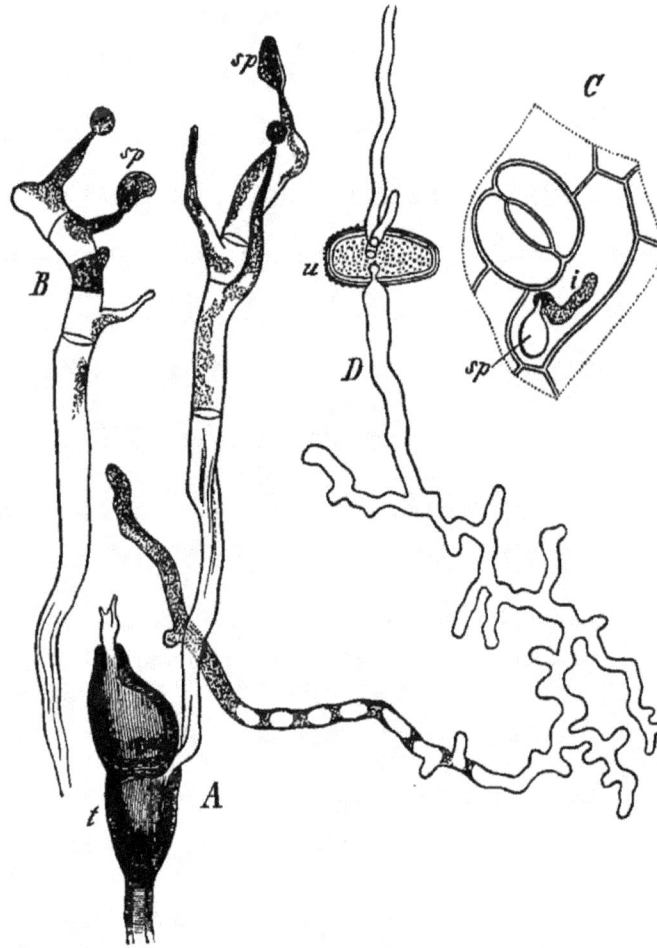

FIG. 55. *Puccinia graminis.* A pair of teleutospores germinating with promycelium and sporidia *sp.* B promycelium detached. C epidermis of the under surface of the leaf of *Berberis vulgaris* with a germinating sporidium, the germ-tube of which has penetrated into an epidermal cell. D uredospore putting out a germ-tube fourteen hours after being placed on the surface of water; four equatorial germ-pores are seen in the empty membrane of the spore. C, D magn. 390 times, A, B more highly magnified. From Sachs' Lehrbuch.

mode to depict, which may be termed *tube-germination* (Schlauchkeimung) and *sprout-germination* (Sprosskeimung).

The characteristic feature in **tube-germination** is that the spore grows out at one or more than one spot in its surface into a tubular process which is of the nature of a fungal hypha. This the first product of germination is accordingly known as the *germ-tube* (Keimschlauch). If the tube receives sufficient nourishment it develops directly in many cases into a mycelium or thallus like that of the parent, and it is therefore the *primordium of the mycelium* (Fig. 55 D). In other cases its normal development soon comes to an end; it elongates for a little while and then abjoints acrogenously at the expense of its protoplasm a small number of spores unlike the mother-spore, and itself soon dies away. The product of germination in this case bears the name of *promycelium* which was given to it by Tulasne, and the abjointed spores are termed *sporidia* (Fig. 55 *A, B,* Fig. 56). Both types of formation of germ-tubes are always peculiar to certain species and to certain forms of spores, on which point further remarks will be found in Chapter V; both behave alike in those first stages of their development with which we are concerned in this place.

Taking the simple non-septate spore-cells first, we find the simplest formation of the germ-tube among swarm-spores. As soon as they have come to rest and are provided with a membrane, they grow out at one or two or even more points into a cylindrical tubular process the membrane of which is the immediate continuation of the membrane of the spore (Fig. 53.) In most of the non-motile spores the proceeding is essentially the same, but with this difference, that the tube is covered only by a delicate continuation of an innermost layer of the spore-membrane. It is uncertain whether in the case of these spores the whole of the spore-membrane is ever prolonged as the covering of the tube; even where the spore-membrane before germination is delicate and without obvious separation into endosporium and episporium, the delicate wall of the germ-tube may often be clearly seen to be continuous with the layer of the inner surface of the spore-membrane, as in Acrostalagmus, Penicillium, &c. Even where an endosporium is stout its whole substance is not extruded to form the membrane of the germ tube, but only its innermost lamella, as in uredospores. The advancing germ-tube breaks through the outer layers where the episporium is strongly developed, causing them to open by

FIG. 56. *Rhytisma Andromedae,* Fr. Asco-spores germinating on the surface of water. The spore *x* is still surrounded by the original gelatinous border, which has disappeared in the other two ; *p* the promycelium, *s* a sporidium. Magn. 390 times.

valves or perforating them, and this either at spots not previously marked out by any special structure, or in places where pits were formed before the spores had matured and which were spoken of above as *germ-pores.* In a few cases the development of the germ-tube and the swelling of the inner part of the spore associated with it cause the layers of the episporium to break up into small pieces, as in Ascobolus sp., Diplodia sp.[1], and some others.

In some spores with a very stout episporium and a narrow germ-pore, as in Coprinus, Sordaria, and Chaetomium, the tube as it passes through the pore is a very slender protuberance, but immediately in front of the outer orifice of the pore it swells into a round and comparatively broad vesicle and after that grows on as a cylindrical tube, which may be branched or unbranched, a mode of growth which looks peculiar and has given occasion to strange misconceptions[2], but is really nothing more than a special case not deserving any particular designation.

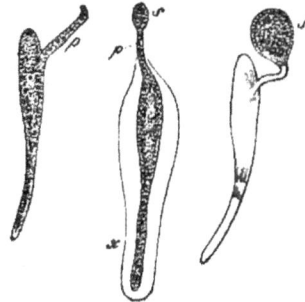

[1] Bauke, Beitr. z. K. d. Pycniden.
[2] Bot. Ztg. 1866, p. 158.

A more noteworthy special case which recalls the formation of swarm-spores is that of the germination of the acrogenously formed spores (gonidia) of the plasmatoparous Peronosporeae (Peronospora densa, Rab. and P. pygmaea, Unger) ; here when a spore is placed in water the whole of the protoplasm suddenly swells and issues from the papilla-like tip of the spore which opens to admit its passage, and assumes the form of a spherical body, which invests itself at once with a new and delicate cellulose-membrane and then puts out a simple germ-tube.

The number of the germ-tubes which proceed from a single spore is usually small, 1, 2, or 3, seldom only a few more. The germination of the spores of the genera Pertusaria, Ochrolechia, Mass. and Megalospora, Mass., first observed by Tulasne[1] and more closely examined by myself[2], is therefore all the more striking. These spores which

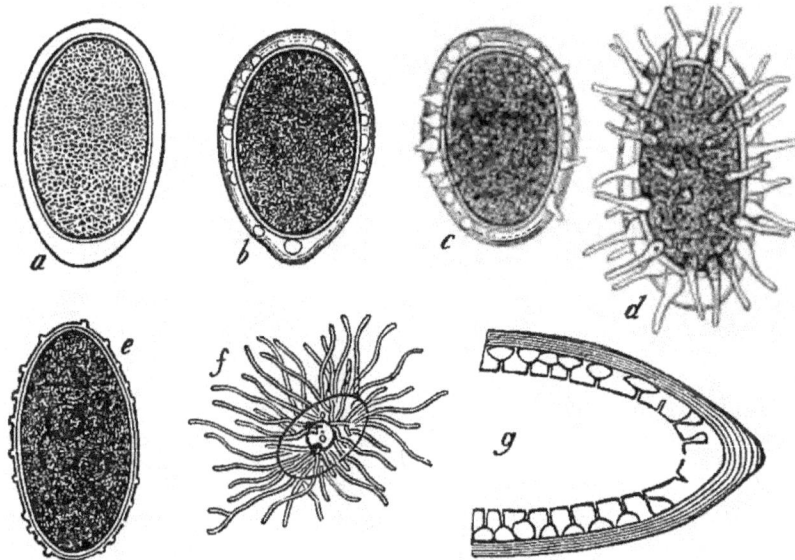

FIG. 57. *a—d. Megalospora affinis*, Kbr. *a* a ripe ejected spore. *b—d* successives tages of germination on a moist microscopic slide ; *b* and *c* optical longitudinal sections only, in *d* the surface also is seen. *e—f Ochrolechia pallescens*, Mass. ; *e* first beginning of germination in optical longitudinal section, *f* with elongated germ-tubes. *g Pertusaria deBaryana*, Hepp. ; optical longitudinal section through the half of a spore which is just beginning to germinate, showing the canals with their enlargements in the wall. The specimen was treated with glycerine ; the wall with its cavities appeared when fresh as in *b* ; the contents of the spore are omitted. *f* magn. 190, the other figures 390 times.

are formed in asci (Figs. 57, 59 A, B) are unusually large (in some species 180 μ and more in length) and ovoid or ellipsoidal, filled with a dense oily protoplasm and surrounded by a thick colourless stratified membrane usually of many layers. Each spore in germination puts out simultaneously a large number, 50–100, of slender tubes, which either spring from all parts of the surface of the spore, as in Pertusaria, or only from the side which is towards the substratum. The tubes when formed have no special peculiarities. The formation of a tube begins with the appearance of a narrow canal running from the inner cavity of the spore in the outward direction, and passing at a right angle through the inner layers of the membrane. The extremity of the canal enlarges in the outer layers of the membrane and at their expense into a lenticular or spherical cavity, in which a homogeneous protoplasm collects, and the cavity at once

---

Mémoires sur les Lichens (Ann. d. sc. nat. sér. 3, XVII).
[2] Pringsheim's Jahrb. V, p. 201.

becomes invested with a very delicate membrane of its own and appears as a small vesicle which elongates in the outward direction as the germ-tube and grows through the episporium. In the thick-walled spores of Pertusaria the tubes often ramify inside the episporium and the ramifications spread in it parallel with the surface of the spore. The canals in the membrane are, as far as can be ascertained, new formations at the time of germination, and are not enlargements of primordial formations. They continue usually so narrow that vesicles and germ-tubes appear at first sight to be completely surrounded. On acccunt of the slight thickness of the episporium in Ochrolechia these project at an early period above the surface of the spore, and, as Tulasne observes, may often be separated with the episporium from the apparently uninjured endosporium. The use of reagents, especially Schulze's solution, shows the state of the case to be everywhere as it has been described, and this is apparent in the large spores of Pertusaria even without their use (see the explanation of Figs. 57 and 59 *A, B*).

In all cases the spore absorbs water before the commencement of germination, and as a consequence of this it swells and forms vacuoles (Fig. 58). If it contains a reserve of food in the form of drops of oil, these are seen to decompose and disappear; the nucleus also becomes indistinguishable. As soon as the germ-tube begins to develope protoplasm moves into it from the spore. In many cases the germ-tube grows exclusively at the expense of the protoplasm and the reserve of food in the spore. Germination of this

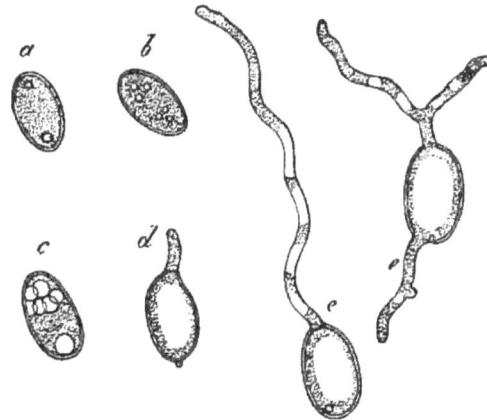

FIG. 58. *Helvella esculenta*, Pers. *a* a ripe ejected ascospore with the two characteristic focal oil-drops. *b—e* germination of ascospores in water. The stages of the development follow the letters. Magn. 390 times.

kind takes place when water only is present, and succeeds best in water; examples of it are to be seen especially in Fungi which lead a purely parasitic life, such as the Peronosporeae and Uredineae (Fig. 55); the large spores of Pertusaria, Ochrolechia, and Megalospora are also of this kind. In these cases the spores do not increase in size after the first formation of germ-tubes; their protoplasm and reserve of food, with the exception of an accidental and unimportant residue, passes over into the tubes in proportion as they develope; water takes the place of these substances in the original spore-cavity, and the spore-membranes thus emptied of their contents soon perish.

Other spores require a supply of nutrient substances as well as water in normal germination, or at least absorb them when present. They then increase considerably in size and their inner cavity is permanently lined with a layer of protoplasm, like a vegetating cell; in other words they are permanently vegetating portions of the mycelium that grows out of the spore, and it is not surprising that transverse walls should also be sometimes formed in them. Many Mucorini, as Mucor stolonifer and M. Mucedo, and the Sclerotinieae, as Sclerotinia Fuckeliana, are excellent examples of this kind. Intermediate forms are found, as might be expected, between the two extremes.

In compound spores each merispore germinates in the same way as a simple spore or has the power of doing so (see Fig. 59 *C*). It is not uncommon to see a germ-tube proceeding from almost every merispore, even where they are many in number, as in Pleospora herbarum and Cucurbitaria Laburni. Sometimes certain merispores only germinate as a rule, and if the cells are arranged in a simple row his is usually the case with one or both the terminal cells of the row, as in Melogramma Bulliardii, Tul., Melanconis, Tul., Aglaospora profusa, Not., Exosporium Tiliae and the stylospores of Cucurbitaria macrospora. The merispores which do not germinate gradually give up their contents to those which do[1], that is, their contents disappear and are replaced by water in proportion as the germ-shoots develope. But their membranes remain uninjured, suffering no perceptible perforations.

The **sprout germination** occurs in single genera and species, as Saccharomyces[2] Exoascus[3], Dothidea Ribesia, Fr.[4], and some species of Nectria[5], not to speak of certain doubtful forms like Dematium pullulans which will be described further on. Small processes with a very narrow base sprout, like commencing germ-tubes, from the surface of the spore, then generally assume an elongated or cylindrical form, and finally are abjointed

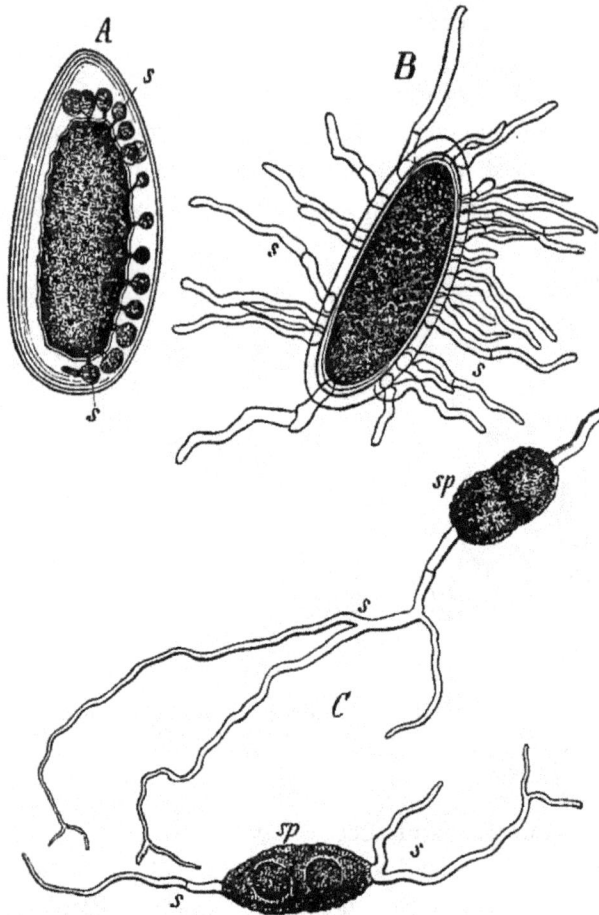

FIG. 59. *A Pertusaria communis.* Optical longitudinal section of a spore germinating on a moist slide after long lying in glycerine. *B Pertusaria lejoplaca,* spore with germ-tubes. *C Solorina saccata,* germinating spores *sp,* after Tulasne from Sachs' Lehrb.; *s* germ-tubes. *B* magn. 390 times.

in the manner described above in the case of the sprouting Fungi. A second and third or more sprouts may follow the first from the same point in the spore, till its protoplasm is exhausted. The sprouts may be formed at any point in the spore (Exoascus, Dothidea) or at fixed points, as at the two extremities of the fusiform dimerous compound spore of Nectria inaurata, or on the whole surface of the spore which is thus thickly covered over with sprouts which stand out from

[1] Tulasne, Carp. II, and I, p. 95. See also Cornn in Comptes rend. 84 (1877), p. 132.
[2] Reess, Unters. ü. d. Alkoholgährungspilze, Leipzig, 1870.
[3] De Bary, Beitr. I.—Tulasne in Ann. d. sc. nat. sér. 5, V. See also section LXXVII.
[4] Tulasne, Carp. II. t. IX.
[5] Janowitsch in Bot. Ztg. 1865.

it at right angles, as in Nectria Lamyi. In some of these forms, Saccharomyces, Exoascus, Nectria Lamyi and its nearest allies, sprouting is the only known mode of germination in the spores in which it is found; in others, as Dothidea Ribesia, spores of a similar character exhibit some of them sprout-germination, while others put out germ-tubes, which may or may not give off sprouts, and this is the case also with the ascospores of Bulgaria inquinans, in which the germ-tubes which give off sprouts swell into vesicles as they issue from the episporium (Fig. 60).

In the last-mentioned cases the proper growth of the tubes often ceases with the formation of sprouts; they resemble therefore the forms described above as pro-mycelia, and are connected with typical promycelia by a variety of intermediate forms which will be dealt with in Chapter V.

Finally it must be remarked that the modes of germination here described take place, as has been always hitherto assumed, after shedding and discharge of the ripe spores; but a number of cases are known among the Ascomycetes in which the spores germinate inside the ascus which has just matured, and form simple germ-tubes (Sphaeria praecox, Tul., Peziza tuberosa), or form sprouts (Exoascus, Peziza Cylichnium, P. bolaris, and especially Nectria). In some species this occurs quite exceptionally, as in Peziza tuberosa; in others it is very frequent, as in Sphaeria praecox, according to Tulasne, in Exoascus and Taphrina; it may even be the general rule, as in Nectria inaurata, Lamyi, and others[1]. As in the cases last described the products of germination are numerous sprouts which are abscised and from which in Exoascus new sprouts are at once

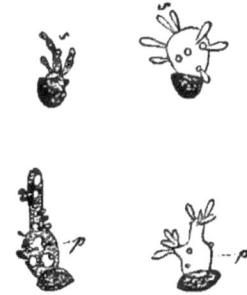

FIG. 60. *Bulgaria inquinans,* Fr. Spores germinating in water. *s* sprout cells. *p* short thick germ-tube to be regarded as a promycelium.

abscised, so in this case the ascus is often densely filled with the sprout-cells, so that the original spores formed after the type of the Ascomycetes (section XIX) are entirely concealed by them, a state of things which has given rise to all kinds of misconceptions.

Historical review of spore-formation. Organs resembling the seeds of Phanerogams and developing into new individuals were up to the times of Tournefort and Micheli (1707, 1729) supposed not to exist in Fungi, or were at least but little sought for, though it is true that a few places are to be found in older writers, which speak of the seeds of Fungi. On this point Ehrenberg, Ep. de Mycetogenesi, and Tulasne, Sel. Fung. Carpologia, Prolegomena, Cap. I, V, should especially be consulted.

The development of the spores was at first examined chiefly in the larger mushrooms. Micheli, Nov. plant. genera (1729), saw the spores grouped in tetrads on the lamellae of the Agarici (l. c. p. 133, tt. 73, 76), but did not perceive the mode of their attachment; but he saw the asci of Tuber plainly and the spores in the asci (l. c. p. 221, t. 102). Bulliard (Champ. de France, 1791) recognised the sterigmata (filets) on which the spores of the Hymenomycetes are placed, and O. F. Müller gave an excellent account of the spore-tetrads of Coprinus conatus in the Flora Danica, Fasc. xiv. in 1780; Hedwig, Descript. &c. Musc. frond. (1788), discovered the eight-spored asci of the Discomycetes, and he and the writers of the succeeding period gradually found these organs in the majority of the orders of Ascomycetes; Persoon

---

[1] Janowitsch, l. c.

especially described them in 1794 in Peziza, Helvella, Morchella, Ascobolus, Sphaeria, and Geoglossum in his epoch-making attempt at a classification of the mushrooms in Römer's Neues Magazin f. Bot., I, p. 62. Persoon's Icon. et descr. Fungorum, I (1798), pp. 6, 25, should also be consulted; also J. Hedwig, Theor. generat. et fructif. plant. Cryptog., Ed. 2 (1798), and from among later writers Ditmar in Sturm's Deutschl. Fl. III, 1, &c. Many of the accounts of these older authors are reproduced in Nees v. Esenbeck's System der Pilze und Schwämme, Würzburg, 1817.

The discovery of the asci in a large number of Fungi led first of all to the mistaken assumption that all the higher mushrooms, and especially Hymenomycetes, are furnished with similar organs. This view is expressed from the time of Persoon's attempt above mentioned, and especially of Link's Observationes in Ord. plant. naturales, 1 (Magazin d. Ges. naturf. Freunde, Berlin, 1809), down to modern times (Fries, Syst. Mycolog.; Epicrisis Syst. Mycolog.), and it is even represented, though somewhat obscurely, in figures (Nees, Syst. d. P.). Further historical details will be found in the writings of Berkeley, Phoebus, and Léveillé, which are cited below.

Vittadini, Monogr. Tuberacearum, discovered in 1831 or rediscovered the basidia of Boletus and Hymenogaster, but it was not till after the appearance of the classical and contemporary works of Léveillé, Recherches sur l'Hyménium des Champignons, (Ann. d. sc. nat. sér. 2, VIII, 1837), and Berkeley, On the fructification of Hymenomycetous Fungi (Ann. of Nat. Hist. I (1838), p. 80), that they became more generally known and more carefully studied, especially in the Hymenomycetes. Others obtained independent results which agreed with those of Léveillé and Berkeley, but were not published for some time after; as—

ASCHERSON, in Wiegmann's Archiv, 1838, and Froriep's Notizen, Band 50.

PHOEBUS, Ueber d. Keimkörnerapparat. d. Agaricinen u. Helvellaceen (Nov. Act. Acad. Nat. Curios. XIX, II), (1842).

CORDA, Icon. Fung. III, p. 40 (1839), in which Corda's earlier observations are noticed.

Berkeley and Tulasne were the first who supplied more exact information with respect to the basidia of the Gastromycetes (see end of section XCIV), and Tulasne more recently with respect to those of the Tremellineae in Ann. d. sc. nat. sér. 3, XIX.

Among later investigators of basidia J. Schmitz should be mentioned, Ueber Thelephora hirsuta, &c. in Linnaea, Bd. 17 (1843).

The asci were simultaneously examined by Léveillé and Phoebus (l. c.), but without adding anything very important to the results of previous observers.

In the more simple Fungi, the Hyphomycetes, Micheli (N. gen. t. 94) represents the acrogenously formed spores of Botrytis and Aspergillus as placed on the extremities of the hyphae. Succeeding writers for some time either gave similar accounts or were unable to satisfy themselves with regard to the origin or insertion of the spores. Corda in his later writings, Fresenius (Beitr.) and Bonorden (Allgemeine Mycologie) were the first to throw more light on the questions relating to the origin of the spores. The reader is referred to their works and to the descriptive literature; no distinct epoch marks the advance in our knowledge of these questions.

As regards more delicate questions of histology and development which have only recently been brought within reach of examination, I was myself the first to make more exact researches into the development of spores in asci in my work Ueber die Fruchtentwickelung der Ascomyceten, Leipzig, 1863, after various previous writers had prepared the way for more precise enquiries, such were—

NÄGELI in Linnaea XVI, p. 257 ;—Id. in Zeitschrift f. wiss. Bot. Heft. I, p. 45, Heft III and IV, p. 23.

SCHLEIDEN, Grundzüge, Aufl. 3, II, p. 45.

CORDA, Icon. fung. III, 38, V, 66, 74, 80.

FRESENIUS in Flora, 1847, p. 11.

SCHACHT, Pflanzenzelle, p. 50 ;—Id. Anat. u. Phys. d. Gew. I. pp. 71, 73, 170.

KÜTZING, Philosoph. Botanik, p. 236.

TULASNE, Fungi hypogaei ;—Id. Selecta fungor. Carpol I.

HOFMEISTER, in Pringsheim's Jahrb. Bd. II, p. 378 (Tuber aestivum).

SOLLMAN'S Beitr. z. Kenntniss d. Sphaeriaceen (Bot. Ztg. 1862 and 1863) made no important additions to our knowledge for reasons which have been already given.

More recent works on the Ascomycetes have confirmed in all important points the views published by me in 1863, while those of Strasburger and Schmitz quoted in section XIX have supplied the corrections that were rendered necessary by the modern doctrine of the cell. Boudier's account of Ascobolus has been confirmed by Janczewski (Bot. Ztg. 1871) in those points in which it differs from that given by myself.

The formation of the spores of the Mucorini has been described by—

CORDA, Icon. fung. II, p. 19.

FRESENIUS, Beitr. p. 6.

SCHACHT, Pflanzenzelle ;—Id. Anat. u. Phys. d. Gew. I.

HOFFMANN in Bot. Ztg. 1856 ;—Id. in Pringsh. Jahr. II.

COHN, Entw. des Pilobolus, N. Act. XIII.

COEMANS, Monogr. du genre Pilobolus, in Mém. prés. de l'acad. Brux. XXX.

DE BARY, Beitr. z. Morph. u. Phys. d. Pilze, p. 83, and in the literature cited in section XVIII.

Corda, Fresenius, Schacht, and Hoffmann consider the formation of the spores in the Mucorini as more or less closely allied to that which takes place in the asci, as a process therefore of free cell-formation within the protoplasm of the mother-cell and at the cost of a portion of it, and this view has been quite recently maintained by Brefeld (Schimmelpilze). In the same way the acrogenous abjunction of spores has also been regarded by later writers as a process of free cell-formation, which differs from that in the typical asci only in the circumstance that the daughter-cells arise in special protuberances of the ascus. Vittadini (Monogr. Tuberac.) goes so far as to make the spore of the Hymenomycetes and Gastromycetes arise inside the basidium and afterwards emerge from it enclosed in a protuberance of the inner layer of the membrane, the sterigma. Montagne takes a similar view in the Esquisse organographique. Schleiden also, in his Grundzüge, Aufl. 3, II, p. 38, and Schacht (Pflanzenzelle, p. 54, and Anat. u. Phys. d. Gew. I, p. 74) are of the same opinion, and H. Hoffmann in the Bot. Ztg. 1856, p. 153 and in Pringsheim's Jahrb. II, p. 303 adopts it in the most decided manner ; he says, ' One fundamental type with many variations occurs again and again ; the spores are formed by free cell-formation inside mother-cells (tubes), which sometimes becomes cemented with them, as in Phragmidium, Agaricus, and Phallus, sometimes only loosely envelope the spore or spores, as in Mucor, Peziza, and Tuber.' Van Tieghem and Le Monnier in the Ann. d. sc. nat. sér. 5, XVII, pp. 332 and 370, and sér. 6, I, p. 37 have recently re-introduced this way of explaining the acrogenously formed spores of Chaetocladium, Piptocephalis, and Syncephalis ; they represent them as produced endogenously and singly or in a simple row in sporangia placed close to one another, like the spores of Mucor or Mortierella, but they do not rest their view on distinct facts in the history of development. These notions are not in harmony with clearly ascertained facts, as Berkeley (Ann. and Magaz. of Nat. History, Vol. IX (1842), pp. 9, 283 note) and Tulasne (ll. cc.) have always contended ; they arose in the case of Schleiden from his erroneous views on the first principles of cell-formation, views which have long been abandoned ; in the other writers above mentioned evidently from a striving after the establishment of homologies, for which purpose, however, they would be superfluous if they were correct. The present state of our knowledge of cell-formation and cell-division, as it is briefly stated on page 61 and fully explained in the third edition of Strasburger's Zell-bildung u. Zell-theilung, admits of our conceiving of all spore-formation as special cases of the process of cell-division without doing violence to the facts, and makes any further discussion of the points in controversy unnecessary.

The historical data respecting the swarm-spores of the Fungi and their development are to be found in the following works :—

PRINGSHEIM, Entw. d. Achlya prolifera, (N. Act. Acad. Nat. Cur. XXIII, p. 1);—Id. Jahrb. f. wiss. Bot. I, 290 ; II, 205 ; IX, 191.

A. BRAUN, Verjüngung, p. 287, and Ueber Chytridium (Abh. d. Berlin. Acad. 1856).

DE BARY in Bot. Ztg. 1852, p. 473 ;—Id. in Pringsheim's Jahrb. f. wiss. Bot. II, 169. On the swarm-spores of Cystopus, Id. Ueber Schwärmsporenbildung b. Pilzen (Ber. d. naturf. Ges. Freiburg, II, 314, and in Ann. d. sc. nat. sér. 4, XIII ;—Id. Sur le développement de quelques Champ. paras. (Ann. d. sc. Nat. sér. 4, XX).

SCHENCK in Verhandl. d. phys. Ges. in Würzburg, IX.

B. PREVOST, Mém. sur la cause de la Carie ou Charbon des blés, Montauban, 1807.

LEITGEB in Pringsheim's Jahrb. VII, 357.

CORNU, Monogr. d. Saprolegniées, Ann. d. sc. nat. sér. 5, XV.

Lastly, the formation and abjection of acrogenously produced spores and the development of the spores of Saprolegnieae, Peronosporeae, and Mucor have been lately reviewed in the works of A. Zalewski, Ueber Sporenabschnürung und Sporenabfallen bei den Pilzen, Diss. and Flora, 1883, and by M. Büsgen, Die Entwickelung. d. Phycomycetensporangien, Diss. and in Pringsheim's Jahrb. XIII, Heft 2 (1882).

The special works quoted in Chapter V should be referred to for details connected with this subject.

DIVISION II.

COURSE OF DEVELOPMENT OF FUNGI.

CHAPTER IV. INTRODUCTION.

SECTION XXXII. There can be no question that the course of development of Flowering plants, Pteridophytes, Mosses, and most of the better-known groups of Algae is very nearly alike, in spite of all the multiplicity of detail and all the variety exhibited by the extreme forms in these great groups of the vegetable kingdom. In Ferns and Mosses where it is most absolutely differentiated two stages alternate periodically in each species; one of these takes its origin from the spore and ends with the formation of antheridia and archegonia, while the other is the regular product of the oosphere of the archegonium and ends with the formation of spore-tetrads. However much in the above-named classes the equivalent chief stages, that which produces archegonia and that which produces spores, may differ from one another in conformation, differentiation, and physiological character, it is quite evident that they agree together in the chief points here briefly indicated and in several matters of detail. Suppose the paths pursued by the development in the species to be laid down as similar geometrical figures, then the several similar members, the archegonia, spores, &c. will occupy corresponding sections in the figures. Members and stages of the development in different species, which thus correspond to one another, are said to be *homologous*. The view that homologous members of different species are the result of modifications of a member of an ancestral type is based on the theory of descent. The phylogenetic or 'natural' affinities of a species are proved by demonstrating the homologies.

It is an accepted truth, which therefore does not require a formal demonstration in this place, that strict homologies may be shown to exist between Ferns and Mosses on the one hand and Phanerogams on the other, and that these homologies imply not only phylogenetic affinity, but also that agreement in the periodic succession of the two main stages, or we may say in the general rhythm of the development in the different species, to which attention has just been called. The same may be said of relations of affinity between most groups of Algae and Mosses. It follows that all these portions of the vegetable kingdom are closely united by bonds of relationship into one large main group with a like course of development, in a word into one main series of the vegetable world which then divides, in accordance with special phenomena, into a few subordinate series[1].

It was for a long time a reasonable matter of doubt and may to some extent be so still, whether Fungi are naturally related to this main group, and at what point they

---

[1] See Bot. Ztg. 1881, 1.

connect with it.   The question was scarcely ripe for discussion at all before the middle of the present century, for scarcely anything was known of the life-history of the Fungi.   Botanists were acquainted with single forms only, and it was known that some of these were reproduced in a like form from spores, as a tree of a distinct species is reproduced from its seeds; hence every distinct form which produced anything that resembled spores was considered to be the complete representative of a species. These *form-species* and the *form-genera* composed of them had for the most part only a distant resemblance to non-fungal plants.   In the comparative simplicity of their structure they approached nearest to members of the lower groups of Algae, the course of whose development was also then very imperfectly known; hence the Fungi were always placed next the Algae, but no points of connection were established to show a closer natural affinity between the two groups.   But while Hofmeister's ' Vergleichende Untersuchungen ' were giving a new impulse to the comparative study of the Mosses and Ferns and of the Algae also, and supplying a number of new points of view, the study of the Fungi was directed into new paths after the year 1851, especially by Tulasne[1].   Starting from a few observations by older writers which had been repeatedly neglected Tulasne undertook to show that the form-species of mycologists up to that time did not in many cases represent a true species, but that such a form-species might be part with others of the cycle of development of a true species. He showed that a regular succession in time takes place in this development between the forms which belong to the species, that, as we now say, the appearance of the successive forms indicates the successive stages in the development of the species. Beginning with the Ascomycetes he extended these views in many publications, which will be cited below, over a large number of different groups of Fungi.

Tulasne rested his views chiefly, if not exclusively, on the proof of the anatomical continuity of the forms which occur in the fully developed state, of the origin, for example, of the sporophores in question from one and the same mycelium.   The sowing and culture of spores under careful management and watching, which were introduced especially by myself, gave us a more accurate acquaintance with the succession of forms.   The application of these two methods of investigation has at present resulted in showing[2] that a certain number of groups of Fungi are associated by the rhythm of their development with the main series of the vegetable kingdom indicated above, and approach nearest to the ovigerous Algae ; the stages of their development are homologous with the stages of those Algae from which the Fungi may be phylogenetically derived.   To these forms I give the name of **Ascomycetous Series** or **Main Series of Fungi**, composed of the Phycomycetes, Ascomycetes, and Uredineae.   Other groups of Fungi, especially the Ustilagineae and the Basidiomycetes, cannot in the present state of our knowledge be ranked with the Ascomycetous Series, though they connect with certain members of the series as phylogenetically derivative through divergent lateral lines.

It follows from what has now been said, that to give an account of the course of development of the Fungi we must submit the groups one by one to a comparative examination. This account should begin with the Ascomycetous Series and include a

---

[1] Comptes rend., 24 and 31 March 1851, and Ann. d. sc. nat. sér. 3, XV.
[2] See Beitr. z. Morphol. u. Physiol. d. Pilze, IV.

comparison with allied forms which are not Fungi. For the latter object it will be well to recall briefly in the first instance the phenomena generally characteristic of the course of development in those groups, which in a comparative account come next to the Fungi, viz. the Algae, Mosses, and Ferns. It is presumed that the details, for which the reader is referred to Sachs' Text-book or Goebel's Outlines of Classification, are already known.

SECTION XXXIII. In all the above groups, with some exceptions, which, however, are so extremely few that they may be at first disregarded, we may set out from a cell which is fertilised by the sexual act of conjugation, or from a cell-group, its equivalent, which has been shown to be everywhere homologous and may therefore be designated by a common name. For this name we may choose the word *archicarpium* (*archicarp*), meaning commencement of fructification, with reference to the evident fact that the body usually known as the fructification very often proceeds from it as the immediate product of development. In the Ferns and Mosses the archicarp is the oosphere in the archegonium; in the two chief classes of flowering plants it is the cells which are the homologues of the oosphere and in some cases bear the same name; in the Florideae it is the procarpium (procarp), which consists of a single cell or a small cell-group; in the ovigerous Algae, as Vaucheria, Chara, Oedogonium, Coleochaete, it is also a single oosphere which in this case is formed in the oogonium; lastly, in the isogamous Algae, as the Zygnemeae, Desmidieae, and Botrydium, it is any gamete-cell which conjugates normally with a similar cell.

If now we follow the course of development onwards from the archicarp, fixing our attention first of all on the simplest cases which are confined as it were to the most necessary members produced by development, we find the archicarp consisting of a single cell become under favourable conditions a *spore*, i.e. a cell capable of growing directly into a vegetative body like the parent, and giving rise to fresh archicarps as the last product of the development; with these archicarps the cycle of development begins anew. This is the history of Vaucheria aversa[1], Chara, Fucus, and Spirogyra. The case is one degree more complicated, when other conditions being the same one spore is not formed directly from the archicarp, but a pluricellular body unlike that which produced the archicarp, all or some of the cells of which have the qualities of spores in the sense indicated above, and so far as this is the case are capable of developing into other bodies bearing archicarps. Such a body the product of an archicarp and essentially serving to the formation of spores is termed a *sporocarpium* (*sporocarp*). The Florideae and the Mosses afford excellent and well-known examples of this kind. But we meet with a certain number of cases of a much less marked character which form a very gradual passage to the phenomena just mentioned as occurring in Vaucheria aversa and Fucus, for the ripe oospheres of Coleochaete, Oedogonieae, Desmidieae, &c., which divide into many or into four or only two daughter-cells developing into spores, are strictly speaking merely sporocarps which have undergone successive stages of simplification in the order here indicated. We may indeed say that if only two cells with the function of spores

---

[1] Walz in Pringsheim's Jahrb. V.

were formed by division from the spore which is the product of the archicarp of Vaucheria, it would be a sporocarp though a sporocarp of very simple form.

By the term sporocarp we designate bodies originating in the manner just described which serve almost exclusively to the formation of spores, and which cease to exist after having once with comparative rapidity developed a certain number of spores. This is distinctly true of the majority of the illustrative cases adduced above and of the simpler ones also. But here too a gradation of difference may be observed, not only in the number of different successive steps in the development which have to be traversed, but also in the number of repetitions of the same step. Compare in respect to the first point the rapid and simple development of the sporocarp of Riccia with the tedious and complicated development of Polytrichum or even of one of the Jungermannieae; in respect to the second point compare Riccia with Anthoceros.

If while other conditions are again the same the differentiation of the sporocarp is carried beyond the conventionally and traditionally determined limit, the formation of spores is confined to definite comparatively small segments or portions of segments of the body developed from the archicarp; and if the formation of these sporogenous segments is repeated in a periodical succession, which is typically and in most cases also actually unlimited, we cease to use the term sporocarp. We may adopt the word *sporophyte* or any other suitable and plain expression in its place. This case occurs when we pass from the Mosses to the Ferns and the classes of the Flowering plants arranged in series with them. Here again it is presumed that the facts on which the term sporophyte is based are already known. In the Ferns the leafy sporiferous plant is the sporophyte developed from the archicarp which is the product of the prothallium; in the Phanerogams the sporophyte is in like manner the entire growth from an oosphere (archicarp), with the exception of the embryo-sacs, each of which is the homologue of a spore and of the products of development other than the embryo which are formed in the sacs[1].

In flowering plants the sporophyte and the body from which it originates, the oosphere, are so related to one another as regards their size and position that the oosphere appears to be a small portion of the sporophyte. The opposite is the case in Vaucheria, Fucus, and Chara, where the sporophyte or sporocarp is reduced, so to speak, to a single spore which forms a small portion of its parent. In Ferns, Mosses, Florideae, &c. the relationships of the two bodies are less unequal; both alike appear as special stages of the development which proceed from one another in turn, and it is for this reason that these plants have served as guides in the determination of the morphology of the rest.

The course of development described above may follow the rule there laid down without any further complication. Instances of this may be seen in Chara crinita, in the apogamous Ferns[2] with some slight limitation which will be noticed hereafter at greater length, and also in Coelebogyne[3]. But these are only isolated exceptions. The almost universal rule, as is well known, is that the development of the

---

[1] It hardly need be observed that it does not affect the questions with which we are at present concerned, whether the strict homology is between fern-spore and embryo-sac or between mother-cell of the fern-spore and embryo-sac.

[2] See Bot. Ztg. 1878, p. 449.   [3] Strasburger, Ueber Polyembonie, Jena, 1878.

archicarp into a spore, sporocarp, or sporophyte is not possible without a sexual act of conjugation, and hereby, to speak generally and without regard to individual details, a morphological complication is introduced. The simplest case of the kind, that of the most typical conjugation, may be described as the union of two similar archicarps. In other cases male sexual organs are formed differing from the archicarps. The production of these organs is an essential distinction; they are essential segments of the stage of the development in which the archicarp is formed, and are absent from the other stage which is developed from the archicarp. The two stages have therefore been naturally distinguished as the *sexual* and the *asexual*; the distinction is quite correct since it corresponds to that which is the almost universal rule, but it is to be observed that the sexual function of the segment in question is of no moment in morphological consideration and comparison; and the distinction by putting forward the sexual function lays stress on that which is not generally essential, because, as we learn from cases of parthenogenesis or apogamy, sexual processes may entirely fail, the segments which have usually sexual functions being functionless or altogether wanting, without causing any essential change in the entire course of development.

It is indeed true that cases of parthenogenesis and apogamy do occur in which the elimination of the sexual functions is accompanied by a change in the course of the development; speaking figuratively, by a partial flaw or displacement in the curve which represents it. This is not always the case, as for example in Chara crinita. Instances of it occur in the apogamous Ferns, when the sporophyte, instead of being developed out of the archicarp, shoots out from the prothallium beside it or without its formation at all. It appears still more strikingly in the formation of adventitious embryos observed by Strasburger in certain Phanerogams. Here the course of development shows a distinct interruption of the strict homology with the nearest allied species, though the homology is at once and completely restored as the development proceeds. In such cases therefore we may speak of *interrupted* and *restored* homologies.

A further complication of the course of development is introduced by the assignment of the sexual functions and the corresponding morphological differences to different individuals, each the product of a single spore; by this means the differences are carried in extreme cases as far back as to the parent-spore, or, as in the Phanerogams, to the sporogenous segments of the sporophyte. This brief notice of the above well-known phenomena is all that is required in this place.

In the cases which we have been considering the ripe spore usually separates from its connection with the parent-organism and developes under favourable conditions into a morphologically and physiologically independent individual, to use the customary phrase, or into a bion. The sum-total of the bions produced from one parent-organism (and from its sisters) is termed a new or the next youngest generation. Solution of continuity and development of the separate segments into a new and independent generation may also occur at other places in the course of the development than that of spore-formation; the embryo of the Ferns and Phanerogams which is developed out of the archicarp is an instance of the kind. If the solution of continuity occurs several times and at dissimilar places in the course of the development, as for example in the Ferns, the course of development is divided into unequal generations of independently living bions proceeding alternately from one another. This phenomenon has been known since Steenstrup's time as *alternation of generations*[1]. This takes place in the Ferns between the sporophyte developed out of the archegonium as the one generation, and the prothallium which grows from the spore and bears archegonia as the other. The two chief stages in the course of the development answer in this case to the two alternating generations, and hence the term generation or alternating generation has been extended to the two relatively homologous chief stages in the course of the development which were distinguished above, and its rhythm as there described has been commonly termed the alternation of generations, without regard to the biontic independence or continuity of the successive stages. This is one meaning of the expression alternation of generations, and it is the one strictly adhered to by Sachs especially (see his Text-book). By this use of the expression the entire course of development is always composed of two alternating generations, the sexual which closes with the production of sexual organs, and the asexual which proceeds from the other and forms spores, or of the homologues of these two generations.

Further complications may make their appearance in the course of the development which we are describing and at very different places in it, inasmuch as portions besides the archicarp and its products may separate as reproductive organs from the body in one stage of the development, and grow into new independent bions resembling their immediate parent. Each of these bions may then under favourable conditions reproduce the other stage of the development, the other alternating generation, and thus return into the typical path. Organs of reproduction of this kind increase the number of single segments of a stage of the development; they may be compared to branches and are in fact connected with them by many intermediate forms, being generally distinguished from them only by the fact that they separate from the parent-form, while what is termed a branch does not. They usually serve in an especial manner for the multiplication and dissemination of the bions belonging to the particular species, and are therefore fitly termed *organs of propagation*. They are always asexually produced and separate in very different states of development from the parent-shoot; they may be highly differentiated shoots, or small tubers composed of a few cells, or single cells—*brood-buds, bulbils, brood-cells* (spores), &c. They are wanting in some species, as in Vaucheria aversa, V. dichotoma, Preissia commutata, many Filices and Dentaria pinnata, while they occur abundantly in their nearest allies, as for example in Vaucheria sessilis, V. sericea, Marchantia polymorpha, Lunularia, some Filices and Dentaria bulbifera. In the latter case they may not only become highly characteristic members of the species, but may actually multiply the species through an unlimited number of generations and always in the same form, and thus divert it from the typical rhythm of the development which is preserved unaltered by the allied forms. External causes often cooperate to a considerable extent in this process, and the possibility of a return to the typical rhythm is still preserved as was stated above. It is easy to ascertain this rhythm in the more highly differentiated forms, in the Phanerogams, Filices, Mosses, and some Algae. To do so in the lower and comparatively simply organised plants requires

---

[1] Ueber d. Generationswechsel, Kopenhagen, 1842.

careful examination and a comparison of allied species. In dealing with such plants the observer is confronted by difficulties arising from the strong preponderance of the propagative forms, from the variety of shapes which they assume in the same species,— a variety sometimes due to external causes, sometimes established by inheritance,—and lastly from the alternation in some species, partly due to external causes, partly also hereditary and constant, between purely propagative and fructificative generations of bions returning again to the typical rhythm. Excellent examples are afforded by the history of the development of Botrydium and Acetabularia[1]; many of the Oedogonieae are connected with Botrydium as presenting a simpler case of what may be termed facultative alternation of bions, while the gynandrosporous Oedogonieae in which this alternation is a necessary part of their course of development connect with Acetabularia[2]. In these instances we have succeeded in ascertaining the typical rhythm from the successive production and reproduction of the forms. In many of the lower species this has not yet been done, and there are still some comparatively highly differentiated species in which the rhythm of the development is not yet fully made out; this is especially the case with the numerous species of the Florideae, in which the propagative organs (tetraspores) are assigned to special plants having no archicarps or antheridia. We have not yet exactly ascertained the genetic relation or possible relation of alternation between these forms and those first mentioned; but we know that the organs of propagation do not occur on special plants in all the Florideae, and that they are not generally necessary members of the course of development in these plants, but are entirely wanting in some species.

Where fructificative and purely propagative generations of bions proceed alternately from one another, it is also quite natural to speak of alternating generations and of alternation of generations, using the latter expression therefore in a less limited sense than Sachs. The practice is adopted in this work in accordance with the original meaning of the word; alternation of generations here indicates every kind of course of development, which is made up of alternating generations of independent bions. In this very obvious sense the word is both necessary and convenient. In every other sense it means nothing more than the rhythm of the development, and the homologous alternating generations are merely homologous stages of the development. To term these stages generations, even where they are not independent bions, may be convenient for comparing them with homologous independent bions; but if the practice is carried beyond certain narrow limits, into the succession of shoots &c., every course of development necessarily becomes an alternation of generations, and the expression ceasing to have a special signification becomes therefore superfluous or objectionable.

It has already been pointed out that the main series of the vegetable kingdom contains species, in which the propagative form of development greatly preponderates, and is in fact the only one which does or can make its appearance through a long succession of generations. In these species therefore a fructification is a comparatively rare occurrence. But the existence of the species is not prejudiced by this morphological defect, for so it may be termed; and experience shows further that the preponderance of the propagative form of development may be carried so far, that no

---

[1] Bot. Ztg. 1877. [2] See Pringsheim's Jahrb f. wiss. Bot. I.

fructifications are ever formed. Clear instances of this are seen in Ulóta phyllantha and especially in Barbula papillosa among the Mosses, and in Allium sativum and some other species among the Angiosperms[1]. In these cases also it cannot be shown that the species has less than the usual power of maintaining its existence. The names show that we are dealing with species whose nearest congeners go through the complete normal course of development which includes the formation of fructifications, and it may be proved, as has been done in another place, that in the former species the fructificative members of the course of development have been excluded and have disappeared in the progress of the phylogenetic evolution, and that the species themselves therefore are in a *reduced* state owing to the loss of the most highly differentiated stages of the development. Homologies with allied and perfectly differentiated species may be followed for a certain distance, and then the homology breaks off, is *interrupted* and is *not restored*; the course of the development as a whole has become different from that of the type.

This interruption of the homologies and the above-mentioned restitution of them occur only in isolated species among the higher forms. If other species following the same general plan of development were to be produced phylogenetically from them, they would form secondary series in the vegetable kingdom which would diverge from the main series and not fit into it. It is possible that such secondary series do exist among the lower groups of plants which are not Fungi, but this is not yet certainly ascertained. It was hinted above that we must assume the existence of such forms among the Fungi, and proof of this necessity will be given in future sections.

The points of view explained in the foregoing paragraphs will enable us to understand fully the phenomena known as the alternation of generations in the vegetable kingdom. Other facts and other points of view may have to be considered in reference to some of the similar phenomena which occur in the animal kingdom, but we cannot discuss them here. Views differing from those here expressed, and as it seems to me unnecessarily complicated, have been advanced on more than one occasion. On this point the reader is referred to Pringsheim and other writers[2].

Section XXXIV. The general phenomena to which attention has been called in the preceding pages recur everywhere in the Fungi also, that is to say, the phenomena observed in the Fungi are only special cases of phenomena which are of general occurrence in the vegetable kingdom, and the Fungi from the purely morphological point of view are 'like any other plants.' This has gradually come to be understood since Tulasne's work of reformation mentioned above. He himself could not possibly have a clear view from the first of the meaning and application of his discoveries in every direction. He therefore named the phenomenon which he had discovered the *pleomorphy* or *pleomorphism* of the Fungi, especially of their reproductive forms, and as these expressions clearly indicate the discovered facts and do not go beyond them, they were good and correct at the time and are so still if we are speaking of that which Tulasne was discussing. If they have ever given rise to misapprehensions, it has not been the fault of their author, but of those who did not understand them.

---

[1] See Bot. Ztg. 1878, p. 481.

[2] See Pringsheim, Ueber d. Generationswechsel d. Thallophyten (Monatsber. d. Berliner Acad Dec. 1876), and the literature there cited.

Tulasne's doctrine of pleomorphy raised indeed a storm of controversy that lasted for some time. On the one hand a conservative opposition was naturally to be expected on the part of those who had already enquired into the species of Fungi, and had become accustomed to find a species in every detached form with some kind of spore-formation, disregarding entirely the question of the course of development and genetic connection. The most aggressive representative of this conservative opposition was perhaps Bonorden [1]; others were more cautious and professed to be open to conviction.

On the other hand a misunderstanding of the word pleomorphy led men of enthusiastic character to extravagant conclusions. 'The explanation of all enigmatical fungal forms has been reserved,' it was said, 'for our decade [2],' and the appearance of two or more different fungal forms in the same spot close to each other, or one after the other, was regarded at once as a case of pleomorphy of one species and proclaimed as loudly as possible. Many experiments in culture were also undertaken, and here too every thing that grew on the spot which had been purposely sown found its place if possible in the cycle of whatever pleomorphous species the observer happened to think of. If the character of the substratum was changed in such cultures, the fungal forms that were obtained were often different; nor could anything be more desired, for the phenomenon of pleomorphy appeared now to have found its physiological explanation in the effect of the various physical and chemical qualities of the substratum. This gave rise to accounts such as the following: Saccharomyces Cerevisiae, the Fungus of beer-yeast, grows in saccharine fluids in the form indicated by the name. If it is eaten by flies it develops in them into the primordia of Entomophthora; these grow in the dead insects or moist substrata either into the form which bears the last name (section XLV) or into Mucor (section XLI), but into Achlya (section XL) if the flies fall into water. Finally Saccharomyces may again be formed from Mucor in saccharine fluids, &c. &c. [3] Bail, Hoffmann, and especially Hallier, were the chief representatives of these pleomorphistic extravagances. Others eager to share the laurels of these discoverers modestly adopted their views. So far as these observations possess any historic interest in special cases they will be mentioned in succeeding sections. Such of them as, like Hallier's especially, only belong to the scientific *chronique scandaleuse* will not be further noticed; this course is justified by a critical notice of these matters which appeared some time ago [4].

A critical examination of these pleomorphistic aberrations shows at once that the different forms were not duly discriminated from one another, that no sufficient notice was taken of the fact that several species are social and are in the habit of appearing together, or of the possibility that one may take the place of another, or be developed as a parasite on another, and that strange spores may come from without and mix with the spores which it is desired to cultivate. It is true that an elaborate provision was made of bell-jars, glass tubes, and similar appliances to prevent the last source of error. But even admitting that the apparatus 'worked with absolute certainty,' it was forgotten in using it that unbidden guests might be introduced into the apparatus when the living objects were placed in it which were intended to be cultivated, and that in fact it was scarcely possible to prevent their intrusion without a more complete control of the conditions than is afforded by a non-microscopical apparatus. The first requisite in a morphological investigation which is also an enquiry into the history of development, namely the proof of the organic continuity necessarily

---

[1] Abhandl. aus d. Gebiete d. Mycologie, Halle, 1864.—Zur Kenntn. einiger d. wichtigsten Gattungen d. Coniomyceten u. Cryptomyceten, Halle, 1860.

[2] Bot. Ztg. 1856, p. 799.

[3] Bail, Ueber Krankheiten d. Insecten durch Pilze (Ber. d. Vers. d. Naturforscher zu Königsberg);—Id. Die wichtigsten Satze d. neuern Mycologie (N. Act. Nat. Curios. 28).

[4] Jahresber. ü. d. Leistungen u. Fortschr. d. Medicin, herausgeg. v. Virchow u. Hirsch, Jahrg. 2, 1867, Bd. II, Abth. 1, pp. 240–252; reproduced in Bot. Ztg. 1868, p. 686. See also a clear historic account by A. Gilkinet, in his work cited at the end of section XLIV (Mucorini).

maintained at every instant of the successive stages of the development, in virtue of which the later-formed member begins as a portion of the one which immediately preceded it, was neglected or even expressly declared to be unattainable[1].

It is obvious that criticism could not alter this required condition; and that it is not at all impossible to carry it out is shown by many investigations, by those especially which have been made since the appearance of the first edition of this work, but also by some of a much earlier date. They have left much it is true still to be done, but they have helped as a rule to clear up the subject, and have done away with extravagant notions. The scientific method in the conduct of these investigations was and is obviously that of strict observation of the whole course of the uninterrupted development, the ascertaining the organic continuity of the parts of the development at each instant of its course; this is simply the method by which we determine that the apple is a product of development of the apple-tree, and that the tree is produced from the apple; the logic does not change with the size of the objects or with our apparatus and manipulation. The technical expedients adopted in applying this method to the Fungi depend on the particular case. The use of the microscope is generally necessary on account of the small size of the objects to be dealt with, and we must often follow the development directly under the microscope, the objects being cultivated on a microscopic slide or in a moist chamber and on a substratum which is at once sufficiently nutrient and transparent. The juice of fruit carefully kept pure, decoctions of fresh animal excrement, saccharine solutions with addition of ash-constituents, or gelatine saturated with the above fluid matters have proved to be suitable substrata for most artificial cultures under the microscope. Which is best in each particular case must be determined by considering the habits of the species. A different method of cultivation to the one here described must be adopted in the observation of most parasitic Fungi, as will be shown in Division III. We are indebted to Brefeld especially for the perfecting of the technical procedure and technical methods in the cultivation of Fungi and particularly in their cultivation under the microscope, after the guiding principles had been already indicated by myself in the first edition of this work.

Mistakes may be made and doubtful questions arise even in the use of the most correct scientific method. Such of these as may be noticed in the sequel must not be confounded by the student with theories of pleomorphism which are now obsolete.

SECTION XXXV. As regards the **terminology** of the subject it may be observed, that in Fungi as in other plants the development of a new bion very often begins with a cell, which is detached from the parent or at least ceases to draw its nourishment from it, and then has the power of further development if the conditions are favourable. The origin and structure of such cells are very various, and are characteristic in each particular case. But the general phenomenon remains the same in all and requires therefore to be indicated by a general expression which has reference only to it. Such an expression is the word *spore* (spora) which was introduced by C. Richard and Link, and has continued to be used ever since in spite of all attempts to supersede it. At the same time there were good reasons for these attempts, which rested on the consideration that spores in the sense of Richard and Link may be produced in a variety of forms on the same species and at dissimilar places in the course of the development, and that it would be well to distinguish these forms according to their structure, development, and homologies. Accordingly A. Braun distinguishes in the chlorosporous Algae, for instance, between spores and

---

[1] See for example Bot. Ztg. 1867 p. 351; also De Bary, Ueber Schimmel u. Hefe, Berlin, 1873.

gonidia; Tulasne reserves the word spore in many Fungi for certain special forms of them, and uses a different word for all others. Sachs' practice is intelligible and precise[1]; he sets out from a consideration of the Ferns and Mosses, and reserves the word spore for the cells produced in the sporophyte or sporocarp and for their homologues in other plants, while for all such as are not homologous with them he uses the words gonidia, brood-cells, or some similar term. This plan may be easily carried out in the very distinct and comparatively limited domain of Ferns and Mosses, even if the formation of free propagative cells other than spores were not reduced in these plants to the very smallest possible amount. For the whole field of botany it is correct but impracticable because the homologies of many free reproductive cells in the lower Thallophytes are still unknown, and we are still in want of a satisfactory general expression to denote a distinct and evident phenomenon.

To meet this need we here conform to the *vox populi* and apply the term spore quite generally to every single cell which becomes free and is capable of developing directly into a new bion, without regard to genesis and homology. This definition excludes from the conception of a spore all oospheres and gametes which require fertilisation or conjugation, and all cells directly endowed with the male sexual function. The forms that come under the definition may for convenience sake be distinguished either by compounds of the word spore or by some special terms, and these may continue to be simply opposed to the word spore, where this is possible, as in Ferns and Mosses. The discrimination of spores will depend on different relationships, according to which the same spore may receive different names, as happens in all other things. For example we should distinguish—

1. According to sexual relations:
   a. spores which are developed from sexually fertilised oospheres, *oospores* as Pringsheim happily termed them; or the product of conjugation of two similar gametes, *zygospores.*
   b. spores not sexually developed.

2. According to structure: *swarm-spores* and *spores that do not swarm* (see above in Chapter III) and many special forms.

3. According to position in the course of development, homology: in many of the Hepaticae, Scapania nemorosa for example, we must distinguish between the *carpospores* (spores simply) which are produced in the fructification and the structures formed on the leaves and usually termed *brood-cells* or *gonidia*, and we may extend these expressions to all members of the vegetable kingdom which show the rhythm of development of the main series. Carpospores would thus be the spores which represent the stage of the development proceeding directly out of the archicarp or typically concluding its further development, such as the oospores of the Algae, the 'spores' of Mosses and Ferns, the carpospores of the Florideae; gonidia would be all other spores, as the tetraspores of the Florideae, most of the swarmspores of the Algae, the spores of many of the Fungi which are the homologues of these swarmspores and will be described in succeeding chapters; for these latter forms the term *conidia* introduced by Fries is in general use.

4. According to the mode of development: thus in the Fungi we have *ascospores (thecaspores), acrospores,* &c.

---

[1] Lehrbuch, Aufl. 4.

In this system of nomenclature the gonidium (3) for example of Oedogonium and Vaucheria would be a swarm-spore (2); the oospore of these genera would be non-motile spores, like the zygospore of Spirogyra; but the zygospore of Acetabularia and Botrydium is a swarm-spore, and all oospores are at the same time carpospores, and so on.   The terminology indicated under 3 apparently comes so near to that of Sachs, that the question may be asked why the latter should not be accepted as the foundation of the system; it will be well therefore to point out once more the fundamental distinction between Sachs' terminology and that now proposed. With Sachs spores and gonidia stand side by side as different things; in this work carpospores and gonidia are special phenomena subordinated to the general conception of the spore.   On practical grounds, which have been to some extent already indicated and which will be further explained below, as under the Basidiomycetes, we cannot do without a word to express this general conception as defined above. The terminology here proposed is therefore by no means unnecessary, as it might appear to be if we confine our attention to Ferns, Mosses, and some types of Algae.

The terminological sketch which has now been given supplies a scheme which is drawn from actual phenomena forming the general rule, and is applicable to them. But the highly complex reality is not designed after a single scheme; there are intermediate forms connecting phenomena which are separated in the scheme, exceptions to the rule, and no terminology can take all these exceptions equally into account. In such cases the systematic terminology must necessarily be modified to suit the particular occasion.   This must not be forgotten either here or in connection with the objects which will be described in the succeeding chapter.

A spore proceeds to a further development if the conditions are favourable. The commencement of this further development is termed *germination*.   Germination considered morphologically consists in a very large majority of cases in the construction of the vegetative body of a unit-bion, as for example when Fungus-spores put out germ-tubes, as described above in section XXXI.   But there are other cases also.   Not only are there those in which the plantlet growing from a spore remains a rudiment morphologically and in this condition is devoted to special functions, as in the germinating androspores of heterosporous Pteridophytes or the pollen-spores of flowering plants, but there are others in which the germinating spore produces other spores directly as daughter-cells and is entirely expended in forming them, as was described above in section XXXI in the case of certain Fungi. In other words in such cases the germinating spore becomes the mother-cell or receptacle of spores of the next higher generation, a *sporangium,* as such receptacles are generally termed.   The same object viewed in the different relations specified above may equally well be called a spore or a sporangium.   The same thing may occur in spores of the most different morphological value, as in the oospores of Oedogonium and Sphaeroplea and in the gonidia of Cystopus.   Spores which are strictly homologous and come from nearly allied species may develope, some as sporangia, others as a rudimentary thallus; for example, the gonidia of Peronospora (section XXXVII), the oospores of Oedogonium on the one hand and on the other those of Vaucheria.   The homology between the spore-forming oospore of Oedogonium and the sporocarp of a Moss has been already pointed out, and the former may accordingly be termed a sporocarp of a very simple kind; here we have the more comprehensive name implying a more comprehensive relation. If we succeed in getting a clear general idea of these modes of expression, which necessarily vary in all cases according to the various points of view, we shall have no difficulty in finding the term that is suitable to each particular case. It will be well to do this before proceeding to the consideration of the Fungi which often have so great a variety of spores.

Historical and critical remarks on the terminology. The word *spore*, together with the related term sporangium and others, is found first in Hedwig (Descriptio musc. frondos. Lips. 1787), who uses it in the same sense as 'semen' and promiscuously with it. It was next employed by L. C. Richard (Démonstr. bot. ou Analyse du fruit, 1808) under the form of *sporula*, and distinctly defined as the small body, functionally corresponding to the seed, in agamous plants, i. e. plants which have no embryo. Link (Elem. Phil. Bot. 1829) again introduced the words spore and sporangium, and adds the word *sporidium* for objects as to which it was not clear whether they were spores or sporangia; in the Fungi for instance he calls (l. c. p. 359) the acrogenously formed spores of Penicillium and Aspergillus sporidia. The construction of the true cell-theory necessarily resulted in the recognition of the fact that spores are reproductive cells. Fries (Syst. Mycol.) generally uses the word sporidium for the spores of Fungi. Berkeley (Introd. to Crypt. Bot. p. 269) terms endogenously formed spores sporidia, using the word spores for those that are acrogenously formed. These attempts to confine the term to special formations have been repeatedly made, but none of them could meet with decided success for the reason already given in the text. The mode of expression adopted in section XXXI, which limits the use of the word sporidium to spores abjointed from promycelia, was introduced by Tulasne in his work on the Uredineae and Ustilagineae. In species in which different kinds of spores had been observed, the difference in their position in the course of development, the difference in their homology as we should now say, has long been clearly recognised and fully appreciated. The expression spore or sporidium was then limited to those bodies which could be regarded as homologous with the spores of Mosses, having or appearing, though without sufficient reason, to have an equal 'dignity' with them. This idea was nowhere distinctly expressed, but was nevertheless everywhere implied. The other spores therefore which made their appearance in the course of development ending in the formation of these 'spores' required another name. Wallroth (Naturgesch. d. Flechten, 1825) called them *gonidia*, on the ground, it is true, of some misinterpreted observations (see Division III), and this term, though occasionally discarded, has maintained itself or been resumed, as has been partly noticed above. Kützing in the Phycologia generalis (1843), where the account of the matter has some of the old obscurity, A. Braun (Verjüngung, p. 143) and other writers may be consulted. In the Fungi Fries (Syst. Mycol. especially Band III, pp. 234 and 263) substituted the older word *conidium* for gonidium, clearly implying that the conidia answer to the gonidia of plants which are not Fungi.

Fries found his especial and most distinct examples of conidia in the Erysipheae (l. c. 234). Since the conidia are formed by acrogenous abjunction in these Fungi Fries seems to consider this as their only mode of origin, and to have chosen the name he gave them from their forming a powder or dust (κονία) on their conidiophores. He did not mean that all acrogenously abjointed spores were conidia, but all conidia are acrogenously abjointed. As this view is often in accordance with the facts, and men are strongly influenced by the word which they have been taught, the expression used by Fries has unintentionally given occasion to confusion, because while conidia were said to be of acrogenous origin, spores acrogenously formed were opposed as conidia to other spores which were formed endogenously, in spite of the fact that the two forms were most certainly homologous. This was the case in the Mucorini ; see Brefeld, Schimmelpilze, IV. 141, and sections XLI–XLIV below.

It is then quite clear that only the homology and not the mode of origin determines the nature of the structure in this case, and that conidia may be formed endogenously and even in an ascus, as well as acrogenously. On these grounds Fries' venture in terminology is not a happy one, and it would be desirable to discard the word conidium and employ gonidium in its stead. The confusion pointed out above would thus be avoided, and homologous structures would be named alike with one generally applicable name in Fungi and in all other plants. We need not at the present day trouble about the special terminology of the Lichens, in which the word gonidium has hitherto had a special meaning, because as there used it is not only superfluous but objectionable, as will be shown in Division III. Moreover its application in the sense explained above is only a restoration of the meaning which its author Wallroth intended to give it, though under a misunderstanding of the facts.

For further details on this subject the reader is referred to Tulasne's Carpologie, I. chapter VI.

SECTION XXXVI. According to the leading points of view here indicated and the present state of our knowledge a review of the course of development of the several groups of the Fungi arranges them in the following manner :—

I. SERIES OF THE ASCOMYCETES.

1. **Peronosporeae** (with Ancylisteae and Monoblepharis).
2. **Saprolegnieae.**
3. **Mucorini** or Zygomycetes.
4. **Entomophthoreae.**
5. **Ascomycetes.**
6. **Uredineae.**

II. GROUPS WHICH DIVERGE FROM THE SERIES OF THE ASCOMYCETES OR ARE OF DOUBTFUL POSITION.

7. **Chytridieae.**
8. **Protomyces** and **Ustilagineae.**
9. **Doubtful Ascomycetes** (Saccharomyces, &c.).
10. **Basidiomycetes.**

Groups 1–4 have been brought together under the name of **Phycomycetes** on account of their close approximation to the Algae.

Groups 7 and 8 in the second category will be considered in connection with the Phycomycetes ; group 9 naturally in connection with 5, and 10 with 6.

CHAPTER V. COMPARATIVE REVIEW OF THE SEPARATE GROUPS.

PERONOSPOREAE.

SECTION XXXVII. Of the Peronosporeae some species of Pythium live in the bodies of dead animals and plants, the greater number as parasites in the tissues of Phanerogams, and chiefly in the intercellular spaces of the host, though there are species which, like Phytophthora omnivora, also spread through the cells. The vegetable thallus consists of copiously and regularly branching tubes, which are at first non-septate, but are divided into chambers at a later time when sexual organs are being formed by comparatively few transverse walls disposed at irregular intervals. Many of the parasitic species put out from the flanks of the intercellular thallus-tubes numerous small branches, which vary in form according to the species and penetrate into the interior of the adjacent cells in the form of haustoria (Fig. 8). The development of the thallus ceases with the formation of sexual organs: female

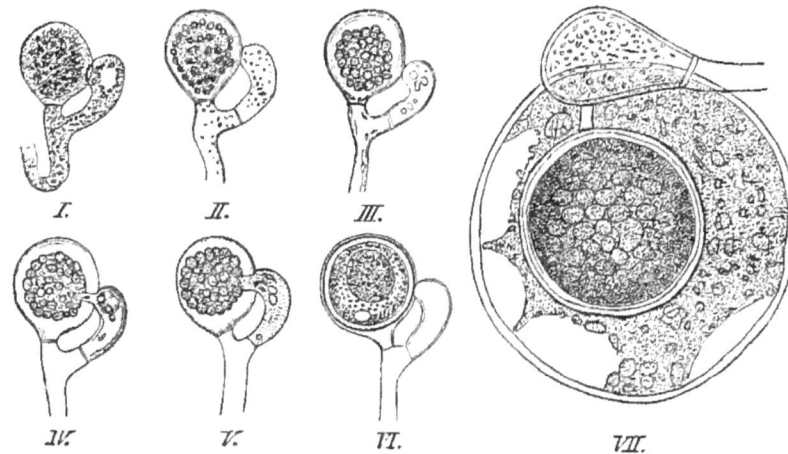

FIG. 61. Formation of oospores and processes of fertilisation in the Peronosporeae. *I—VI. Pythium gracile.* Successive states of an oogonium. *I* mature oogonium; to the right of it is an antheridial branch formed but not yet delimited. *II* antheridium delimited by a transverse wall. *III* the oosphere has rounded itself off in the oogonium, and a thin zone of periplasm lies between the oosphere and the wall of the oogonium. *IV* the antheridium has put out the fertilisation-tube, and a clear receptive spot is visible on the oosphere. *V* passage of the gonoplasm from the antheridium into the oosphere. *VI* ripe oospore surrounded by a thick membrane and almost entirely filling the cavity of the oogonium. *VII. Peronospora arborescens;* an oogonium with antheridium attached which has put out a fertilisation-tube. The oosphere is already invested with a thick membrane; outside it is a comparatively broad zone of periplasm, which is contracting to form the exosporium round the oospore. *I—VI* magn. about 800, *VII* 600 times.

organs, oogonia, in each of which one oosphere is formed, and male organs, antheridia, by which fertilisation is effected in the oosphere which developes into the oospore. The behaviour of these organs in the process of fertilisation varies considerably in the different genera. It shall be described first in the genus Pythium.

The *oogonia* of this genus (Fig. 61) are terminal or intercalary spherical

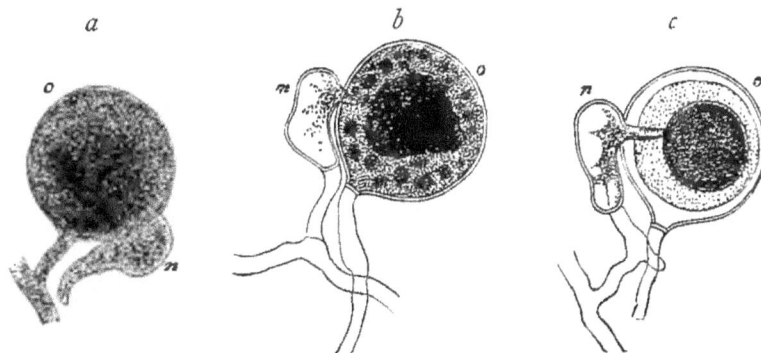

FIG. 62. Sexual organs of *Peronospora Alsinearum.* Casp. *a* young state. *b* formation of the oosphere and fertilisation-tube. *c* after fertilisation; periplasm somewhat contracted by the preparation, and the fertilisation-tube unusually thick in this specimen. *n* antheridium, *o* oogonium. Magn. nearly 350 times.

swellings of the thallus-tubes. Their surface remains smooth as it enlarges, or is rendered spiky in some species by projecting points. Their wall of cellulose is moderately stout. When they have reached their full size they are delimited from the thallus by one, or, if their position is intercalary, by two transverse walls, and are then filled with dense protoplasm containing numerous drops of oil (Fig. 61, *I*).

A slowly progressing separation begins in the protoplasm soon after the delimitation; the denser protoplasm with the oil-drops moves from the wall to the centre of the oogonium and aggregates there into a smooth globular body, the *oosphere*, which is surrounded by a delicate hyaline pellicle. The space between the oosphere and the wall of the oogonium continues to be filled with a slightly granular hyaline protoplasm, the *periplasm*, which may easily be overlooked (Fig. 61 *II, III*). About the same time that delimitation of an oogonium takes place the formation of at least one *antheridium* begins close to it (Fig. 61; see also Fig. 62 *n*). The antheridium is in the simplest case a younger sister-cell of the oogonium formed by delimitation of the contiguous portion of its parent-filament; terminal oogonia are therefore placed upon it as on a stalk-cell, which may either remain straight (Fig. 67) or have a characteristic curvature. In another series of cases the antheridium is the terminal cell of a special branch of the thallus, which grows towards the oogonium and attaches itself firmly to it; and this branch either springs from the same filament as the oogonium for which it is provided and close to it, bending over towards it, or it proceeds from some other branch of the thallus lying near the one which bears the oogonium. These circumstances of form and insertion vary sometimes in different species, sometimes in different individuals; so also the number of antheridia attached to one oogonium, which are often two, but in P. megalacanthum may be as many as six.

The cell-wall of the antheridium is not thickened, and its protoplasm is at first parietal and granular and not distinguishable from that of the thallus. When the oosphere has been formed in the oogonium, the antheridium sends a delicate cylindrical or conical tube-like process, the *fertilisation-tube*, from the part of its surface which is in contact with the oogonium into the interior of the latter; the tube grows till it reaches the surface of the oosphere, attaches its apex firmly to it and subsequently opens at that spot. Soon after the formation of the tube the protoplasm becomes differentiated in the antheridium; the larger and denser granular portion moves into the centre of the cavity, forming there an irregular and somewhat indistinct strand, the *gonoplasm*, while a thin layer (periplasm) remains on the wall of the cell. Then the gonoplasm passes slowly in most cases through the fertilisation-tube which has opened in the meanwhile into the oosphere, the transference of the whole of the gonoplasm being completed in from one to two hours. If there is more than one antheridium, they all usually, but not always, empty their gonoplasm, one after the other, into the oosphere. At least one antheridium discharging its contents in this way is invariably present on normal specimens, that is to say, on those which develope an oosphere. The function of the antheridium is fulfilled when it has discharged its contents. The oosphere at once becomes invested with a thick cellulose-membrane and ripens into the oospore (Fig. 61 *III–VI*). In the genus Phytophthora the process is in every respect the same as in Pythium, except that only a very minute quantity of protoplasm passes over into the oosphere through the fertilisation-tube, and this portion is not distinctly separated beforehand. Phytophthora omnivora has one noticeable peculiarity; the oogonium and its antheridium are formed almost simultaneously beside one another and develope in connection with one another. In Peronospora again such phenomena of development and fertilisation as can be observed are essentially the same as in the two preceding genera, but the passage of the protoplasm into the oosphere cannot be directly seen, and the periplasm in the oogonium is much denser and more copious (Figs. 61 *VII* and 62). Lastly the species of Cystopus behave like Peronospora in every point that can be distinctly seen, but the details in this genus require further investigation.

In Pythium, Phytophthora, and Peronospora the constituents of the protoplasm of the oosphere are rearranged as it begins to mature, and the structure of the ripe oospore is of the following kind (Fig. 61 *VI*; see also section XL, Fig. 69 *c*). The membrane becomes thicker and is composed of two main layers, an episporium and an endosporium; the latter usually swells up into a jelly in water in Peronospora, and both show the reaction of cellulose. The original oil-drops are dissolved, but their substance has collected into a comparatively large sphere pale on the outside and of weak refringent powers occupying the centre of the spore-cavity. The space between it and the membrane is either filled with protoplasm containing fine uniformly disseminated granules, or a layer of protoplasm of this kind lines the wall, but is separated from the central sphere of fatty matter by a limpid zone. When the spore is ripe a round or elongated and perfectly pellucid spot makes its appearance in the parietal layer close to the membrane. The nature of this spot has not been clearly ascertained; it is perhaps a nucleus, which is visible as a transparent central body when the oosphere is first formed and afterwards disappears from that position.

In Pythium and Phytophthora the periplasm cannot be seen to take part in the maturing of the oospore; it surrounds it in the form of an inconspicuous sparingly granular mass. In Peronospora (Figs. 61 *VII*, 62) on the contrary it develops into a thick membrane usually of a deep brown colour, the *exosporium*, which forms a closely fitting envelope round the ripe oospore, and has a structure and surface-sculpturing characteristic of each species. In this point Cystopus closely resembles Peronospora, but the structure of the ripe oospores still wants a more thorough examination. The wall of the oogonium undergoes no special changes as the oospore matures; in most cases it becomes decomposed; in some of the Perono-sporeae it becomes strongly thickened at the time of the formation of the oosphere, and may persist after the latter is ripe.

The oospores remain dormant for some time and then germinate under water; in some species they put out a germ-tube, which on a favourable substratum grows at once into a thallus which produces oospores in the same way as the mother plant; this is the only form of germination in Pythium de Baryanum, Artotrogus, and Peronospora Valerianellae. In other species the whole germinating oospore always becomes a sporangium; its protoplasm divides simultaneously into from several to many spores which swarm out of a previously formed tubular process, and after a short period of movement each developes like the spores of the first-mentioned species into a thallus that produces oospores. Cystopus candidus (Fig. 63) is an instance of this kind. In a third group of species some oospores show the one, some the other mode of germination, sometimes apparently under the influence of external causes; this is the case with Pythium vexans and P. gracile. A fourth group containing Pythium proliferum and Phytophthora omnivora exhibits a mode of germination which is to some extent intermediate between the first and second kind and which will be described further on.

The above phenomena close the cycle of development and in some species the course of development is actually limited to them; many observations at least have failed to detect anything further, for example in Pythium vexans and Artotrogus. We may therefore conclude that the essential points in the life-history of the whole group are confined to these phenomena.

But in most species there is this difference, that the course of the development is extended by the intercalation of numerous propagative cells, *gonidia*. In some cases indeed the formation of gonidia is actually a necessary part of the entire development; the germ-tube which proceeds from the oospore developes into

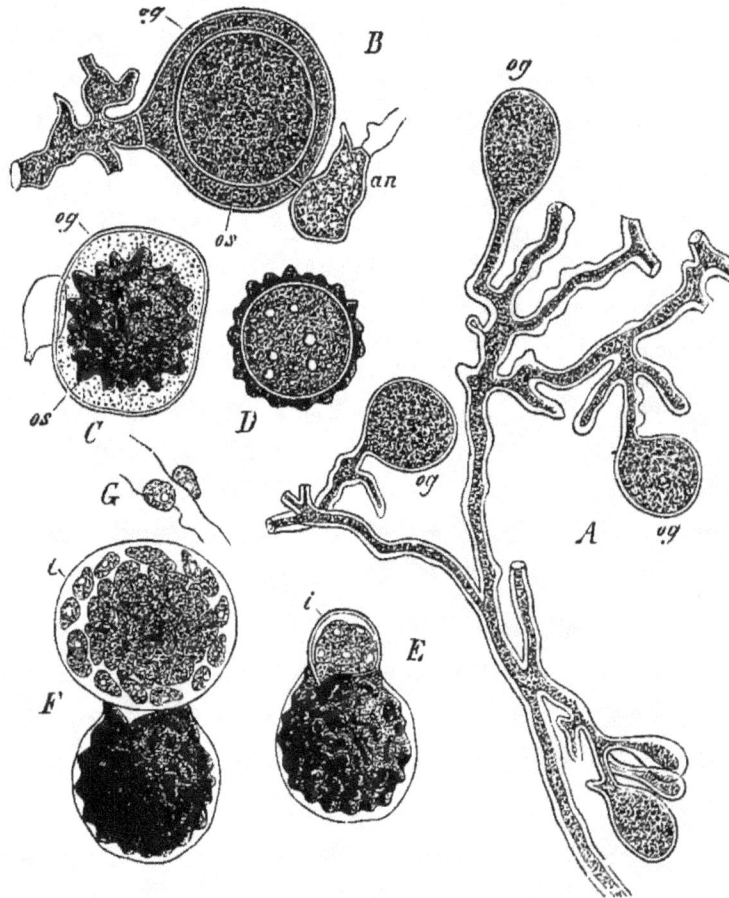

FIG. 63. *Cystopus candidus. A* mycelium with young oogonia *og. B* oogonium *og* with oosphere *os* and antheridium *an. C* mature oogonium *og*, oospore *os. D* ripe oospore in optical longitudinal section. *E, F, G* formation of swarm-spores from oospores; *i* endosporium. Magn. 400 times.

a small rudimentary plant, which we may call a promycelium (see on page 111), and this produces a few gonidia and then dies, while the gonidia give rise to new perfect fertile plants. This is the case with Phytophthora omnivora and Pythium proliferum which represent the intermediate mode of germination mentioned above. A very large majority of species, the two last-named among the number, form gonidia, not or not only in the way just described as terminal members of a short-lived alternate generation, but as accessory products of every normally developed thallus; the gonidia are usually produced in such large quantities as to further the propagation of the species to an enormous extent by their germination, and they have such characteristic forms, that the characters of species, and especially of genera, may be taken chiefly from the gonidial formations, while the few species which have no gonidia are not easily classified.

The main features in the formation of gonidia in the genera and subgenera of the Peronosporeae are as follows :—

Pythium. A persistent cell, usually the terminal cell of a branch, is delimited by a transverse wall and becomes a spore-mother-cell (sporangium). The gelatinously thickened wall at its apex suddenly expands into a thin-walled spherical vesicle, and into this at the same moment the whole of the protoplasm of the cell, which is hitherto undivided or has only shown transitory beginnings of division, streams rapidly, within a few minutes' time at most; there it breaks up at once into a number of swarm-spores, which issue from the delicate swelling vesicle and finally germinate. The sporangia in some species are of the same form as the gonidia of Phytophthora (Fig. 64),—round or ovoid vesicles prolonged at the extremity into a neck or beak, in the apex of which the swarm-spores are formed ; in others any portion, and often a very long portion, of the cylindric filamentous thallus-tube is delimited to form the sporangium, and its apex in which the swarm-cells are formed is then a small knob-like enlargement at the extremity of a branch, but is not otherwise distinguished by any particular form. There is usually no strict regularity discernible in the arrangement of the sporangia. In species with vesicular sporangia the filament which bears them often grows on into the empty sporangium from the point of its insertion, or through its entire length, and then forms a new terminal sporangium as in Saprolegnia (see p. 46). A certain kind of regular arrangement and succession occurs in one species only, Pythium intermedium, and in

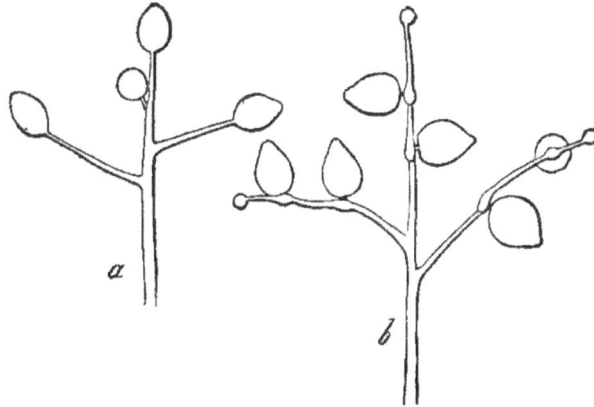

FIG. 64. *Phytophthora infestans.* Extremities of two simple sporophores. *a* formation of the first gonidia on the tip of each branch. *b* two ripe gonidia on each branch, a third beginning to form. Magn. about 200 times.

two ways, sporangia being either formed sympodially and successively on a branch of the thallus and separated by elongated portions of the thallus, as is described on pages 47 and 65, or by serial successive abjunction (see p. 66). In the same species the sporangia with their wall of delimitation are easily and abundantly shed from their sporangiophore, so that they may equally well be called spores, and the name is still further justified by the fact that under certain circumstances they may put out a germ-tube immediately without forming swarm-spores. P. de Baryanum also often forms spores (gonidia), which have the same form as the sporangia, in place of sporangia, but germinate directly by the emission of a tube.

Phytophthora (Figures 64, 65). Branches of the thallus either solitary or growing side by side in small tufts con-

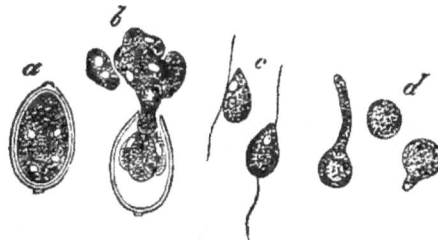

FIG. 65. *Phytophthora infestans.* *a* sporangium in water after the division is completed. *b* escape of the ten swarm-spores from the sporangium. *c* spores in the motile state. *d* spores come to rest and beginning to germinate. Magn. 390 times.

stitute peculiar gonidiophores and form gonidia. They generally send out a few branches monopodially disposed, and each of these secondary branches or the unbranched gonidiophore forms a number of gonidia sympodially and successively at elongated intervals; accidental and unimportant deviations from this plan are sometimes but rarely observed. The gonidia are usually abjointed as spores, and in germination, which takes place under water, become swarming sporangia. The swarm-cells are not formed in the same way as in Pythium but inside the original membrane of the sporangium, from which they issue through an aperture at the apex. As a frequent exception the germination of the spores abjointed from the gonidiophores is direct by the growth of a germ-tube.

Peronospora. Gonidiophores, disposed as in Phytophthora and regularly and usually copiously branched, give off by abscision a single spore at the extremity of every branch and then die away. The gonidia germinate in water and the process varies in different species.

(a) In a number of species they behave as in Phytophthora.

(b) In the majority of species they send out a germ-tube directly either from their extremity or laterally, and never form swarm-spores.

(c) In the plasmatoparous species mentioned on page 112 there is an intermediate form between (a) and (b) which is especially allied to the variety of (b) in which there is an apical germ-tube.

Cystopus. Gonidia are abjointed in a long simple chain from the apex of club-shaped branches arranged in tufts (Figs. 66 and 33). The branches are set close together and parallel to one another and form broad hymenia (see section XII). The gonidia in germination form swarm-spores only; the terminal member alone of each chain, except perhaps in C. candidus, may behave otherwise; it has a thicker wall than its younger sisters and is poor in protoplasm and incapable of germination. Tulasne's statement[1] regarding the peculiar form of the terminal members in C. Portulacae and its sending out a germ-tube has not been confirmed. As the gonidial chains of Cystopus begin to develope beneath the epidermis of phanerogamous plants and afterwards force their way through it, the thick-walled terminal members evidently serve to protect their younger sisters behind them.

The other genera of the Peronosporeae as at present determined entirely agree with those described above in the points with which we are here concerned.

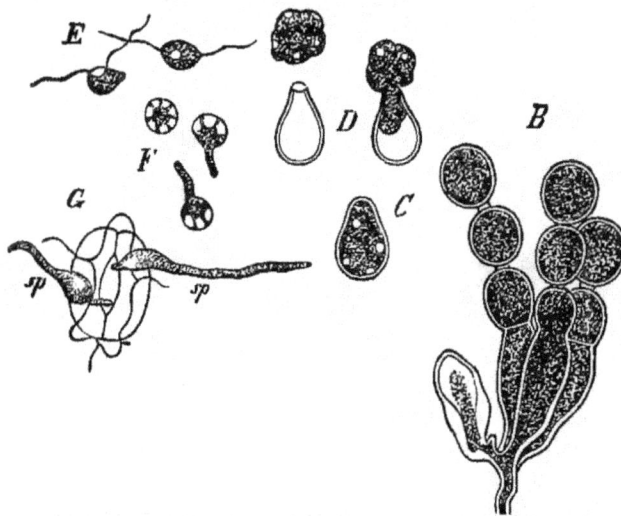

FIG. 66. *Cystopus candidus*, Lév. *B* gonidiophores. *C* gonidia in the act of germinating, i. e. forming swarm-spores; the protoplasm is already divided. *D* escape of the swarm-spores. *E* swarm-spores in the motile condition. *F* swarm-spores come to rest on a microscopic slide and germinating. *G* two germ-tubes which have penetrated into a stoma of *Lepidium sativum*; it is the inner surface of the stoma or epidermis which is seen, and the spores from which the germ-tubes come are on the outer surface at the stoma. Magn. 400 times.

In all the above cases each germ-tube proceeding from a gonidium either directly or through the intervention of swarm-cells developes on a suitable substratum into a thallus, which first forms gonidia and ends with the production of oospores. If the conditions are less favourable the plant forms often a luxuriant thallus, and abundance of gonidia through an unlimited number of generations proceeding from

---

[1] Second Mémoire sur les Uredinées in Ann. d. sc. nat. sér. 4, II, p. 110.

gonidia, but does not produce oospores. This may occur in all species; and among those which have been more thoroughly examined there are two, Phytophthora infestans and Pythium intermedium, which, as far as accurate observation extends, have lost the power of forming oospores, and produce gonidia only (see p. 125).

Literature of the Peronosporeae.

DE BARY, Recherches sur le développement de quelques Champignons parasites (Ann. d. sc. nat. sér. 4, XX, where the older literature, especially Tulasne's works, is cited;—Id. Beitr. z. Morph. u. Phys. d. Pilze, II;—Id. Beitr. IV, Unters. ü. d. Peronosporeen u. Saprolegnieen, &c.;—Id. Zur Kenntn. d. Peronosporeen (Bot. Ztg. 1881). Further notices of works are given in the two last-mentioned treatises.

PRINGSHEIM in Jahrb. f. wiss. Bot. I (Pythium).

M. CORNU, Monographie des Saprolegniées (Ann. d. sc. nat. sér. 5, XV, 1872);—Id. in Comptes rendus, XCI and XCII, 1880, 1881;—Id. Observations sur le Phylloxera et les parasitaires de la vigne. Étude sur les Peronosporées, II. Le Peronospore des vignes, Paris, 1882 (Acad.).

SCHRÖTER, Peronospora obducens (Hedwigia, 1877, p. 129);—Id. Protomyces graminicola (Hedwigia, 1879, p. 83).

FARLOW, On the American Grape-vine mildew (Bullet. of the Bussey Institution, 1876, p. 415).

A. MILLARDET, Le Mildiou, Paris, G. Masson, and in Journal d'Agriculture pratique, 1881, I, No. 6, and 1882, II, No. 27.

A. ZALEWSKI, Zur Kenntn. d. Gattung Cystopus (Botan. Centralbl. 1883, No. 33).

ANCYLISTEAE.

SECTION XXXVIII. The Ancylisteae are a small group of small plants, parasites on fresh water Algae and nearly related to the Peronosporeae and especially to the genus Pythium. The tubular thallus is at first unsegmented in all the members of the group and is divided into cells by transverse walls for the purpose of forming oospores; some of these cells swell into vesicles and become oogonia, the rest continue narrow and answer to the antheridia of Pythium. The two kinds of organs either lie side by side in the monoecious thallus, somewhat as in Fig. 67, as is the case in Myzocytium globosum of Cornu; or they are dioeciously distributed, some plants forming only oogonia, others only antheridia, and the antheridia send out a tubular filament by means of which they unite with the oogonia, as in Lagenidium Rabenhorstii of Zopf, which lives in Spirogyra, and in Ancylistes of Pfitzer, a parasite on Closterium. In these plants there is no distinct differentiation of oosphere and periplasm previous to the formation of oospores; it appears, on the contrary, that the partition-membrane between the antheridium and oogonium is first perforated, and the whole of the protoplasm of the antheridium passes into the oogonium, and it is not till then that the combined protoplasm withdraws from the wall of the oogonium and develops into the spherical thick-walled oospore. The more special structure of the oospore varies with the species. Their germination has not been observed. No organs are known in Ancylistes Closterii except those just mentioned, the propagation of the plant being effected by the thallus-tubes growing from one closterium-cell to another and forcing an entrance into it. The other forms named above produce zoospores after the manner of Pythium, and from sporangia which originate like the oogonia by transverse segmentation of the thallus. The group requires more thorough investigation; Pringsheim's Pythium endophytum[1] and Schenk's Achlyo-geton[2] are certainly very near it.

Literature of the Ancylisteae.

CORNU, Monogr. des Saprolegniées in Ann. d. sc. nat. sér. 5, XV, and in Bull. soc. bot. de France, XVI (1869), p. 222;—Id., cited in Sachs, Traité de Bot. translated by Van Tieghem, p. 328.—See also Schenk, Ueber contractile Zellen, Würzb. 1858, p. 9.

ZOPF in Bot. Ztg. 1879, p. 351.

PFITZER in Monatsber. d. Berlin. Acad. 1872, p. 351.

MONOBLEPHARIS.

SECTION XXXIX.   The aquatic genus Monoblepharis with three species, which at present has been examined only by Cornu[3], is related to the Peronosporeae. These plants, according to Cornu's short and somewhat incomplete description, resemble the Pythieae in their vegetative structure and in their mode of life.   They form on their thallus sporangia with swarm-spores, and the latter originate and escape, not in the manner of Pythium, but in that of Phyto-phthora, &c. and have only one cilium.   Oogonia and anther-idia are terminal or intercalary on the branches of the thallus; their disposition, which varies in the different species, agrees with that of some species of Pythium; Fig. 67 shows it in Monoblepharis sphærica.   The points in which M. differs from Pythium appear in its further development.   Firstly, the whole of the protoplasm of the oogo-nium with its numerous oil-globules is transformed into the oosphere without separation of

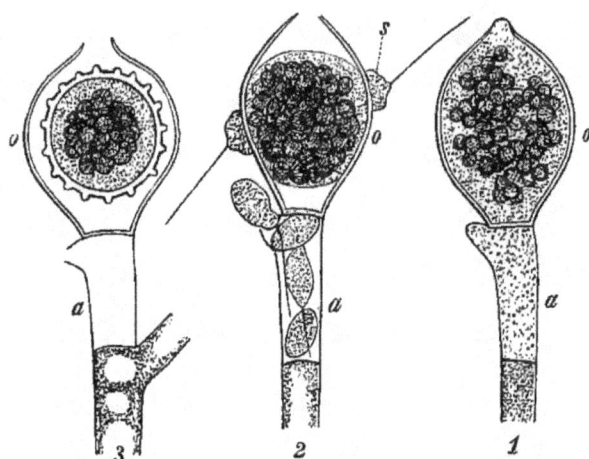

FIG. 67.   *Monoblepharis sphaerica*, extremity of a filament bearing an oogonium *o* and an antheridium *a*.   1. before the formation of the oosphere and spermatozoids.   2. oosphere formed, oogonium opened, spermatozoids *s* escaping from the antheri-dium.   3. ripe oospore in the oogonium which is borne on the empty antheridium. After Cornu.   Magn. 800 times.

periplasm and with diminution of volume, and as this takes place the wall of the oogonium opens at its upper end.   Secondly, a few swarm-cells (spermatozoids) are formed by division of the protoplasm in the antheridium; the spermatozoids escaping through an aperture in the wall of the antheridium move with a gliding motion over the wall of the oogonium, till one of them finds its way through the aperture to the oosphere and coalesces with it.   The oosphere is thus fertilised and an investing membrane is formed, which subsequently becomes strongly thickened and rough with warts on the outer surface; in this state the body is a resting oospore, the further development of which is unknown, but can hardly differ from that of the Peronosporeae.

---

[1] Jahrb. f. wiss. Bot. I.   [2] Bot. Ztg. 1859, p. 398.
[3] Ann. d. sc. nat. sér. 5, XV, 1872.

SAPROLEGNIEAE.

SECTION XL. These plants, which live on dead organic bodies in water, closely resemble the Peronosporeae in the course of their development and to some extent also in habit; they are most of them of large growth, with tubular hyphae 1–2 cm. in length standing out from the substratum and slender rhizoids spreading through it (Fig. 68). They differ from the Peronosporeae chiefly in the development of the oosphere and in the circumstance that in all the better-known species the antheridia though existent do not perform their fertilising function, or are entirely wanting. The oogonia are formed on branches of the thallus-tubes as in the Peronosporeae, and the whole of the fatty protoplasm is transformed into a single oosphere, or divides into several portions which become so many round oospheres, without any separation of periplasm. The number of oospheres varies in both species and individuals. Most species have as a rule several oospheres in an oogonium, some have as many as 30 or 40 or more, feeble specimens often only 2–4. The oospheres ripen into oospores, which in most cases have the same structure as those of the Peronosporeae, especially the Pythieae (Fig. 69 *C*). A few species are unlike the rest in this respect.

Like the Peronosporeae many Saprolegnieae have antheridia; in a few species only (Saprolegnia hypogyna, Pringsh.) the antheridium is the stalk-cell which bears the oogonium, as in Monoblepharis (Fig. 67): usually it is the obliquely club-shaped or cylindrical terminal cell of slender branches which grow one or more in number towards the oogonium and apply themselves closely to it. These lateral antheridial branches spring, according to the species, either from the branch of the thallus that bears the oogonium to which it attaches itself, and then usually close to it

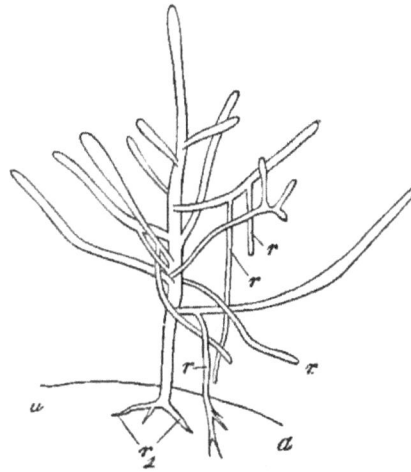

FIG. 68. *Achlya prolifera.* A germ-plant twenty-four hours old, about 1.5 mm. in height, growing on the surface of the larva of a gnat indicated by the line *aa*; *r¹* indication of branches of the primary rhizoid which have penetrated into the substance of the larva; *r* secondary rhizoid-branches growing towards the substratum from the erect and subsequently fertile branches.

(androgynous forms; Fig. 69 *A, B*); or from special branches of the thallus which do not bear oogonia (diclinous forms); cases intermediate between the two extremes are of comparatively rare occurrence. The antheridial branches apply themselves to the oogonia. The first appearance of the lateral branches and the delimitation of the antheridia take place before the formation of the oospheres. When these are formed, the antheridia usually send out 1, 2, or 3 delicate tubular processes which, like the fertilisation-tubes of Pythium, grow into the oogonium and in the direction of the nearest oosphere and apply their apex firmly to it; but they do not open, and nothing like a discharge of their protoplasm has ever been observed (Fig. 69 *B*); on the contrary they generally continue to elongate after their first contact with an oosphere and grow over its surface and not unfrequently beyond it. When there are several oospheres present the tubes often grow from one to another, and even form branches which grow up to and past different oospheres, and sometimes even pierce through the wall of the oogonium and pass outside it; but they always remain closed and die in the course of 1–2 days while the oospheres are maturing. The short tubes of Aphanomyces scaber are the only ones which I have examined which never showed this luxuriance of growth; they apply their apex firmly to the oosphere which in this species is solitary, and remain in that position without change till the oospheres ripen in 2–3 days' time. These facts show that the antheridia and fertilisation-tubes of the Saprolegnieae must be considered to be homologous with those of Pythium and that they may be called by the same name: but there is no ground for regarding them as really fertilising organs; for while in some species, as Achlya polyandra and Saprolegnia monoica, their mode of formation is always such as has just been described, there are other species in which the same plant may have these antheridia with their fertilisation-tubes and at the same time antheridia without tubes, or oogonia without any antheridia (Aphanomyces scaber, Saprolegnia hypogyna); and lastly there are species or races, extremely like those named above in other respects, which very seldom or never form antheridia. In all these cases the formation of oospheres and oospores does not vary in the smallest detail.

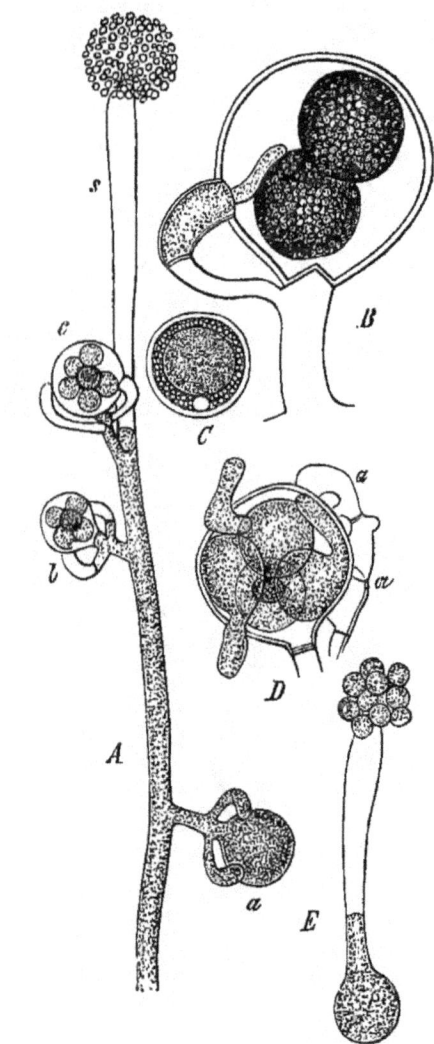

FIG. 69. *A—C. Achlya racemosa*, Hildebr. *A* end of a fertile branch with an empty sporangium at *s* surmounted by a head of gonidia out of which most of the spores have already swarmed away. Beneath it on short lateral branches are three oogonia with antheridial branchlets; *a* before the delimitation of the oogonium and antheridia; *b* and *c* in the same stage as *B*; both oogonia have two antheridia, and *b* has six, *c* seven oospheres. *B* oogonium with two oospheres and an antheridium resting on it; a fertilisation-tube from the antheridium has reached the surface of the nearest oosphere. *C* a ripe oospore. *D—E. Achlya polyandra*. *D* an oogonium with three germinating oospores about five weeks after maturity. The short germ-tubes of two of the oospores are protruding from the oogonium, the third lies bent inside it. The oogonium contains also two oospores which have not yet germinated, one of which is shown in the figure. *E* a germinating oospore which has formed a small sporangium with a head of spores. *A* magn. 145, *B* and *C* 375, *D* and *E* 225 times.

The ripe oospores remain dormant during a time which varies from some days to some months according to the species, and germinate in the same forms as in the Peronosporeae (Fig. 69 *D, E*). All the different forms of germination described in the Peronosporeae have also been observed in most of the species of Saprolegnieae which have been examined, the form varying according to the external conditions, especially those of nutrition; but in some species certainly only one or the other form of germination has been observed. In a new species, Aplanes Braunii, the whole course of the development, as far as my observations go, is usually limited to the direct development of a thallus from the germ-tube which proceeds from the oospore, and to the formation of oogonia, oospores, and antheridia on it. There is usually no formation of gonidia in this species.

In all other cases the fully-grown thallus which forms oospores also produces gonidia; the production is comparatively scanty and uncertain in Achlya spinosa, abundant usually in all other species. The gonidia are formed first, the oospores appearing during a later period of the development, partly on the same main branches of the thallus as the gonidia, partly on special branches. Here, as in the Peronosporeae, and evidently as the result of external causes, the gonidia are produced in much greater quantities than the oospores, and they are themselves the most effective in the propagation of the species. At the same time no species in the Saprolegnieae is known to be without oospores. The gonidia in all species, except the Aplanes mentioned above, are normally swarm-spores and are formed either in the germinating oospore, or in sporangia, which are usually of some size and borne on branches of the thallus. The species with swarming gonidia have occasionally resting gonidia also, but their appearance is accidental and exceptional. The genera of the Saprolegnieae, like those of the Peronosporeae, are chiefly distinguished by the sporangia and the formation of the swarm-cells in them.

The genera Saprolegnia, Achlya and Dictyuchus, when well developed have club-shaped sporangia, the protoplasm of which divides into numerous spores arranged in many rows (Fig. 70 *A*). Very feeble specimens form only one row of spores, and there is scarcely ever more than one row in the long narrowly cylindrical sporangia of Aphanomyces. (See section XVIII *a*, p. 74.)

The distinguishing mark of **Saprolegnia** is that the spores are in the motile state as they issue from the sporangium, and that the branch of the thallus which bears the sporangium grows through it when it has discharged its spores. **Achlya** and **Aphanomyces** are known by the discharged spores collecting into little heads forming the hollow spheres described on page 108, which they subsequently leave when they begin to swarm (Figs. 69 *A*, 70 *B*). Another exceptional fact besides the forming of these heads is observed in some species of Achlya and according to Sorokin[1] in Aphanomyces also; the spores are invested with a membrane of cellulose in the place in which they are formed inside the sporangium, and afterwards burst through this membrane and through the lateral wall of the sporangium to swarm. A number

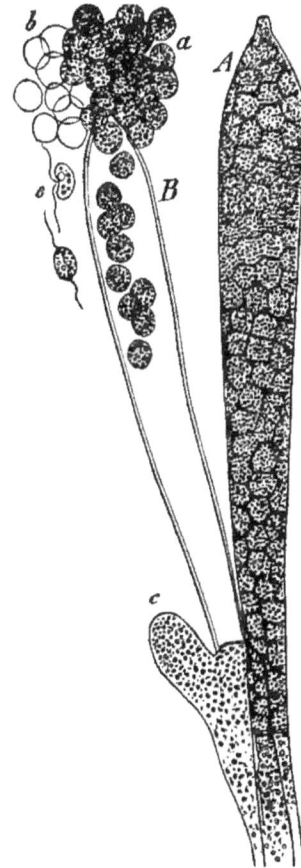

FIG. 70. Sporangia of a species of *Achlya*. *A* after formation of the spores (gonidia) but still closed. *B* after ejection of the gonidia, a few only remaining in the sporangium. The larger number are grouped at its mouth in a hollow sphere *a* and have become invested with a cell-wall; at *c* they begin to swarm away leaving their cell-walls *b* behind them. Magn. about 300 times.

---

[1] Ann. d. sc. nat. sér. 6, II (1876), p. 46.

of species which agree with Achlya in other respects have the last named peculiarity alone, and their spores do not collect into heads; these are included under the generic name of **Dictyuchus**. Some histological details will be found in section XVIII. The formation of swarm-spores in the germination of the oospores agrees in every known instance with the character assigned to the genera, but has not yet been observed in Dictyuchus.

Aplanes, a new genus, identical probably with Reinsch's Achlya Braunii, very rarely forms gonidia on the developed thallus, but more often in the germination of the oospores. The gonidia are formed as in other genera; in the germination of the oospores either directly in the cavity of the oospores, or in a single row in sporangia on short-lived dwarf plants. They put out short germ-tubes at once in the place where they are formed, nor have I ever been able to perceive any appearance of swarming.

Resting gonidia. In old tufts, those of Saprolegnia especially, it not unfrequently happens that the thick thallus-tubes become broken up by transverse walls into cylindrical, barrel-shaped, or inflated spherical cells, which are sometimes thick-walled and always rich in protoplasm. In some species, and especially according to my own observation in Achlya prolifera, these cells may be very large and spherical, unusually full of protoplasm, and abjointed serially and successively at the end of a tube[1]. All such cells may under favourable circumstances—in pure water containing free oxygen, and when supplied with suitable food—either develope directly into new thallus-tubes or become swarm-sporangia (the resting sporangia of Pringsheim). They are not, as far as we know, characteristic of particular species, but simply resting states which frequently make their appearance under the influence of external causes.

The deviation in the structure of some oospores from the ordinary type has been described by me at length in another work[2]. It occurs in a few species of Achlya (A. polyandra and A. prolifera), in Dictyuchus clavatus and in an undescribed Saprolegnia, and is not generally characteristic of any of the genera in question.

Leptomitus lacteus and **L. brachynema** are imperfectly known forms, belonging probably to the Saprolegnieae, with their thallus-tubes constricted at intervals and with swarm-spores formed in a similar manner to those of Saprolegnia; with respect to Cornu's genus **Rhipidium** the author's short preliminary description leaves it uncertain whether its place is here or with the Peronosporeae near Pythium.

Controverted points. Pringsheim has recently put forth some views which if correct would require a modification to some extent of the account here given of the Saprolegnieae, but only so far as concerns the possibility of a fertilisation of the oospheres by the antheridia.

Pringsheim claims the office of fertilisation for some small portions of protoplasm with amoeboid movements, which are supposed to make their way through the closed wall of the fertilisation-tube and to pass into the oosphere. Pringsheim has never seen this take place; he only suspects it on the evidence of some stained preparations, which seemed to show a possible open communication between the protoplasm of the oosphere and that of the antheridium, and of a peculiar phenomenon observed in the antheridia of Achlya racemosa, the account of which must be read in the original publication, but which has certainly nothing to do with the process of fertilisation. His observations moreover refer to other species than those specially described above. If there is really an open communication in those species between the protoplasm of the antheridium and oosphere, which, as has been said, is extremely doubtful, we must admit a fertilisation in their case, and in the mode already described in Pythium

---

[1] Walz in Bot. Ztg. 1870, t. IX, Fig. 20.　　　　　[2] Beitr. IV, p. 69.

and Phytophthora. All beside remains untouched; especially the absence of sexuality in the species, forms, and individuals, which have no antheridia producing fertilisation-tubes.

For the details of these disputed points the reader is referred to the publications, which are cited at the close of the following list and which have appeared since the year 1882.

Literature of the Saprolegnieae.

N. PRINGSHEIM, Entwicklungsgeschichte d. Achlya prolifera (N. Acta Acad. Leopoldin. Carolin. 23, I, pp. 397-400).

A. DE BARY, Beitr. z. Kenntn. d. Achlya prolifera (Bot. Ztg. 1852, p. 473).

Both these works contain also a record of the earlier copious literature of the subject.

PRINGSHEIM, Beitr. z. Morph. u. Systematik d. Algen, II ;—Id Die Saprolegnieen in Jahrb. f. wiss. Bot. I, 1857, p. 284 ;—Id. Nachträge z. Morphol. d. Saprolegnieen (Jahrb. f. wiss. Bot. II, 1860, p. 205) ;—Id. Weitere Nachträge, &c. (Jahrb. f. wiss. Bot. IX, 1874, p. 191).

DE BARY, Einige neue Saprolegnieen (Jahrb. f. wiss. Bot. II, p. 169) ;—Id. Beitr. z. Morphol. und Physiol. d. Pilze, IV (1881).

HILDEBRAND, Mycolog. Beitr. I (Jahrb. f. wiss. Bot. VI, 1867, p. 249).

LEITGEB, Neue Saprolegnieen (Jahrb. f. wiss. Bot. VII, 1869, p. 357).

K. LINDSTEDT, Synopsis d. Saprolegniaceen (Diss. Berlin, 1872).

M. CORNU, Monographie der Saprolegniées (Ann. d. sc. nat. sér. 5, XV) ; see page 139 above.

P. REINSCH, Beob. ü. einige neue Saprol. (Jahrb. f. wiss. Bot. XI, 1878, p. 283).

M. BÜSGEN, Entwickelung d. Phycomycetensporangium in Diss. and Pringsheim's Jahrb. XIII, Heft 2, 1882.

N. PRINGSHEIM, Neue Beobacht. ü. d. Befruchtungsact v. Achlya u. Saprolegnia (Sitzungsber. d. Berlin. Acad. 8 Juni, 1882) ;—Id. in Jahrb. f. wiss. Bot. XIV, Heft I.

DE BARY in Bot. Ztg. 1883, Nr. 3.

See also Zopf and Pringsheim in Bot. Centralblatt, 1882, Nr. 49, 1883, Nr. 25 and 31. Some smaller treatises have been already noticed in the text and in my publication of 1881 cited above.

MUCORINI.

SECTION XLI. The Mucorini or Zygomycetes agree very nearly both in structure and in the course of their development with the Peronosporeae and Saprolegnieae, but there is this essential difference between them, that, instead of the oospores which have just been described in the two latter groups, the Mucorini form *zygospores*, which are typically produced by the coalescence, conjugation, of two nearly or perfectly similar cells of separate origin (gametes). The formation of zygospores is not the only form of propagation in any known species of the group. On the contrary they all produce gonidia as well, and the gonidia in some species have considerable variety of form owing to complicated life-conditions and adaptations, and in all are much more abundant than the zygospores. Even in species where the zygospores are comparatively abundant the gonidia predominate to such an extent, that they propagate the species unaided through many successive generations, the zygospores on the whole rarely arriving at their full development. In some species the zygospores are great rarities; in a whole series of forms, which it cannot be doubted are in other respects closely related to the species which produce

[4] L

zygospores, these organs have never yet been observed ; and since some of the species have been frequently and carefully examined, it may perhaps be conjectured that they do not at present produce zygospores, but only gonidia.

The members of the group of Mucorini, with the exception of the doubtful genus Zygochytrium, which will be considered in a later page, are plants of the dry land, and grow most of them on dead organic bodies (especially animal excrements), some being parasitic on other Mucorini.

A spore gives rise to a mycelium having the form of a much-branched unicellular tube, as may be seen most readily by cultivating the plant in drops of fluid or in mucilage on a microscopic slide (Fig. 71 *B*), and it is not until gonidiophores begin to be produced that the tube is in a condition to form the transverse septa which then usually appear in it. The *typical gonidiophores* (Fig. 71 *g*) begin after one or more days' time to shoot upwards from the mycelium which has spread in the substratum ; they appear in the form of branches which are usually erect, and, like the parent-tube, are at first without transverse septa. In some species, as Mortierella and Syncephalis, they are almost microscopically small ; in most cases, however, they are of considerable size, from one to several centimetres in length, in Phycomyces 10–30 cm. Older or imperfectly nourished mycelia may subsequently produce a fresh crop of *accessory gonidial forms*. But in all species which have been thoroughly examined the mycelium under favourable

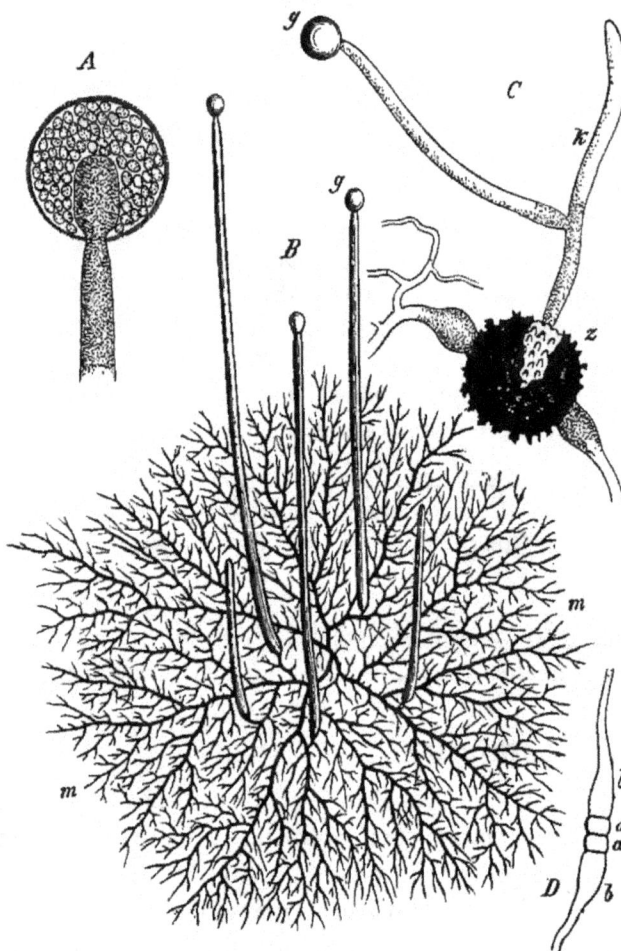

FIG 71. *B. Phycomyces nitens.* Plants three days old grown from a gonidium in gelatine with decoction of plums ; the mycelium *m* has spread horizontally, *g* a gonidiophore. *A, C, D Mucor Mucedo* highly magnified. *A* sporangium in optical longitudinal section. In *C* the germinating zygospore *z* is borne on suspensors ; *k* germ-tube, *g* sporangium. *D* conjugation ; *aa* gametes, *bb* suspensors. *B* slightly, *A, C* and *D* more highly magnified. After Brefeld from Sachs' Lehrbuch.

circumstances completes its development by forming zygospores. Finally the ripe zygospore, after remaining dormant for several months, puts out one or more strong germ-tubes, which develope at once without mycelial formation into the typical gonidiophores which are characteristic of the species (Fig. 71 *C*). The gonidia of every species invariably produce, if the conditions are suitable, a mycelium which displays the characters and follows the course of development here described. Hence a regular alternation of generations is observed in the life-history of these plants; typical gonidiophores are formed directly and normally from the germinating zygospore, and their products germinate and form a mycelium which produces gonidia again and ultimately zygospores. This course of development may, it is true, be so far disturbed, that the germ-tube which proceeds from the zygospores, if it is hindered from forming gonidia directly but is well supplied with food, as when it is artificially sunk for instance beneath the surface of a nutrient solution, may develope into a mycelium which does not become sporiferous till after the lapse of some time; but this fact does not alter our determination of what is the normal behaviour of these plants. No other phenomena of regular alternation of generations have been observed; I have seen indeed the mycelium formed from gonidia produce zygospores first and then gonidiophores in Sporodinia grandis, the species which most frequently produces zygospores and grows on fleshy Hymenomycetes; but the reverse order of events is quite as common, and sometimes no zygospores are formed at all. The mycelium developed from zygospores may in this species produce zygospores directly, omitting the formation of gonidia [1].

In all other known species the formation of zygospores is always preceded by a copious development of gonidia on the same mycelium, and most specimens never get further than the gonidia. It may be a question therefore whether possibly a generation forming zygospores succeeds with some regularity to a large number of successive generations which only form gonidia; experiments in artificial cultivation have only given negative answers [2].

SECTION XLII. The **formation of zygospores** is known only in the smaller number of species, but these belong to almost all the chief genera of the group. It begins in the case of each zygospore with the appearance of a pair of archicarps (section XXXII), and these are formed in Sporodinia on special erect dichotomously branching hyphae resembling the sporangiophores; in all other known cases they appear singly on sporangiophores (Chaetocladium, Absidia), or as immediate branches of the mycelium spread in or on the substratum; the latter case occurs also in Sporodinia, if the drawing in the vignette which precedes the tables in Tulasne's Carpologia is correct. The archicarps in each pair spring either from spots in the hyphae which are morphologically close to one another, as in Phycomyces [3] and Sporodinia [4], or from spots which are only locally adjoining, places of contact of branches of morphologically remote origin, as in Piptocephalis [5] and Mucor stolonifer [6]. Both conditions of origin correspond to those which have been described for the antheridia and oogonia of the Peronosporeae and Saprolegnieae, and appear as in them to vary from species to species and in individuals of the same species. The archicarps of a pair (Fig. 72 *a*, *b*) are small branches of the hyphae,

---

[1] Brefeld in Sitzgsber. d. Ges. naturf. Freunde z. Berlin, 15 Juli, 1875.

[2] Brefeld, Schimmelpilze, IV.

[3] Van Tieghem, I, t. 20, Fig. 4.

[4] De Bary, Beitr. I.

[5] Brefeld, Schimmelpilze, I, p. 48.

[6] De Bary, Beitr. II.

and are at first of the same breadth as the hyphae, and the even surfaces of their extremities become firmly united together, in some species when their length is not yet greater than their breadth, in others, it is said, not till a later stage of their development. While thus attached to one another, and while dense oily proto-plasm constantly. passes into them from the hyphae, they grow into bodies which are broadly club-shaped towards the surface of attachment and straight or curved according to the species; the two together therefore are spindle-shaped. When they have reached a certain size, which in some species may be as much as 1 mm., the further development follows two different paths according to the species.

In the **Mucoreae** (Fig. 72) and the **Chaetocladieae** each archicarp is divided by a transverse wall into a nearly cylindrical terminal cell, the conjugating-cell or *gamete* which remains connected with the other of the pair, and into a larger portion the *suspensor* (c), which adjoins the gamete and continues club-shaped or conical. The two gametes are at first separated from one another by their original membranes which form a tolerably thick partition-wall; but these are soon

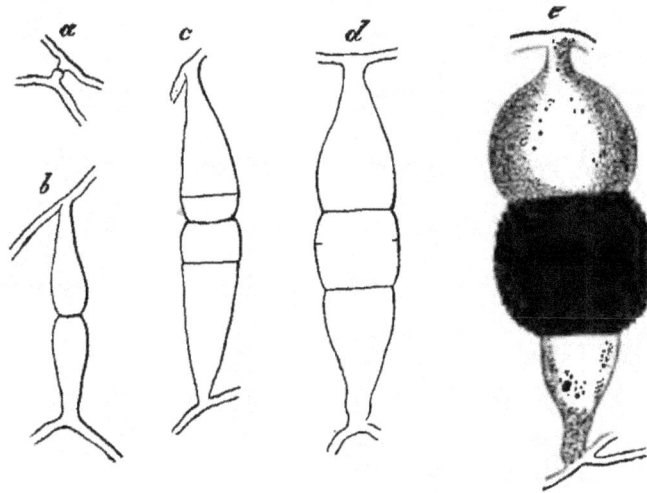

FIG. 72. *Rhizopus nigricans*, Ehr. (*Mucor stolonifer*, Ehr. Silv. myc.). Formation of a zygospore. Stages of the development according to the letters. *e* a nearly ripe zygospore magn. 90 times. The other figures reduced to about the proportion of *e* from larger drawings.

dissolved, the dissolution beginning from the centre, and thus the two protoplasmic bodies conjugate and coalesce into a *single zygospore d*. After conjugation the zygospore still increases in volume at the expense of the protoplasm of the suspensors, swells into the shape of a barrel or of a ball with its surfaces abutting on the suspensors flattened, and assumes the characteristic structure which will be described presently (Figs. 72 *e*, 71 *C*). During these proceedings the two gametes of a pair behave in some species, as in Sporodinia, precisely alike, excepting inconstant variations of form in individual plants. In other species tolerably constant dissimi-larities make their appearance with the delimitation of the gametes. In Mucor stolonifer (Fig. 72), where the point has been more exactly investigated, the one gamete is almost always only half the height of the other; and after conjugation the suspensor belonging to the smaller gamete grows into a large stalked spherical vesicle, sometimes divided again by a transverse wall, while the other retains its original size and conical form.

In the group of **Piptocephalideae** the archicarps are curved and so disposed that the pair has very nearly the form of an Ω or of an inverted Ω. The surface of union lies in the apex of the bow (Figs. 73, 74 Z). In Syncephalis nodosa the archicarps are coiled spirally round one another. Up to the time when the conjugation of the slender gametes is completed the development is essentially the same as in the first-mentioned cases, but then the product of conjugation swells at the place of coalescence into a spherical vesicle, which bulges on the convex side of the bow of the Ω. Protoplasm passes into it in proportion as it increases in size. When it has reached the limit of its growth, it is delimited by a partition-wall from each limb of the bow and becomes a nearly spherical zygospore; it may at least be so called for the sake of clearness and simplicity, though it is plain that it is not the strict morphological equivalent of the zygospore of the first case, but is a daughter-cell of the zygospore, if the zygospore is the cell which results directly from the conjugation of the pair of gametes. If we adopt the proposed terminology, the spherical zygospore of Syncephalis is placed at the apex of the bow formed by the pair of suspensors, and each suspensor is divided by a transverse wall.

The behaviour of the zygospores, while they are maturing, is essentially the same in both cases, apart of course from specific differences. The fatty matter in the protoplasm, which continues to be dense and darkly granular, usually collects into several large round drops. A further and exact insight into more delicate points of structure in the protoplasm is scarcely attainable owing to its opaqueness.

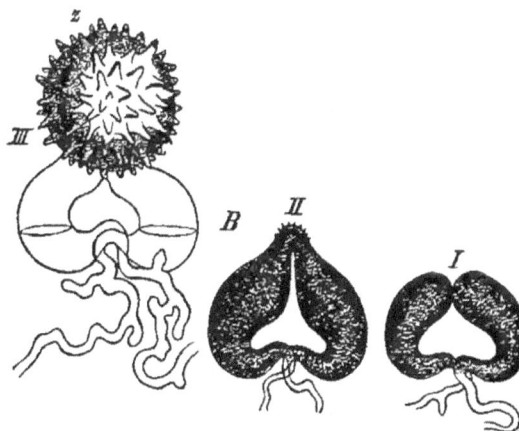

FIG. 73. *Piptocephalis Fresemiana*, conjugation and formation of zygospores, the development in the order of the numbers. *Z* a ripe zygospore on its suspensors. After Brefeld from Goebel, Grundzüge. Magn. 650 times.

The wall forms usually wart-like or conical projections on its free outer surface, in Piptocephalis even before the delimitation of the zygospore, only those parts remaining smooth where it is in contact with the suspensors, and becomes differentiated into a stout episporium the colour of which varies from brown to black, and a thick stratified endosporium formed of more than one layer (Figs. 72 e, 71 C, 73 III, 74 Z). The latter has either its outer surface quite smooth, as in Chaetocladium, and the projections belong entirely to the episporium; or it is furnished with stout solid projections which fit into corresponding depressions in the episporium (species of Mucor, Sporodinia). Mortierella is the only exceptional case, in which the episporium is not rough on its outer surface and owing to its close adherence to the investing envelope is but little developed.

In many species, and in the majority of those which have been named above and represented in Figs. 71–74, the ripe zygospore thus constituted lies naked and without any further covering between the suspensors, which ultimately wither and decay. But in some species an *envelope* is added to the parts already described in the zygospore. Short branches shoot out in a simple or multiple circlet in the more simple cases after conjugation from each suspensor near their line of junction with the ripening zygospore; these branches, which have stout dark-coloured walls, complete their growth as the zygospore ripens and form a loose envelope round it. In Phycomyces[1] the branches of the envelope divide dichotomously and their ramifications spread in every direction; in Absidia[2] they either do not branch again or they have short tooth-like lateral branches, and bend crosier-wise from the two suspensors over the zygospore. The formation of the envelope is more complicated in Mortierella, in which the ripe zygospore is encapsuled by a number of layers of compactly woven hyphae. The accounts which we possess show that the development of the envelope varies in the different species. In M. nigrescens[3] the hyphae of the enveloping tuft grow out at first after conjugation from the suspensors and later also from the filaments bearing them, and enclose the zygospore as it ripens in their copious ramifications. In M. Rostafinskii[4] the hyphae do not spring from the suspensors, but only from the point of origin of the adjacent mycelial hyphae, and envelope the conjugating pair with their branches before conjugation is effected.

Azygospores. It happens not unfrequently in the Mucoreae that zygospores which are otherwise normally developed contract a permanent constriction at the place of coalescence, and the coalescence is therefore to outward appearance less perfect than it usually is. But it may also happen that the gametes in a pair, after being formed in the normal manner, do not coalesce, do not even come into contact with one another, and the gametes may even make their appearance singly and not in pairs (Absidia, Mucor tenuis), and yet have the normal structure of zygospores and their normal capacity of germination. Spores thus formed and resembling zygospores have been named *azygospores*. They are not produced in all species; they have not been found, for instance, after long continued search, in Mucor stolonifer or in Chaetocladium; they have been found in Absidia septata, A. capillata[5], Mucor fusiger[6], and most frequently in Sporodinia. In this species they are associated with the zygospores. Mucor tenuis according to Bainier only forms azygospores, and perhaps Fries' Azygites[7], a plant which requires more precise investigation, is also one of the Mucorini which have this peculiarity.

Zygospores or azygospores have up to the present time been described in Sporodinia grandis; Mucor Mucedo, M. racemosus, M. tenuis (Bainier), M. fusiger, M. stolonifer; Pilobolus (Pilaira) anomalus (Brefeld, Van Tieghem); Phycomyces nitens, Ph. microsporus (V. Tieghem); Absidia capillata, A. septata (V. Tieghem); Mortierella 2 species (vid. sup.); Choanephora 1 spec.; Chaetocladium Jonesii and Ch. Fresenii (Brefeld); Piptocephalis Freseniana (Br.); Syncephalis Cornu (V. Tieghem), S. nodosa, S. curvata (Bainier); altogether 19 species, very few compared with the large number of existing species. In many they may have been overlooked, being rare and concealed in the substratum, and may yet be found. In others, as Pilobolus crystallinus, P. oedipus, and Thamnidium elegans, they have been so often sought for in vain that it may be assumed that they never occur.

---

[1] Van Tieghem, I, t. 20.
[2] Id., III, t. 11.
[3] Id. III.
[4] Brefeld, Schimmelpilze, IV, t. V.
[5] Van Tieghem, III.
[6] Id. II, p. 73.
[7] See Tulasne, Sel. Fung. Carpol. I, p. 64.

The species of the Mucorini which have zygospores, but no other organs of propagation, have since Ehrenberg's time[1] received the generic name of Syzygites. Hildebrand[2] has described two such forms, one of which may belong to Chaetocladium, the other (Syzygites ampelinus) is of uncertain affinity.

SECTION XLIII. **Typical gonidiophores.** It has been already stated that the spore when germinating normally produces gonidiophores as the first result of development in the form of hyphae (simple sporophores) which are characteristic of each species. Every strongly developed mycelium of the species produces similar gonidiophores, which are at least the primary and often the only organs of propagation, if we disregard the deviation from the usual course of development in Sporodinia, which was mentioned above and scarcely requires to be considered. The species that are thoroughly known, those, that is to say, in which zygospores have been observed, have accordingly gonidiophores of a distinct form, which are shown by more than one consideration to be constant and necessary members of the normal development and also to be strictly homologous with one another, and it is these which are here designated typical gonidiophores. There are besides many species in which the zygospores and their germination are not known, but which have gonidiophores agreeing so entirely in all essential points with those of typical and thoroughly known species, that they must be regarded as homologous with them and always have been so regarded. These therefore come under the category which we are now considering.

The gonidia themselves, apart from the doubtful genus Zygochytrium, which may be put aside for the present, are in all cases simple motionless comparatively thin-walled spores, in other respects possessing the usual structure.

The genera and subdivisions of the group are chiefly distinguished as in the Peronosporeae by the mode of their development, the structure of their sporangia, and the structure and segmentation of their gonidiophores; to these are of course added characters derived from the zygospores, mycelium, &c. The arrangement in the following brief review of this group will be founded on these particulars. I go with Van Tieghem in the definition and naming of the several genera, but merely for the sake of brevity and convenience and without fully approving his minute divisions of the forms; I differ from him as regards the primary division. Three primary groups may be distinguished according to the manner of forming the gonidia or spores :—

The **Mucoreae**; spores formed endogenously in terminal sporangia by division of the protoplasm into several or many portions (see section XVIII).

The **Chaetocladieae**; spores abjointed acrogenously one by one (see section XVI).

The **Piptocephalideae**; spores formed acrogenously and serially by cross-septation (see section XVI).

1. MUCOREAE. The sporangia (Fig. 71 *g, A*) are nearly spherical, rarely club-shaped terminal vesicles on the primary stalk or on the branches of the sporangiophore. In **Mortierella** they are placed with a narrow insertion on the slender extremity of the branch that bears them, and are delimited in the plane of insertion by a horizontal

---

[1] Verhandl. d. Ges. naturf. Freunde zu Berlin, I, p. 98.
[2] Jahrb. f. wiss. Bot. VI, p. 270.

basal transverse wall ; their wall, except at the plane of insertion and the parts immediately adjoining, deliquesce in water when it is mature.   In the other genera (Fig. 71)
the basal wall of all or of the larger sporangia expands into a conical dome-like or
ellipsoid form and projects into the cavity of the sporangium, and from its shape is
termed the *columella*.   Among these genera is **Pilobolus** : its sporangium thick-walled
with a basal expanding layer (p. 83), solitary terminal on an unbranched turgescent
vesicular sporangiophore, from which it is abjected when ripe or (Pilaira, V. Tieghem) removed by swelling.—**Mucor** : outer wall of sporangium deliquescing when mature, usually
incrusted on the outside with short spreading needles of calcium oxalate ; sporangiophore narrow filiform, with or without single branches arising monopodially (**Phy-**
comyces, Spinellus = Mucor fusiger but differing from Mucor in the formation of the
zygospores, **Sporodinia** differing in its dichotomous sporangiophores and suspensors).
The sporangia in the genera Rhizopus, Circinella, and Absidia have the same
structure as in Mucor, but the filiform sporangiophores branch sympodially.   In
Rhizopus and especially Rh. nigricans (= Mucor stolonifer) the sporangiophores rise
like stolons from the mycelium, ascending at first in a curve and then letting their
points drop downwards.   In this way they may reach a length of from one to several
centimetres.   If their apex touches a firm substratum, the previous growth in length
ceases, and a number of incipient branches make their appearance behind the
apex.   Some of these branches develope with the apex of the stolon itself into shortly-
branched rhizoids closely appressed to the substratum ; others, usually from three
to five in number, raise themselves above the substratum and become rigid branchlets
2–3 mm. in height, each of which terminates in a sporangium ; lastly, one or a few
of them develope as stolons which may repeat the same course of development[1].
Absidia has according to Van Tieghem a similar mode of growth to Rhizopus, but
with the difference that the stolons of the successive orders describe very regular
curves, and that a tuft of sporangiferous branches spring above from the convexity
of the curve.   On the erect sympodial sporangiophores of **Circinella** see Van Tieghem,
as cited in end of section XLIV.   While the above genera have sporangia of only one
kind, those which constitute the Thamnidium group have two kinds : large ones
quite like those of Mucor, and terminal like them on the primary stem of the sporangiophore which is monopodially branched or not branched at all; and small ones,
sporangiola, usually formed on the extremities of the many ramifications of lateral
branches of the sporangiophore which ends itself in a large sporangium.   In all the
species of this group sporangiophores occur which only produce one large sporangium, and others which produce only sporangiola.   The sporangiola are distinguished
from the sporangia not only by their small size and the reduction of the number of
their spores to 2 or 1, but also by their level or slightly convex basal wall, which does
not rise high into a columella, and the comparatively thick smooth outer wall without
incrustations, which persists till after the ripening of the spores and is in many cases
only burst by their germination.   There is no difference in the germination of the two
kinds of spores or in their products.   Van Tieghem distinguishes the genera **Thamni-**
dium, Chaetostylum, and **Helicostylum** by the shape and ramification of the branchlets that bear the sporangiola.

2. CHAETOCLADIEAE.   The genus **Chaetocladium** which forms this division
contains two species very like one another, which are usually parasitic on the larger
Mucoreae in the manner described in section V.   The filiform erect gonidiophores
produced from the zygospore usually terminate in a slender subulate hair-point,
beneath which they form a whorl of usually 2–5 short branches standing out at a
right angle, which again branch and form further whorls of a second or third order.
The branching axes which become shorter with each successive order end each in its

---

[1] On deviations from this rule of growth see de Bary, Beitr. II, and Wortmann in Bot. Ztg.
1881, p. 368.

turn in a hair-point; the lateral branches of the last order swell into irregularly capitate basidia, from the short slender sterigmata of which 8–20 spherical spores are simultaneously abjointed. Similar sporiferous structures with hair-points are formed on the terminal ramifications of copiously branched gonidiophores, which rise in a curve into the air from well-fed mycelia in a similar manner to the stolons of Rhizopus. Many variations occur in the number and disposition of the whorls and the successive orders of branches; a small cluster of spores may take the place of the terminal hair-points, especially in weakly specimens. It is hardly possible to detect any strict rule in the primary branching of the stolons, the typical form of which appears to be indiscriminately sympodial and monopodial. Some of the primary branches terminate in spore-clusters; others seize on sporangiophores of Mucor and adhere to them, and send out new stolons from the points of adherence which are swollen. Zygospores also are formed on the stolons. Cunningham's **Choanephora** may also belong to this division, with a creeping endophytic mycelium and straight erect simple sporophores ending in umbellately arranged heads of basidia, from which many spores are simultaneously abjointed.

3. PIPTOCEPHALIDEAE. In **Syncephalis** the very delicate mycelium gives rise to short erect unicellular usually unbranched gonidiophores with a circle of rhizoids at their base, and with one bifurcation in S. furcata. A dense umbel of simple or dichotomously branching spore-rows is formed at the capitate swollen apex of the gonidiophore by the cross-septation of cylindrical mother-cells (see section XVI). **Piptocephalis** (Fig. 74) differs from Syncephalis in its repeatedly dichotomously branched and septate simple sporophores which are often large and tall, and in the circumstance that the capitate summit which bears the spore-rows falls off with them when they are ripe. The mycelium of the species of both genera is parasitic on the larger Mucoreae, making its way into their cells by means of delicate haustoria (section V).

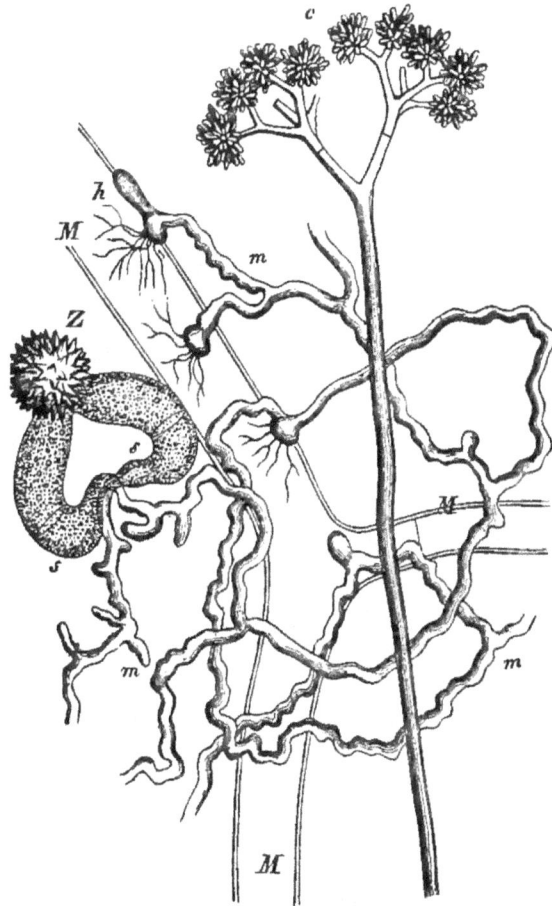

FIG. 74. *Piptocephalis Freseniana.* M piece of a mycelial tube of *Mucor Mucedo*; *m* mycelium of *Piptocephalis* with haustoria *h* applied to *M* and sending slender filaments into it, *c* gonidiophore, *Z* zygospore on the two suspensors *ss*. After Brefeld from Sachs' Lehrbuch. *c* magn. 300, the other figures 630 times.

The formation of the gonidia of Chaetocladium and the Piptocephalideae was described and named above in section XVI in accordance with the facts observed in Chaetocladium, Piptocephalis Freseniana, and Syncephalis. Van Tieghem's view of the process is different from mine. He considers that the spore-chain of Syncephalis and Piptocephalis is formed, like the gonidia of Mucor, simply by division of the protoplasm of the mother-cell, and is then set at liberty by the disappearance (resorption) of its membrane; and he regards the acrogenously formed spores of Chaetocladium as monosporous sporangia completely filled by the spore, and most comparable with the sporangiola of the Thamnidieae. These interpretations, which are not in accordance with the facts, have evidently arisen from a perception of the truth, that all the gonidial formations in question are homologous, coupled with the erroneous notion that homologous spores must necessarily originate by exactly the same mode of cell-formation. We naturally gain nothing more from such interpretations in this case than in others of older date which were noticed above on page 116. The homologies are brought out quite simply in our case, as has been shown above, without these artificial helps; and conversely the case before us shows with special distinctness that members and cells may be homologous, though all the facts observed with regard to them do not belong exactly to one and the same category of the scheme of cell-formation which is accepted at the time.

Section XLIV. **Accessory gonidia.** In a number of species, in Sporodinia grandis for example, Brefeld's Mucor Mucedo, Rhizopus nigricans and Chaetocladium, the cycle of forms is exhausted by the appearance of the members which have now been described; but in other species gonidial formations occur in addition to the typical ones, some of which by their form and structure are eminently characteristic of species and genera. It would seem that they are always found on starved or old mycelia, but the conditions of their formation have not been in every case clearly ascertained. It would be difficult to find any other general name for them than the one here chosen. Many of them have been called by Van Tieghem *chlamydospores* and *stylospores*, others *gemmae*, &c. It will be most convenient to choose a suitable name for each case as it presents itself.

Characteristic accessory gonidia are the *acrogonidia* of **Mortierella** and some species of **Syncephalis.** They are solitary and acrogenously abjointed on the summit of slender cylindrical branchlets of the mycelium, and are spherical in shape and usually of considerable size in Mortierella (as much as 20 μ or more in diameter), but small in Syncephalis (6 μ); they have a thick episporium with its surface marked with warts or spikes in a manner characteristic of the different species, and they emit germ-tubes which may develope into the normal mycelium. They are formed in Mortierella on erect branches of the mycelium which are either isolated or are united into small umbels; in the species of Syncephalis from short stalks which are arranged in dense racemes springing at right angles from a fusiform swollen portion of a mycelial filament. A very remarkable instance of the formation of accessory gonidia is that described by Cunningham in his genus **Choanephora,** in which the heads of basidia mentioned above must, from the mode of their occurrence on the normally developed mycelium, be certainly regarded as the typical gonidiophores, though the germination of the zygospores has not yet been observed, and cannot therefore confirm this view. But erect simple sporophores make their appearance on old and starved mycelia by the side of meagre heads of basidia, and form at their summit a spherical mucor-sporangium having a warted outer wall, a slightly convex basal wall, and containing a number of ellipsoid smooth-walled spores, which can germinate and produce a normal mycelium.

Another form of accessory gonidia is known under the name of *gemmae* or *brood-cells.* Their ordinary mode of production is that short pieces of a mycelial tube or gonidiophore, which is rich in protoplasm, become delimited by transverse walls to form cylindrical or nearly spherical or ovoid or pear-shaped or similarly shaped cells, which often acquire thick membranes and then under favourable conditions of vegetation and in many cases after a long resting period develope into normal mycelial tubes. These cells are densely filled during their resting period with tolerably homogeneous protoplasm; but sometimes, and especially in starved or dying or dead specimens, numerous large drops of oil may become separated in the protoplasm, a circumstance which once gave occasion to some passing misapprehensions[1]. On old mycelia especially, and in species of Mucor on sporangiophores also, the protoplasm of which is usually employed in forming gonidia of other kinds, these gemmae occur frequently as cylindrico-ellipsoid cells inserted like stoppers at irregular intervals in hyphae which are otherwise without protoplasm[2]. They occur frequently in many species, especially in those of Mucor; in some, as Chaetocladium, Piptocephalis, and Phycomyces, they have not yet been observed. In old cultures of Pilobolus they make their appearance on the slender branches of the mycelium especially, and not unfrequently in thick swellings in them, which if vegetation is continued develope into typical sporangiophores; they resemble in fact incipient sporangiophores which have remained stationary at the first stage of their development and have become dormant, and have a thick yellowish brown outer membrane and dense reddish yellow protoplasm; they are liable to be mistaken for zygospores, and grow when they germinate into typical gonidiophores. The smooth-walled ‘chlamydospores’ of Mortierella, described by Van Tieghem, also belong to this class. Those which are not intercalary but terminal in the mycelial hyphae are obviously transition-forms to the acrogonidia mentioned above.

Under certain conditions of vegetation some of the Mucoreae also form gemmae, which may be distinguished from the preceding by the names of *chain-gemmae* and *sprout-gemmae*. These forms have been most thoroughly studied in Mucor racemosus and have been described by Brefeld[3], but they occur also, according to Van Tieghem, in other and very different species. The formation of these gemmae begins when the mycelium is submerged in a nutrient fluid, especially in a saccharine solution capable of alcoholic fermentation, and is thus cut off from the free access of oxygen. Numerous transverse walls divide the entire mycelial tubes into short segments, which may be narrowed into mere disks and are swollen with protoplasm. The segments may, as Berkeley[4] said in 1838, remain united together in a confervoid manner or separate from one another, and, if the conditions remain the same, they often sprout luxuriantly after the manner of the Sprouting Fungi described on p. 4. If spores are submitted to the same conditions, they first swell into large spherical vesicles and then sprout directly without first forming germ-tubes. The sprouts which proceed from them have almost, if not quite always, the form of spherical vesicles which grow to more than 40 μ in diameter and repeatedly put forth new generations of sprouts by copious development at all parts of their surface; the new sprouts either continue united together or some of them separate from one another. This sprouting form of Mucor was formerly known as

---

[1] See Bot. Ztg. 1868, p. 765.

[2] Bail in Flora, 1857, t. XIII.—Zabel in Mélanges biolog. Acad. de St. Pétersbourg, III.—Brefeld in Thiel's Landw. Jahrb. V (1876), p. 282, t. 1. See also the works cited below passim.

[3] Landw. Jahrb. V (1876).—Id. in Flora, 1873.

[4] Mag. of Zoology and Botany, II, p. 340.

'sphere-yeast' or 'mucor-yeast,' partly from its resemblance in shape to the Saccharo-mycete of yeast and partly from its power of exciting alcoholic fermentation. Under certain conditions the sprouts behave like spores of Mucor.

All the gemmae which have now been described are capable of developing by germination into normal mucor-mycelium, either at once or after a prolonged period of rest, if properly fed and supplied with air. Each gemma of Mucor racemosus, cultivated in moist air without supply of food, is capable of developing at the expense of its protoplasm into a typical but very minute sporangiophore with a sporangium containing in extreme cases not more than eight spores (Brefeld).

Doubtful Mucorini. 1. Sorokin[1] has described under the name of **Zygochy-trium aurantiacum** a Fungus which grows on dead insects beneath the water, and which if that writer's observations are confirmed is a small species of this group adapted to growing in water. The entire plant consists of an erect tube with two bifurcations, alto-gether scarcely 0.1 mm. in height, attached to the substratum by short lobes at its base without any proper mycelium. Globular sporangia producing numerous swarm-spores with one cilium as in the Chytridieae (see section XLVI) are formed at the extremities of the branches; and a zygospore is also produced at the first bifurcation in perfect specimens in the manner which has been described in the case of Mucor. The course of development is in other points like that of Mucor. These observations have yet to be confirmed.

2. Van Tieghem[2] describes under the names of **Dimargaris cristalligena** and **Dispira cornuta** two Fungi growing on dung, which have very peculiar gonidiophores, but otherwise agree with Piptocephalis in being parasitic on the Mucorini, in the mode in which they enter the cells of their host, in the presence of transverse walls in the gonidiophores and in the chains of gonidia. The genesis of the gonidial chains is not indeed fully described, but the drawings would seem to show that it is similar to that of Piptocephalis. It is as yet by no means certain that these Fungi belong to the Piptocephalic group, but this affinity is highly probable, since we know of no species outside the Mucorini which they approach in form. The point can only be decided by the discovery of zygospores or some structure homologous with them.

3. The same may be said of a small group, which may be called the Coemansieae, consisting of Coemans' **Kickxella** and **Martensella** and **Coemansia** of Van Tieghem and Le Monnier. The mode of life of these Fungi, of Kickxella at least and Coemansia, is the same according to Van Tieghem as in Piptocephalis. In the structure of their gonidiophores, the only part at present known beside the mycelium, they differ materially from the acknowledged members of the group of the Mucorini. Their common and chief peculiarity consists in the possession of basidial branches of somewhat fusiform appearance, falcately curved, and divided by transverse walls into several cells from which on the concave side of the branch numerous spores, placed close together in two or more comb-like longitudinal rows, are abjointed both simultaneously and one by one. The spores themselves are narrowly fusiform, pointed at both ends, and emit their germ-tubes at right angles to their own longitudinal axis. Basidial branches of this kind form in Kickxella a whorl of 6–14 members with their concave sides upwards on the apex of the erect septate filament that bears them, which is elsewhere usually unbranched and is scarcely 0.3 mm. in height. In the other genera they stand singly in racemose arrangement on the dichotomously ramifying branches of the gonidiophore. No zygospores are known; but Coemans and afterwards Van Tieghem and Le Monnier found small ascomycetous sporocarps in the neighbourhood of the gonidiophores; whether they belonged to each other is a very doubtful question.

---

[1] Bot. Ztg. 1874, p. 305.   [2] Van Tieghem, II.

Historical Remarks. The course of development in the Mucorini was first observed throughout by myself in my examination of Sporodinia or Syzygites megalocarpus carried out in 1860 and published in a complete form in 1864, and next by Tulasne in 1867 in Mucor fusiger, after he had already shown in 1855 that the fungal forms designated by the generic names above given are parts of the same species. Schacht's observations on Sporodinia [1] were published at the same time as my own and not before them, and his results agreed with mine. For the further enlargement of our knowledge of this rich group we are indebted mainly to Brefeld and Van Tieghem. In the work [2] which I brought out in conjunction with Woronin in 1865 I gave a fresh account of the development of Rhizopus nigricans, but it was imperfect, as it did not contain the full history of the germination of the zygospores; and some confusion was caused in the same work by our introducing Chaetocladium, a parasite on Mucor, into the cycle of forms of Mucor Mucedo on the evidence of cultures, which, through my fault not Woronin's, were not perfectly regulated. It is not true indeed that Thamnidium elegans was also introduced into the same cycle, but it would be no serious fault if it had been, because it can form gonidiophores without sporangiola which are then scarcely to be distinguished from those of Mucor Mucedo, and the separation or non-separation of two species nearest to one another is almost without effect on the determination of the course of development of the Mucoreae. Then Brefeld at my instigation undertook a revision of 'the apparently irregular pleomorphy' of the collective supposed Mucor Mucedo and succeeded in making out the true state of the matter. Van Tieghem and Le Monnier confirmed and added first to our incorrect, and then to Brefeld's correct results.

Other views on the course of development in the Mucorini differing from those given above, especially the idea of a genetic connection between Mucor and Saccharomyces, which will be noticed again in section LXXVIII, belong to the history of the pleomorphy craze (page 126). The special literature already cited and to be cited below contains references to it. The reader is referred to the same source for an account of the strange controversy maintained at an earlier time on the subject of the structure of the sporangium of Mucor. In the case of a group so much discussed as the Mucorini we can only give the main sources of information, in which some further and more particular references on points of detail will be found.

Literature of the Mucorini.

DE BARY and WORONIN, Beiträge, I and II.

TULASNE, Note sur les phénomènes de Copulation, &c. (Ann. d. sc. nat. sér. 5, VI, 1867).

O. BREFELD, Bot. Unters. ü. Schimmelpilze, I and IV.

P. VAN TIEGHEM et G. LE MONNIER, Recherches sur les Mucorinées (Ann. d. sc. nat. sér. 5, XVII, 1873); cited in the text as Van Tieghem, I;—Id., Nouvelles Recherches sur les Mucorinées (Ann. d. sc. nat. sér. 6, I, 1875); cited in the text as Van Tieghem, II;—Id., Troisième mémoire sur les Mucorinées in Ann. d. sc. nat. sér. 6, IV (1878); cited in the text as Van Tieghem, III.

G. FRESENIUS, Beitr. z. Mycologie, I (1850), III (1863).

E. COEMANS, Spicilège mycologique No. 3, in Bull. Soc. Bot. Belg. I (Kickxella);—Id., Quelques Hyphomycètes nouveaux (Mortierella, Martensella), in Bull. Acad. roy. de Belgique, sér. 2, XV;—Id., Recherches sur le polymorphisme et les différents appareils de reproduction chez les Mucorinées, I et II (Bull. Acad. roy. de Belgique, XVI);—Id., Monographie du genre Pilobolus (Mém. couronné de l'Acad. roy. d. Belg. XXX).

---

[1] Sitzungsber. d. Niederrh. Ges. z. Bonn, 7 Apr. 1864.
[2] Beitr. II.

H. HOFFMANN, Icones analyt. Fungor. IV, 1865 (Mucor, Rhizopus).

ZIMMERMANN, Das Genus Mucor, Chemnitz, 1871.

J. KLEIN, Zur Kenntn. d. Pilobolus (Pringsheim, Jahrb. VIII, p. 305).

A. GILKINET, Mémoire sur le polymorphisme des Champignons (Mém. couronné de l'Acad. roy. de Belg. XXVI, 1878).

O. BREFELD, Ueber Gährung, III, in Landw. Jahrb. ed. Thiel. V, 1876 (Mucor racemosus).

D. D. CUNNINGHAM, On the occurrence of conidial fructification in the Mucorini, illustrated by Choanephora (Linn. Soc. Trans. London, sér. 2, I, 1878).

BAINIER, Observations sur les Mucorinées et sur les zygospores des Mucorinées (Ann. d. sc. nat. sér. 6, XV, 1883).   First brought to my notice while this work was being printed.

ENTOMOPHTHOREAE.

SECTION XLV.   We proceed to give a brief account of this small group to which we may apply the terminology in use for the Mucorini; the Fungi which supply the material for our description penetrate into the cavities of the bodies of living insects and there develope, forming their gonidiophores on hyphal branches, which make their way through the body of the insect after its death and complete their development on its outer surface.

In a certain number of species, as Empusa Muscae and E. macrospora, Now., numerous detached and at first spherical cells are formed by repeated sprouting from the germ-tube which has penetrated through the skin into the interior cavities of the insect, and each cell developes as the insect dies into a long tube containing much protoplasm.   In other species, as Entomophthora radicans, E. ovispora and E. curvispora, the germ-tube in the insect's body produces a mycelium composed of copiously branched hyphae, divided by transverse walls and often connected together by anastomosing branches.   In most instances the Fungus forms its gonidia on the surface and after the death of the insect.   In the Empusae one extremity of each separate tube pierces the insect's skin, grows outside it into a short cylindrico-club-shaped body, and then forms acrogenously a single spore, which is abjected by the mechanism described on page 72.   The protoplasm and other contents of the tube are expended for these purposes, and the tube itself then perishes.   In the Entomophthoreae numerous branches of the entozoic mycelium appear on the surface of the body of the dead insect and there ramify in so copious a manner that they soon wrap it in a close felt.   Much the largest part of this felt consists of branches set very nearly at a right angle to the surface of the body, and their last ramifications, which end at about the same level and form a compact hymenium, are cylindrical in shape and unicellular, and ultimately one spore is abjointed and flung off in the manner just described.   Hyphae and tufts of hyphae also appear before the sporogenous hymenium at certain spots on the ventral side of the dead insect, especially on caterpillars attacked by Entomophthora radicans, and develope into rhizoids which secure the body to the substance on which it lies.   In both cases the whole of the protoplasm of the Fungus is expended in forming the spores; as they are produced one after another on the extremities of the tube or its branches, these shrink in size and with them the body which they occupy; at last there remains only a shrivelled drying mummy, surrounded by a circle of spores or gonidia which have been thrown off. The latter are capable of immediate germination, and do germinate if duly supplied with moisture, either emitting a germ-tube which penetrates at once into the body of a suitable insect and goes through the process of development above described, or only producing short tubes from the extremity of each of which a new secondary gonidium with the same qualities as the primary is abjointed. The gonidia soon lose the power of germination; in Empusa Muscae, for example, it does not continue beyond about fourteen days.

The Fungi are limited to the above course of development in most of the insects which they attack. In some cases, however, few or no gonidia are formed and *zygospores* or *azygospores* are ultimately developed, and in most species inside the body of the insect. Nowakowski gives the following description of the formation of zygospores in Entomophthora ovispora and E. curvispora. The cells of adjacent hyphae develope an H-shaped union by means of the necessary processes, and establish an open communication at the point where the processes are in contact with one another. Then a spherical protuberance makes its appearance near the point of union, either on the cross-bar of the H or close to it, and receiving as it grows the entire protoplasm of the conjugated pair of cells is finally delimited by a membrane, and becomes a zygospore in the sense in which the word is used in the case of Piptocephalis.

No azygospores have been observed in these species. But Entomophthora radicans and the species of Empusa examined by Nowakowski have only azygospores which are produced without conjugation as lateral outgrowths on the mycelial tubes or acrogenously like the gonidia, as in the species named by Nowakowski Lamia culicis. It would appear therefore that there is a difference in the matter of conjugation in the different species similar to that which is found in the Saprolegnieae as regards the presence or absence of the antheridial branches. Both zygospores and azygospores become resting spores in the same way. The membrane becomes much thickened and is differentiated into a thick usually bright yellow episporium with a smooth surface in most species, and a thinner endosporium, while a large, globular, nearly central drop of oil separates in the protoplasm from the general finely granular turbid mass. The mycelium is dissolved and disappears after the formation of the resting-spores, which are therefore the only remains of the plant in the mummified body of the insect. Their germination was observed by Nowakowski in Empusa Grylli, and consists in the emission of a short tube, the promycelium, which forms a gonidium at its apex; the gonidium is abjected, in the same way as gonidia from the gonidiophores described above.

The above account is a summary of Brefeld's and Nowakowski's observations. The differences in the statements of the two writers with regard to conjugation are to be explained by the different behaviour of different species. The genetic connection once frequently assumed between the Entomophthoreae and other Fungi, especially Saccharomycetes and Saprolegnieae, is a subject for the history of botanical errors. Information on the point will be found in the works cited below.

Cohn has described as **Tarichium** a species which preys on ground-caterpillars; it has only resting-spores with thick walls roughened on the outside with warts, and may perhaps belong to this section. In Nowakowski's opinion both the development and affinity require to be more closely examined.

The Entomophthoreae do not live exclusively on insects. **Completoria complens,** which according to Leitgeb's observations is one of the group, is a small parasitic Fungus in the cells of the prothallia of ferns. Leitgeb's account of its structure, of the formation of its gonidia on the extremities of branches which have emerged from the cells of its host, and of its resting-spores, agrees almost exactly with these features in Empusa and Entomophthora. Brefeld also[1] has recently reported an allied species, which he found parasitic on some Tremellineae, and names **Conidiobolus utriculosus.**

Literature of the Entomophthoreae.

F. COHN, Empusa Muscae und d. Krankheit d. Stubenfliegen (N. Act. Acad. Leop. XXV, pars I, 1855).

S. LEBERT, Die Pilzkrankheit d. Fliegen (Verh. d. Naturf. Ges. z. Zürich, 1856).

G. FRESENIUS, U. d. Pilzgattung Entomophthora (Abh. d. Senkenb. Ges. II, 1858).

O. BREFELD, Unters. u. d. Entw. d. Empusa Muscae u. E. radicans (Abh. d. Naturf. Ges. z. Halle, XII, 1873).

F. COHN, Ueber eine neue Pilzkrankheit d. Erdraupen (Beitr. z. Biolog. d. Pflanzen, I, 1874, p. 58).

L. NOWAKOWSKI, Die Copulation einiger Entomophthoreen (Bot. Ztg. 1877, p. 217).

BREFELD, Schimmelpilze, IV (1873), p. 97 ;—Id., Hefepilze, l. c.

H. LEITGEB, Completoria complens, ein in Farnprothallien schmarotzender Pilz (Sitzungsber. d. Wiener Acad. 84, Abth. 1, 1881).

M. SOROKIN, Zwei neue Entomophthora-Arten in Cohn's Beitr. z. Biol. II, Heft 3.

A. GIARD, Deux espèces d'Entomophthora, &c. (Bull. sc. du départ. du Nord, sér. 2, année 2, No. 11, p. 253).

L. NOWAKOWSKI, Entomophthoreae (Abh. d. Acad. d. Wiss. z. Krakau, 1882, (Polish). Report on the same in Bot. Ztg. 1882, p. 560.

CHYTRIDIEAE.

SECTION XLVI. This is the name for what has gradually become a very varied series of small microscopic forms, which spend their entire vegetative period, or at least a definite stage of their spore-producing time under the water, and agree morphologically in forming swarm-spores in sporangial cells of a fixed and definite shape ; each swarm-spore has usually one cilium, and developes either directly or through inconspicuous transition states into fresh sporangial cells. Resting-spores are also known in certain species, and these develope directly in germination into sporangia or produce them after a short intermediate stage. There is such great similarity in the formation of sporangia and swarm-spores, that the species composing the group have always, one may almost say instinctively, been considered to be closely related to one another. Yet our knowledge of the several species is so unequal, and the course of development in the best-known forms is so different in extreme cases, and these extremes are so imperfectly connected by the intermediate forms with which we are acquainted, that we must at present feel in doubt whether we are dealing in this case with a series of objects naturally related to one another, or with a number of groups of similar adaptation and therefore of similar form, but of different natural

---

[1] Hefepilze, p. 10.

though close affinities. With this reservation we will first proceed to describe the points of form which are common to them all. See the Figs. 75–77 below.

The sporangia are cells of varying shape and diameter in the different species, and are often furnished with one or more wart-like or neck-like processes, which finally discharge the spores from their swollen and projecting apex, or which, as in Chytridium Olla, are cast off like a lid for the same purpose. They are furnished when fully grown with a moderately thick wall of cellulose, and densely filled with a uniformly finely-granular fatty protoplasm, which at length divides simultaneously into numerous spores. In most species the division is preceded by separation of the protoplasm-granules, which are colourless or coloured yellow, orange, or rose, according to the species, into as many groups divided by narrow hyaline streaks, in each of which the granules then coalesce to form bodies of successively larger size, and ultimately a single sphere consisting chiefly of fatty matter. This sphere of fatty matter then lies, usually excentrically, in the body of the swarm-cell, which otherwise consists of hyaline protoplasm and allows a nucleus to be seen in the larger forms[1]. Such a swarm-cell when set at liberty is roundish or elongated in shape, and is furnished with one cilium which is several times longer than the diameter of the body. The sphere of fatty matter is much more frequently, but not always, brought near the point of insertion of the cilium; exceptions to this structure occur in most of the species only as monstrosities. But in some there is no fatty sphere (Chytridium macrosporum, Ch. roseum, &c.). The spores of Olpidiopsis Saprolegniae, Woronina, and Rozella have according to A. Fischer always two cilia.

The spores are discharged from the sporangia by the process of swelling described in section XX, and in some species are at first held together in a mass by mucilage, from which they are afterwards gradually set free one after another; in other species they leave the cavity of the sporangium one by one. Where the dimensions and the speed of their movements allow of exact observation, the cilium is usually seen to follow the body in the process of release from the sporangium. The movement in the water is described as being clearly in many species a hopping movement; a progression by hops in no strictly determinate direction alternates in longer or shorter periods with a state of quiescence and each hop is associated with a stroke of the cilium like the stroke of a whip. But this kind of movement is not found in all the species. The spores of Nowakowski's Chytridium Mastigotrichis and the highly phototactic spores of Polyphagus Euglenae and of Strasburger's Chytridium vorax move forward with moderate speed and uniform rotation round their longitudinal axis and with the extremity that has no cilium in front, while the cilium follows passively behind. The time that the movement continues varies in each case, being seldom more than an hour, often much less; in a few cases, as in Synchytrium Taraxaci, it is considerably more. Towards the end of the period of movement an undulating change in the outline of the body often takes the place of the phenomena which have been described, together with an amoeboid creeping on a firm substratum, in which the cilium is dragged behind.

The *resting-cells* or *resting-spores* of the Chytridieae are on an average nearly or quite as large as the sporangia, and are distinguished from them by thick, often

---

[1] See above, pp. 107–109.

very thick, and many-layered membranes, the outer layers of which, the epi-sporium, are coloured and sclerosed in many species, in some are furnished with wart-like or slender spike-like prominences; they may also be known by their very dense protoplasm containing a large quantity of fatty matter uniformly distributed in small drops or granules, as in species of Synchytrium, or aggregated into a few drops or into one comparatively large round drop, as in Polyphagus, Chytridium Brassicae, Wor., Rhizidium mycophilum, A. Br., Chytridium Olla. The resting-spores remain dormant for some time before germinating.

If we next endeavour to form an idea of the course of development in the Chytridieae, we shall find that our present knowledge permits of our distinguishing four types, which might perhaps be combined by pairs into two main types. Each of these has one or more chief representatives, and each of these again has a crowd of imperfectly known forms doubtfully associated with it. There are no distinctly intermediate forms between the two types.

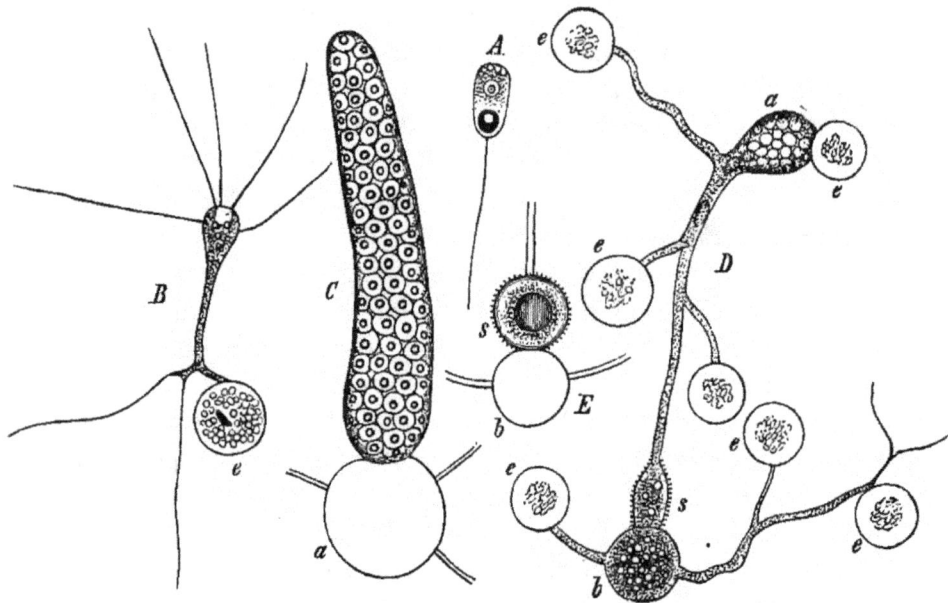

FIG. 75. *Polyphagus Euglenae.* *A* swarm-spore with sphere of fatty matter and nucleus. *B* young plant grown from a swarm-spore with a branch of the rhizoid attached to a resting *Euglena e.* *C* zoosporangium with formation of spores just completed and resting on the empty mother-vesicle *a* (prosporangium) from which it has proceeded; on the vesicle are three rhizoid-branches. *D* conjugation; *a* the receptive individual, *b* the supplying individual, *s* the swollen end of the tube of conjugation connecting *a* and *b*, which end is becoming the rudiment of the resting spore, *ee* the *Euglenae* attacked by the *Polyphagi. E* a portion of the pair shown in *D* 5½ hours later than *D*; *b* and *s* indicate the same parts as in *D, b* empty, *s* mature. After Nowakowski. *A* magn. 550, *B, D, E* 350, *C* about 400 times.

SECTION XLVII. 1. **Rhizidieae.** One species belonging to this section, **Polyphagus Euglenae,** a parasite upon resting Euglena viridis, has become the best-known of the Chytridieae through Nowakowski's beautiful investigations (Fig. 75). The swarm-spore when it has come to rest in the water becomes spherical in shape, and at once puts out hair-like tubular-rhizoid processes in indefinite directions (*B*). If one of these encounters a resting Euglena (*e*), it penetrates into its body, destroying and exhausting it to supply food to the parasite. The parasite then begins to increase in size, the rhizoid-tubes become larger and thicker, and new ones are formed which throw out branches, and attack and destroy any new

Euglenae which they encounter. In this way a much-branched plant is formed with hair-like terminal branchlets, which connect with the larger main stems and through these with the body of the original spore; the latter has grown in the meantime into a large round or elongated vesicle at the expense of the Euglenae, which have been exhausted by the rhizoids. When it has reached a certain size, varying according to the food which has been supplied to it, it shows itself in many specimens to be a sporangium, or, if the term is preferred, a *prosporangium*. It grows out at one spot into a bluntly and irregularly cylindrical thick tube with a delicate membrane, into which the whole of the protoplasm passes, and is at once divided into swarm-spores (Fig. 75 *C*). This process of development may be repeated for many generations, and leads to an immense multiplication of individuals if there is a sufficient number of Euglenae within reach. When this has taken place, the course of events changes. The young plants remain for the most part small and become *gametes*, which conjugate in pairs, each pair forming a *zygospore*, and these behave as *resting-spores*. The two conjugating gametes of a pair (Fig. 75 *D*) have no definite position or distance with respect to one another, and are similar in form to the non-conjugating plants. The one (*b*), which from the processes to be described may be termed the *supplying gamete* (abgebende Gamete), has usually a round and larger body, but shows no other apparent difference before contact with the other (*a*), the *receptive gamete* (aufnehmende Gamete). The latter usually continues to be smaller, and often very small, and puts out rhizoid branches, and if one of these, after longer or shorter growth, encounters a supplying gamete it applies its extremity to it as a conjugating tube (*s*) and increases in thickness while it ceases to increase in length. The membrane between the conjugating tube and the supplying gamete disappears at the point of attachment, and an open communication between them being thus established, the whole of the united protoplasm of both gametes passes into an enlargement of the conjugation-tube close to the point of attachment; the swelling gradually expands into a spherical vesicle, and being delimited by a membrane after receiving the protoplasm becomes a thick-walled zygospore (*E*, *s*). The outer wall of the zygospore assumes a pale yellow colour, and in some cases continues smooth, in others is covered with short spikes, which begin to form at the same time as the enlargement in the tube. The whole process of forming a zygospore from the attachment of the conjugating tube to the maturation of the zygospore was completed, in the case observed by Nowakowski, in about 6–7 hours.

A few instances are known of the conjugation of 2–3 receptive with one supplying gamete and of the consequent formation of 2–3 zygospores. The zygospore, as has been already said, is a resting-spore. It germinates when its resting time is over and produces a zoosporangium like the non-conjugating plants.

Polyphagus therefore is essentially characterised by the gametes with their rhizoids, the mode of forming the zygospores, and the production of the zoosporangium or of swarm-cells from it. It may be assumed to be possible for these swarm-cells to develope directly into gametes; but an indefinite number of generations of non-conjugating plants are in fact interposed between two successive gamete-generations. The gametes in each pair behave differently in conjugation, as has been shown, and the species is dioecious. Which of the two should be called the male and which the female is not easy to determine, and must not be further discussed in this place.   It is evident that we have before us an intermediate case between the ordinary forms of oogamous and isogamous conjugation.

Besides a second species of Polyphagus, exactly like P. Euglenae, which attacks Conferva bombycina and is called by Nowakowski P. parasiticus, there are a number of Chytridieae which *appear* to belong to the type of P. Euglenae.   Among the first of these, according to Schröter's brief communication (1882), is **Physoderma pulposum**, Wallr., a plant endophytic in the Chenopodiaceae, and with some highly peculiar characters about which we await fuller information.   Another in all probability is **Rhizidium mycophilum** which inhabits Chaetophora elegans, and, according to Nowakowski's account, and putting aside some variations in form which we may at present disregard, is entirely like Polyphagus Euglenae in all important points of structure and development ; we do not indeed know the formation of the resting-spores, but their germination is the same as in Polyphagus.   Then there are a number of forms described under the names **Rhizidium** and **Rhizophydium** (with Obelidium of Nowakowski), most of them parasitic on larger Algae, in which as in Polyphagus the body of the germinating swarm-spore puts out rhizoids and grows into a swarm-sporangium, and in which mature resting-spores of unknown origin have been found here and there.   These Rhizophydieae also, from the account which we have of them, require further examination.   Finally we may class with those already mentioned a number of A. Braun's typical **Euchytridieae** and **Phlyctidieae** which live in Algae, since they resemble them in structure, and their rhizoids were overlooked by the older observers on account of their great delicacy.   How far this is the true view of their affinity must be determined by further research, which, if I am not misled by imperfect observations, requires to be extended even to the most typical Euchytridieae such as Chytridium Olla.

I add what follows in justification of this remark. **Chytridium Olla**, A. Br., is parasitic, as Braun tells us, on the unripe oospores of Oedogonium rivulare and kills them.   Its swarm-spores settle according to Kny's description on the orifice of the oogonium which is filled with mucilage, put out slender rhizoid-processes from thence towards the oospore suspended in the oogonium, and then convert their body which is outside the oogonium into a sporangium; the rhizoid-process becomes a thick cylindrical stalk by means of which the sporangium is attached to the oospore, and which is delimited from the sporangium by a transverse wall when the development is complete.   I am inclined to doubt whether the rhizoid-process or stalk of the sporangium proceeds from the cilium of the swarm-spore, as Kny states, because this is never the case in other species of similar growth, though there is often the appearance of it ; I have never examined into this point in Chytridium Olla.   The swarm-sporangium is ovoid in shape when fully grown, and when it discharges its spores it throws off the apical portion of its membrane like a small lid, as is shown in Fig. 76 *A*, *B*.   According to the reports of observers the stalk is attached firmly to the surface of the oospore by its obtuse extremity only ; but when attempting to detach it I often saw the extremity prolonged into a little point which seemed to pierce through the membrane of the oospore, but could not be followed into its interior. My material was very old when I undertook the investigation, and the sporangia of the parasites on the oogonia were already emptied of their contents, and it was quite possible that rhizoid-processes might have penetrated into the oospore at an earlier stage but have disappeared in the decomposed contents of the oospore at the time of the investigation.

In older cultivated plants of Oedogonium which have been attacked by Chytridium Olla we often find a great many oospores of the Alga, which were killed by the parasite while still young and when their walls were still thin, and inside them in their decomposed cell-contents colourless glistening round bodies usually in large numbers.   These when carefully examined (Fig. 76 *A*) and isolated (*C*) prove to be thick-walled spherical cells containing a dense strongly refringent central sphere of fatty matter in a finely granular protoplasm tightly packed in a vesicular receptacle, which was evidently an intercalary member of a very slender branch filament. The spherical cells rich in fatty matter are resting-spores of Chytridium. After a long resting period (about four months in the cases which have been observed) they germinate and put forth a cylindrical germ-tube, which as it grows takes the shortest way to reach the outside, piercing through the membranes of the oospore and the oogonium. When its extremity has reached the outer surface of the oogonium, it swells into an ovoid sporangium, which resembles in every respect that which is ascribed to Chytridium Olla. It is developed at the expense of the protoplasm of the resting-spore, which passes through the germ-tube into the sporangium after the dissolution of the fatty sphere; before this transference has come to an end a transverse wall makes its appearance in about the middle of the tube, and the sporangium when fully grown is also delimited from the tube by a transverse wall (Fig. 76 A, B).

So much is matter of observation. The gaps in our knowledge of the details are readily descried; speaking in general terms we may say that the question of conjugation and fertilisation is still unsettled, and that the continuity of the development between the germinating swarm-spore and the filaments which form the resting-spores has not been satisfactorily established. This latter defect would be of very small importance in presence of the perfect similarity between the sporangia developed from the swarm-spores of Ch. Olla and those from the resting-spores, if another form, also a Rhizidium, had not been observed in the plants examined under cultivation, which at least resembles Chytridium Olla in the formation of the sporangia. The sporangia of this species are not as like the sporangia developed from the resting-spores as are those of Ch. Olla, but they always resemble them enough to make it necessary to be careful in interpreting the observed facts.

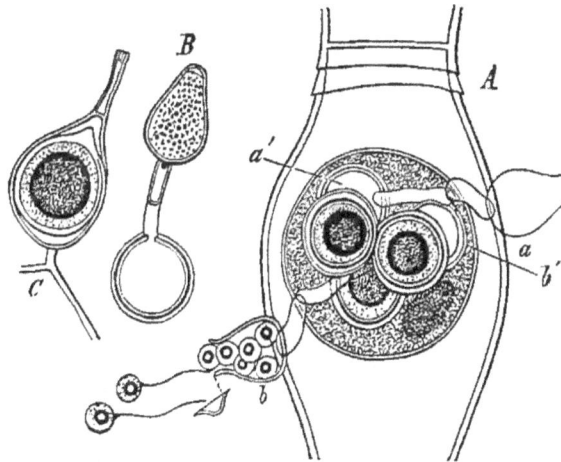

FIG. 76. *Chytridium Olla*, A. Br. (?). *A* oogonium of *Oedogonium rivulare* with an immature oospore killed by the parasite; the oospore contains several resting-spores of *Chytridium* which ripened in October; three of these spores are seen still unchanged, two have germinated. By turning the specimen round it was seen distinctly that the empty sporangium *a* was formed from the resting-spore *a'*, and the sporangium *b*, which is ejecting its contents, from *b'*; near the mouth of *b* are the cast-off lid and two zoospores. *B* an isolated resting-spore after germination with the sporangium still undivided, from another specimen. *C* resting-spore in a receptacle attached to the branches of the filament on which it was formed, prepared from a dead oospore like that shown in *A*. *A*, *B* magn. 375, *C* 600 times.

In conclusion we refer once more in this place to the genus Zygochytrium mentioned on p. 156.

SECTION XLVIII. 2. **Cladochytrieae.** A delicate much-branched and widely spreading mycelium resembling the rhizoids of the Rhizidieae forms numerous terminal and intercalary sporangia on its branches, and the germinating swarm-spores give rise to a mycelium resembling the original one. The sporangia of many species are only known in the form of resting-spores.

The species first determined by Nowakowski in this group inhabit the decaying tissues of marsh-plants and the jelly of Chaetophora. The course described above has been directly observed in them from the germination of the swarm-spores to the production of the next sporangiferous generation. The sporangia are formed in large numbers on the vegetating mycelium and proceed to form their spores at once without passing through a period of rest.

A second series of forms are intracellular parasites in the living and otherwise sound foliage of some of the marsh plants infested by the first mentioned group. They form brown spots or pustules on the leaves, and spread from cell to cell, and often produce a large number of sporangia in each cell without coming out to the surface of the plant. To this group belongs the form which has been described as Protomyces Menyanthis, and another which may be named provisionally **Cladochytrium Iridis**. All the sporangia in these two species are converted, as the mycelium disappears, into thick-walled often brown ellipsoid resting-spores (Fig. 77), which, as far as is known, germinate only after hybernation and then form swarm-spores. This was observed by Goebel, as he informed me by word of mouth, in Cladochytrium Menyanthis, and by myself in C. Iridis. The germination of the swarm-spores has not yet been observed.

The occurrence of members of both these groups in the tissue of the same plant, as for example in Iris Pseudacorus, might lead to the supposition that the forms of the first group belong to the same species as those of the second which form resting-cells; but there is no fact to show this, and it is opposed to the observation that a large number of swarm-spores of Cladochytrium Iridis will not germinate even on dead tissue of Iris Pseudacorus. They seem to require living cells for their further development, but I could not see that they made their way into them.

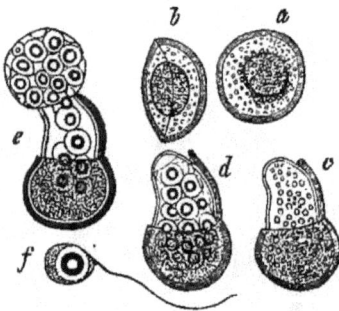

FIG. 77. *Cladochytrium Iridis.* *a* a resting-spore with a brown membrane seen from the broadside. *b* the same rotated through 90°; in its centre a large sphere of fatty matter. *c—e* successive stages of germination of a resting-spore; the inner cell developes into a tubular swarm-spore-receptacle, when the brown outer membrane has opened by a lid. *d* completion of the formation of the spores. *e* exit of the spores. *f* a single swarm-spore. *a—e* magn. 375, *f* 500 times.

No act of conjugation and no sexual organs have been observed in these plants. In the formation of intercalary sporangia and resting-cells it is often observed, that an intercalary swelling of a branch of the mycelium first appears and is then divided by a transverse wall into two halves; one of these halves swells into a sporangium, while the other does not enlarge and loses its protoplasm. But it cannot be seen that the protoplasm has passed over, as might be supposed, into the half which is increasing in size, it seems on the contrary to travel into the growing mycelium; and on the other hand the whole of the part that swells at first often becomes a sporangium or resting-cell without previous transverse division. The conjugation of the swarm-spores, which might also be expected from analogy, has not been observed.

The forms of **Physoderma**, Ph. maculare, Wallr. inhabiting Alisma graminifolium, Ph. Heleocharidis, Fuckel, Schröter's (1882), Ph. Butomi and Ph. vagans, the latter of which is found in different Phanerogams (Potentilla anserina, Ranunculus flammula, &c.), are very like Cladochytrium Menyanthis and C. Iridis as regards the intracellular development and the structure of their resting-spores in the inner layers of the parenchyma of the leaves of their host. But these species according to Schröter have no mycelium; the single resting-spore is formed in the same way as in Synchytrium. We are not told how their primordia find their way into the interior of the cells.

SECTION XLIX. 3. **Olpidieae.** We are indebted to A. Fischer for a complete account of the development of Olpidiopsis Saprolegniae and O. fusiformis, Cornu. The former plant causes a pouch-like swelling in the tubes of the Saprolegnieae which it inhabits, and in its full-grown state is an ellipsoid or round cell which forms neither rhizoids nor mycelium, but ultimately becomes a sporangium and discharges its zoospores through one or more cylindrical necks which pierce through the wall of the tube of the Saprolegnia. Each sporangium is the product of a swarm-spore which forces its way into the interior of the tube of the young Saprolegnia, and there lives in amoeboid form in and at the expense of the protoplasm of the tube, and finally after 3–5 days' time is invested with a membrane and becomes a sporangium. The sporangia appear in two forms; either they are smooth-walled and as soon as their growth is completed produce swarm-cells or else die and disappear; or they have the surface of their membrane beset with delicate spikes and form swarm-cells at once from their dense protoplasm, or first spend a period of at least several weeks' duration as resting-cells, and then produce swarm-cells. The spores of both kinds of sporangia are alike in origin and structure. But the spores formed in the smooth-walled sporangia generally develope into spiky sporangia, and those in the spiky sporangia into smooth ones; if the conditions are unfavourable to vegetation spiky sporangia are produced in greater numbers than the smooth. Every smooth and every spiky sporangium is formed, as has just been said, directly from a single swarm-spore. Former statements or conjectures to the contrary have not been confirmed by direct and more complete observation. Olpidiopsis fusiformis, which lives in species of Achlya, agrees with O. Saprolegnieae according to A. Fischer in all the details of its history.

We may consider not only the other parasites of the Saprolegnieae which were placed by Cornu in the genus Olpidiopsis, but A. Braun's Olpidieae also and other forms that are without rhizoids, to be nearly related to those which have just been described; their affinity is shown by the facts observed, especially the entire absence of rhizoids on the sporangia and the accompanying phenomenon of the resting-cells with spiky membranes. It is not at all easy to see why Cornu separates Olpidiopsis from Olpidium; but this group of species is imperfectly known and needs further investigation.

Section L. 4. The group of **Synchitrieae** is marked by the entire absence of rhizoids and by the circumstance that an *initial cell* proceeding from one swarm-spore breaks up by simultaneous division into a heap (*sorus*) of polyhedric sporangial cells which produce swarm-spores. The number of sporangia in a sorus varies from two to more than a hundred.

The typical forms of this group are gall-forming parasites which live in the epidermal cells of phanerogamous land-plants, where they produce swollen vesicles; the sporangia which ripen or the resting-cells which germinate if supplied with water burst the membrane which encloses them in order to release the spores or sori, if these have not been previously set at liberty by the decay of their investment.

Two subordinate groups may be distinguished according to their special course of development, **Eusynchytrium** and **Pycnochytrium**, the latter embracing Schröter's Chrysochytrium and Leucochytrium. Pycnochytrium has the simpler development. The swarm-spores which have made their way into the young cells of their host are converted into very large thick-walled resting-cells, one usually in each cell. The resting-spores remain dormant for some time, usually for about a year, and then germinate in the manner shown in Fig. 78; the colourless endosporium emerges through a small orifice in the layers of the episporium in the form of a slender papilla, which then slowly developes into a spherical vesicle on the outer surface of the episporium and receives the whole of the fatty protoplasm of the resting-cell (*a–c*). While the protoplasm is gradually becoming finely granular, it divides into the numerous sporangia which form the sorus (*c, d*).   The swarm-cells (*g*) discharged from the sporangia become new resting-cells, if they find a host.   This is the history, for example, of Synchytrium Mercurialis, S. Anemones, &c.   The course of development in the Eusynchytrieae is somewhat different; an unlimited number of generations of sori may be and actually are interpolated between every two successive hybernating generations of resting-cells; these sori are formed from swarm-spores without there being any resting period and produce again new swarm-spores.   The spore which has penetrated into the swelling cell of its host developes into the thin-walled initial cell, which divides into the members of the sorus as soon as it has reached a definite stage of development.   After a number of such generations have been produced, partly perhaps owing to external causes, the formation of resting-cells recommences.   The behaviour in germination of these also varies in different species; in some, as for instance Synchytrium Stellariae (Fig. 78) and S. Oenotherae, they form sori like those of the Pycnochytrieae; in others each resting-cell becomes a single sporangium directly without forming sori, as in S. Taraxaci.

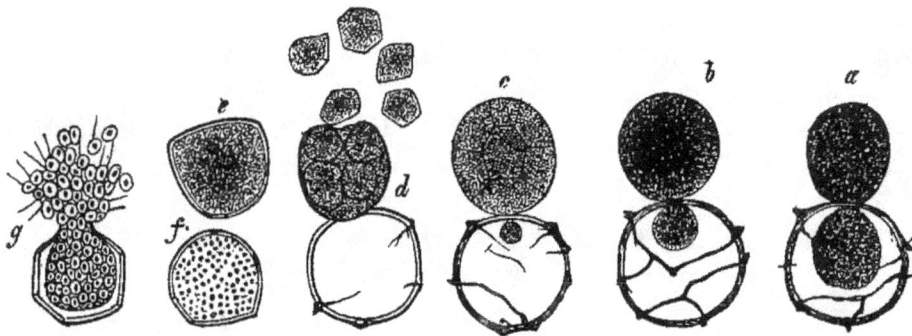

FIG. 78. *Synchytrium Stellariae*, Schröt., from *Stellaria media*. *a—d* germination of the resting-cells when placed in water after hybernating in the dry state. *a—c* successive states of the same specimen, *b* four hours later than *a*, *c* seven days later and five days after division into the cells of the sorus. *d* a mature sorus with the mature sporangia falling asunder. *e—g* a single sporangium in water. *e* immediately after being taken from the sorus. *f* the same two hours later shortly before formation of spores, the small sphere of fatty substance being divided into parts of equal size. *g* forty-five minutes later with the swarm-cells escaping. *a—d* magn. 145, *e—g* 375 times.

It is obvious that the difference between the Synchytrieae and the group of the Olpidieae lies in the formation of sori.   Species like Synchytrium Taraxaci with resting-cells which become sporangia without dividing are to some extent intermediate forms.   Neither conjugation nor any sexual process has been observed in either group. From the facts before us we can only gather that a sorus or resting-cell proceeds directly from a swarm-spore.   Influenced by the analogy of the isogamous Algae and Protomyces, I sought carefully in Synchytrium Stellariae and S. aureum for the conjugation of free swarm-cells, but, as was the case in my examination of Cladochytrium, I never found an instance of it, either between spores of one or of different sporangia belonging to the same or to different sori.   The abnormally large swarm-cells which are observed occasionally with two or more spheres of fatty matter and additional cilia are monstrous forms and not products of conjugation.   These negative results do not preclude the possibility of conjugation; it might occur inside the cell of the host if more than one swarm-cell had found its way there.   Cornu is inclined to account in this way for the formation of the resting-cells of Synchytrium, but he adduces no decisive observations in support of his view, but only an arbitrary comparison of states of S. Stellariae found side by side.   The objections to this are so obvious that they need not be specified.   Direct observation of the development is in this case indispensable; it is not easy to manage but still it is not impossible. Observation should be directed also to the origin of the initial cells of the sori as well as to that of the resting-cells, for there is no apparent reason why the latter should arise without and the former with conjugation. It should also be extended to more than one species, for the oogonia and antheridia of the Saprolegnieae show that what is the rule in one species is not so necessarily in another. If we were prepared to draw arguments for the formation of sori by conjugation from isolated facts, we might find them in the one which was first observed by Schröter, that there is a broad empty brownish-looking membrane adhering to one side of every full-grown sorus in Synchytrium Stellariae and S. Succisae inside the swollen cell of the host. Schröter, however, gives a different and probably the true explanation of this circumstance; he supposes that a vesicle which takes up the protoplasm issues out of the original membrane of the initial cell shortly before the formation of the sorus, as in the germination of the resting-cells of Pycnochytrium, but this has not however been directly observed.

The course of development in the genera **Woronina** and **Rozella**, Cornu, is very like that of the Synchytrieae; in outward appearance, as parasites of the Saprolegnieae, they are certainly sufficiently unlike the Synchytrieae which inhabit Phanerogams. A. Fischer's complete investigations have distinctly proved the absence of conjugation and sexuality in these genera; for a detailed account of them the reader is referred to Cornu and A. Fischer. Their very peculiar mode of life will be noticed in Division III.

SECTION LI. If we compare the facts of development in these four groups with those in other Fungi, we see that the chief points of connection with the latter are to be found in the first group. Polyphagus may at once be put side by side with the Mucorini as a very small form adapted to a submerged life by the formation of swarm-spores; the connection is still more perfectly effected by Zygochytrium, which ranks with the Mucorini as truly as with Polyphagus, if Sorokin's statements are confirmed (see page 156). The homologies are sufficiently brought out by the terminology employed. There are also obvious points of connection with the Ancylisteae (page 139). Of the other groups the Cladochytrieae come nearest to the Rhizidieae, even if sexual processes are really wanting in them and are not merely undiscovered. It is possible that some of the Rhizidieae themselves may be in the same condition. With a similar supposition necessary at present, we may regard the two other groups as nearly related to the Rhizidieae, if we consider them to be forms in which the formation of rhizoids or mycelia has fallen into disuse, or been entirely lost, in consequence of special and intimate parasitic adaptation; the Olpidieae will then come nearest to the Rhizidieae, and there is no difficulty in connecting the Synchytrieae with the Olpidieae. In this view of the subject the whole division of the Chytridieae would be regarded as a lateral branch either of the Mucorini or of the Ancylisteae, which has been gradually simplified in correspondence with its submerged parasitic life and has reached its most specialised condition, in which it differs most from the preceding divisions in Woronina and Rozella, genera of the Synchytrieae.

On the other hand the resemblance of the simple Chytridieae without rhizoids to unicellular Algae, especially the Protococcaceae, Characium, Chlorochytrium, &c.[1], has always been recognised. The question of course arises whether this resemblance is the expression of close phylogenetic affinity, or only of analogous adaptations; for ·apart from the presence of chlorophyll, the conjugation of swarm-cells which prevails

---

[1] See Klebs in Bot. Ztg. 1881, p. 249.

among these algal forms is a fact which separates them at present from the Chytridieae. At the same time the first view is not one which can be entirely set aside ; we are not sure, as has been said already, that phenomena do not occur in the Chytridieae which are at least very near akin to the conjugation of swarm-cells, namely, the supposed conjugation of the young plants just formed from swarm-cells, which have made their way into cells of the host. But if the simple Chytridieae are really related to the Protococcaceae and form a natural series of affinities with all the rest, this series linking with the Protococcaceae must also extend to Polyphagus and the Ancylisteae and Mucorini.

At the foundation of all such considerations lies the supposition, that the four different groups of the Chytridieae really form a genetic series. This supposition is usual and admissible, but is not necessary. There may quite as well be two or even more than two series before us of separate affinities connected together by certain similarities of adaptation ; on one side the Olpidieae and Synchytrieae without mycelia, and on the other side the Rhizidieae with Cladochytrium ; the latter would unite with the Mucorini or Ancylisteae and could be derived phylogenetically from them ; the former would come from other forms, for instance the Protococcaceae. These questions and considerations cannot lead at present to any certain decision, they can only point out the direction for further enquiries.

SECTION LII. **Doubtful Chytridieae.** 1. Along with the above-mentioned Zygochytrium Sorokin[1] describes a **Tetrachytrium triceps**, which forms four swarm-cells in each sporangium ; the swarm-cells conjugate in pairs after their escape from the sporangium, then round themselves off and germinate. Before conjugation they resemble those of the Chytridieae. Those which have not conjugated are incapable of germination. Germination gives rise to a small sporangiferous plant, which is attached by a short rhizoid-process to the substratum (vegetable substances decaying in water), and consists of a tube-like cell with four branches formed by bifurcation, three of which terminate in a sporangium and the fourth is sterile, having the form of a horn-like process.

2. The same observer[2] found on submerged rotten wood in Venice a plant, **Haplocystis mirabilis**, consisting of a pear-shaped cell, about 110 μ in diameter, the protoplasm of which divides by successive bipartitions into 32 parts ; the parts round themselves off, and conjugate in pairs after rotating inside the mother-cell. Then the cell opens by a lid and the products of conjugation issue from it in the form of swarm-cells provided with two cilia ; these cells come to rest in about a quarter of an hour and develope into a cell like the mother-cell. These two remarkable accounts have still to be confirmed. If they are correct, we have to do with organisms which can scarcely belong to the Chytridieae, but must be related to the Protococcaceae.

There are other and probably many forms of similar habit to the Chytridieae, some of which are very imperfectly known, while others would seem to be certainly distinct from the Chytridieae, but not referable at present to any other group. We cannot go further into the subject of these forms in this place ; reference to some accounts of them will be found in Schenk, Algol. Mittheilungen (Amoebidium parasiticum, see below), and Cienkowski, Bot. Ztg. 1861, p. 169.

---

[1] Bot. Ztg. 1874, p. 308.   [2] According to Just's Jahresbericht, 1875, p. 190.

Literature of the Chytridieae :—

A. BRAUN, Ueber Chytridium, eine Gattung einzelliger Schmarotzergewächse (Monatsber. d. Berlin. Acad. Juni, 1855, and Abhandl. derselb. Acad. 1855, p. 21, tt. 1–5);—Id., Ueber einige neue Arten v. Chytridium u. d. damit verwandte Gattung Rhizidium (Monatsber. d. Berlin. Acad. 1 Dec. 1856).

F. COHN, Ueber Chytridium (Nov. Act. Leop. Car. 24, 1, p. 142).

BAIL, Chytridium Euglenae, Ch. Hydrodictyi (Bot. Ztg. 1855, p. 678).

CIENKOWSKI, Rhizidium Confervae glomeratae (Bot. Ztg. 1857, p. 233).

A. SCHENK, Algol. Mitth. (Verhandl. d. Phys. Med. Ges. z. Würzburg, Bd. VIII);—Id., Ueber d. Vorkommen contractiler Zellen im Pflanzenreiche, Würzburg, 1858 (Rhizophydium).

DE BARY u. WORONIN, Beitr. z. Kenntn. d. Chytridieen in Ber. d. naturf. Ges. zu Freiburg, III (1863) and Ann. d. sc. nat. sér. 5, III (Synchytrium).

DE BARY, Beitr. z. Morph. u. Phys. d. Pilze, I, in Abhandl. d. Senkenb. Ges. Frankfurt, 1864 (Cladochytrium Menyanthis).

WORONIN, Entwicklungsgesch. v. Synchytrium Mercurialis (Bot. Ztg. 1868, p. 81);—Id., Chytridium Brassicae (Pringsheim's Jahrb. Bd. XI, 1878, p. 557).

KNY, Entwicklung v. Chytridium Olla (Sitzungsber. d. Berliner Naturf. Freunde). See also Bot. Ztg. 1871, p. 870.

M. CORNU, Chytridinées parasites des Saprolegniées in Ann. d. sc. nat. sér. 5, XV (1872), p. 112 (Olpidiopsis, Rozella, Woronina, &c.).

J. SCHRÖTER, Die Pflanzenparasiten aus d. Gattung Synchytrium (Cohn's Beitr. z. Biol. I, 1, 1875);—Id., Ueber Physoderma (Berichte d. Schlesischen Ges. 1882).

L. NOWAKOWSKI, Beitr. z. Kenntn. d. Chytridiaceen in Cohn's Beitr. z. Biol. II (1876), p. 73 (Chytridium, Obelidium, Rhizidium, Cladochytrium);—Id., Polyphagus Euglenae (Cohn's Beitr. z. Biol. II, p. 201);—Id., Ueber Polyphagus, in Polish (Abh. d. Krakauer Acad. 1878).

A. FISCHER, Ueber d. Stachelkugeln in Saprolegniaschläuchen (Olpidiopsis) in Bot. Ztg. 1880);—Id.. Unters. ü. d. Parasiten d. Saprolegnieen (Habilitationsschrift, Berlin, 1882, and Pringsheim's Jahrb. Bd. XIV).

PROTOMYCES AND THE USTILAGINEAE.

SECTION LIII. **Protomyces macrosporus** (Fig. 79) is a parasite which lives in the intercellular spaces of umbelliferous plants, especially Aegopodium; it has a branching filiform mycelium with transverse segmentation, which produces numerous intercalary resting-spores and then dies. The resting-spores, which may be more than 60 μ in diameter, are irregularly ellipsoid in shape, filled with dense fatty protoplasm and provided with a very thick many-layered cellulose-membrane. They germinate after hybernation in water when set free by the decay of the umbellifer. In germination the endosporium with the protoplasm which has become finely granular swells into a spherical vesicle, bursts the thick layers of the episporium, and issuing from it becomes a sporangium. A large number of rod-like spores only about 2.2 μ in length are formed in a sporangium in the manner described in section XIX and are ejected from it; they are thin-walled and without independent movement, but show slight oscillations caused by currents in the water; they are thus brought into close proximity with one another if this has not happened directly by ejection, and conjugate in pairs by means of slender processes usually in the form of the letter H. Scarcely anything but pairs thus united are to be seen in a few hours after ejection; the two halves of each pair are in open communication, and in most cases also swollen to a larger than their original size. Each of these double spores is capable of germinating under conditions which will be described in section XCVI; one of the halves puts forth a tube which takes the whole of the protoplasm of the pair, and penetrating into a suitable host developes there at once into a new mycelium, which forms resting-spores. No further stage of the development is known.

The plant in question with its peculiar and simple course of development is at present an isolated form; only the resting-spores are at present known of a second species, Protomyces pachydermus, Thümen[1], which grows on the Cichoraceae and has probably a similar history.

SECTION LIV. The **Ustilagineae** are endophytic parasites in phanerogamous plants. Their mycelium, which usually spreads through the intercellular spaces of the host, consists of slender hyphae with comparatively long segments. In some species it has haustorial branches each with a tuft of sinuous branchlets

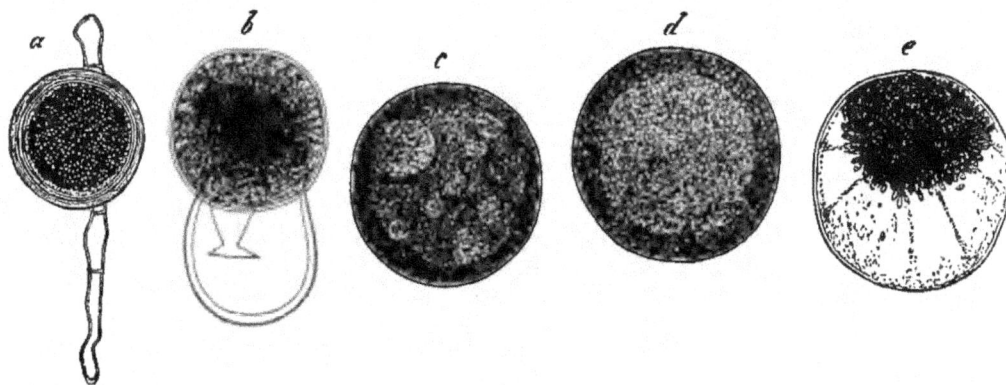

FIG. 79. *Protomyces macrosporus,* Unger. *a* a mature resting-spore in the dormant state with the remains of the hypha which bears it. *b* further development of the same when grown in water; the protoplasmic body surrounded by the innermost membranous layer (the inner cell) is swelling and escaping from the ruptured outer membranous layer. *c—e* development of spores in the escaped inner cell, the sporangium. *c* the parietal protoplasm. *d* the same divided into spores. In *e* the spores are rounded off and separated from the remainder of the parietal protoplasmic layer. Magn. 390 times.

which penetrate into the cells of the host (see page 20). *Resting-spores* or more shortly *spores*, are formed in or on the host, either, as in Entyloma, in all parts of the mycelial hyphae, or in special branches different from those first formed. These *sporogenous branches* of the Fungus form very numerous ramifications, and are usually woven together in great numbers into compact masses of definite shape and occupying definite spots; they are generally found inside the plants attacked by the Fungus, especially in the flowers and fruit or in parts of them that are diseased and swollen, and there they in great measure destroy and consume the tissue; less often (for instance Sorosporium Saponariae, Ustilago Tragopogonis, U. hypodytes) on the surface of the part attacked which they closely cover up. We may therefore in these cases speak of spore-producing bodies, *compound sporophores*, in the sense of section XII. The Entylomeae, on the other hand, are simple hyphomycetous forms, and other species, those especially which live in leaves (species of Tilletia and Urocystis), are intermediate between the two extremes.

---

[1] Hedwigia, 1874 and 1878, p. 124.

It is true that most of these compound sporophores have no independent form of their own, for they merely fill cavities in the substance of the host, though these cavities owe their existence and particular shape to the Fungus; or they are flat layers something like a Thelephora covering superficial portions of the plant. Yet the growth of many of these bodies evidently follows a single common plan; it is continued for a time at fixed spots by the constant formation of fresh sporogenous branches which form a weft—in the superficial layers, for example, on the whole of their basal surface,—and on removal from these spots comes to an end or passes on to formation and maturation of spores.

Most of these bodies consist only of sporogenous hyphae which are gradually used up to form spores, so that heaps of spores only remain at maturity. But there are species whose compound sporophores form other organs as well, which accompany or envelope the spores. Rudimentary forms of this kind are perhaps to be seen in the pellucid vesicles which Meyen discovered between the spores of Ustilago longissima and which require to be further investigated, also in part at least in the filaments in the spore-masses of U. olivacea which are still unexplained. Cornu's **Doassansia**, one of the Ustilagineae which is parasitic on water-plants and approaches Entyloma in the structure and germination of its spores, forms its spores inside a round closed receptacle, the wall of which consists of compact palisade-cells belonging to the Fungus. Ustilago Hydropiperis has a much and distinctly differentiated spore-receptacle, and might on that account be made a separate genus under the name of **Sphacelotheca**. But the greatest amount of differentiation in the compound sporophore is to be seen, according to Ed. Fischer's researches, in the genus **Graphiola,** if it should prove really to belong to the Ustilagineae.

We have said enough to show that the compound sporophores of the Ustilagineae deserve more attention than has been bestowed upon them since Tulasne's first memoir. Cornu's examination of Doassansia is only known to me from a preliminary and private communication from the author, but a more detailed account of it is to be expected shortly. The reader is referred to Ed. Fischer's own paper for his investigation of Graphiola; and we may proceed to give a description of Sphacelotheca, the parasite of the flowers of Polygonum Hydropiper mentioned above, as an example of a highly differentiated compound sporophore, though the investigations of the year 1854, from which we must draw our material, are old and in some points defective (Fig. 80).

Sphacelotheca forms its compound sporophore in the ovule of its host. When the ovule is normally and fully developed in the young flower, the parasite, which always grows through the flower-stalk into the place of insertion of the ovary, sends its hyphae from the funiculus into the ovule, where they rise higher and higher and surround and penetrate its tissue to such an extent as almost entirely to supplant it, and thus an ovoid Fungus-body of densely interwoven hyphae takes the place of the ovule. The micropylar end of the integuments alone escapes the change, and remains as a conical tip (Fig. 80 *C, o*) on the apex of the Fungus-body and gradually turns brown and dries up. The Fungus-body is at first colourless and uniformly composed of much-branched hyphae, which are woven together into a compact mass and have the gelatinous walls of the simple sporophore of Ustilago to be described below. If it has retained its ovoid form as it steadily increased in volume, differentiation begins first in the apical region into a comparatively thick outer *wall* which is closed all round, an axile columnar cylindrical or club-shaped body, the *columella*, both parts remaining colourless, and a dense *spore-mass* which fills the space between the two and becomes of a dark violet colour (Fig. 80 *C, D*). The lower part which corresponds to the funiculus and chalaza of the ovule remains undifferentiated, and an abundant formation of new hyphae is constantly taking place in it. This new formation is so added from below to the differentiated portion, that the latter constantly increases in height without becoming materially broader, and maintains therefore the form of a cylinder pointed at the upper end. Where the parts below approach the wall, columella, and spore-mass, they assume their structure and colour. In other words, each of the three portions grows from its base by addition of new tissue-elements, which are constantly being produced and pushed onwards from a basal formative tissue, and are differentiated and assume their ultimate form in the order in which they are produced (Fig. 80 *C* and *D*). The development and mature structure of the spore-mass are the same as those of Ustilago,

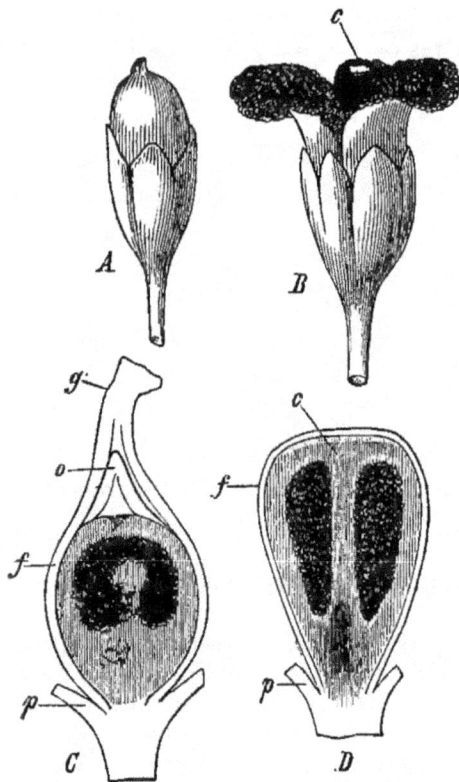

which will be described presently. The wall in its fully developed state is a thick coat formed of many irregular layers of small round cells not very firmly united together. These cells are formed in the same way as the spores from the hyphae of the primary tissue, and are of about the same size as the spores with a delicate colourless membrane, and for the most part with watery hyaline contents. The columella has the structure of the wall, but it usually incloses in its tissue evident brownish fragments of the tissue of the ovule, and consists at its uppermost extremity of much larger, firmer hyaline cells, the origin of which I am unable to explain. I may also observe that the upper extremity in young specimens always ends blindly in the spore-mass (*C*), but in some older ones reaches to the apical portion of the wall and passes into it (*D*); it is still uncertain whether this is a difference in the individual plants or a difference of age.

The spore-receptacle which has now been described is formed only from the ovule. The perianth and stamens of the flower continue in their normal state. The wall of the ovary and the style are also not attacked by the Fungus; they do not follow the growth of the spore-receptacle, and as this advances the lateral wall is distended and at length bursts transversely; the style with the upper portion of the wall dries up into a

FIG. 80. *Spacelotheca Hydropiperis* in the flower of *Polygonum Hydropiper*. *A* ripe compound sporophore of the Fungus projecting from the perianth of *Polygonum*. *B* the same with the mass of spores emerging from the sporophore. *C* median longitudinal section through a young fructification and its environment. *D* longitudinal section through an older sporophore. *c* the columella, *p* the perianth, *f* the wall of the ovary, *o* the integument (mycropyle) of the ovule, *g* the style. In *C* and *D* the sterile or young tissue of the Fungus is shaded by longitudinal lines, the mass of ripening spores is darker. Further explanation in the text. Slightly magnified.

small point at the apex of the receptacle, which is borne by the latter as it grows out of the perianth (*A*). The wall of the spore-receptacle, especially where it is covered above by the withered remains of the wall of the ovary, is very fragile and tears asunder at the slightest touch to discharge the spores (*B*).

SECTION LV. The **development of the resting-spores** commences in **Entyloma** (Fig. 81 *a*), as in Protomyces, with intercalary abjunction at any spot in the hyphae; in **Tilletia** the resting-spores are terminal, being formed singly from the extremities of branches of sporogenous hyphae. In **Geminella Delastrina** they originate by serial abjunction ; all the sporogenous hyphae divide, according to Winter, by transverse walls arising from the extremities of their curved terminal branches in basipetal succession into short members, which develope into spores united together in pairs.

The development of the spores of **Ustilago** may be briefly described in about the same words, only in this case the ripe spore-cells are·not united in couples, but are isolated and free. The terminal ramifications of the sporogenous hyphae are in some species (U. Ischaemi according to Winter, U. hypodytes according to my old account) slender and filiform, but they are usually furnished with many tufts of short branches, so that they have a peculiar clustered and lobed appearance (Fig. 82). Ustilago has the further peculiarity that the membranes of the hyphae swell strongly into a jelly synchronously with or even before the basipetal segmentation by cross walls into short isodiametric members, which also takes place in them. The protoplasm of each individual member is therefore soon surrounded by a broad hyaline

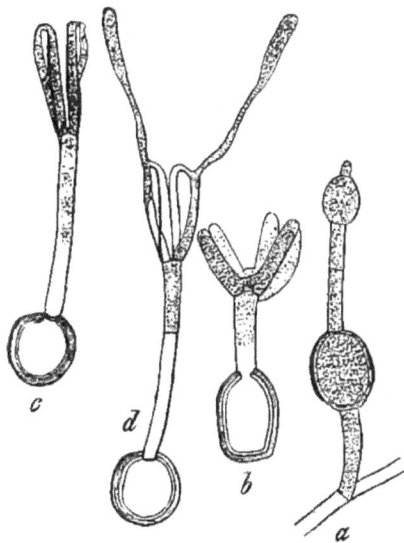

FIG. 81. *a* and *b Entyloma Calendulae. a* mycelial filament with two young resting-spores. *b* resting-spore germinating ; the front pair of primary sporidia in the whorl shows conjugation at the base. *c* and *d Entyloma Ungerianum,* De Bary. *c* a germinating resting-spore; four primary sporidia conjugating by pairs at their apices. *d* the same specimen seven hours later ; commencement of the abjointing of a secondary sporidium (gonidium) on each pair. Magn. 600 times.

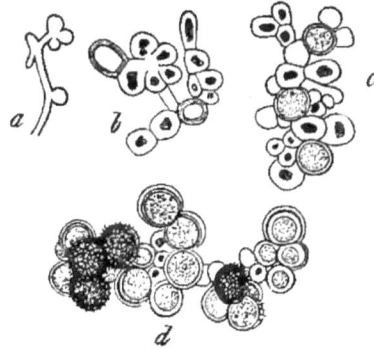

FIG. 82. Development of the spores of *Ustilago Tragopogonis;* successive stages of the development according to the letters. *a* sporogenous branch, just appearing on the surface of the young corolla of *Tragopogon pratensis* and beginning to form a tuft of branchlets. *d* spore-tuft with several ripe spores, the episporium of which is coloured a dark violet and furnished with reticulate thickenings. Magn. 300 times.

gelatinous sheath (Fig. 82 *b, c*), and forms inside it a relatively small almost homogeneous and strongly refringent nucleus-like body. The definitive membrane of the spore is then formed on its outer surface inside the gelatinous envelope, and developes with the protoplasm into the comparatively large spore. As growth proceeds the gelatinous sheath becomes more delicate and paler and disappears altogether when the spore has reached maturity (*b–d*).

In the genera Urocystis, Sorosporium, and Tuburcinia the ripe spores are joined together from two up to a considerable number in a cluster and are each provided with a persistent or temporary special envelope. The development of the cluster is not clearly known in all points. In **Urocystis**, according to Wolff and Winter, the first beginning of a cluster is represented by a turgid and often curved branch from the sporogenous hypha, in place of which a group of 2–4 delicate cells subsequently makes its appearance after some intermediate stages which have not been clearly ascertained; these cells are firmly united together and develope into spores, and the group in its early stages has slender curved hyphal branches growing close round it and forming an envelope for it. The hyphae of the envelope are divided by transverse walls into short cells, most of which disappear as the spores ripen, while a small number of cells varying with the individual develope into persistent half lenticular envelope-cells, which lie close upon the group of spores as it matures. The envelope appears to be formed at earlier or later periods in the development according to the species, and the division of the spore-primordia to be sometimes omitted, as only one ripe spore is not unfrequently found inside a group of envelope-cells. According to Prillieux the process is much more simple; a number of sporogenous hyphal branches become felted into a knot, and one or more branches form a spore acrogenously, as in Tilletia, while the cells of the other branches develope more or less perfectly into an envelope; but the envelope may be omitted and then the spore is formed as in Tilletia. My older observations agreed pretty well with these statements of Prillieux, but it would be desirable to have other confirmation of them.

The commencement of a spore-cluster in **Sorosporium** and **Tuburcinia**, according to Woronin, is also a short turgid lateral branch, sometimes perhaps two branches, lying close upon one another. Every such primordium is then wrapped up in a tangle of the many ramifications of slender hyphal branchlets, which become woven together into a round compact coil. No further differentiation is at first perceptible in the coil. When it has grown to a certain size its central portion consists of a compact group of delicate polyhedric cells, which then ripen all together into perfect spores without further perceptible division by multiplication. The group is at first still enclosed in a dense many-layered web of hyphae, but this disappears as the spores mature, in Sorosporium with a previous conversion into mucilage. The first formation of the group of spores in the coil is not clearly ascertained. Frank's statement, that each spore comes from a turgid cell of one of the original filaments of the coil, and that the other parts then form the temporary envelope, is plausible but requires more distinct proof. We learn from Fischer von Waldheim that spores sometimes occur even in Sorosporium Saponariae, which are abjointed singly and without envelope on the extremity of a hypha as in Tilletia. If these spores really belong to Sorosporium, as they must be supposed to do without proof to the contrary, and not to a parasite of the plant, Frank's view finds in them material support.

The development of the spores has not been closely followed in the genera **Thecaphora** and **Schizonella**, Schröt., which certainly come near to Sorosporium.

SECTION LVI. **The single ripe spore** of the Ustilagineae is usually round or polyhedric and has the ordinary structure of firm Fungus-spores,—a delicate colourless endosporium enclosing the protoplasm, and a stout episporium, which in most species, and also in a form of Entyloma distinguished as Melanotaenium, is dark-coloured, and in many has a delicacy of structure and superficial marking which is characteristic and useful for the discrimination of species; in some cases it has a distinct germ-pore. It is only in the majority of the Entylomeae that the thick stratified episporium is almost or altogether colourless. The envelope-cells which adhere to the spores in Urocystis have likewise pale-coloured membranes at the time of maturity, and their contents are almost entirely watery.

The **germination** of the resting-spores takes place when they are sufficiently supplied with moisture and have absorbed it to a sufficient degree. The form differs according as only water is supplied to the spores or nutrient matter also dissolved in the water.

In the first case a short germ-tube is emitted, which takes the protoplasm of the spore and developes into a *promycelium* in the sense of section XXXI (p. 111). In most species the promycelium remains in connection with the membrane of the spore, in a few, as Ustilago Vaillantii, it soon separates from it. To this rule of development there are a few individual and specific exceptions which will be considered presently. The further development is as follows :—

1. Shoots are put forth acrogenously or laterally at the cost of the protoplasm and are abjointed; these in the terminology which has been proposed would be *sporidia* (of the first order). The particular forms which their development assumes vary much in the several species; the chief ones, with which isolated intermediate or divergent kinds can easily be connected, are these :—

a. A whorl or circle of narrowly cylindrical or subulate sporidia ('Kranzkörper') shoots out from the obtuse and comparatively broad apex of the promycelium. They appear simultaneously; their number in a whorl is different in different species and individuals, varying between 4 and 10. This mode is characteristic of most species of Entyloma (Fig. 81), and for Tilletia (Fig. 83), Tuburcinia, and Urocystis.

b. The promycelial tube is divided by transverse walls into a series of two or more short cells, and from these are abjointed, usually at their acroscopic extremity, a number

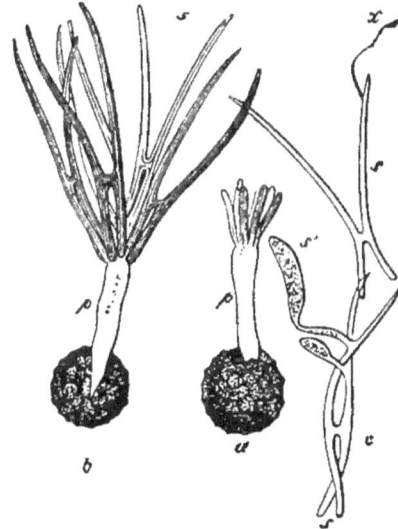

FIG. 83. *Tilletia Caries*, Tul., germinating. In *a* primary sporidia beginning to shoot out from the promycelium *p*. In *b* the primary sporidia *s* of the promycelium *p* coojugated in pairs. In *c* a germ-tube *x* proceeding from a pair of primary sporidia *s*; *s'* a secoodary sporidium or gonidium. After Tulasne. Magn. 460 times.

of elongated or rod-like shoots as sporidia. This is the process in Tolyposporium Junci, Woronin, and in many species of Ustilago, for instance in U. Tragopogonis (Fig. 84 *B*), U. flosculorum, U. utriculosa, U. Cardui, U. Kühniana, and some others.

c. The spore emits a simple, or in a very few cases a branched, short, slender promycelial tube, and from this is abjointed one cylindrico-fusiform sporidium or several in a row one after another (Ustilago longissima, Fig. 84 *A*, Thecaphora Lathyri according to Brefeld).

2. The slender promycelial tube, which sometines puts out one or two branch-like protuberances, is divided by transverse walls into a few cells from which single sporidia are abjointed or sometimes none. This is the case with Ustilago Carbo (Fig. 84 *C*), U. destruens, and, according to Woronin, with Thecaphora hyalina.

[4]   N

The abjointed sporidia then unite in pairs in many species, either before or after their separation from the promycelium.   They unite by means of short transverse processes, which may be placed at the point of insertion or at the apex or in the middle, and form double cells, as is shown in Figs. 81, 83, 84 *B*.   Coalescence by pairs takes place also in the cases which come under No. 2 between the segment cells of the promycelium.   Adjacent cells unite with one another by means of short processes arising near the partition-wall and forming with one another a short lateral excrescence (Fig. 84 *C* left), through which the protoplasm of the two cells coalesces after disappearance of the separating membrane.   Communication is established between cells which are not next neighbours by larger lateral branches, which may form curved tubes or loops on a promycelium or even transverse bridges between two promycelia.

The simplest form in which the development proceeds is by the production at a spot in the conjugated pair of a slender *germ-tube* with acropetal growth, which gradually receives the whole of the protoplasm of the two cells (Fig. 83 *x*); it has been shown that in many species this tube can penetrate into the phanerogamous plant which is its proper host, and there develope a mycelium which produces new resting-spores; it may therefore be shortly described as an *incipient mycelium* or mycelial primordium.

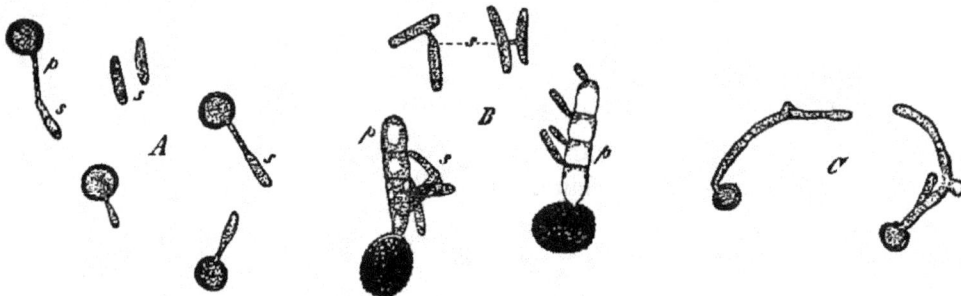

FIG. 84.   *A Ustilago longissima*, Tul., germinating.   *B U. Tragopogonis*.   *C U. Carbo*, Tul.   *p* promycelium, *s* primary sporidia.   Further explanation in the text.   *A* magn. nearly 700, *B* 390, *C* more than 390 times.

The process is more complicated in Tilletia, in many species of Entyloma, in Tuburcinia Trientalis, and as a general rule in Urocystis Violae, where a (secondary) sporidium is acrogenously abjointed on a short lateral branch from the conjugated pair (Figs. 83 *s'*, 81 *d*), which then gives rise to the incipient mycelium.

Deviations from the above course of development occur under otherwise similar conditions in individuals of the species in which this development is the almost invariable rule.   Firstly, the germ-tube which proceeds from the resting-spore may assume the characters not of a promycelium as described above, but of an incipient mycelium with acropetal growth but not producing sporidia.   Secondly, the conjugation of sporidia of the first order in pairs may be omitted, but the sporidia may still put forth tubes which are similar in conformation to the incipient mycelium; this may happen to all or to a majority of the primary sporidia of a whorl in some species, for instance, of Entyloma, or, when the number of the members of a whorl is uneven, to one sporidium, while the others conjugate in pairs.

These phenomena, which are individual exceptions in the cases which we have been hitherto considering, are the prevailing and even the invariable rule in another series of forms. Here we may have formation of sporidia according to one of the above modes or one like them, but without their conjugation, as in Ustilago Maidis, U. Vaillantii, U. longissima, and in Entyloma Magnusii, Wor.; or there is no formation of sporidia, but production of a germ-tube from the resting-spore which shows the growth of an incipient mycelium without conjugation of any of its segments, as in Sorosporium Saponariae according to Woronin.

The germination of the resting-spores in the nutrient solutions mentioned on p. 127 has been recently studied by Brefeld in great detail. In some species it is easier and more rapid in these solutions than with only a supply of pure water, and generally so far differs from the process as described above, that luxuriantly vegetating forms take the place of the short-lived products developed at the expense of the protoplasm of the spore, and that these can continue to grow on without change of shape and without limit if supplied with sufficient nutriment. The form which they assume depends on the species; either the promycelium which begins in the manner above described continues to develope like the Sprouting Fungi (see page 4), or the germ-tube of the resting-spore grows into a branched filamentous mycelium from which spores are then abjointed in the fluid itself or on branches which rise into the air. Ustilago antherarum, U. Carbo, U. Maidis, and U. Kühniana are beautiful examples of the sprouting form; U. destruens of the sporogenous mycelia.

There are species intermediate between the two forms, a detailed account of which will be found in Brefeld. U. destruens itself may to some extent be regarded as an intermediate form, since some of its spores germinate by sprouting in the nutrient fluid, while others develope a mycelium which vegetates and ramifies in the fluid, and sends up erect branches into the air from which elongated spores serially developed as branched chains of sprouts are abjointed (see p. 66). In the case of Tilletia Caries, Entyloma, and Thecaphora Lathyri the resting-spores have not been seen to germinate in a nutrient fluid, or the product of the commencing germination soon dies away; on the other hand when primary or secondary sporidia from plants which have germinated in water are placed in nutrient solutions, there is a copious formation of mycelium and acrogenous abjunction of spores on erect aerial branches of the mycelium; these spores in Thecaphora are like the primary sporidia of germination in water, those of the other species mentioned are like the secondary.

Geminella Delastrina differs from the rest, as Schröter found, in the acrogenous serial abjunction of roundish sporidia on the short promycelial tube in water; the further development of the spores was not observed. The same species produced large mycelial bodies in Brefeld's nutrient solutions, but their further history also was not ascertained.

SECTION LVII. Finally, in some Ustilagineae spores of another kind are produced on the mycelium forming the resting-spores and inhabiting the host-plant; these may be termed provisionally *gonidia*. Schröter found that in a number of species of Entyloma, E. Ranunculi for example which is common on Ranunculus sceleratus and R. Ficaria and especially E. serotinum on Symphytum officinale, the mycelium which lives in the leaves sends a large number of short branches often closely crowded together into the air, partly through the stomata, partly through the lateral walls of the epidermis, and that from the extremities of these branches single spores (or perhaps several successively) are abjointed. The spores are narrowly fusiform in shape, like the secondary sporidia of the promycelium, and it may be assumed, though it has not been observed, that their mode of germination is the same. They are formed on the mycelium before the resting-spores, and appear to the naked eye as a slight sprinkling of flour or mould on the spot which is attacked in the leaf.

The most remarkable case of this kind was observed by Woronin in Tuburcinia Trientalis. The resting-spores of this Fungus germinate in late autumn. The mycelial primordia produced from the secondary sporidia penetrate into the young subterranean shoots of Trientalis destined to pass the winter, and develope a mycelium which also hibernates in their parenchyma. Next spring the mycelium spreads through the whole of the shoot as it unfolds, and at first forms gonidia on the under side of its leaves and afterwards the cluster of resting-spores, but more in the stem than in the leaves. The gonidiophores are branches of the intercellular mycelium and come out to the air through the stomata and lateral walls of the cells of the epidermis, covering the whole of the surface of the leaf with a white down. They are unbranched and subulate in form, and from the apex of each several spores (gonidia) are successively abjointed one after another, which are pear-shaped and not fusiform like the sporidia of the promycelium. Germ-tubes are put forth by the gonidia on the moist surface of a leaf of Trientalis and penetrate into it, and develope mycelia which are confined to small spots in the leaf, where they form clusters of resting-spores but not gonidia.

Section LVIII. According to the above well-ascertained facts the course of development in the Ustilagineae (in species of Entyloma, in Ustilago Carbo, and Urocystis occulta), where the process is the simplest, appears to be, that the resting-spore produces a promycelium, and the promycelium the conjugating pairs, and the mycelial primordia which proceed from them enter and develope in the host-plant and produce fresh resting-spores. The first complication is caused by the production of mycelial primordia, not directly from the conjugating pairs, but from secondary sporidia proceeding from them, as happens almost invariably in some species of Entyloma and in Tuburcinia Trientalis. Then comes in Entyloma Ranunculi, E. serotinum, and in Tuburcinia Trientalis the interpolation of *gonidia* which are formed on the endophytic mycelium and can produce similar mycelia. If we regard the resting-spores as carpospores in the sense of section XXXIV, a point which will be discussed further on, the secondary *sporidia* must be termed *gonidia* which have been interpolated in the course of the development. Lastly, a mycelium can also be produced on any dead substance which yields suitable nourishment and gonidia on the mycelium, and it may at least be assumed that the *gonidia* are capable of reproducing a mycelium forming carpospores. I have purposely avoided the general use of the word gonidium in the preceding pages, though it will have to be used from this time, in order to show by one example how the same object may and must have different names when considered from different points of view and in different connections.

The facts that have been certainly ascertained are only these: that the species, whose course of development has been really followed throughout, Ustilago Carbo, U. destruens, Tilletia, Tuburcinia Trientalis, Entyloma and Urocystis occulta, all show conjugation of the segments of the promycelium, or primary sporidia, and that in Urocystis Violae, whose development was also fully observed by Kühn, conjugation must be supposed to take place from Prillieux' figures, though it is not expressly described. On the other hand there are species, which, according to our present observations, do not conjugate. But we have no complete observations of their development; only from the data which we possess, and the positive result of Kühn's extended researches into the infection of host-plants with Urocystis Maidis, it must be supposed that the endophytic sporogenous mycelium is in this case developed directly from the resting-spore or from interpolated gonidia which do not conjugate. Accordingly the course of development, apart from certain special formations and complications which vary in the different species, would be the same in its main features in the entire group of the Ustilagineae up to the difference which lies in the presence or absence of conjugation.

What importance is to be assigned to this difference is open to discussion. The pairing I have formerly called conjugation and do so still, thus giving expression to the view, that it may or ought to be considered as analogous with a sexual process. Brefeld does not admit this and explains it to be an analogous phenomenon to the coalescence of vegetable cells, especially young germ-tubes (see on page 2). In such cases as these no decisive argument for or against is to be found in the phenomenon in itself, but we must look about for indirect grounds of probability. Mine are as follows, if I confine myself to cases that are thoroughly known. First, the almost invariable occurrence of pairing under the normal conditions of germination; the conditions, I mean, to which the species is actually adapted in nature. The conditions are those for germination in water in the case of Tilletia, the species of Entyloma, Urocystis, and Tuburcinia Trientalis, as has been already shown; and here pairing takes place so promptly that it is not easy to keep primary sporidia from uniting, if we disregard some special exceptional cases mentioned above. But these cases are of the kind which prove the rule. Secondly, the great predominance of union in pairs. The sporidia which are placed close to one another in whorls in Tilletia, Entyloma, Urocystis, and other genera unite, almost without exception, in pairs only, and where there is an odd sporidium it usually does not conjugate, though its union with some pair would be easy, one might almost say would be very natural. The segments of the promycelium of Ustilago Carbo conjugate when they are immediate neighbours in a mode similar to the ' clamp-connection ' of page 2; under different local relationships pairing is effected in a different way, as by loop-unions between two segments of a promycelium which are separated by a pair which have already united. These facts show that a change usually takes place in a pair after conjugation which renders a second union difficult or impossible, while it introduces the further development. All these are phenomena which find their analogy, as far as we know at present, only in sexual processes, or, as it may be briefly expressed, in sexual processes of conjugation, and must be interpreted by them, till we obtain further knowledge. The case is different in the coalescence of germ-tubes, which are the nearest comparable. A glance at Fig. 1 shows that in that case coalescence may take place between any number of spores or may even be omitted, while the further development is the same in kind, but shows different degrees of strength according to the number of the germs which have coalesced. It is true that conjugation of more than two primary sporidia has been observed in the plants which we are considering; in Tilletia, for instance, and Ustilago Tragopogonis; but these exceptional cases are comparatively rare and would only confirm the rule, even if similar exceptions in typical sexual conjugation, as in Zygnema and Acetabularia, were unknown.

The reasons which Brefeld alleges against my conception of the pairing, as far as I understand them, are these. Pairing occurs in germination in water; it does not occur when the primary sporidia of Tilletia, for example, are placed in nutrient solutions before pairing can begin; it also does not occur in Ustilago Carbo and similar germinating species, if they germinate and sprout in a nutrient solution. But when the supply of food is exhausted, pairing sets in in many species and is followed by the formation of incipient mycelium. A fresh supply of food again sets up gonidial sprouting. This shows that the uniting cells are not sexual organs in themselves incapable of germination, for 'according to our present ideas with respect to the essence of sexuality' no such change in the properties of sexual cells could be brought about by changes in the nutrition. The answer to this is, that, according to our present *knowledge*, the only processes of conjugation which can be here brought into comparison, and which are to be distinctly regarded as simplified analogues of sexual processes, consist solely in the characteristic coalescence in each case of two cells into one capable of further development. What further 'essence' is involved in the matter we do not know. Many of these conjugating cells are not capable of further development without conjugation, the gametes, for example, of many Chlorophyceae; but this is not the case with all. I will not mention the Phaeosporeae, which supply examples of the opposite behaviour, because the circumstances in their case are not quite clear and are to some extent open to question. But Rostafinski and Woronin found that the swarm-spores developed from recently matured resting-spores of Botrydium granulatum[1] do typically and necessarily conjugate, and that the same swarm-spores germinate without conjugation when they proceed from resting-spores which are more than two years old. A change therefore takes place in these cells, affecting those properties in them with which we are at present concerned, through the influence of external causes which must be closely connected with processes of nutrition. Even highly differentiated sexual cells may be capable of further development when fertilisation is prevented, as is shown by the oogonia with unfertilised oospheres of Pythium megalacanthum[2] which send out strong germ-tubes. We see therefore that the exceptional cases to the pairing which is the general rule in the Ustilagineae, whether artificially produced or spontaneous, are not opposed to phenomena known to occur in organs which are undoubtedly sexual.

That there are species which do not pair cannot affect our decision with regard to those which do; moreover it is well known, that in other groups of Fungi organs strictly homologous have a distinct sexual function in some species and may be asexual in others, as we learn by comparison of the Peronosporeae and those Saprolegnieae which are without antheridia.

---

[1] Bot. Ztg. 1877, p. 662.
[2] Bot. Ztg. 1881, p. 543.

The strongest argument which could be advanced at the present time in support of Brefeld's view is, that, according to his statements, species like Ustilago longissima show pairing in the 'gonidia' which are produced in nutrient solutions and subsequent formation of mycelial primordia under the given conditions, while pairing is not observed when germination has taken place in water. But the course of development is not perfectly known in these species, and they cannot be a standard for judging of those which are thoroughly known. And what we know about these species, especially Ustilago longissima, gives ground for suspecting that even in them pairing is a necessary preliminary to the formation of mycelial primordia capable of infection, and that it occurs in opposition to the case of Tilletia only under special conditions of nutrition which have yet to be ascertained.

In whatever way the question of the sexual value of the pairing may ultimately be decided, it is at all events a characteristic fact in the cases in which it occurs and cannot be disregarded.

The homologies in the course of development of the Ustilagineae are quite clear within the group itself, and require no further discussion. It is moreover obvious, and the fact is expressed by the terminology which has been partly anticipated, that this development corresponds in general to that of the preceding groups, and resting-spores therefore may be compared with the resting-oospores, *oospores* or *carpospores* of the Peronosporeae and Entomophthoreae, &c.; the comparison of the simpler forms, especially Entyloma and Tilletia, confirms this view. If this comparison is accepted, we have at once a justification of such terms as gonidia, &c. The resting-spores or carpospores, as they may be presumed to be, of the Ustilagineae are formed, it is true, *asexually*, while the contrary is the case with those of the Peronosporeae; but this, as we have learnt from the Saprolegnieae, affords no criterion for the determination of the homologies. It is not easy to see why it should be the Entomophthoreae particularly to which the Ustilagineae must be considered to approach nearest, as Brefeld maintains[1]; it might very well be the Peronosporeae. But while the agreement between the groups makes itself thoroughly felt, a *near* approximation of them is in most points impossible; this can be brought about in many places, if we put one thing forward and disregard another, but we gain nothing by these arbitrary proceedings. If, on the other hand, we look for the points which are distinctly characteristic of the Ustilagineae, the most prominent is that of the conjugating pairs of cells. This phenomenon recurs, so far as is known, only in one form not belonging to the Ustilagineae at the same place in the course of the development and in a quite similar form, namely, in Protomyces macrosporus. There is only one important difference between the two cases; in Protomyces the conjugating cells are of endogenous origin, in the Ustilagineae they are acrogenously abjointed, but this difference is no real objection to the homology, for it occurs in a similar form among the undoubtedly homologous gonidia in the Mucorini (section XLIII); the pairs of cells in Protomyces are formed endogenously, like the gonidia of Mucor, those of the Ustilagineae acrogenously, like the gonidia of Chaetocladium. Hence Protomyces macrosporus appears to be in every respect very nearly related to the Ustilagineae, and its resemblance in habit to species of Entyloma

---

[1] Brefeld, Schimmelpilze. IV, p. 165.

is in harmony with this. We should be able to connect the Ustilagineae more closely with other groups, if we could do the same with Protomyces, and for this purpose the approximation of Cladochytrium is important. In these two forms indeed there is complete agreement except in two points. First, the propagative cells formed in the resting-spores are swarm-cells in Cladochytrium, but not in Protomyces; but this difference is quite unimportant for the present question, because strictly homologous spores of most closely allied species may behave differently in this respect. Secondly, and this is important, there is no conjugation or pairing, as far as we know, in Cladochytrium. If further researches only confirm this fact, then there is a gap in the chain of connection which is not bridged over, but this has yet to be determined. According to the data before us it would appear that the best line of connection of the Ustilagineae with other groups must lead through Protomyces and Cladochytrium to the Chytridieae or to the group of the Chytridieae to which Cladochytrium belongs (see page 165). The Ustilagineae therefore, considered phylogenetically and in connection with what was said above of the Chytridieae, may be regarded as a more highly developed group proceeding from the Chytridieae, whose development has advanced from the simpler forms like Entyloma in two diverging directions, and has reached its highest point on one side in the cluster-forming Sorosporia and Urocystes, and on the other in the compound sporophores of Sphacelotheca and other forms. The same considerations arise moreover if we connect the Ustilagineae with other groups, such as the Entomophthoreae, which again can only be effected through Protomyces and Entyloma. And whichever view we adopt, if we accept our present data, we must look upon *conjugation* or *pairing* in Protomyces and the Ustilagineae as a special new phenomenon, that is, as one which has *nothing homologous with it* in the groups supposed to be allied to them. If it has its *analogue* in them in sexual processes or processes of conjugation, if, in other words, it is a sexual process, this process makes its appearance in the Ustilagineae at a place in the course of the development *other than* that which it occupies in groups before considered; in these the resting-spore is always the direct result of the sexual process, while its homologue in the Ustilagineae is of asexual origin.

It is superfluous perhaps to say in conclusion that the Ustilagineae cannot at present be brought into closer connection with the succeeding groups, as the following sections will show.

Historical remarks. Brefeld (Schimmelpilze, IV) and myself (Beitr. IV) were the first who attempted recently to settle the homologies and the position of the Ustilagineae in the system in the sense indicated in the text. The old writers placed them next the Uredineae on the ground of external resemblance. I pointed, out as early as 1853 (Brandpilze, p. 28), that this position was inadmissible. When I attempted, seventeen years ago, in the preface to the first edition of this work, to give a short survey of the groups of Fungi according to the knowledge then before us, I followed the old custom of placing the Uredineae and Ustilagineae together, under Fries' name of Hypodermii, expressly quoting that writer, because the state of our knowledge at the time did not allow of my substituting a new and safer view for the old one. I write this in correction of statements which make me the author of Fries' division Hypodermii, or which even seem to make it a reproach to me that I retained that division when nothing else was possible, and abandoned it as soon as our increased knowledge permitted me to do so.

Literature :—

Protomyces.

DE BARY, Beitr. z. Morph. u. Phys. d. Pilze, I, where the older literature is cited ;—Id. in Bot. Ztg. 1874, p. 81.

Ustilagineae.

LÉVEILLÉ in Ann. d. sc. nat. sér. 2, XI, 1839.—Article Urédinées in d'Orbigny's Dict. d'hist. nat.

TULASNE, Mémoire sur les Ustilaginées comparées aux Urédinées (Ann. d. sc. nat. sér. 3, VII, 1847);—Id., Second Mémoire sur les Urédinées et les Ustilaginées (Ann. d. sc. nat. sér. 4, II).

DE BARY, Unters. ü. d. Brandpilze, Berlin, 1853 (this work and Tulasne's memoirs give the older literature);—Id., in Flora, 1854, p. 648;—Id., Protomyces microsporus u. seine Verwandten in Bot. Ztg. 1874, p. 81 (Entyloma).

J. KÜHN, Die Krankheiten d. Culturgewächse, Aufl. 2 unaltered, Berlin, 1859;—Id. in Fühling's Landw. Zeitsch. 1879, p. 81 (according to Just's Jahresber.).

A. FISCHER v. WALDHEIM, Beitr. z. Biol. u. Entw. d. Ustilagineen (Pringsheim, Jahrb. VII, 1869). This work contains copious accounts of the older writers and literary minutiae. See also another work by the author not properly belonging to this place, Les Ustilaginées et leurs plantes nourricières (Ann. d. sc. nat. sér. 6, IV).

R. WOLFF, Beitr. z. Kenntn. d. Ustilagineen (Urocystis occulta) in Bot. Ztg. 1876, p. 657;—Id., Der Brand d. Getreides, Halle, 1874;—Id. in Bot. Ztg. 1874, p. 814.

G. WINTER, Einige Notizen ü. d. Familie d. Ustilagineen (Flora, 1876, Nr. 10).

J. SCHRÖTER, Bemerkungen und Beobachtungen ü. einige Ustilagineen (Cohn's Beitr. z. Biol. II, pp. 349 and 435).

E. PRILLIEUX, Quelques observations sur la formation et la germination des Spores des Urocystis (Ann. d. sc. nat. sér. 6, X, 1880).

A. B. FRANK, Die Krankheiten d. Pflanzen, Breslau, 1880, p. 419.

M. WORONIN in Beitr. z. Morph. u. Phys. d. Pilze, V, Frankf. 1882 (Tuburcinia u. Keimungen).

M. CORNU, Contributions à l'étude d. Ustilag. (Bull. soc. Bot. de France, Aug. 1883);—Id. in Ann. d. sc. nat. sér. 6, XV (1883).

ED. FISCHER, Beitr. z. Kenntn. d. Gattung Graphiola (Bot. Ztg. 1883).

BREFELD, Bot. Unters. ü. Hefenpilze, Leipzig, 1883.

ASCOMYCETES.

GENERAL CHARACTERS. SPOROCARPS.

SECTION LIX. The group of the Fungi known as the Ascomycetes contains an enormous number of species, but all its members have their bodies composed of branched hyphae which are always septate, and they are all distinguished by the formation of their spores in asci, as has been described in section XIX, from which peculiarity they derive their name. Putting aside the species which will be noticed at the conclusion of the division as doubtful Ascomycetes, the asci of this group are *sporocarps* in the sense of section XXXIII, or parts of sporocarps, and in the latter case are often collected several or many together into hymenia (section XII).

The sporocarps are formed and borne in some species on an inconspicuous filiform mycelium, and then exhibit the characters of *compound sporophores* described in sections XII and XIII. In other species they are produced and borne on larger compound sporophores, each called a *stroma*, which take the form in the different species of flat expansions, crusts, foliaceous or erect and branched shrublike bodies. The general structure and conditions of growth of these formations, which are often of considerable size, have also been described above in section XII. Well-known examples of them among the Pyrenomycetes are the cushion-like or membranous expanded stromata of the 'Sphaeriae compositae,' i. e. of the genera Hypoxylon, Diatrype, Ustulina, Epichloë, and many others, and the erect sporophores of the Xylarieae, Claviceps, Cordyceps, Thamnomyces with its many bifurcations, and others. Of the Discomycetes which belong to this group the most important are the Lichen-fungi with their disk-shaped and alveolate sporocarps; next, Rhytisma and its allies with flat disk-shaped stromata, and perhaps too the remarkable South American Cyttarieae —spherical or club-shaped or gelatinous bodies more than an inch in diameter with the broad upper half covered with deeply alveolate hymenia, regarding which it is doubtful whether each is in itself a sporocarp or a portion of the sporocarp formed by the whole club-shaped stalk.

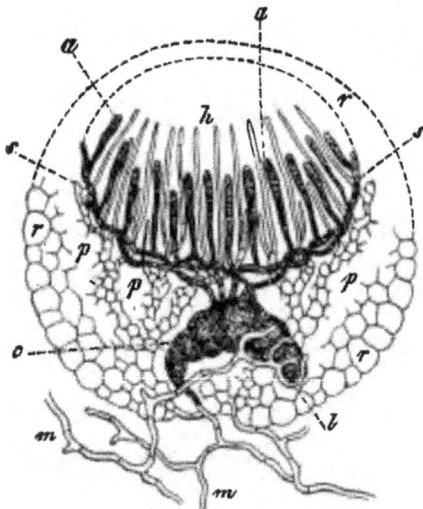

FIG. 85. *Ascobolus furfuraceus.* Young sporocarp in median longitudinal section; *m* mycelium, *h* hymenium, *c* archicarp with the ascogenous hyphae *s* in the subhymenial layer and the asci *a* shaded, *i* antheridial branch, *p—r* tissue of the envelope giving rise to the paraphyses. Diagrammatic representation from Sachs after Janczewski.

Two chief constituents may be distinguished in almost all the better-known sporocarps of the Ascomycetes (Fig. 85); one (*c, s, a*) is the *ascus-apparatus* and consists of asci together with the hyphae or cells from which they are immediately derived, the *ascogenous hyphae* or *ascogenous cells*; the other is the *envelope-apparatus* which is formed of all the other parts of the sporocarp. The two parts are necessarily in the closest relation to one another from the purely morphological and also from the physiological point of view, since the envelope-apparatus bears, protects, and feeds the asci. The elements of the two parts may be most intimately united and interwoven with one another in the mature sporocarp and in fact may be difficult to separate or distinguish. Nevertheless the origin and growth of the two parts are usually distinct from their first inception or at least in a very early stage of the sporocarp, so that only similar parts, asci, and not elements of the envelope, spring from an ascogenous hypha or ascogenous cell, and no asci from the elements of the envelope. Exceptions to this rule are according to accurate observations at least very rare. Former statements to the effect that asci and elements of the envelope, especially paraphyses, spring directly from the same hyphae are in most cases certainly incorrect; even some recent statements, like those regarding Pleospora and Ascodesmis, which will be noticed again in a later page, require a more searching investigation. But even if there are some exceptions to the rule, and a good many cases require further examination, the prevailing rule undoubtedly is that the two constituents of the sporocarp though close to one another are independent in their origin and growth.

Eremascus, a typical Ascomycete of the simplest form, recently described by Eidam, is remarkable for the entire absence of an envelope-apparatus.

Three chief forms of sporocarp in Ascomycetes may be distinguished according to the arrangement and conformation of its two constituent parts, and the coarser structure which results therefrom : the *discocarp* or *apothecium* and the *pyrenocarp* or *perithecium*, which are used to distinguish the Discomycetes from the Pyrenomycetes, and the *cleistocarp*, a form which remains closed. I use the words apothecium and perithecium in the sense introduced by P. A. Karsten in his Mycologia fennica.

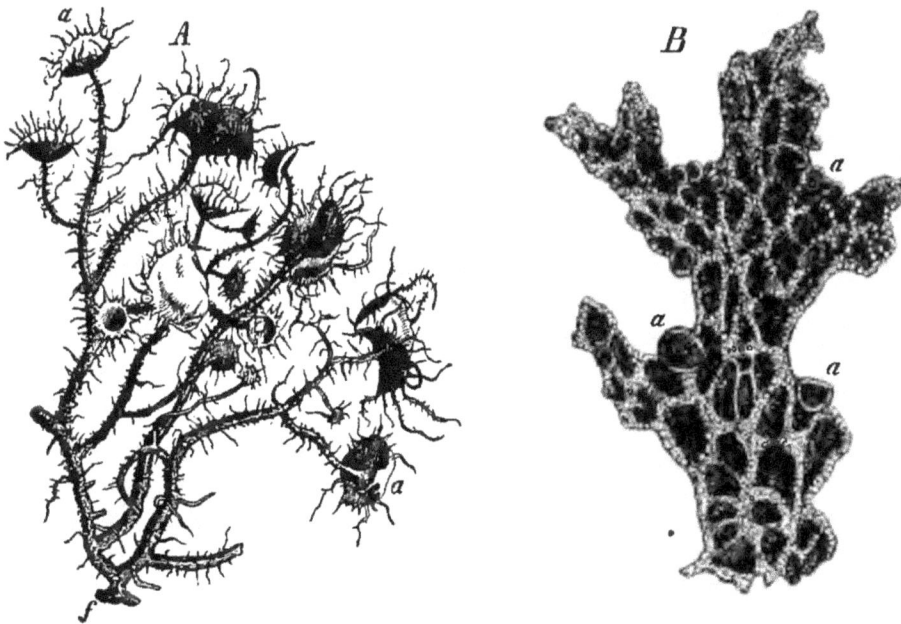

FIG. 86. *A Usnea barbata. B Sticta pulmonacea.* Portions of thallus. *a* apothecia, *f* point of attachment to the substratum. After Sachs. Natural size.

There are a few forms or small groups which cannot in strictness be arranged under one of these three types, but must be placed along with them as peculiar and exceptional cases. The accounts which we possess of the history of the development, though imperfect, are sufficient to show that the above types are distinguished more by habit in their advanced states than by any more fundamental distinction.

SECTION LX. The distinguishing feature in the **apothecium** (Figs. 86–89, see also Figs. 19, 22, 85, and 99) is, that the hymenium lies exposed on the surface of the sporocarp while the spores are forming and maturing. The hymenium itself, the *discus, lamina proligera* and *sporigera* of the old terminology, consists first of the *asci*, secondly of capilliform hyphal branches, the *paraphyses.* The latter are placed perpendicularly to the surface of the hymenium, and terminate at a uniform height on this surface, being crowded together in great numbers and usually giving the hymenial layer its characteristic tint from the colouring of their walls, or contents: not infrequently, especially in the hymenia of the Lichen-fungi, they are united laterally and without gaps by the gelatinously thickened walls, so that their lumina appear to be set in a homogeneous structureless jelly. The paraphyses spring by their inner or lower extremities, that is, those turned away from the outer surface as branches from a dense hyphal tissue beneath the hymenium, the *subhymenial layer* or *hypothecium*, which is then continued further downwards into the more or less largely developed *receptaculum* or stipe of the apothecium, or at least into an outer envelope, the *excipulum*, which belongs to it, though it is not greatly developed.

The paraphyses, together with the elements of the hypothecium which produce and bear them and the receptaculum or excipulum, belong to the envelope-apparatus of the apothecium. But the ascogenous hyphae take their course in the hypothecium,

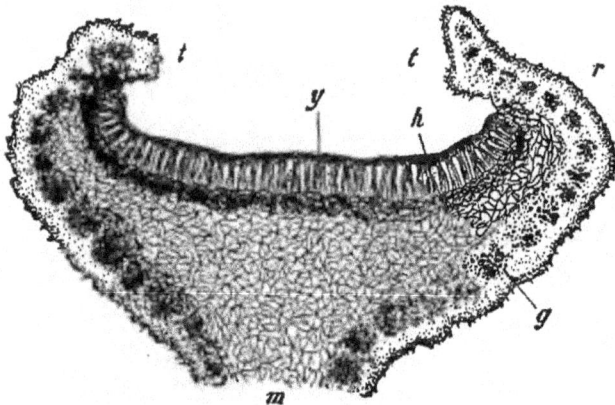

being interwoven with the elements of the envelope; they grow up in the commencing apothecium from points of origin, which will be described more exactly in section LXIII, in the direction of the hymenial layer, and afterwards spread themselves out near the under surface of the disk with copious branching which follows the progressive growth of the whole . apothecium, and thrust the extremities of their branches of the last order successively in between the paraphyses to

FIG. 87. *Anaptychia ciliaris.* Médian section through an apothecium; *h* hymenium, *y* subhymenial layer and excipulum. All beside belongs to the thallus which forms a rim round the excipulum at *t*; *m* medullary layer, *r* rind, *g* its Algae. After Sachs. Magn. about 50 times.

form the asci. In many of the species belonging to this division the ascogenous hyphae can scarcely be distinguished from the elements of the envelope which surround them by anything but the formation of asci which belongs to them only. In others, especially in many but not all the Lichen-fungi, in which Schwendener first discovered them, they are distinguished from the surrounding tissue by their greater thickness, by the abundance of their protoplasm, and by their membrane turning blue after treatment with potash. The development of the envelope-apparatus is always in advance of the ascus-apparatus, though the latter may have begun to be formed at the same time; the paraphyses are therefore always the parts which are first present in the hymenium. The asci make their appearance after them and grow vertically upwards between them in the direction of the surface, reaching it usually or rising above it only as the spores become ripe (see section XXII). Only a few asci appear at first, then their number increases through successive branching of the ascogenous hyphae, often to such a degree that the paraphyses are displaced and become indistinguishable.

The Ascomycetes which bear apothecia are well known under the name of **Discomycetes** and **Gymnocarpous Lichens**. The apothecia in the largest species are compound sporophores of considerable size with limited growth in the direction of the apex or margin, club-shaped or cochleariform in Geoglossum, Spathulea, &c., a stalked cap in Morchella, Helvella, Leotia, Verpa, and others. The early stages of the development of these bodies are little known, but they may be ranked as sporocarps with the forms which will be mentioned directly on account of similarity of structure, and the presence of intermediate forms, especially the large stalked Pezizae. The most characteristic and frequent form is that of the roundish or oblong disk-shaped

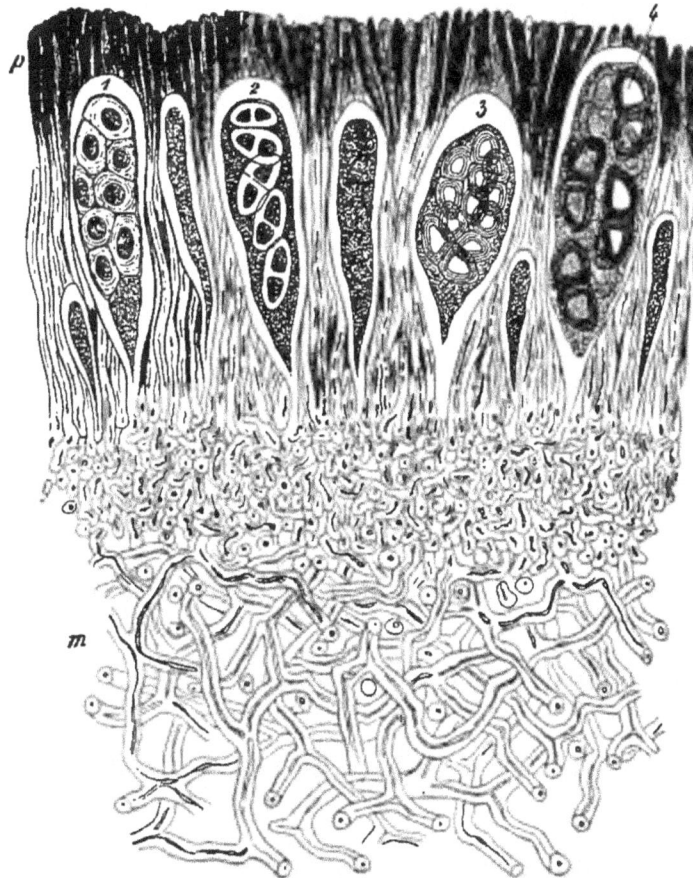

FIG. 88. *Anaptychia ciliaris.* Small piece of a vertical section through an apothecium ; *m* medullary layer of the thallus, *ỹ* subhymenial layer, *p* paraphyses with asci between them. The numerals 1—4 represent successive stages in the development of the spores. After Sachs. Magn. 550 times.

hymenia, which are plane, convex or concave, and in the latter case usually like a bowl or cup, on a stalked or sessile receptaculum or excipulum, as in the Pezizae and in most gymnocarpous Lichen-fungi. The usual mode of growth by gradual advance towards the apex or margin does not prevent the appearance of intercalary surface-growth, which does in fact occur very often and with very varying distribution of the preferred places of growth ; and this may produce a variety of changes in the original shape of the hymenial surface, such as splittings and prolifications, the latter producing a very peculiar and characteristic form in Gyrophora [1]. For the details of these phenomena, which have yet to be more certainly ascertained in many points, we must refer the reader to

---

[1] See Krabbe in Bot. Ztg. 1882, Nr. 5–8.

the special literature. The most important and most general phenomenon of intercalary growth in the surface of the hymenium consists in the introduction of new asci already mentioned, which goes on for a long time at all points. This is the cause of the long continued superficial growth of many hymenia.

Some smaller disk-shaped apothecia, those for example of **Ascobolus** and **Pyronema**, show no marginal progressive growth or only a trace of it. In this respect and in some others also their development approaches that of the Pyrenomycetes.

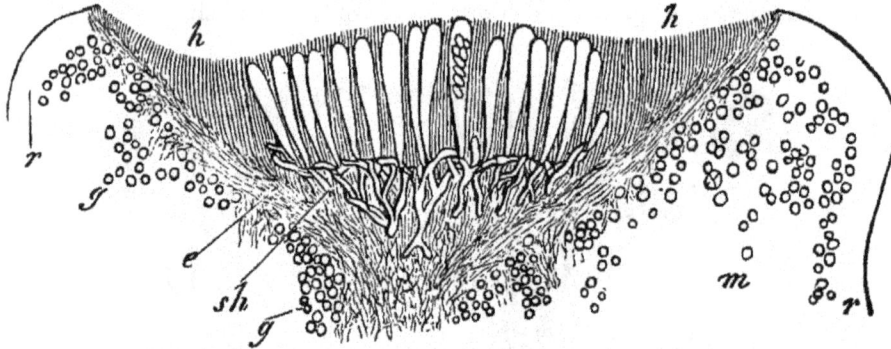

FIG. 89. *Lecanora subfusca.* Median section through a young apothecium, swollen up in ammonia, somewhat diagrammatically represented; *h h* hymenium, *e* excipulum from which spring the paraphyses represented by strokes running vertically towards *h, sh* ascogenous hyphae giving rise to the asci, *r* rind, *m* medullary layer of the thallus which forms a rim round the excipulum. The round bodies are the algal cells contained in the thallus. Magn. 190 times.

This is the case in a still higher degree with the ascocarps of the **Hysterineae** and **Phacidiaceae**, the structure and development of which have been but little examined. According to Hartig's account of Hypoderma macrosporum and H. nervisequum and my own imperfect observations on some species of Rhytisma and Phacidium, the hymenia in these groups are formed in the interior of flat sclerotioid Fungus-bodies (*xyloma*, see p. 43), and become exposed at the time of maturity, when the layer of tissue over the surface of the hymenium separates from it

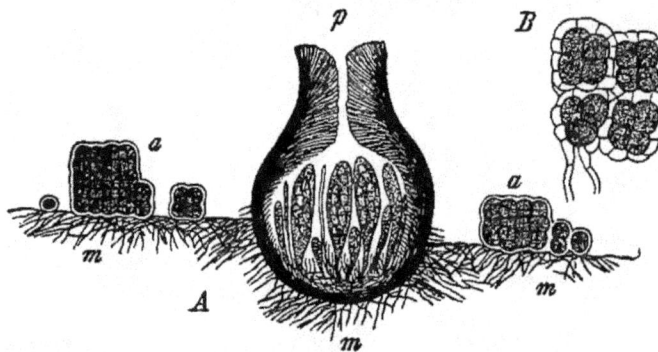

FIG. 90. *Thelidium minutulum. A* perithecium borne on the thallus : *a* group of Algae, *m* the part of the thallus in the substratum which has no Algae, *p* the perithecium in median section diagrammatically represented and slightly magnified. *B* a group of Algae with hyphae winding round them. After Stahl. Magn. 480 times.

and tears asunder in the mode which has often been described in the accounts of the different genera. In this case there are at first only paraphyses present in the hymenia ; the origin of the asci which are introduced between them has yet to be exactly ascertained.

SECTION LXI. The **perithecia** of the Pyrenomycetes (Fig. 90, see also Fig. 44), may be described as cup-shaped discomycetous sporocarps with the margin incurved so as to form a narrow-mouthed cavity. They are generally roundish or flask-shaped, and they are seldom more, usually less, than 1 mm. in height. They are bounded on the outside by the *wall* which encloses an asciferous hymenium, and are furnished in the full-grown state with a narrow aperture or ostiole (*ostiolum, pore*), which is a *canal* passing through the wall and serving for the discharge of the spores (section XXIII). The ostiole usually lies at the apex in reference to the point of origin of the perithecium, seldom at the side (Pleurostoma, Tul.) or at the base (Melanospora parasitica). In many species the orifice is drawn out into a conical or cylindrical *neck* (tubulus of Tode), through the middle of which the canal passes, and which in some cases, as in the Valseae, Sordarieae, and Melanospora, may be more than 1 mm. in length. The asci are inside the perithecium on the side away from the ostiole, and are borne on the ascogenous hyphae or cells from which they successively shoot out, as in the Discomycetes. With this mode of formation it is possible for some hundreds of asci to be formed in the small space of some perithecia, new ones continually appearing in place of the ripe ones which have discharged their spores. The asci together either exclusively form the hymenium or are at least its most essential parts; the hymenium either occupies a narrow bit of surface opposite the ostiole of the perithecium, on which the asci grow as a small tuft parallel and erect towards the ostiole, or is spread over a larger, sometimes over the largest portion of the inner surface of the wall, and the asci then converge radially towards the middle line of the perithecium. Here too the asci and ascogenous organs alone form the ascus-portion of the perithecium; all the other parts belong to the envelope-apparatus; in structure therefore they are on a smaller scale than the envelope-apparatus and in their main features they are like cup-shaped apothecia.

The *wall* is formed of a dense hyphal weft or pseudo-parenchyma, in Pleospora, according to Bauke, of actual parenchyma-like tissue. It is usually differentiated into a stronger often sclerosed outer layer, the thickness of which varies in the individual, and which can produce rhizoids and various other hair-structures on perithecia which stand detached on the mycelium, and an inner layer with delicate and often large cells. In some perithecia inserted in stromata, for instance in Tulasne's Dothidea, in Claviceps, species of Cordyceps and Polystigma, this differentiation does not occur or it is not very distinct, and the wall is formed almost throughout of delicate cells. In its quite young state the wall is entirely closed round the initial organs of the ascus-apparatus and its ultimate investment, which latter will be described below. The ostiole is not formed till the development is more advanced, and it appears as an intercellular passage in the originally close tissue; it is partly schizogenetic by the separation of persistent tissue-elements in consequence of unequal growth, partly lysigenetic by the dissolution of a strip of tissue lying originally in the direction of the canal. The two processes seem to be often combined, and it is not easy to decide whether there is a solution of small strips of tissue along with the schizogenetic formation, or not. The ostioles in Sordaria, Melanospora, Claviceps, Epichloë, in Eutypa also according to Füisting[1], and Stictosphaeria are very largely schizogenetic; Füisting finds the lysigenetic formation

---

[1] Bot. Ztg. 1868, pp. 369, 641.

in Diatrype, Verrucaria, Endocarpon, Pyrenula, and in a very striking manner in Massaria and its nearest allies. In the latter the perithecia are formed inside a flat stroma which lives in the rind of trees, and the orifice is due to the disorganisation and ultimate disappearance of a comparatively thick strip of tissue outside and over the apex of the perithecium, together with the tissue of the rind enclosed by it.

The *neck* is of course a prolongation of the wall, chiefly of its outer layers; this appears very distinctly in some perithecia developed in the interior of the thallus, as in Xylaria, the Valseae, Verrucaria, Endocarpon, and species of Pyrenula[1], where it pierces through the surface of the thallus and comes out to the air. Its development is either rapidly completed at an early period, or it is capable under certain conditions of a long-continued apically progressive or intercalary growth in length, and while this is proceeding it is, especially in Sordaria[2], in a high degree heliotropic; this variety in the mode of development depends on the species and genus.

The *asci* are inserted in the places in the delicate tissue of the inner wall which have been already indicated, and the ascogenous hyphae or cells are thrust in between the elements of the wall or lie immediately upon them. The asci fill the inner space of the perithecium, or at least the largest part of it, excepting the neck. All the space not occupied by them is filled with branches of the hyphae, which grow out from the inner layer of the wall toward the median line of the perithecium. Some of these branches lie between the asci, and are then termed *paraphyses*, as in the Discomycetes, and stand in the same relation as regards their development to the asci as in the Discomycetes, since as parts of the envelope they are formed before the asci which are afterwards introduced between them. Others may cover the portion of the perithecium which is without asci, and even the canal, and then they are called by Füisting *periphyses*. In the canal they are like small closely set hairs of uniform height, which converge from all sides and are directed obliquely upwards towards the median line of the canal where their extremities almost touch. Below the inner (lower) entrance of the canal, in the part of the perithecial cavity where there is no hymenium, their direction and arrangement either remain the same as in the canal, as in Chaetomium and Sordaria fimiseda, or they point downwards towards the median line and the hymenium (Fig. 90). Periphyses would appear to be seldom entirely wanting, though this is sometimes the case, according to Füisting, as in some species of Massaria.

It more frequently happens that there are no paraphyses between the asci, and then the asci alone constitute the hymenium; this is the case in Sordaria, Melanospora, Claviceps, Epichloë, Chaetomium (Zopf), Sphaeromphale, and species of Dermatocarpon, Endocarpon, and Verrucariae (Winter, Füisting). Further details will be found in special treatises, though the accounts there given must often be received with caution, for the paraphyses and periphyses and other organs are often delicate and perishable, or easily overlooked or confounded if the observations are not conducted with sufficient care, and thus mistakes may often be made, especially if the material is not in a favourable state and the observer is wanting in experience.

---

[1] Bot. Ztg. 1868, p. 641.   [2] Woronin, Beitr. II.

It has already been said that the wall or outer wall of the perithecium is usually thick, and often firm and hard. That which is enclosed by it is on the contrary comparatively soft, and has usually great capacity for swelling in water; it comes out from the firm wall, when a perithecium is opened, like a soft kernel from its shell. Hence the old expression *kernel* or *nucleus* of the perithecium, which included all the soft parts described above, the asci first of all, and then the paraphyses, periphyses, hypothecium, and the soft layers of the wall; it had therefore no strict morphological foundation.

SECTION LXII. The **cleistocarps** of the Cleistocarpous Ascomycetes are, as their name imports, surrounded by a wall, which remains closed and without an ostiole even when the spores are ripe, and the latter are only released by external influences which cause the rupture of the wall or by its decay. A great variety of special forms are included under these general characters. A number of these are in other respects scarcely anything else than simple pyrenomycetous perithecia without an ostiole. Among the Chaetomieae which have been carefully studied by Zopf there is one species, Chaetomium fimeti, which is distinguished from all the species nearest to it by this want of an ostiole. Others are removed from typical perithecia by further peculiarities of structure, the Erysipheae, for instance, Eurotium and Penicillium, which can only be briefly noticed here, as they will have to be described at greater length in a subsequent page. The sporocarps also of Sphaerophoron may be mentioned in this connection; their structure is given in Tulasne[1], but their development has yet to be ascertained. All these sporocarps may be regarded as perithecia with a greater or less amount of deviation or simplification. The structure, on the other hand, of the compound sporophores of **Elaphomyces**, the **Tuberaceae, Onygena**, and **Myriangium** is quite different. The early stages of their development are still too little known, and we can count them among the sporocarps of the Ascomycetes only because they form asci and on the ground of some other analogies and resemblances; whether they can rightly be regarded as homologous with the others must for the present be left undecided. With this reservation a short description of these forms may be inserted in this place, as it will be scarcely possible to recur to them while subsequently relating the histories of development.

1. **Elaphomyces.** The sporocarps, which become of the size of a hazel-nut as they ripen, are round hollow bodies with a perfectly closed wall, usually known as the *peridium*, and enclosing the sporiferous tissue or *gleba*. The wall is some millimetres in thickness and consists of two firmly connected concentric layers. The inner of the two, the peridium in Vittadini's narrower sense, is a dense and massive tissue of hyphae which are sometimes very thick-walled. The outer layer, the cortex of Vittadini, is thinner and of a consistence which varies with the species, and may be smooth, warted, hairy, or spiky. Its structure also varies with the species, and in most of them has not yet been described with exactness. In Elaphomyces granulatus it is hard and brittle and thickly beset with warts; the centre of each wart is formed of a conical group of irregularly shaped cells with their bright yellow walls everywhere strongly thickened. The bases of these cones are immediately on the inner layer, and touch each other by their sides. The intervals between the cones and the summits are occupied by a tissue without interstices composed of many layers concentric to

---

[1] See the figures in his Mém. sur les Lichens.

the surface of four-sided prismatic cells; the cells in each layer are arranged in rows which radiate from each cone and meet those which proceed in the same manner from the neighbouring cones. A section parallel to the surface is consequently made up of delicate roundish facets composed of radiating rows of cells, each of which has in its centre a group of thick-walled bright yellow cells.

A loosely felted weft of slender hyphae with elongated segments proceeds on all sides from the inner surface of the peridium and traverses the gleba; here and there, especially in younger specimens, the hyphae are more closely combined into larger plates or strands, but there are no distinctly closed chambers. The gaps in the weft of slender filaments are everywhere loosely filled with *ascogenous hyphae* twice or thrice the thickness of the other hyphae, which are divided into short cells and are often coiled into knots and bear asci on the extremities of their branches. As the spores ripen the whole of the ascogenous tissue becomes mucilaginous and disappears, while the tissue with the slender filaments remains behind as the delicate '*capillitium*' between dry masses of spore-dust. On the asci and on the ripening of the spores, see above, pp. 80 and 97. The spores are large and globular and remarkable for the enormous thickness of their walls, which may be more than two-thirds of the semidiameter of the spore; the wall consists chiefly of a thick, gelatinous, stratified, colourless membrane, covered on the outside with a thin but firm and darkly coloured episporium. For further details see Tulasne's Fungi hypogaei and De Bary's Frucht-entwicklung der Ascomyceten. The germination of the spores has not been observed.

A few observations only have been made on the early stages of the development of Elaphomyces.

The youngest sporocarps of E. granulatus which have come under my notice are small round bodies from $1\frac{1}{2}$ to 2 mm. in size in the interior of a compact dirty-yellow filamentous mycelium. Their outside is covered by a cortical layer of the same thickness, colour, and warty surface as in full-grown specimens, and consists of a thin-walled pseudo-parenchyma, the elements of which are often in continuous connection with the mycelial hyphae. The cortical layer encloses a tissue-mass of delicate closely woven hyphae, which fills the whole of the interior and shows the same structure, but is differently coloured in different parts; a small central portion is of a whitish colour, and is surrounded by a dirty-violet layer, between which and the cortex is a narrow white zone. Later states show that the whitish central mass becomes the gleba, the rest the peridium. This structure and the relative size of the separate portions continue unchanged till the sporocarp has reached the size of a large pea. Larger specimens show the gleba proportionately more enlarged than the peridium, and the ascogenous hyphae beginning to be developed between the slender filaments of its original tissue; soon after the gleba forms the chief mass of the sporocarp, which grows by degrees to the size of a hazel-nut. While the peridium consequently increases greatly in circumference, its absolute thickness increases at the same time or does not diminish. The structure of the inner layer and especially the thickness of its hyphae remain all this time unchanged; the cells also of the cortical layer become only about one half larger than in the first observed stage, while the warts multiply so that without altering much in size they always closely cover the surface of the sporocarp, their multiplication being effected by the splitting of one wart into two or more. All this shows that growth must take place up to late stages in the development by simultaneous and continued formation of new cells in all parts. Tulasne's figures agree with the above account, except in the statement that young specimens must at first be hollow; the difference arises perhaps from variations in the species examined. All the facts seem to show that we have in Elaphomyces a sporocarp with an extremely thick envelope-apparatus and formed of all the described parts except the ascogenous hyphae. The whole recalls the sporocarp of Penicillium, as will be seen from the description of that genus below.

2. The sporocarps of the **Tuberaceae** have the forms of tubers, which either have

an evident basal portion resting on the mycelium, as in Terfezia and Delastria, or are entirely enveloped when young in the mycelium and are connected with it, as in Tuber, the mycelium disappearing when the sporocarp is mature and leaving it naked and detached in the soil. Its surface, if we disregard the frequently occurring warts and roughnesses, is either smooth and marked only with quite irregular and so to speak accidental large unevennesses, as in Tuber aestivum, T. melanosporum, &c. and in Terfezia, or it shows typical pit-like depressions or narrow deep sinuous furrows, as in Hydnobolites and Genabea.

The sporocarp in its simplest form, as in **Hydnobolites**, consists of a fleshy tissue formed of closely woven hyphae, in which numerous asci on the extremities of the branches are everywhere imbedded ; the outermost layer of tissue only forms a kind of wall or peridium, a delicate down composed of sterile hyphae.

In a second series of forms we can distinguish between a sterile fundamental mass and a large number of groups or nests of fertile tissue, i. e. tissue containing asci imbedded in it. The fertile tissue is a more or less compact hyphal tissue in which asci springing from the ends of the branches are distributed irregularly and in large numbers. This tissue fills the spaces between the fertile groups in the form of broad bands constituting much the larger part of the sporocarp, as in **Genabea**, or comparatively narrow plates which show in section as veins with many fine ramifications, as in **Terfezia** and **Delastria**. The sporocarp is surrounded on the outside by a layer of sterile tissue of varying thickness, forming a peridium from which the veins and bands in the interior take their rise ; the hyphae of the fertile groups originate in the adjacent sterile hyphae.

A third type is represented by the genus **Balsamia.** The outside of the compound sporophore is a thick perfectly closed peridium, and the interior is divided by means of thick plates of tissue springing from the peridium into many narrowly sinuous air-conducting chambers. The wall of each chamber is covered with a hymenial layer the elements of which are placed at about a right angle to the wall.

A similar structure is found in the genus **Tuber**, or at least in several species of that genus in the young state (T. rufum, T. mesentericum, T. excavatum, &c.[1]), only the chambers are very narrow and very much coiled and branched. But nevertheless hyphae from the adjacent tissue grow into the cavity of the chambers at an early stage in the development, and fill it quite full with a dense tissue which contains air in its interstices and is therefore white. At the same time the hymenial layer on the walls of the chambers increases considerably in thickness, and assumes the character of a massive irregular tissue which everywhere bears asci. The middle layer of the wall of the chamber retains its original condition in some species. It is these relationships which produce the characteristic marbled appearance of a section through a ripe or ripening truffle (Fig. 91), in which two kinds of branched veins run through a dark-coloured fundamental mass, the fertile tissue ; the one kind dark-coloured and therefore less striking to the sight, which answer to the walls of the chambers and contain no air (venae lymphaticae, veines aquifères of Tulasne, venae internae of Vittadini), the other white and conveying air (veines aërifères, venae externae). The former always originate in the inner surface of the peridium. The latter and probably the previous cavities, which they are formed to fill, extend at certain points to the outer surface of the peridium, and form a kind of opening there to the outside ; this takes place either at spots irregularly distributed over the surface, or in such a way that the veins from all parts unite into a chief trunk with an orifice at a fixed spot in the circumference. In some species of Tuber, T. dryophilum, for instance, and T. rapaeodorum, air-veins only can be distinguished in the fundamental mass, which is traversed uniformly in all parts by asci; this is the case at least in all the states of development in which they are at present known.

---

[1] Tulasne, Fungi hypog. tt. XVII, XVIII.

With respect to the more minute anatomical structure of the Tuberaceae, it may be further added, that the peripheral layer, known as the peridium, is usually a stout, thick mass of pseudo-parenchymatous tissue. The outer cell-layers are in most cases furnished with thickened walls corresponding in colour to the surface, which varies in shade from brown to black; in a few cases they are thin-walled and have their surface covered with spreading hairs, as in Tuber rapaeodorum, &c. Except in Stephensia, in which the layers of the peridium are distinctly separated from each other, the outer cell-layers pass gradually into the inner and these in like manner into the sterile veins and bands which spread between the fertile tissue, and which either show the same pseudo-parenchymatous structure as the peridium (Genabea) or, as in most cases, have their hyphae disposed in a course which follows that of the veins. Here too ascogenous hyphae appear in the tissue known as the fertile tissue interwoven with but strictly distinct from other hyphae which may be termed paraphyses. Moreover it often occurs, both in Tuber and Elaphomyces, that a young ascus is placed on a knee of the hypha which bears it in such a manner that it seems to be borne on two small stalks, somewhat as in Eremascus which will be described below. Tulasne has given representations of this phenomenon, and Dr. Errera has recently called my attention to it. It may at

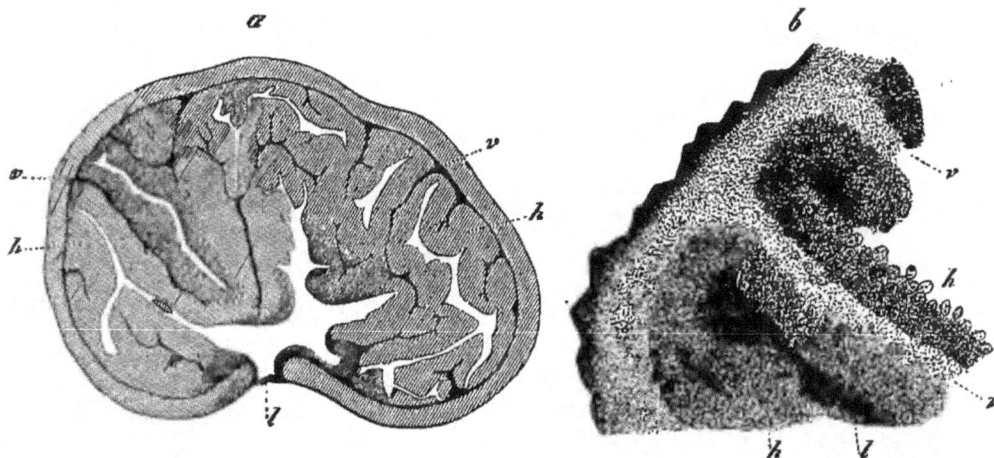

FIG. 91. *Tuber rufum.* *a* small specimen divided in half in reflected light: the white veins *l* contain air, the dark ones *v* fluid, *h* the hymenial tissue. *b* a thinner section through a young specimen in transmitted light; lettering as in *a*, light and dark appearance of the veins reversed. *a* magn. 5 times, *b* 15 times.

least be a question whether the development of the ascus in these cases is the same or similar to that of the ascus in Eremascus; the whole subject requires investigation.

The genera **Hydnocystis**, **Hydnotria**, and **Genea** are not noticed here because a full consideration of them would lead us too far into descriptive details, and we must be satisfied with remarking that they are intermediate in their whole structure between Tuberaceae and typical Discomycetes, especially the Pezizae; they are evidently closely related to both groups.

We are indebted to Tulasne for the little that we know of the origin of the sporocarps in Tuber, and from this they would appear, as has been already stated, to be formed inside a mycelial weft. The different regions and tissues are differentiated in them while they are still quite young; the surface of specimens of Tuber mesentericum of the size of hemp-seed has the same structure and the same black colour as in those which are fully grown. Very little more is known than this. We shall have to wait for a complete knowledge of the history of development in these subterranean plants till we have succeeded in cultivating them.

3. **Onygena corvina**, A. S. grows on the feathers of birds of prey and the mycelium which spreads in them produces stalked spherical sporocarps. The stalk is 7–10 mm. in length and about 1 mm. in thickness, and consists of longitudinally parallel

hyphae closely united together. It bears at its apex the spherical spore-receptacle which is about 2 mm. in diameter and has its dense hyphal weft differentiated into a wall or peridium composed of many layers of an unevenly floccose loose pseudo-parenchyma, and a spore-forming gleba within the peridium. The general form of the gleba is that of a flattened sphere; it consists of closely interwoven and very copiously branched hyphae, which produce everywhere countless asci in dense tufts on the extremities of their branches. The asci are comparatively small ovoid bodies containing each eight delicate ellipsoid spores, which are released when ripe by the disappearance of the membrane of the ascus. When the entire mass of spores is ripe, the whole structure dries up, the peridium opens by a circular fissure and becomes detached like a cap, and the cinnamon-coloured spores are scattered like dust from the flocculent remains of the ascogenous hyphae. Onygena equina, P. a more robust form has, according to Tulasne, a precisely similar structure.

4. **Myriangium Durieui** grows on the rind of trees and forms a large, flat, black thallus from one to a few millimetres in size composed of a tolerably uniform narrow-celled pseudo-parenchyma with brown cell-walls. The sporocarps are protuberances on the outside of the thallus, and consist, as we know from Millardet's researches, of a pseudo-parenchyma similar to that of the thallus, but of finer texture, between the cells of which spherical asci are everywhere distributed usually at some distance from one another. The sporocarp thus constituted occupies the middle of a round protuberance of the thallus in the earliest stages that have been examined. It then grows by constant formation of new cells in a meristematic tissue which lies on its inner side, the side which is turned to the substratum. From this tissue new layers of ascogenous parenchyma are thrust one after another towards the outside. In consequence of the pressure thus produced the original tissue of the thallus in this part is ruptured, and the exposed ascogenous tissue breaks away as new tissue is pushed forward. The great capacity for swelling in the membranes of the ripe asci in the older tissue-layers favours their removal. The youngest asci lie near the layer of meristem between the cells of the pseudo-parenchyma, from which they are chiefly distinguished by their greater abundance of protoplasm. As they move towards the outside they grow to 8–10 times their original diameter and produce eight pluricellular compound spores; the germination of the spores is as little known as the further details of their development.

Origin of the Sporocarp.

Section LXIII. The first steps in the **formation of the sporocarp (fructification)** of the Ascomycetes have up to the present time been closely studied in comparatively few species; the search for and clear preparation of the first commencements is in most cases difficult from the small size of the objects, and there is difficulty also in unravelling the hyphal weft. The subject has been investigated in a certain number of species in all the chief divisions of the group, and important differences, recurring in all cases in their most essential points, are found to exist even between species which resemble each other very closely in the more advanced state. Intermediate forms are found between the several cases which differ extremely from one another. The whole series may be arranged in the following order.

1. **Eremascus albus** is the name given by Eidam to a small Mould which may be cultivated in nutrient solutions, and which has a filiform septate many-celled mycelium. To form the sporocarp (Fig. 92) two adjoining cells of the mycelium put out each a lateral branch (*a*) close by the transverse wall which divides them. The two branches are from the first in contact with one another; they are exactly alike, and they grow coiling spirally round one another to a length exceeding the

transverse diameter of the mycelial hypha 10 times or more (*b*).   Then they cease to lengthen, are delimited from their parent-hyphae and conjugate at their summits, the cell-walls disappearing at the point of contact and the two protoplasmic bodies coalescing into one (*c*).   The place of conjugation then swells into a spherical vesicle, which is delimited when the protoplasm of the pair of cells has passed into it, and having thus become an ascus forms 8 spores capable of germination (*d–f*).   The spores are formed, as far as can be gathered from Eidam's somewhat superficial description, in the manner described in section XIX.   No further complications have been observed in the formation of these sporocarps.

2. Distinct archicarps are formed as branches on the mycelium or on vegetating hyphae in the thallus singly, or rarely in groups, as in Pyronema and Physma.   It depends on the species whether the archicarp is a single cell or, as is more commonly the case, a cell-row, and whether it is spirally coiled or of some other shape. The whole ascus-apparatus of the sporocarp is derived exclusively from the archicarp.   In Podosphaera a single ascus borne on a short stalk-cell is formed by transverse division of the unicellular archicarp; in other species ascogenous hyphae sprout as branches from the archicarp, or the cells of the archicarp divide into ascogenous daughter-cells, that is, into daughter-cells which sprout out into asci.   The archicarp takes no part in the formation of the envelope-apparatus, that is, of the wall, receptaculum, excipulum, paraphyses, &c.   This has its origin in the hyphal branches which arise in the neighbourhood of the archicarp, usually

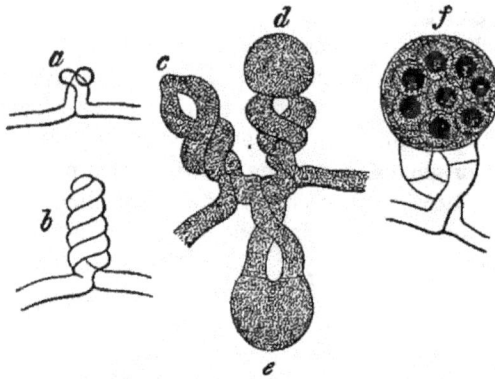

FIG. 92.   *Eremascus albus.   a* inception of the sporocarp.   *b—f* further development in the succession of the letters.   In *f* the ascus is matured and the spores formed in it.   After Eidam.   Magn. 900 times.

at its base, and grow round the ascus-apparatus in a way which is determined by the species.   From this specific ascogenous function the archicarp may in this case be termed an *ascogonium*.   It has also been called a *carpogonium*.   Of cleistocárpous and pyrenocarpous forms the **Erysipheae, Eurotium, Penicillium, Sordaria (Hypocopra)**, and **Melanospora parasitica** belong to this section; of gymnocarpous and discocarpous forms **Gymnoascus, Pyronema, Ascobolus**, and the **Collemaceae** which were examined by Stahl (Collema, Synechoblastus, Leptogium, Physma, &c.).

In a number of the species of this division an *antheridial branch* makes its appearance in characteristic form in connection with the archicarp before the commencement of the formation of asci.   This is the case especially with Pyronema, the Erysipheae, Hypocopra, Gymnoascus, and Eurotium.   In Pyronema, before further development begins, *conjugation*, the union of the two protoplasmic bodies into one, takes place between antheridial branch and archicarp, by means of a special apparatus belonging to the archicarp, the *trichogyne*, and the same thing happens in less striking form in Eurotium.   In the Collemaceae the antheridial branches are formed separately from the archicarps in special layers or receptacles, the *spermogonia*, and give off small spore-like cells, the *spermatia*, by abscision. The spermatia are conveyed to the archicarp and to a special receptive process of it, the trichogyne, and conjugate with it. These phenomena correspond in part, and, excepting in some points of detail which will be described further on, to those observed elsewhere in distinct sexual organs and processes, and without them the sporocarp is not developed. The organs here described in the Ascomycetes are therefore to be regarded in the above cases as sexual organs, the archicarps as the female, the antheridial branches or spermatia as the male organs.

In the Erysipheae, Penicillium, Sordaria, and Gymnoascus conjugation has not been observed, but the union of the two kinds of organs is as firm as it is invariable. Their sexual function therefore has not been certainly proved, but it may be assumed to be highly probable.

The antheridial branches are less constant and less distinct in Ascobolus and Melanospora. According to present observations they are not to be clearly distinguished from the first envelope-filaments that grow round the archicarp; their *sexual function* must therefore be considered to be undetermined. The question of the *homologies* is not hereby prejudged, as will be explained in section LXVI.

3. An archicarp is formed in the compact thallus of **Polystigma rubrum** and **P. fulvum** very similar to that of Collema, and in this case also the archicarp alone produces the ascogenous hyphae at a later period. Spermogonia and spermatia are likewise present, but the union of the latter with the archicarp has not been certainly observed, perhaps owing to their extreme delicacy. Moreover the archicarp here makes its appearance inside a delicate (pseudo-parenchymatous) hyphal coil produced at first as a new formation in the thallus, which may be termed the primordium of the sporocarp, and from it the envelope-apparatus of the sporocarp is subsequently developed under conditions of new formation and resorption. The archicarp is a long, coiled row of many cells. In this respect it is like the archicarp of the Collemaceae, and one extremity of it projects as in that group in the form of a trichogyne above the surface of the thallus, while the lower coils are concealed in the primordium. Before the formation of asci commences, these coils are found to be divided into portions containing from one to several cells and distributed in the future hypothecium, and from here they put out the ascogenous hyphae in the form of branches; but the portion which protrudes as the trichogyne perishes without taking any direct part in the formation of asci. All these phenomena are exactly similar to those observed in the Collemaceae, as will appear from the special description of a subsequent page, with the exception of the union of the spermatia and the presence of the primordium from the first concealing the archicarp, neither of which has yet been ascertained.

4. The processes observed in **Xylaria** again are similar to those in Polystigma. First the appearance of a delicate primordial hyphal coil in the thallus; then inside that of a coiled row of large cells similar to the archicarp of Polystigma (named by Füisting *Woronin's hypha*); finally of cell-groups distributed in the hypothecium from which the asci sprout, while Woronin's hypha is to a great extent at least disorganised and disappears. But a piece of Woronin's hypha projecting from the primordium, a trichogyne, has not been observed, and therefore no visits of spermatia to it; nor is there any proof of genetic connection

between the ascogenous cells distributed in the hypothecium and Woronin's hypha. Considering the difficulty of getting at the young states of these sporocarps the data before us leave it open to possibility that both Polystigma and Xylaria do really behave like Collema, only certain initial and intermediate states being at present unknown, and that we shall at length discover organs in Xylaria equivalent to spermatia. But if what we at present know is the full account of the matter, then Xylaria is distinguished from Polystigma, and of course still more from all the forms mentioned in number 2, by the fact that the ascogenous cells and hyphae do not spring from a distinct archicarp, but, like the paraphyses, from parts of the primordium, while the archicarp, unmistakeably present in form as Woronin's hypha, perishes without taking part morphologically in the formation of asci.

5. The difference from the first case is still more distinct in certain **discocarpous Lichen-fungi** which have been examined by Krabbe (Sphyridium, Baeomyces, Cladonia), in **Sclerotinia** and in a number of **Pyrenomycetes**.

Ascogenous and envelope-hyphae are everywhere inserted between one another and are closely interwoven in the hypothecium of the long-stalked cup of **Sclerotinia Sclerotiorum,** but a direct genetic connection, an origin of the two from a common source, can nowhere be shown; the lowest extremities of both pass into the uniform sterile tissue which ascends from the stalk. It is nevertheless highly probable that the two kinds of elements have a separate origin from the commencement of the formation of the cup, for small round coils of very delicate hyphae are formed in the sclerotium beneath the rind before the cup emerges from it. The commencement of a cup always rises from above a coil of this kind as a comparatively thick bundle of hyphae, the innermost of which are branches of the hyphae of the coil, while the much more numerous peripheral hyphae originate in the surrounding tissue of the sclerotium. It is probable that the latter represent the envelope-apparatus, and those from the coil the ascogenous portion of the cup, and that the coil therefore is a kind of ascogonium. But a distinct proof of this has never been forthcoming, because, as the stalk elongates, it is no longer possible to show a morphological distinction between the two elements and by this means to establish the connection between the later ascogenous hyphae and their supposed primordia. Neither antheridial branches nor spermatia were observed during these developments.

In the **Lichen-fungi** mentioned above, the ascogenous hyphae may be seen, according to Krabbe, distinctly marked between the hyphae of the envelope-apparatus at a very early stage in the development of the sporocarp. But no initial organ has been observed from which they originated, and it must be presumed that they both have a common origin in the hyphae of the vegetative thallus or of the primordium of the sporocarp, and without the co-operation of spermatia or antheridia.

In the Pyrenomycetes, **Claviceps, Epichloe, Pleospora,** and perhaps also **Nectria,** no co-operation of the above-named organs has been observed, and no distinct ascogonium. The young perithecium, as at present known, is a body consisting of similar hyphae or parenchymatous cells, and its elements are gradually fashioned and differentiated into the parts of the perithecium; in this process a cell-group occupying the position of the hypothecium undertakes the formation of the asci, and in Pleospora and Nectria the paraphyses are even formed from the same group. Hartig's conjecture with regard to Nectria may certainly hold good of Claviceps and also of Epichloe, that special ascogenous initial organs are really present on the very young stroma, but up to the present time have been overlooked; as regards Pleospora we have only Bauke's somewhat imperfect preliminary communication. With the accounts at present before us our knowledge is limited to the alleged mode of differentiation. If we choose to speak of ascogonium or archicarp in these genera, we must apply the term only to the initial organs which are late in forming and not very distinct in their differentiation. Van Tieghem's discomycetous genus **Ascodesmis** would belong to this series, if that writer's not very complete account of it is correct.

We now proceed to give the details which are necessary for the full understanding of what has been said above, and to add some supplementary observations. The arrangement of the material is for perspicuity's sake somewhat different from that adopted in the foregoing account.

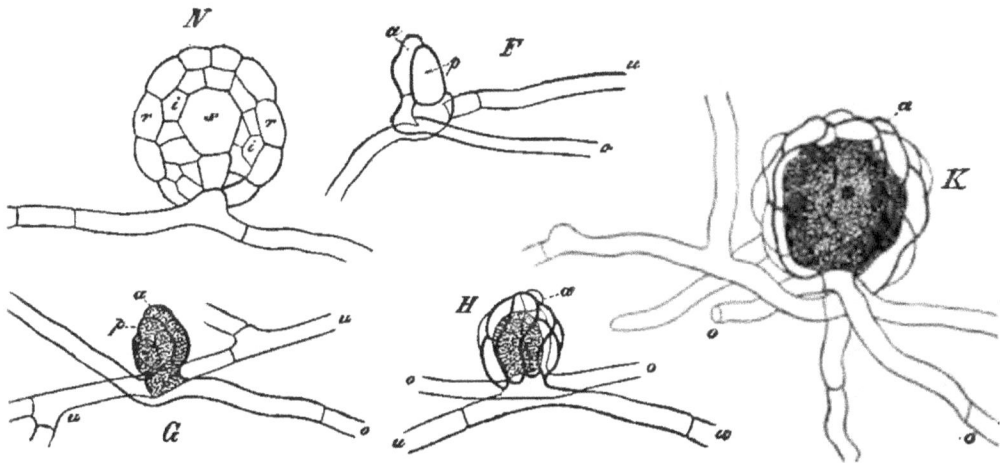

FIG. 93. *Podosphaera Castagnei* on *Taraxacum*. Development of the sporocarp. Stages of the development according to the letters F—N. *o* superior, *u* inferior mycelial filament, *a* antheridial branch, *p* archicarp. In *G* the envelope is beginning to form, in *H* the outer wall is complete. *K* a young sporocarp quite transparent. *N* a similar one in optical longitudinal section; *s* an ascus, *r* the outer wall, *i* the cells of the inner wall formed from the outer. Magn. 390 times.

SECTION LXIV. 1. **Erysipheae** (Fig. 93; see also Fig. 107). The mycelium of these epiphytic parasites is composed of branched septate hyphae which spread over the surface of the host, being attached to its epidermis by the haustoria described in section V, and frequently touch and cross one another. The formation of a sporocarp begins at the point of contact or crossing of two branches. The process is of the simplest kind in **Podosphaera.** Two branches put out short protuberances at the same time, which rise erect from the surface of the epidermis and are soon delimited by a transverse wall. The one which proceeds from the lower of the two branches where they cross takes the form of an elongated ellipsoid cell, 2–3 times the length of the transverse diameter of the parent-branch, and is the *archicarp*. The other, the *antheridial branch*, remains cylindrical in shape, being of the same breadth as the mycelial hypha from which it springs or a little narrower; it is always closely applied to the archicarp, and its upper extremity bends over and covers the apex of the archicarp, and it is presently delimited by a transverse wall, forming a short nearly isodiametric cell, the 'antheridium,' which is borne on the lower part of the branch as on its stalk.   The archicarp now developes into the sporocarp, being usually divided by a transverse wall into two cells, an upper which becomes the solitary ascus and subsequently produces eight spores, and a lower which as a stalk-cell bears the ascus.   Tulasne has found two asci in a sporocarp as a very rare and individual exception; they were probably caused by the formation of two transverse divisions in the archicarp-cell.   The envelope-apparatus also begins to be formed at the same time as the ascus.   From 7–9 tubular outgrowths appear close round the base of the archicarp on the hyphae which bear it and the antheridial branch, and grow up round it and in close contact with it and in close lateral contact with one another and with the antheridial branch, till they all meet together above its apex.   Each tube then divides by one or two transverse walls, so that the incipient sporocarp is surrounded by an envelope formed of a single layer of many cells.   These cells then increase in size in the surface-direction, their walls thicken by degrees and assume a dark brown colour, and they thus form the *outer wall* of the sporocarp.   During this time they form no further divisions, but those nearest the substratum send out rhizoid branches which spread over the substratum; and in some, but not all, species some of the cells at the apex of the sporocarp form a number of hairs with delicate ramifications which are described under the name of *appendiculae*.   Branches shoot out at an early period from the inner surface of the cells of the outer wall, which insinuate themselves between it and the growing archicarp, and ramify and develope into a dense parenchyma-like weft without interstices formed of two or three or more layers of cells according to the species; this weft has been termed the *inner wall* of the sporocarp and compared from its origin and arrangement with the paraphyses of more highly differentiated sporocarps.   With these formations the sporocarp in its envelope is complete in all its parts, and they are followed by further considerable increase in size only, which in the end chiefly affects the ascus and leads to a partial displacement of the cells of the inner wall.   The antheridial branch separates from the archicarp when the branches begin to be formed from the inner wall; it takes part with less increase in size of its parts and less considerable change of shape in the formation of the outer wall, between the other cells of which it remains laterally enclosed.

The development of the sporocarp of the species of **Erysiphe**, under which genus I include all the Erysipheae which do not belong to Podosphaera, agrees with the above description except in certain points, the chief of which only will be pointed out in this place; the reader is referred for further details to the account given in another work[1].   The archicarp has the form of an elongated club-shaped cell, curved spirally round a hooked antheridial branch.   The two organs are surrounded and enclosed by the tubes of the envelope, which give rise, as in Podosphaera, to the outer wall of the sporocarp and to the inner wall which is much more largely developed in these species.   The antheridial branch enclosed in the inner wall soon disappears from observation.   The archicarp on the other hand, lying in the basal portion of the sporocarp, grows into a curved tube and

---

[1] Beitr. z. Morph. u. Phys. d. Pilze, III.

divides by transverse walls into a row of several cells, from which a number of broadly club-shaped erect asci are formed by each cell of the row growing out directly into an ascus, or putting out a few short branches which terminate in asci and are therefore ascogenous.

In both Erysiphe and Podosphaera the formation of the envelope is at first in advance of that of the asci, and is nearly finished when the asci or the cells which produce them are still quite small; and it is not till the last stage of the development that the growth of the asci advances, chiefly at the expense of the tissue immediately

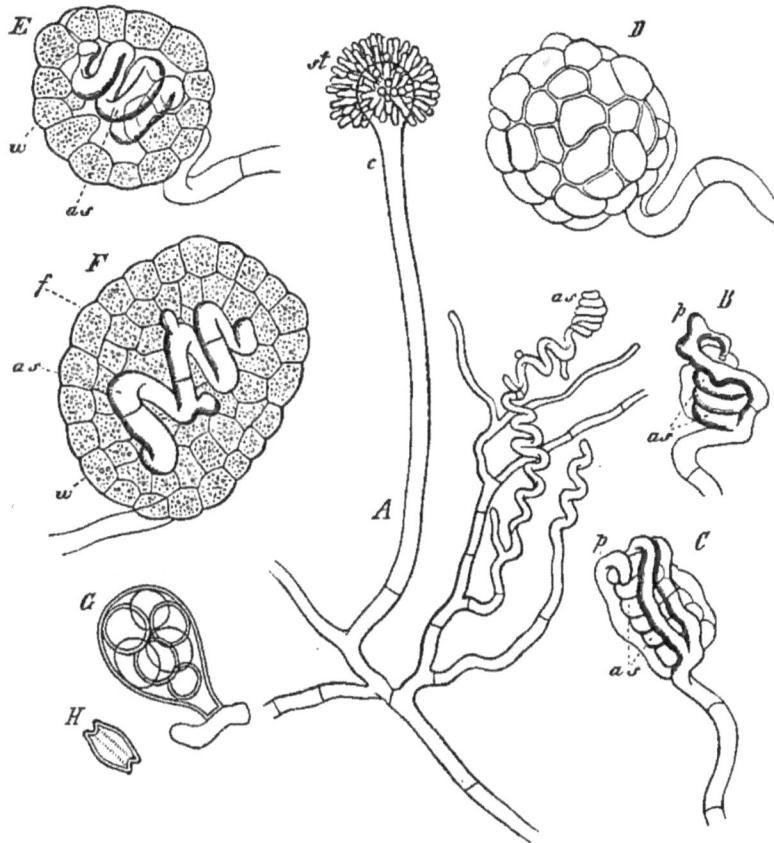

FIG. 94. *Eurotium repens.* *A* branch of the mycelium with a gonidiophore *c* and young archicarps *as*; *st* sterigmata. *B* spirally twisted archicarp *as* with the antheridial-branch *p* and an envelope-branch. *C* older specimen with a larger number of envelope-branches growing round the archicarp; *p* antheridial branch. *D* young sporocarps seen from without. *E* and *F* other young sporocarps in optical longitudinal section. In *E* the inner wall is beginning to be formed; *w* the outer wall, *f* the inner wall-cells and the cells filling the space between the ascogonium and the wall. *as* the ascogonium. *G* ascus with spores. *H* ripe ascospore of *E. Aspergillus glaucus* isolated. *A* magn. 190 times, the rest of the figures 600 times.

surrounding it. The spores are in most cases formed as soon as the asci have reached their full size; but in some species, as Erysiphe Galeopsidis and E. graminis, (Wolff), there is a pause in the development before the formation of the spores, and further progress only takes place under favourable conditions of temperature and moisture after a resting period of some duration which happens to fall in the winter-time; the protoplasm of the tissue of the inner wall is evidently employed to form the spores.

2. The archicarp of **Eurotium** (Fig. 94) is produced by the gradual basi-petal coiling of the extremity of the upper end of a branch of the mycelium into

the shape of a hollow spiral with four or five turns which lie close to one another. The spiral is divided by transverse walls into about as many cells as there are turns. Then two or three slender branchlets grow from the lowest turn in the direction of the apex, and are closely applied to the surface of the spiral; one of these gets in advance of the rest and is the first to reach the apex; there it lays its upper extremity on that of the spiral filament, and, if we may trust our observations, the two filaments conjugate, that is, their protoplasmic bodies unite by the disappearance of the intervening membranes. Sometimes the branch which anticipates the rest is seen to grow up inside the spiral, and then the conjugation cannot be so certainly ascertained; from its behaviour it must be regarded as the antheridial branch. When it has reached the apex of the spiral it is followed by the rest, and now all of them put out new branches which become so interlaced and divided by transverse walls that the spiral is soon covered by a compact layer of isodiametric cells. The lowest turn of the spiral itself participates in the formation of this layer, which surrounds the rest of the spiral, the ascogonium-hypha, as the perfectly closed *outer wall of the globular sporocarp*. The cells of the outer wall do not divide again; but while the sporocarp increases considerably in volume they grow in the direction of the surface into a tabular form, and secrete on their outer membrane, which continues thin and colourless, a golden-yellow substance readily soluble in alcohol in the form of a thick brittle pellicle. Branches shoot out, as in Erysiphe, from the inner surface of the cells of the outer wall, as soon as these have united, and ramify and become interlaced and soon form an inner wall of many layers, while fresh branches from them push in between the loosening turns of the spiral, and fill the space between it and the outer wall with a tissue composed of thin-walled cells rich in protoplasm and without interstices. The growth of this tissue causes the sporocarp at first to increase in volume in every direction and constantly forces the coils of the spiral ascogonium further apart. When it has reached a certain point, the spiral begins to put out numerous branches, the ascogenous hyphae, which thrust themselves in between the inner wall-cells in every direction, and replace them, and the many extremities of their numerous ramifications become ovoid eight-spored asci. The continuity of the ascogenous hyphae is more and more lost as the asci are formed, so that as the spores begin to ripen the outer wall encloses only asci and the remains of the hyphae and the cells of the inner wall, and at length the walls of the asci themselves disappear and the sporocarp contains scarcely anything but ripe spores.

3. According to Brefeld's researches, the development of the sporocarps of **Penicillium glaucum** also begins with the appearance of a spirally twisted hyphal branch. But here we find in the first stages that are open to observation two similar branchlets surrounded by felted mycelium, which always arise close to one another and are spirally twisted round one another in one or two turns; whether they are morphologically and physiologically of equal or unequal value cannot be directly determined, and the further development gives no certain information on this point, so that we can only speak of a distinction between archicarp and antheridial branch of like form with it from the analogy of the otherwise nearly related Eurotium. Then from the spirally twisted body—whether from one only of its component parts or from both is not ascertained—short ascogenous hyphae grow out as branches

in every direction, and at the same time numerous branches begin to be formed on the neighbouring mycelial hyphae which grow rapidly round the others, and inclose them in a compact envelope composed of from 8–16 layers of cells which leave no interstices. The elements of this enveloping weft thrust themselves everywhere in between the ascogenous hyphal branches; in their early states they are much narrower than these and therefore easily distinguished from them in section. The spherical sporocarp which in this state is about 0.05–0.09 mm. in size now increases to an average size of 0.5 mm. and more, and this increase is chiefly due to the enlargement of the cells of the envelope. The great mass of the inner substance is the part most strongly affected, and its cells become irregularly polyhedral and colourless, and are provided with much thickened pitted cellulose-membrane and hyaline cell-contents which turn dark yellow with iodine. The membranes of the cells of two or three peripheral layers become yellowish brown in colour and form a thick persistent outer wall, while a few layers on the outside do not share in the thickening and are cast off when the sporocarp is ripe. With the commencement of these changes the ascogenous hyphae elongate and force themselves in between the growing tissue of the envelope in irregular courses in every direction. In doing this they do not appear to form many new branches or to increase much in breadth, and the latter is the case also with the cell-layers of the envelope which are in contact with them. In sections through an older sporocarp we therefore find in the interior in the large-celled tissue of the envelope ascogenous hyphae cut through in various directions, transverse, oblique or longitudinal, accompanied by small-celled tissue. The cell-walls of these ascogenous hyphae also become thickened, and when this thickening has reached a certain point in them and in the envelope, there is a pause in the development, a *resting condition*. This lasts 7–8 weeks, if the sporocarp is placed as soon as it is ripe in moist surroundings which are favourable to further development; but the resting state cannot last much longer, according to Brefeld's observations, if the surroundings are dry, for dry sporocarps 3–4 months old proved incapable of further development. If the sporocarps are placed within the time stated on a moist substratum in a suitable temperature, they recommence their development; the ascogenous hyphae begin to branch copiously, the branches grow at the expense of the colourless tissue of the envelope which is by degrees entirely dissolved, and again branch, and at length a large number of small eight-spored asci connected together in rows and resembling those of Eurotium are formed on branches of the last order. There only remains at last of the whole sporocarp the pores and the brownish yellow outer wall, which forms a loose envelope round them. Of the details of these changes which may be obtained from Brefeld, it will be sufficient to mention here, that the branches formed on the ascogenous hyphae are of two kinds; comparatively slender ones which penetrate between the cells of the tissue of the envelope, ramify copiously in it and are evidently used to effect the dissolution of that tissue and to take up the products of the dissolution, but do not form asci; and secondly, thicker much curved forms with short branches of their own, from the ramifications of which the asci are ultimately produced. These facts recall the two forms of hyphae in the ripening sporocarp of Elaphomyces (see page 193). The entering of the sporocarp on a resting period and the change from this to the formation of asci at the expense of the tissue of the envelope has its analogue in the processes described in Erysiphe graminis and E. Galeopsidis, only that there the resting-time begins after the formation of the asci, and the spores only have to be formed when vegetation reawakens. Van Tieghem[1] describes a **Penicillium aureum**, which is distinguished from P. glaucum, as regards the processes in question, by having no resting-period; he asserts, that in this species both the initial branches which are spirally wound round one another are ascogenous, that is, that they both send out branches, the last ramifications of which are asci.

The sporocarps of **Aspergillus (Sterigmatocystis) niger** and **A. purpureus** have, according to Van Tieghem[2], essentially the same development and structure as those of Penicillium; and Eidam's new **Sterigmatocystis nidulans**, a remarkable plant in many respects, should also find its place here.

4. The sporocarps of **Gymnoascus** and **Ctenomyces** which live on animal excrements usually appear, according to Baranetzky, Eidam, and Van Tieghem, as small coils of felted tissue crowded together in heaps on the mycelium, the largest somewhat more than 1 mm. high, but most of them of much smaller size. Their formation commences with the union of two unicellular segments of the mycelial hyphae, one of which is wound round the other in a close spiral; the two cells arise close to one another as lateral branches of one hypha, or spring from different hyphae, or only the one which is wound round the other is a lateral branch and the other is an intercalary member of a hypha. The member that winds round the other is an archicarp or ascogonium. It ceases to grow in length when it has formed a varying number of turns, and sends out branches instead, forming numerous ramifications, which are woven together into a coil and have finally at their extremities round eight-spored asci like those of Eurotium. The member, round which the ascogonium is coiled, and which, from the analogy of the cases previously described, must be called the antheridial branch, is cylindrically club-shaped and either coiled like the other cell or straight, and can no longer be distinguished by direct observation when it has undergone a moderate increase in size and become divided by a few transverse walls. It is not till the ascogonium begins to send out branches that the young sporocarp becomes loosely enveloped by a number of hyphal branches, which spring partly from the base of the ascogonium itself partly from adjacent branches of the mycelium, and have their membranes thickened and coloured yellow or brick-red; the peculiar antler-like ramifications of these filaments loosely intertwined form a lattice-like envelope composed of several layers round the ascogenous coil as it advances to maturity. Rhizoid branches also spread over the substratum.

5. The sporocarps of the species of **Ascobolus** (Fig. 95) when fully matured have the typical discomycetous form. They have the form of a broad short cone, the obtuse point of which is seated upon the filamentous mycelium while the broad basal surface is covered by the hymenium. Their development would appear to be alike in all the species; the following account of it refers especially to Janczewski's exact observation of the best-known species A. furfuraceus. The development begins with the appearance of an archicarp in the form of a comparatively thick arched lateral branch from a mycelial hypha; this branch becomes by successive divisions

---

[1] Van Tieghem in Bull. Soc. Bot. de France, XXIV (1871), p. 157.
[2] Loc. cit. pp. 96, 203.

a series of simple apparently similar cells rich in protoplasm which grow to be about as long as broad, and then a preliminary cessation of this growth takes place.   Slender branches which spring from the mycelium near the archicarp, and also branch themselves, now grow in the direction of the archicarp and apply themselves and their branches closely to its free extremity (Fig. 95 *l*).   They behave in this respect like the antheridial branches of Eurotium and Erysiphe and may therefore receive the same name.   Their contact with the archicarp is followed at once by the formation of a large number of fresh branches on the hyphae which produced them and on adjacent mycelial hyphae, and all these later branches grow closely interlaced round the archicarp and the first antheridial branches, which from this time cease to be distinguishable.   The archicarp is thus at once inclosed in a compact hyphal coil, the envelope, which then grows considerably, partly by the introduction of new hyphal branches partly by the increase in size of those previously formed, the cells of which

become vesicular and for the most part continue united together into a dense pseudo-parenchyma. A few peripheral layers of these cells form a thick-walled rind, which is yellow in Ascobolus furfuraceus from the colour of the membranes but is differently coloured in other species, and which sends rhizoid-hyphae into the substratum at the points of contact, while in many species, but not in A. furfuraceus, it produces spreading hairs of peculiar form and arrangement.   With all these changes the sporocarp assumes a spherical shape, and the course and direction of its growth are such that the archicarp remains inclosed in the *basal* portion of the sphere where it rests on the substratum.   The formation of paraphyses begins at the same time as the differentiation of the rind in the opposite *apical* region, and their first

FIG. 95. *Ascobolus furfuraceus*. Median longitudinal section of a young fructification ; *m* mycelium, *c* archicarp, with the ascogenous hyphae *s* spreading in the sub-hymenial layer and the asci *a* dark, *l* antheridial branch, *p—r* tissue of the envelope from which the paraphyses *h* spring. Diagrammatically represented by Sachs after Janczewski.

beginnings appear as branches from a zone of narrow cells still rich in protoplasm in the tissue of the envelope, *the subhymenial zone*, which passes across the apical region beneath the rind. The paraphyses are formed as slender hyphal out-growths in this region ; each of them can send out new and similar branches of more than one order near its point of origin.   But the ends of the branches of every order grow into slender long-celled filaments which constitute the paraphyses, and converging at first have all a direction from the subhymenial zone towards the apical region of the rind and end below it.   In proportion as they elongate and increase in number by the introduction of new branches from the subhymenial region and whilst the surrounding parts follow these processes of growth, the space between the rind and the subhymenial layer grows broader, but is always being filled up by the paraphyses which are arranged close beside one another as the first commencement of the *hymenium*. It may be observed here by anticipation that the growth of the hymenium continues

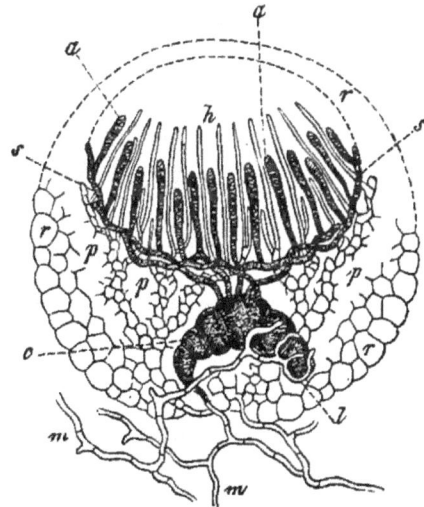

longer in the surface-direction, while the rind covering it does not grow correspondingly, and the rind is therefore ruptured above the hymenium, which is thus exposed as a discus.

Up to the time of the inception of the hymenium beneath the rind no changes of importance take place in the ascogonium.   But now all its cells except one are seen to become thick-walled and poor in protoplasm, and in this state they continue permanently; but that one cell, the third or fourth uppermost cell, becomes the initial cell of the formation of asci, the *ascogenous cell.*   It is full of protoplasm and swells considerably, and then sends out twelve or more strong cylindrical branches from its free outer surface.   These are the ascus-forming branches, the *ascogenous hyphae,* and they thrust themselves in between the elements of the tissue of the envelope as they grow in the direction of the subhymenial layer, into which they send their many branches and spread abroad between the points of insertion of the paraphyses.   Finally the asci appear as lateral branches of the last order on these subhymenial spreading hyphae, which, as has been already said, grow between the paraphyses and in the same direction with them towards the outer surface of the hymenium.   The long-continued successive formation and introduction of new asci at all points is the chief cause at least of the surface-enlargement mentioned above, and of the exposure and often even of the convexity outwards of the hymenium.   Janczewski's observations have been confirmed in the case of several species by Borzi, who has also described an allied form, a species of **Ryparobius,** in which every shoot from the ascogenous cell becomes an ascus directly.   Borzi's view respecting the fertilisation of the archicarp is not supported by any other case.

6. The development of the sporocarp of **Pyronema confluens** (Peziza, P.) was described by myself, but imperfectly, in 1863.   Tulasne then added something to my statements.   Kihlman's recent examination of the species gives the following results (Figs. 96–99).   The Fungus spreads the stout filaments of its mycelium over wide spaces of ground, especially where charcoal has been made or fires have burned. The inception of the young sporocarp is preceded by the formation of groups of obliquely erect curved branches, which in their turn put forth many branchlets.   Some of these, usually two in each group, swell strongly and form a few short bifurcations, which grow in a direction vertical to the substratum and then cease their longitudinal growth.   The bifurcations form together an erect loose tuft or rosette (Figs. 96, 97 *A*), and some of them terminate in a short roundish cylindrical cell which remains sterile.   The extremities of others become either *archicarps* or *antheridia* (Fig. 97 *A, B*).   The former are broadly club-shaped bodies consisting of a much inflated and usually somewhat curved cell, densely filled with protoplasm and borne on one or two disk-shaped stalk-cells; the antheridia are the club-shaped terminal cells of the branches of the bifurcations, about the same height, but only half as broad as the archicarps.   Several, at least two or three, organs of both kinds are present in each rosette, and no other relations than those stated between the points of origin of each pair of dissimilar organs have ever been observed.   When the two kinds of organs have reached the shape and length which have been described, each archicarp puts out a broad protuberance near its apex, which grows rapidly into a blunt cylindrical tube filled full with protoplasm; and the tube becoming bent like a bow in a plane differently disposed in different individuals, grows on towards a neighbouring antheridium, and

embraces its apex and presses its obtuse extremity firmly against it. When this has taken place, seldom before, the tube is delimited by a firm transverse wall from the inflated portion of the archicarp, and, as soon as the wall is formed, the membrane in each of the connected organs is dissolved at the point of contact of the tube with the antheridium, and the protoplasmic bodies of the two organs unite together through a broad aperture (Fig. 97 *B*). The bent tube is therefore an organ which affects an

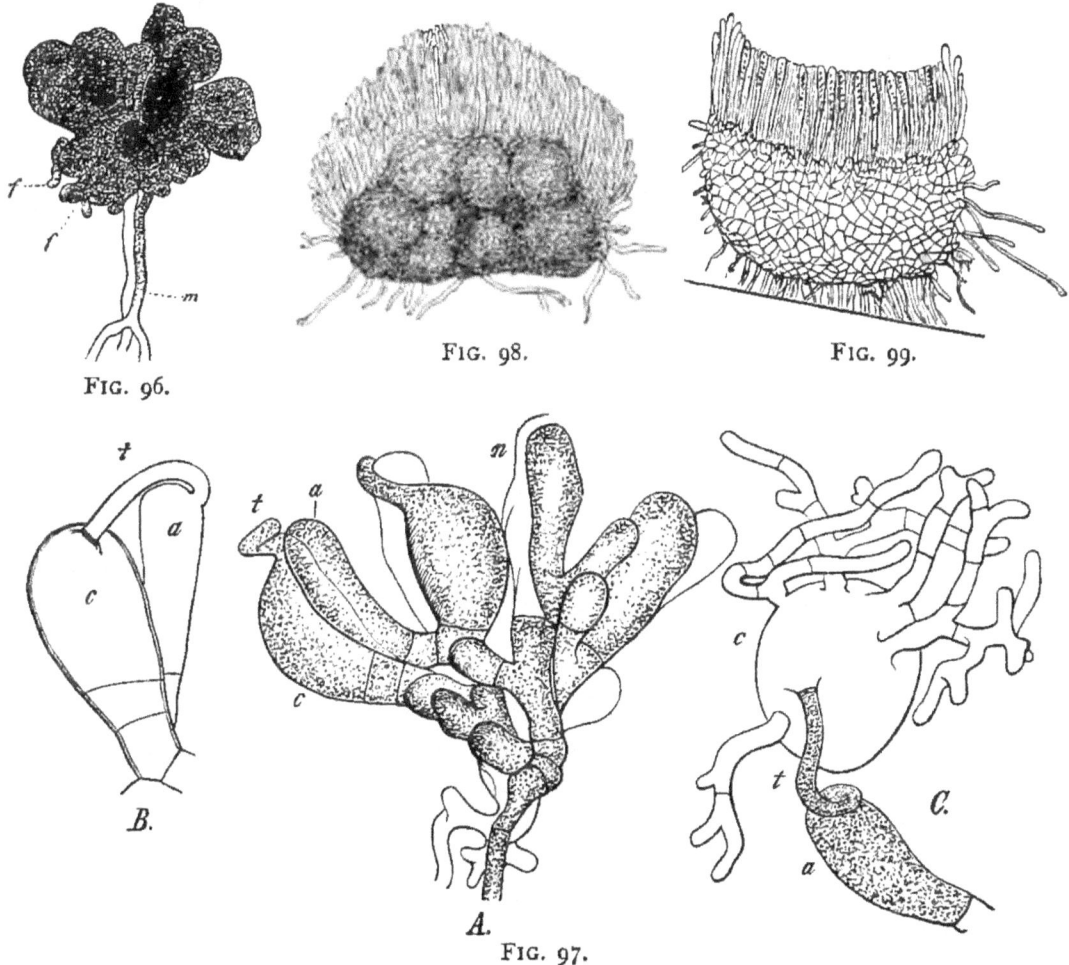

FIG. 96.

FIG. 98.

FIG. 99.

B.

A.

C.

FIG. 97.

FIGS. 96—99. *Pyronema confluens*, Tul.

FIG. 96. Rosette of antheridia and archicarps on the mycelial filaments *m*; *f* first beginnings of the filaments of the envelope. Magn. 190 times.

FIG. 97. *A* a small rosette of incipient sporocarps; *c* archicarps, *a* antheridia, *t* a trichogyne which has not yet entered into union with *a*. *B* from an older rosette; the trichogyne *t* proceeding from the archicarp *c* and cut off by a transverse wall is in open communication with the antheridium *a*. *C* a pair of organs isolated, from a young sporocarp in about the same stage as Fig. 98; *a* antheridium in communication through *t* with an archicarp *c*, which is much swollen and has put out branched ascogenous hyphae from its surface. After Kihlman's preparations and drawings. Magn. about 300 times.

FIG. 98. Young sporocarp in water showing through the cover-glass. The group of antheridia and archicarps is densely overgrown by hyphae of the envelope which have formed erect paraphyses above; the archicarps appear through the envelope-weft as large vesicles. Magn. 90 times.

FIG 99. Median longitudinal section through a sporocarp in the process of maturing. Archicarps and antheridia can no longer be distinguished, and many asci have been formed between the paraphyses. (See Fig. 39.) Magn. about 45 times.

union between the archicarp and the antheridium and, in accordance with the terminology which is in use in other cases and which will be further considered below, may be termed a *trichogyne*. Conjugation is followed by increase in size in the archicarps, and by the formation of protuberances in a dozen or more places scattered over the surface of each archicarp, which develope into thick short-celled *ascogenous hyphae* (Fig. 97 *C*). Simultaneously with this, or even before it, copiously branched

[4]   P

hyphae begin to shoot out from the sterile sister-branches of the archicarp and from the whole of the rest of the basal region of the rosette to form the envelope-portion of the sporocarp (Fig. 98), consisting of a receptaculum enclosing the group of archicarps and antheridia and the hypothecium with the paraphyses, which latter always rests free on the receptaculum without an enveloping wall. The distribution of the ascogenous hyphae and asci and the gradual multiplication of the latter between these elements of the envelope are essentially the same as in Ascobolus. The paraphyses form at first a conical tuft on the hypothecium, which generally broadens out into a disk through the introduction of new elements. The receptaculum becomes a comparatively large thick large-celled pseudo-parenchymatous disk covered with rhizoids, and between its elements those of the primary rosette are inclosed, and are at length indistinguishable (Fig. 99). The antheridia continue longest visible and indeed almost unaltered, being very full of protoplasm, and take no part in the formation of the envelope.

7. In **Sordaria** among the Pyrenomycetes especially S. (Hypocopra) fimicola and in **Melanospora parasitica** the course of development in the perithecium, according to Gilkinet's and Kihlman's researches, is essentially the same as in Ascobolus, of course with certain specific differences, and with differences of conformation corresponding to the difference between Discomycetes and Pyrenomycetes. The archicarp is a spirally coiled cell-row, though in Melanospora it is sometimes almost straight. Antheridial branches are less plainly seen in Melanospora, or at least are not sharply distinguished from the incipient filaments of the envelope, which here too grow close round the archicarp soon after its formation. In Sordaria the growing archicarp divides by transverse walls into numerous cells, and ascogenous hyphae sprout from the great majority of these cells; whether any portion of the archicarp takes no part in their production could not be determined owing to the early gelatinous disorganisation of the walls of all its cells. In Melanospora only one or two cells in the middle of the large archicarp become ascogenous, the rest being disorganised and afterwards partly ejected in this state from the orifice of the young perithecium. These one or two ascogenous cells develope by successive bipartitions in varying directions into a parenchymatous body, the many cells of which are full of protoplasm and subsequently produce the asci. These are arranged in Sordaria nearly parallel to one another in a thick tuft, in Melanospora they form a nearly spherical body, and their apices converge towards its middle. In both cases the many-layered pseudo-parenchymatous wall of the perithecium is formed from the weft of hyphae of the envelope. It is a spherical body at first closely surrounding the future group of asci on every side, the neck and the canal of the ostiole being formed in it later. There are no paraphyses standing between the asci; these are placed separate and beside each other on a surface of insertion which covers a part of the interior of the sphere. On the side left free from the paraphyses there is a narrow empty space between the wall and the ascus-group, but this is soon filled with a large number of closely packed hyphal branches, which grow into it as *periphyses* from the wall, converging radially till they touch one another. The group of these which is mostly turned away from the asci then grows vertically towards the wall in the direction of the neck which is to be, and so on to the outside through a hole in the wall, and thus forms the commencement of the neck, which may lengthen out considerably, and in Sordaria is

covered on the inside with periphyses which converge towards the median line. All the periphyses, those of the neck as well as those beneath it, converge till their extremities touch, but without becoming firmly united, so that asci or spores can pass between them to the outside when they are mature. In Melanospora parasitica the future canal of the ostiole is to some extent marked out from the first, for the non-ascogenous basal cells of the spiral archicarp, that is the cells turned towards the place of insertion, remain in their place as large vesicles, together forming a strand outside which the periphyses which converge towards it afterwards spring from the wall. Then the neck also grows in the direction of the strand outwards at the place of insertion of the perithecium or archicarp, while the vesicles are ejected as disorganised masses of mucilage. In Sordaria it would appear that the canal of the ostiole and the neck are formed on the side diametrically opposite to the *place of insertion* of the perithecium ; the first simply in consequence of corresponding surface extension of the young wall, and as an intercellular space which is at once filled with the periphyses ; the formation of the neck has been less exactly studied.

According to Van Tieghem's observations **Chaetomium** is nearly related to the forms which we have been considering. The conclusions of this writer have, it is true, been stoutly assailed by Zopf, but on the other hand they have been recently confirmed by Eidam, and rightly as far as I can see. Some particulars are still doubtful, and should be submitted to further examination

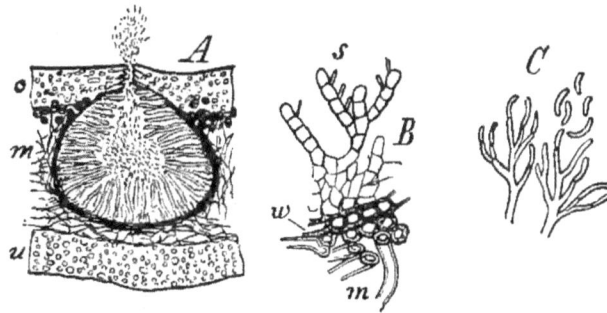

FIG. 100. *A, B Gyrophora cylindrica. A* a vertical median section through a spermogonium imbedded in the thallus ; *o* upper, *u* under rind, *m* medullary layer of the thallus. *B* portion of a very thin section from the base of the spermogonium ; *w* its wall from which proceed sterigmata with rod-like spermatia *s*, *m* medullary hyphae of the thallus. *C Cladonia Novae Angliae*, Delise ; sterigmata with spermatia from the spermogonium. After Tulasne. *A* magn. 90, *B* 390 times, *C* highly magnified.

ination with due reference to the works of these observers. The latter remark will apply also to Bainier's short description of some species of **Ascotricha** examined by him and *perhaps* belonging to this place ; I have not been able clearly to understand his account of them.

8. The development of the apothecia in the Discomycetes which are included in the group of the **Collemaceae** is in all points similar, according to Stahl's observations, to those which have just been described. But it is preceded by fertilisation of the archicarp by spermatia formed at a distance from it, and this causes the following modifications.

The Collemaceae form a gelatinous Lichen-thallus with lobe-like branches (see section CXVI. 5). Sections through the thallus show much-branched hyphae loosely distributed and interwoven in the thick gelatinous membrane-substance ; the branches are also closely united together at certain spots in the fertile thallus in order to form the receptacles which produce the spermatia, and which were first clearly distinguished by Tulasne as *spermogonia*. These organs, of which Fig. 100 below will give some idea, though taken from other species, are in the mature state small bodies, but visible to the naked eye, having very much the shape of the flask-shaped perithecium of the

Pyrenomycetes protuberant below and with a short neck; they are sunk in the thallus but have the free extremity of the neck on a level with the outer surface.   The neck is traversed throughout its length by a canal open at both ends, the canal of egress. The wall of the inner ventral portion, which is formed of a close weft of hyphae, bears a hymenium on its inner surface composed of delicate hyphal branches of uniform height closely packed together, and converging in the direction of a central space which is free from them and is in open communication with the neck-canal.   These hyphal branches behave like basidia or sterigmata, and abjoint serially and successively numerous *spermatia* at their apices in the form of small cylindrical rod-like cells. The apparently homogeneous protoplasm of the spermatia has a sharply defined contour, outside of which is a hyaline jelly, which swells and deliquesces in water and forms probably the outer layers of their membrane.   In this gelatinous envelope the spermatia when abscised lie at first in the central cavity of the spermogonium and remain there as long as it is kept dry.   If the water finds its way in, the jelly swells

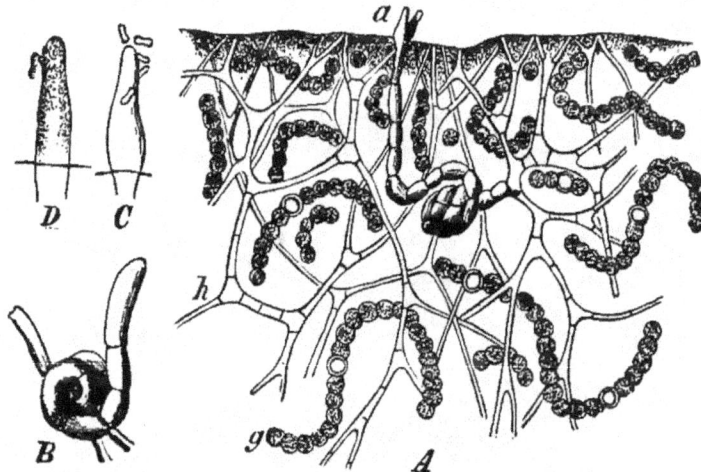

FIG. 101. *Collema microphyllum. A* transverse section through the thallus; *h* the hyphae, *g* the Algae (see section CXVI), *a* the trichogyne projecting above the surface of the thallus; the spirally twisted archicarp which is imbedded in the thallus terminates in the trichogyne. *B* a younger archicarp drawn separately. *C, D* summits of trichogynes with spermatia attached projecting above the surface of the thallus which is shown by the horizontal lines.   After Stahl. *A* magn. 350, the rest 750 times.

and forces them out of the neck on to the surface of the thallus, over which they are distributed if there is sufficient moisture, as can be seen by transferring them to a drop of water on a microscopic slide, where they move in a slow uncertain manner, due no doubt to the currents in the dissolving jelly.

The formation and discharge of the spermatia usually precede the appearance of the commencement of the apothecium, as will be noticed again below.   These, the *archicarps* or *carpogonia*, are formed in most species of the group, in **Collema**, for example, beneath the surface of the thallus, and singly as lateral branches from hyphae which have nothing else to distinguish them.   They are rather broader than the parent-hyphae and coil up near their point of origin, forming usually two or three narrow turns of a spiral, and then lengthen at the free extremity into a straight or slightly curved filament, which grows towards the surface of the thallus and out beyond it into the open air; here it often forms a narrow flask-shaped expansion and ceases to grow in length when the portion outside the thallus is about from four to six times as long as broad (Fig. 101).

The further development shows that the spirally coiled portion of the archicarp is the *ascogonium* or place from which the formation of the asci proceeds, while the elongated part with its point projecting above the surface of the thallus is the *trichogyne*[1], serving as the receptive-organ in fertilisation and conveying its influence to the ascogonium. Both parts are divided as they grow by transverse walls into cylindrical cells, of which there are about twelve in the ascogonium and as many or more in the trichogyne; they are all at first thin-walled and filled with a homogeneously hyaline protoplasm. The formation of archicarps generally takes place under the same external conditions as the discharge of the spermatia; namely, in the cool, damp, rainy period of the year. If the spermatia which have reached the surface of the thallus encounter the top of a new-formed trichogyne they attach themselves firmly to it often in great numbers, and notwithstanding the difficulty of minutely observing such small bodies when adhering to the comparatively thick trichogyne, Stahl was repeatedly able to assure himself that some of them put out a short process by means of which their protoplasm becomes united with that of the apex of the trichogyne (Fig. 101 *D*). The effect of this union with the spermatia is seen in peculiar changes in the trichogyne, advancing from the apex to the base, and in the ascogonium and surrounding parts, and these changes are not observed if no union takes place between trichogyne and spermatia. The changes are these; the cells of the trichogyne lose their turgidity and shrink into slender threads; their transverse walls only maintain their former breadth and at the same time swell strongly in the direction of the axis of the trichogyne, and thus form knot-like protuberances in the collapsed cell-row. The cells of the ascogonium continue turgescent, thin-walled, and rich in protoplasm, and increase in size and number by transverse divisions. Finally a large number of branches begin to be formed in the neighbourhood of the ascogonium on the adjacent hyphae of the thallus, and these ramifying repeatedly and intertwining grow round the outside of the ascogonium and also push in between the turns of its spiral and force them apart; the ascogonium is thus rapidly inclosed in a mass of closely coiled filaments, the first-state of the envelope of the fructification. The coil is at first round and occupies the place in the interior of the thallus in which the ascogonium inclosed by it was first formed. When it has reached a certain size (Fig. 102), the hyphae on the side towards the adjoining surface of the thallus send out branches which again branch repeatedly, and the final ramifications are the first *paraphyses* which grow straight towards the surface of the thallus, pushing aside the tissue of the thallus which stands in their way, and terminate in it. At the same time the envelope-coil which was originally round increases in breadth in the direction of the surface of the thallus—chiefly no doubt by centrifugal formation of new hyphal branches on its lateral margin, so that it assumes the form of a concave disk cutting the surface of the thallus with its edges; and thus it developes into the ultimately pseudoparenchymatous *excipulum*, which continues for some time to increase in breadth at the margin which rests on the surface of the thallus. As the development of the excipulum advances, new paraphyses similar to the first shoot out one after another from its side towards the surface of the thallus, pushing aside the tissue of

---

[1] The term was introduced by Bornet and Thuret to denote the analogous organs in the Florideae. See Ann. d. sc. nat. sér. 5, VII, p. 137.

the thallus which is in their way, till at length the space between the excipulum and the surface of the thallus is filled by an incipient hymenium consisting of paraphyses standing side by side with no gaps between them. The ascogonium has meanwhile at first slowly followed the growth of the excipulum by intercalary growth accompanied by a loosening of the turns of the spiral. As the development proceeds the *ascogenous hyphae* sprout from it, and spread their abundant ramifications through the zone of origin of the paraphyses, the *subhymenial layer,* in essentially the same manner as in Ascobolus, and thrust the asci as branches of the last order one after another in between the paraphyses (Fig. 102).

The species of **Physma** also examined by Stahl agree with Collema except in the following peculiarities. The archicarps here spring from the hyphae which form the protuberant base of the spermogonia, from four to eight on each spermogonium. The ascogonia are but slightly curved and are inclosed in the hyphal

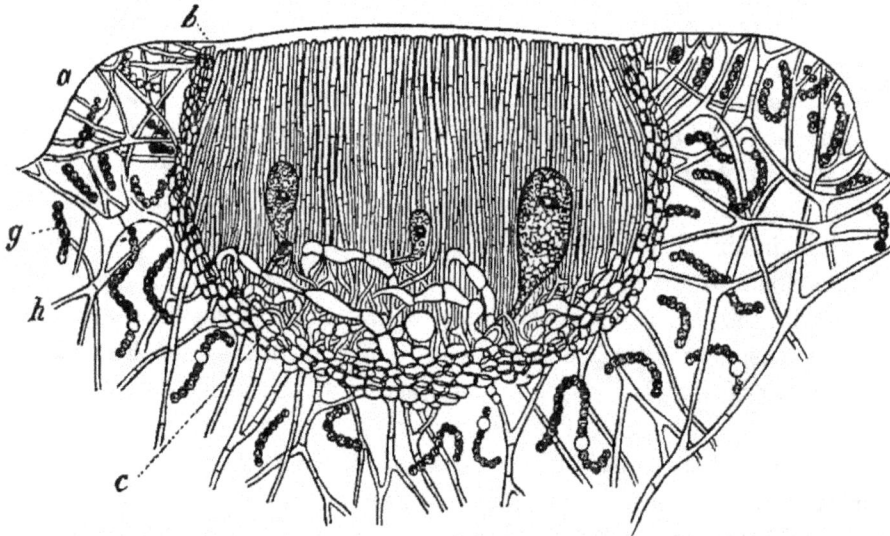

FIG. 102. *Collema microphyllum.* Median section through a young apothecium imbedded in the thallus; *h* and *g* as in Fig. 101, *b—c* excipulum and hypothecium; from the latter proceed crowded upright paraphyses, between which asci are beginning to be formed on the ascogenous hyphae above the hypothecium. After Stahl. Magn. 530 times.

weft of the wall of the spermogonium, and the trichogynes protrude beyond the outer side. The discharge of the spermatia of a spermogonium coincides as a rule with the completion of the trichogynes which belong to them, and these become covered with the spermatia adhering to them. Then paraphyses grow out from the wall of the spermogonium into its now empty cavity, displacing the sterigmata, and soon fill it up in the form of a tuft of filaments which converge towards the former orifice and are so closely packed as to leave no spaces between them. Into the subhymenial zone, which is thus defined, ascogenous branches shoot out from the archicarps and push the asci in between the paraphyses; the spermogonium is thus converted into the apothecium.

Borzi has repeated Stahl's observations on other species of the Collemaceae with confirmatory results.

9. Incipient sporocarps or archicarps, of doubtful character and requiring a fresh examination, have been assigned by Woronin to Sphaeria Lemaneae, Sordaria

fimiseda, Peziza granulata, and P. scutellata, and by Tulasne to P. melanoloma; by R. Hartig to Rosellinia quercina and Nectria.

10. **Polystigma rubrum and P. fulvum.** The thallus of these Fungi forms compact disk-shaped stromata in the living tissue of the leaves of species of Prunus. It forms spermogonia, which resemble those of Collema in structure and produce curved filiform spermatia. At the same time or soon after, the primordia of the perithecia make their appearance in its interior. These are small coils of closely interwoven hyphal filaments, which are perfectly similar in structure in their early stages and show no manner of differentiation. The individual cells of the coil are distinguished from those of the thallus by their small dimensions and especially by turning yellowish brown, not blue (see page 9), with solution of iodine. As the growth of the coil proceeds the cells in its interior assume a more delicate appearance from diminution in thickness of their walls and become densely filled with protoplasm. A spirally coiled filament composed of broad and rather short cells, which comes out very beautifully when coloured with iodine, now becomes conspicuous among them. Its two or three turns extend through the entire space of the primordium, which in this stage is usually of an elongate ovoid form. The extremity of the filament rises above the primordium and passing through the cells of the mycelium penetrates to the surface of the leaf, where it finds its way through a stoma into the outer air, and has a perfect likeness to the apex of the trichogyne in the ascogonium of the Collemaceae as described by Stahl. The point is generally accompanied by more slender mycelial hyphae which grow out through the stoma when the trichogyne has decayed, and form a penicillate tuft of companion hyphae. Spermatia have been frequently seen to adhere to the summit of the trichogyne, but in no case has a more intimate connection, especially conjugation, been proved to take place. After some time the cells of the trichogyne-filament begin to die away from above downwards and to be no longer distinguishable in the tissue of the thallus, while the young perithecium enlarges at the same time throughout, the cells of its outer layers becoming elongated to form the wall. The remaining portion of the spiral filament, the ascogonium, also enlarges its cells considerably and now appears as a thick highly refringent strand of cells. In Polystigma rubrum the young perithecium remains in this state during the winter, but the development goes on without interruption in P. fulvum, and consists in the upward growth of the envelope in a conical form and the flattening of the basal portion of the young perithecium, while the inner tissue swells at the same time into a jelly. The ascogonium lies irregularly distorted on the base of the young perithecium. The hyphal weft of the base of the perithecium sends in paraphyses between the cells of the ascogonium in the form of thick cell-rows with walls which readily swell into a jelly, while the upper part of the wall of the perithecium is covered with periphyses. The cells also of the ascogonium, all of them apparently, form protuberances which elongate into slender filaments with abundance of protoplasm and then branch, and soon become a mass of interlacing threads amongst the basal tissue of the paraphyses; these are the ascogenous hyphae, and their last ramifications grow upwards as asci, and as they increase in size they displace the paraphyses, which are then dissolved. The periphyses also disappear, and the basal tissue of the asci and paraphyses swells up and can

no longer be recognised, so that the perithecium in the mature state is broadly ovoido-conical with an indistinct ostiole. The wall is formed of three or four layers of not much thickened elongated cells.

Phyllachora Ulmi appears to show similarity to the process here described.

11. The club-shaped stroma of **Xylaria polymorpha** (Fig. 103) consists in the young state, according to my earlier observations, of a white medulla surrounded by a firm black rind. The former is composed of an air-containing tissue of colourless hyphae; the rind of the portion bearing perithecia consists of small-celled pseudo-parenchymatous tissue, which is overlaid on the outside by the hymenium which bears gonidia (see section LXXI) and ultimately disappears. The primordia

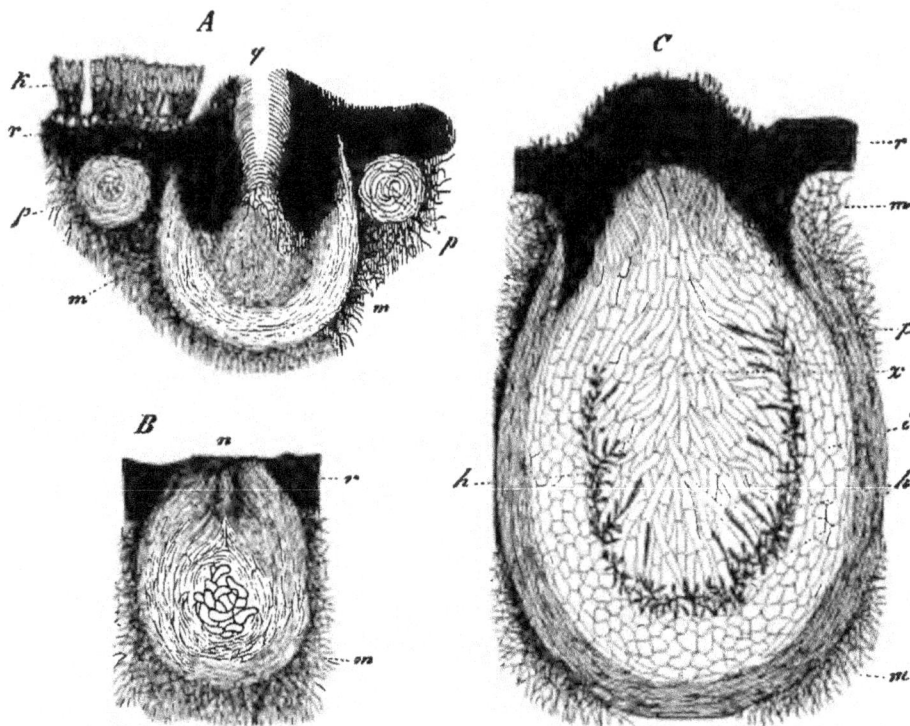

FIG. 103. *Xylaria polymorpha. A, B, C* transverse sections through young stromata with perithecia divided more or less exactly in half, all three magn. 90 times. *r* rind, *m* medullary layer of the stroma. *A, p* very young perithecium cut through the middle, *p* a similar one cut through near the median plane, *q* older perithecia, *k* gonidial layer. *B* perithecium with the mouth *n* bursting through the rind. *C* a nearly fully developed perithecium; the section passes close to the mouth, which is fashioned as at *q* in *A*, elsewhere through the median plane; *p* the outer, *i* the inner wall of the perithecium, *x* the large-celled paraphyses filling the centre of the perithecium having entirely displaced the short-lived inner tissue, *h* the inner surface of the wall with the insertions of the paraphyses and asci.

of the perithecia (*A, p*) make their appearance in the form of small spherical bodies which lie in the medulla close beneath the black rind, and are at once distinguished from the medullary tissue by containing no air and therefore being transparent. They are formed of a closely woven mass of slender hyphae, which are much thinner than the hyphae of the original tissue and must therefore be a new formation in it. In somewhat older specimens an irregular large-celled coil of tissue is found lying in the middle of the sphere. The spheres now increase in size in the direction of the medulla, the shape, structure, and position remaining the same. Then a dense tuft of straight hyphae, in the shape of a broad truncated cone, shoots forth from the part which abuts on the rind, and elongates in the direction of the rind, which is first bulged out a little and then gradually pierced through, so that the extremities of the hyphae project above the surface (*B, n*). The young perithecium has meanwhile become egg-shaped, its broader portion lying in the medulla being the future basal part, while the narrow end which is wedged into the rind is the future neck with the ostiole. The canal lined with periphyses is formed at an early period in the median line of the neck in a way which has not been exactly ascertained, and the elements at its circumference become thick-walled and black, and hence the neck is soon surrounded by a black outer wall which is continuous with the cortex (*q*). The process of turning black progresses very slowly towards the base of the perithecium and is only completed when the perithecium is ripe. After the origin of the neck the basal portion of the perithecium extends itself further into the medulla. Its circumference is all the time occupied by a layer of slender firmly woven hyphae running parallel to its surface, and this layer is the outer wall, which also becomes afterwards thick and black. This encloses a tangled mass of filaments filling the whole inner space; the component hyphae, with the exception of the large cells just mentioned, remain delicate and slender and swell strongly in water. The following is the result of Fisch's investigation into the further development.

The peripheral portion of the delicate hyphal weft last-named takes an active part in the further growth, and developes into the thin-walled hyaline pseudo-parenchymatous inner wall or subhymenial layer, which is about 6–8 layers of cells in thickness. The whole of the inner surface of the wall gives rise to slender hyaline large-celled paraphyses which appear at first singly but are afterwards closely crowded together and converge towards the middle having walls that can swell gelatinously, and also to small-celled ascogenous hyphae abundantly supplied with protoplasm, which stand between the points of insertion of the paraphyses and are everywhere in connection with the elements of the subhymenial layer. These hyphae do not reach their full development till the inner space is quite filled with paraphyses, and they develope at the cost of them. With the beginning of the formation of paraphyses, and in proportion as it advances, the primordial tissue, or as much of it as has not been expended on the construction of the inner wall, becomes spongy and gelatinous, and then dissolves and the paraphyses take its place. The same thing happens also to the coil of large-celled tissue which was merely alluded to above and must now be noticed again. In its very early state, in which it was first mentioned above, it may often be easily recognised as one simple row of comparatively large cylindrical cells very full of protoplasm and irregularly rolled up together. In some cases it is uncertain whether it is composed of one or more than one row of cells. It has an exact resemblance, especially in the first case, to the ascogonium of Polystigma, but with this difference, that it is always entirely inclosed in the spherical primordium, and does not send out a trichogyne-process outside it. It takes no direct part in the formation of the asci but, as the primordium increases in size, its coils are drawn asunder and then separated into pieces by the intrusion between them of branches of the transitory primordial hyphal weft. The entire coil remains inclosed in this weft and swells into a jelly and is dissolved with it; it is rare to see small portions of it taken up by the subhymenial layer and withdrawn from their inevitable fate. The transverse walls in the cells of the coil resist the dissolving influences longest, and even swell for a time into highly refringent plates like those of the trichogyne of Collema; but at length they entirely disappear, the last remains being still recognisable as the paraphyses begin to form.

From Füisting's[1] many researches it is more than probable that the development of the perithecia inside the stroma, not only in all the species of Xylaria and in Ustulina where I observed it some time ago, but also in the genera Diatrype, Stictosphaeria, Eutypa, Nummularia, Quaternaria, and Hypoxylon, runs essentially the same course as has now been described in Xylaria polymorpha, with the exception of course of specific differences of shape and especially of the formation of the wall and the orifice.   In all cases there appears, especially in the delicate coil of hyphae which forms the primordium, the row of broad cells irregularly rolled up and full of protoplasm, which Füisting terms *Woronin's hyphä*; in all cases Füisting observed the gradual disappearance of these cells without being able to prove a direct connection between them and the ascogenous hyphae, which sprang finally with the paraphyses from the wall of the perithecium in similar relations to them of place and time as in Xylaria. It is true that the views and objects of observers have so far changed since his investigations that his statements cannot be regarded as certainly infallible, and fresh examination might not be superfluous.

12. The sporocarps of **Sclerotinia Sclerotiorum**, the conformation of which was shortly described on page 52 (see Fig. 106), show, as the cup begins to expand, the first asci between previously formed paraphyses, and new paraphyses are added one after another with the growth of the margin of the cup, and then more asci are interposed between them, at first singly, but afterwards in greater numbers and crowd out the paraphyses.   Asci and paraphyses are of course branches or the extremities of branches of the hyphae of which the original bundle was composed; but the asci when once found cannot be referred back with the paraphyses to common parent hyphae; only hyphae are found, as Brefeld also states[2], which terminate either in paraphyses or in numerous asci; the latter hyphae penetrate deep into the subhymenial layer.   In this layer the ascogenous hyphae cannot be distinguished from the others, and where their first origin is to be found, remains uncertain.   The examination of the first beginning of the cup in the sclerotium leads to a conjecture on the subject.   Certain bodies are formed in sclerotia, if kept moist, before there is any external appearance of sporocarps.   These bodies, which here too may be termed primordia (Fig. 104), appear in large numbers in the periphery of the sclerotium, either close beneath the black rind or a little further in, as round transparent objects about 70–100 µ in diameter.   They consist of a coil of very narrow tangled hyphal branches with gelatinous membranes; their cavities, which are filled with protoplasm, appear to run through a homogeneous jelly.   They originate in single stout medullary hyphae of the sclerotium which are not distinguished from the rest in any other way; branches from these hyphae form the coil of filaments, and its development is accompanied with displacement and gelatinous disorganisation of the adjacent medullary hyphae.   The bundle of hyphae of which a cup is formed always bursts from the sclerotium above a primordium of this kind (Fig. 105), and consists of a smaller central portion which branches off directly from the primordium, and a larger

---

[1] Bot. Ztg. 1867.   [2] Schimmelpilze, IV.

peripheral mass the elements of which originate in the periphery of the primordium as branches of the stout medullary hyphae. The small central bundle is short, the peripheral hyphae are longer in proportion as they are nearer to the circumference, and, like the periphyses of the Pyrenomycetes, their extremities converge towards the median line, and thus a narrow depression is formed at the apex of the whole which has been noticed before on page 52. No other decided difference of structure is to be observed even at this time between the two kinds of hyphae, and during the subsequent growth

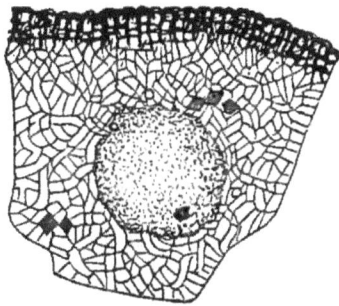

FIG. 104. *Sclerotinia Sclerotiorum.* Thin vertical section through the periphery of a sclerotium which has been kept moist and is ready to develope; beneath the black rind is the primordium of a sporocarp. The dark angular bodies are calcium oxalate. Magn 150 times. See also Fig. 14.

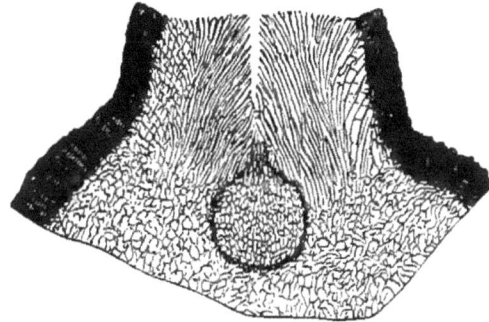

FIG. 105. *Sclerotinia Sclerotiorum.* Median section through a young sporocarp which is bursting through the rind. Magn. 90 times, but completed from higher enlargements.

of the whole body all possibility of distinguishing them ceases. But it is not improbable that the difference reappears with the formation of the ascus, in other words, that the hyphae which have proceeded from the primordium are the ascogenous hyphae and the primordium is therefore an ascogonium, while the envelope-apparatus of the sporocarp with the paraphyses comes from the peripheral hyphae; and thus the young sporocarp contains from the first the two elements side by side, though they are not anatomically different. The original structure of the primordium is obscured after the emergence of the sporocarp, but its place usually continues to be distinctly marked by the brown colour of the walls of the medullary cells at its circumference; this however may often ultimately spread to the primordium also. The number of primordia in a sclerotium is always much larger than that of the sporocarps which are matured; many are obliterated by their peripheral cells turning brown or are destroyed by the emergence of neighbouring sporocarps.

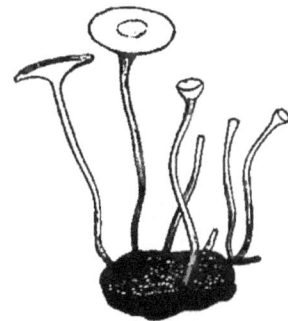

FIG. 106. *Sclerotinia Sclerotiorum.* Sclerotium with eight sporocarps of different ages; natural size.

Sclerotinia Fuckeliana shows phenomena of development quite similar to those which have been described; but there is one difference which adds greatly to the difficulty of observation: the primordia are not formed inside, but on the surface of the sclerotium. A thin bundle of hyphae from the medullary tissue bursts through the rind and developes on its outer surface into a dense round coil, the central part of which is like the primordium of S. Sclerotiorum and is surrounded by the peripheral hyphae as a large-celled envelope. The round coil then developes into the at first cylindrical sporocarp by the aid of branches from the primordium and from its envelope, which have the same relation therefore to each other in this respect as the primordium and medulla in S. Sclerotiorum. The incipient sporocarp is therefore now seated on the rind of the sclerotium in the same form as the one represented in Fig. 105 beneath it. Meanwhile more branches from the elements of the medulla have grown through the rind up to the envelope, so that the rind is pierced by a strand of hyphae as broad as the sporocarp and passing into the envelope, a condition of things which continues in the form represented in Fig. 19. Whatever could be observed of the final maturing of the sporocarp is the same as in S. Sclerotiorum.

13. According to Gibelli's and Griffini's researches confirmed by Bauke the development of the perithecium in **Pleospora herbarum** differs to some extent from those described above, the perithecium being formed by differentiation at a late period of growth of a spherical primordium originally composed of a uniform pseudo-parenchyma. This arises from one or two adjacent mycelial cells which are converted into the spherical primordium by active cell-division in every direction. The initial cell was previously constituted by one, rarely several, hyphal branches, which show no fixed arrangement and no peculiar changes in their further development. Then a bundle of slender paraphyses springs from the basal region into the parenchymatous body, displacing and dissolving its original central tissue, and grows on into the inner space; and after that, in the observed cases after a winter's rest, the asci are formed 'in the middle of the paraphyses as branches from their basal cells;' the paraphyses swell into a jelly and disappear as the asci mature.

Similar proceedings are perhaps to be observed in Sphaerella Plantaginis according to Sollman's statements [1], but these are not to be relied upon.

14. In **Claviceps purpurea**, according to Fisch, the formation of the perithecia begins with the differentiation of a few cells in the periphery of the young capitate end of the stroma which proceeds from the sclerotium (see page 38 and also section LXV). Two or three hyphal cells become filled with strongly refractive protoplasm and begin to form by divisions in all directions a very small roundish or elongate-ellipsoid cellular body, which is clearly distinguished from the pseudo-parenchyma of the capitulum by the small size of its cells and the nature of their contents. The mode of formation of the cavity of the perithecium could not be certainly ascertained; but in all probability it is effected by the mutual separation of the cells in the interior, either by the simple parting of the walls or by dissolution of a cell-layer; in this way a cavity would be formed, the roof of which would be the greater portion of the wall of the perithecium, and its floor become the incipient hymenium. The young perithecium as a whole soon acquires the form of an elongated cone by growth in the direction of the radius of the capitulum, and this change is accompanied by an elongation of the whole peripheral cell-layers of the body, thus plainly delimiting the wall of the perithecium. The point of the young perithecium elongates above into a cone, and forms a canal which is beset all round from below upwards with periphyses, while small protuberances grow out of the upper cell-layers of the young hymenium and lengthen

---

[1] Bot. Ztg. 1864, p. 281.

and form asci. No paraphyses are formed. The whole process shows great similarity with that described by Bauke in Pleospora.

The formation of the perithecia in the stroma of **Epichloë** is undoubtedly very like that of Claviceps; so probably, according to Fisch, is the same process in **Cordyceps** (C. militaris, C. ophioglossoides, and C. capitata). The perithecia also of **Nectria,** according to Janowitsch's earlier observations and of Cucurbitaria according to Bauke's report, are formed without initial archicarps, antheridia, or spermatia, but simply by late differentiation of portions of the stroma which were at first uniformly parenchymatous or consisted of closely woven hyphae; in both genera there is also a dissolution of the original central pseudo-parenchyma to make the inner cavity and a formation of paraphyses between the asci. But these older investigations require to be repeated at the present day. R. Hartig's account of Nectria mentioned above (page 217) is specially deserving of attention; he suspects that the perithecia in N. ditissima are produced from archicarps which are formed originally superficially on the stroma under a covering of gonidia-forming hyphae, and are then inclosed by branches from adjacent hyphae, and that in conjunction with these latter hyphae they then give rise to the primordial pseudo-parenchymatous formations from which Janowitsch's investigation sets out. The possibility of similar processes is not entirely excluded by the accounts which we possess in the case also of Epichloë.

15. Van Tieghem gives the name of **Ascodesmis** to two small Discomycetes which in the full-grown state look like small Ascoboli and are distinguished by reticulate thickenings of their spore-membranes. He describes the development of their apothecia from specimens cultivated on microscope-slides in the following manner. A slightly bent lateral branch rises from a cell of a filament of the mycelium, and, after a short increase in length, branches in a pseudo-dichotomous manner; this mode of branching is repeated through several orders in planes which intersect each other alternately, and the successive branches have a similar curvature. At length they all become woven together into a cushion of compact pseudo-parenchyma, which is attached on one side to the mycelial filament by a short stalk. Then closely crowded paraphyses shoot out from the superficial cells on the opposite side, and then the asci one after another from the same surface and between the paraphyses. We are not told whether any difference appears between the ascogenous cells and those which form paraphyses, or if the asci at least which follow one another spring from distinct ascogenous hyphal branchlets.

16. The apothecia of **Sphyridium fungiforme, S. placophyllum** and **Cladonia Papillaria** consist in the mature state of close-set paraphyses and of asci inserted between them, and the asci arise from distinct ascogenous hyphae in the hypothecium. According to Krabbe[1] the commencements of these apothecia are peripheral shoots from the outer surface of the thallus, and the layer of paraphyses first makes its appearance and afterwards the ascogenous hyphae with the asci. No trace was observed of a distinct carpogonium or archicarp, as the source of the ascogenous hyphae, or of any co-operation of spermatia; on the contrary, the ascogenous hyphae are branches of 'ordinary' hyphae, hyphae, that is to say, which are not distinguishable from vegetative hyphae and from those which form

---

[1] Bot. Ztg. 1882.

paraphyses. The sporocarps of **Baeomyces roseus**, which are very like those of Sphyridium in shape, first appear as coils of hyphae in the interior and deep beneath the surface of the thallus, and are there differentiated into layers of paraphyses and ascogenous hyphae. Subsequently they emerge from the thallus as long-stalked bodies in consequence of the elongation of their basal portions. But Krabbe was unable to arrive at any more positive conclusion with respect to the origin of the ascogenous hyphae than in the case of Sphyridium. According to the same observer **Sphyridium carneum** exhibits a curious variation from the genera with which it has hitherto been associated. Its sporocarps would appear to be only pseudo-sporocarps, sporocarp-like shoots from the thallus, which form neither paraphyses, nor asci, nor even spores, only coils of hyphae beneath the surface, in appearance like the ascogenous hyphae of allied species, but never forming asci.

We learn from Krabbe's latest 'preliminary' communication that in the Cladonieae, except Cladonia Papillaria, the whole of the large body known in descriptions as *podetium*, for instance the well-known cup in C. pyxidata and the branching shrubby form in C. rangiferina, is by its mode of origination an apothecium. It is formed as a primordial hyphal coil in the interior of a crustaceous or foliaceous thallus and forces its way through the rind outwards and then arrives by progressive or intercalary growth at its final form. The differentiations into ascogenous hyphae, distinguishable from the rest by turning blue with iodine, and the paraphyses is effected without a distinct archicarp and virtually in the same way as in the species of Sphyridium and Cladonia already mentioned; taking place either when the body is just emerging from the thallus, as in C. decorticata, or not till it has acquired in separate parts the final cup-like or shrubby form. Ascogenous hyphae and even asci may in some species revert to the vegetative form; and certain · kinds which have paraphyses in the normal manner either do not produce perfect asci, or only do so exceptionally. We must wait for the author's more detailed accounts, and we shall return to the anatomical character of the podetium in section CXVI.

From the agreement found among the sporocarps of the Ascomycetes in the mature state, it may be considered to be certain that they may all be included as respects their origin in one or other of the types above described or come very near them; which type it should be must be inquired into in each particular case, and cannot be decided with certainty from the mature condition. The many careful researches of Schwendener and Füisting into the formation of sporocarps in Lichens may still be adduced in support of this view; these observers failed to account for the first beginnings only of the ascogenous hyphae which are differentiated at a very early period. Füisting reports the presence of Woronin's hypha in the young sporocarp of Lecidea formosa, and Stahl[1] says of Parmelia stellaris and P. pulverulenta and Endocarpon miniatum: 'It is not difficult, especially in layers of Parmelia stellaris with many sporocarps, to find the extremely delicate apex of the trichogyne in young sporocarps; I succeeded in some successful preparations in showing the connection between these processes and the ascogonia which were distinguished by the abundance of their protoplasm.' Since spermogonia and spermatia are present in all these forms, as they are in the Collemeae, it is natural to suppose that there is a near agreement between them and the Collemeae; but this is not yet demonstrated. Füisting says that he has not found his Woronin's hypha in species of Verrucaria,

---

[1] Beitr. z. Entw. d. Flechten, p. 41.

Pyrenula, and Polyblastia, and Krabbe's results given above make it highly probable that the processes in the Lichen-fungi with which he was dealing are of a different kind.

With regard to the rest of the phenomena observed in the sporocarps of the Lichens of which we are speaking, it may be remarked that apothecia as well as perithecia are not formed on the surface as in Collema and most of Krabbe's species, but inside the thallus, as in Xylaria, in the shape of coils of delicate primordial hyphae which only come to the surface in the course of further development, pushing aside the tissue of the thallus above them in a manner which varies with the species. In species of Placodium, Lecanora, Zeora, Callopisma, Lecidea, Blastenia, Bacidia, and Pannaria, which have been examined and which form apothecia, a dense tuft of slender branched filaments, growing towards the outside, shoots out at an early period from the whole of the upper side of the primordial coil which is turned towards the upper surface of the thallus; these filaments are the first paraphyses. An outermost layer of similar filaments, open above and of varying thickness in each case, surrounds the tuft of paraphyses and runs to the surface of the primordial coil; this layer is the excipulum, though not exactly in the sense in which that word has been used hitherto in descriptive Lichenology. The excipulum is either formed at the same time as the first paraphyses, so that the outermost rows of the tuft become the hyphae of the excipulum, as in Placodium, Lecanora, Lecidea, &c., or before the paraphyses as in Blastenia ferruginea, Huds. according to Füisting. While the filaments of the primary tuft of paraphyses increase in length and form new branches, which insert themselves vertically between the first ones, and while the excipulum enlarges its surface in every direction by the formation of new interposed hyphae and grows by the appearance of new elements at its margin and of new hyphal branches continually behind the margin, which are like the primary paraphyses and in contact with them on the outside,—while all these processes are going on simultaneously, the young sporocarp by accession of new elements increases in height and thickness. The introduction of new branches continues for some time in the lower portion of the original tuft of paraphyses, and in such a manner that the filaments which were at first parallel become irregularly woven together, forming a tissue which cannot be distinguished from the primordial coil. The formation of new elements is followed directly by increase of size through expansion of the previous elements. The whole growth is first completed in the middle of the sporocarp; it continues longest, and often a long time after the sporocarp has appeared on the surface of the thallus, in the upper margin of the excipulum and close underneath it, where new constituents are being constantly and progressively added by apposition. The ascogenous hyphae also make their appearance with the first paraphyses.

The formation of the perithecia in Lichens from the primordial coils of hyphae follows in general the same course as that which has been given for Xylaria and Polystigma, &c.; but the first origin of the ascogenous hyphae is unknown. Peculiar features and deviations from rule may be studied in the authors named above, especially Krabbe and Füisting.

COURSE OF DEVELOPMENT OF THE ASCOMYCETES.

SECTION LXV. The **life-history of the Ascomycetes** has been thoroughly studied in the same species as the development of the sporocarps; there are many besides in which it is sufficiently known to allow of our comparing them with the others and judging of them with certainty.

The simplest case is where under normal conditions the germinating spore develops directly a mycelium or thallus, which also directly produces sporocarps in the modes described above, without the appearance of any other organs of propagation not belonging to the sporocarp. This is the case in Eremascus albus, Hypocopra fimicola, Ascobolus furfuraceus, Pyronema, Gymnoascus, species of Collema, Endocarpum pusillum, Thelidium minutulum, and very many, if not all, Lichen-fungi; a similar course of development is the almost invariable rule in Sclerotinia Sclerotiorum, where a filamentous mycelium grows from the germ-tube of the spore, the mycelium producing sclerotia and the sclerotia sporocarps. Single resting-cells, which occur accidentally in the mycelia and then resume active growth, are no more to be taken into consideration in this connection than the soredia of Lichens which will be described in section CXVII. More distinct formations of gonidia have not been observed in this course of development.

There is a second case in which the development *may* proceed as in the first; but it commonly happens that a formation of distinct gonidia is introduced, and the products of their germination resemble those of the ascospores. A good example of this kind is to be found in Sclerotinia Fuckeliana[1]. A primary mycelium grows from the germinating ascospore, and may in the simplest case produce sporocarps directly and without going through any intermediate state. I have observed this once and in one specimen only cultivated artificially in grape-juice on a microscopic slide. The sporocarp was produced directly from a tuft of mycelial branches resembling in appearance the primordium of a sclerotium, but its initial states were not further examined. The general rule is that sclerotia are formed (p. 34) on the primary mycelium; then either sporocarps alone proceed from the sclerotium, as in Peziza Sclerotiorum, or filamentous gonidiophores, which are known under the name of Botrytis cinerea, Pers. One only of the two forms appears on a sclerotium, never both together or one after the other. These gonidiophores may also grow directly from the mycelium which has proceeded from ascospores, without prejudice to future formation of sclerotia, but this certainly does not very often occur. Finally, the germinating gonidia produce a mycelium with all the characteristics of one that has proceeded from the ascospore and giving rise to the same products, but with this difference that it inclines much more to the formation of gonidiophores. To these phenomena is to be added the occasional formation of special abortive gonidia or doubtful spermatia, which must however be noticed again in section LXXIV.

A third case is that of a number of species in which it must be allowed that the course of development proper to the first category is *possible*, but is never actually observed. More usually the primary mycelium or thallus formed from the ascospore always produces gonidia. Strictly speaking, two subdivisions may here be distinguished:—

a. The primary mycelium which proceeds directly from the ascospore is reduced to a promycelium (section XXXI) which produces sporidia; these give rise to the definitive thallus, which then behaves as in the first category or as in the cases placed under *b.* Of this kind is Polystigma rubrum and Rhytisma Andromedae also, to judge from the germination of the ascospores.

In this case also the formation of gonidia is as a rule a necessary part of the development, for a perfect fertile thallus is not produced without its interposition.

---

[1] See Pirotta in N. Giorn. Bot. Ital. XIII, p. 130.

b. A richly vegetating primary mycelium or thallus proceeds from the ascospore, and its complete development closes with the formation of sporocarps, but as a matter of fact it always first produces gonidiophores with gonidia. The germinating gonidia again give rise in all cases to a mycelium or thallus of the same qualities and capabilities as those which sprang from the ascospore. Here therefore the formation of gonidia is not necessary to the development, but it is, as a matter of fact, an invariable occurrence. It precedes the formation of sporocarps in the history of the individual, and the gonidiophores were therefore often termed the precursors of the sporocarps.

The formation of gonidia is usually extremely copious in this third category, often much more copious than that of the sporocarps, and then generally owing to external causes which can be demonstrated; it may be the only mode in which the species is propagated during many successive generations, while sporocarps appear exceptionally and under special conditions. Some species have more than one kind of gonidia, which may then be distinguished according to size as *micro-gonidia, megalogonidia, macrogonidia*, or by special names in particular cases according to other characters.

Again, the gonidia are formed in certain cases on the free surface of the thallus on single hyphae or in crowded hymenia; or else in receptacles resembling perithecia. These latter have been termed by Tulasne *pycnidia*, and the spores or gonidia formed in them *stylospores*—not very happy expressions; the first, however, may be retained here, the latter replaced by the words *pycnospores* or *pycnogonidia*.

All known gonidia in the Ascomycetes are acrogenously abjointed after one or other of the forms described in section XVI, and none are swarm-cells.

Examples of species which have been fully examined. The germinating ascospore in the **Erysipheae** (Fig. 107) puts out a short germ-tube, which on a suitable substratum, namely the living epidermis of certain Phanerogams, sends first of all a haustorium (Fig. 6) into a cell of the epidermis, and then developes into the filamentous branched thallus which spreads over the whole surface. Short erect branches of this thallus then serially and successively abjoint large colourless cylindrico-ellipsoid gonidia, and each of these gonidia yields in germination under favourable conditions the same product as the ascospore. Every thallus which proceeds from these germinations ends, when it has reached its full development, with the production of archicarps and antheridia, that is, of sporocarps. But it need not always arrive at this conclusion, but may only form gonidia and propagate itself by means of them through an unlimited number of generations. This imperfect development may usually be traced to obvious external causes, such as climatic conditions or the absence of the nutrient substance required for perfect development, that is, the proper species of Phanerogam. The Erysiphe of the grape-vine is the best example of the group[1]. From the circumstances attending its first appearance and its diffusion in Europe, it may be safely assumed that it suddenly migrated, and was transferred to our vines from some other species of Phanerogam. It most probably came from America. In spite of its destructive diffusion over the whole of wine-growing Europe the most careful examination has never detected any sign of a sporocarp; the invasion was entirely carried out by vast numbers of gonidia, the shape of which procured the plant the name of **Oidium** (O. Tuckeri, Brk.). The sporocarps are probably found in N. America on native species of Vitis and described as Erysiphe (Uncinula) spiralis, Brk. and Curt., but this is not certain.

---

[1] Beitr. z. Morph. u. Phys. d. Pilze, III, p. 50.

The course of development in **Eurotium** and **Penicillium** may be described in the same words as in Erysiphe, making allowance for differences of form and for the circumstance that the species in the two last genera are not epiphytic parasites, but (for the most part) inhabit dead organic bodies ; here too we find frequent absence of sporocarps where the vegetative conditions are not altogether favourable. The gonidiophores of Eurotium (Figs. 94 and 35 *b*) are erect usually unicellular hyphal branches inflated and bladder-like at the apex, where closely crowded radiating sterigmata of uniform height are developed, and from these sterigmata spores are serially and successively abjointed. The gonidiophores in Penicillium (Fig. 36) are narrowly filiform, septate, and cymosely branched, and at their extremities, which are erect parallel and close to one another and terminate at nearly the same height, spores are formed by serial successive abjunction. The sporocarps of Penicillium glaucum have at present been found only in dark or imperfectly lighted places where the supply of oxygen is small, and chiefly on bread (Brefeld) ; I myself found them in abundance on a heap of grape-skins, both growing naturally and after the spores had been sown by hand.

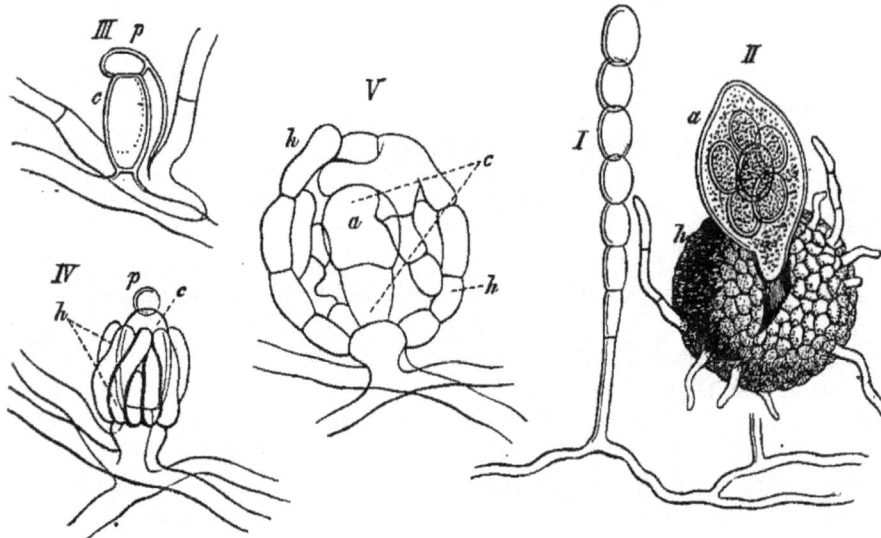

FIG. 107.   *I, II Podosphaera pannosa. I* chain of gonidia on the gonidiophore and mycelium. *II* ripe sporocarp ; the ascus *a* is emerging through the wall of the sporocarp *h* which has been ruptured by pressure. *III—V Podosphaera Castagnei. III* archicarp *c* with antheridial branch *p* on the mycelium. *IV* older state ; *c* archicarp invested by the hyphal branches of the wall, *p* antheridial branch. *V* still older state in optical longitudinal section ; *a* ascus with its pedicel-cell, the product of *c*, *h* the wall. *I, II* after Tulasne. Magn. 600 times.

The course of development of **Melanospora parasitica** again is on the whole very like that of the above species, except that the gonidiophores—short verticillately branched hyphae with whorls of secondary branches, from which spores are acrogenously and serially abjointed—are very rarely produced, the work of propagation falling chiefly to the ascospores. The peculiar parasitism of this Fungus will be considered in Chapter VII.

The ellipsoidal ascospores of **Polystigma rubrum** ripen in spring. They put out a short tube on a moist substratum, and the extremity of the tube, which also swells into an irregularly ellipsoidal form, receives the whole of the protoplasm, and is then abjointed and forms a thick-walled spore-cell (gonidium, sporidium). This cell readily germinates on a moist substratum, and on the epidermis of a foliage-leaf of Prunus the germ-tube penetrates at once into the nearest cell and there puts out branches, which then grow rapidly through the wall of the cell into the parenchyma of the leaf. Here they grow at the cost of the tissue of the leaf and displace its elements, but always covered by the epidermis, and in a few weeks' time they have formed closely woven thallus-structures (see page 43), which cause roundish red spots about 1 cm. in diameter in the still living green leaf; spermogonia and archicarps make their appearance in the spots in the course of the summer. The Fungus does not go beyond the complete development and subsequent fertilisation of the archicarps during the summer, but falls to the ground in autumn with the leaf, and there the further development of the perithecium takes place, if the conditions are favourable, at the expense of the reserve of food in the thallus; the spores become ripe in the ensuing spring. The process can of course be to some extent hastened or retarded in plants under artificial cultivation by changes in the temperature and in the supply of moisture.

The sclerotia of **Claviceps purpurea** and its nearest allies (see section VlII) mature during the summer and remain dormant all the winter; in the next spring each

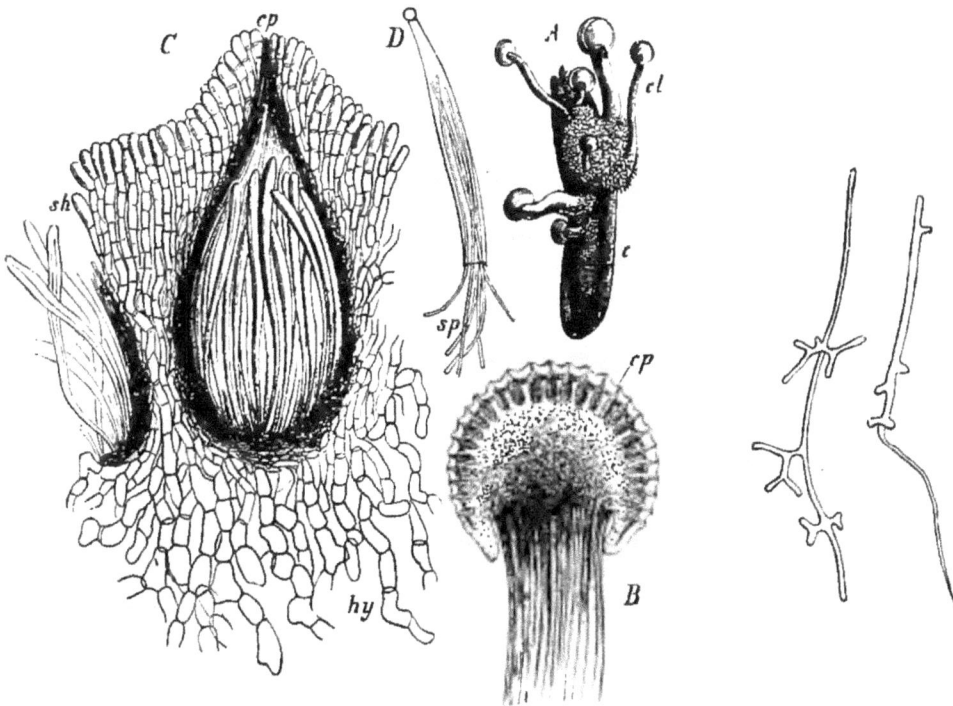

FIG. 108. *Claviceps purpurea*, Tul. *A* sclerotium which has given rise to seven stromata. *B* upper portion of a stroma in median longitudinal section; *cp* perithecia. *C* highly magnified perithecium divided through the middle with the surrounding parts; *cp* orifice, *sh* cortical tissue, *hy* inner tissue of the stroma. *D* ascus isolated; *sp* ascospores issuing. After Tulasne from Sachs' Lehrbuch. *A* natural size. *B* slightly, *C* and *D* highly magnified.

FIG. 109. *Claviceps purpurea*, Tul. Ascospores germinating 48 hours after being scattered on water. Magn. 375 times.

sclerotium, if it happens to lie on moist ground, usually produces several spherical stalked stromata (Fig. 108 *A*), the upper spherical portion of which is thickly covered with perithecia sunk half-way beneath the surface (*B, C*). The ejected cylindrical filiform ascospores (*D*) swell in different parts under the influence of moisture, and put out germ-tubes at several points (Fig. 109). If the ascospores of Claviceps have found their way into young flowers of the Gramineae (Secale in artificial cultivations) under conditions favourable to germination, their development begins in the pistil, according to Kühn's observations, and doubtless after the germ-tubes have penetrated into the pistil, though this has never been directly observed. The young pistil concealed between the paleae is first of all traversed in every direction and enveloped by a luxuriant growth of the hyphae of the Fungus, as has been already described : a white hymenium, Léveillé's Sphacelia, then forms on the whole of the furrowed surface, and from cylindrical sterigmata on it gonidia are abscised (Figs. 110, 111 *a*). During the formation of the hymenium the well-known saccharine fluid is secreted, which oozes out from between the paleae in thick drops rendered turbid by countless gonidia, and thus betrays the presence of the parasite. This juice is eagerly sought by insects, which thus carry away the gonidia. Soon the formation of the sclerotium begins in the basal portion of the gonidia-forming body in the way already described. The sclerotium reaches maturity by the time that the grass is ripe and passes into the dormant state which lasts till the next spring. The gonidia readily put out germ-tubes as soon as they become free, and the tubes sometimes produce small upright branches on the microscope-slide, from which fresh gonidia are then abscised (Fig. 111 *b*). Kühn informs us that new gonidiophores and sclerotia are developed in the manner described above from the germ-tubes of

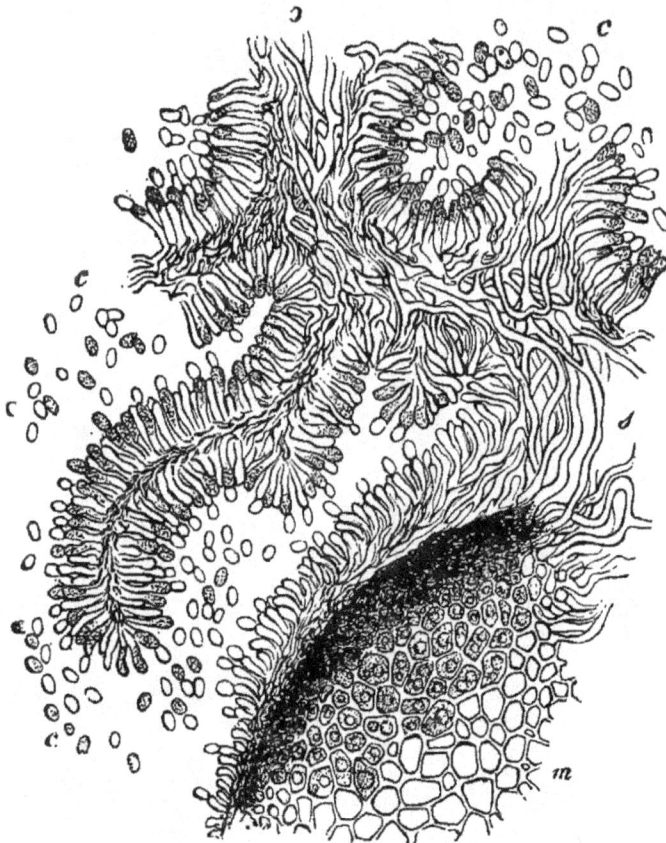

FIG. 110. *Claviceps purpurea*, Tul. Portion of a thin longitudinal section on the boundary line between the gonidiophore *ss—cc* and the young sclerotium *m*. See Fig. 17. After Tulasne, from Lürssen's Handbuch, highly magnified.

gonidia, which have found their way to the flowers of a grass.

Nectria ditissima may be given from R. Hartig's description[1] as an example of a species furnished with more than one kind of gonidium. The mycelium lives in the rind of leafy trees, and causes the disease known as 'canker.' It forms a small cushion-like pseudo-parenchymatous thallus beneath the surface of the rind; the thallus eventually bursts through the rind and produces first gonidia and then perithecia on its outer surface. A sufficient account has already been given of the perithecia, which in the primordial state are concealed beneath the gonidia and the structures producing them, but these are displaced and thrust aside by the perithecia. The gonidia are now formed acrogenously outside the cushion on short slender filiform sterigmata arranged side by side and parallel to one another, so as to

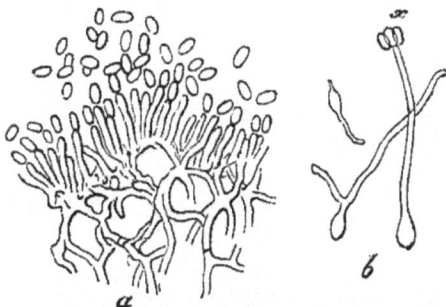

FIG. 111. *Claviceps purpurea*, Tul. *a* thin transverse section through the layer from which gonidia are being abscised, *b* gonidia germinating and producing by abjunction a small group of secondary gonidia at *x*. *a* after Tulasne, highly magnified, *b* after Kühn.

---

[1] Unters. a. d. forstbot. Instit. München, 1. See also Tulasne, Carp. III, and R. Göthe, Der Krebs d. Apfelbäume in Thiel's Landw. Jahrb. IX (1880).

produce a dense hymenium. The gonidia in the highest state of development are abjointed successively in long rows forming a close colourless mass which covers the cushion, each gonidium having the form of a bent cylinder 60 μ in length and divided by transverse walls into several members (spore-cells), which may be as many as eight in number. Besides these there are some much smaller, but of similar origin, in which the number of members sinks to two. Each segment-cell of these gonidia may develope in a moist atmosphere into a branched hypha, from single branches of which fresh smaller gonidia are abjointed. If the mycelium vegetating in the tissue of the rind of the tree comes to lie exposed in a moist atmosphere, it sends out numerous branches into the air, and from these also countless small gonidia are abjointed. The size of all these small gonidia sinks by regular gradations to 1.5 μ; they are all cylindrical and rod-like; those of medium size are divided by a transverse wall into two segments which often separate from one another; the smallest are undivided. All down to the size of 2 μ can put out germ-tubes or may also multiply by sprouting. The very smallest have never been observed to put out germ-tubes, but they appear to multiply by division and fission and by sprouting. Lastly, gonidia of the smallest kind are also formed in great numbers by slender branches of the mycelium in the interior of the tissue attacked by the Fungus. All germ-tubes, whether from gonidia or from ascospores, can develope new fertile mycelia in the proper substratum, that is, in the living rind of a tree; it is doubtful whether the smallest gonidia can produce mycelia (see section LXXIV). There is also some doubt as to the true nature of certain acrogenously abjointed spores which occur on the sporocarp-bearing portions of the Fungus which we are considering; they appear to belong to parasites of Nectria, and it will be sufficient therefore in this place to refer the reader to Hartig and Tulasne.

The development of **Cordyceps** with its great variety of forms will be described subsequently in Division III.

The cycle of development of the one or perhaps two species included under the name of **Pleospora herbarum** is particularly rich in forms. These are found in dead and rotting herbaceous plants. The results which will be given here were obtained from plants cultivated in nutrient solutions on the microscopic slide. The mycelium forms (1) the *perithecia* mentioned above with pluricellular compound ascospores : (2) *gonidia* of three kinds produced acrogenously on filiform gonidiophores, namely,— (*a*) bicellular or pluricellular spores resembling the ascospores, in general shape shortly cylindrical to roundish, with dark-brown thick membranes rough with fine points on the outside, named by Berkeley as a form-species in 1838 Macrosporium Sarcinula and therefore called by Gibelli and Griffini the **Sarcinula**-form; they are usually produced singly at the end of the gonidiophore; (*b*) the **Alternaria**-form, classed with the old form-genera Alternaria, Sporidesmium, Mystrosporium, and Polydesmus; these are conically pear-shaped pluricellular compound spores having a smooth light-brown membrane, and arising at the extremities of the hyphae in long and often branched rows (Fig. 34); (*c*) a form said by Bauke to be a microgonidial form, but of which he gives no further description; it is not the one known as Cladosporium herbarum and placed by Tulasne with Pleospora herbarum, for this, according to all later investigations, does not belong to this place at all and its genetic connection is uncertain: (3) *pycnidia* (see below, section LXXI, Figs. 118, 119) which appear as intercalary formations on branches of the mycelium. A piece of the hypha consisting of one or several cells swells in the same way as when a perithecium begins to be formed, and its cells divide at the same time meristematically and irregularly by walls inclined in every direction. By this process of growth a small-celled parenchymatous body is formed of many layers, which is round or irregularly elongated in shape and seldom more than 0.2 mm. in size, often much less. This body is at first uniformly dense, but towards the end of its growth a central cavity is formed in it surrounded by the many-layered wall; this cavity is produced by the cells of the central part ceasing to follow the growth of the outer parts in the direction of the surface and therefore separating from them.　The cell-rows which bound the cavity project into it from the first with radial convergence, and fresh ones are formed resembling these and between them. The wall of the cavity is thus lined with slender converging rows of short cells, which begin at once to form numerous *pycnospores* terminally and laterally on all their cells by successive abjunction.　These spores are elongate-cylindrical, very thin-walled, 2.8–4 μ in length and about half that breadth, and are surrounded by a hyaline gelatinous or gum-like substance, perhaps the outer wall-layer. ´Imbedded in this gelatinous envelope they are heaped up in large numbers in the interior of the pycnidium.　When their formation begins the outer wall-layer of the pycnidium, which has hitherto been colourless, is thickened and becomes brown.　At the same time a narrow opening is formed in the wall usually at one, rarely at more than one, place by the retreat of the cells from one another, and the orifice is usually surrounded on the outside by a circle of projecting cells like short papillae.　In this state the pycnidium has reached its full development ; addition of water causes violent swelling of the mucilage which envelopes the spores, so that they collect into a gelatinous mass and are squeezed out of the narrow opening in countless numbers, forming a tendril-like body or a round gelatinous drop according to the degree of moisture ; they separate at once in water.

All the vegetative cells of this Fungus and its allied forms show a tendency to form thick brown membranes like those of the perithecia, pycnidia, and gonidia. With these walls they can pass eventually into a long period of rest and return from it under favourable conditions to vegetative activity and the formation of spores. These *resting mycelia*, and detached single cells or portions of hyphae (*gemmae*), may appear in numbers at the close of their culture and add to the variety of forms.

As regards the genetic relations between all these forms, we are only told of the microgonidia, that they occur in company with other gonidial forms.　Observers also agree in saying that a mycelium is produced from the germinating ascospore, and can first produce gonidia and then perithecia and pycnidia ; it behaves therefore in the same manner as the rest of the gonidia-bearing plants described above as regards the successive formation of organs of propagation, except in the matter of the pycnidia. Gibelli and Griffini saw a mycelium proceed also from the Sarcinula-gonidia, which first produced similar gonidia and then perithecia ; sometimes in this case also the mycelium which has grown from the two kinds of gonidia does not proceed beyond the formation of gonidia.　But the pycnospores, which swell strongly in a nutrient fluid and even form transverse walls and then put out germ-tubes, produce a mycelium from them which *up to the present time has never borne anything but pycnidia* in cultures, even when these have been continued through a large number of generations.　It would appear probable that the mycelium could produce perithecia when growing on its natural substratum, but this has not been proved. While observers are agreed up to this point, the views of the Italian botanists differ in other respects from those of Bauke.　They are of opinion that two similar but always distinct species are confounded together under the name of Pleospora herbarum, one Pleospora Sarcinulae, which is constantly characterised by the presence of the Sarcinula-gonidia and by larger ascospores, the other Pleospora Alternariae, which as constantly has the Alternaria-form of gonidia.　They found pycnidia only in P. Sarcinulae.　On the other hand according to Bauke a mycelium proceeds from the ascospores of the same perithecium, which forms *either* pycnidia in company with Alternaria-gonidia *or* perithecia with the gonidia of P. Sarcinula.　The mycelium from each of the gonidial forms always produces gonidia of the same kind as the one from which it proceeded. Further observation must determine which of the two views is correct, but the analogy of other Fungi makes that of the Italian writers the more probable.　The decision of this question is not important for our present purpose, because, as was intimated above, in the one case we are dealing with one species with a great variety of forms, in the other with two species with fewer forms and in other respects like one another.

Section LXVI. The remarks which have now been made render it unnecessary to say anything more concerning the **homology** of members of the same name in the Ascomycetes which we are here considering; nor need we further discuss the fact, that the mature sporocarp contains in some cases only the products of the development of one archicarp, in others, as in Physma and Pyronema, the results of the development of several archicarps. We could if necessary establish sub-forms on this distinction. The question of the homology of the spermatia and spermogonia is not so readily settled; but even here the difficulties are not great. It may be conceded that the consideration of the function of the spermatia of the Collemaceae puts us on the right track. That function, as will be shown in the following section, is the same as that of the antheridia in other and allied species. Hence arises the consideration whether the spermatium with the spermatiophore, the sterigma, is not the homologue of an antheridial branch from which portions are abjointed in the form of spermatia, according to the arrangements of particular species, in order to be capable of the fertilising function. Forms like Collema, in which spermatia and archicarps are formed at a distance from one another, may not afford any sure ground for an answer to the question; but the case is different with Physma, where spermatia and archicarps spring close together from branches of the same hyphal coil, like the antheridia and archicarps of Pyronema. If the spermatia in Physma remained fixed to their spermatiophores in order to conjugate with the archicarp, the only difference between the two forms would be that of conformation. The actual differences, it is true, go farther than this, since the spermatiophores are combined into a spermogonium from which the spermatia are discharged, and the archicarps are outside of it, and send up the trichogyne to the place where it encounters the spermatia. But we can understand all these phenomena as adaptations to suit the origin of the two organs inside a dense thallus which impedes their direct meeting, and still maintain the homology with Pyronema. Even the excessive numbers of the spermatia or antheridial branches will be quite intelligible in view of the very general rule that the number of male sexual cells in a species increases with the difficulties in the attainment of its physiological aim. But homology of the spermogonia and archicarps of Physma with those of Collema is quite obvious; the latter agree perfectly with the former in every respect except in their diclinous and monoecious distribution, which in some forms[1] inclines to dioecism. But this arrangement is no difficulty in the question before us, since diclinism may appear everywhere and is actually observed in many species, in which sexual cells are endowed with free motion whether active or passive.

It follows from these comparisons and considerations that in Collema also the spermatia with their spermatiophores may properly be considered to be homologues of the antheridial branches and antheridia of more simple forms, and the peculiarities of their development, and the excessive numbers in which they are produced

---

[1] Stahl, Beitr. z. Entw. d. Flechten, pp. 30, 38.

in special receptacles, as adaptations to special circumstances of development, the further consequence of which is the establishment of diclinism.

It must be admitted that comparisons of this kind are always somewhat insecure and readily afford too much room for the play of fancy, if they are not supported by a more complete series of well-known intermediate forms than can at present be produced in this case. The comparison in the text may be offered with this reservation attached to it. It appears to me to fit in quite naturally with ascertained facts; nor is it too much to expect that the intermediate forms that are wanted will be found, if we consider how few of the whole number of Ascomycetes have been examined up to the present time.

If we compare the whole course of development of the Ascomycetes in which this course is fully known with that of the groups of Fungi described in previous sections, we become distinctly aware of a parallelism between Eremascus and the Ascomycetes, which are provided with archicarps and antheridial branch on the one side, and the Mucorini, Peronosporeae, and Saprolegnieae on the other. A thallus proceeds from the carpospores (ascospores, oospores), and terminates its development with the formation of an archicarp and antheridial branch and of the new carpospore which is formed from them. To this the whole course of the development is limited in many cases, as in Eremascus, Pyronema, and species of Ascobolus on the one side, and on the other in Pythium vexans and Artotrogus; in most cases it is interrupted by the formation of other spores, the gonidia. The gonidia of a species are sometimes all of one kind, as in Erysiphe and Peronospora, sometimes of more than one kind in the same species. The parallelism extends even to close resemblance in form in organs of the same name in certain groups. Eremascus might, to judge from Eidam's description, be almost ranked with the Mucorini, especially the Piptocephalideae; on the other hand, none of the essential developmental characters of an Ascomycete are wanting in it. In the form of its archicarps it is perfectly like Penicillium, Gymnoascus, Eurotium and other genera. There is also a great amount of agreement in respect of the thallus, formation of the gonidia, archicarp, and antheridial branch between the Erysipheae, especially Podosphaera, on the one hand, and some Peronosporeae on the other[1]. These groups therefore establish a nearer connection between the Ascomycetes in question and the Peronosporeae; a convergence of the two groups which amounts to actual contact and may be regarded as phylogenetic affinity. This connection with the Peronosporeae is closest in Podosphaera, because there *caeteris paribus* not only the archicarp and antheridial branch are like the same parts in the Peronosporeae, as, for example, in Phytophthora omnivora, but the ultimate development also of the archicarp only goes a little further; the eight-spored ascus with its stalk proceeds from division of the cell twice repeated, while in Phytophthora the archicarp becomes the oogonium with the oospore. Erysiphe is closely connected with Podosphaera, in which the archicarp by repeated cell-division and branching gives rise to a number of asci, and all the rest of the Ascomycetes in question with Erysiphe, as appears plainly from the details which have been given

---

[1] For a more extended comparison see Beitr. IV, p. 109.

above. These comparisons show that archicarps, antheridial branches, and all other parts of the same name in all the Fungi which are here compared, are homologous.

The homologies go as far as the archicarp; they cease with its further development, unless we may perhaps compare the oospore of Cystopus which forms swarm-spores directly in germination (page 135) with an ascus; the ascus of Podosphaera and Eremascus is an organ which does not appear in the Peronosporeae, and this may be said still more of the sporocarp of Erysiphe and the series of forms which follow it. The series of the Ascomycetes and that of the Mucorini and Peronosporeae set out on divergent routes from Podosphaera and Eremascus, as the members which touch one another in the two series. It must not be forgotten that *in this comparison* of sporocarps, the parts spoken of above as the ascus-apparatus are *alone* to be taken into consideration, and have been considered here. The envelope-apparatus, important as it is in other connections, does not enter into the question. For the case is exactly the same if there is no envelope-apparatus, as really happens in Eremascus, and would be the same if there were Peronosporeae with their oogonia in envelopes; this, it is true, has not been observed, but it is quite possible, and in Mucorini (see section XLII) the zygospores are provided with envelope-apparatus of great variety of form.

When once the homology between the archicarps of the two groups is proved, that of all the spores, which have been termed gonidia in the preceding pages, is also established. The expression was anticipated above throughout in the case of the Ascomycetes, so far as it was supposed to have exactly the same meaning as in the Peronosporeae and their nearest allies.

In the Peronosporeae the antheridial branch and the archicarp function as *sexual organs*. But homologous *members* need not also function in all cases as exactly similar *organs*, as appears at once from the case of the Saprolegnieae with doubtful sexuality and with sexuality undoubtedly wanting. Hence when the homology has been established it is still an open question, whether the members of the Ascomycetes in question are sexual organs or not. To understand clearly this much discussed question[1] we must first of all remember that, in our imperfect knowledge of the nature of sexuality and the sexual process of fertilisation, we have no simple mark or reagent by which we can recognise the sexual quality of an organ. We learn from the facts before us that in every process of fertilisation there is a material union of one peculiar male or fertilising cell or at least of a portion of its protoplasmic and nuclear substance with one other, a female cell, which is to be fertilised, or, as in the Florideae, with a pluricellular female apparatus[2]. The result of this union is that the female portion is rendered capable of further development: the development does not take place without this union, and union with the male portion is necessary that the female may become capable of it. In a doubtful case therefore the determination will depend first on the observation of the union of the protoplasm or nucleus, and secondly on the experimental proof of the necessity of this in order that the presumed female portion may become

---

[1] See Beitr. IV, pp. 74, 111. We cannot enter here into a discussion of the general question of sexuality; the beginner is referred to Sachs' Text-book, 2nd Engl. ed.

[2] See the work cited on page 213, and Fr. Schmitz in the Monatsbericht d. Berliner Acad. 1883.

capable of development. We may derive assistance also from analogies of cases which have been certainly ascertained; but these have only a subordinate value, for experience has certainly shown that sexuality is a phenomenon of irregular occurrence in the greater part of the vegetable kingdom, varying sometimes from species to species even in the higher plants, and hence homologies and analogous functions are here also not necessarily coextensive.

According to these criteria the sexual organs of Pythium, for example, are really sexual, for the union of protoplasms is evident, and its necessity, though not demonstrable in strictly experimental manner by artificial separation and conjunction of the two parts, is all but absolutely proved by the fact that the union always takes place: but if we apply the same criteria to the homologous organs of the Saprolegnieae the sexuality will at least be very doubtful.

Similar results are obtained by the same method of determination in the case of the Ascomycetes in question. The protoplasms unite in Pyronema, and there is no exception to this rule; and as strict experiment is impossible, we may conclude, as in Pythium, with almost perfect certainty from the constancy of the phenomenon, that the union is necessary. The phenomena in Pyronema are so far different from those in Pythium that the archicarp grows to meet the male organ by means of a special organ, the *trichogyne*, and unites its protoplasm with that of the male organ; this union takes place after the trichogyne has previously been permanently delimited as a special cell by a transverse wall. These facts could hardly be understood but for the exact analogue presented by the majority of the Florideae. But these make it plain that the trichogyne is an organ of conception which first receives the fertilising matter, and that the effects of fertilisation are conveyed from it, in a manner which cannot be further considered here, to other parts of the female apparatus, which in the present case is the ascogonium.

We are made acquainted with quite similar phenomena to those in Pyronema, though different certainly in form and more complex, from Stahl's observations on the Collemaceae which have been described above. The most important complication arises from the fact that the male organs are spermatia detached by abscision, not antheridial cells formed beside the archicarp. The union with the trichogyne and the changes which proceed from the place of conjugation and affect the female apparatus which ultimately forms asci are very apparent. The necessity of the union for the further development of the female apparatus is in this case also not shown by strict experimental proof on account of technical difficulties; but it is as good as certainly proved by the observation that not only the union of the spermatia with the trichogyne precedes those characteristic changes and developments, but that these do not take place if the spermatia for any reason have not made their appearance.

We are met by the same arguments and results in the case also of Eurotium, though the facts observed in this case are not so striking at first sight as in Pyronema; and lastly Eidam's observations on Eremascus show that the union of the protoplasms is as evident as it can be in organs which are extremely similar to those in question in Eurotium, Penicillium, &c. In these cases we may therefore affirm, that, according to our present criteria, the antheridial branches, or the spermatia which must be supposed to be abjointed portions of such branches, are in function *male sexual organs*, and that the archicarps are *female sexual organs*, and that the Fungi which produce them possess the power of sexual propagation.

This statement is not admissible in the case of all the other forms furnished with homologous members. Resting on what we know, we may suspect that in Polystigma the trichogyne and spermatia behave in the same way as in Collema, but we have no proof of the necessary material union. In Gymnoascus and the Erysipheae, especially Podosphaera, the two organs appear with the same constancy, one might say with the same morphological necessity, as in Pyronema, and the possibility of a material union of the protoplasmic bodies is not excluded by the known facts. The antheridium, it is true, always remains separated from the archicarp by a membrane which, as far as we can see, is not perforated, but it is closely attached to it, and dissolved or very finely comminuted substance may pass through the membrane, as must be assumed in the case of the fertilisation of Angiosperms. But after all nothing is proved about the matter; the constant contact of the antheridial branch proves nothing; the envelope is constantly in the same position; we cannot get beyond probabilities and possibilities. Beneath the level of probability we arrive at last at species like Melanospora parasitica and Ascobolus, which needs revision however in this respect, with a beautifully developed carpogonium, but with the attachment of the antheridium not constantly or certainly observed. The conclusion on the whole is, that some of the forms in question have sexual organs which can be shown to fulfil their functions, others have organs perfectly homologous with the first, but with the sexual function not certainly ascertained or certainly wanting.

We have secondly to enquire after the homologies of the Ascomycetes, in which there is no distinct archicarp, as far as we at present know, when the sporocarp begins to appear. We will consider first the extreme cases, Pleospora and Claviceps. Here the question is, are the parts in these species to be considered as really homologous with those of the same name in the other series which has archicarps, or only as very similar to them in form and function; or, expressed in terms of the phylogeny, do these Ascomycetes belong to a *single* series of forms descended from the same stock, or to *at least two* series descended from different stocks only with analogous ultimate construction? We can only advance probable arguments in deciding between these alternatives, but these are against the second of the two and in favour of the unity of the Ascomycetes. First of all the difference alleged is the only one, while they agree together in all other points of importance to a degree which is elsewhere found only in allied forms, and not in those which are merely analogously developed. Secondly, no other close affinity can be found for the Ascomycetes which have no archicarps than that with the others; and they must have some relation of the kind, some connection with other forms. Thirdly, the extremes are evidently connected together by intermediate forms. The first of such forms is to be found in Melanospora parasitica with its beautifully developed carpogonium but inconspicuous or absent antheridium; other like phenomena appear to occur occasionally[1] in the series of the Sordarieae, and these therefore claim the attention of observers. Sclerotinia also

---

[1] See also Zopf in Sitzgsber. d. Brandenb. bot. Ver. 1877.

is one of this kind. On the other hand, forms like Xylaria, which have Woronin's hypha as a transitory formation only, offer transitions connecting them with Polystigma. Amongst the former cases which connect with Melanospora are forms in which at one end of the series distinct archicarps are present functioning as certainly asexual (par-thenogenetic) ascogonia, and along with them distinct elements of an envelope; towards the other end of the series the difference between ascogonium and envelope-formation diminishes till it disappears, and it is only in a more advanced stage of development of the sporocarp that the formation of ascus and of envelope is undertaken by separate elements, which up to that time were apparently uniform with the rest (Pleospora, Claviceps). Sexuality therefore is not developed, and in the extreme cases there is entire disappearance of primordia of the sporocarp which are homologous with sexual organs. In the other series of cases Woronin's hypha in the Xylarieae can be under-stood if we compare it with the archicarp of Polystigma or Collema. It occupies the same morphological position, but takes no active part, as far as can be seen, in the formation of the sporocarp, and then disappears apparently without having had any function to perform, while the formation of the asci is undertaken by neighbouring hyphae belonging to the envelope. Here then there is an archicarp or ascogonium present in form, but it remains functionless in the sense expressed by these names, and the formation of asci falls to the lot of other organs not strictly homologous with it.

These facts all lead to the result, that in these extreme cases we are in presence of phenomena, which were spoken of above on page 123 as interruption and restoration of the homology. Such a conception would perhaps be rash, if we had not before us the clear cases above mentioned of the occurrence of this phenomenon in Ferns and Angiosperms; but, from our acquaintance with it there, we are led naturally to this assumption by arguments which have now been stated. The species with an aborted Woronin's hypha and a formation of asci at the same time are parallel to the apogamous Ferns with functionless archegonia and to Angiosperms with an aborted egg-apparatus replaced by adventitious embryos; the rest approach nearer to simply parthenogenetic apogamy, as seen in Chara crinita and the Saprolegnieae, but with the peculiarity that the παρθένος itself entirely disappears in the extreme cases.

In the above discussions the simple forms which have distinct archicarps, such as Eremascus, Erysiphe, and Eurotium, have been treated throughout as forming an indivisible group of closely related species, and it has been sought to bring the rest into connection with them. More is not at present possible. It ought not by any means to be affirmed that all these forms which serve as points of departure are connected with the same species outside the Ascomycetes, and that subordinate parallel or diverging series do not proceed from individual forms among them inside the circle of the Ascomycetes. It was shown above that Eremascus comes near the Mucorini, and perhaps special groups of other Ascomycetes connect with Eremascus. Podosphaera is nearer on one side to the Peronosporeae, on the other to the main body of the Pyrenomycetes, and so on. But at present we are not in possession of the empirical material necessary for the enquiry into these details, and the main results, as here represented, are not decidedly affected by it. The resemblance also in the development of the sporocarps which we have been considering to that of the Florideae has necessarily always attracted attention, but whether it points to an actual closer affinity must for the present remain undecided; other connection than that to which attention has been called above cannot in my opinion be established.

A nearer approximation of the Ascomycetes generally, or of some of them, to the Florideae would not materially affect the conclusions at which we have arrived above with regard to the main questions connected with the homologies.

In my first investigations (Beitr. III) into the development of the sporocarps in Erysiphe, Eurotium, Pyronema, and others I called the archicarps and antheridial branches generally sexual organs; and from the great amount of agreement between the sporocarps when fully formed I expressed the *conjecture* that all the Ascomycetes have homologous and analogous organs for the production of these sporocarps. Others followed me in this view, especially when they became acquainted with individual cases which confirmed it. The investigations which are here communicated have had the result of showing that my generalisation was incorrect, and that the mistake arose not merely from want of consideration of facts not then known, but more especially from not distinguishing sufficiently between morphological and phylogenetic homology and physiological analogy. I trust that I have taken this distinction sufficiently into account in my last special treatise (Beitr. IV) and in this work.

Van Tieghem has been one of the chief opponents of my view, for he takes his ground on forms that have no distinct archicarps and does not allow of 'sexuality' in the Ascomycetes. His opinion briefly stated is this, that the differentiation of the ascogenous hyphae and their envelopes takes place at stages in the development of the sporocarp which vary according to the species, and that it occurs at the earliest possible stage in the species which are supposed to have sexual organs. The supposed female organ is only an ascogenous hypha differentiated at a very early period, the supposed male organs are simply part of the envelope-structures. The facts on which Van Tieghem originally founded his opposition were certainly not happily chosen. But if he is content to rest his case on Pleospora for instance, or even on the actual condition of things in Sclerotinia with which he has never been acquainted, he must be allowed to be quite in the right as against my original conjectural generalisation literally taken; and if he objects that the actual sexual function of these organs has not been proved in the case of Eurotium and Podosphaera, he will find that this is allowed in my work of 1870. But Van Tieghem makes no enquiry into the homologies and extends his negation beyond the limits allowed by the facts. If he had duly considered the indisputable fact of the constant presence in Podosphaera of my antheridial branch, that is of an organ distinctly different from the later-formed structures of the envelope and accompanying the proper commencement of the sporocarp, he would have been led to those true subjects of enquiry which have been discussed in the foregoing pages, and which it has been attempted to elucidate; and the facts at present known about Pyronema, Eremascus and other forms should have led him to a different answer to his own enquiries. We need not go into his positive views with regard to the function of the organs in question, that, for instance, the antheridial branches serve to support the ascogonium and that the trichogyne in Collema is an organ of respiration, before it has been shown to be to some extent probable that the ascogonia are in danger of falling without this support, and that the particular organ in Collema is obliged to have an apparatus of its own to get air, and cannot respire without it quite as well as the elements of the interior of the thallus near which it is placed. Such fancies must certainly deserve the name of gratuitous hypotheses quite as much as the views which I have here explained.

Another opponent of my ideas is Brefeld. He wavers between Van Tieghem's views on the one hand[1], and certain others, which, when stripped of some accessories which do not strictly belong to the question, agree with those of the present work[2]. I have therefore no reply to make, apart from the corrections of some matters of fact

---

[1] Bot. Ztg. 1876, p. 56, Abs. 23, and Schimmelpilze, IV, p. 142.
[2] Bot. Ztg. 1877, p. 371, and Schimmelpilze, IV.

which have been made in previous sections. I shall return in a later page to the matters which, as I have said, do not belong to this place.

DETERMINATION OF IMPERFECTLY KNOWN ASCOMYCETES.

SECTION LXVII. The facts detailed in the foregoing pages were established at first from a comparatively small number of species, but they nevertheless enable us to pass judgment with tolerable certainty on all the varied phenomena which have been observed in the countless forms of the Ascomycetes, especially the Pyrenomycetes and the Discomycetes; they are a frame in which the latter may be set. It must at the same time be remembered that very many of these phenomena were known, named, and provisionally disposed of according to the best knowledge of the time before a secure basis was laid for a decision respecting them, and that it was by starting from single phenomena that this basis was gradually reached. Especially should it be remembered (see section XXXII) that at first every distinct *form* was supposed to represent a distinct *species*; the gonidiophores of Sclerotinia Fuckeliana were made a species under the name of Botrytis cinerea, the sclerotia another species as Sclerotium echinatum, while the sporocarps by themselves were made a species of Peziza; in Erysiphe the gonidiophores were supposed to be a species of the genus Oidium, and only the perithecia were assigned to Erysiphe. The researches of Tulasne first led gradually to an understanding of the real condition of things to which he gave the name of pleomorphism, and to him we are chiefly indebted for the distinguishing and naming of the possible forms in the development of a species. These researches rested on the broad foundation of the comparative observation of numerous forms, of their cohabitation, of their anatomical connection, and their succession in time. Pursued in this way they arrived on the whole at the truth, and it is a small diminution of their merits that they should have given rise to some erroneous views on special points, or that they occasionally made a too extensive application of *schemes* drawn from a number of observations. This latter proceeding led indeed to more important mistakes in the hands of some less careful followers. The task of critical examination could only be satisfactorily performed after more profound investigation, aided especially by complete experiments in artificial cultivation; and this has resulted in showing that, owing to the great number of the species, the differences in the course of development between the homologous and analogous terminal points which are often very important, and the frequent symbiotic relation social or otherwise between several species, the complications may be much greater than would appear at first sight and than can be expressed by one scheme. Various controversies also have arisen, as appears from the case of Pleospora described above, out of all these labours and efforts which are still far from their final conclusion. Much that belongs to this subject is only of interest in connection with the individual cases and must be referred to in the descriptive literature. We can here call attention only to the chief points of view, but it will be well first of all to enumerate briefly the chief phenomena and members of the development observed in the species above described.

1. From the ascospore is developed a thallus which only produces ascogenous sporocarps, or *archicarps* which produce these sporocarps together with

antheridial branches and sometimes *spermogonia* with *spermatia*, as in Pyronema, species of Ascobolus, and Collema. This may be called the **simple course of development** in the Ascomycetes.

2. The **dimorphous or pleomorphous course of development.** Similar to the simple one in the terminal points being represented by the ascospores, but *formations of gonidia are intercalated* between them. These formations make their appearance sometimes as a *transitory intermediate generation* (Polystigma), sometimes as *precursors* of the ascocarp on the same thallus, and capable under favourable conditions of uniform reproduction through an unlimited number of generations. Excellent examples are Erysiphe, Eurotium, Penicillium, Sclerotinia Fuckeliana. The gonidia are usually acrogenous, seldom intercalary also, in their abjunction, and are produced—

(*a*) On solitary simple sporophores, or sprouting cells.

(*b*) On the exposed upper surface of compound sporophores, as in Claviceps.

(*c*) In peculiar receptacles, pycnidia (pycnogonidia, pycnospores, 'stylospores').

Each species can only produce one of these gonidial forms, as is the case with Erysiphe, or under favourable conditions more than one, as Pleospora and Nectria.

In all cases that have not been thoroughly examined and are therefore more or less doubtful, an organ or member of the development must be determined and named according to the agreement of its observed characters with those of thoroughly known forms. The correctness of the naming will be more or less certain according to the degree of agreement, and will vary from the extreme of probability to entire uncertainty. The result of this examination of the separate parts and organs is as follows :—

SECTION LXVIII. 1. There is nothing to add here to what has been already said of the **archicarps** and **antheridial branches.**

2. The **sporocarps** (apothecia and perithecia) with the asci agree so entirely in the essential points of structure, development, and moment of appearance in the general course of the development, as they are known to us and have been described above, that, as has already been pointed out, they may or must be regarded as generally homologous in the sense and with the modifications above indicated. In by far the largest number of species, as far as our experience goes, they are the most constant members in each species in their structure and especially in the structure of the asci and ascospores. Exceptions to this rule, in which the number or size of the spores is strikingly unequal in different asci, are comparatively rare, and some instances have been mentioned above on page 79. Similar cases are recorded in Pleospora and some other forms. Calosphaeria biformis, Tul. and Cryptospora suffusa, Tul. are said to have two kinds of perithecia, one of which has asci with a large number of small spores, the other asci with from four to eight much larger spores[1]. How far this is really a case of difference within the same species, and not also of the mixing up of two similar or associated species, should be enquired into, and the investigation is rendered more necessary by the question which has arisen in the case of Pleospora noticed on page 230.

---

[1] Tulasne, Carpol. II.

SECTION LXIX.   3. **Spermatia, spermogonia.**   Organs in every respect extremely like those which are thus named in Collema, Physma, &c. (page 211) are found in almost all the rest of the Lichen-forming Ascomycetes; the genus Solorina may be mentioned as an exception among those in which this point has been carefully examined.   These organs occur also in many species which do not form Lichens both among the Discomycetes and especially among the Pyrenomycetes.   On the ground of these points of resemblance the organs in question are entitled to the names given to the corresponding organs of the Collemeae and Polystigma, and are at least to be regarded as homologous with them.

All these organs agree first of all in the formation of *spermatia*, small ellipsoid, or more commonly narrowly rod-shaped, bodies, which are often also bent, as in Rhytisma, Diatrype (Fig. 114), and Polystigma.   Their absolute size varies much in the different species; those that have the form of narrow rods are according to Tulasne 6 or 7 μ in length in species of Diatrype, and as much as 30 μ in Polystigma rubrum, or less than 6 μ in some species of Gyrophora (Fig. 100); the ellipsoid spermatia of Peltigera have a length of 12–22 μ.   Their structure, as far as it can be ascertained, is similar to that of very small and delicate spores with homogeneous

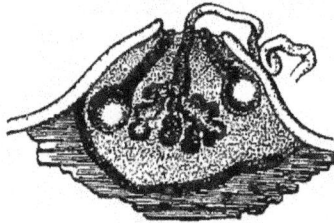

FIG. 112. *Valsa nivea*, Tul. Vertical section through a stroma; in the centre a spermogonium ejecting spermatia; on each side a perithecium.   After Tulasne. Slightly magnified.

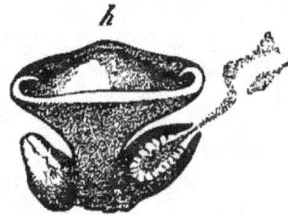

FIG. 113. *Tympanis conspersa*, Fr. *h* a shortly stalked apothecium with two spermogonia at its base, in median longitudinal section.   Spermatia are escaping from the spermogonium to the right.   After Tulasne. Slightly magnified.

protoplasm, and they are formed in the same way as acrogenously produced spores, being abjointed singly or in rows from short and narrow ends of filaments (sterigmata, basidia); the latter organs vary in the different species and genera, being either elongated and cylindrical, unsegmented or with indistinct septa, and forming spermatia at their apex only (sterigmata in the narrower sense of Nylander), or they are many-membered cell-rows in which the cells are little longer than broad, and form each of them lateral spermatia close to their upper end (Fig. 100 *B*, the arthrosterigmata of Nylander).   This latter form has been chiefly, if not exclusively, observed in certain genera of Lichen-fungi.

These spermatia are always formed side by side in large numbers, and are imbedded, as in Collema, in a jelly which becomes hard and brittle as it dries, and dissolves and disappears in a super-abundance of water.   If they are placed with the jelly in a comparatively large quantity of water, they exhibit a gently tremulous oscillating movement; but since this movement appears in spermatia which have been killed by boiling or by being treated with absolute alcohol as well as in those which are alive and fresh, it must be considered to be a purely physical phenomenon due to the motion which is caused in the water by the swelling and partial dissolution of the jelly, and which must necessarily be communicated to such small and light bodies.

With these characteristics the spermatia cannot be certainly distinguished from small spores. The distinction however is, that, like those of Collema or Polystigma, they are all, as far as has been hitherto observed, *incapable of germination*.

Secondly, these organs all agree in having the spermatiophores collected together into close hymenia in the *spermogonia*. These spermogonia are usually, as in Collema and Polystigma, hollow receptacles like perithecia sunk in the tissue of the thallus, with the cavity smooth and pitcher-shaped, or, as is very often the case, repeatedly and very irregularly folded into sinuous depressions and projections, so that where the folds are narrow the receptacle has the appearance in section of being divided into a

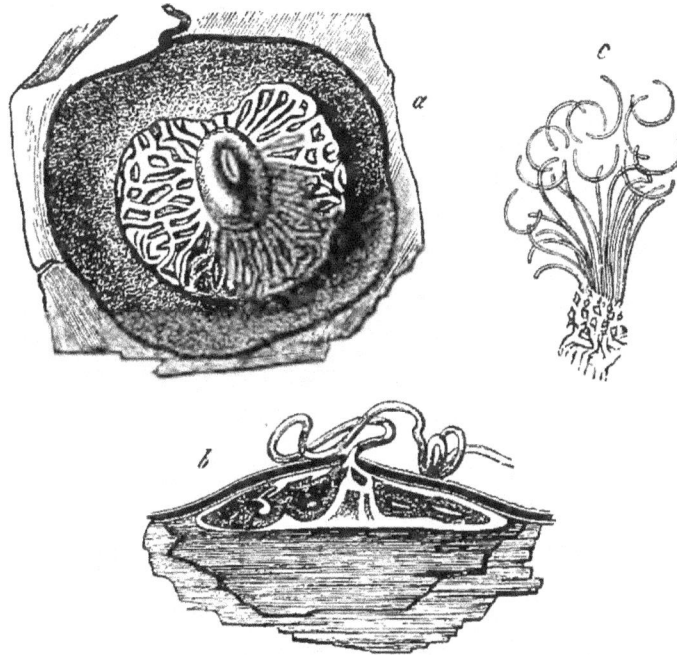

FIG. 114. *Diatrype quercina*, Fr. *a* a spermogonium on a piece of bark, laid open by removing the periderm. The conically pointed upper surface which is folded in coils bears the hymenium of the spermatia. *b* vertical longitudinal section through a spermogonium; a tendril-like mass of spermatia is issuing from an opening in the overlying periderm. *c* fragment of a thin section through the surface of the spermogonium with sickle-shaped spermatia and their sterigmata. After Tulasne. *a* and *b* slightly magnified, *c* magn. 360 times.

number of compartments. The cavity is everywhere lined with the hymenium which produces the spermatia, and the spermatia when mature are imbedded in jelly and occupy the centre of the cavity, and when the jelly swells in water they issue crowded together in drops or long strings from the narrow orifice of the spermogonium (Figs. 112, 113).

Some Pyrenomycetes which live in the rind of trees form layers agreeing in every respect with the spermatiophores, except that they are not inclosed in receptacles altogether belonging to the Fungus; on the contrary they are disk-shaped or cushion-shaped bodies with the spermatiogenous surface folded into deep sinuous depressions, as in species of Diatrype (Fig. 114), Quaternaria, and Stictosphaeria, Tul., or else smooth as in Calosphaeria princeps, Tul., and in both cases covered only by the peripheral layers of the rind. The spermatia escape through a narrow fissure in the rind, usually formed above a conical projection in the Fungus-body.   Since the agreement is otherwise so complete we may certainly consider these bodies as spermogonia which have no outer wall of their own, but are covered over instead by the rind of the tree which they inhabit.   But this affords reason for a further concession and allows us to give the name of *open spermogonia* to such cushion-shaped or club-shaped bodies formed on the surface of the substratum as Tulasne has described in Bulgaria sarcoides and Peziza fusarioides; for the outer surface of these bodies is covered with a hymenium, which, together with its products, behaves in the same way as the hymenium of the closed spermogonia previously mentioned.

The third and last point of agreement to be observed in all the formations of which we are speaking lies in their relation in place and in the time of their development to the production of the ascocarps.   In all cases which have been fully investigated we find the same arrangement as in Collema and Polystigma; the formation of spermogonia and spermatia always precedes that of the sporocarps or coincides with the first appearance of their primordia.   At the same time the formation of spermatia may continue beyond the period of the orientation of the sporocarps, and both kinds of organs may be repeated more than once on a long-lived thallus; but this makes no essential difference.   The two kinds of organs usually occur close to one another on the same thallus; dioecious distribution, to which attention has been called above in the case of Collema, has been observed a few times in some Lichen-fungi (Spilonema, Bornet, Ephebe pubescens).

We do not certainly know the true *function* of all the bodies which are spoken of in this section as spermogonia and spermatia.   What we are able to conjecture on the subject may be gathered from previous sections, and will be considered also below in section LXXIV.

SECTION LXX.   We have insisted in the foregoing remarks on the invariably small size of the spermatia, and on their simplicity of structure and incapacity of germination, or, to speak more correctly, the absence of observed capacity of germination; and these conditions make it difficult to determine a number of other cases, which may for the present be placed together under the head of **doubtful spermatia.** We learn from a series of observations that there are small rod-shaped or spherical cells in the Ascomycetes, which have all the known positive and negative characters of spermatia, but are abscised at other places in the thallus than in or on distinct spermogonia.

Firstly, such cells are said to occur in the sporocarps themselves, between or near the asci.   Gibelli[1] states that many Verrucarieae, especially those with simple spores and without paraphyses in the hymenium, have no proper spermogonia, but that the lower portion of the perithecium is covered with asci, the upper with spermatia-forming sterigmata; but other observers[2] do not corroborate this statement.   We learn from Tulasne that slender branched hyphae, from which countless small rod-shaped 'spermatia' are abscised, are found between the asci at the places where paraphyses otherwise occur in some but not all the apothecia in

---

[1] Sngli org. reprod. del gen. Verrucaria (Mem. soc. ital. di sc. nat. I).
[2] Stahl, Beitr. z. Entw. d. Flechten, I, p. 40.

his Peziza benesuada (Fig. 115); similar organs occupy the margin of the platter-shaped tube-bearing hymenia of Cenangium Frangulae, Tul. Small round cells incapable of germination, which will be noticed again in a subsequent page, are said by Brefeld[1] to be sometimes abscised from the ramifications of the paraphyses in Peziza Sclerotiorum.

The second place where these doubtful 'spermatia' occur is in the pycnidia of certain species, in which spores as well as spermatia are produced; such species, according to Tulasne, are Cenangium Fraxini, Tul., Dermatea carpinea, Fr., D. Coryli, Tul., D. dissepta, Tul., where the spermatia-forming hyphae also occupy chiefly the margin of the hymenium, also in D. amoena, Tul., Peziza arduennensis and Aglaospora.

Thirdly, small short-lived cells, which do not germinate and may be compared with spermatia, are abscised in many species from filiform branchlets of the mycelium and from the germ-tubes, or even directly from the germinating spores.

Brefeld[2] found a multitude of such formations on the mycelium of artificially grown plants of Peziza (Sclerotinia) tuberosa. From short branches, often with tufts of branchlets as in Penicillium, are abjointed successively and serially at the extremities of their ramifications small cells, each containing a small sphere of a highly refringent perhaps fatty substance, and these are cemented together by a jelly and thus collected in heaps on the parent-filaments. Tulasne[3] found just such formations on the germ-tubes of the same species and on those of Peziza bolaris and P. Durieuana when the spores were sown in water. A similar phenomenon occurs sometimes on old cultures of the mycelium of P. Sclerotiorum, as Brefeld states and I can myself confirm; but, as far as my experience goes, only in isolated cases which cannot be more precisely defined. I observed it also on the young germ-tubes of this species, but only in a few and these poor and evidently weakly plants. The small cells mentioned above which are abscised

FIG. 115. From the hymenium of *Peziza benesuada*, Tul. Ascus surrounded by paraphyses which are giving off spermatia by abscision. After Tulasne. Highly magnified.

in the cups of P. Sclerotiorum are similar, according to Brefeld, to those which we are now describing. The same formations appear also not unfrequently on old luxuriant mycelia of P. Fuckeliana grown from ascospores in fruit-juice on a microscopical slide (Fig. 116); and Zopf found quite similar structures, the narrowly flask-shaped sterigmata, singly or in a tuft according to the luxuriance of the individual, on the mycelium of species of Chaetomium, especially on starved specimens, and also in species of Sordaria (S. curvula, S. minuta, S. decipiens), Woronin having seen them before in S. coprophila.

Small bodies of the kind here described sprout out directly from the cells of the multicellular compound spores of Tulasne's[4] Peziza Cylichnium when sown in water.

---

[1] Schimmelpilze, IV, p. 121.
[2] Schimmelpilze, IV, p. 113.
[3] Ann. d. sc. nat. sér. 3, XX, p. 174, and Carpol. III, t. XXII.
[4] Ann. d. sc. nat. sér 3, XX, and Carpologia, III, pp. 200, 202.

The small rod-like cells which sprout from the cells of the spores of Nectria inaurata and N. Lamyi [1] while still inside the ascus, filling it quite full and giving rise to strange misunderstandings, may also be mentioned in this place, though it is not very probable that they are of the same significance. The point of agreement between all these forms lies in their outward resemblance and in the absence of any certain knowledge as to their morphological and physiological value.

SECTION LXXI. 4. **Gonidia.** The course of development in the few forms mentioned above on page 238, 1, is shown with certainty by our observations up to the present time to be that which is there termed simple; and almost all Lichen-fungi also are without gonidia unless we count among them the *soredia*, which will be described in section CXVII, as there is certainly good reason for doing; other gonidial formations are described in a few species only, as exceptional cases therefore, and in these are not beyond doubt.

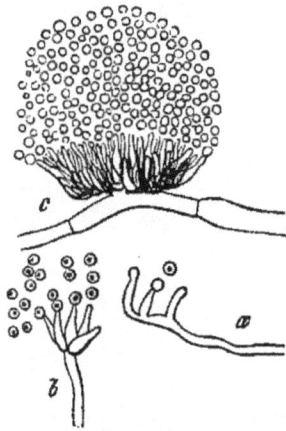

The course of development in the larger part of the Ascomycetes with which we are acquainted, and especially in the Pyrenomycetes, is pleomorphous with copious production of gonidia of more than one form. All the gonidia are unicellular or pluricellular compound spores formed by acrogenous or intercalary abjunction, as in the examples which have just been described. Anatomical investigation and observation of different portions of the development show that they usually appear as precursors of the ascocarps, whether their development comes to an end when the formation or at least the completion of these begins, or they make their first appearance before the latter but continue to develope simultaneously with them. Claviceps, which has already been described, is an excellent example of the first case, for gonidia and perithecia follow one another in that genus in successive periods of vegetation. The development of species of Stigmatea according to Tulasne [2], and probably of some other small Pyrenomycetes that live in leaves, follows a similar course, but without forming sclerotia; and this is the case also with Epichloe, which was described in a former page, and with Tulasne's Xylarieae (Xylaria, Poronia, Ustulina, Hypoxylon) and some species of Nectria, especially N. cinnabarina (= Tubercularia vulgaris, P.), which all behave in a similar manner to Epichloe. The compound sporophores of these forms are at first covered by a hymenium which produces gonidia, but this ceases to grow and is cast off as soon as the development of the perithecia formed within its plane of insertion begins to advance.

A second case is exemplified in the Erysipheae mentioned above, in Fumago salicina [3], Cucurbitaria macrospora (Fig. 117), Pleospora polytrichum, P. Clavariarum [4],

FIG. 116. *Peziza Fuckeliana.* From a specimen grown on a microscopic slide in grape juice examined under water after being treated with alcohol. *a* three young sterigmata forming 'spermatia' at the top of a branch of the mycelium, the abjunction shown plainly ou the middle one. *b* group of five sterigmata before all the spermatia are shed. *c* view in profile of a stout dense tuft of sterigmata springing from more than one mycelial filament, the apex of which is formed of a mass (merely outlined) of spermatia abscised and imbedded in jelly. Magn. 375 times.

---

[1] See Janowitsch in Bot. Ztg. 1865, p. 149.
[2] Carpol. II.
[3] Tulasne, Carpol. II.
[4] Tulasne, Carpol. II.

and some others. Ripe and ripening perithecia may in these plants stand side by side on the same mycelium or stroma with developing gonidiophores.

It is scarcely necessary to remark that the succession of events does not always proceed in these examples with perfect regularity; and the instance of Peziza Fuckeliana described above on page 224 shows that considerable deviations from it may and do occur.

As regards the *point of origin of the gonidiophores* and their *arrangement* and *structure*, it would seem that they occur in particular species in the ascocarps themselves, like the doubtful spermatia noticed on page 242. According to Berkeley[1] single paraphyses are found between the asci in Sphaeria oblitescens B. et Br., in which one or two of the cells are enlarged into somewhat elongated septate 'spores;' the terminal cells of such paraphyses in Dothidea Zollingeri, Berk.[2] are like simple ellipsoid spores. Berkeley[3] makes a similar statement in the case of a species of Tympanis and for Lecidea Sabuletorum[4] or an allied form; but these points require re-examination, as Tulasne has intimated[5], because we are still ignorant of the qualities of these *spore-like* bodies.

Putting aside these few doubtful and possibly exceptional cases, all gonidial formations conform to the examples described above which have been thoroughly examined. The following special forms may be enumerated:—

(a) *Free filiform gonidiophores*; often very characteristic in their conformation, as in Penicillium, Eurotium, Erysiphe, &c., and in such cases formerly assigned to established form-genera. Thus species of Hypomyces were assigned to the form-genera Verticillium, Sepodonium, and Mycogone of the old descriptions, and Fusisporium Solani[6]; species of Nectria to Fusisporium and Spicaria of the old descriptions, and many other cases might be cited. To these may be added some other forms, in which the distinction between gonidia and gonidiophores on the one hand and portions of the mycelium on the other is less sharply defined, and may even be arbitrary in each individual instance up to the extreme cases in which each cell of a hypha or a hyphal strand first performs the part of a mycelium and then assumes the characters of a spore. The latter is in extreme cases naturally termed the *formation of resting mycelium* and has been elaborately studied by Bauke and Zopf, especially in saprophytic Pyrenomycetes, though older observers often mentioned it incidentally. It occurs in old and specially in starved mycelia for example of Pleospora, Fumago, and Cucurbitaria, in which the cells of the mycelium acquire thick and usually brown walls, store up reserve food material, and pass into a dormant state, and subsequently under suitable conditions germinate as spores. Changes of form, especially swelling of the individual cells into a spherical shape, may or may not accompany the changes which characterise the state of rest, and hence the resting states differ in very various degrees from the vegetative mycelial forms.

---

[1] Ann. Mag. Nat. hist. ser. 3, III, p. 373, t. XI, 32.
[2] Hooker's Journ. III (1844), p. 336.
[3] Introd. Crypt. Bot. p. 244.
[4] See Ann. Mag. Nat. Hist. ser. 2, IX, and Crypt. Bot. p. 391.
[5] Mém. s. les Lichens, p. 110.
[6] Reinke u. Berthold, Die Zersetzung d. Kartoffeln durch Pilze, 1879.

(*b*) *Dense hymenia giving off gonidia by abscision on the free outer surface of compound sporophores.* Examples of this kind are Claviceps (page 227), Epichloe, the Nectrieae before mentioned, Xylarieae (Fig. 103 *A*), Cucurbitaria macrospora (Fig. 117), and many others. The form of the separate gonidiophores which together constitute the hymenium, the special mode of abjunction of the gonidia, and the structure and form of the gonidia themselves, all vary extremely according to the species. And again it depends on the species whether the formation of gonidia is entirely, or almost entirely, confined to these hymenia or to the stromata which bear them, as is the case in Nectria cinnabarina and the other genera last named, or whether gonidia-forming hyphae of like structure occur either united into hymenia or appearing singly on a filamentous mycelium as in the Hyphomycetes, as happens in Nectria Solani and Hypomyces Solani[1].

Whether we have always to do with gonidia in the cases which have been given as examples, especially in the Xylarieae, or sometimes also with non-germinating spermatia, is often uncertain and must be determined in each separate case.

(*c*) *Pycnidia:* receptacles (conceptacles) of more or less similar character to those described in Pleospora, and producing gonidia which are known as pycnospores or pycnogonidia or more commonly as *stylospores.* They are wanting in many or most species of Ascomycetes, in all forms, for instance, mentioned under *b* and in most of those mentioned under *a*, and in almost all Lichen-fungi. They were said indeed to have been found by Lindsay in Bryopogon jubatus, Kbr., Imbricaria saxatilis and I. sinuosa, Kbr.; by Gibelli in 'Verrucaria carpinea, Pers.,' Sagedia carpinea, Mass., S. Zizyphi, Mass., S. callopisma, Mass., S. Thuretii, Kbr., Pyrenula minuta, Näg., P. olivacea, Pers., Verrucaria Gibelliana, Garov.; by Füisting in Opegrapha varia, Acrocordia

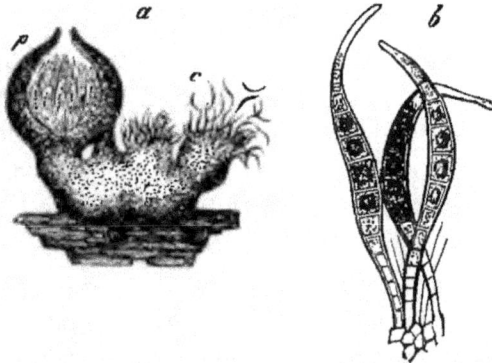

FIG. 117. *Cucurbitaria macrospora*, Ces. and de Not. *a* stroma. in longitudinal section ; *p* developed perithecium, *c* layer of gonidia. *b* gonidia on the gonidiophores. After Tulasne. *a* slightly magnified, *b* magn. about 200 times.

gemmata, Mass., A. tersa, Sagedia netrospora, Hepp., and S. aenea. Lindsay's account also of two kinds of spermogonia in Roccella Montagnei, Bel. and Opegrapha vulgata, Ach. may be mentioned here since some of the receptacles which he termed spermogonia may be pycnidia. But in all these cases we know so little concerning the development of the organs in question that it is still uncertain whether they belong to the species named above or to parasites living in their thallus.

The pycnidia, like the perithecia, are according to the species either formed singly from the filamentous mycelium, or are placed in or on compound pycnidiophores (stromata), as in Cucurbitaria Laburni, Dothidea Melanops, &c.[2] Their development proceeds, in several cases that have been observed, in the manner

---

[1] Reinke u. Berthold, Die Zersetzung d. Kartoffeln durch Pilze, 1879.
[2] Tulasne, Carpol. II.

described above on page 229, in the case of Pleospora; an intercalary portion of a mycelial filament grows by successive divisions which arise without fixed order in every direction, and the cells thus formed are subsequently differentiated, while branches from adjoining hyphae usually grow up round the new body and thus help to form its wall (see Fig. 118). This is the mode of formation according to Gibelli and Griffini, Eidam and Bauke not only in Pleospora herbarum, but also in Cucurbitaria elongata, Leptosphaeria doliolum and two other species not precisely determined, and according

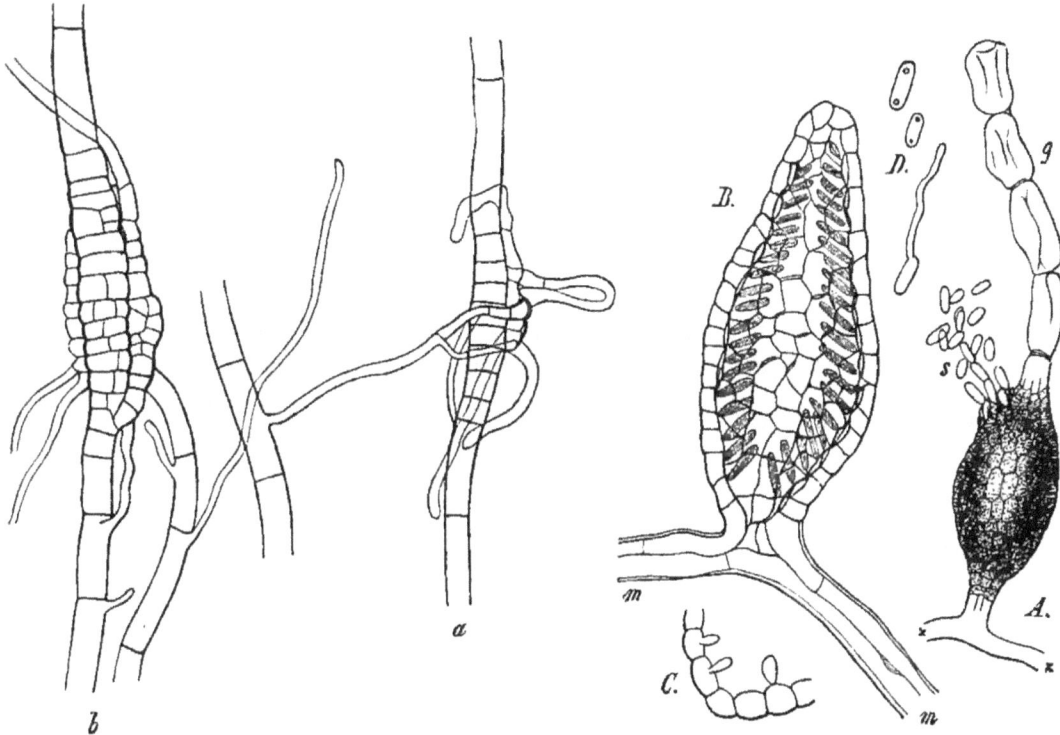

FIG. 118. *Pleospora Alternariae*, Gibelli. (Determination not certain from the absence of perithecia.) Young stage of development of pycnidia. *a* commencement of the swelling and rapid transverse division of the intercalary portion of a hypha, which is developing into a pycnidium and has branches from itself and from an adjacent hypha attached to it. *b* older stage of the development. The mature structure of these pycoidia closely resembles that represented in Fig. 119, only the wall is formed of several layers. Magn. 600 times.

FIG. 119. *Cicinnobolus Cesatii* (De Bary, Beitr. III), parasitic on *Erysiphe*. A ripe pycnidium (seen from without) open above oo the left and discharging its spores *s*, having developed in a gonidiophore of the *Erysiphe* which is attached to the mycelial filament x x and bears four dead gonidia *g* oo its summit. *B* a small and nearly ripe pycnidium, formed on a branch of a mycelial hypha *m m* of the *Erysiphe*, in which the slender mycelial hyphae of the *Cicinnobolus* may be seen. The figure shows the upper surface and the optical longitudinal section of the transparent peridium; the section shows the young spores growing inwards from the one-layered wall. *C* transverse section through the wall of a ripe pycnidium with three primordial spores sprouting inwards. *D* two ripe spores just discharged from the pycnidium and a germinating spore. *A* magn. 380, *B*, *C* 600, *D* 300 times.

to Zopf in some pycnidia of Fumago; Brefeld's 'Pycnis sclerotivora' must also be added to the number. Other pycnidia are not meristogenetic but symphyogenetic formations, that is, they are produced by union and interweaving of hyphal branches; such are those known under the name of Cicinnobolus, some in the genus Fumago, and the Diplodia-form examined by Bauke. The formation of the pycnidia in Pleospora polytricha, according to Bauke, is of an intermediate kind, the inner portion being meristogenetic and the numerous outer wall-layers symphyogenetic. The structure of pycnidia formed in these different ways may be quite alike in the matured state, as is seen by comparing Cicinnobolus (Fig. 119) and Pleospora.

The shape of the pycnidia is in general the same as that of perithecia or spermogonia, and the internal cavity, like that of the spermogonia, is according to the species either simple or divided by projections from the wall into usually irregular narrow compartments communicating towards the orifice. The pycnospores of the different species exhibit the usual variety of modifications in the structure of spores. Two extreme forms may if necessary be distinguished: *small-spored* pycnidia (corresponding to the old form-genus Phoma) with very small, colourless, somewhat elongated spores, which are imbedded in mucilage and are discharged in masses from the orifice of the pycnidium, as in Pleospora and Cucurbitaria elongata, and *large-spored* pycnidia with comparatively large either simple or compound spores, the walls of which are often thick and of a brown colour.

Section LXXII. In the species of which we are now speaking, as well as in those previously described and which have been thoroughly examined, the different forms of **gonidia and gonidiophores occur according to the species in the greatest possible variety of combinations with the perithecia and sometimes with one another.** Examples of this, such as may be considered to be well ascertained, are to be found in the works here quoted, and especially in Tulasne's Carpologia. Some of the accounts contained in this book, and still more those in the more recent descriptive writings, must be accepted with caution. We may here call attention more distinctly to the fact, which may be gathered indeed from some of the previous remarks, that each species has the faculty of forming asci and gonidia within broader or narrower limits as the result of its inherent inherited qualities; external causes, especially the quantity and quality of the food at its disposition, then determine in a variety of ways the phenomena which are actually observed. A few examples will now be given in illustration especially of the latter point, which will be further discussed in section LXXIII.

The genera **Ustulina, Poronia,** and **Hypoxylon** among the Xylarieae may be considered, as far as we at present know, to be plants of one form, since, like Epichloe (and Claviceps), they produce gonidia of *one* definite form on the young stroma, and then perithecia.

Cucurbitaria Laburni[1] forms in the dead rind of Cytisus Laburnum large flat roundish cushion-like stromata which reach a breadth of some millimetres and finally issue from the ruptured periderm covered with numerous black round spore-receptacles. Some of these are perithecia, some gonidial receptacles, pycnidia, with a single cavity and a narrow canal at the orifice; a stroma according to Tulasne may bear only perithecia or only pycnidia, but usually has both and more than one kind of pycnidium. In the latter case the receptacles make their appearance on the stroma, which increases in size while they are forming, in *about* the following order in time and centrifugal succession:—

1. In the middle of the stroma one or a few comparatively large colourless pycnidia, producing colourless thin-walled unsegmented cylindrico-elliptical spores 5–10 μ in length, on short sterigmata.

2. Numerous pycnidia with thick black walls in which are abscised from short sterigmata

   (*a*) Colourless spores of very unequal size,

---

[1] Tulasne, Carpol. II, p. 215, t. XXVII.

(*b*) Spores with dark brown walls, nonseptate or with one transverse wall, closely resembling the spores of No. 1 in form and size,

(*c*) Brown spores like the last but pluricellular compound, 20–30 μ in length and 7–10 μ in breadth.

Each of these spore-forms generally occurs separately in a special receptacle, so that four kinds of pycnidia may be distinguished; but combinations are also found especially of (*a*) and (*c*) in the same receptacle.

3. The perithecia.

The formation of germ-tubes has been observed in all the forms of spores produced in the pycnidia. No other gonidial forms are found in Cucurbitaria Laburni.

Species of **Hypomyces** are characteristic examples of the regular formation of two kinds of gonidia; these plants, like Hyphomycetes, live on the larger Fungi, especially the Hymenomycetes, Hypomyces rosellus for instance, H. chrysospermus, and other species[1], or on parts of dead plants, as H. Solani on rotten potatos[2]. Besides the perithecia which are comparatively rare, and are always later in their appearance, the mycelium produces

(1) Microgonidia, comparatively thin-walled and colourless, but tolerably large, ellipsoid, cylindrical, or fusiform unicellular or compoundly pluricellular spores, which are abjointed successively and form small heads at the extremities of the ramifications of verticillately or irregularly branched gonidiophores, and were assigned to the old form-genera Verticillium, Dactylium, Fusisporium, and others.

(2) Megalogonidia or macrogonidia, sometimes also called chlamydospores, acrogenously formed and solitary or more rarely appearing a few together one behind another on branches of the hyphae which produce the microgonidia; their formation generally begins later than that of the microgonidia, from which they are distinguished by their thick membranes, which are often rough with warts and usually coloured, and in most species by their greater size; they are unicellular or compound pluricellular according to the species. The thickness of their membranes shows that they are adapted to a persistent state of rest.

Reinke and Berthold have shown that in Hypomyces Solani mycelia are formed from the germ-tubes of both kinds of gonidia as well as of the ascospores, which can again produce both kinds of gonidiophores, and Tulasne's less complete account agrees with theirs. More complete knowledge of the general course of development has not yet been obtained.

Zopf's **Fumago**[3] may be adduced here in conclusion as an example of a cycle of forms still more copious and varied than that of Pleospora or Nectria ditissima which have already been described. Though Zopf found no ascocarps in his Fungus, yet I place it here with the Ascomycetes, because according to Tulasne[4] the very similar species Fumago salicina has ascocarps, and because we are specially considering at the present moment the Ascomycetes which are not yet quite perfectly known. The species of Fumago are the soot-dew which is found in the form of black fuliginous coatings covering parts of living plants. Zopf studied his plants chiefly in pure cultures on microscopic slides in nutrient saccharine solutions of various degrees of concentration, and ascertained the agreement of the cultivated forms with those which occur in nature.

1. The mycelium of the Fungus is composed of hyphae with short segments, which, like the cells of the gonidiophores, acquire usually at an early period a brown coloration of the walls accompanied by a gelatinisation of an outer colourless layer, while the

---

[1] See Tulasne, Carpol. III.

[2] Reinke u. Berthold, as cited on page 245.

[3] Die Conidienfrüchte von Fumago (N. Act. Leopold. XL).

[4] Çarpol. II.

cell-contents develope much fatty matter. The gonidia are formed acrogenously on distinct gonidiophores, or in gonidial receptacles, and may be termed *acrogonidia*. They are when first matured small thin-walled colourless ellipsoidal cells with a gelatinous outer covering and a small drop of fatty matter in the focus; in size they are 4–5 μ in length and 2 μ in breadth, and are formed in a variety of ways.

(*a*) In specimens grown in a poor solution (containing at most five per cent. of nutrient matter) and kept as dry as possible the poorly developed mycelium produces small slender filiform erect gonidiophores composed of a few cells, which form their acrogonidia successively like heads at their apex, or in whorls beneath the upper end of their subterminal cells (Zopf's formation of microgonidia).

(*b*) Stronger mycelia are developed with a more generous supply of food, and these produce erect tufts of many-celled hyphal branches on which acrogonidia are formed. From 2–12 of these branches grow close together, at first closely parallel to one another, but afterwards diverging at an acute angle, and may be nearly a millimetre in height; the branches in a tuft are of a nearly uniform height. The lower cells of each hypha are elongated and cylindrical, the upper short, scarcely longer than broad, and from them proceed other branches with short cells usually placed on one side only of the cell from which they spring, and rising in the same direction and to nearly the same height as the primary branch, thus resembling to some extent the tuft of branchlets at the extremity of the gonidiophore of Penicillium. From the short cells of all these tufts of branchlets acrogonidia are abscised, terminal ones at their apex, the rest near their upper bounding wall, and all usually on one side and in the same direction.

(*c*) The tufts of gonidiophores formed on the mycelium in the way described under division (*b*) may be firmly united into a bundle along their whole length, while their other characters remain the same. This bundle is at first nearly cylindrical, but its apex spreads out into the shape of a funnel as the terminal ramifications are being formed which will produce gonidia, and their extremities spread out from one another in a penicillate manner; the abscision takes place only inside the funnel-shaped enlargement, the outer side remains sterile, and the pointed barren extremities of its hyphae extend with a slight divergence a short distance above the tuft from which the gonidia are abscised.

(*d*) The sterile extremities of these hyphae may unite laterally and firmly into a narrowly conical tube open at the top, and grow out far beyond the region which supplies the gonidia; in other words they form a symphyogenetic gonidial receptacle, a more or less elongated flask-shaped pycnidium. The ventral portion of the flask is the region from which the gonidia are abscised, and Zopf found that the process was always carried on from the cells of the wall which continues to be of one layer, and not from other hyphal branches which project into the interior of the tube.

(*e*) Lastly, pycnidia of essentially the same definitive structure as those in (*d*) only of roundish less elongated form, with a wall that is usually two-layered, may also be formed meristogenetically.

Intermediate forms are found, as might have been expected, between the formations described above from (*c*) to (*e*) inclusively.

2. The acrogonidia when sown in a dilute (five per cent.) nutrient solution sprout and form successive roundish ellipsoid sprout-cells like those of Saccharomyces Cerevisiae [1] if the supply of atmospheric air is restricted; with free admission of air the sprouts are frequently elongated cylindrical shoots (the 'Chalara-' and Mycoderma-form).

3. All the parts and forms of the Fungus which have now been described may pass, if the supply of food diminishes slowly, into *resting states* under a great variety of particular forms, while the cells swell, acquire a brown colour and store up fatty

---

[1] But they incite no alcoholic fermentation.

substances ; thus we get *resting gonidia, resting gemmae,* and *resting mycelia,* the latter often appearing as torulose filaments or crust-like masses. All the *gonidial and resting forms* are capable of germination under favourable conditions, and *all the forms included in divisions* 1, 2, *and* 3 *may be changed into one another if the conditions of growth are suitably varied.*

We must not enter here into the question whether Fumago has other organs of propagation besides the perithecia which are said to occur and those which may possibly occur, though it is one which may very well be asked after Tulasne's account of F. salicina[1].

SECTION LXXIII. It is almost to be expected in the case of **pleomorphous species of Ascomycetes,** that only separate members of their form-cycle should often be found on a substratum at any given time, whether ascocarps or gonidiophores or gonidial receptacles. The frequency of this occurrence in a particular species will depend on the ease with which it spreads as a rule over different kinds of substrata under a great variety of external conditions, and on the other hand, also on the strictness with which it is tied to special conditions of vegetation, in order to produce the particular member which closes the cycle. Examples are seen in the species of Fumago, Pleospora, and Penicillium, and in Sclerotinia Fuckeliana which have been described above ; the mycelia of these Fungi bearing simple gonidiophores are of universal occurrence in the form of Moulds, and they are propagated in the same form, while the sporocarps are much more rarely found ; the conditions under which they occur in the first two genera are not yet precisely ascertained[2]; in Penicillium they are produced on plants artificially grown on bread and spontaneously on the skins of grapes that have been pressed for wine, and in Sclerotinia Fuckeliana only on sclerotia which have developed and arrived at a certain degree of maturity on particular foliage-leaves (Vitis, Castanea, Quercus); a large number even of the sclerotia of this species produce only new gonidia, as is especially the case with those which are so common on dead cabbage-stalks, the Sclerotium durum of old writers. The majority of known species which form gonidia behave in a similar manner ; the converse proceeding, a comparatively copious formation of sporocarps with scanty production of gonidia, is comparatively rare, though it does sometimes occur, as in Melanospora parasitica. Many or indeed most of the gonidial forms of species that are now better known were for these reasons described as form-species, before their genetic relations had been ascertained, and were distributed into corresponding groups, the pycnidia and spermogonia being arranged in the Sphaeropsideae, Cytisporeae, and Phyllosticteae[3], the simple hyphal gonidiophores and open hymenial layers in the Hyphomycetes, Haplomycetes, and Gymnomycetes of Fries. Proof of this statement is to be found in the special descriptive literature, to which reference here is unnecessary, since the historical facts are of the same kind as those to which attention was called above in the case of the Mucorini, Peronosporeae, and other forms.

Another fact which was noticed in the description of those groups is also repeated in the Ascomycetes, namely, that there are forms which strongly resemble the members of the development of thoroughly well-known species, some even

---

[1] Tulasne, Carpol. II.   [2] Ibid.   [3] Fries, Summa Veget. Scand. II.

exhibiting the same comparatively minute specific distinctions, but in which the formation of an ascocarp such as belongs to the particular development has never been observed, while at the same time there is no reason for considering that they belong to any group outside the Ascomycetes. We are compelled by this condition of our knowledge to regard these isolated forms as homologous with those which are like them and the position of which is known in the course of development of other species, and to call them accordingly spermogonia, gonidiophores, pycnidia, or the like; it is true that this practice is founded only on probabilities, but it has already found its justification in many cases in the strict proofs which have been subsequently obtained. Most of the Haplomycetes, Gymnomycetes, Sphaeropsideae, &c. of the old systems, one might indeed say all of them that do not belong to the groups described in the previous sections, fall in this way and in accordance with the present state of our knowledge into the class of Ascomycetes, some connecting immediately with well-known ascomycetous species, others through the forms first mentioned; the grouping of the very large number of species and particular forms is necessarily attended with practical difficulties of a different kind from those which are met with in dealing with the few dozen Mucorini or Peronosporeae.

While referring the reader to the descriptive literature of this subject, it may be well to mention a few names in illustration of the above remarks; most species for instance of the old form-genera Naemaspora, Cytispora, Libertella, Septoria, Lepto-thyrium, Phyllosticta, Cheilaria, Gloeosporium, Spilosphaeria, Ascochyta, Phoma, Diplodia, Myxocyclus, Hendersonia, Sporocadus, Sphaeropsis, Cicinnobolus, Ehr. and some others must be classed with the pycnidia and partly also with the spermogonia; species of the form-genera Cylindrosporium, Oidium, Dematium, Conoplea, Periconia, Cladosporium, Helminthosporium, Macrosporium, Dendryphium, Mystrosporium, Brachycladium, Sepedonium, Mycogone, Aspergillus, Verticillium, Polyactis, Botrytis, Fusisporium, Alternaria, Torula, Isaria, Stilbum, Atractium, Graphium, Melanconium, Stilbospora, Steganosporium, Coryneum, Exosporium, Vermicularia, Tubercularia, Sphacelia, and many others, whose affinity to undoubtedly typical Ascomycetes is either certain or very probable, go with the filamentous gonidiophores and open gonidia-bearing hymenia. To these may be united with the needful reservation a large number of forms, in which the mycelium and the formation of spores, which must be considered homologous with the gonidia, are all that is at present known. Some of these forms do actually belong to the above form-genera, for the determination of these genera was made to rest on certain particulars of conformation which in some cases appear to our present knowledge to have been very superficially examined, and which, as we have since learnt, may occur in very various genetic connections. For instance Oidium leuco-conium, Desm. and O. erysyphoides, Fr. are names for the gonidiophores of Erysipheae. Oidium fructigenum, Kze. and O. lactis, Fres. are somewhat similar forms which do not belong to the Erysipheae and whose genetic affinities are quite unknown; Botrytis cinerea is the name of Sclerotinia Fuckeliana when it produces gonidia; B. Bassii denotes an isolated gonidial form by no means closely related to the last-named Botrytis. Other forms of this series are so widely different from those named above that the old describers gave them distinct generic names; thus they called their Hyphomycetous forms Arthrobotrys, Gonatobotrys, Haplotrichum, Cephalothecium, Stysanus, &c. &c.

These forms are for the present arranged with the Ascomycetes, because from what we know of them they appear to have more connection with that division than with other Fungi; but they are only known to us under one form, which may be considered to be that in which they produce gonidia.

It may be affirmed of the majority of the species which have just been considered, that they are imperfectly known, because no attempt has been made to ascertain the entire course of their development. But some among them have not only been observed in the mature state or occasionally grown from the spores, but have been repeatedly submitted to long and careful observation and cultivation; and yet they are only known to produce the same supposed gonidial forms, and there is no sign of ascocarps or of any other member of the development, which the analogy of very similar forms in well-known species would have led us to expect from them. The large species, Aspergillus clavatus[1] for instance, has never been known to produce anything but gonidiophores; the sporocarps like those of Eurotium and Penicillium, which were to be expected from the structure of the gonidiophores and of the mycelium, never made their appearance either in Wilhelm's many experiments in the artificial cultivation of the plant, which were made for the purpose of determining this point and were conducted under a variety of conditions, nor in many others which I have myself often repeated in the course of years. Botrytis Bassii[2] is a very common insect-destroying Fungus, which in the character of its vegetation is very like Cordyceps militaris, but resembles another Pyrenomycete, Hypocrea rufa[3], in the way in which it forms its exposed gonidia; hundreds of cultivated specimens of the plant have produced only the same organs bearing gonidia, never a sign of perithecia; my conjecture[4], which has become an assertion in Brefeld[5], with regard to the latter has proved therefore to be incorrect. The same must be said now of another insect-killing form which agrees very closely with Cordyceps militaris in the mode also of forming its gonidia, and which I have described under the name of Isaria strigosa[6]. Another instance which may be noticed here is the universally distributed and repeatedly cultivated Oidium lactis; this plant never produces anything but the mycelium with cylindrical serially abjointed gonidia[7]. The common Cladosporium herbarum, Lk. also should not be forgotten in this connection. Further instances of this kind have been discovered in the course of the investigations which have been made into the pycnidia. I refer to Zopf's account of Fumago of which a *résumé* has just been given. Brefeld[8] cultivated a pycnidia-bearing form, a not uncommon parasite on the sclerotia of Sclerotinia, under very varied conditions through more than a hundred successive generations, without ever obtaining anything but the pycnidium-form. Similar results are recorded for other species in Bauke's work on pycnidia; Ehrenberg's Cicinnobolus, a parasite on Erysipheae[9], may also be quoted here, and it may be added that the pycnidial forms mentioned here are so like others which undoubtedly belong to typical Ascomycetes, that they may be readily mistaken for them.

In view of these facts of experience the question again arises which was discussed above in connection with the Mucorini and Peronosporeae whether we have before us species which are only imperfectly known to us, and which under

---

[1] Desmazières in Ann. d. sc. nat. sér. 2 (1834), II, tab. II Fig. 4.—K. Wilhelm, Dissert. p. 62.

[2] See Bot. Ztg. 1867.

[3] See Tulasne, Carpol. III.

[4] Bot. Ztg. 1869, p. 590.

[5] Schimmelpilze, IV, p. 136.

[6] Bot. Ztg. 1869, p. 590.

[7] See above at p. 67.

[8] Schimmelpilze, IV, 122.

[9] Beitr. III, and above, Fig. 119.

certain conditions do really fill up the gaps in our knowledge, that is, are able to produce the typical ascocarp of an Ascomycete; or whether there are species which in their known characters do come near to typical Ascomycetous genera and may even be included in them, but do not at present actually produce the ascocarp of an Ascomycete. In the latter case a further question arises, whether and how far the problematic defect comes from the loss of the capacity or from its having never been possessed. The attempt to examine and answer these questions leads us necessarily into the domain of conjecture, and forces upon us the reservations here proposed. Every new unexpected fact may change the grounds of the decision. If we begin with the *second* question from the stand-point of our present knowledge, we must conclude that the plant has *lost the capacity*, as long as the views on the unity and affinity of all the Ascomycetes and on their homologies and connection with the Peronosporeae, which were developed above in section LXVI, are not shown to be incorrect. This is the necessary result of what has been already stated and does not require to be again explained. Whether the loss has in every case directly befallen the species *b* under observation, or another *n* from which *b* is descended, must remain an open question and is not an essential element in the matter. The assumption of such a loss would also be in harmony with other known facts in the Fungi which imply retrogression in the development of a species with excision of certain members (see below in section LXXXII). Moreover there are distinctly observed facts within the group in question, which enable us to form a clear idea of the way in which this loss may be occasioned; but these very facts themselves compel us to be cautious and to leave it to time to give a decided answer to the second as well as to the *first* of our questions. These facts are, first, that some species of Ascomycetes, as has been already pointed out more than once, form their ascocarps only under fixed and narrowly limited conditions, while they reproduce themselves by forming gonidia under very varied conditions. Of this fact Penicillium, Peziza Fuckeliana, Zopf's Fumago, and the species of Hypomyces described above, are examples. Secondly, there are pleomorphic species which also show a marked tendency to the reproduction of like forms when the conditions remain unchanged, that is, each spore-form chiefly produces its own form again, more rarely a form of another kind. Peziza Fuckeliana is an excellent example of this. If the gonidia of this Fungus, the spores of 'Botrytis cinerea,' are sown in a good nutrient solution, grape-juice for instance, the product is always a filamentous mycelium with copious formation of gonidia. If the ascospores are sown in the same solution under exactly the same conditions, a mycelium is developed with sclerotia, but gonidiophores never or scarcely ever appear; wherever they have appeared they have appeared singly, and cases of the kind are highly exceptional and not free from suspicion on the score of the purity of the sowing. If similar sowings are made on suitable dead leaves of Vitis or Castanea which have been boiled to free them as far as possible from foreign spores, sclerotia are generally developed; if ascospores are sown, sclerotia only are formed without filamentous gonidiophores; if gonidia are sown, the sclerotia are accompanied by an abundance of gonidia which spring from the filamentous mycelium. In cultures of the latter category single gonidiophores may certainly be produced from a sowing of ascospores, just as they may be developed, as is well known, from the sclerotia (see on page 224). The general result which is the important point here, the tendency to the reproduction of like forms, is not the less manifest.

Similar relationships appear to exist in Pleospora herbarum (or Gibelli's P. Alternariae), but I cannot speak decidedly respecting them. By corresponding relationships I should now be inclined to explain the facts, observed by myself in connection with the reproduction of like forms in the gonidial form of Cordyceps known as Isaria farinosa[1], which led to a controversy respecting its connection with the cycle of development of Cordyceps militaris, for which Tulasne contended.

Moreover it must easily happen, that in species which show the tendency in question, this tendency and the external conditions work together in the same direction, and the possible consequences of this co-operation can be readily conceived. The long-continued effect of the combined conditions might ultimately result in the permanent separation of the originally connected forms, each preserving its existence as a species. This means that each loses the other out of its cycle of development, whether the other continues a separate existence or for any reason disappears. It is quite conceivable that species of Ascomycetes producing gonidia only like those described above might originate in this way.

On the other hand, we have to consider that the two sets of causes determining the production of forms in these Fungi, the external and the internal, also work in opposite directions, and that the external causes may eventually overcome the internal tendency and lead to the reproduction of the other form from the one first produced. We may also imagine *a priori* that there are cases in which special conditions must combine to produce this result, cases in which the few observations that we possess do not acquaint us with the real conditions, though they are perhaps very simple. Ten experiments may be made with a similar result under different conditions, and the eleventh may all at once give a totally different result. Experiences of this kind are quite common in this portion of the field of research. These facts admonish us to be cautious, and require that the suggestions in the text should once more be expressly declared to be only very guarded conjectures.

SECTION LXXIV. No less caution is advisable in determining some of the **organs which have been described above as doubtful**, and we must say a word more about them in this place. The attempt has more than once been made to remove the doubt which exists as to their real nature by declaring them to be *rudimentary* (rudimentär). It will be well therefore to remember that organs or members are said to be rudimentary which do not reach the height of development attained by their homologues but are stunted in their growth, that is, remain stationary at a stage in their development in which they are in every respect immature; such are the rudimentary stamens of Salvia and of some diclinous flowers. It is true that the term rudimentary has been used in another sense, when an *organ* is highly developed, but is not properly adapted to the *function usually discharged* by its homologues, being applied to some other purpose through a necessity resulting from its high organisation, as, for instance, the median staminode of Cypripedium. In some cases this mode of expression may be obvious and therefore admissible, especially as there

---

[1] Bot. Ztg 1867, 1869.

are many gradations in the scale of arrestment and of perfection. Still it is more correct in such cases to speak not of arrest of development, but of different *adaptation*, or the *metamorphosis* of members, as we call foliage leaves, tendrils, and anthers in their various adaptations metamorphosed leaves or phyllomes, but do not call foliage-leaves rudimentary anthers, or anthers rudimentary foliage-leaves. At present we will keep to the correct and customary usage and apply the term rudimentary to those parts only which are arrested in respect to their development as members or structures, and also as regards their capacity in every respect for discharging their proper functions.

There is still another phenomenon which is allied in many points to unusual metamorphosis and rudimentary development of members and yet is distinct from it, namely, the occurrence of organs which are well developed and *capable of performing their function*, but which, as far as can be ascertained, are *actually functionless*. The phenomenon is of course rare; but the antheridia and archegonia of the apogamous Ferns[1] afford an example of its actual appearance among the organs of reproduction, which, from the facts recorded elsewhere, may be considered as beyond doubt. The occurrence therefore of this phenomenon must not be forgotten when we are engaged in deciding about doubtful formations.

The doubtful formations which we have to consider here are first of all the 'doubtful sporocarps' of the Aspergilli (and Sterigmatocystis of Van Tieghem), and secondly most spermatia and spermogonia. The Aspergilli show exactly the course of development of Penicillium. The doubtful sporocarps (p. 206) are bodies that look like sclerotia and resemble the perithecia of Penicillium, but differ from these, according to some observers at least, in that they show no development of asci. Brefeld's statements[2] to the contrary have been silently withdrawn or ignored by himself[3] since the appearance of Wilhelm's profound treatise on the subject. The question then naturally arises, what was the reason for the negative results respecting the formation of asci which were all that were obtained during so many years? Brefeld answers the question by declaring these bodies to be 'rudimentary primordia of perithecia.' This may be so with the bodies which Brefeld found in Aspergillus flavus and which he describes as undifferentiated tuber-like structures. But Wilhelm repeatedly obtained from A. flavus as well as from the other species a large number of well-developed bodies of the nature of sclerotia which in A. flavus had a black rind; the development therefore goes in this case also beyond the undifferentiated rudiment.

These bodies in view of their structure can no more be called rudimentary than the sclerotia of Penicillium. The difference in structure which seems actually to be found in all cases, namely, that there are no distinct ascogenous hyphae such as appear in the sclerotia of Penicillium, cannot be of any importance, for this difference exists in like manner up to the commencing formation of asci between other primordial sporocarps, the perithecia of Claviceps and Pleospora for instance, on one side and those of Melanospora and others on the other. It may be allowed that these objections would be merely a playing with words, if there were sufficient grounds

---

[1] See above, p. 122.   [2] Bot. Ztg. 1876, p. 265. ᴜ
[3] Schimmelpilze, IV, 134.

for believing that these bodies are *functionless* or discharge a very subordinate function. But the known facts which may be summed up in the words structure of a sclerotium and evident homology with Penicillium seem to be in favour of the contrary view, and to show that the sclerotia in question are capable of development, and are stages in the formation of ascocarps. The single argument for their being without function, and therefore, as I am prepared to allow, for their being in a rudimentary state, is to be drawn from the fact that no further development has been observed in them during the two years that they have been known, and under the conditions of cultivation which have been hitherto employed. On the other side it would be well to remember our experience with other sclerotia and resting states, to simply acknowledge our present ignorance and to give two more years to investigation before attempting to decide the question. I have purposely referred here only to Brefeld and Wilhelm. But Van Tieghem now asserts that he has actually obtained asci from Aspergillus niger, in just the same way as from Penicillium. If this is confirmed, the whole dispute is set at rest as far as that species is concerned, and the same thing will happen probably in the case of the others also.

Many doubts, uncertainties, and controversies have also arisen in determining the **spermogonia and spermatia**. I do not include among these the many separate cases more or less imperfectly investigated, in which it remains uncertain whether an organ described as a spermogonium and assigned to a particular species really belongs to that species or to some other which perhaps lives as a parasite in or with the first species, and similar cases. To questions of this kind we may almost always apply the remarks which will be found in a previous page respecting the determination of the genetic connection between forms occurring together, and which are really self-evident; we will only touch here on those cases in which the genetic connection of the organs in question is certainly or as good as certainly ascertained.

We will fix our attention first of all on those spermatia only which are produced from spermogonia and on the spermogonia themselves, remembering always that the formation of spermatia differs in no essential point from that of acrogenously formed spores, and that the one and always recurring distinction between the two organs is that the spores germinate, while, as far as our observations go, the spermatia do not. The spores germinate by the extrusion of a germ-tube, and it has either been actually observed or it is assumed from analogy that the tube can grow into a mycelium. Tulasne, the original discoverer of the spermatia and spermogonia, suggested in the year 1851 that they were male sexual organs; he rested his view partly on the non-germination of the spermatia, partly on the fact that their formation usually precedes that of the ascocarps, and in these points they certainly agree with the male organs of other plants. It was clear from the mutual connection of the observed facts, that it was the ascocarps and not the gonidial forms which must stand in special and direct relation to the supposed fertilisation. Beyond this no certain idea existed at the time of these first discoveries respecting the mode of fertilisation, nor was anything known of the presumptive female organ which was supposed to be fertilised. The discoveries which were needed to clear up these points have been made from the year 1863 onwards.

Tulasne in his first works, and others after him, had described, under the name of spermatia, some cells which resembled spermatia in their small size and in their origin, and had termed their receptacles or the organs that carried them spermogonia; but it was shown by further observation that these cells were germinating spores or gonidia, and that their receptacles should be termed gonidiophores or pycnidia. Of a similar kind were the gonidia of Claviceps and other forms mentioned in Tulasne's Carpologia, which germinate very readily. As these observations multiplied, the question naturally arose whether there really are spermatia which are absolutely without the power of germination, or whether the absence of germination in the alleged cases did not arise from defects in the mode of conducting the experiments, since some spores only germinate under fixed conditions, and the conditions may not always have been properly secured in the artificial cultivation of the plants. A work of Cornu[1] endeavours to give an answer to this question, and a further answer is to be found in Stahl's treatise on Collema which appeared almost at the same time. The two are very different.

Stahl's work shows that there are spermatia which are not spores but fertilising organs, and describes the mode of fertilisation and the organ to be fertilised (see above on page 211). It does this, it is true, in a limited number of cases only; but what is known of the rest of the Lichen-fungi, and is not disputed by Cornu, proves further that by far the larger part of them possess spermatia which show no more signs of germination than those of the Collemeae, and that these spermatia are homologous with those of Collema. This is sufficient to distinguish the spermatia and spermogonia from spores and their receptacles in this long series of cases, even though nothing certain is yet known as. to the function of most of these spermatia. That exactly the same condition of things is to be found also outside the group of Fungi which form Lichens, is evident from the case of Polystigma described above on page 215.

Cornu, on the other hand, simply does not allow that the spermatia are special organs, but would have them regarded as spores with the power of germination, while retaining the name which they have hitherto borne. His arguments for this view are not convincing. He saw first of all the 'spermatia' of certain species, which hitherto perhaps had been considered to be incapable of germination or had not been examined (for example those of Massaria Platani), produce germ-tubes when sown in nutrient solutions; a few more therefore to be added to the previously known cases of pseudo-spermatia. He also saw other known spermatia, those, for instance, of Stictosphaeria Hoffmanni, Tul. and Valsa ambiens, Tul., undergo changes of form, also in nutrient solutions, and swell up, but without showing further signs of germination. He gives no other new facts; the cultivation of the spermatia of Lichens gave him only negative results, and he can scarcely be said to have advanced the subject even in a single minor point. His treatise was published before the results of Stahl's profound investigations were given to the world.

From the facts which have been established we now know of spermatia or spermogonia in certain species or genera as organs with a definite function different from that of spores. We can also form a plausible view as to the homology of these bodies with the antheridial branches or functioning antheridia in other species which have no spermatia, as was attempted to be done above on page 231. Lastly, we are acquainted with a large number of species in which the homology of the spermatia

---

[1] Reproduction d. Ascomycètes (Ann. d. sc. nat. sér. 6, III)

and spermogonia with those first mentioned is indubitable ; but, on the other hand, we know nothing certain in almost all these latter cases about the function of the spermatia. We may certainly assume that they are male fertilising organs in all the species which possess an organ (trichogyne or ascogonium) which may be intended to be fertilised. The swelling too of the spermatia in a nutrient solution, or even the extrusion of a germ-tube, would not be an objection to this assumption, since processes of growth might make their appearance, as Stahl has already remarked, in this mode of cultivation, which in the natural course of development would only be set up after contact with the organ to be fertilised, just as pollen-tubes are formed in saccharine solutions. But, as was shown above, the female organs to be perhaps fertilised are actually known only in a comparatively small number of species, and in the rest, which are the large majority, the functions of the spermatia must therefore be declared to be doubtful. If we suppose them to occur in species which have no female organ, they cannot there have a sexual function. Yet we can hardly call them rudimentary organs; and the enormous numbers in which they are produced are opposed to the view that they are entirely without function : their function therefore remains for the present undetermined.

Supposing after what has now been said that we have satisfied ourselves as to the distinction between spermatia and small spores, and as to the mode of naming them and their receptacles and so on, and can find our way in the practical description of them, an interesting subject of enquiry still remains in the homological relations between the two kinds of organs, for there is a striking agreement between them, not only in form and structure, but also in that which is of much greater importance, namely, the place or moment of their appearance in the course of the development. In the latter respect, for instance, the commencing small-spored pycnidium of Cucurbitaria Laburni agrees with the spermogonium of Polystigma or Physma. Here we may say, without exaggeration, that the only difference lies in the germination. In this and similar cases we only know the first products of germination, the germ-tubes; we do not know what is developed from them. We might therefore consider them to be formations which are incapable of further development, like the pollen-tubes in a saccharine solution, or we might at least enquire if they are not of this nature. However, we may put this point out of consideration and assume that they are in all cases capable of producing a mycelium. This assumption does not prevent us from maintaining their homology with true spermatia. On the contrary, we may very well conceive that we are dealing with homologues of *different adaptation or metamorphosis*, and this different adaptation would arise in correlation with the absence of the female organ capable of fertilisation; for, as far as we can see at present, the phenomenon in question occurs exactly in those forms which have no female organs, having lost them probably in the course of the phylo-genetic development, as we have already endeavoured to show. It is no sufficient objection that this metamorphosis of the spermatia is not found in all species that have experienced this loss, for phenomena of this kind vary continually from species to species.

The hypothesis that there are such metamorphosed spermatia would make many facts more intelligible than they have hitherto been. It will depend on the results of further special investigations how far this hypothesis can be applied to small-spored pycnidia and similar formations, and we must not here enter further into the details of the subject than the reader can at any time do for himself by comparing the examples described in the preceding pages. But we may just observe that it is possible that metamorphosis may be found in cases where it has been quite unsuspected. The hymenium which has already been repeatedly described is found on the young stroma of the Xylarieae, Claviceps, Epichloe, &c. either before or together with the first appearance of the sporocarp, and gives off small cells, which in structure, origin, and size might be spores (gonidia) or spermatia. There can be no doubt that they are homologous in all these species. They were called gonidia in a previous page, because they germinate in Claviceps and Epichloe, and in Poronia and Ustulina among the Xylarieae ; but, as far as we know, they do not germinate in Xylaria—a fact which may be added here to complete our former observations on the genus. These phenomena find their explanation in the hypothesis here proposed, and may therefore be brought forward in support of it ; but it is obvious that the hypothesis is not hereby made a certainty.

In conclusion we must recur again to the objects which were included in page 242 under the name of doubtful spermatia. The word 'doubtful' must still be repeated of many of them, for we possess only brief accounts of them, and portions of these are disputed. I confine myself therefore to the cases of Sordaria, Chaetomium, and Sclerotinia, which we know more in detail through the labours of Zopf and Brefeld. Here, according to these observers, the organs in question agree as much in their characteristic development and structure as in their power of germination under the conditions to which they were submitted, and they have no other function, as far as could be ascertained, than that of spores. They are therefore organs whose function is unknown to us. It is *possible* that they have no function at all; at any rate as they do not function as sexual or otherwise reproductive organs, they can scarcely have any important duties to fulfil, for they are usually few and small; and if the case is otherwise in the specimens of Sclerotinia tuberosa grown by Brefeld, we must not forget that in this instance the Fungus was growing under conditions quite foreign to its usual circumstances. Having regard to the known facts and to the analogies and homologies which may be applicable, there is the alternative proposed by Zopf for the determination of these bodies, that they are either functionless spermatia, or spores or gonidia not capable of germination. This is not a matter of indifference in reference to the question of the homology. But gonidia without the power of germination, according to all trustworthy data, are things which nothing but extreme necessity can allow us to assume ; and in the case of the Chaetomieae, where, according to Zopf, almost every cell of the mycelium may become a gemma or gonidium with power of germination, and in Sclerotinia Fuckeliana with its characteristic and highly reproductive gonidiophores, it would be an absurd thing that such well-furnished appliances should be occupied solely in producing sterile gonidia. But if we suppose these bodies to be homologous with spermatia, the whole matter becomes intelligible from the points of view which have now been discussed. Only one objection has been brought forward to this view. Brefeld[1] calls attention to the difficulty of accounting for the concurrence in 'Sordaria' of

---

[1] Schimmelpilze, IV, p. 143.

spermatia with or without special functions and an antheridial branch which conjugates with the archicarp. But the 'Sordaria' in which the antheridial branch has been observed is Sordaria or Hypocopra fimicola described by Gilkinet, and spermatia are not said to have been found in it. The Sordarieae in which the spermatia in question have been found are different species, S. curvula, S. minuta, and others[1], and it has not been shown that their young sporocarp is like that of S. fimicola. On the contrary, it has been already intimated on page 235 that there appear to be important differences in the matter of the inception of the sporocarp in the group of the Sordarieae. Brefeld's objection therefore rests on a misconception, and is at present at least not justified; and if it is removed, it is no longer necessary to consider at length the question of the connection between these cases and those discussed above in which the spermatia were supposed to be without sexual function.

Literature of sections LIX–LXXIV.

VITTADINI, Monogr. Tuberacearum, Mediolani, 1831.

TULASNE, Fungi hypogaei, Paris, 1851 ;—Id., Selecta Fungorum Carpologia, I–III, Paris ;—Id., Recherches sur l'organisation des Onygena (Ann. d. sc. nat. sér. 3, I, 1844) ;—Id., Note sur l'appareil reproducteur des Lichens et des Champignons (Ann. d. sc. nat. sér. 3, XV, 1851) ; see also Compt. rend. XXXII, p. 470 ;—Id., Mémoire pour servir à l'histoire organographique et physiologique des Lichens (Ann. d. sc. nat. sér. 3, XVII) ;—Id., Discomycètes (Ann. d. sc. nat. sér. 3, XX, p. 128) ;—Id., Mém. sur l'Ergot des Glumacées (Ann. d. sc. nat. sér. 3, XX, p. 5) ;—Id., Note sur l'appareil reproductive d. Hypoxylées et d. Pyrénomycètes (Ann. d. sc. nat. sér. 4, V, p. 108) ;—Id., Nouvelles obs. sur les Érysiphées (Ann. d. sc. nat. sér. 4, I, 299, and Bot. Ztg. 1853, p. 257) ;—Id., Notes sur les Isaria et les Sphaeria entomogènes (Ann. d. sc. nat. sér. 4, VIII, p. 44) ;—Id., De quelques sphéries fongicoles (Ann. d. sc. nat. sér. 4, XIII, p. 5) ; see also Comptes rendus, 41, p. 615 and 50, p. 16;—Id., Note sur les phénomènes de copulation dans les Champignons (Ann. d. sc. nat. sér. 5, V, p. 216).

CURREY, On the fructification of certain Sphaeriaceous Fungi (Philos. Trans. Royal Soc. London, 147, 1858).

DE BARY, Ueber d. Fruchtentwicklung d. Ascomyceten, Leipzig, 1863 ;—Id., Eurotium, Erysiphe, Cicinnobolus, nebst Bemerkungen ü. d. Geschlechtsorgane d. Ascomyceten (Beitr. z. Morph. u. Physiol. d. Pilze, III, Frankf. 1870). See also Beitr. IV, p. 111.

S. SCHWENDENER, Ueber d. Entw. d. Apothecien von Coenogonium (Flora, 1862, p. 224); —Id., Ueber d. Apothecia primitus aperta u. d. Entw. d. Apothecien im Allgemeinen (Flora, 1864, p. 320).

FÜISTING, De nonnullis Apothecii Lichenum evolvendi rationibus (Diss. inaugur. Berol., 1865) ;—Id., Zur Entwicklungsgesch. d. Pyrenomyceten (Bot. Ztg. 1867, 1868) ;—Id., Zur Entwicklungsgesch. d. Lichenen (Bot. Ztg. 1868).

WORONIN, Entwicklungsgesch. d. Ascobolus pulcherrimus u. einiger Pezizen (Beitr. z. Morph. u. Phys. d. Pilze, II). Id., Sphaeria Lemaneae, Sordaria, &c. (Beitr. z. Morph. u. Phys. d. Pilze, III).

JANCZEWSKI, Morph. d. Ascobolus furfuraceus (Bot. Ztg. 1871, p. 257).

---

[1] Zopf, Chaetomium, p. 237.

J. KÜHN in Mittheil. d. Landw. Instit. Halle, I, 1863 (Claviceps).

O. BREFELD, Bot. Unters. ü. Schimmelpilze, II (Penicillium) IV.

VAN TIEGHEM, Comptes rendus, 81, 1875 (Chaetomium) ;—Id., Nouvelles Observ. sur le développement du fruit, &c. des Ascomycètes (Bull. Soc. Bot. de France, 23, 1876, p. 99) ; see also Bot. Ztg. 1876, p. 165;—Id., Sur le dével. du fruit des Ascodesmis (Bull. Soc. Bot. de France, 23 (1876), p. 271 ;—Id., Nouvelles Obs. s. l. dével. du périthèce des Chaetomium (Bull. Soc. Bot. de France, 23, 1876) ;—Id., Sur le dével. de quelques Ascomycètes (Aspergillus), (Bull. Soc. Bot. de France, 24, 1877).

GILKINET, Rech. s. l. Pyrénomycètes (Sordaria), (Bull. Acad. Belg. 1874).

BARANETZKI, Entw. d. Gymnoascus Reesii (Bot. Ztg. 1872).

EIDAM, Beitr. z. Kenntn. d. Gymnoasceen in Cohn's Beitr. z. Biol. III, 271 ;—Id., Z. Kenntn. d. Entw. d. Ascomyceten in Cohn's Beitr. z. Biol. III, 377 ;—Id., Ueber Pycniden (Bot. Ztg. 1877).

E. STAHL, Beitr. z. Entwicklungsgesch. d. Flechten, I, Leipzig, 1877.

A. BORZI, Studii sulla sessualità degli Ascomicete (N. Giorn. Bot. Ital. X (1878), p. 43).

BAINIER in Bull. Soc. Bot. de France, 25, 1878.

C. FISCH, Zur Entwicklungsgesch. einiger Ascomyceten (Bot. Ztg. 1882).

O. KIHLMAN, Zur Entwicklungsgesch. d. Ascomyceten (Pyronema, Melanospora) in Act. Soc. Sc. Fennicae, XIII, Helsingfors, 1883.

W. ZOPF, Zur Entwicklungsgesch. d. Ascomyceten (Chaetomium), (N. Act. Leopold. XLII, 1881) ;—Id., Die Conidienfrüchte v. Fumago (N. Act. Leop. XL, 1878).

GIBELLI e GRIFFINI, Sul polymorphismo della Pleospora herbarum (Archiv. del Laborat. di Bot. Crittogam. in Pavia, I (1873), p. 53).

H. BAUKE, Zur Entwickelungsgesch. d. Ascomyceten (Bot. Ztg. 1877, 313) ;– Id., Beitr. z. Kenntn. d. Pycniden (N. Act. Leop. XXXVIII, 1876).

K. WILHELM, Beitr. z. Kenntn. d. Pilzgattung Aspergillus (Diss. Berlin, 1877).

O. MATTIROLO, Sullo sviluppo e sullo sclerozio della Peziza Sclerotiorum, Lib. (N. Giorn. Bot. Ital. XIV, 1882), p. 2.

B. PIROTTA, Sullo sviluppo della Peziza Fuckeliana, &c. (N. Giorn. Bot. Ital. XIII, 1881), p. 130.

G. KRABBE, Entw., Sprossung u. Theilung einiger Flechtenapothecien (Bot. Ztg. 1882, Nr. 5–8) ;—Id., Morphol. u. Entwicklungsgesch. d. Cladoniaceen (Ber. d. deutsch bot. Ges. 1883).

REINKE u. BERTHOLD, Die Zersetzung d. Kartoffel durch Pilze, Berlin, 1879.

R. WOLFF, Beitr. z. Kenntn. d. Schmarotzerpilze (Erysiphe) in Thiel's Landw. Jahrb. 1872 (?).

M. CORNU, Reproduction d. Ascomycètes (Ann. d. sc. nat. sér. 6, III).

R. HARTIG, Wichtige Krankh. d. Waldbäume, p. 101 (Hysterium) ;—Id., Unters aus d. Forstbot. Inst. z. München, I (Rossellinia, Nectria).

W. LAUDER LINDSAY in Trans. Roy. Soc. Edinburgh, I, p. 101. (Spermogonia and 'stylospores' of Lichens.)

GIBELLI, Sugli org. reprod. del gen. Verrucaria (Mem. Soc. ital. di Scienc. nat. I).

A. MILLARDET in Mém. de la Soc. d'hist. nat. de Strasbourg, VI, 1868. (Myriangium, Naetrocymbe, Atichia.)

See also the works quoted in the foot-notes to the text.

The **reader** is also referred to descriptive and phytopathological literature. Nylander's Synopsis is specially valuable among works on the Lichen-fungi; others will be found fully given in Von Krempelhuber's Geschichte u. Literatur d. Lichenologie.

Doubtful Ascomycetes.

Section LXXV. There are certain small groups of Fungi which, as far as we know them, show a greater amount of agreement with the Ascomycetes than with any other Fungi, and must therefore be classed with the Ascomycetes. Some, like the Laboulbenieae and the group formed of Exoascus and Saccharomyces, have asci, but are so widely separated by structure and course of development from typical Ascomycetes, that there may be some scruple about uniting them directly with this division of the Fungi; others greatly resemble certain typical Ascomycetes in all that is ascertained of their life-history, but are hitherto only known to produce peculiar small cellular bodies, 'bulbils,' without power of germination, instead of sporocarps with asci.

To the latter category belong the forms **Helicosporangium parasiticum**, Karst., and **Papulaspora aspergilliformis**, Eid., which have recently been described by Eidam. We can only mention them thus briefly in this place, referring the reader to Eidam's publication; the plants themselves should be further investigated. We proceed to give a short account of the other species.

Most of the **Laboulbenieae** grow on the outer surface of beetles which live in or near water, but some are found on other insects, as the species especially of Eastern Europe, Stigmatomyces Baeri, Peyr., which is common in Vienna on house-flies. They appear like small brushes on the surface of the insect, either singly or often, like Stigmatomyces, forming a thick fur on it. Each of these

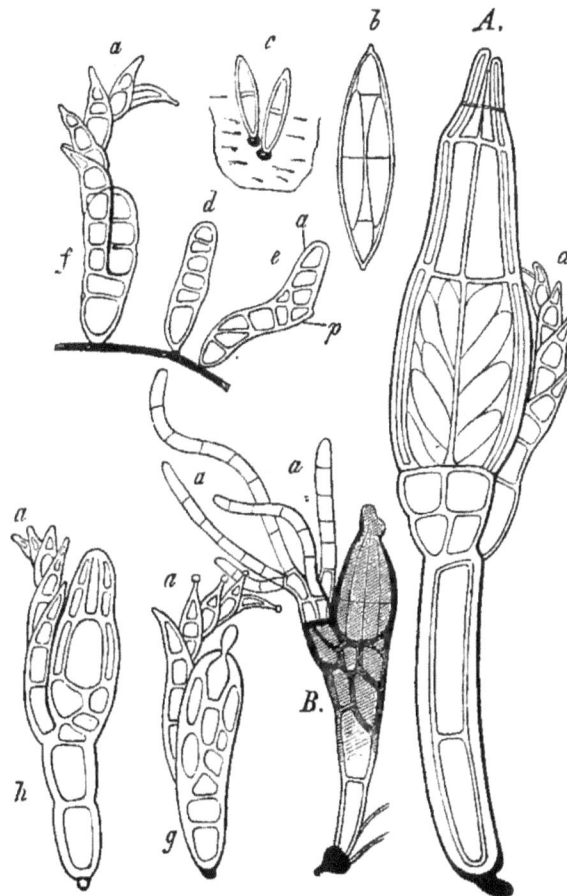

FIG. 120. *A, b—h Stigmatomyces Baeri*, Peyritsch (*St. Muscae*, Karsten). *A* ripe specimen with its black organ of attachment released from the skin of the fly, showing the surface and an optical longitudinal section; the asci are seen through the wall of the perithecium. *a* everywhere the appendage. *b* an isolated ascus with ripe spores. *c—h* development of the perithecium and appendage; successive stages of development according to the letters. *c* two double spores fastened to the wing of a house-fly. *d, e* older states on the chitinous membrane cut through perpendicularly. *p* commencement of the perithecium. *g* the delicate projection (? trichogyne) from the apex of the perithecium, with the small round swellings on the extremities of the branches of the appendage. *h* after the formation of the perithecium is completed. *B* full-grown specimen of *Laboulbenia flagellata*, Peyr. from the wing-cover of *Bembidium lunatum*. The stalk-like base of a second specimen is indicated, with the same black organ of attachment. *a* the appendage. All the figures after Peyritsch. *A c, g, h* magn. 350, *b d, e, f* 450, *B* 125 times.

small brush-like bodies is a separate plant. The entire length in the largest known species, Laboulbenia Nebriae, is about 1 mm., in most species little or not more than 0.5 mm. The phenomena observed in them most nearly resemble those known in the Ascomycetes, and are named accordingly. The small plant (Fig. 120) is attached to the substratum by a filiform or clubshaped stalk, consisting usually of two cells one above the other; at the apex of the stalk is a perithecium and a body which may be here briefly termed an *appendage* (*a*). The perithecium is narrowly conical in form or flask-shaped and in some species oblique, and consists when mature of a wall formed of a few cells disposed in two layers at the base and one layer at the sides with a narrow orifice at the apex; the group of asci which rises erect from the base of the perithecium is closely surrounded by its wall. The number of the asci and the way in which the spores are formed in them are not exactly ascertained. The number of spores in an ascus is said to be 8 and 12; the ripe spores are fusiform and colourless, and being divided by a transverse wall into two equal cells are therefore compound and bicellular; they escape singly and one after another through the orifice of the perithecium, no doubt in consequence of the gelatinous deliquescence of the wall of the ascus. The appendage springs from close to the base of the perithecium in the form of a segmented hair or filament, varying, according to the species, in length and number of cells and presence or absence of branches, which in some species are very peculiar in their form and arrangement. All the cells in the mature Fungus except the asci, the spores, and the extremities of the branches of the appendage, have very thick membranes of a deep and often a dark brown colour.

The Laboulbenieae have no mycelium; the ripe double spore attaches itself by one extremity to the chitinous covering of the insect, and sends into it a small short point, which sometimes enlarges into a knob at its extremity and with the surrounding chitin soon assumes a brown colour; this point is its only organ of attachment and of nutrition. Thus firmly planted it developes at right angles to the substratum and reaches its mature state by the necessary successive cell-divisions and differentiations. Most of the details of these formations can be seen at once in the accompanying figure for the case which it represents, but some important points have still to be cleared up. I select the following for special notice, and refer the reader to Peyritsch's treatises for further details. The appendage is developed from the cell of the double spore which is the *upper* one in reference to the point of attachment; it is therefore originally terminal and is completed before the perithecium. The stalk and the perithecium are formed from the *lower* cell of the double spore; the perithecium shoots out laterally from beneath the point which is afterwards that of insertion of the appendage, and as it increases in breadth it thrusts the appendage to one side. In its earliest stage it is unicellular; as it grows it divides by successive transverse divisions into three tiers of one cell each, and each tier in acropetal succession then separates by longitudinal division into an axile and several parietal cells. But before the longitudinal division begins (Laboulbenia vulgaris), or before it has reached the uppermost tier-cell (Stigmatomyces), it is observed that this cell puts out a short protuberance at its apex, which is either very thin-walled or seems to have no membrane, and disappears again at a later stage of the development (Fig. 120 *g*, *h*). Simultaneously with the formation of this protuberance on the primordium of the perithecium small thin-walled swellings are seen on the apex or on the tips of the branches of the young appendage, and these also subsequently disappear. According to Karsten these small swellings, which are spherical in Stigmatomyces, separate in that plant from the cells on which they are formed and attach themselves to the protuberance on the young perithecium, as spermatia in the Fungi or Florideae attach themselves to the trichogyne, and then the spores or asci are developed from an axile cell. If this were so, the organs in question would have to be termed sexual organs and their homologies with the sexual organs of the Ascomycetes would be evident enough. But it appears from Peyritsch's careful observations that no such supposed abscision of spermatia really takes place. We know nothing more than has been stated above, and Peyritsch himself does not think very highly of his own attempt to save the trichogyne, which might be fertilised by contact with a young branch from the appendage. It is not yet quite certainly ascertained whether the asci are formed by division or by sprouting from one or more initial cells. With these data only to guide us, it will be best for the present to allow the remarkable little group to remain *next* the Ascomycetes, being marked as doubtful, till further information is obtained concerning them.

Section LXXVI. The species of **Taphrina**, Fr., the **Exoascus** of Fuckel in Sadebeck's sense[1], are parasites developing on the surface of parts of living plants, which are more or less deformed by them; Exoascus Pruni, for example, grows on the young fruit of species of Prunus, and produces swellings in them which are known as pockets, more rarely on the leafy shoots, while E. aureus is found on the leaves and ovaries of Poplars and Aspens and E. alnitorquus on the deformed fruits and upon leaves of the Alder.

The Fungus when fully developed is composed chiefly of a single palisade-like layer of asci standing close beside one another, which breaks through the cuticle and covers the outer surface of the epidermis of the part attacked. The species which live on the Amygdaleae, Exoascus Pruni for example and E. deformans, develope this layer from a filiform mycelium, which first spreads in the inner parenchyma of the part and then thrusts its branches in between the outer walls of the epidermal cells and the cuticle. In this situation the branches ramify copiously, and spread out in the direction of the surface, the ramifications, which grow alongside and between one another, forming a single layer and then becoming divided into isodiametric cells. Each of these cells next swells into a vesicle, and breaking through the cuticle elongates in a direction perpendicular to the substratum and becomes club-shaped, and at length divides by a transverse wall into a lower cell, the short *stalk-cell*, which rests on the substratum, and an upper cell, the club-shaped *ascus*. The connection of the ascus-layer thus formed with the intramatrical mycelium can be seen even when the asci are mature.

Other species, Exoascus alnitorquus for instance and E. aureus, according to Sadebeck's and to some extent also of Magnus' earlier investigations, spread their mycelium only between the cuticle and the epidermis. Then, as the plant developes, all the hyphae become divided into ascogenous cells, and these proceed as in E. Pruni; consequently asci only are to be seen when the fructification is mature, and these are either borne on a stalk-cell (E. alnitorquus) as in E. Pruni, or have no stalk

---

[1] In Winter, Pilze, II.

(E. aureus). In the latter case especially and in the last-named species the outer extremity of each ascus bursts through the cuticle, while the inner extremity grows into a narrowly conical process, which becomes deeply and firmly fixed between the lateral walls of the epidermal cells. The forms of a third series, represented by Sadebeck's Exoascus epiphyllus, which grows on Alnus incana, and E. Ulmi, also spread their hyphae between the cuticle and the wall of the epidermal cells, but form their asci from a part only of their cells, while the other part remains sterile; hence the asci are here less crowded.

The structure of the asci, the formation of the spores in them, and the ejection of the spores by mechanical contrivance, are essentially the same, as far as is at present known, as in other organs of the same name. The number of the simultaneously formed ascospores is also usually eight in Exoascus Pruni; other numbers will be mentioned further on. All the spores are small, simple, ellipsoidal cells with a delicate colourless membrane.

The spores of Exoascus Pruni are ejected when ripe and germinate at once in water or a nutrient solution, sprouting repeatedly and perfectly and forming many orders of sprouts. Those of the first orders are of much the same shape and size as the mother-spore, those of the higher are often much smaller. If the ripe spores are detained in the ascus, they often form their germ-sprouts in it, and the ascus becomes filled with countless sprouts of different orders and sizes, which readily separate from one another and escape as individual 'spores' when the ascus opens.

The spores germinate in a very similar manner in the other species. In many of them, those for instance which live on Poplars and Alders, a very large number of small sprouting spores are found in the ripe ascus. Sadebeck states that these are always sproutings from eight primary ascospores; according to my earlier researches and Brefeld's investigations, repeated quite recently, the original spores in Exoascus Populi may be less than eight; Brefeld says that there are usually four, and I remember to have seen only two and three. Short germ-tubes, which soon give off sprout-cells, are occasionally formed from the spores, as, for example, in E. alnitorquus.

Sadebeck has noticed that the products of the germination of the spores of Exoascus alnitorquus and E. bullatus penetrate into young leaves of Alnus glutinosa or Pyrus communis, and there develope directly into ascogenous hyphae. The mode of penetration is not stated. This observation would justify our assuming a similar behaviour in the other species, with the addition that in some of them at least the mycelium vegetates and maintains itself for a long time in the plant which it attacks. For instance, it is found early in spring in the rind of the branches of Prunus and spreads from them into the young twigs and fruits; and in E. deformans, which inhabits the cherry-tree, it is perennial and lives for years in the rind of the branches, where it causes the 'witches' brooms,' and sends branches every year to form asci in the leaves, which are disfigured in a similar manner.

There is another doubtful Fungus which Reess has named **Endomyces decipiens**, and which must for the present be placed near Exoascus. It grows in old lamellae of Agaricus melleus and consists of septate hyphae, which are often constricted at the septa and produce small ellipsoid asci arranged in lateral clusters.

Four hemispherical spores are formed in each ascus, which escape when mature by the dissolution of the wall of the ascus, and put out germ-tubes in water. Nothing more is known of this species; the controversy which has arisen in the attempt to determine it will be noticed again in section XCIII.

Section LXXVII. The chief representatives of the genus **Saccharomyces** are the Yeast-fungi which excite alcoholic fermentation and are known as Saccharomyces Cerevisiae, S. ellipsoideus, S. Pastorianus, &c. These names, according to E. Hansen's recent investigations, denote form-groups, which will no doubt have to be otherwise distributed. Besides these there are the Flowers of wine, S. Mycoderma, Reess, Cienkowski's Chalara Mycoderma, and the Fungus of thrush (apthae), S. albicans, Reess, which grows as a parasite on the mucous membrane of the human digestive organs, but also thrives in saccharine fluids, where it excites a slight fermentation. The rest are found in quantity in or on fluids which are fermenting or have undergone fermentation. S. Cerevisiae is added intentionally to the wort of beer and is cultivated largely for this purpose. Other kinds and indeed S. Cerevisiae as well appear of themselves in must, finding their way into it chiefly from the surface of the juicy fruits which yield the must. They are conveyed to these fruits along with dust from the surfaces of other bodies (see below, section C).

The larger part of these Fungi vegetate, as far as we know, only by sprouting (Fig. 121). Continuous branching hyphae with long segments are found only in Saccharomyces albicans, in S. Mycoderma and Cienkowski's Chalara; these grow directly, according to Cienkowski's observations on S. Mycoderma, from cells formed by sprouting, produce fresh cells from their sides by sprouting, and ulti-

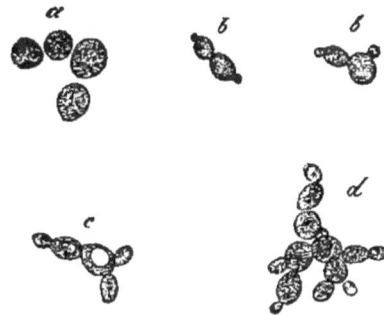

FIG. 121. *Saccharomyces Cerevisiae. a* cells before sprouting. *b—d* cells sprouting in a fermenting saccharine solution. Successive stages of development according to the letters. Magn. 390 times.

mately divide transversely into short cells which then vegetate simply by sprouting. Other species, especially Reess' S. Pastorianus, show a certain approach to this growth-form in that they are frequently chains of elongated sprout-cells, from which short cells are abscised laterally. The shape moreover of the single sprout varies generally between spherical and elongated cylindrical, with certain rules and limitations in each species. The cells that are formed one after another by sprouting in the fermenting fluid are usually at once separated from one another; connected strings of cells are to be seen when the plant vegetates on still surfaces, such as a microscopic slide especially, and the length and number of the chains vary according to the species. The cells of S. Cerevisiae, when developed in large quantities, often adhere together irregularly and form largish lumps, being attached to one another apparently by the muci-laginous outer lamellae of their membranes (see on page 9). The structure of the sprout-cells is that of other vegetative fungal cells, but their membranes are comparatively thin and colourless.

The Saccharomycetes in the sprouting form may be said to be capable of unlimited growth and multiplication if supplied with sufficient food. This is shown by the hundreds of thousands of pounds weight of yeast which are produced year by year and which consist entirely of the sprouts of S. Cerevisiae.   But a certain number of known species also forms *spores in asci* under certain conditions ; this fact was discovered by De Seynes in Saccharomyces Mycodermain, 1868, was afterwards examined more thoroughly especially by Reess, and has now been certainly established in S. Mycoderma and the forms included under the names of S. Cerevisiae, S. ellipsoideus and S. Pastorianus.   It occurs most readily and most frequently in the S. ellipsoideus of wine-yeast.   It has been studied by Reess, Hansen, and other observers in S. Cerevisiae, but it is often difficult to induce this cultivated form to produce spores. Spores begin to be formed if well-fed specimens, protected as far as possible from invasion by other Fungi and Schizomycetes, are kept without food or restricted to the least possible amount of it in presence of water and air containing free oxygen and in a suitable temperature ; if, for example, yeast is spread in a thin layer on moist surfaces, such as succulent parts of plants, plaster of Paris, or a microscopic slide, or kept in a little distilled water.   At first new cells are formed by sprouting in such cultures at the expense of the old ones, which may become exhausted and sometimes die.   Then spores are formed in cells which are not distinguished by their origin, shape, or any other particular, sometimes in a few isolated cells, at other times in all or most of the cells of a chain.   Two or four, or some-

FIG.122. *Saccharomyces ellipsoideus*, R. (Wine-yeast). Formation of spores in sprout-cells, taken from fermenting must and spread out for thirty-six hours on a microscopic slide in distilled water. The spores are not yet fully formed. Magn. about 600 times.

times three, seldom more than four spores are formed in a cell according to its size.   The stages observed in the formation of the spores correspond to the processes known to occur in asci (see section XIX).   The young spores appear simultaneously as delicately circumscribed round bodies of homogeneous protoplasm collected into a group inside the protoplasm of the mother-cell, in which a parietal layer of protoplasm remains at first everywhere unbroken (Fig. 122).   The spores soon form a membrane which always remains thin, and increase in volume while the protoplasm more or less completely disappears.   When full-grown they may just fill the cavity of the mother-cell, but generally they do not quite fill it ; if they are four in number they are disposed tetrahedrally as quadrants of a sphere or in a row according to the shape of the cell.   They are now arrived at maturity.   In older specimens the membrane of the mother-cell often collapses and disappears ; it is ruptured according to Cienkowski's account in Saccharomyces Mycoderma and releases the spores.   The ripe spores can germinate as soon as they enter the nutrient fluid.   In germination they swell slightly and form vacuoles, and then begin to sprout in the manner proper to each species ; the membrane of the mother-cell is broken through as the first sprouts are extruded.

E. Hansen found in the species which he examined sprout-cells which were divided by thick flat partition-walls into 2–4 daughter-cells, and these cells germinated in the same way as the ascogenous spores, but he did not see the formation of these septa.   Meanwhile, judging from the figures, we should be inclined to suspect that the formations in question are simply asci with their walls much collapsed after ripening and with the spores closely pressed one against another.

The foregoing brief account of the formation of the spores of the Saccharomycetes is taken from Reess' earlier statements and a recent revision of them in examining

Saccharomyces ellipsoideus; it would appear to be an evident case of partial division or free cell-formation (see page 61), in which the observed facts perfectly correspond to what is known of the formation of spores in smaller asci (Exoascus, Eurotium). The term asci is accordingly chosen or retained. It is true that there are differences of opinion with regard to the process in question. Cienkowski suspects that in S. Mycoderma the whole of the protoplasm of the mother-cell is divided into spores, and Brefeld speaks in the same way as regards other species, ('wine-yeast') in so far as he considers the sporogenous process in Saccharomyces to be like that in Mucor, regarding the latter indeed as a case of partial division. It is otherwise in the Saccharomycetes examined by Reess and myself. The continued presence of the parietal layer of protoplasm after the formation of the spores is decisive in their case even now, when the distinction between 'free cell-formation' and (total) division is less sharp than it once was. The spores are not formed in the sporangia of Mucor in the same way as in Saccharomyces (see page 74). Lastly, Van Tieghem has proposed a view which differs entirely from any other[1]. He thinks that the spores of Saccharomyces are produced by the division of the whole of the protoplasm, but that they are pathological formations induced by the assaults of Bacteria; this idea was suggested by the behaviour of spores of Mucor in presence of Bacteria, but it is at once refuted by the observation of a good specimen grown beneath the microscope in distilled water and free from Bacteria, and appears to have been recently abandoned by its author[2].

SECTION LXXVIII. There can be no doubt, from what we know of the history of development in the ascogenous Saccharomycetes, that they are immediately connected morphologically with the Exoasci. The differences in form between them and the Exoasci, whose hyphae are broken up into asci, would even allow of the two groups being united into one genus. The two genera therefore together form a natural group, which may be called here the **Exoascus-group.**

If we enquire further into the connection of this group with other Fungi, we can only take morphological arguments into consideration in determining the question. No decisive argument of the kind is to be drawn from the simple vegetative structure; the tendency to vegetate by sprouting or the actual occurrence of this mode of vegetation in Saccharomyces cannot determine anything, for this phenomenon occurs in the most heterogeneous fungal groups, as has been already pointed out and will be again noticed below. But our group forms asci, and this peculiarity it shares only with the Ascomycetes, if we disregard Protomyces, which, however, is much further removed from it (see p. 171), and this must be decisive at present for its connection with the Ascomycetes. Brefeld's early opinion, expressed in the year 1876, that Saccharomyces belongs to the Mucorini was disposed of when it was shown that the chief argument in its favour drawn from the similarity in the mode of spore-formation cannot be maintained.

The connection with the Ascomycetes rests entirely on the resemblances which have been pointed out between the two groups. It remains uncertain how far these resemblances are the expression of natural and phylogenetic affinity. That they are the results of such an affinity is rendered highly probable by the great agreement between the hymenia of the more highly differentiated Exoasci and typical Ascomy-

---

[1] Ann. d. sc. nat. sér. 6, IV, p. 9.
[2] Traité de Botanique.

cetes. Objections to this view are drawn from the entire absence of archicarp and ascogonium and differentiation of ascus-apparatus and envelope-apparatus. But we have constantly seen variations in the amount of differentiation in these organs within the group of the Ascomycetes, and can therefore conceive of an extreme simplification; we must perhaps wait for the discovery of more thoroughly intermediate forms.

If without waiting for more exact proof we assume a phylogenetic connection between the Exoascus-group and the Ascomycetes, two views are open to our adoption. Either the former group represents the simple *starting-point* of the series of Ascomycetes, the Saccharomycetes containing the simplest forms from which Ascomycetes have been gradually developed; or its members are greatly *reduced* Ascomycetes (see page 125), with extensively interrupted homology which is only restored with the appearance of the asci. The latter is the only admissible assumption if the account given above of the relation of the Ascomycetes to the Phycomycetes by affinity and descent is the true one, and it must be maintained with the necessary reservation so long as these relations, which are the natural conclusion from known facts, are not set aside by the discovery of new ones.

Most of the species of **Saccharomyces** which have been mentioned above are the most common and practically the most important inciters of alcoholic fermentation; they are the yeasts of fermentation and are for this reason generally termed **Yeast-fungi**. Some of them, beer-yeast for example, are purposely added to the liquid which is intended to ferment; the juices of certain fruits ferment spontaneously, the yeast-plants appearing in them without artificial assistance. This led the earlier observers to the notion, which afterwards reappeared from time to time, that the Yeast-fungi are formed without parentage from the organisable material contained in the juices of fruit. Karsten[1] in 1848 derived them from organised 'vesicles,' which had been normal parts of the cells of living plants and continued to vegetate independently after their death. But ideas of this kind have been given up for many years. We now know that the Yeast-fungi are derived from parent-forms as members of the development of a normal species of Fungus, and that their germs find their way from without into fermentable fluids; this latter point will be noticed again in a subsequent chapter. But those who upheld this view carried on a lively controversy respecting the systematic or morphological relations of the Yeast-fungi, especially before the asci of Saccharomyces had been observed by De Seynes and Reess in 1868 and 1870. Some writers regarded them as independent representatives of distinct species which always appear in the form of sprouting Fungi (Schwann, Pasteur, De Bary). Others, on the contrary, considered them to be special forms of species of Fungi which are generated in suitable fluids, and which appear in some other form, chiefly that of the Hyphomycetes, outside these fluids; they thought that the characteristic sprouting form of the plants and the yeast-fermentation depended on the nature of the medium, and that the Fungus could be brought back again to the other form, the Hyphomycetous form for example, by changing the medium. Either some particular species were indicated as capable of this transmutation, chiefly the common Moulds, species of Mucor for example, Sclerotinia Fuckeliana (Botrytis cinerea) (Bail) or Penicillium (Berkeley), or the capacity was supposed to exist in a great variety of Fungi, though here again the commonly diffused species just named stood at the head of the list (H. Hoffmann).

Continued investigations have brought to light the reasons for this variety of

---

[1] Bot. Ztg. 1848, p. 457.

opinion. They have shown that the earlier observers obtained uncertain results from having different and imperfectly distinguished forms mixed up together in their impure cultures, and have revealed another source of obscurity in their belief that every form of sprouting Fungus must be regarded as an inciter of fermentation or 'Yeast-fungus,' and conversely that all alcoholic fermentation was caused by the vegetation of a sprouting Fungus resembling Saccharomyces. We know now that this is not so.

But there are *first* of all many species of Fungi in which the only mode of vegetation is by sprouting or which vegetate in this way under certain circumstances or in certain stages of their development. Foremost among these are the ascogenous species of Saccharomyces. Connected with the latter are the forms which resemble them exactly in their vegetative construction, but in which asci and distinct spores are not known, or it should be said perhaps are not *yet* known. These are usually, and for the present rightly placed in the genus Saccharomyces; whether they really belong to it has yet to be ascertained; among them are S. apiculatus which has been so thoroughly examined by E. Hansen, and 'Pasteur's Torulae' recently investigated by the same observer. To these must be added Exoascus, also the plants mentioned above on page 114 as examples of germination by sprouting, and certain Mucorini (see page 155) with further instances in the Ustilagineae (see page 179), Tremellineae, and Exobasidium recently supplied by Brefeld (see section XCII). Lastly, we must mention Fumago (see page 249) on Zopf's authority, and a form most probably nearly related to Fumago or Pleospora and at present imperfectly known, which I formerly described as **Dematium pullulans.** It is very common on the surface of plants; and for this reason and because its sprout-cells are very like those of some species of Saccharomyces the two forms have no doubt often been mistaken for one another by earlier observers, who did not distinguish different forms very acutely. It is probable that a similar confusion is at the bottom of Pasteur's statement, that certain

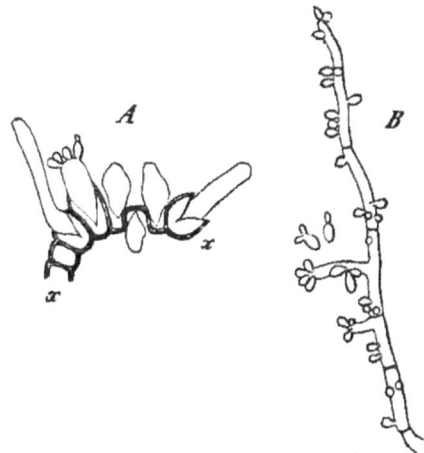

FIG. 123. *Dematium pullulans. A, x x* portion of a row of cells with brown membranes forming tubes and occasionally sprouts in a saccharine solution. *B* portion of a filament vegetating in a saccharine solution and covered with sprout-cells. *A* magn. 390, *B* nearly 200 times.

brown-walled cells which are found on succulent fruits are the resting-states of species of Saccharomyces which excite fermentation. It will be well therefore to repeat here my former description of the plant which has been confirmed by Löw (see Fig. 123).

Small ellipsoid cells sprout in large quantities in a saccharine solution or in water from the colourless branched and septate mycelial hyphae of Dematium, some from the extremities of short branches, some from their sides. They are abscised and multiply in exactly the same way as the cells of Saccharomyces. Finally, when the available food is exhausted, the mycelial hyphae divide by transverse walls into cells which are as long as broad and then swell into a roundish shape; their membranes also become thick and two-layered and of a brown colour, and they secrete small drops of oil in their interior. The free sprout-cells show the same changes under similar conditions. When they are again placed in a suitable fluid, each brown cell goes through a period of rest, and then puts out a germ-tube, which proceeds to abscise fresh cells either at once or after it has developed into a branched hypha. The sprout-cells of Dematium attain a considerable size and then become of an elongate cylindrical form, and not unfrequently grow into long septate hyphae. But many remain of a smaller size, and these together with their secondary sprouts may easily be mistaken for Saccharomyces.

The other Fungi enumerated above, with the exception of those which are classed with Saccharomyces, agree with Dematium under favourable conditions in forming hyphae; for this fact the names of the observers given above are a sufficient guarantee.

Secondly, among these sprouting Fungi some species have the power of exciting alcoholic fermentation, others have not. Both kinds are found even in the genus Saccharomyces; S. Mycoderma, the flowers of wine, usually belongs to the second category, the other ascogenous forms to the first; of the species, which as far as we know at present do not form asci, S. apiculatus belongs to the first category, S. glutinis, Cohn[1] to the second. Mucor racemosus is an inciter of fermentation when it developes the sprouting form. E. Hansen[2] has recently studied two forms, not yet certainly determined but resembling Saccharomyces, which are powerful fermenting Fungi and are developed from the hyphae of a mycelium. Most of the other sprouting forms mentioned above either excite no alcoholic fermentation or only a trace of it. The former is the case, for instance, with the sprout-cells of Exoascus Pruni, Dematium, and Fumago, which are very like those of Saccharomyces, and may be affirmed of most of the other species in the absence of any statements to the contrary. Sadebeck found signs of the formation of alcohol in the case of Exoascus alnitorquus.

Thirdly and lastly, alcoholic fermentation is not confined to the sprouting form; some species can excite it in the hyphal form also. Mucor racemosus is an instance of this and Eurotium Aspergillus glaucus also, according to Pasteur[3].

If we consider that these various matters were not originally kept separate from one another, and that, as has been already said, really distinct species were often enough mixed up with one another, we see an explanation of the protracted disputes mentioned above, and we need not enter upon any long account or critical examination of them in this place; these will be found in the publications cited on page 127, and in the works of Reess, E. Hansen, and Pasteur on Saccharomyces and yeast mentioned below, as also in the first edition of the present book. We shall arrive even in the present day at very different conclusions, according as we understand the word Yeast-fungus to mean the sprouting form, or the exciting cause of fermentation, or Saccharomyces in particular; but then our disputes become only a contest about words. From this point of view we must criticise the confusion which Brefeld has recently tried to introduce into the history of the 'Yeast-fungi.' He has done good service by proving that the sprout-form is more general than was supposed, but then he proceeds to mix all the sprout-forms together under the name of yeast; moreover since some sprout-forms, those for example of Exoascus, may according to their position in the course of development be conveniently termed the spores or gonidia of Fungi which form hyphae, he transfers this designation under the name of 'conidia-fructification' to all his Yeast-fungi, that is to his sprout-forms. In doing this he forgets or misunderstands the facts at present known about Saccharomyces of which an account has been given in the preceding pages, and to which we must adhere so long as no new ones are put in their place.

Literature.

　　1. Papulaspora and Helicosporangium.

Eidam in Cohn's Beitr. III, p. 411, t. 13.

---

[1] This name may be given to a rose-coloured form growing on boiled potatoes, which answers to Cohn's description (Beitr. I, p. 187).

[2] Bot. Centralblatt, 1884.　　　　　　　　[3] Études sur la bière, p. 100.

2. Laboulbenieae.

J. PEYRITSCH, Ueber d. Laboulbeniaceen, &c. (Sitzungsber. d. Wiener Acad. 64, Abth. 1, 1871 ; 68, Abth. 1, 1873 ; and 72, Abth. 3, 1875), in which a full account is given of the literature of the subject.

Ch. Robin (Végét. parasites) was the first to distinguish the Laboulbenieae, and Karsten first gave a more detailed description of Stigmatomyces in his Chemismus d. Pflanzenzelle, 1869.

3. Exoascus.

M. J. BERKELEY, Introd. to Cryptog. Bot. p. 284.

L. FUCKEL, Enumeratio fungorum Nassoviae, Wiesbaden, 1861, p. 29.

DE BARY, Exoascus Pruni (Beitr. I, p. 33).

L. R. TULASNE, Super Friesiano Taphrinarum genere (Ann. d. sc. nat. ser. 5, V, p. 122).

REESS, Bot. Unters. über d. Alcoholgährungspilze, Leipzig, 1870, p. 77 ;—Id. (Kutsomitopulos), Zur Kenntn. d. Exoascus d. Kirschbäume (Sitzungsber. d. Phys. Med. Ges. zu Erlangen, 11, Dec. 1862), with some notices of the literature.

P. MAGNUS in Hedwigia, 1874, p. 135, and 1875, p. 97, with a figure.

M. SOROKIN, Quelques mots sur l'Ascomyces polysporus (Ann. d. sc. nat. sér. 6, IV, 1876).

E. RATHAY, Ueber d. Hexenbesen d. Kirschbäume (Sitzungsber. d. Wiener Acad. 83, Abth. 1, März, 1881).

SADEBECK in Verhandl. d. Ges. f. Bot. zu Hamburg, I, 1881 ;—Id. in Tageblatt d. 55 Vers. d. Naturf. zu Eisenach ; see also Bot. Centralblatt, 1882, XII, p. 179 ; —Id. in Winter, Pilze von Deutschland, &c., I, Abth. 2, p. 3.

4. Saccharomyces, 'Yeast-fungi' and their allies.

L. PASTEUR, Mémoire sur la fermentation alcoholique (Ann. Chim. et Phys. LVIII, 1860) ;—Id., Études sur la bière, Paris, 1876.

M. REESS, Bot. Unters. ü. d. Alkoholgährungspilze, Leipzig, 1870, with an enumeration of the older literature ;—Id., Ueber den Soorpilz (Sitzungsber. d. Phys. Med. Ges. z. Erlangen, 9 Juli, 1877, und 14 Jan. 1878).

ENGEL, Les ferments alcooliques, 1872.

L. CIENKOWSKI, Die Pilze d. Kahmhaut (Mélanges Biol. Acad. St. Pétersbourg, VIII, p. 566).

O. BREFELD, Mucor racemosus u. Hefe (Flora, 1873) ;—Id., Ueber Gährung in Thiel's Landw. Jahrb., 1875, 1876 ;—Id., Bot. Unters. ü. Hefepilze, Leipzig, 1883.

EMIL C. HANSEN, Oidium lactis, Saccharomyces, colorés en rouge, &c. (Meddelelser fra Carlsberg Laboratoriet, I, p. 235, Résumé, p. 75) ;—Id., Unters. ü. d. Phys. u. Morph. d. Alkoholfermente in Meddelelser fra Carlsberg Laboratoriet, p. 293 or 159 (Saccharomyces apiculatus) ;—Id., Unters. ü. d. Organismen welche sich z. verschiedenen Jahreszeiten in d. Luft finden, &c. (Meddelelser fra Carlsberg Laboratoriet, p. 381 or 198) ;—Id., Unters. ü. d. Phys. u. Morph. d. Alkoholgährungspilze (Meddelelser fra Carlsberg Laboratoriet, II, p. 30 or 13) ;—Id., Bemerkungen ü. Hefepilze (Allg. Zeitschr. f. Bierbrauerei u. Malzfabrication, 1883, p. 871). See also Bot. Centralblatt, XVII, Nr. 6, 1884.

E. Löw, Ueber Dematium pullulans in Pringsheim's Jahrb. VI.

Other treatises are cited by the above authorities, especially by Hansen ; see also those mentioned above in the notes to the text.

UREDINEAE.

SECTION LXXIX.   The Uredineae, as far as we are acquainted with them, are all parasites on living Phanerogams and Ferns.   Many species are able to complete their development on one host, as will be described more at length in section CX; others are obliged to migrate from one host to another in order to arrive at certain stages of their development.

The mycelium consists of delicate copiously branched septate hyphae, and the cells thus formed, especially when they are young, contain a large quantity of orange-coloured oil-drops.   The mycelium spreads chiefly in the intercellular spaces of the parenchyma of the host, but often sends into the interior of the cells short branchlets which themselves have usually short branches, and are then considered to be haustoria (see section V).   In Hemileia only haustoria were found by Ward which have the form of unbranched vesicles on slender stalks, not unlike those of Cystopus.   We are at present able to divide the Uredineae into two groups according to the course of their development, **aecidia-forming Uredineae** and the **tremelloid Uredineae.**

The development in the former group, which will be first considered separately from the other, agrees so nearly with that of the typical Ascomycetes, that certain stages in each group may be regarded as homologous with one another, though it must be allowed that the proof of the homologies is not quite perfect.

The *sporocarps* are termed *aecidia*, and are developed in the subepidermal parenchyma of the host; the mode of development, except in one case which will be noticed in the sequel, is as follows (see Fig. 124 *I*).

The first beginnings, primordia, are found in the intercellular spaces of the parenchyma of the host and are composed of a dense weft of felted mycelial hyphae with air-containing interstices, at first scarcely larger than a cell of the parenchyma, but gradually increasing in size and displacing the surrounding elements of the tissue.   The separate cells of the originally narrow cylindrical hyphae of the primordium increase in circumference, the change proceeding from the centre to the periphery of the primordium, and it gradually acquires the appearance of a body of pseudo-parenchyma with round or ellipsoid thin-walled pellucid cells, and narrow interstices containing air.   This body, which corresponds in all important points to the one to which I formerly gave the name of *perithecium*, continues to be enclosed in a weft of ordinary mycelial filaments, which are directly continuous with the elements composing its exterior surface.   It lies with its apex just beneath the epidermis of its host, while its base reaches some way down into the parenchyma (Fig. 124 *A*). In form it is spherical or flattened vertically.   The *hymenium* now makes its appearance at the base of the aecidium, on the flat surface which abuts on the surrounding mycelium; it is composed sometimes of an irregularly shaped, but much more usually of a circular and continuous layer of short cylindrically club-shaped basidia directed vertically towards the apex; each basidium abjoints a single long row of spores in basipetal succession one after another (Figs. 124 *a*, 125) with temporary intermediate cells (page 71).   The surface-diameter of this layer and the number of its basidia are at first comparatively small, and some time is required before the ultimate breadth of the mature hymenium is reached; whether the new basidia are inserted between those which are first formed, or grow up outside them, has not been ascertained. The spores are between round and polyhedric, more rarely ellipsoid and filled with a dense protoplasm, which is sometimes colourless, but is more commonly coloured by a reddish yellow oil; the walls of the spores are colourless or of a brownish colour, and often show the prismatic structure described on page 100.

The hymenium and the rows of spores which proceed from it are enclosed in a membranous *envelope* composed of a single layer of cells (the peridium, pseudo-

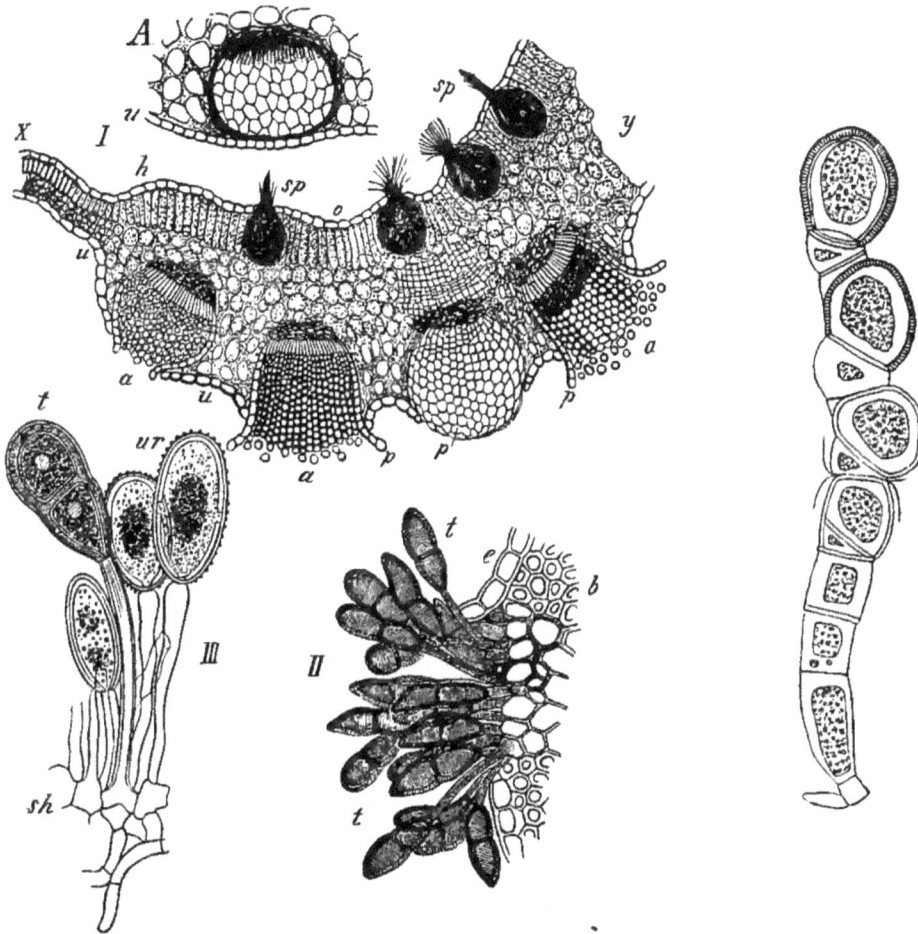

FIG. 124. *Puccinia graminis.* *A* portion of a thin transverse section of a leaf of *Berberis vulgaris* with a young aecidium beneath the epidermis *u*. *I* section through a spot in a *Berberis* leaf containing aecidia. At *X* is seen the normal structure of the leaf, the part *u—y'*, which contains the Fungus, monstrously thickened ; *h—o* upper surface of the leaf. *sp* spermogonia. *a* aecidia opened by a section through the middle, *p* their peridium. The specimen marked with *p* only without an *a* shows the peridium exposed by the section in a surface-view. *II* group of ripe teleutospores bursting through the epidermis *e* from the tissue *b* of a leaf of *Triticum repens*; *t* teleutospores. *III* teleuto-spores *t* and uredospores *ur*. See above, p. 62. From Sachs' Lehrbuch. *I* slightly magni-fied, *II* magn. 190, *III* 390 times.

FIG. 125. *Chrysomyxa Rho-dodendri.* Basidium from an aeci-dium bearing a chain of spores. Magn. 600 times. For the explan-ation of the figure see page 71.

peridium or paraphyses-envelope); the cells of the envelope are in rows like the spores, and it grows *pari passu* with the chains of spores and by the constant addition of new elements from the base. This growth is effected by a compact annular row of cells, like basidia, which occupy the margin of the hymenium, and it advances therefore exactly in the same way as that of the spore-chains, but without forming intermediate cells. The cells of the envelope are united laterally to one another without interstices, the uppermost inclining towards one another and closing over the apex of the spore-chains; and as this is done from the first, before the hymenial layer has reached its definitive breadth, the envelope is formed at least at the same time as the first basidia, and perhaps before them. All cells of the envelope are polyhedric in form, and are distinguished from the spores by their larger size, by their thicker wall which often shows a very delicate prismatic structure, and by their slightly granular or quite pellucid contents, ultimately often containing air. The spore-layer consisting of the spore-chains with the envelope increases in circumference by the constant introduction of new elements from the base and their subsequent enlargement, and encroaches on the surrounding pseudo-parenchyma. Its increase in breadth squeezes the adjacent cells together till they can often be no longer recognised. By its growth in length the summit of the perithecium is first pierced, the epidermis of the host is ruptured, and the spore-layer is raised with the summit of the perithecium above the epidermis and grows, if protected from injury, by constant additions from below into a tube filled with spores and upwards of 1 mm. in length. After it has burst through the epidermis, the cells of the envelope separate from one another at the apex, and the envelope itself opens out into the shape of a cup (Fig. 124 *I a*), or in some species (Gymnosporangium Sabinae) is split longitudinally into narrow lobes; the uppermost ripe spores fall out, and this disintegration of the envelope and the spore-chains advances in the direction of the base, more rapidly in exposed specimens and when the moisture of the surrounding atmosphere varies than in cultivated and carefully protected plants.

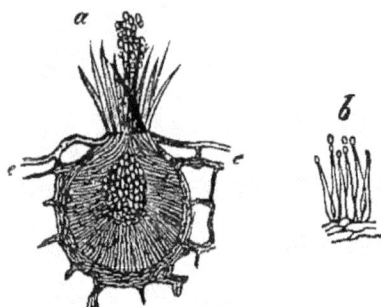

FIG. 126. *Puccinia graminis. a* median section through a spermogonium in the leaf of *Berberis vulgaris,* bursting out from the epidermis *e—e. b* sterigmata from a similar spermogonium with young spermatia. *a* magn. 200, *b* about 350 times.

The exceptional case mentioned above is exemplified in the aecidia of the genus **Phragmidium**, which are distinguished from all others in the mature state by having no compact tubular envelope. In place of an envelope a circle of club-shaped unicellular hairs, paraphyses, occupies the margin of the hymenia, which often spread out into broad cushion-like layers; their earlier development has not been investigated.

Except in a very few cases the aecidia are always accompanied by *spermogonia,* which are in all points very like the more simple organs of the same name described above in Collema and other Lichen-fungi (Figs. 124 *I sp,* 126). In most species they are small roundish to flask-shaped receptacles, looking to the naked eye like dots sunk in the subepidermal tissue, and have a thin smooth wall formed of several layers of closely woven hyphae which at length opens in the epidermis. The wall encloses a single cavity, and closely packed *sterigmata* springing from the whole of the inner surface of the cavity converge towards its centre leaving a narrow space free, which is afterwards densely filled with *spermatia.* A few rows of narrowly subulate pointed *paraphyses* or *periphyses* grow from the mouth of the receptacle instead of the sterigmata, and pierce through the epidermis and issue to the open air, forming a slightly divergent tuft, in the middle of which is the very narrow canal of the mouth.

The sterigmata are shortly subulate unicellular hyphal branches. They abjoint at their apex in basipetal order a succession of narrowly ellipsoidal spermatia wrapped in a mucilage and forming a single row. The morphological characters of these bodies which have been ascertained are the same as those described above in Collema, &c. They collect in the cavity of the spermogonium and are discharged through its mouth, when the investing mucilage swells by accession of moisture. Like the spermatia of the Ascomycetes, they have never been known to germinate.

Small deviations from the above arrangements have been observed; in some species the spermogonia are formed on the outer surface of the epidermis, only covered by the cuticle, through which they ultimately force their orifice, as in Puccinia Anemones, Peridermium elatinum and Phragmidium, and hence they have a flatter basal surface ; in others the paraphyses at the orifice do not protrude beyond the epidermis, as in Chrysomyxa, or are entirely wanting, as in Phragmidium. The spermogonia dry up when they have discharged their spermatia, and their colour, which is at first yellow or reddish yellow, changes to brown.

The spermogonia, where they occur, are always the precursors of the aecidia which belong to them, that is, they are always present in a fully developed state when the first beginnings of the aecidia make their appearance in their neighbourhood. All these facts point to a physiological relation of the two organs to one another similar to that which exists between the spermogonia and sporocarps of the Collemeae. But a more certain proof of this relation is still wanting; the organs in question must therefore for the present be classed with those in which the physiological import is doubtful.

It was conjectured that in the Uredineae as in the Ascomycetes the spermatia were sexual fertilising organs, as soon as it was understood that the spermogonia are parts of the species which form aecidia, and do not represent, as their discoverer Unger supposed, distinct species of Fungi. The arguments for that conjecture were and are the same as those which have been discussed in connection with the Ascomycetes; this is sufficiently apparent from the foregoing statements, and they need not be repeated. Further reflection suggested that only the sporocarps known as aecidia and no other bodies could be the products of the presumed fertilisation, because it is the aecidia which are almost always preceded by the formation of spermogonia and spermatia. But there have always been minor objections to this view. In some cases spermogonia occur as the companions or precursors, not of the aecidia, but of gonidiophores, the layers of the uredospores and teleutospores which will be described presently. This is the case in Puccinia fusca (Anemones), P. suaveolens and some others, but it occurs only in some particular species and the objections founded on it are easily met by the suggestion that in these species we have to do with the male gonidia-forming plants of dioecious species, while the female plants occur perhaps elsewhere or for some reason are not known. Again, I have given[1] a description of a case in which Endophyllum sempervivi, which usually forms spermogonia in the ordinary way, produced aecidia with normally developed and fertile spores on plants cultivated in perfect isolation, without any spermogonia or spermatia being found upon them. Schröter found the aecidia on some cultivated specimens of Puccinia Alliorum always unaccompanied by spermogonia; in Puccinia Galiorum on Galium Aparine, and in Uromyces Viciae Fabae on Ervum hirsutum, spermogonia were constant in the spring, but in the later part of the year there was a copious formation

---

[1] First Edition, p. 169.

of aecidia, but no spermogonia. But these may be exceptional cases connected with parthenogenetic development, and would be no valid objection to the presence of sexuality as a rule, if the latter is supported by other and sufficient reasons. But these are at present wanting. We know of nothing in the Uredineae, which, like Woronin's hypha in Xylaria, can be regarded with any probability as even a rudimentary archicarp, much less therefore as a distinct female sexual organ or even carpogonium. Yet it cannot be maintained with absolute certainty that there is no such organ, for we are still unacquainted with the first beginnings of the hymenium with its envelope ; there is a gap in the observations, and technical difficulties which are not yet overcome stand in the way of the completion of our knowledge. In young groups of aecidia a phenomenon is observed not unfrequently and without great difficulty, which seems to be in favour of the supposition that there is an archicarp duly equipped for conception ; short obtuse hyphal branches project from some of the stomata like the tips of the trichogyne in Polystigma (see on page 215) and may be traced here and there to a young perithecium. But the chief point, the continuity of such a possible trichogyne with the supposed archicarp on the one side and on the other its distinct relation to the spermatia, has not yet been shown ; there is nothing in the phenomena observed which compels us to speak of trichogynes and not simply of branches of the mycelium which may as well grow outwards from a stoma as in an inward direction into an inter-cellular space.

In presence of this uncertainty we cannot at present deal with the recently published statements of Rathay, that the spermogonia of the Uredineae allure insects to visit them by their fragrance which has long been known, by the saccharine contents of the mucilage which surrounds the spermatia, and by their colour which is often supplemented by the bright red or yellow tints of the part of the host where they occur, and thus make them involuntary agents in the dispersion of the spermatia.

The *spores*, as they may be shortly termed, which are produced in the aecidium are mature and capable of germination from the moment that they are set free from the hymenium, and retain their vitality for some weeks under favourable circumstances. In a moist environment they put out—usually from pores previously formed in their wall (see page 100)—thin-walled germ-tubes which receive the proto-plasm of the spore-cavity together with the oily colouring matter ; each spore forms as a rule only one tube. The next stages in the growth of the tubes and in connection with them the general course of development in the species exhibit in the better known cases two chief sets of phenomena.

In *the first case* observed only in the genus **Endophyllum** the germ-tube, when it has become about ten times longer than the diameter of the spore, ceases to lengthen and **assumes the characters of a promycelium** (see on page 111). It then divides at once by transverse walls, and each of the cells thus formed, generally four or five in number, with the exception usually of the lowest, sends out a short subulate lateral sterigma, and on its apex abjoints a thin-walled curved ovoid spore (Fig. 127). This is called a *sporidium* to distinguish it from other spores. Regarded from another connection it belongs to the category of gonidia, if the view of the nature of an aecidium with which we set out is the true one. The promycelium dies away when the sporidium has been abscised. The sporidia germinate under favourable conditions immediately after abscision, sending out a germ-tube which penetrates through an epidermal cell into the parenchyma of the proper host, and there developes into a mycelium which ultimately produces new spermogonia and aecidia. This completes the life history of the plant.

SECTION LXXX. In *the second* and much most common case **the germ-tube of the aecidiospores does not become a promycelium**, but forces its way by acropetal growth through a stoma into the interior of the host, and there develops directly into a mycelium. The mycelium ultimately forms *gonidia* on distinct hymenia and always by acrogenous abjunction ; and the gonidia, which can germinate either directly or after a period of rest, develope a promycelium with sporidia like that of Endophyllum. The sporidia are usually obliquely ovoid like those of Endophyllum, but in some species are round. The germ-tube of the sporidia when it has penetrated into the host gives rise to a new mycelium producing spermogonia and aecidia. These gonidia which form promycelia have been named *teleutospores* (Fig. 124 *II, III*, Fig. 128 *A*, and Fig. 129).

The species which produce aecidiospores and teleutospores are again distinguished into two subordinate groups according to the stages in the course of the development. In some the stages of the development in the order of their appearance are mycelium, aecidium, teleutospore, promycelium, sporidium, and there are no others. This is the case in **Gymnosporangium**, Puccinia section **Hemi-puccinia**, as for instance in P. Falcariae.

In other species the mycelium formed from the aecidiospore produces other gonidia besides the teleuto-spores, which are called *uredospores* (Tulasne's stylospores). These too arise by acrogenous abjunction (Fig. 124 *III, ur*), and as regards the time of their development they may be said to be always the precursors of the associated teleuto-spores. They are formed according to the species either in the same hymenia as the teleutospores, or apart from them in special ones which are simply styled uredo or uredo-layer. Being short-lived they are able to germinate immediately after abscision, and put out a germ-tube (Fig. 128 *D*) which penetrates through a stoma into the tissue of the host like the aecidiospore, and develops directly into a mycelium. The mycelium forms only new uredospores and teleutospores, and since the process of germination, the development of the mycelium, and the formation of the uredo on

FIG. 127. *Endophyllum Sempervivi,* Lev. Germinating spore with the pro-mycelium and a sporidium almost fully formed on the uppermost sterigma. Magn. 200 times.

it are very rapidly performed, the sowing of an uredospore is followed in from six to ten days under favourable circumstances by the ripening of the uredo of the next generation ; thus they are especially effective organs for the dissemination of the species which produce them abundantly.

To complete our account it must be added that the mycelium which forms aecidia may in exceptional cases subsequently produce gonidia, that is teleutospores with or without uredospores. Sometimes individuals of a species, as in Uromyces appendiculatus, sometimes single species are distinguished from the majority by the regular occurrence of this phenomenon, as Uromyces Behenis, U. Scrophulariae, U. Cestri, and Puccinia Berberidis. It is moreover not unusual for the germ-tubes of sporidia, when they cannot penetrate at once into the host, first to abscise a secondary sporidium at their apex, which has the characters of the original sporidium.

It has been already said that both kinds of gonidia are produced on hymenia on the outer surface of somewhat cushion-shaped bodies, which are formed by the interweaving of mycelial hyphae immediately beneath the epidermis of the host, more rarely at a greater depth, and burst through it when they form spores. Both are formed acrogenously on crowded sporiferous cells (sterigmata, basidia), which cover the outer surface of the hymenium, either alone or in certain species mixed with or surrounded by paraphyses of peculiar structure; in a few cases, as the uredo of Melampsora populina, and of Cronartium, they are enclosed in a one-layered

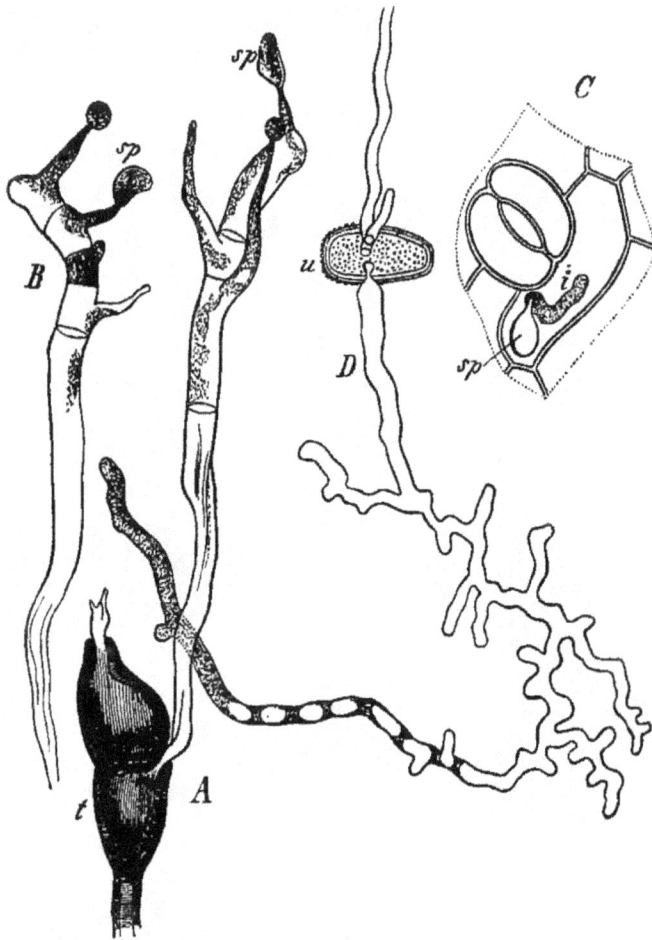

FIG. 128. *Puccinia graminis*. *A* a pair of teleutospores *t* germinating with pro-mycelium and sporidia *sp*. *B* a promycelium with sporidia *sp* torn from the spore. *C* epidermis of the under surface of the leaf of *Berberis vulgaris* with a germinating sporidium *sp*, the germ-tube from which has penetrated at *i* into an epidermal cell. *D* uredospore putting out a germ-tube fourteen hours after being placed on water. Four equatorial germ-pores are seen on the empty spore-membrane. *C, D* magn. 390 times, *A, B* somewhat more highly magnified.

FIG. 129. *Puccinia Rubigo vera.* Pair of teleutospores, the lower not having germinated, the upper germinating; *p* promycelium, *s* sporidium. Magn. 390 times.

envelope like that of the aecidia, the development of which has yet to be more closely examined.

The uredospores are formed in some species in successive rows and with intervening cells like the aecidiospores, in others singly on slender stalk-like sterigmata, from which they always separate by abscision as they ripen with a view to their dispersion. Their form and structure agree essentially with those of the aecidiospores, Hemileia being the only exception. The teleutospores of most species are formed in the same way as the uredospores, only there are no intervening cells (Figs. 129, 130); but in Triphragmium three spores are formed on the same plane by simultaneous (?) division in the solitary acrogenetic mother-cell, and in Melampsora the mother-cell is divided by longitudinal walls into several (four) spores placed side by side. The teleutospores are generally distinguished from the uredospores by two peculiarities as well as by their mode of germination. In the first place they do not separate from the sterigma, but in most species remain with it at their place of origin until they germinate; more rarely, as in some species of Uromyces and Puccinia (Uromyces Phaseolorum, Puccinia fusca, and some others), they become disengaged, though not from the sterigma, since they carry with them a portion of the sterigma, which is severed from the hymenium by a transverse rupture. Secondly, all teleutospores are filled when they are mature with a finely granular protoplasm which is either colourless or coloured by a very finely comminuted reddish yellow fatty matter surrounding a comparatively small spherical cavity, the cavity itself being filled with a weakly refringent substance. Whether this is a nucleus or a space containing a nucleus or something of another kind is not yet certainly ascertained. In other respects, such as their special mode of formation, their connection with each other and with the hymenium, their configuration and the structure of their walls, there is considerable variety among the teleutospores, and the distribution of the Uredineae into genera has been founded chiefly on these differences. The mode of germination is very similar in them all; Coleosporium only differs from other genera in the circumstance that the teleutospores, which are placed one above the other in rows, generally four together, each put out one sterigma only in germination, and the sterigma abscises a sporidium, so that a row of germinating spores resembles in outward appearance the promycelium of other genera.

It appears from the foregoing review that the simplest known course of development among the Uredineae which form aecidia is found in the Endophylleae. But even here between the aecidiospore and the next generation which forms aecidia there is always an intervening alternating generation which forms gonidia, the promycelium with its sporidia, though it is only a transitory state. In the two known species, Endophyllum Sempervivi and E. Euphorbiae, a year elapses between the formation of two successive aecidium-generations, and this time is employed in the complete development of the mycelium in the tissue of the perennial host. There may possibly be species in which the course of development is still further simplified by the absence of sporidia, the germ-tube from the aecidiospore developing directly into the mycelium with aecidia ; but these the simplest conceivable species are not known. The simple tubes which may develope from the spores of Endophyllum when they are placed under water do not become promycelia, and are, as far at least as we at present know, incapable of further development.

The intercalation of teleutogonidia and uredogonidia in the course of development in the other species, either as necessary or at least as regular members of it, makes it more complicated, and there are various degrees of complication in different species according to the greater or less abundance or the entire absence of the uredo. Besides these purely morphological gradations there are biological differences also, which give an extraordinary variety to the actual living forms in the plants by frequent change of combinations from species to species. The chief difference here is that the teleuto-spores of some species germinate as soon as they are ripe, those of others are obliged to pass through a period of rest ; the former mature in the period of vegetation, the latter mature at the end of one period and germinate at the beginning of the next.

The first species that were more thoroughly studied belong to the latter category. It was natural therefore to begin the account of their life-history with the formation of sporidia, and then the teleutospores form its close ; this is the origin of their name.

Since the special forms of these teleutospores serve to characterise form-genera which are still preserved in an altered sense, they still bear special names derived from these genera, as Puccinia spores, Triphragmium spores, Phragmidium spores, &c. Tulasne called them simply spores, to distinguish them from the rest. The *terms* uredo and aecidium at present in use were originally the names of *form-genera*, and were retained, but with an altered meaning. The fact that almost all the aecidia and uredines, unlike the teleutospores, were included in a single form-genus points to a very great similarity in the forms of the same name, and there is this similarity in fact between most of the aecidia. More minute investigation has already discovered greater variety among the uredines, and especially two chief types in the abjunction of the spores, namely, solitary abjunction on filiform sterigmata, as in Puccinia, Uromyces, and other genera, and successive serial abjunction, as in Coleosporium and Chrysomyxa. A more considerable departure from all others is found in one genus only, Hemileia.

There is no need to enter here into the details of the combinations which arise from the points of view indicated above and of their application to the determination and grouping of species and genera, since they can be seen quite well in Fungus-floras ; the reader is especially referred to Winter's Pilzflora.

SECTION LXXXI. Beside the many species of aecidia-forming Uredineae which are now known to us in the whole course of their development, there are a number of forms which resemble them so closely in certain portions of it, that they are regarded without hesitation as their homologues, though their complete life-history has not yet been ascertained. The further development of the stage with which the investigation begins has been followed for a certain distance, but without returning again to the starting-point. It is therefore also still uncertain what kind of organs actually belong to the course of development in each species. Examples of this are found in abundance in the Floras, and include every kind of known organs. In Peridermium elatinum for example, Phelonites strobilina, Aecidium Sedi, and many others we know only the aecidia, in Phelonites strobilina not even the germination of the spores ; this is known in the two other species and agrees perfectly with that of other aecidia which do not form a promycelium, but it is not known how and where the further development of the germ-tubes takes place. The teleutospores only are at present known in some Puccinieae, in Uromyces and Triphragmium echinatum, &c., and their germination has been partly observed, but we have not yet learnt what proceeds from the sporidia. A Fungus, for instance, in which the uredo only is known, is described as Uredo Symphyti, the name Uredo being still inconsistently but intelligibly employed to denote a form-genus ; in the case of some forms included under the form-genus Caeoma (Tulasne), as C. Mercurialis, C. Euonymi, and some others, it is doubtful whether they are to be considered as uredines or as naked aecidia like those of Phragmidium. Species like Melampsora Salicina and M. populina, Coleosporium Campanularum, Hemileia vastatrix, &c. form uredospores and teleutospores, and are reproduced with these spores in unlimited numbers from the germ-tubes of the uredospores. The teleutospores form promycelia with sporidia, but no one has been able to find out what becomes of the sporidia.

Cases like the one last mentioned on the one hand, and on the other the occurrence of aecidia with fertile spores outside a known typical cycle of development, would in themselves be sufficient ground for supposing that we are here, in the main at least, simply dealing with Uredineae whose course of development is the same as that described above, but is only imperfectly known to us because we have not yet ascertained the conditions for the further development of the particular organs on which the investigation turns. Experience, and especially the discovery of the phenomena of a change of host (section CX), have to a great extent confirmed this supposition; fresh typical species forming aecidia are being constantly completed out of the separate stages known to us, and we may therefore say with confidence that many of the gaps still existing are only gaps in our knowledge and not in the development of the species. This, it is true, does not exclude the *a priori* possibility that our scheme does not fit every case. There are perfectly well-known species coming under it, for instance Puccinia Rubigo vera of our corn-fields, which are reproduced year after year in frightful quantities by the uredo only. They produce also millions of teleutospores which germinate but without result, because the sporidia seldom meet with the conditions necessary for developing aecidia. Aecidia certainly are developed if the conditions are favourable; but the instance shows that the species can multiply abundantly without the interposition of aecidia, and hence there is nothing to prevent the assumption that there may also be species in which the aecidia are not only rare but are altogether wanting; perhaps they once were there but have been lost, while the other members of the development, uredospores, teleutospores, and sporidia have maintained their existence without changing their former characters. Other species may then have also lost the teleutospores which had become useless to them and become confined to the uredo only. These are at least possibilities which may be cautiously weighed and tested by further investigations. We are led up to them by another path also, which will be discussed in the next paragraph.

Section LXXXII. There are Uredineae of which it may or must be acknowledged that the cycle of their development, which is known to us throughout, is different from that of the Uredineae which form aecidia. The species here alluded to are known as the **tremelloid Uredineae,** to which the **Leptopuccinieae** and **Leptochrysomyxa** also belong.

Organs are known in the tremelloid Uredineae which in structure, development, and germination with formation of sporidia resemble in all essential points the *teleutospores* of species of Puccinia which form aecidia, and may therefore be called by that name, being distinguished at most from the teleutospores of most of these species by having their spore-membranes as a rule softer and more gelatinous. But this is not always the case; the aecidia-forming Puccinia Berberidis from Chili, for example, which unfortunately is only known to us from old specimens in collections, agrees in this respect with the Leptopuccinieae which have the softest spores. In this species of Puccinia and in all Leptopuccinieae this character is connected with another peculiarity; the teleutospores, or at least the great majority of them, germinate as soon as they mature and while they are still crowded together on the comparatively long stalks by which they are attached to their nutrient substratum. Schröter found that in some species, as Leptopuccinia Circaeae, L. Veronicae, and L. annularis, teleutospores with thicker walls were also formed in addition to the others, and that these cannot germinate without a winter's rest. The further development of the sporidia abscised from the promycelium is perfectly known, at least in L. Dianthi, L. Malvacearum, and L. Circaeae; the germ-tubes from the sporidia penetrate at once into the proper host, either through the wall of an epidermal cell, as in L. Malvacearum, or through a stoma, as in L. Dianthi, and develope a mycelium which again produces only teleuto-spores. Aecidia and uredines have never been observed in these species.

A similar relation to this between the Puccinieae which form aecidia and Leptopuccinia exists between the Chrysomyxae and Leptochrysomyxa. The former (Chrysomyxa Ledi and Ch. Rhododendri) form aecidia, teleutospores, and uredines in the regular succession. The teleutospores (Figs. 130, 131) are cylindrical cells with soft colourless membranes forming simple or branched pluricellular rows, which stand densely crowded together parallel to one another and perpendicular to the surface of the hymenium. They germinate as soon as they are ripe, and where they are formed.

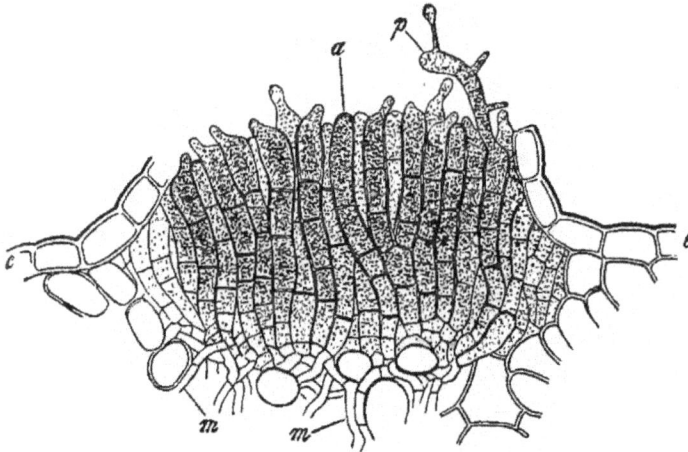

FIG. 130. *Chrysomyxa Rhododendri* in a leaf of *Rh. hirsutum.* Vertical section through a teleutospore-layer. *e—e* epidermis of the under surface of the leaf. Adjoining the spores is the tissue of the leaf traversed and distorted by mycelial filaments *m* of *Chrysomyxa*; *a* a row of teleutospores which have not yet germinated; *p* a similar row in which the uppermost teleutospore has formed a promycelium, and on this sterigmata and sporidia are beginning to be formed in basipetal succession. Most of the other rows show the first commencement of the formation of promycelia on the uppermost teleuto-spore. Magn. 140 times.

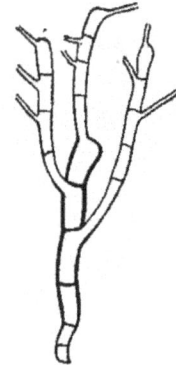

FIG. 131. *Chrysomyxa Rhodo-dendri.* A single isolated row of teleutospores after completion of germination and shedding of the sporidia.

Leptochrysomyxa (Chrysomyxa Abietis, Unger) forms exactly the same teleutospores with promycelium and sporidia, which are scarcely distinguishable from those of the two former species. But the germ-tubes from the sporidia produce mycelia which only form teleutospores; no aecidia or uredines have ever been observed on them.

It is evident that the provisions for the life of these species are different from those of the species first described. This however might be the case while the morphological conditions remained the same; but here there is the important difference, that the whole aecidium-generation is passed over and struck out of the cycle. It may be that it does occur in some Leptopuccinieae which have not been thoroughly examined as much as in Puccinia Berberidis, and that we merely have not yet learnt the place and the conditions of its occurrence. We may allow the possibility of this even in the four last-named species, which have been very often and very thoroughly examined and in which there is no indication of any formation of aecidia of their own.   But this makes no change in the facts observed in the course of development in the tremelloid Uredineae.

Schröter distinguishes by the name of **Micropuccinia** a small group, in which also teleutospores only are known, for instance Puccinia Pruni, P. Aegopodii, and P. Asari.   These forms differ from the Leptopuccinieae in the circumstance that the teleutospores drop off when they are ripe and only germinate after a long rest. We have no complete knowledge of their development.   It is possible from these data that it belongs to the tremelloid type, but we must wait for further investigations to decide this point.

SECTION LXXXIII. The **homologies within the group of the Uredineae** are at once apparent from the foregoing account.   The organs of the same name are in all of them undoubted homologues in the morphological sense.   Starting from Endophyllum, the species which are furnished with teleutospores and uredo have a number of members connected by a variety of gradations in the segment between two successive sporidia-generations; but in the tremelloid forms the homology, as far as we know is interrupted and not restored (see page 125).   If, on the contrary, we make the tremelloid forms the starting-point in the comparison, we then have in the case of the species which form aecidia an addition, an acquisition of a new and generally important member of the development, the aecidium with its attendant spermogonia.

It was pointed out at the beginning of these remarks, that the rhythm of the development in the species which form aecidia is very closely allied to that of the typical Ascomycetes.   There is the most complete agreement between the course of development in Polystigma (see page 226) and Endophyllum up to the difference between perithecium and aecidium.   At present we know of no intermediate forms between these two kinds of sporocarps.   But the difference between them is sufficient to make the homology appear questionable.   Here is the gap indicated at the outset, for all other peculiarities in the Uredineae which form aecidia are nothing more than special cases of the general course of development in the Ascomycetes.   But there are no facts recorded at present which would lead us to infer a nearer connection of the aecidia on any other side than that of the sporocarp of the Ascomycetes, and this circumstance, which is in harmony with the agreement in the general rhythm of the development, to which attention has been repeatedly drawn, turns the scale in favour of assuming that the two are homologous; anything more than an assumption is of course out of the question.   But if this assumption is correct, the Uredineae which form aecidia belong to the series of the Ascomycetes, as a special subordinate or collateral group distinguished by special peculiarities of the sporocarp.   Further, if the view taken above of the connection of the Ascomycetes with the Peronosporeae and Mucorini, and through them with the general system, is correct, the Uredineae which form sporocarps cannot have been developed from the tremelloid.   For the latter therefore there remains only the assumption that they show a retrograde development by the loss of the aecidia, being descendants of aecidia-forming species and apparently homologous with certain segments of their development, sometimes even so like them that they might be mistaken for them.

Literature.

UNGER, Die Exantheme d. Pflanzen. Wien, 1833.
The rest of the older writers are noticed in the works enumerated below.

LÉVEILLÉ, Sur la disposition des Urédinées (Ann. d. sc. nat. sér. 3, VIII ; and Article Urédinées in D'Orbigny, Dict. hist. nat.).

TULASNE, Mém. sur les Ustilaginées et les Urédinées, Ann. d. sc. nat. sér. 3, VII, and especially the Second Mémoire s. l. Urédinées et les Ustilaginées (Ann. d. sc. nat. sér. 4, II).

KÜHN, Krankheiten d. Culturgewächse, Berlin, 1859.

DE BARY, Rech. sur l. Champignons parasites (Ann. d. sc. nat. sér. 4, XX, page 64) ;— Id., Unters. ü. d. Brandpilze, Berlin, 1853;—Id., Ueber Caeoma pinitorquum (Monatsber. d. Berl. Acad. Dec. 1863) ;—Id., Neue Unters. ü. Uredineen (Monatsber. d. Berl. Acad. Jan. 1865 and Apr. 1866) ;—Id., Ueber d. Krebs u. d. Hexenbesen d. Weisstanne (Bot. Ztg. 1867) ;—Id., Aecidium abietinum (Bot. Ztg. 1870).

SCHRÖTER, Die Brand- u. Rostpilze Schlesiens (Abh. d. Schles. Ges. f. vaterl. Cultur, 1869) ;—Id., Entwicklungsgesch. einiger Rostpilze in Cohn, Beitr. I, Heft 3, p. 1, and III, 1, 51 ;—Id., Ueber einige amerikanische Uredineen (Hedwigia, 1875) ;—Id., Aecidium Euphorbiae u. Uromyces Pisi (Hedwigia, 1875).

M. REESS, Die Rostpilze d. deutschen Coniferen, Halle, 1869.

R. WOLFF, Aecidium Pini u. s. Zusammenhang mit Coleosporium Senecionis Lév. (Festschrift, Riga, 1876).

A. S. OERSTED, Om Sygdome hos Planterne, &c., Kopenhagen, 1863 ;—Id., Ueber Podisoma resp. Roestelia (Bull. de l'Acad. Roy. d. sc. de Copenhague, 1866, 1867, and k. Danske Vidensk. Selskab. Skrifter. ser. 5, VII, 1863).

WORONIN, Puccinia Heliantbi (in Russian), St. Petersb. 1871.

R. HARTIG, Wichtige Krankheiten d. Waldbäume, Berlin, 1874.—Id., Lehrbuch d. Baumkrankheiten, Berl. 1882, p. 49.

W. G. FARLOW, The Gymnosporangia or Cedar-apples of the United States (Memoirs of the Boston Soc. of Nat. Hist. Boston, 1880).

E. RATHAY, Unters. ü. d. Spermogonien d. Rostpilze (Denkschr. d. Wien. Acad. 46, Wien, 1882).

H. MARSHALL WARD, Researches on the life-history of Hemileia vastatrix (Linn. Soc. Journal (Botany), XIX) ;—Id., On the morphology of Hemileia vastatrix Berk. (Quarterly Journ. of Micr. Sc. New Series, XXI).

G. WINTER, Die Pilze Deutschlands, I.

BASIDIOMYCETES.

SECTION LXXXIV. This group is composed of two large sub-groups, the **Hymenomycetes** and the **Gastromycetes**, and derives its name from the *basidia* which are proper to all members of the group and abjoint spores acrogenously. These organs, which were described above on pages 63, 64, are closely crowded together and usually stand parallel to one another, forming *hymenia* on or in the compound sporophore; they are not unfrequently accompanied by sterile hyphal branches, which here as elsewhere may be included in Montagne's term[1] *paraphyses* (see page 53). The basidia appear almost always, it may be said indeed uniformly, as the

---

[1] Esquisse organographique, &c. des Champignons.

club-shaped bodies described above producing two or four, seldom a larger number of spores. No great deviations from the ordinary rule occur except in the Tremellineae and a few exceptional genera, as Tulostoma (section XC) and Kneiffia [1], in which latter the basidia are said to produce only one spore. The ripe spore is always abjointed as a single cell, except in a few Tremellineae and the doubtful case of Agaricus rutilus [2], and exhibits every gradation in form from spherical to narrowly fusiform.

Hymenomycetes and Gastromycetes are distinguished from one another by the structure of their compound sporophores, which we will now proceed to describe.

HYMENOMYCETES.

SECTION LXXXV. This division differs from the other in having the hymenium on the free outer surface of the compound sporophore at or more commonly before the time of the abjunction of spores. In the simplest cases, such for example as are presented by **Corticium, Dacryomyces, Exobasidium,** and some species of **Hypochnus,** the compound sporophores do not vary essentially in form and differentiation from the layers of teleutospores in the Uredineae, such for instance as the layer shown in Fig. 130, if basidia are substituted for the teleutospores. They are therefore flat or cushion-shaped bodies which bear the hymenial layer on the free surface and are attached by the opposite surface to the mycelium or substratum. From these which are the simplest forms there is a passage into more-highly developed forms and chiefly in two directions. In the one case the substratum is vertical and the margin of the compound sporophore which points upwards raises itself from the substratum and continues to grow nearly at right angles to it; in this way fan-shaped, mussel-shaped, or horse-shoe-shaped sporophores are formed, bearing the hymenium on the surface which looks towards the ground and sterile on the opposite side. In the other case the compound sporophore rises in a vertical, erect position from the usually, if not always, horizontal substratum and takes the form of the **Cap-fungi** and **club-shaped Hymenomycetes.** The former are obconical bodies or funnel-shaped or umbrella-like and borne on a stalk; in a few cases the hymenium is on the surface which is turned away from the substratum, and therefore on the upper, inner surface in the funnel-shaped form, while the rest of the surface remains sterile, as in Cyphella, Guepinia, Tul. and Exidia. Generally the opposite is the case, and the hymenium is localised on the side towards the substratum, on the surface of the cone and the lower surface of the stalked umbrella, as in Gyrocephalus of Persoon (= Guepinia helvelloides, Fr.) and most of the Hymenomycetes in the narrow use of the word. In compound sporophores of the latter kind the portion to which the hymenial surface belongs has the special name of *cap* or *pileus*, and is placed on a more or less distinct *stalk*, the *stipes*. The term cap or pileus has been extended for convenience sake from this its original signification to the compound sporophore generally in which the hymenial surface looks towards the ground, and therefore

---

[1] Fries, Hymenomyc. Europ. p. 628. Berkeley and Broome in Ann. and Mag. of Nat. Hist. ser. 4, VII, p. 429.

[2] Léveillé in Ann. d. sc. nat. sér. 2, VIII, 328.

also to those which have the shape of a horse-shoe or fan. The club-shaped Hymenomycetes are erect, club-like or cylindrical, unbranched or branched shrub-like
bodies, and have their upper part covered all round with the hymenial surface.
Calocera, Dacrymitra, and the Clavarieae are well-known examples of this subdivision.

A very extensive and almost unbroken series of intermediate forms connects
the chief forms specified above, and is most complete in the group of the Thelephoreae.

The *hymenial surface* shows considerable variety of formation, to some extent independently of the general form of the compound sporophore, and this variety has been
the chief means employed to distinguish the groups and genera of the Hymenomycetes
in the narrower sense. It is smooth or in large irregular folds, and sometimes has
also slight prominences, hair-structures, and similar formations, as in most of the
Tremellineae, Clavarieae, and Thelephoreae. In other groups it is considerably
enlarged by peculiar projections of definite shape; by teeth or regularly conical sharp-
pointed spikes in Tremellodon and the Hydneae; by plates, *lamellae*, something like a
knife-blade in shape, which radiate towards the margin of the pileus in the Agaricineae, or
are concentric with the margin in the genus Cyclomyces; lastly, in the Polyporeae
by folds or plates connected together into a reticulation, which are either shallow
(Merulius, Favolus), or grow to such a height that the meshes become comparatively
long narrow tubes (*tubuli, pori*) united laterally to one another and bearing the
hymenium on their inner surface (Polyporus, Boletus). Between these forms also,
which in their extreme condition are very characteristic, there is no lack of intermediate states. A larger or smaller number of transverse connections between
lamellae or teeth may make a species or even individuals of the same species
occupy a doubtful position on the border-line between Agaricineae and Polyporeae,
or between Hydneae and Polyporeae, and so on. The genera Irpex, Lenzites,
Daedalea, Cantharellus, &c. supply numerous instances of the kind, and should be
studied in descriptive works.

In by far the great majority of cases the construction of these compound sporophores
is by progressive growth of a primordial bundle of hyphae in the direction of the margin
or apex. There may at the same time be intercalated areas where growth continues
longer or recommences, but this point is not certainly ascertained. The instances of
progressive growth of the compound sporophores given in the general description
at page 50 are chiefly taken from the Hymenomycetes, and the reader is referred to
them for further information. Where the hymenial surfaces have projections, it is a
general rule that they retain on the average the same shape, and especially the same
breadth and distance from one another over the whole extent of the surface. The
absolute number of the projections must therefore increase in proportion to the growth
of a fan-shaped or cap-like sporophore at the periphery. This is effected in lamellae
which radiate towards the margin either by bifurcation of the original lamellae, as in
Cantharellus, Daedalea, and some species of Lenzites, or all lamellae when once
formed grow radially marginwards without branching, and new ones are formed
between them from the point where the distance between the old lamellae exceeds a
certain measure; this is what happens in most Agaricineae. Hymenial surfaces
therefore of this kind come to have successively formed lamellae, which, starting from
the margin, either extend as they were first formed to the point of insertion of

the pileus, or come to an end on the surface of the pileus at different distances from the insertion; they may be shortly spoken of as lamellae of different and successive orders.

Each projection on the hymenium, lamella for example or spike, rises above the hymenial surface, which in its earliest state is always level, in consequence of the stronger growth perpendicular to the surface of the hyphae of which it is composed, as compared with that of the portion of surface between the projections. Each projection also when once begun grows perpendicularly to the surface and advances towards the margin. This growth is quickly completed in the many short-lived forms; it may last for years in less transitory species with periodical alternations of cessation and recommencement. See page 57.

Attention has been already called at page 56 to the epinasty and hyponasty which alternate according to the age, especially in rapidly growing forms with a pileus, to

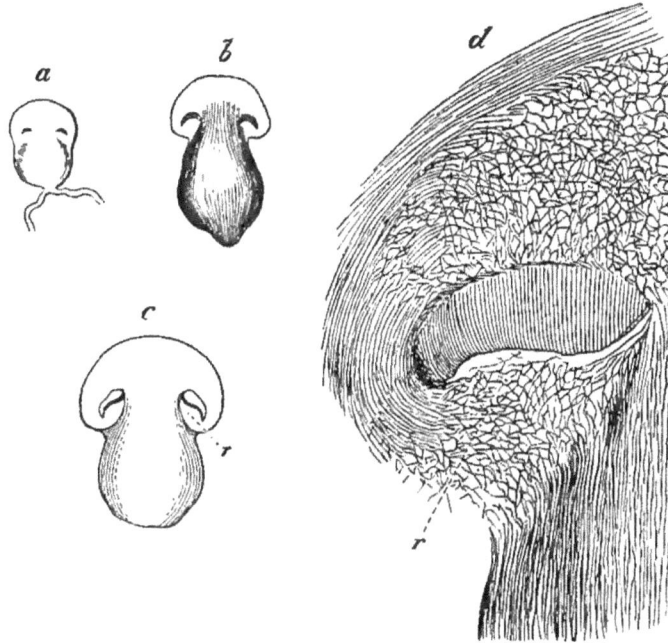

FIG. 132. *Agaricus campestris*, L. *a—c* three stages of the development of a pileus in vertical radial longitudinal section slightly magnified, *a* 6 mm., *b* 16 mm. in length. Successive stages of the development according to the letters. *d* thin section of *b*, showing the course of the hyphae, enlarged and somewhat diagrammatically represented. *r* the veil (annulus).

the consequent primary involution of the margin of the pileus or application of the hymenial surface to the stipe, and lastly to the expansion of the pileus.

Hymenomycetes with compound sporophores of purely marginal or apical progressive growth are termed *gymnocarpous*, because the hymenium lies from the first on the free surface and is not covered by any special envelope, though it is protected when young by adjacent hairs and by epinastic curvatures.

SECTION LXXXVI. Many Agaricineae and some Boleti differ in one respect from true gymnocarpous forms, since the growth of their stalked pileus or of a portion of it goes on within a special envelope, which was termed by Persoon the involucrum, by Fries the *velum* or *veil*; the latter name is still generally

[4]   U

adopted. The veil is rent by the upward extension of the pileus, and often, but not always, in such a manner that a portion of it remains behind on the stipe as an annular frill (*ring* or *annulus*).

The veil appears in two principal forms; first, as a membrane running from the margin of the pileus to the surface of the stipe, and therefore does not enclose much more than the hymenial surfaces, but leaves all the other part free; in this case it may be called a marginal veil, or with Fries *velum partiale* (Fig. 132). Secondly, as a sac which encloses the entire sporophore from its base, and is ruptured at the apex by the unfolding pileus; in this state it is the *velum universale* or *volva* (see below, Fig. 135).

1. As regards the development of the **forms which have the marginal veil** only the following special facts have been established by observation. Up to the first formation of the pileus on the summit of the stipe-primordium the phenomena are the same in essential points as in the gymnocarpous forms (Fig. 24 *a*). The young pileus is entirely delimited from the stipe by a transverse annular furrow running along its future hymenial surface. But then the superficial hyphal layers of the stipe and of the young pileus send out numerous branches towards one another from the edges of the furrow; these unite into a close weft, the marginal veil, which bridges over the furrow and closes it on the outside (Fig. 133). The body of the pileus then developes from the inner hyphal layers of the primordium, those which are nearest the furrow, and chiefly by uniform growth in the direction of the margin and by alternate epinastic and hyponastic growth, as in the species which have no veil. The veil, together with the portion of the stipe inclosed by it, follows the superficial increase in size of the pileus by intercalary growth, till the hyponastic expansion of the latter commences.

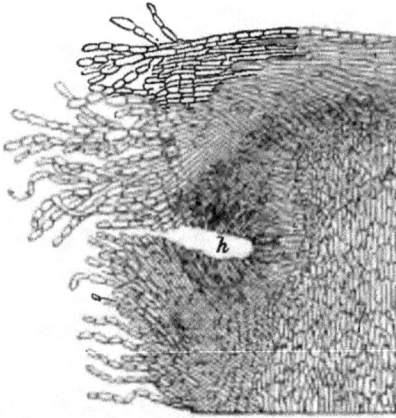

FIG. 133. *Agaricus melleus.* Half of a thin median longitudinal section through a young pileus before the closing of the veil; *h* the annular furrow between the pileus and the stipe bounded above by the commencement of the hymenium, elsewhere by the veil which has begun to develope. After R. Hartig, Wichtige Krankh. d. Waldbäume, t. II. Magnified.

The veil therefore in these cases is, as Bonorden described it, a continuation of the outermost row of cells of the stipe, which grows with the stipe for some time by intercalary growth and passes into the margin of the pileus, and conversely a continuation of the outermost hyphae of the pileus passing into the surface of the stipe; it is composed in fact of hyphae of this twofold origin. The structure is differently formed in each separate case, the difference depending on whether it begins to be formed in a somewhat later or earlier stage of the development, and in connection with this whether the hyphae which take part in its formation and run downwards from the pileus have their origin at a greater or less distance from the centre of the pileus, while those which run upwards from the stipe reach more or less of the way to the apex of the pileus; and also on whether the margin and hymenial projections of the pileus are closely applied to the stipe or are more or less distant from it in consequence of the special localisation of the epinasty. Coprinus lagopus [1] on the one hand, and Agaricus melleus [2] on the other, represent the extreme cases which have been observed. In the former the hyphae from the periphery of the stipe reach to the apex of the young pileus from the first commencement of its formation, and hyphae pass downwards from the whole outer surface of the young pileus, except the extreme apex, and become interwoven with those from the periphery of the stipe to form the veil. The margin of the pileus and the edges of the lamellae moreover lie so close to the stipe up to the time of the hyponastic upward extension, that the veil has an extremely narrow slit to cross. In Agaricus melleus the formation of the veil is comparatively late in beginning, the young pileus being already delimited from the stipe by a deep annular furrow, and the only hyphae which take part in it are the exterior hyphae of the margin of the pileus and those of the stipe which bound the furrow. Moreover the margin of the pileus is at some distance from the surface of the stipe, almost at right angles to it, and the veil therefore spans from the first a comparatively broad channel, and grows as the pileus increases in size into a collar which in vigorous specimens may be more than a centimetre in breadth. Most marginal veils are formed in the same way as that of Agaricus melleus, and Fig. 132 of A. campestris will serve to illustrate these remarks. Something will be added further on with regard to its first origin. Fig. 134 shows the phenomena in most species of Coprinus in the state succeeding the very early ones and in more advanced stages of development.

The veil continues to grow in all these cases with the general growth of the pileus until the time of the hyponastic upward expansion of the latter; then it is ruptured and in more than one way. In most species of Coprinus it separates smoothly in the direction of the surface of the stipe without leaving any conspicuous trace of itself behind either on the stipe or on the margin of the pileus. In other cases, as Hypholoma and Cortinarius, it separates from the stipe and remains hanging to the margin of the pileus as a membrane which usually tears into irregular perishable shreds, the veil or velum in the narrower sense of Persoon, the *curtain* (*cortina*) of Fries. Thirdly, the separation takes place at the margin of the pileus, the veil remains attached to the stipe as a ring running down its surface, the *annulus* or *annulus inferus*, as in Agaricus (Armillaria) melleus above described, A. (Psalliota) campestris, and many others.

Lastly, in Coprinus ephemerioides [3] hyphae of the veil grow downwards at first from the margin of the pileus between those which ascend from the stipe, while the margin of the pileus and the lamellae lie close to the stipe, as in the other species of the genus (see Fig. 134). When the elongation of the upper portion of the stipe and the upward expansion of the pileus begin, the veil is first of all torn away in consequence of the stretching from its connection with the stipe, and continuing to be united to the margin of the pileus on the stipe is carried upwards with it. Then when the upward expansion of the pileus begins, it separates from its margin and remains behind on the stipe, forming an annular sheath broader above and capable

---

[1] Brefeld, Schimmelpilze, III, t. VII.
[2] R. Hartig, Wichtige Krankheiten d. Waldbäume, t. 2.
[3] Brefeld, Schimmelpilze, VI.

of being moved up and down on the stipe in consequence of the first separation, the *annulus mobilis.*

2. In the species of the groups of Amanita and Volvaria which are **furnished with a volva,** the development of the pileus is essentially different from that of the rest of the Hymenomycetes, as far at least as has been ascertained by my older and Brefeld's recent examination of the Amaniteae. In this group the compound sporophore in its earliest state is a small tuber-like body produced on a mycelium consisting of a weft of hyphae uniformly capable of development. This body grows in every direction, and not by advance on one side only, to the size of a hazel-nut or even larger, and stipe, pileus, and lamellae are formed inside it by differentiation of the tissue, by being moulded as it were out of the originally homogeneous fundamental mass. The youngest roundish tubers, somewhat more than 1 mm. in diameter, which Brefeld saw in Amanita muscaria, are close wefts of hyphae, composed in the larger portion of the tuber of slender cylindrical cells which are

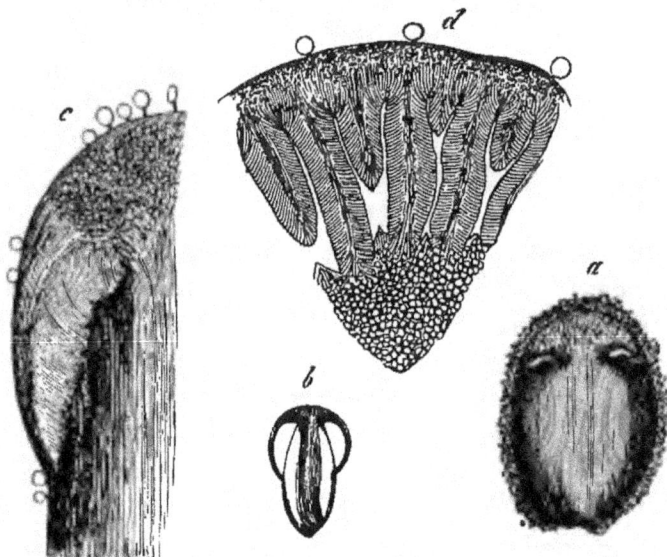

FIG. 134. *Coprinus micaceus,* Fr. *a* a young specimen 2 mm. in length in radial longitudinal section; the annular furrow beneath the future hymenial surface is bridged over on the outside by the veli. *b* a specimen 3·5 mm. in length in radial longitudinal section. *c* a thin radial longitudinal section through a somewhat younger specimen than *b.* *d* transverse section through the middle of the pileus of *c.* *a* and *b* slightly magnified, *d* magn. 90 times. *c* somewhat less highly magnified.

here and there dilated into vesicles. In a small peripheral section, which, from its position with respect to the substratum, may be called the apex, the hyphal tissue is exclusively formed of slender delicate filaments, and its elements are evidently in the act of branching copiously. This apical portion is the primordium of the pileus of the larger part of the stipe and of the volva which covers the summit of the pileus. All beside is the young base of the stipe, which is always swollen and tuber-like and is therefore termed the *bulbus.* A median longitudinal section through tubers of about twice the size shows that the tissue of the apical region is differentiated into a peripheral portion formed of many layers, the volva, and a bluntly conical body inclosed by the volva, the primordium of the pileus and stipe; both volva and primordium of pileus and stipe are lost below in the bulbus. In the next stage the flat umbrella-like commencement of the pileus appears at the apex of the primordium, being covered by the volva and borne on the broad and stoutly conical stipe-primordium (Fig. 135 *a*).

At first only the upper surface of the pileus appears distinct and separate from the volva; then the commencement of the hymenial layer or the lamellae is seen in the general shape of a narrow ring beneath the upper surface of the pileus and separated from the volva by a layer of tissue, the future body of the pileus (*b*, *c*). All the principal parts of the structure are now commenced; the whole tuber has reached a size of 10–20 mm., being irregularly ellipsoid in shape and having towards the apex a deep depression with thickened and rounded edges, in the middle of which the young pileus with the volva forms a regularly shaped cushion-like prominence. All these parts are shown successively and sharply defined in section, most sharply in a median longitudinal section, but none of them with distinct edges bounded by ever so small a gap. The definition is entirely due to the fact that the structure is different in the different parts; the young hymenium, the outer surface of the pileus, and a strip which runs vertically from this into the middle of the stipe are formed of slender hyphae and have little air or none in the interstices of the weft; they thus have an aqueous transparence and are contrasted with the tissue round them, which is white in colour from the presence of broader interstices filled with air and has inflated hyphal cells among its narrowly cylindrical elements. If the air is removed the sharpness of the definition at once disappears.

Fig. 135. *Amanita rubescens*, Fr. *a—d* radial longitudinal section through compound sporophores of different ages. Successive stages of development according to the letters. *d* a small and nearly mature specimen. *f* transverse section through *d* in the direction of the dotted line. *g* thin tangential longitudinal section through the young pileus and lamellae of *b*, slightly magnified. The inner portion only of the volva is shown. In all the figures *v* is the volva, *r* the annulus (armilla), *h* the substance of the pileus, *l* lamellae. *a—d* very slightly magnified. *a* 9 mm. long and the rest in proportion, *g* slightly magnified.

It should be particularly observed that the hymenial layer, which received a passing notice above, is seen to be separated into the future lamellae from the moment that it first becomes visible; these appear as plates of tissue stretching from the inner surface of the pileus to the outer surface of the stipe. Brefeld observed in Amanita muscaria that there was never anything more than a narrow space containing air between each pair of plates; according to Woronin's and my own previous investigations, especially into A. rubescens, it is shown by tangential longitudinal sections that a narrow plate of air-

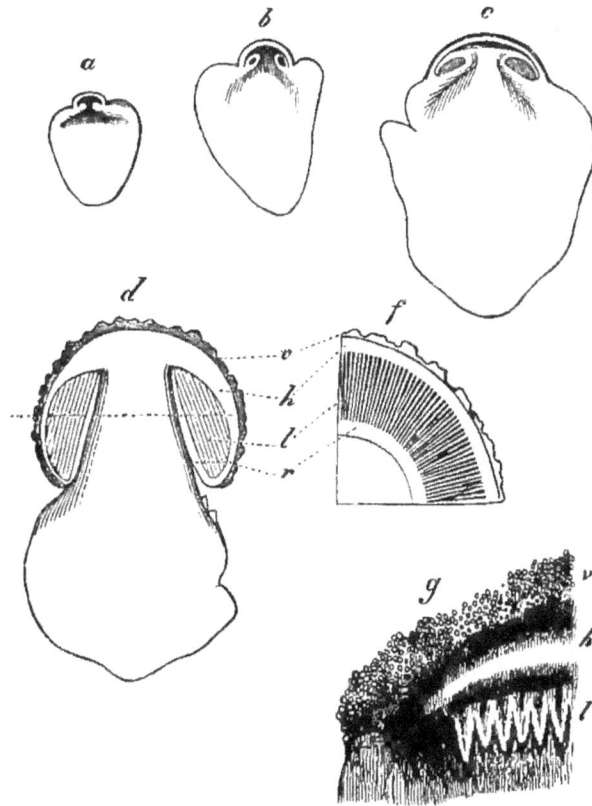

containing tissue lies originally between each pair of future lamel ae, which does not
however grow with them, but disappears as the general structure developes and thus
completes the delimitation of the surfaces of the lamellae (Fig. 135 *g*).　The dispute
is not important and arises perhaps from specific differences.　But there is no
question that the primordia of the lamellae have no free lateral margin; their
longitudinal margin on the side of the stipe passes into the tissue of the surface
of the stipe as continuously as the margin that looks towards the pileus passes into
the substance of the pileus.

The commencement of all the principal parts is followed by a general vigorous
growth throughout, and from this time the pileus and the stipe grow more rapidly
than the bulbus, the contrary having been the case in former stages.　The portion of
the stipe above the bulbus, hitherto a low narrow disk surrounded by the commencing
hymenium, grows into a stout body slightly conically pointed at the apex, and the
growth of the young pileus and of the hymenium keeps pace so exactly, that the
attachment of the lamellae to the surface of the stipe, which has been already noticed,
is not interfered with; the pileus increases in size by superficial extension in the
direction of the margin, and the lamellae which are first formed increase
correspondingly in radial length in the same direction, and in breadth in the
direction of the stipe.　The increase in breadth, however, is comparatively small, so
that the marginal portion of the pileus is somewhat curved over and bent downwards
(epinastically) in towards the surface of the stipe (Fig. 135 *d*).　Lastly, as the general
surface of the hymenium becomes broader towards the margin in consequence of
this growth, new lamellae make their appearance between the original ones, being
distinguished from them only by the fact that the edge towards the apex of the
pileus does not extend as far as the apex, while that towards the margin lengthens as
the margin advances; the later they are formed the greater the distance from the apex
of the pileus at which they terminate.　All this growth of pileus and lamellae
marginwards takes the same course as regards its direction as in the rest of the
Agaricineae; but it does not result in the formation of free margins or even of
margins bounded by gaps.　The various margins are bounded by undifferentiated
hyphal weft with intercalary growth and pass into it without any distinct boundary
line, and at the zone of transition the differentiation is constantly spreading over
newly formed sections of this weft.

These processes of growth and differentiation at length reach their limit, and
are followed by the final completion of all the parts, the last acropetal elongation of
the now tall cylindrical stipe and the hyponastic upward expansion of the pileus
—phenomena which again follow in their main features the same course as in the
species previously described.　There is only one more peculiarity to be observed in
the Amaniteae, namely, that the peripheral layer of the stipe into which the edges of
the lamellae pass, a thick loosely felted layer, takes no part in the elongation of the
inner portion of the stipe which it encloses.　It separates everywhere as the
elongation proceeds from the surface of the stipe, retaining its connection with it
only at the point of insertion of the pileus (Fig. 135 *d*); it also continues at first in
connection with the edges of the lamellae, forming a continuous membrane which
extends over the whole surface of the hymenium up to the margin of the pileus.　As
the pileus expands the connection is dissolved, beginning from the margin of the

pileus, until at length the entire membrane is attached only at the uppermost end of the stipe and hangs down from it like a frill, being cone-shaped and broadening downwards, and plaited in delicate folds corresponding to the former lines of contact with the lamellae; in this state it is known as the *annulus superus, frill* or *armilla*.

It remains to remark that the volva slowly follows the growth of the pileus, partly by increase in size of its cells, partly also during a certain period by a not very copious formation of new cells. This growth continues longest in the inner layers next the surface of the pileus, while the outer, which cease to grow earlier, are consequently torn and rent in pieces. At length it ceases altogether, and when the stipe begins to elongate greatly the volva tears all round where the margin of the pileus and the bulbus meet and is split up over the top of the pileus into the angular pieces which lie loose on the ripe pileus of Amanita muscaria, and look like white warts.

Amanita vaginata and some species of **Volvaria** appear, from the few data with which we are acquainted, to have the same or very nearly the same development as the forms just described, except that they form no armilla, the lamellae and the margin of the pileus separating smoothly from the stipe at the time of upward expansion, and that the volva is ruptured at the apex and hangs together at the base of the elongated stipe as a sac opened by lobes.

The following remarks are added to the above account by way of illustration.

1. The account given above of the development of the species which are furnished with a marginal veil is founded, wherever it departs from my former statements, chiefly upon the facts discovered by Hartig and Brefeld. It so far differs from the interpretation of these facts expressed by Brefeld, that it acknowledges the existence of only a marginal veil in the **Coprini**, Coprinus stercorarius for example and C. lagopus, whereas Brefeld thinks that these species have a volva like that of Amanita. This difference of opinion arises from the fact not mentioned above, that the apex of the pileus in these species is covered with hair-formations, with which the hyphae of the marginal veil are so interwoven that the two together form a dense *covering* over the young pileus, and also that in some species the veil itself and the surface of the young stipe are overlaid by a similar covering of hairs. The hairs begin to be developed in the earliest stage of the formation of the pileus; in etiolated specimens[1] on the top of the primordium before or without the downward growth of the hyphae which build up the margin of the pileus. They grow out from the pileus and multiply, that is, the members of the cell-rows of which they are composed are developed in progressive succession towards the future definitive surface of the pileus, and from it new cells are added one after another to the hairs already formed, and new hairs are inserted between the original ones. To this must be added in many cases the resemblance in form between the hairs of the pileus, the constituents of the veil, and the clothing of the stipe. All are in the young state cylindrical hyphae; all may retain this form during their existence, as in Coprinus lagopus[2]; or the segments swell as they mature into round vesicles, and the hair assumes the appearance of a rosary and the cells part from one another as they grow old. A desquamating covering of hairs of this kind clothes pileus and veil and stipe, as in C. ephemeroides and C. micaceus. The covering of hairs generally ceases to grow before the upward expansion of the pileus and final elongation of the stipe, and

---

[1] Brefeld, Schimmelpilze, III, t. IV, Fig. 4.
[2] Ibid. t. VII.

is ruptured and torn by these processes in different ways according to the species and thrown off by the more or less sharply defined and more persistent tissue of the definitive surface of the pileus. These are all facts which remind us of similar ones in the volva of Amanita. Here too no clear distinction can be drawn between hairs and hyphae. This is the case with the Fungus-bodies generally ; when we speak of hairs in the Fungi we always mean parts of hyphae of special form and arrangement. The peripheral elements in the species of Amanita may if necessary be also termed hairs. But the difference already pointed out in the first stages of the formation appears to me decisive in favour of the account given above. In Coprinus *the apex of the pileus is first formed*, and then the hairs sprout out from it. Brefeld's figures of C. lagopus show this clearly, and it strikes the eye especially, when, as in C. stercorarius, the hairs of the surface of the pileus have an entirely different form from those of the stipe and marginal veil[1]. In Amanita, on the contrary, the parts in question separate from one another *in the previously formed uniform compact primordial hyphal tissue* by differentiation, that is, by difference in the further development. The distinction remains unaltered, though the effect, the practical use if we may say so, is the same in both cases, namely, the formation of a *protecting envelope* for the young growing pileus. Still less is it affected by Brefeld's query, why the pileus of C. lagopus should indulge in the luxury of forming hairs from which it gains nothing but the trouble of getting rid of them again. For the situation remains the same in this respect if instead of hairs we say volva, or if it really were a volva ; and hairs or oftentimes a close felt of hair are found as a protecting covering in an infinite number of cases, not only on the pilei of Fungi, but on all possible parts of plants during a certain portion of the young state, and are thrown off before the parts are unfolded.

2. In the case of **Amanita** Brefeld's and my own previous statements complete one another without any important differences of opinion. Our direct observations on the first separation of pileus and volva would not in themselves be sufficient to establish with entire certainty the marked distinction which has been pointed out above between this genus and the forms with a marginal veil ; it would still be possible that the volva is formed in the same way as the envelope in the Coprini or in Agaricus melleus, and that the contrary conclusion arose from the unsuitableness or minute examination of the scanty material at the disposal of the observer, which must be laboriously collected in our forests. But that the processes are really different may certainly be gathered from the development of the lamellae, for which it is much less difficult to obtain suitable materials and preparations. Ocular inspection also in conjunction with the facts thus established scarcely leaves a doubt as to the correctness of the view here proposed with respect to the first separation of the volva.

3. The final result of the development as exhibited in the expanded pileus is always the same in the truly gymnocarpous forms whether they have a marginal veil or a volva. This circumstance makes it probable that intermediate forms are to be found in the long series of the Agaricineae connecting the three types here distinguished. The conjecture finds support in a comparison of the forms described above ; Agaricus melleus with its marginal veil comes nearer to the gymnocarpous forms than the closely enveloped Coprini, which rather approach the Amaniteae in the character of the envelope, though the difference before insisted on still remains. More extended and searching investigations may be expected to add to the number of intermediate forms. According to former statements of mine a further phenomenon of an intermediate character occurs in some species with the marginal veil, in which the compound sporophore is at first a small undifferentiated tuber as in Amanita, and then a narrow space containing air, a horizontal annular slit, appears in the interior of the

---

[1] Brefeld, Schimmelpilze, III, t. II, Figs. 2-4, &c.

part which is afterwards the upper extremity of the stipe; the part above the slit becomes the pileus, that below it the stipe, and that which bounds it on the outside the marginal veil. The further development is the same as that of the species with marginal veils described above. So far as these statements related to Coprinus, they have been shown by Brefeld's researches to be incorrect; my own did not pay sufficient regard to the earliest stages of the development. I will not even maintain that they are quite correct for Agaricus campestris and A. praecox, but readily allow that the facts in their case are always the same as in A. melleus and that the first extension of the marginal veil over the hymenial surface which was originally exposed had there also been overlooked. At all events the question should be further investigated. I bring forward these earlier statements chiefly to show how we may picture to ourselves the phenomena which would be intermediate between Amanita and species with a marginal veil, the modes in which differentiations may be combined with subsequent progressive growth of free margins bounded by gaps in the pileus and lamellae.

4. With the exception of certain special cases arising out of the circumstances just described, all, or very nearly all, the veiled forms here mentioned may be arranged under the chief types specified in the text. The origin also of peculiarities like the annulus mobilis in the comparatively few cases in which it occurs, as in Agaricus (Lepiota) procerus, can scarcely be different from that in Coprinus ephemeroides.

Information as to the presence or absence of the volva in the several species or groups will be found in descriptive works. It appears to be found only in the groups Amanita and Volvaria and the peculiar genus Montagnites, a form of the Agaricineae which is distinguished by the absence in the mature state of a proper pileus and which requires investigation. The lamellae in the latter case are radially disposed round the upper and somewhat broader end of the cylindrical stipe which projects from out of a volva[1].

Examples of marginal veils are supplied by the Coprini which have been described above, by the groups Lepiota, Armillaria, Pholiota, Hypholoma, Psalliota, &c. in the Agaricineae, and by Boletus luteus, B. elegans and their nearest allies. All species not belonging to the Agaricineae, the non-fleshy species among the Agaricineae and the fleshy ones of the divisions Mycena, Clitocybe, Omphalia, Pleurotus, Paxillus, Gomphidius, Lactarius, Russula, Cantharellus, Nyctalis, are, as far as can be ascertained, truly gymnocarpous. Other groups or genera which are at present considered distinct contain gymnocarpous species and species with a marginal veil. The group of Boletus luteus is an instance of the kind. Among the Coprini, which otherwise agree so closely together, Coprinus ephemerus is distinguished, according to Brefeld, from the species described above by the entire absence both of a veil and of the dense covering of hairs; in place of the latter only short, scattered, conico-cylindrical hairs are found on the surface of the pileus and stipe.

Some species of the section Collybia for instance, Agaricus dryophilus, A. tuberosus, and A. cirrhatus are gymnocarpous, while others, according to Hoffmann, as A. velutipes and A. fusipes, have marginal veils. Similar differences are found among the Cortinarii, Hygrophori, and others. We are still without comprehensive and certain knowledge of all these circumstances, nor have the various formations described from time to time as vela ever been critically examined; a more thorough investigation of these points is therefore to be urged now as it was twenty years ago.

SECTION LXXXVII. The **structure of the mature compound sporophores**, excluding from consideration the hymenial layer which will be specially noticed hereafter, in perhaps all non-fleshy and many fleshy forms is always 'hyphal' ('fädiger'), and the general rules specified above in section XIII are applicable to it. Exceptions and

---

[1] Corda, Icon. VI, t. XX; Explor. sc. d'Algérie, t. 21.

details will be found in special descriptive treatises and in Hoffmann's, R. Hartig's and Brefeld's anatomical works. Most Tremellineae are distinguished by the circumstance that their whole body, except a thin cortical layer at most, consists of gelatinous tissue capable of the utmost degree of swelling. See above, pages 9 and 12.

The structure of some fleshy species is partly pseudo-parenchymatous, and this is associated with other and peculiar appearances. The characteristic composition of almost all parts of the Amaniteae out of slender confusedly woven hyphae and inflated bladdery cells, which are segments of the hyphae, has been already mentioned. The volva also has the same structure, and is thus distinguished from the surface of the pileus, which continues to be formed of slender closely compacted hyphae, and in A. muscaria assumes the character of a gelatinous felt. Further instances of a structure which departs from that of simple hyphal filaments are seen in the Russula, Fr. and Lactarius, Fr. groups of the Agaricineae, and a description of them may be reproduced in this place.

The compound sporophores in these Fungi are round umbrella-shaped pilei with a thick central stipe, and they are of a firm fleshy consistence.

Sections in different directions through the pileus and stipe of **Russulae**, which were examined in the case of R. integra and R. olivacea by Bonorden, and in R. integra, Fr., pileo rubro and R. adusta, P. by myself, show two different kinds of tissue in all parts, except in the outermost surface, which is formed of a close weft of slender hyphal filaments sometimes and in some parts of a gelatinous consistence, as, for example, in the pileus of R. integra. There are large groups of broad, pellucid, roundish cells, and riband-like strands composed of slender branched hyphae which contain an abundance of protoplasm. The pellucid groups of roundish cells have an irregularly elongated shape in the stipe with rounded or pointed extremities, the longer diameter being parallel to the stipe. In the fleshy substance of the pileus they are roundish and irregularly disposed. The slender filaments of the plates and strands grow round them in every direction and in such a manner that the former appear on a cross section as an irregular network, the meshes of which are filled by the round-celled tissue. The arrangement of the cells in the latter shows no fixed order in the middle of the stipe and in the pileus; towards the surface of the stipe they are placed in irregular horizontal rows or layers. The size of the groups and that of the individual cells gradually diminishes from the middle to the surface of the sporophore, while the thickness of the bands of slender filaments increases. The fibrillation in the stipe is chiefly longitudinal, but follows no rule in the pileus. Branches separated from their hyphae may be seen to enter the round-celled tissue from all parts, and branch irregularly and spread in it. Closer observation of sections in which the filaments have been unravelled makes it quite plain that the round cells are connected in such a manner with the hyphae that are woven round them and spread among them, that they must be members of greatly enlarged branches of these hyphae with the form of a rosary; transition-forms also may be seen between the truly round cells and the narrow cylindrical members of the hyphae which have grown between them and around them. Bonorden was the first who called attention to this point; in others the connection and development of the two kinds of tissue still require more careful examination.

The structure of **Lactarius** has been investigated by Bonorden principally in Lactarius pallidus, by Hoffmann in L. mitissimus, and by myself in L. subdulcis, L. chrysorrhoeus, and L. deliciosus. Here, as in Russula, there are groups of broad roundish cells which appear to be set in a weft of slender cylindrical hyphae (Fig. 136). The superficial or cortical layer is formed of the latter only. The large-celled groups have much the same shape as in Russula, but they are usually narrower and more sharply defined than the strands of slender hyphae; they are often much elongated in the stipe and not unfrequently branched in the longitudinal direction or anastomose with others. In transverse sections, especially in the stipe, the cells of many of the large-celled portions are ovoid or wedge-shaped, and are so arranged, usually five or six together, round a centre that their narrow ends converge towards it, and they thus form a rosette on the transverse section. The cells thus arranged either form the large-celled group by themselves, or they are surrounded by one or more irregularly concentric layers of roundish cells ; other groups show two rosettes in the transverse section, others again show no indication of arrangement in rosettes. The small circular centre of the rosettes is formed by the transverse section of a narrow, cylindrical, thin-walled hypha with limpid cell-contents, which runs longitudinally and usually in a very winding course through the groups of large-celled tissue, as appears in a longitudinal section. The laticiferous tubes which are characteristic of Lactarius run through the strands of fine hyphal tissue, both close beside the large-celled groups and at a distance from them, but without ever entering them. These tubes have a large diameter as compared with that of the surrounding hyphae and a

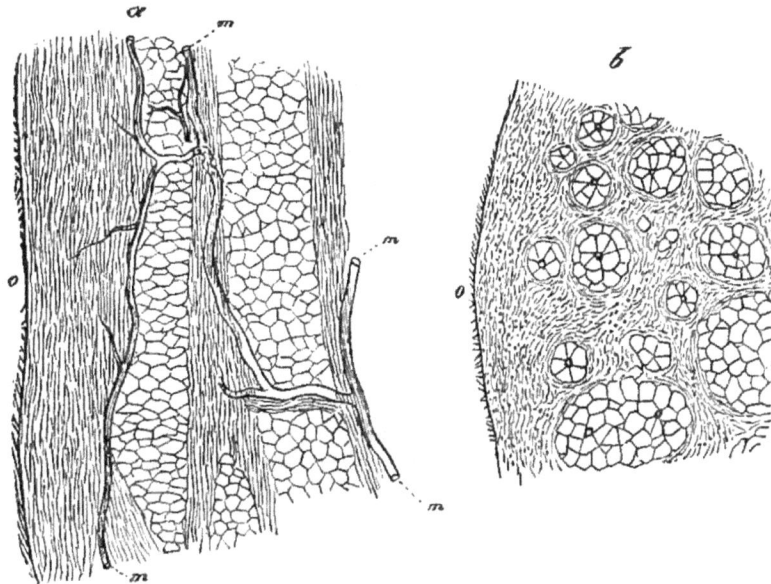

FIG. 136. *Lactarius subdulcis,* Fr. Outer region of the stipe. *a* longitudinal section. *b* transverse section ; *o* surface ; *m* laticiferous tubes. Magn. 90 times.

very soft and extensible membrane, and they are filled to overflowing with a finely-granular turbid latex which is differently coloured in the different species and oozes out in large drops from the injured Fungus. The latex coagulates at the temperature of boiling water and when treated with alcohol. It is therefore advisable in examining the course of the tubes in sections to place the Fungi for a short time in alcohol ; in order to make preparations through the isolated tubes, the parts of the Fungus should be previously boiled a short time in water. Such a preparation shows that the tubes send out numerous strong branches in every direction, which often form H-shaped connections between two primary tubes, but never give rise, as far as I have observed, to a net-work with narrow meshes. Here and there the stronger branches put out short and delicate branchlets with very slender closed blind extremities. In older specimens especially the laticiferous tubes are not unfrequently divided by single septa placed at a considerable distance from one another. These organs traverse the hyphal weft of the whole of the compound sporophore and their finer branches extend to close beneath its surface.

The foregoing account agrees in the main with Hoffmann's statements, only that observer describes an intercellular passage in the centre of the rosettes instead of the winding h'ypha, which it resembles in shape. In the species which I have examined the intercellular passage has evidently a wall of its own, which is even separated here and there by intercellular spaces from those of the adjoining cells; this is shown with more than usual distinctness in Lactarius subdulcis and L. deliciosus; but it is possible that the species differ from one another in this respect. The structure in Russula foetens var. lactiflua, Corda[1] is certainly very like that here described. I do not attempt to determine what is the cause of the difference between Corda's account and the one in the text above. Schleiden's statement that the latex of Agaricus deliciosus is 'decidedly contained in small groups of parenchymatous cells' is quite unfounded, and the same may be said of Kützing's[2] confused account of these parts.

It may not be out of place to state here, that laticiferous tubes or long-membered hyphae have also been found by De Seynes in striated hyphal tissue of the sporophore of Fistulina hepatica, which otherwise is uniform in structure. Similar organs, that is, long tubes filled with dense and often glistening cell-contents, occur in some other fleshy mushrooms, especially Agaricineae, as in Agaricus praecox and A. olearius according to Tulasne; they require further investigation.

Formations of a peculiar kind were found by Hoffmann[3] on the armilla of Amanita muscaria, but have been searched for in vain in allied species. The outer surface of this organ is covered by a thin layer of a yellowish greasy structureless substance. If this is placed in water a number of microscopically small bodies, with a glistening oily appearance and with the shape of small cylindrical rods ending usually in a knob, emerge quickly from it. They display a lively undulating and trembling movement, and changes of form such as elongation and abbreviation, forming of loops, and so on. If left to the effects of the water they cease to move after twenty-four hours or some longer period, and generally, but not always, assume the form of hollow spheres with a wall glistening like oil and watery contents. These bodies consist of a fatty or resinous substance soluble in alcohol and ether mixed with a small quantity of another substance not soluble in those fluids and turning yellow with iodine. They resemble in appearance the motile formations observed in Beneke's myelin (protagon-mixtures)[4] when placed in water. There is no reason to suppose these rods to be special organs of Amanita muscaria. The substance from which they are developed may perhaps be a product of decomposition of elements of the tissue, which were destroyed by the expansion of the pileus.

SECTION LXXXVIII. As regards the **structure of that part of the sporophore which bears the hymenium,** and which sometimes receives the special name of *hymenophorum,* it is to be observed that the hymenium itself consists of the basidia and the cells which accompany them, and that both are the terminal members of copiously branching hyphae and are placed in a vertical position on the hymenial surface. The description in section XII of the construction of the sporophore generally applies equally to the structure, interlacing, and direction of these hyphae at the commencement of their course. Their ramifications become more numerous, delicate and compact the nearer they approach the hymenium; immediately underneath it they are excessively abundant, more rich in protoplasm than the rest of the

---

[1] Corda, Icon. IV, t. X.

[2] Phil. Bot. I, p. 247.

[3] See Bot. Ztg. 1853, p. 857, and 1859, p. 212. De Bary in Flora, 1862, p. 264. Fr. Darwin in Quart. Journ. of Micr. Sc., new series, XVIII (1878), p. 74.

[4] See Beneke, Studien ü. Gallenbestandtheile, &c., Giessen, 1862.

tissue of the sporophore, and closely interwoven and united with one another. In more simple forms, as Hypochnus centrifugus, Tul. (Fig. 137) and the Tremellineae, they may still be separated from one another for considerable distances; but they usually form a delicate tissue very difficult to unravel, which is distinguished by the name of the *subhymenial layer* or *subhymenial tissue*.

Where the hymenial surface is furnished with projections of definite forms, these projections and the spaces between them are covered uniformly by the hymenium and the subhymenial tissue. Only the outermost free margin of the projections, the edge therefore of the lamellae, the orifice of the pores, the tip of the spikes, is in many species not covered by the hymenium. The inner portion of the projections which bears the subhymenial layer is named the *trama* (also dissepiment or intralamellar tissue). The trama in by far the largest number of cases is distinctly

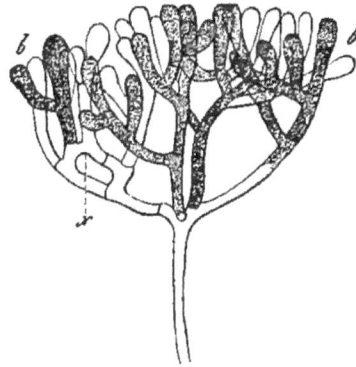

FIG. 137. *Hypochnus centrifugus*, Tul. *b, b* young basidia, the terminal cells of the ramifications of a hypha, the branches of which form a tuft; at *x* an H-shaped anastomosis. Magn. 390 times.

hyphal in structure, and consists of a hyphal mass of the form of the projection, in which the hyphae arise as branches of those of the sporophore along the whole line of insertion of the projection, enter it at its base in a straight or curved line, and run from thence to its free margin in a course parallel to the surface. The trama therefore usually exhibits a distinctly marked fibrillation running from the line of insertion to the free margin, as in many Agarici (Fig. 138), Lenzites, species of Polyporus, Trametes Pini, Hydnum zonatum, H. cirrhatum, H. gelatinosum, and in Boletus edulis. The separate hyphae in the trama pursue a straighter or more undulating and winding course according to the species. It is more unusual for the trama to be composed of a tangled hyphal-weft without definitely directed fibrillation, as in Polyporus hirsutus and P. annosus (see on page 57). The structure, consistence, colour, &c. of the constituents of the trama are either the same as those of the rest of the sporophore or they may be different from them, as a glance at the generic characters in the Hymenomycetes is sufficient to show. Sub-

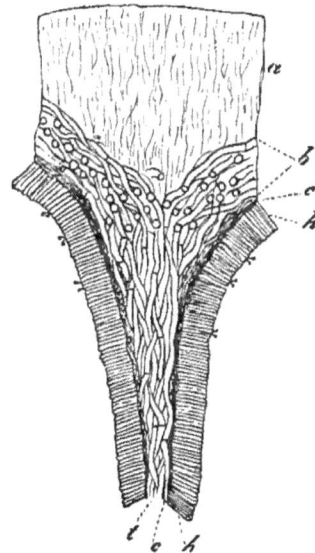

FIG. 138. *Agaricus vulgaris*. A semi-diagrammatic representation of a tangential section through a pileus which has just reached its full growth; *a* outer substance of the pileus consisting of soft gelatinous tissue, *b* lower substance of the pileus formed of stout hyphae, *c* subhymenial layer, *h* hymenium, *t* trama of the lamella. Magn. 70 times.

hymenial tissue and hymenium spring from the trama in the manner specified above; the elements of the hymenium are everywhere perpendicular to its surface.

The trama of the lamellae in the group or genus **Lactarius** shows the structure just described, at least in the case of L. subdulcis and L. chrysorrhoeus which I have myself examined. The groups of large-celled tissue become suddenly fewer and smaller towards the under surface of the pileus or hymenophorum ; this consists itself of numerous layers of hyphae which run in an undulating course from the middle of the pileus to its margin and give off the hyphae of the trama as branches. The subhymenial tissue is composed of small isodiametric cells, which show plainly by their arrangement that they are members of the delicate short-celled interwoven hyphae. The laticiferous tubes appear both in the hymenophorum and in the trama ; in the former they usually run parallel to the principal hyphae, while in the trama they spread in abundance in every direction and are much branched. The structure of the trama in Russula, and according to Hoffmann in Lactarius mitissimus also, is for the most part pseudo-parenchymatous.

Schizophyllum, a genus of the Agaricineae, differs from the rest in the structure of the lamellae. The lamellae are formed in this case, as in other genera, as projections from the hymenial surface, but they split from the edge to the middle parallel to the surfaces into two plates which curve away from one another as they continue to grow. The dorsal surfaces which are convex to one another are sterile and hairy with spreading hyphal branches ; the concave surfaces answering to the surfaces which bear the basidia in undivided lamellae with the interstices between them bear the hymenium. **Fistulina** so far agrees with Schizophyllum that its hymenium is beset with small tubes, which bear the hymenium on their inner face and are barren without ; only the tubes are formed separate from the first, according to the accounts which we possess, and not by the splitting of the trama of reticulately connected projections.

The hymenial layer itself consists of the terminal members of the subhymenial hyphae closely packed together and placed vertically to the surface. The larger number of these cells develope into basidia. Others may remain sterile and then they surround the basidia as the paraphyses surround the asci and may therefore be called by that name. They appear to be the hymenial tissue of Léveillé which he distinguishes from the basidia in his fundamental researches into the hymenium of the Mushrooms. It is uncertain whether they can be entirely wanting ; yet it has been expressly stated by more recent observers that in some cases the hymenium contains only basidia, for example by De Seynes in the case of Fistulina hepatica, and many other descriptions say the same thing by implication inasmuch as they speak only of basidia. The fact is that in many cases amongst undoubted spore-abscising basidia there are present only structures that may possibly prove to be paraphyses, but which are so very like the basidia, differing from them at most in their somewhat smaller size, smaller amount of protoplasm and present sterility, that it is not possible to say when they are examined in prepared specimens whether they might subsequently have assumed the function of basidia or not. This is the case with Agaricus melleus[1]. From these circumstances the organs in question have often been termed *sterile basidia*, a name descriptive of the appearance in all cases, but requiring critical examination.

On the other hand, cases are known in which the elements between the basidia differ from them entirely in their character. The bowl-shaped compound sporophore of Corticium amorphum, Fr. may be mentioned especially in support of the expression paraphysis, in which a few long club-shaped basidia are found between a large number of narrowly filiform branched hairs often constricted into the form of a rosary, so that the sporophore would be thought at first sight to belong to a Peziza rather than to a Hymenomycete[2]. It has long been known that the larger portion of the surface of

---

[1] Hartig, Krankheiten d. Waldbäume, t. II.   [2] Hartig, l. c. t. V.

the hymenium in Coprinus, while maturing and when mature, is covered with irregularly 3–5 angled prismatic almost isodiametric cells of uniform height and with pellucid contents. The much narrower basidia are inserted without interruption between the corners of these paraphyses-cells, alternating therefore with them, and it is only rarely that the corners of two paraphyses meet together (Fig. 139).

Other formations occur not unfrequently, different from these paraphyses-forms and in certain cases at least close beside them, which are generally distinguished from them by the circumstance that they stand out as large unicellular structures far above all others on the hymenial surface. As they are often inflated in appearance in the fleshy species Léveillé named them *cystidia*; Phoebus called them specially paraphyses.

The **cystidia** according to present accounts have been found in species of all the groups except the Tremellineae, Clavarieae, and Hydneae, but so distributed that their presence or absence and their relative frequency of occurrence vary according

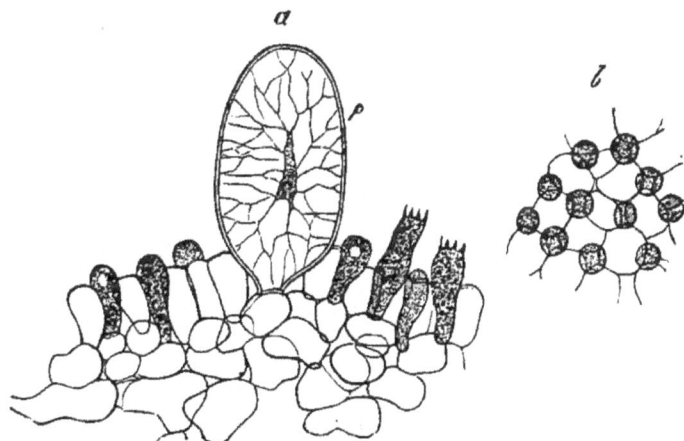

FIG. 139. *Coprinus micaceus*, Fr. *a* a thin longitudinal section through the upper surface of a lamella, the basidia distinguished by their turbidly granular contents and springing from the subhymenial cells between pellucid inflated paraphyses; *p* a cystidium. *b* surface-view of the hymenium. The intercellular space between two paraphyses to the left above appears by a mistake in the woodcut, but was not shown in the drawing from which the woodcut was taken. Magn. 390 times.

to the species within narrow cycles of affinity. While they are wanting, for instance, in most of the non-fleshy Polypori, they are found in Polyporus igniarius and Trametes Pini; while they are abundant in most Coprini they occur rarely or not at all, according to Brefeld, in Coprinus ephemerus. They originate like the basidia in the subhymenial tissue and their position is the same as theirs; but sometimes, as in Trametes Pini and Lactarius deliciosus, they terminate special branches of hyphae which ascend to the hymenial surface from the interior of the trama without directly bearing basidia also. They are sometimes scattered without order over the hymenial surface, more frequently they are found on the free margin of the hymenial processes, especially on the edge of the lamellae in the Agaricineae. Their number is usually small compared to that of the basidia, often very small; but in species of Stereum which Léveillé named Hymenochaete[1] (S. rubiginosum and S. tabacinum) the hymenium, owing to their presence, has the appearance of being covered with bristly hairs.

[1] Ann. d. sc. nat. sér. 3, V (1846), p. 150.

Their shape and size vary in the different species; they are usually constant and characteristic in each species, less so in genera and subgenera.   The large, ellipsoid or elongated, obtuse vesicles of the Coprini, which strike the naked eye, are remarkable forms which require to be especially mentioned (Fig. 139).   In a number of other species they are cylindrical, club-shaped or flask-shaped, blunt at the extremities in Polyporus umbellatus according to Corda, and in Agaricus viscidus, L. according to Phoebus, or pointed, or with a knob in Lactarius, Russula, and Boletus according to Corda; Agaricus fumosus, P. and A. laccatus, Scop. &c. have simple or branched cylindrically hair-shaped cystidia according to Hoffmann; in A. Pluteus, P. they are flask-shaped and furnished at the upper extremity with several short sharp projections which are bent a little backwards into the shape of a hook[1].   In most of the leathery or woody species in which they occur they are narrowly conical in form with sharply pointed extremities which stand out prominently 'like lance-points' from the hymenium, as in Stereum, species of Corticium, Trametes Pini, Polyporus igniarius, &c.

The structure of the cystidia is as follows.   In the species with a juicy flesh a thin and usually colourless membrane encloses the colourless cell-contents which are either protoplasm with vacuoles or quite transparent.   I found in cystidia from half-matured hymenia of Coprinus micaceus (Fig. 139 *p*) a central irregularly elongate protoplasmic body, from which numerous branched anastomosing threads with active amoeboid movement radiated to the wall; older cystidia in the Coprini are almost perfectly transparent.   In Lactarius deliciosus and allied species the cystidia are filled with densely granular opaque contents.   In this respect they resemble the laticiferous tubes, and in thick sections it often looks as though they were branches of these tubes, especially as they extend in this species far below the subhymenial tissue into the interior of the trama.   But I always observed that they sprang as branches from non-laticiferous hyphae of the trama.   The cystidia are of a deep purple-red colour in Agaricus balaninus, Berk.[2]   In the conical cystidia of the non-fleshy species the membrane, especially in the projecting extremities, is strongly thickened and coloured to correspond with the rest of the tissues.

Further details will be found in special works on the Hymenomycetes, especially those of Corda, and in a separate work on the cystidia by H. Hoffmann, and in R. Hartig's writings.

According to Corda and the doubtful statements of earlier observers, the cystidia of the fleshy Fungi discharge their contents in the form of drops through the apex which is open in the figure.   Neither I nor Hoffmann nor Brefeld were able to satisfy ourselves that this was done spontaneously; it was but rarely that I saw the cystidia burst when placed in water, and according to Hoffmann this takes place quite irregularly.   The moist surface, which often bears small drops of liquid, is a phenomenon which it has in common with all free Fungus-cells that are rich in cell-sap.

It is plain that the formations above described in the non-fleshy forms belong to the same category as those in the fleshy ones, for there is no other more general difference between them than that between fleshy and non-fleshy species.   It is as little to be disputed that the cystidia belong morphologically to the category of hair-formations, one may say indeed that they are prominent hymenial hairs.

---

[1] Ditmar, in Sturm D. fl. III, 1, t. 28.

[2] Montagne, Esquisse org. et phys. de la classe des Champignons.

What function they have as hairs is still a matter of enquiry and is perhaps different in different cases. Brefeld's suggestion is best worth considering, that they serve to protect the sporogenous basidia, and perhaps take part in the Agaricineae in loosening the appressed lamellae from the stipe. That they and especially the strikingly large vesicles of the Coprini were taken by the old observers[1] for male sexual organs, and that this notion once put into print was repeatedly being discussed for more than a hundred years is a matter now only of a certain historical interest. The terms antheridia, anthers, *pollinaria* owe their origin to these views. Further details respecting them will be found in older treatises on the formation of spores in the Basidiomycetes (see page 116), especially in Phoebus, in Tulasne[2], and in the first edition of this book at page 170.

The **basidia** themselves, and the formation of spores on them, have been already described in chapter III, pages 63, 64. We have only to add here that the club-shaped form of basidium putting out 2–4 sterigmata at its upper end, such as is represented in Figs. 28 and 30, and spores varying from round to fusiform, are found in all the Hymenomycetes which have been examined, except the Tre-

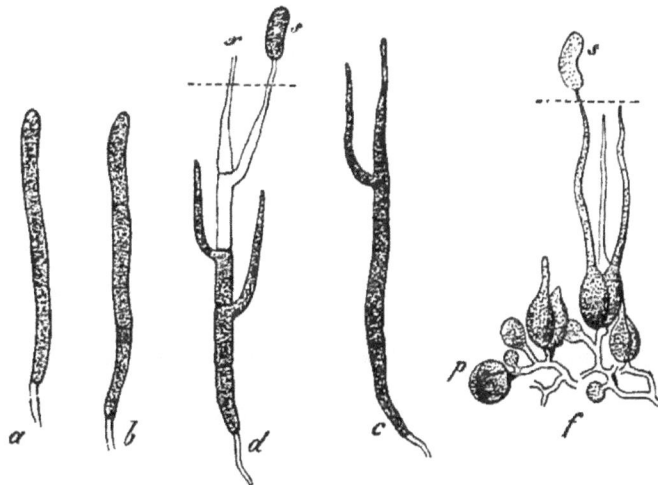

FIG. 140. *a—d Auricularia Auricula Judae.* Basidia and formation of spores. Successive stages of the development according to the figures. *a* a cylindrical terminal cell of a hypha, from which, *b*, several definitive basidia are formed by transverse divisions; each of the basidia sends out a long narrowly conical sterigma, *c, d,* from its upper extremity, and the swollen apex of the sterigma is abjointed to form a spore; *x* a sterigma from which the spore has dropped. *f Exidia spiculosa,* Sommerf. Development of basidia. Four basidia have been formed from the cell *p* by cruciform division. Younger and later stages of development are shown in the other parts of the figure: *s* a spore. The dotted lines indicate the surface of the hymenium. *a—d* magn. 390 times, *f* after Tulasne, highly magnified.

mellineae. The members of the latter group are distinguished by their variations from the general rule (Fig. 140). But there are intermediate forms. Dacryomyces, Calocera, Dacryomitra, Guepinia, and other genera have basidia which produce two spores and are distinguished from typical Hymenomycetes (Fig. 28) only by the sterigmata; these spring from a comparatively broad base and are so finely drawn out and to so great a length that the basidia appear to have long pointed bifurcations. But these forms afford no ground for a more decided separation of the Tremellineae from the other Hymenomycetes. Tremella, Exidia, and Tremellodon

[1] Micheli, Nova plant. genera.—Bulliard, Champ. de France, I, pp. 39–50.
[2] Carpologia, I, p. 163.

have another form of basidium, as was first observed by Tulasne. Subhymenial hyphal branches swell in these genera to a spherical or ellipsoid cell filled with protoplasm, the initial or primary basidium. This cell then divides by vertical longitudinal walls usually into four cells arranged as the quadrants of a sphere, secondary or definitive basidia, each of which then puts out a long sterigma and abjoints a spore upon it which absorbs the whole of its protoplasm. Slight variations occur in the number of the secondary basidia formed from a primary. But more important than these are the differences in the separation of the basidia, which may go so far that before the spore-formation each of the three sister-basidia becomes separated down to the base, or, on the other hand, all may remain united as at first; or again the division of the primary basidium may be incomplete or be omitted, and the basidium become imperfectly chambered or lobed, each division answering to a sterigma; this is the case, according to Tulasne's account, in Tremella violacea and T. Cerasi and in Sebacina incrustans, and according to Brefeld in Tremella violacea [1]. Cases of the latter kind form the transition to the two-lobed or four-lobed basidia of Dacryomyces and to ordinary Hymenomycetes.

In Auricularia Auricula Judae (A. sambucina, M.) the primary basidium is long and cylindrical and very like the basidia of Dacryomyces and Calocera. It divides by transverse walls into a row of four or five daughter-cells, each of which sends out an erect subulate sterigma, which rises above the surface of the hymenium and abjoints a spore. The sterigma issues from the apex in the uppermost basidium of a row, in the rest from the side close beneath the upper wall; the formation of sterigmata and the abjunction of spores begins in the uppermost basidium of a row; the rest follow it in order from above downwards. Exactly similar phenomena are described by Tulasne in Hypochnus purpureus, only the end of the row is in this case hooked and the terminal cell itself is sterile. The shape of the ripe spore also is peculiar, being *reniform* in most of the Tremellineae. These characters necessitate the separation of the Tremellineae as a special division of the Hymenomycetes. The gelatinous constitution of the sporophore is a convenient character in many cases, but would not be of sufficient importance by itself, the more so as Sebacina incrustans and Hypochnus purpureus, the latter of which I have not myself examined, do not appear to have gelatinous membranes.

By far the greater number of the Hymenomycetes form only one hymenial layer on each sporophore, whether it is of only brief duration or lasts longer and even for many years. The course of its development, the maturation of its parts, exhibit in general the same progressive advance towards the margin and apex as has been already described in connection with the growth of the whole apparatus which bears the hymenium. A few slight deviations however occur. On the one hand, the definitive completion and differentiation of the originally uniform elements of the hymenium are effected, according to Brefeld, in Coprinus simultaneously at all points of the hymenial surface; it has long been known that in C. micaceus and C. comatus the ripening of the spores, as shown by the lamellae turning black, even begins at the margin of the pileus and the edges of the lamellae and advances to the middle of the pileus and the base of the lamellae. On the other hand, indubitable

---

[1] See particularly the Ann. d. sc. nat. XV (1872), p. 234.

basidia have been found close to one another in very different stages of spore-formation, or apparently younger basidia among full-grown ones on a small portion of the surface of hymenia of moderate age in many Hymenomycetes. It would seem as if the basidia which appear to be younger were of later formation than those which are maturing, and had been subsequently introduced between them; in other words, that an intercalary growth of the hymenium had taken place by insertion of new elements, and the fact first mentioned of the juxtaposition of basidia in different stages of maturity supports the idea of such intercalation. It is true that the first formation of the basidia may have advanced strictly from the middle to the margin of the hymenium, and only the last differentiation and maturation have proceeded in a different direction; and the apparently young basidia may only have been paraphyses which look like basidia. More thorough investigation of this point is desirable.

In many long-lived Hymenomycetes the hymenial surface, which remains simple, enlarges in each successive period of vegetation by the growth both of the whole pileus and of the separate projections of the hymenium, the growth being in the direction of the margin, as was described above on pages 56, 57. In old Polypori which come particularly into consideration in this connection (P. fulvus and P. igniarius and Trametes Pini) the older parts of the tubes formed in the earlier years of the growth of the plant become stopped up, according to Hartig's detailed account, by a dense formation of hyphae in proportion as the tubes themselves increase in length at the margin. The hyphae proceed from subsequent ramifications of the hyphae of the wall of the adjoining tube, on which the peculiarly short and very short-lived basidia performed their office a long time before and then disappeared.

But a certain number of long-lived Hymenomycetes renew their hymenium in successive periods of vegetation even on the same portion of its surface. A part only of the parallel extremities of the hyphae which form the hymenium in the first period developes into basidia or into cystidia also. Others, which do not differ in structure from young commencing basidia, do not arrive at the formation of spores, but remain capable of development and ramify and grow out in the next period of vegetation beyond the first hymenial surface to form a new hymenium upon it, which is similar to the first and covers it everywhere. The old basidia and even their ripe spores, and in a very striking way the acutely conical hairs or cystidia where there are any, are overgrown by the new layer and enclosed in it. The process may be repeated from period to period, from year to year. In this way distinct layers corresponding to the years, or to shorter spaces of time periodically repeated, are formed in hymenia by the remains of the basidia and hairs of each period; a few only, according to Hartig, in Trametes Pini; the same observer counted 5–8 in Hydnum diversidens, in Thelephora perdix as many as 20; I myself found as many as 6 on not particularly old specimens of Corticium quercinum.

The question of '*gonidial formations*' reported in some Hymenomycetes will be deferred till section XCII in order to avoid repetition.

GASTROMYCETES.

SECTION LXXXIX. The Gastromycetes include the chief groups of the **Hyme-nogastreae, Lycoperdaceae, Nidularieae**, and **Phalloideae**; to these are joined a few smaller divisions composed partly of forms intermediate between them and partly of divergent genera and some small groups.

The compound sporophores in these Fungi spring from a simple filamentous or from a compound mycelium (see page 22). They are for the most part large, often very large, bodies; Fig. 141 represents a beautifully small specimen of a small species. When forming their spores they are all, with the exception of **Gautieria**, a genus of the Hymenogastreae, receptacles or sacs entirely surrounded by a closed wall of dense texture, the *peridium* or uterus, and usually divided inside by plates of tissue springing from the peridium into chambers in which the hymenium and the spores are formed.

Gautieria has no peridium; the chambers in the periphery lie on the free surface and are open to the outside. In all other forms the peridium varies in thickness according to the species and is often extremely thick; in many cases it is largely and peculiarly differentiated, partly into persistent, partly into temporary parts, as will be set out at greater length presently. It is a general occurrence in the course of this differentiation that the peridium becomes strongly, often very strongly, thickened at the base, though there are exceptions to this rule, as in Hysterangium and the Nidularieae. The thickened portion either projects outwards in the form of a stipe which bears the chambered portion, as in Lycoperdon and Octaviania, Fig. 141; or it projects inwards forming a cushion, as in Hymenogaster, Rhizopogon, Geaster hygrometricus (Fig. 146), or as an elongated vertical central column, as in most species of Geaster (Vittadini), the Phalloideae (see pages 316, 322) and others. The point of origin of the central column is termed the base, because in all cases in which the earliest states are known it corresponds to the point where the compound sporophore springs from the mycelium, and generally also to the place of insertion of the full-grown peridium. In some forms, as Rhizopogon and Geaster, mycelial strands run into the peridium at very various and often at many points in the outer surface, and the first commencements are not known; here therefore the expression base is only applicable by analogy.

Except in the Nidularieae and some divergent genera which will be specially considered at a future time, the chambers inclosed by the peridium are narrow irregularly curved and branched cavities, just large enough or too small to be seen with the naked eye, and separated from one another by thin curved plates of tissue

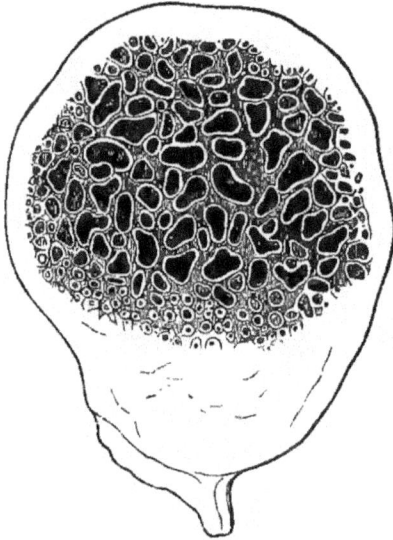

FIG. 141. *Octaviania asterosperma*, Vittad. Median longitudinal section through a nearly ripe compound sporophore. From Tulasne, Fungi hypogaei. Magn. 8 times.

which anastomose with one another in every direction and pass on one side into the tissue of the peripheral peridium, on the other it may be into that of the central column, seeming as if they radiated from it. The chambers of Polysaccum are larger, of the size even of a pea, and are less irregular.

The chambers are in most cases in countless numbers; they form altogether a mass of tissue which is distinguished from the adjoining tissue by its chambered structure and by the formation of spores and is known as the *gleba*.

As regards the more minute structure, we may distinguish a middle layer or *trama* in the walls of each chamber, and a *hymenial layer*, on both surfaces of the trama. The two parts (Fig. 142) resemble in all essential points the parts of the same name in the Hymenomycetes. In the cases which have been most thoroughly examined (the Hymenogastreae, Phalloideae, Lycoperdon, Bovista, Scleroderma, Geaster) the trama is formed of a weft of copiously branched hyphae, which chiefly run parallel to the surface of the walls, and pass without interruption from one chamber-wall to the adjoining ones and into the tissue of the peridium. Numerous closely packed branches from the hyphae of the trama run towards the interior of the chambers, and there form the hymenial tissue. In some cases they are comparatively short, of uniform height and placed palisade-like side by side and perpendicularly upon the surface of the trama; thus they form a sharply defined hymenial layer which lines the empty cavity of the chamber and is exactly like that of the Hymenomycetes (very many Hymenogastreae (Fig. 142), species of Geaster, Lycoperdon, Phallus). In another series of cases (Melanogaster, Scleroderma, Polysaccum, Geaster hygrometricus), all the hyphae which enter a chamber from the hymenium are elongated, copiously branched, and woven together into a weft which fills the chamber.

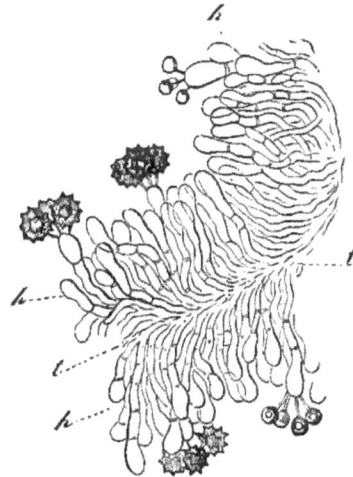

FIG. 142. *Octaviania asterosperma* Vitt. Thin section of the wall of a chamber of the gleba; *t* trama, *h* hymenium with five basidia forming spores. From Tulasne, Fungi hypogaei. Magn. 180 times.

The special formations also of basidia and paraphyses in most Hymenogastreae scarcely differ at all from those of any Hymenomycetes except the Tremellineae. The basidia of some Lycoperdaceae and Phalloideae, described already on page 63, vary a little more from them, but only in points of special conformation. Those of Geaster tunicatus and Tulostoma are strangely shaped objects; the former are ellipsoid to flask-shaped vesicles with a narrowly conical neck, the apex of which puts out about six sterigmata which abjoint spores; the latter (Fig. 143) are narrowly club-shaped cells which abjoint four almost unstalked round spores on their lateral faces.

1. There is not much to add to what has been already said of the **Hymenogastreae**. The gleba retains the structure which has been described from the commencement of its formation to its perfect maturity (Fig. 141). Its tissue is all the while either fleshy and formed of thin-walled juicy cells, the interstices conveying air or fluid, as in Hymenogaster Klotzschii, Tul. and Octaviania carnea,

Corda; or it is composed of a tough gelatinous felt, as in Hysterangium and Melanogaster.

The peridia show no marked peculiarities of structure, having a close weft like that of the walls of the chambers formed of hyphae which run chiefly in the direction of the surface. They decay after the spores have ripened and while the gleba is gradually becoming disintegrated.

The **Secotieae**, or at least the genus **Secotium** and **Cauloglossum transversarium**, are Hymenogastreae in structure with a stipe and a thick central column which traverses the entire peridium up to the apex in the line of prolongation of the stipe (Fig. 144).

2. The young compound sporophores of the **Lycoperdaceae**, with whose development we have any acquaintance (**Lycoperdon, Bovista, Geaster**), display the structure of the Hymenogastreae in all important points up to the time of the formation of spores, only the young peridium is developed on a much larger scale. One point of difference makes its appearance at an early period; two kinds of hyphae are formed in the trama when still young: slender delicate segmented hyphae rich in protoplasm, which make up the chief mass of the trama and by their branches form

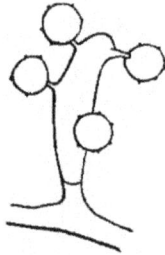

FIG. 143. *Tulostoma mammosum,*
Fr. Basidia with fully formed spores
highly magnified. After Schröter.

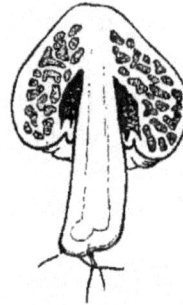

FIG. 144. *Secotium erythrocephalum,*
Tul. Sporophore divided in half, of
the natural size. After Tulasne.

the constituents of the hymenium; and stouter tubes usually non-septate, which are members or branches of the same hyphae as the delicate elements and run for the most part in the trama, but may also, as in Lycoperdon and Bovista, send branches transversely through the chambers from one wall to the wall opposite. When the spores begin to ripen the delicate hyphae and the elements of the hymenium become dissolved with copious effusion of water and entirely disappear. The thick tubes on the contrary persist and grow, and acquire a shape and structure which vary in the different genera and species, and their membranes become thickened and usually assume a lively colour, yellow passing into brown. Together they form in the compound sporophore, which ultimately becomes dry by the evaporation of the water produced in it, a woolly mass of loose texture, the *capillitium*, the interspaces of which are filled with large quantities of a dry powder, the ripe spores.

I include here under the name Lycoperdaceae all the species in which the structure of the ripe gleba points to the above course of development, excepting only for the present the genus Tulostoma, which will be considered below. They are distinguished from other groups by the capillitium between the masses of spore-dust in the ripe sporophore.

It has been already said that the peridium of the Lycoperdaceae is often of considerable thickness and is much differentiated at the time of the formation of the spores. The most important differentiation, which recurs in a variety of forms in the different species, is the separation into an inner layer directly enclosing the gleba, and an outer one which opens in different ways and becomes detached from the inner, the *inner* and *outer peridium* of authors. The genus Geaster supplies excellent examples of this. In **Batarrea** after the spores have ripened an axile strand of tissue, beneath the middle of the inner peridium about a centimetre in thickness, developes into a stout stipe, which may be as much as 2 decimetres in length and raises the closed inner peridium above the outer which has opened irregularly.

The genus **Scleroderma** agrees in the structure of its sporophore up to the formation of the spores with the Lycoperdaceae and Hymenogastreae, especially with those in which the chambers of the gleba are filled with a tangled mass of hymenial elements. Here too the hymenial tissue is dissolved and desiccation takes place when the spores are fully ripe; the chambers remain filled with the dry powdery masses of spores and the trama is disorganised, but persists as a dry fragile network in which the original structure is indistinctly shown. No capillitium with its characteristic structure is formed, at least not in the species examined by Tulasne and myself. It is evident therefore that Scleroderma is intermediate between the Lycoperdaceae and Hymenogastreae.

As regards the mode in which the Lycoperdaceae and Scleroderma form their spores, the question raised by Berkeley in 1841[1], whether the spores always attain their full development and maturity while still attached to the basidia, or whether they do not mature till after the disappearance of the constituents of the hymenium and at the expense of a portion of the products of disorganisation, as is the case in Elaphomyces (see page 97), requires further investigation; Sorokin has recently pronounced in favour of the second alternative.

3. If we imagine the entire number of the chambers in the compound sporophore in Fig. 141 reduced to from twenty to thirty and each chamber comparatively large regularly lenticular in shape and furnished with very thick walls, we get the plan of the young sporophore in **Nidularia**. When ripe the outer wall-layers of the peridium, except at the apex, and those which directly surround the cavity of each chamber with its hymenium, persist as thick membranes consisting of many layers. The tissue between these persistent layers becomes transformed into a jelly, and the wall of the peridium also disappears over the entire surface of the apex. The ripe sporophore therefore is an open bowl, in which the separate chambers as closed lenticular receptacles (*peridiola*) are imbedded in mucilage and are at length set at liberty by its disappearance. The genera **Crucibulum** and **Cyathus** exhibit the same phenomena with a still greater diminution in the number of the chambers and an augmentation of the transitory gelatinous tissue-mass; there is also a further complication, inasmuch as each peridiolum remains attached inside to the persistent wall of the peridium by a strand of tissue of complicated structure, which is likewise

---

[1] Annals and Magaz. Nat. Hist. VI, p. 431.

persistent.   This separation of the peridiola in the opened bowl-shaped or cup-shaped wall of the peridium is the distinguishing mark of the **Nidularieae**.   Each peridiolum is lined with a single hymenial layer which almost entirely fills the inner cavity; the basidia disappear after abscision of their spores, which are usually four in number.

4. The **Phalloideae**, according to our present knowledge, have a comparatively small gleba within a very massive peridial wall which is characterised by a gelatinous middle layer.   The structure of the gleba is, as in the Hymenogastreae, up to the formation of the spores; it has a great number of narrow chambers, and (except perhaps in Ileodictyon?) is traversed more or less completely by a thick central column, from which the walls of the chambers radiate.   Its entire tissue, with the exception of the column, is disorganised when the small narrowly cylindrical spores are ripe, and becomes a perfectly structureless mass of jelly which dissipates in water and, like the spores, is of a dark-green colour.   As the gleba developes a certain portion of the peridial wall is transformed into a *receptaculum in the narrower sense*, which remains at first in connection with the gleba, but becomes suddenly and greatly elongated at maturity and carries up the gleba above the wall of the peridium which has opened at the apex, and there allows it to deliquesce.

The conformation of the receptaculum is unusually different in the extreme cases. One extreme is represented by **Clathrus cancellatus** and **Ileodictyon**.   Here it is formed in the inner layer of the peridial wall surrounding the gleba as a large hollow body pierced so as to form a net-work or lattice-work.   The gleba with the central column which shares the gelatinous disorganisation of the gleba adheres to the receptaculum at its final elongation and then breaks up and drops off.   The other extreme appears in **Phallus** and the allied forms; in these the receptaculum is a simple fusiform body formed in the middle of the central column, which ruptures the ripening gleba above its apex so as to form a conical cap and finally carries it up, in consequence of its own elongation, above the ruptured apex of the peridium.   The two extremes are connected by a series of intermediate forms, the important points in which are occupied by **Clathrus (Colus) hirudinosus**[1], **Aseroe**[2], **Calathiscus**[3], and **Aserophallus**[4].

Phallus and its nearest allies are evidently the furthest removed from the rest of the Gastromycetes.   Comparison of early states, on the other hand, show an unmistakeable and close agreement and affinity between forms such as Clathrus and Ileodictyon on one side and Geaster, a genus of the Lycoperdaceae, on the other.   A nearer connection seems to be established through the intermediate genus Mitremyces, but this requires further proof.

SECTION XC.   There are many undesirable lacunae in the **history of the development of the compound sporophores** of the above four groups, chiefly owing to the difficulty in procuring them in their early states.   There is more than one cause of this difficulty; most species pass the first period of their existence beneath the

---

[1] Tulasne, Explor. sc. d'Algérie, Fungi, p. 435, t. 23, ff. 9–22.
[2] See Corda, Icon. fung. VI.
[3] Montagne in Ann. d. sc. nat. sér. 2, XVI.
[4] Montagne et Léprieur, in Ann. d. sc. nat. sér. 3, IV (1844).   See also Corda, Icon. Fung. VI.

surface of the ground, while the attempts which have been made to cultivate them have not been hitherto successful, and many of the most remarkable species grow in countries which are very inaccessible to botanists.

However it has been ascertained that as a general rule every compound sporophore consists at first of a close uniform weft of primordial hyphae, in which the changes which produce its definitive condition are effected by internal differentiation and new intercalary formations. The case may possibly be to some extent at least different in Gautieria.

The differentiation begins chiefly with the separation of the gleba and the young peridium and varies, as might be expected, in its further details according to the species or genus and group. The most important of these details will be given in the succeeding paragraphs together with the most remarkable peculiarities that have been observed in the structure of the mature compound sporophores, and some additional and critical remarks will be appended.

1. **Hymenogastreae. Hymenogaster Klotzschii** appears not unfrequently along with **Octaviania carnea** during the winter months on heath-mould in flower-pots in conservatories, growing at first beneath the surface, but soon coming above it. Its compound sporophore, in the earliest stages observed by Hoffmann and myself, is a small spherical body attached by one side to the substratum and the mycelium, and consisting of closely interwoven hyphae with narrow interstices which in part contain air. In quite small specimens 1 mm. in diameter, a median vertical longitudinal section shows a fibrillation radiating from the point of attachment, in older ones there is no apparent arrangement in the weft. The surface has from the first the same close felt of hairs as the mature peridium. Still older individuals show the chambers in the interior of the gleba in the form of narrow, air-conducting, and very sinuous cavities ; that part of the walls which adjoins the cavities contains no air and shows the structure of the hymenial layer. The chambers themselves are filled at first with loosely woven filaments which run from one wall to the wall opposite and gradually disappear.

These data prove that the parts are formed by the splitting and differentiation of the originally uniform mass of tissue. The formation begins, as far as I could determine, in the periphery and advances in the direction of the base, where a portion of the original tissue (the basal portion) remains unaltered. As the development proceeds the folds in the walls of the chambers become more and more smoothed out and the chambers are thus enlarged. The expansion of the cells of the trama has no doubt much to do with this change. What is known of other Hymenogastreae agrees essentially with the above account. There is nothing of general importance to be added to the remarks which have been made in a previous page on the structure of the mature plant, and the same may be said of the forms included in the group of the **Secotieae**. As regards other forms which have been placed by authors with **Secotium**, especially Berkeley's remarkable **Polyplocium**, it must be left to further investigations to determine whether they belong to this or to some other place.

2. **Scleroderma** and **Lycoperdaceae.** Specimens of Geaster hygrometricus of the size of a pea consist of a uniform soft tissue containing air and formed of delicate segmented hyphae ; it is of a whitish colour in the inside and brown at the circumference, and is attached to a felted mycelium which often spreads in the soil for the distance of an inch all round. Older specimens, which may be of the size of a hazel-nut in strongly developed plants, show the fibrillose layer of the peridium to be described below, in their periphery; in the interior the hyphae part from one another to form the chambers of the gleba, into which the hymenial hyphae shoot out; the layer of collenchyma, the description of which must also be deferred, is not yet present; I have never observed its formation. Here too the facts point

to a splitting and differentiation of an originally uniform hyphal weft, and the accounts which we possess would seem to show that this may be assumed of the other genera. It is true also of Scleroderma verrucosum, according to Sorokin's recent observations, except that the hymenial coil in each chamber of the gleba is formed, if that observer is correct, by the branching of a single hypha which grows from the wall into the chamber at a very early period of the development.

The *ripening of the gleba* begins in Geaster hygrometricus (Fig. 146) at the apex and advances from thence towards the base. According to Bonorden's and Tulasne's statements it begins in Lycoperdon and Scleroderma in the middle line and proceeds centrifugally; according to Sorokin the points where the ripening commences in Scleroderma verrucosum vary in different individuals, but are always situated inside the gleba.

The fully formed *capillitium* consists in most of the forms first to be considered of a countless number of single tubes or portions of hyphae, which are only woven, not grown, together and can therefore be easily isolated without tearing. Form, size, and structure in these capillitium-threads are different in different genera and species, and serve admirably to distinguish the latter. The threads are generally unsegmented and unicellular;

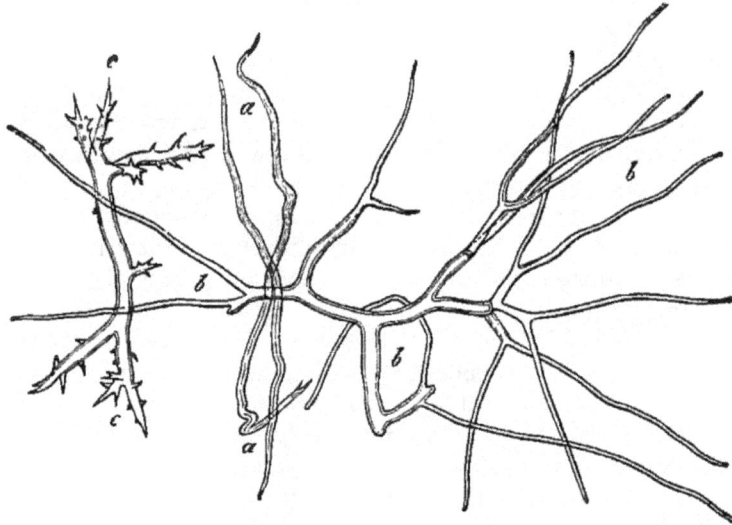

FIG. 145. Isolated threads of the capillitium *a* of *Geaster coliformis*, P.; *b* of *Bovista plumbea*, P.; *c* of *Mycenastrum Corium*, Desv. *a* magn. 190, *b* and *c* 90 times.

they are simple or very rarely branched and shortly fusiform tubes in **Geaster coliformis** (Fig. 145 *a*), elongate fusiform and usually unbranched and with the extremities very finely tapering and the membrane thickened till the lumen disappears in **G. fornicatus**, **G. fimbriatus**, and **G. mammosus**, &c. In the species of **Lycoperdon** the threads are elongated, curved, and sometimes torulose, and simple or divided into single branches which are quite irregular in their disposition; the extremities have long acuminations or are closed by a broad transverse wall which indicates the point of union of the thread with the previously formed delicate tramal hypha. Sometimes, especially in L. Bovista and L. giganteum, they have a transverse septum here and there, and their greatly thickened lateral walls are furnished with pits, which is not the case in the other genera. The threads in **Bovista** (Fig. 145 *b*) have no transverse walls and have the appearance of a many-rayed star; a short thick primary stem, which often shows plainly the former place of junction, sends out short branches in several directions; the branches form on the average four bifurcations, which increase in length but diminish in thickness with each successive order, those of the last order being prolonged into tapering hairs. **Mycenastrum** (Fig. 145 *c*) has short thick unicellular threads with the primary stem fusiform and simple or divided into a few branches and beset, especially at the extremities, with short pointed branches like spikes.

Geaster hygrometricus is the only one of the species in question with which I am acquainted in which the mature capillitium forms a connected net-work. The much branched, often torulose and unusually thick-walled threads wind in and out confusedly among one another, and their extremities are often swollen into a head and adhere firmly together. Like the earlier observers I found no true capillitium in Scleroderma, only the walls of the chambers dried up and disorganised. Sorokin speaks otherwise on this point, but it is possible that he was not examining the same species.

The *peridium* of Scleroderma is thick and leathery, but it has much the same fibrillose structure as the Hymenogastreae described above.

In **Lycoperdon, Bovista, Mycenastrum, Geaster, Sclerangium,** &c. the wall of the peridium is differentiated into two concentric separable layers, peridium interius and exterius. The inner layer is usually a thin membrane of the consistence of paper, but in Mycenastrum it is cork-like and more than two millimetres in thickness. In Bovista, Geaster, and species of Lycoperdon, the latter of which require more exact investigation, it consists of several layers of stout hyphae which run in the direction of the surface and are firmly intertwined, having generally the structure and appearance of capillitium-threads. In Geaster hygrometricus the elements of the peridium are exactly like those of the capillitium, and are continuous with them, and the capillitium-net is therefore concrescent everywhere with the peridium. In all the species of Lycoperdon which have been examined and in Geaster fimbriatus and G. fornicatus the hyphae are distinguished from the capillitium-threads by their smaller size and clearer colour, but they send countless branches into the interior of the peridium, which, as long as they are free, have all the characters of capillitium-threads. The inner peridium of Bovista plumbea has a similar structure and the same dense woolly covering on the inner surface formed of threads which have their origin in the weft of the peridium ; but these do not resemble the capillitium-threads, nor are they connected with them, but are long fine tapering unbranched hyphae. The inner peridium of Mycenastrum Corium is a dense irregular air-containing weft of hyphae with brown membranes, more finely fibrillose and denser in the outer region than in the inner. The hyphae end on the inner surface in pointed branches which resemble but are always more slender than the capillitium-threads.

We know from descriptions that the inner peridia of most forms, Geaster for instance especially, ultimately open in a fixed manner at the apex and discharge the spores. The anatomical conditions which lie at the bottom of the process have not been thoroughly examined. Where the inner peridium passes into a central column or a sterile basal portion of the gleba, these parts, so far as they have been examined, are found to have essentially the structure of the trama. The stalk-like projecting basal portion of Lycoperdon has chambers like the gleba, but the walls of the chambers are sterile or furnished only with insignificant traces of basidia.

The *outer peridium* in **Mycenastrum** is a whitish soft thin membrane covering the inner peridium and composed of a loose weft of colourless thin-walled cylindrical hyphae ; when mature it peels off in lobes and at length leaves the inner peridium uncovered. The peridium externum is more highly developed in **Lycoperdon** and **Bovista.** In these genera it is formed of large-celled usually pseudo-parenchymatous tissue, which in some cases, as Bovista plumbea, is seen to consist of several layers and often has external projections in the shape of warts, spikes, &c. In the young state it adheres firmly to the inner peridium, the elements of the two passing into one another. At maturity, when the copious excretion of water already described takes place in the compound sporophore in forming the spores, the inner layer of the outer peridium becomes disorganised and is changed into a slimy or fluid mass, which soon dries up or is absorbed, and hence the outer peridium often peels off from the inner and falls away (Vittadini). In some, perhaps in most species, as for instance in Bovista plumbea and Lycoperdon perlatum, according to Tulasne and Vittadini, the disorganisation affects the whole of the outer peridium and it becomes changed into a slimy mass, which turns as it dries into a brittle and almost structureless membrane.

The structure of the peridium is more complicated in **Geaster**. G. hygrometricus is up to the period of perfect maturity a roundish body, which may be of the size of a hazel-nut and remains beneath the surface of the ground (Fig. 146). Six layers may be distinguished in the peridium in a vertical longitudinal section a short time before the compound sporophore is mature. The outermost layer is of a brownish colour, flaky and fibrous, and is continued on one side into the mycelial strands which spread through the soil and on the other passes into the second layer; a thick stout brown membrane entirely covering the compound sporophore. This is followed towards the inside by a white layer, which is more largely developed at the base of the compound sporophore than elsewhere and is immediately continuous at that spot with the inner peridium and the gleba. Both of these last-mentioned layers are formed of stout closely-woven hyphae running in the direction of the surface, and may be combined under the name of the *fibrillose layer*. The inner of the two is lined on the inside by the *collenchyma-layer* (Fig. 146 *c*), except where its basal portion passes into the gleba. This layer is cartilaginously gelatinous and consists of hyphal branches of uniform

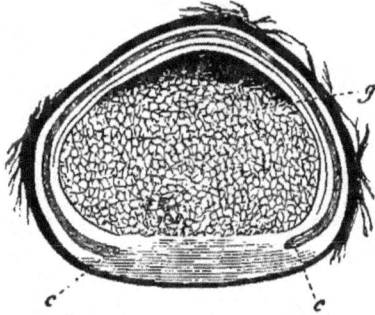

FIG. 146. *Geaster hygrometricus.* Vertical median longitudinal section of a fully grown nearly mature specimen, very slightly magnified. *c* collenchyma-layer, *g* gleba, the apex of which is beginning to assume a dark colour as the spores ripen.

height connected together without interstices, which are placed palisade-like vertically to the surface and are bent as they spring from the hyphae of the fibrillose layer. The strongly thickened stratified walls of the cells of this layer have great capacity for swelling. Inwards from the collenchyma is a white layer, the innermost region of which is the *inner peridium*, while the outer, which may be called the *split-layer*, consists of soft loosely woven hyphae which pass at many points into the inner peridium. When the Fungus is quite matured, the outer peridium, through the influence of moisture and through the swelling of the collenchyma-layer, bursts outwards from the apex in a stellate manner, forming several lobes which turn back, so that the upper surface which is covered by the collenchyma becomes convex. The split-layer is by this means so torn to pieces that its constituent parts remain hanging as perishable flakes, some to the collenchyma, some to the inner peridium. It is known that the collenchyma-layer retains its hygroscopic qualities a long time, and the outer peridium remains a long time lying on the soil, stellate in shape, spreading out its rays in moist weather and bending them inwards in dry. The flaky investment of the outer peridium is often more strongly developed in Geaster fimbriatus and G. fornicatus than in G. hygrometricus, and in G. fornicatus it is composed of the finest of hyphae; it tears away from the fibrillose layer when the peridium is ruptured and lies on the ground beneath the peridium as an open empty sac. The extremities of the lobes remain for the time firmly united to the margin of this sac, and as the collenchyma-layer expands greatly, the star formed by it and the fibrillose layer, especially in G. fornicatus, becomes convex upwards, and carries the inner peridium on the apex of the convexity. The fibrillose layer is comparatively thinner in these and other species than in G. hygrometricus and is not divided into two layers. The collenchyma-layer consists of large-celled transparent pseudo-parenchyma, which swells up strongly in water and causes the opening of the peridium and the convexity by its expansion, whether this is due to swelling only, or perhaps to growth also. In G. fornicatus,

G. fimbriatus, G. coliformis, and others its cells are delicate, and it becomes full of fissures soon after the peridium has opened, and unable to assist in the bending of the rays. In G. mammosus and, according to Tulasne, in G. rufescens on the contrary it has the same persistent hygroscopic qualities as in G. hygrometricus.

3. The differences between the genera **Batarrea** and **Podaxon** and the typical Lycoperdaceae which have been hitherto under consideration are sufficiently striking to require a special description. The early stages of the development are not known in either of them.

A half-ripe specimen of Batarrea Steveni from the south of Russia examined by myself was in shape a cushion-like body (Fig 147 *a*) with a regularly convex upper surface and a diameter of nearly seven centimetres. The vertical median section shows a structure, which may be roughly compared with that of a nearly ripe Geaster. An inner peridium shaped like the plano-convex blunt-edged pileus of an Agaric with an average thickness of one centimetre incloses the almost ripe gleba, which has a scleroderma-like structure, except that many of the stronger walls of the chambers run vertically from the upper to the lower surface; isolated capillitium-threads occur amongst the spore-dust. The threads are short and obtusely fusiform

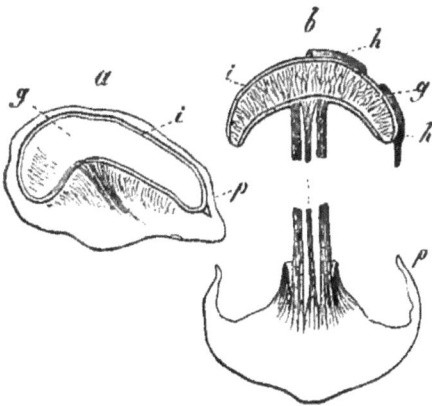

FIG. 147. *Batarrea Steveni*, Fr. Vertical median longitudinal sections. *a* a younger specimen but with most of its spores already ripe. *b* a mature specimen (in the latter the apex only and base of the stipe are depicted). *p* and *h* the outer, *i* the inner peridium, *g* the gleba; the lines in the gleba show the position of the stronger remains of the trama. One third the natural size, semi-diagrammatically represented.

FIG. 148. *Batarrea Steveni*, Fr. Isolated threads of the capillitium, magn. 390 times.

and irregularly curved, and show at their extremities or on their sides evident traces of their former attachment; the inner surface of their thin membrane, smooth on the outer surface, has delicate brown spiral or annular thickenings, which Berkeley[1] was the first to describe (Fig. 148). The outer peridium is a stout membrane about 1 mm. in thickness, which lies everywhere close upon the upper surface of the inner peridium; its lower portion is a massive cushion-like body more than 2 cm. thick in the middle. In later stages of the development an axile portion of the basal cushion beneath the centre of the inner peridium developes ultimately into an erect stipe which may be a foot in length and 1–1.5 cm. in thickness with scales on its upper rough and fissured surface, and carries the inner peridium up with it (Fig. 147 *b*). The apical portion of the outer peridium is torn from the basal by the elongation of the stipe, and remains hanging in shreds from the upper surface and margin of the inner peridium, while the basal portion

---

[1] Hooker's Journ. II, 1843.

surrounds the lower extremity of the stipe like the volva of Amanita. Ultimately the wall of the inner peridium parts by a circular fissure beneath the margin, the upper piece falls away from the lower, which remains connected with the stipe, and from the gleba, and the spores are dispersed as dust. The tissue of Batarrea Steveni is entirely formed of stout hyphae with interstices which usually contain air; the hyphae in the wall of the hollow stipe are vertical and parallel to one another, as is the rule in most of the Hymenomycetes. There is no gelatinous tissue. In specimens with the outer peridium still closed the stipe is indicated only by the denser structure and darker colour of the tissue of the basal portion at the spot where the stipe subsequently originates.

Queletia[1] of Fries seems to approach nearest to Batarrea, and perhaps also to Tulostoma.

Podaxon has long-stalked ovoid or elongated peridia more than an inch in thickness, with a lateral wall of the thickness of a stout sheet of paper, which ultimately opens by lobes or scales and has the stout fibrillose structure of the stipe. The same structure is seen in the central column which traverses the peridium to the apex as a prolongation of the stipe. The space between the column and the wall is filled in the ripe compound sporophore with a connected capillitium, formed of long spiral tubes with a transverse wall only here and there. These tubes shoot out in numbers from the parallel peripheral hyphae of the central column, and run in the young plant into the outer wall of the peridium, which is torn away from them when the plant is matured. They are but little branched, and are rarely found with blind ends and these always on one side only. They are held firmly together and form a net-work after maturity, being interlaced and wound round one another in every direction. They have rather thin yellow walls in Podaxon pistillaris and flatten into ribbons in the mature and dry condition. Some of them behave in the same way in P. carcinomatis, but others have thick yellowish brown membranes, which often have fine spiral striations and readily tear into spiral bands in the line of the striae, as Berkeley[2] has recorded. See Fig. 149. The spaces between the capillitium-threads are filled with spores and the dried remains of the membranes of the basidia, which have become more or less of a brown colour and partly adhere to the threads. Specimens of Podaxon pistillaris or an allied species, which were younger but had reached their full size[3], showed the cavity of the peridium filled with a gleba containing an extremely large number of narrow and very sinuous chambers, very thin trama-plates, and a dense hymenial layer consisting entirely of stout four-spored basidia. The capillitium-threads were already discernible as broad but thin-walled hyphae passing on one side into the wall of the peridium, on the other into the columella, and in the gleba running, as in Lycoperdon, partly in the trama-plates, partly transversely through the chambers. If a statement of Corda[4] may be applied to this plant, the early development of the stalked peridium described above takes place inside a peridium externum which is subsequently broken through, just as in

FIG. 149. *Podaxon carcinomatis,* Fr. Piece of a tube of the capillitium, from a specimen in the herbarium of the University of Leipzig. Magn. 390 times.

---

[1] K. Vet. Acad. Förhandl. &c. Stockholm, 1871, No. 2.
[2] Hooker's Journ. IV, p. 292.
[3] In the Herbarium at Berlin, marked Schweinfurth, Iter 2, No. 275.
[4] See Icon. VI, t. III, f. 44, and the text belonging to it.

Batarrea. At any rate the Fungus described by Corda under the name of Cauloglossum in the place cited is very near Podaxon, and must be placed with that genus among the Lycoperdaceae. It is plain, from the remarks recently made above on the Secotieae, that their arrangement in one special group along with Podaxon is founded only on superficial resemblances and is not tenable, and that the group of Podaxineae as hitherto constituted must be broken up.

4. The ripe compound sporophores of the larger **Nidularieae** were described by earlier observers as delicate open cups about 1 cm. in diameter containing usually 10–20 lenticular seed-like bodies. The cup is now known as the peridium, and the seed-like bodies as peridiola or sporangia; the latter should be called the chambers of the gleba if our present terminology is strictly adhered to. The structure and development of these Fungi have been investigated by J. Schmitz, Tulasne, Sachs, Eidam, and Brefeld. **Crucibulum vulgare,** Tul. may be taken as an example of the group, and the description of it will be founded in the main on the works of Sachs and

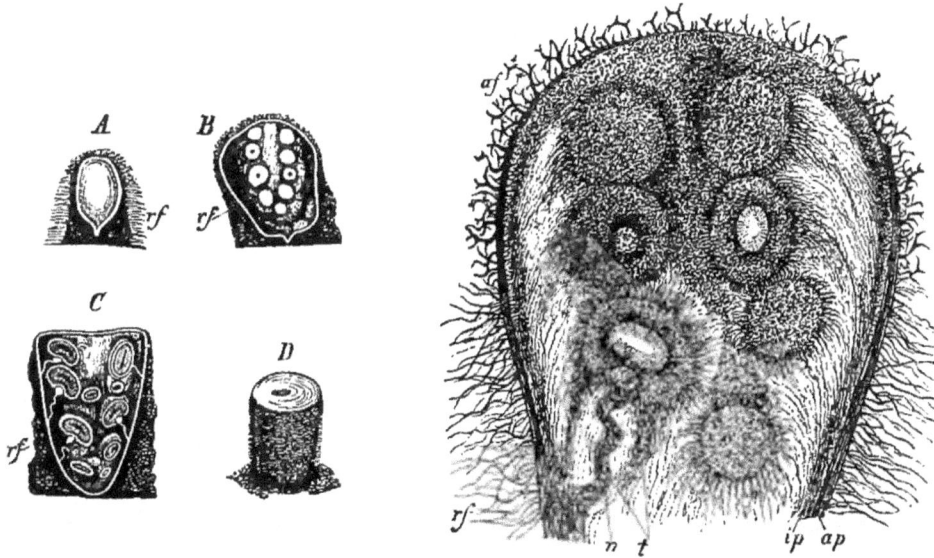

FIG. 150. *Crucibulum vulgare. A—C* median longitudinal section through ripening sporophores; successive stages of development according to the letters. *D* sporophore just ripe in which the epiphragm is beginning to disappear, seen from without. *A—C* slightly magnified. *D* natural size.

FIG. 151. *Crucibulum vulgare.* Thin median section through the upper part of a sporophore of about the same age as *B* in Fig. 150, more highly magnified and seen in transmitted light, the dark parts containing air; *ap* outer, *ip* inner layer of the wall of the peridium, *rf* and *af* its hairs, *n* funiculus, *t* the layer which forms a sheath round it and belongs to a peridiolum which is divided through the middle. After Sachs.

Brefeld. The first beginnings of the compound sporophores are small spherical bodies produced by the interweaving of copiously branched mycelial hyphae. The weft of primordial hyphae thus formed is at first close and colourless and contains air ; but the branches at the periphery soon develope into stout hairs with tooth-like branches and brown membranes, which cover the surface as a brown felt. With this covering the small sphere grows by constant formation of new elements in the interior of the hyphal weft into a thick cylindrical or obconical body about 6 mm. in height. The differentiation into the parts which are found at maturity begins before the body has reached half its ultimate size and advances with the general growth. The first separation is seen in the internal primordial tissue which is originally uniformly white in consequence of the air contained in it, in other words in the opaque primordial tissue, and the result is that a zone between the periphery and the middle becomes a gelatinous felted tissue free from air and therefore transparent. The differentiation of this zone begins above the base of the body; the zone itself, which follows the form of the surface of the body, is concave upwards and thickest in the middle, thinning out towards the upper end, where the margin at length reaches nearly to the apical surface of the peridium (Fig. 150 *A*). By the formation of this gelatinous layer the dense non-gelatinous tissue outside it is separated from the rest and becomes the lateral wall of the peridium, which is then differentiated into two concentric layers, an outer permanent layer of brown felt and an inner whitish layer. Meanwhile no separation takes place in the middle and beneath the apex, but the dense still undifferentiated primitive tissue extends from the latter into the interior space as a thick round sac. At length the differentiation spreads further into the interior of the originally uniform mass, while all the parts increase simultaneously in volume. Portions which are at first round in form become more dense and are separated from one another by the spread of the gelatinisation from without inwards successively between them (Figs. 150 *B*, 151). These are the primordia of the chambers of the gleba or *peridiola*. During their further independent growth, the gelatinisation goes on also beneath the apical surface, which is now covered only by a thin continuation of the inner wall of the peridium. The apex is at first clothed with brown hairs ; as it increases in size with the general growth the hairs are pushed aside and no new ones are formed ; the summit is therefore destitute of the covering of hairs, and is a thin white membrane, the *epiphragm* (Fig. 150 *D*), which becomes torn up and disappears at maturity. The gelatinous tissue round the peridiola also disappears, and the latter lie heaped up at the bottom of the now open cup-shaped sporophore.

The peridiola grow from their first beginnings into a lenticular form, and become inclined obliquely at an acute angle opening upwards to the lateral wall of the peridium (Figs. 150 *C*, 152). A central cavity appears in them at an early period, filled at first, it is said, by a felted gelatinous tissue which afterwards disappears, but this perhaps is uncertain ; the cavity has the same shape as the peridiolum, and continues relatively small, and is densely filled at the time of maturity with somewhat long spores. Two to four spores are formed on each of the basidia, which, with the paraphyses, form a dense hymenial layer lining the cavity. The constituents of the hymenial layer become thick-walled after they have produced the spores and form a stout palisade layer round the cavity, and this is again invested by the still harder thick outer wall of the peridiole, into the structure of which we must not enter further in this place.

From the accounts which we possess it seems certain that the course of the development of the compound sporophore is similar in the genera **Cyathus** and **Nidularia** to that here described in the case of Crucibulum, the only exception being that in

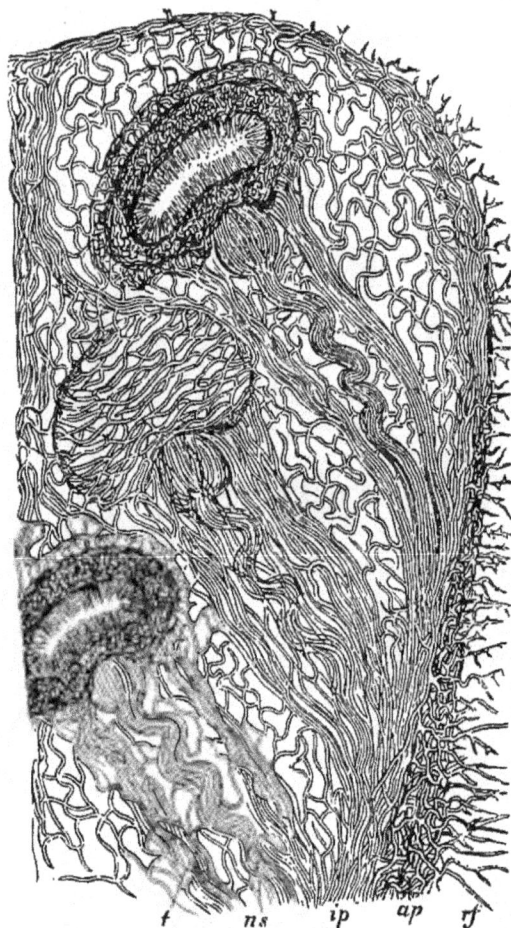

FIG. 152. *Crucibulum vulgare.* Thin median section through a sporophore like that in Fig. 151, more highly magnified. Two peridiola with their funiculi are cut through the middle ; a third lying between them is cut through on the outside only. Meaning of the letters as in Fig. 151. After Sachs.

Nidularia the shape and opening of the peridium is less regular. As regards Nidularia there is nothing to be added to what has been already said. It may be specially noticed that the surface of the peridiola has the same structure in all cases, at least in the stages of the development which have been examined. The earliest states are not known, but peculiar appendages are found on the surface of the mature peridiola of Crucibulum and Cyathus; these appendages are produced, like the other parts, from the primitive tissue by the process of differentiation and are not subjected to gelatinisation; but the history of their development is not yet clearly understood, and some parts of it are matters of controversy (Figs. 150 *C*, 151 *n, f*, 152). The mature peridiolum of Crucibulum has an umbilical depression in the middle of the surface which is towards the peridium, and in the depression is a body which in the intact state is smooth and round and projects towards the outside. This body consists of a compact strand of very slender hyphae rolled up and bent into a dense coil, which may therefore be termed the umbilical coil or tuft. The hyphae are surrounded by a colourless mucilage; and this mucilage and the whole tuft swell when moistened and the hyphae become soft, so that they may be easily drawn out into a filiform strand 3–4 centimetres in length. The hyphae of the tuft are inserted at one end in the outer layer of the wall of the peridiolum, and where they are most perfect they continue their course closely united and parallel to one another as a smooth and somewhat sinuous strand visible to the naked eye as a fine thread, the *funiculus*, which is attached to the inner surface of the wall of the peridium. In the young state, before the differentiation, both strand and coil are loosely enveloped in a layer of hyphae which passes through the gelatinous felt like a bag stretched in the direction of the funiculus. As the development is completed the bag becomes gelatinised, but remains of it are still to be seen as a thin covering especially over the coil. The existence of the funiculus is denied in the case of Crucibulum by Brefeld as against Tulasne and Sachs; but it is certainly often there. It is true that I have been myself unable to find a funiculus in some peridioles, but the coil always showed a small point which answers to it; it would appear therefore that in such cases it has itself finally suffered gelatinisation.

In Cyathus under otherwise similar conditions coil and strand and bag are persistent, and thus more complicated phenomena arise, which vary in particular points according to the species.

In Cyathus striatus, for example, the funiculus when intact has an average length of more than 2 mm. It is nearly cylindrical in shape, and is divided by a deep transverse constriction in the middle into an upper and a lower portion. The lower portion and the slender middle piece are formed of a weft of much-branched and thick-walled but slender hyphae; this weft in the dry state is brittle, but when moist is tough and tenacious and may be stretched to more than double its length. The upper portion is a bag which reaches from the lower portion to the wall of the peridiolum and passes into it; in this bag is a filiform strand of slender parallel hyphae about 3 cm. long and disposed therefore in numerous coils inside the bag, which is only 1 mm. in length. The upper end of the strand is attached to the peridiolum, the lower passes into a coil, which, like the umbilical coil of Crucibulum, is enveloped in mucilage and enclosed in the somewhat swollen lower end of the bag. The wall of the bag has essentially the same structure as the lower portion of the funiculus. The whole body is rather brittle in the dry state. It swells by absorption of water and becomes soft and flexible; the coiled strand may be drawn out to the length mentioned above when the bag is torn up, but not much beyond it; the coil behaves exactly as in Crucibulum; by a little manipulation its hyphae may be drawn out together till the whole strand reaches to a length of 8 cm. The hyphae of the extensible tissue of the funiculi are slender and usually have their walls thickened till the lumina disappear; they are divided into long cells resting one on another with their swollen extremities, where they show peculiar clamp-formations.

It is probable, as Brefeld suggests, that these coils and funiculi, thus provided with slimy mucilage and capable of being drawn out into strands, are contrivances for the dissemination of the peridiola by animals and for furthering the germination of the spores. That there is no spontaneous dehiscence of the peridiola is in favour of this view, but the course of the events which would in that case come under consideration is not yet known.

(5) The development of the compound sporophores of the **Phalloideae** has been studied exactly in **Phallus** (P. impudicus and P. caninus) and **Clathrus**. They first appear in **Phallus** (Fig. 153) as ellipsoid swellings about 1 mm. high on the mycelial strands, and consist at first of a uniform compact weft containing air and formed of very delicate (primordial) hyphae. In specimens of larger growth (Fig. 153 *u, v*) this is differentiated into a dome-shaped central column rising vertically from the point of insertion, a bell-shaped layer of felted gelatinous tissue enveloping the column, *gelatinous layer*, and a white membrane, *outer wall of the peridium*, which surrounds the gelatinous layer and passes at the point of insertion into the central column. The two last named parts consist of primordial tissue. While the whole sporophore as it increases in size becomes more narrowly ovoid in form, and the outer wall and the gelatinous layer grow in circumference and thickness, though their structure remains the same, the central column acquires the form of a globular head supported on a cylindrical stipe. Its primordial tissue at first homogeneous is at the same time differentiated into the *gleba*, the *receptaculum of the gleba*, which is peculiar to the Phalloideae, and in the present case is a simple fusiform *stipe*, and a white membrane surrounding the other two parts (Fig. 153 *w, x*). This membrane forms the *innermost layer of the wall of the peridium*, which consists therefore of three concentric layers, the outer and inner white membranes which unite at the base, and the much broader gelatinous layer lying between the two. The gleba lies in the upper capitate portion of the central column in the form of a thick horizontal ring, which is semicircular on the transverse section and is surrounded on the outside by the inner wall of the peridium, while its inner surface rests on a conical axile portion of the central column. This portion, which may be briefly termed the cone, passes through the whole of the gleba up to the summit of the column. The structure of the gleba is the same as in the Hymenogastreae and Lycoperdaceae but without coils; its chambers are very numerous and narrow, and the trama when somewhat advanced in its development consists of soft gelatinous tissue with laminae which spring on one side from the inner wall of the peridium, on the other from the cone. The outermost zone of the cone bordering on the gleba separates early in Phallus impudicus from the inner tissue, forming a distinct layer, and becoming ultimately the free conical 'pileus' which carries the gleba. In Ph. caninus this separation does not take place. The stipe is at first a very narrowly, afterwards a more broadly fusiform body which runs through the longitudinal axis of the entire central column from its apex to its base. In its earliest condition it is a transparent band and is distinguished from the white primordial tissue, which contains air, solely by the absence of air from its interstices. As growth proceeds, the uniform hyphal tissue becomes differentiated into an axile tissue-strand and a peripheral layer, the wall of the stipe. The latter consists of laminae of a round-celled pseudo-parenchyma, which unite together, like those of the gleba, to form one layer (Ph. caninus) or several layers (Ph. impudicus) of closed chambers. The chambers are large, but so much compressed from above downwards, that the breadth of their interior space is scarcely equal to the thickness of their walls; the walls themselves are much twisted and folded. The chambers are filled with a soft gelatinous felt and the axile strand of the stipe is formed of a like felt. The wall of the stipe at its upper extremity has only deep pit-like folds in its surface, no chambers being formed in the interior. The stipe when once formed increases greatly in size, the parenchyma of its wall enlarging by simple expansion of its cells from the moment that it is distinctly differentiated. The growth of the

two outer layers and of the inner wall of the peridium, so far as it surrounds the gleba, advances *pari passu* with the enlargement of the stipe. On the other hand, the tissue in the cone and in the portion of the central column beneath the gleba diminishes in proportion as the stipe enlarges, till at length in Ph. caninus (Fig. 153 *y*) it is only a thin white membrane ; in Ph. impudicus it remains of larger size underneath the gleba, forming a cup-shaped basal piece which supports the lower extremity of the stipe (Fig. 154) ; in the cone it stretches into a thin white membrane, as in the other species. The gleba, in which the formation of spores is completed or nearly completed during the expansion of the stipe, dilates in Ph. caninus into a thin conical cap covering the upper part of the stipe close beneath the extreme summit; in Ph. impudicus it diminishes less in thickness in proportion to the extension of its surface, and the hyphae of the trama even manifest an active growth by the expansion of their cells. No change in structure worth mentioning here occurs in the parts surrounding the stipe during the time that it is enlarging beyond an evident increase in the size of the hyphae. The cells of the parenchyma of the stipe itself continue thin-walled and

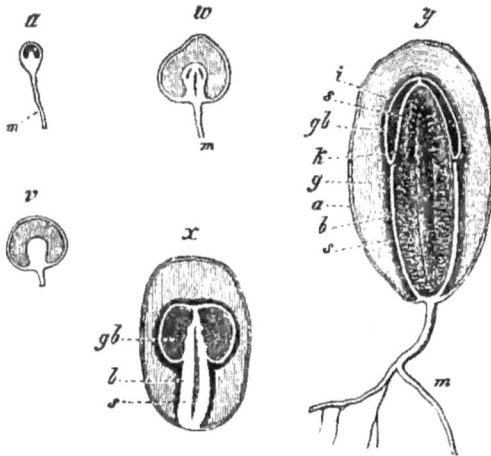

FIG. 153. *Phallus caninus.* Young sporophore partly attached to the mycelium *m* in median longitudinal section and natural size. Succession of stages of development according to the letters *u—y*. *y* a specimen not fully grown but with ripe spores; *a* the outer wall, *i* the inner wall, *g* the gelatinous layer of the peridium, *b* the basal portion, *k* the cone, *s* the stipe, *gb* the gleba.

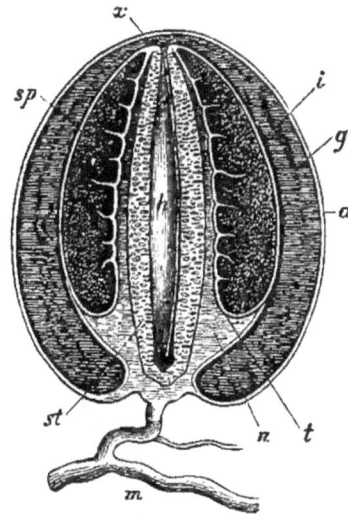

FIG. 154. *Phallus impudicus.* A nearly mature example before the elongation of the stipe, in median longitudinal section ; *m* mycelium, *a* outer, *i* inner wall, *g* gelatinous layer of the peridium, *st* stipe, *h* its cavity filled with mucilage, *t* lower margin of the pileus, *sp* gleba, *n* the cup-shaped basal portion, *x* the spot where the peridium is ruptured by the elongation of the stipe. After Sachs. Two-thirds of natural size.

filled with a watery fluid. At length all growth by expansion of cells already formed or by the formation of new ones comes to an end in all parts, and then comes a sudden increase in the length of the stipe, which thrusts the gleba which is attached to its summit against the apex of the peridium and bursting through it raises the gleba high above it. The elongation is brought about simply by the raising and smoothing out of the folds in the plates of parenchyma in the wall of the stipe, like the folds in a paper hand-lantern, to repeat the comparison of the old botanist Schäffer, till the length of the chambers is at least equal to the breadth ; the elongation of the chambers is caused by accumulation of air within them and their consequent inflation. The gelatinous felt which fills them at first is torn up and disappears, and the axile gelatinous strand is also torn and replaced by air. This process takes place simul-taneously at all points in Ph. impudicus; in Ph. caninus it begins above and advances slowly towards the base. The elongation of the stipe causes the inner peridium of Ph. caninus to separate into two parts by an annular fissure beneath the gleba ; the upper portion with the remains of the cone is carried upwards while the lower continues attached round the base of the stipe. In Ph. impudicus the inner wall of the peridium is also rent at its apex, and the gleba is detached from it and rises clear above it. An annular transverse rent in the lower portion of the cone separates the cup-shaped basal portion which remains round the base of the stipe from the upper portion, which is torn into shreds, and the pileus which bears the gleba is thus separated from the stipe except at its upper margin which adheres firmly to the summit of the stipe.

The gleba when exposed to the air drops from its stalk as a slimy mass containing the spores in consequence of the deliquescence of its gelatinous tissue. In Phallus caninus the cone and the part of the inner wall of the peridium which covers the gleba go through this process of disorganisation and become indistinguishable before deliquescence. For further particulars and differences among the species the reader is referred to the more elaborate works which will be cited below, and the descriptions in systematic treatises.

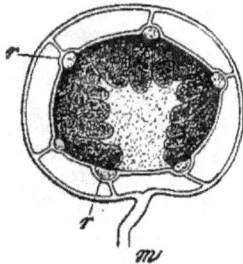

FIG. 155. *Clathrus cancellatus.* Young compound sporophore in median longitudinal section. *m* mycelium, *r, r* sections through the strands of the receptaculum, surrounding the gleba, which is shaded. Further explanation in the text. The figure is diagrammatic after Tulasne's and Berkeley's drawings. Natural size.

FIG. 156. *Clathrus cancellatus.* Mature specimen ; the receptaculum with comparatively narrow fissures has issued from the ruptured peridium. Sketched from a photograph by Bornet. Half the natural size.

Clathrus cancellatus, as we have known since Micheli's time, agrees with Phallus in the character of the gleba and peridium. But the receptaculum which raises the gleba from out of the peridium has the form of a net or lattice-work with large meshes surrounding the outer surface of the gleba. We know from Tulasne[1] especially that the development of the parts begins here, as in Phallus, with a separation of the uniform tissue of the young compound sporophore into central column, gelatinous layer, and outer wall of the peridium (Fig. 155). Lamelliform processes which anastomose and form a net-work pass from the peridium to the surface of the central column, passing like septa through the gelatinous layer. The lowest of these converge towards the point of insertion of the column or of the whole body and unite there. Then the column becomes further differentiated into the inner wall of the peridium, the gleba, and a roundish axile gelatinous body of a cartilaginous consistence. The latter occupies the entire central portion; its base rests on the peridium and passes into it, the whole of its surface, with the exception of the point of insertion, being covered by the thick gleba. The lamellae of the trama in the gleba spring on all sides from the central body ; hence it appears in a transverse

---

[1] Explor. sc. d'Algérie.

section to be furnished at the circumference with numerous unequal processes and projecting points radiating into the gleba. In the net-work of stripes, where the septa from the outer wall touch on the inner wall, the white (primordial?) tissue of the latter is more largely developed than in the interspaces. It is in the region of these stripes that the receptaculum originates after the first formation of the gleba. When the compound sporophore is mature the receptaculum expands greatly, and while it remains attached at the point of insertion it rises far above the ruptured peridium (Fig. 156). The gleba meanwhile, with its gelatinous support which is separated from the point of insertion, is attached to the uppermost part of its inner surface; the whole of the gelatinous tissue soon deliquesces, as in Phallus, into a slime which drops off with the spores. The structure of the perfect receptaculum, as appears especially from Corda's accounts, is so like that of the same organ in Phallus, that we can hardly doubt that the development of the tissues and the mechanism for expansion are the same in both genera, though we cannot appeal to direct observations.

We are acquainted with a considerable number of very varied and sometimes strange forms in the group of the Phalloideae, but most of these have been examined only in single mature specimens. A large collection of them are found in Corda[1]. All that is known of them, as regards the shape and structure of the peridium, gleba, and spores, and especially the structure of the receptaculum, agrees in the main points so fully with the descriptions here given of Phallus and Clathrus, that there can be no doubt that the development is alike in all. This view is confirmed by Tulasne's researches into the young compound sporophores of **Colus hirudinosus** and by Corda's account of **Ileodictyon**.

The species chiefly differ from one another in the shape of the receptaculum, which, as in Phallus and Clathrus, always emerges from the walls of the peridium. Some are more like Phallus in this respect, others more like Clathrus; others again depart considerably from the type of both, as for example **Aseroe**, which has a star borne on a thick erect stipe with dichotomous rays which are either expanded horizontally or rise obliquely upwards, and has the gleba placed originally in the focus of the rays above the stipe. From the material in our possession, imperfect as it is owing to the rarity or perishable nature of most extra-European species, we may arrange the majority at least of the forms in a gently graduated series according to their affinities, with Clathrus (Fig. 156) at one end and Phallus (Fig. 153), Lysurus and similar forms at the other. Fries[2] has already drawn attention to this connection. **Colus hirudinosus** (Fig. 157) agrees very nearly with Clathrus cancellatus in its development. But while in the latter the base of the receptaculum consists only of a few short stripes which converge into a network and are connected at the extreme base, it is developed in Colus into a hollow stipe, open above and below, in form cylindrical-conical and nearly a third of the entire length of the whole sporophore. The stipe divides above into from six to eight riband-like arms, which ascend like the meridians of longitude on a globe and unite their upper extremities into a small terminal plate which is pierced with holes like a coarse sieve. Altogether therefore it forms a net-work, and the position of the gleba in it is the same as in Clathrus, only the shape of the net is different and much more regular. If the meridian arms were set free, and did not end in the plate above, we should have the same plan as in Aseroe (Fig. 158) and in Calathiscus with some modification, where the gleba is borne on the extremity of the stipe, which is split up into diverging arms.

Lastly, **Aserophallus**[3] (Fig. 159) has a comparatively long stipe, which divides beneath the round gleba at its apex into four short arms which embrace it. Here is at once a very near approach to the shape of Phallus, especially if we frame an idea

---

[1] Icones, V, VI. [2] Summa veget. Scand. 434.
[3] Montagne et Leprieur in Ann. d. sc. nat. sér. 3, IV (1844).

for ourselves of the only possible form of the parts when enclosed in the peridium. The cavity of the stipe is in this case also opened wide between the lobes; the somewhat more persistent outer layer of the gleba with its thin walls makes no difference in this respect. To arrive at the form of Phallus caninus or **Simblum**, we must have the stipe closed at the apex and projecting in a conical shape into the gleba, and the lobes must entirely disappear. We are not at present acquainted with any intermediate forms. But if none are found, we have sufficient material to show the family connection. Other genera than those already named are not taken into the comparison, because they are not clearly known or because it is unnecessary to adduce them. The reader is referred for them to the works which describe this group.

Phallus and its nearest allies are the members of the series of the Phalloideae which depart farthest from the rest of the Gastromycetes. Clathrus is a link of connection between them and the Gastromycetes through the Lycoperdaceae. To see this we have only to compare its compound sporophores when ripening but still closed with those of Geaster, and the collenchyma-layer of most species of this genus, at any rate that of G. hygrometricus, with the receptaculum of Clathrus. The relation is

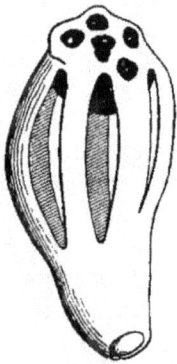

FIG. 157. *Colur hirudinosus.* Sketch of a mature isolated receptaculum which has issued from the peridium. The black gleba hangs down under the sieve-like terminal plate. After Tulasne in Explor. Scientif. d'Algérie. Twice the natural size.

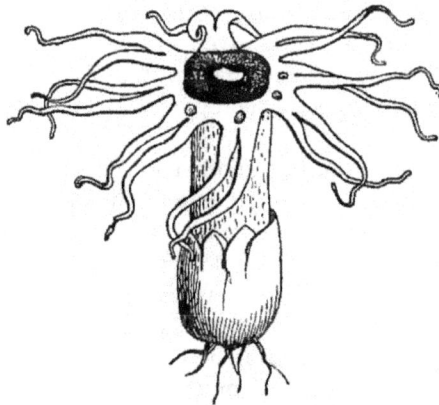

FIG. 158. *Aseroe rubra.* Sketch of a mature receptaculum which has issued from the peridium attached to it below, and bears the gleba in the middle of the radiating expansion. After Berkeley in Hooker's Journ. III, Tab. V. Half the natural size.

FIG. 159. *Aserophallus.* A ripe compound sporophore. After Montagne and Leprieur, l. c., in note on p. 325. Natural size.

brought out very clearly by the circumstance that there are individuals of C. cancellatus, in which the bars of the lattice-work, even in the ripe compound sporophore, are excessively broad and the interstices only narrow slits; here therefore the receptaculum is a hollow body only slightly perforated.

The connection between Clathrus and Geaster appears to me to be still more completely established by the genus **Mitremyces**, which is chiefly American and still far from being thoroughly known. But I do not attempt to describe it here, for I have no sufficient account before me of the history of its development[1].

6. A description of the genera **Tulostoma**, **Polysaccum**, and **Sphaerobolus** may be given here by way of appendix, because they depart in a remarkable manner from the type of the groups which we have just been considering. But no special proof is required to show that they approach very near to them and especially to the Lycoperdaceae, and in this position may be regarded as the representatives of distinct and at present small divisions which have to be co-ordinated with the others.

---

[1] For the facts as at present known, see E. Fischer in Bot. Ztg. 1884.

The peridia of **Tulostoma** are formed, according to Schröter, on subterranean mycelial strands, which are flat sclerotia and may be 6 mm. in breadth; they are probably shoots from these sclerotia, and are round bodies about 4 mm. in diameter composed of a uniform weft of primordial hyphae; the superficial ramifications of the hyphae form a floccose envelope which attaches itself to the grains of sand in the surrounding soil. The differentiation into gleba and peridium may be observed when the compound sporophore has reached the size of 6–8 mm. The peridium is a relatively thick hyphal layer entirely surrounding the gleba and having a conical thickening at the upper end as the primordium of the papilla which afterwards appears at the opening of the peridium; there is also a broadly obconical thickening below. The peridium separates further into an axile cylinder situated beneath the middle of the gleba and into a part which surrounds the cylinder like a sheath. The cylinder elongates at maturity into the cylindrical stipe which attains a length of 3–6 cm. and raises the peridium above the ground; the elongation causes the sheath to separate by a transverse fissure into a lower portion, the base of the stipe, and an upper, which surrounds the upper end of the stipe, according to Vittadini (see Fig. 160), and both portions then dry up. There is no separation into inner and outer wall in the peridium. The gleba is at first a reniform body which afterwards becomes spherical in form, and is distinguished by the absence of chambers. It is formed of a uniform tangled mass of hyphae about 2 μ in thickness, the branches of which produce the strangely-shaped basidia described above on page 309. The abscision of the spores is over and the basidia deliquesce before the stipe begins to enlarge; then the spore-membranes turn brown, the process advancing, according to Schröter, from the centre towards the circumference of the gleba. A large number of the hyphae of the gleba begin to develope shortly before the disappearance of the basidia into the close net-work of stout threads, which form the capillitium and grow all over the wall of the peridium, as in Geaster hygrometricus.

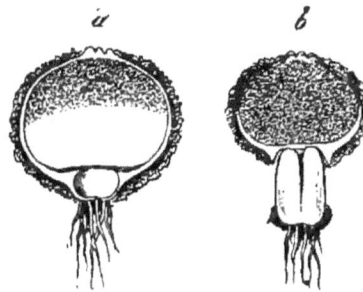

FIG. 160. *Tulostoma mammosum.* Median longitudinal section. *a* before the elongation of the young stipe, the gleba at the apex beginning to assume the dark colour of ripeness. *b* after the stipe has begun to elongate. After Vittadini. Natural size.

Polysaccum is another strange object and deserves further examination; it is a large somewhat elongated or club-shaped body which has chambers throughout except in the thin outermost fibrillose layer; a few concentric layers of chambers in the periphery are smaller and sterile and together represent a peridium; the chambers in the interior which are polyhedric, and may be as large as a pea at maturity, are about 1 mm. in diameter when the spores begin to be formed, and are filled with hyphae closely woven together into a hymenial coil and either sterile or producing basidia. These hyphae have very soft gelatinous membranes, and the whole coil may be removed uninjured from the brown trama. The chambers at maturity contain only a chocolate-brown spore-powder, the tramal laminae with the peridium being disorganised and desiccated and crumbling to pieces. No capillitium is formed. The formation of the chambers begins at the apex and advances, very slowly as it would appear, towards the base of the peridium which is sunk deep in the sandy soil. Specimens are found in which the upper half is quite ripe while in the lower all stages of the development may appear in unbroken succession one above another. The earlier states are not known. The same thing appears to occur in Berkeley's Phellorinia. The agreement with Scleroderma is evident; it is possible that Polysaccum, Phellorinia, Scleroderma, and Melanogaster form a distinct group marked by the hymenial coil which fills the chambers.

The mature peridium of **Sphaerobolus stellatus** is white and about as large as a grain of mustard-seed, and is found either on the surface of the substratum, which is usually decaying wood, or sunk in it. Its differentiation may be compared with that of Geaster. A thick outer wall encloses a spherical peridium internum about 1 mm. in diameter, which in this genus is also termed a sporangium, and at maturity consists to a great extent, and even in its outer surface, of a tenacious mucilage, in which numerous basidiospores and other cells accompanying them are imbedded. When young it contains a gleba formed of delicate hyphae surrounded by a thin hyphal weft next the outer peridium, and imperfectly and irregularly divided into chambers by narrow tramal lamellae containing air. The chambers are filled with hyphal branches closely woven together and bearing basidia, which usually produce from five to seven spores each. Ultimately the hyphae in the compound sporophore are to a great extent disorganised and form the tenacious sticky mucilage mentioned above, which has been carefully studied by E. Fischer. The ripe spores are retained in the mucilage, and with them, especially in the periphery, isolated one-celled or few-celled pieces of the trama or the basidia-bearing hyphae. Some of these pieces become vesicular mucilage-cells several times larger than the basidiospores, and are incapable of further development, but in their turn suffer disorganisation; others become spores or *gemmae* of a different kind, which are originally short cylindrical thin-walled cells rich in protoplasm occurring singly or a few together in a row, different in shape from the basidiospores and germinating very readily. They will be noticed again on page 330. The outer wall is composed of two chief concentric layers; a peripheral, white and floccose on the outer surface, which is again divided into two layers, and an inner layer. The latter separates again into two layers, a peripheral dense fibrillose layer next within the outer layer, and a collenchyma-layer united to the fibrillose layer, and consisting chiefly of relatively large cells placed radially to the whole sphere in the manner of a palisade. The whole of the outer wall when fully mature bursts at the apex into 6–7 lobes in a stellate manner, and the inner peridium which is filled with spores is not at first moved from its place. The rupture is caused by the greater superficial extension of the collenchyma layer. Growth continues in this layer in the direction of the surface after the rupture of the peridium, and as the other layers do not grow with it, it is in a state of positive tension. Ultimately it breaks loose with the fibrillose layer which adheres firmly to it from the peripheral layer, continuing united to the latter only at the tips of the lobes, and carries the inner peridium on its centre, which is arched strongly outwards. If these processes are made to go on slowly, the peridium remains attached though slightly to the collenchyma-layer, as in Geaster fornicatus. Usually and normally the positive tension reaches a high degree before the separation, which occurs suddenly in consequence of fluctuation in the turgescence, the collenchyma-layer becomes convex outwards with a jerk and a crackling sound, and the inner peridium which is only loosely attached is flung away, in favourable cases to a distance of more than a millimetre, and adheres to any object on which it alights.

The development of the structural conditions which are at the bottom of these processes, the details of which are to be found in Pitra and E. Fischer, is the result of the differentiation of a coil of hyphae originating in the branching of the mycelium and at first perfectly uniform.

COURSE OF DEVELOPMENT AND AFFINITIES OF THE BASIDIOMYCETES.

SECTION XCI. The whole **course of the development of a Basidiomycete** was first followed by Woronin in 1867, when he examined the genus **Exobasidium**, a parasite on living species of Vaccinium. This very simple form consists in the mature state of a basidial layer, which comes to the surface of the epidermis of the host, and a mycelium which spreads widely in the part of the plant which it attacks, deforming it more or less, and producing the basidial layer directly from its hyphae. The basidia are as usual club-shaped, and abjoint four fusiform spores. The spores divide as soon as they are ripe by transverse walls into four cells, and the two end-cells germinate at the expense of the protoplasm of the other two as soon as they reach a moist surface. If this is the surface of a young foliage-leaf of Vaccinium, the germination consists in the putting out of a tube, which at once grows through the young epidermis into the inner tissue of the leaf and there developes a mycelium. The mycelium forms the hymenium which bears the basidia directly; the time required in plants cultivated artificially is about 14 days from the sowing of the spores. If the spores germinate on some other substratum not so favourable as the leaf of Vaccinium, the germ-tube after attaining a short length begins to sprout and produces rather elongated fusiform sprout-cells which sprout only at their extremities. The formation of sprouts may be repeated many times. Brefeld saw it continue for a year in specimens cultivated in nutrient solutions, and the power of production appeared to be unlimited. It can scarcely be doubted that the sprout-cells can give rise to a mycelium like that of the primary spores in suitable substratum; but this has never been actually ascertained.

The development of the **Tremellineae**, the species of **Dacryomyces** for example especially, would appear from the investigations which have been made to be the same as in Exobasidium. I refer partly to the statements of Brefeld[1], partly to an unpublished series of researches conducted by Klebs. A mycelium is developed from the germinating basidiospore under suitable conditions of nutrition, and the branches from the mycelium combine to form compound sporophores which produce new basidia. Under different conditions, not in all cases exactly defined, the germ-tubes which proceed from the spores remain short and abjoint (successively ?) small secondary spores or develope by sprouting. We shall return to these phenomena in a subsequent page. There is moreover a further point of agreement between Dacryomyces and Exobasidium; the ripe basidiospore at the time of abjunction divides by transverse walls into short disc-like cells, daughter-spores, which are usually four in number, but each of these is capable of germination in Dacryomyces in one of the forms stated above. In other Tremellineae the basidiospores do not divide or they divide in a different and peculiar form.

We are acquainted with the whole of the life-history of the Hymenomycetes proper in Typhula and species of Coprinus through the researches of Brefeld; in Agaricus melleus through those of R. Hartig and Brefeld; in Crucibulum and Cyathus, genera of the Nidularieae, from the labours of Eidam and Brefeld, and in Sphaerobolus from those of E. Fischer.

The germinating spore of **Agaricus melleus** and **Coprinus stercorarius** puts out a germ-tube, which in all the Coprini is swollen into a vesicle at its point of origin; the germ-tube developes directly into a mycelium and the compound sporophores are formed directly on the mycelium from the ramifications of the hyphae. No organs beside those described in the preceding pages, which can even be supposed to have any connection with the propagation of the plant, nor any rudiments of them, have

---

[1] Hefepilze, p. 198.

ever been found.   Every piece of the mycelium, in Coprinus every bit even of a mutilated sporophore, may under conditions favourable to growth form the hyphal branches which become a sporophore.   The filiform mycelia, the sclerotia which may be produced on them, and the strands of Agaricus melleus (see page 23) behave alike in these respects.

In the rest of the Coprini which have been examined and in Typhula also the mode of development is the same as in the above species, but a complication occurs in them more or less frequently, which will be considered presently.

In **Crucibulum** again and **Cyathus** the course of events is entirely the same as that in the Hymenomycetes above mentioned.   In the proper nutrient solutions, especially decoctions of dung, and a not too low temperature (15°–18° C., according to Eidam 25° c.), the basidiospores put out germ-tubes, which under favourable conditions develope directly into a mycelium, and its branches form the coils of primordial hyphae which develope into peridia without the interposition of any intermediate members.

The history of **Sphaerobolus** is very peculiar according to E. Fischer's account of that plant.   It was stated on page 328 that the sporangium or peridium when flung off has the basidiospores mixed with the mucilage-cells and the gemmae. These are all so firmly held together by the tenacious mucilage, that it is very difficult to isolate them, and a spontaneous separation of the several elements has never been observed, and from what we know is scarcely conceivable.   If a sporangium is flung to a spot where there is no moisture it dries up into a firm hard body, which retains its vitality for months and swells when moistened, and then exhibits the same phenomena of germination as a recently ejected specimen.   But if a sporangium falls at once on a moist substance, it puts out numerous germ-tubes from its entire surface, so that it may be densely covered in a day or two with white hyphae radiating in every direction.   On a substratum which supplies sufficient nourishment these develope without interruption into a mycelium which produces new peridia in a few weeks or months without the aid of any other organs of reproduction.

It may be said therefore that the whole sporangium thus remaining in connection produces by germination the commencement of a mycelium of many hyphae, and that hardly any other mode of germination could occur in the natural course of development, though it is easy to show that portions of the sporangium artificially separated and even individual cells, if naturally capable of germination, may be made to germinate when isolated in water or a nutrient solution.

A closer examination of the incidents in the entire sporangium discloses however the curious fact, that the germ-tubes which spring from the surface and develope into a mycelium proceed, as far as can be observed, from the gemmae and *not* from the basidiospores.   As the germination of the whole body proceeds the basidiospores lose their protoplasm and their membranes become more delicate and paler, and are pierced with holes in places, and they themselves at length entirely disappear; the products of their disorganisation are evidently employed with the mucilage which envelopes them as food for the germ-tubes.   Isolated gemmae germinate quickly and readily in water, and normal mycelium is obtained from their germ-tubes, if proper nutriment is supplied.   Isolated basidiospores rarely produced germ-tubes in Fischer's experiments; as a rule they did not germinate; and in the normal germination of sporangia no germinating spore was observed among the multitude of gemmae with their germ-tubes, only the successive disorganisation of the spores.

Section XCII. If putting Sphaerobolus on one side we consider the course of development in the Basidiomycetes which have been described above and which have been carefully examined, we find that in the simplest case, which is sometimes the one exclusively observed, the germinating basidiospore gives rise directly to a mycelium, the branches of which become sporophores producing basidia without the interposition of any special intermediate members at all comparable with archicarps.

The circumstances which, as was intimated above, may produce **complications in this scheme** are of the following kind. In describing the germination of the basidiospores of **Exobasidium** and **Dacryomyces** in pages 328 and 329, it was stated that the spore can under certain circumstances produce a great abundance of sprouts, or may abjoint small cells acrogenously from short germ-tubes. These small cells are round in Dacryomyces and about 2 μ long, and in nutrient solutions they give rise to germ-tubes, which may develope into filamentous and often large mycelia. These mycelial hyphae and those also which proceed from basidiospores may abjoint small cells in crowded tufts, which differ from the first in being ellipsoidal or rod-like in form, but agree with them in giving rise to a mycelium with similar products. We learn from Brefeld especially that similar phenomena have been observed in many Tremellineae in forms which vary according to the species. The mycelium from which these cells are abjointed may then produce the basidia-bearing compound sporophores of Dacryomyces. The cells in question are, according to this description, *spores*, but we may also call them by analogy *gonidia*, if we wish to maintain the traditional use of the word spore for basidiogenetic cells.

Gonidia in the same sense of the word and with essentially the same characters occur also in some Tremellineae on the sporophores which produce basidia; such are the cells capable of germination in form like a bent rod which are abjointed at the slender extremities of hyphae in depressions with thickened and rounded margins in the compound sporophores of Tremella Cerasi, Tul. The small round cells also 2 μ in diameter, which according to Tulasne are abjointed like small heads on the extremities of much branched hyphae in the hymenial layer of Tremella mesenterica, may be of the same kind, though their germination has not yet been observed. Another form of gonidium, distinguished by the name of *gemma* and produced no doubt by external causes which have yet to be ascertained, is found in the compound sporophores of Dacryomyces deliquescens. The presence of these cells is shown from without by their making the parts of the sporophores where they are formed turn from their normal clear amber-yellow colour to a dark orange. The hyphae swell and their cells become filled with a densely granular substance of a dark orange colour. The cells separate from one another when placed in water, and each may then give rise to a mycelium with the characters which have been described above. Klebs especially has obtained compound sporophores producing basidia from gemmae cultivated on microscopic slides.

The mycelia produced in nutrient solutions from the basidiospores of most of the **Coprini** which have been examined and of Typhula may form small **rod-like gonidia**, like those just described in the Tremellineae, before they arrive at the formation of typical compound sporophores. These are long slender filiform cylindrical bodies formed, often several side by side, in a tuft at the extremities or sides of hyphal branches, and are divided transversely into shorter rods before or after abscision (Fig. 161). These small rods are very abundantly and frequently produced in some species, as for instance in Coprinus lagopus, but not in all the individuals which form basidia. In other species, as C. ephemeroides, they are few and rare; in C. stercorarius, as may be gathered from what has been said above, they do not occur at all. In Brefeld's careful cultivations the rods always perished without germinating, in C. lagopus sometimes after a doubtful commencement of germination; Van Tieghem's statement with regard to their power of germination must therefore be accepted with caution. From the facts in our possession we must speak of them as *gonidia the germination of which has not been observed.*

A formation of gonidia of a certain kind has also been observed in the **Nidularieae**, where young mycelial hyphae, if imperfectly fed, break up by transverse divisions into cylindrical cells, which germinate under favourable conditions and develope new normal mycelia with peridia.

It appears therefore that in the best-known species a *formation of 'gonidia' may be introduced* into the section of the development between two successive generations of basidiospore-generations, in most cases as a facultative occurrence depending on external causes, in others perhaps as a necessary or at any rate a very regular process. More thorough investigation is necessary on this point especially in the Tremellineae; in the Nidularieae the accidental nature of the occurrence and its dependence on external and occasional causes is evident.

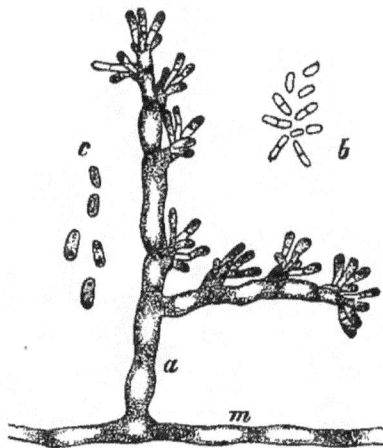

FIG. 161. *Coprinus lagopus*, Fr. Mycelial hypha *m* with a branch *a* from which rods are being abjointed. *b* rods or gonidia detached, some still connected together in rows. *c* the same isolated. After Brefeld, from Lürssen's Handbook. *a* and *b* magn. 400, *c* 600 times.

Finally, the observations on **Sphaerobolus**, to which we must now revert, show when compared with the foregoing that the gonidia described above as gemmae in this Fungus have a prominent position in the development, since the task of propagation devolves almost exclusively on them; at least the basidiospores are much behind them in this respect and in the natural course of things need scarcely germinate at all.

For some time the course of the development in the Basidiomycetes was supposed to be quite different from that which has now been described, and efforts were made to show that it was different; it was thought that the compound sporophore, like the sporocarp of the Ascomycetes, was developed from a fertilised archicarp, which became surrounded in various ways, as in the Ascomycetes, with an envelope of hyphae. Karsten[1] has thrown out some doubtful hints in this direction since 1860, in connection with Agaricus campestris. More decided suggestions of the kind arose from the discovery of the development of the sporocarps of Erysiphe, and these were probably the occasion of the paper of A. S. Oersted[2], which described 'oospheres' formed

---

[1] Geschlechtsleben d. Pflanz. p. 50, and Bonplandia, 1862, p. 63.
[2] Verhandl. d. k. Dän. Ges. d. Wiss. 1 Jan. 1865.

as small branches on the mycelium of Agaricus (Crepidotus) variabilis, P., on which 1–2 slender adjacent branches supposed to be antheridiae usually did *not* lay themselves, while the oosphere, without experiencing any further changes, became surrounded by a hyphal tissue which developed into the pileus.

The question was subsequently resumed in connection with specimens of Coprinus cultivated on microscopic slides, especially by Reess[1], who thought he found spermatia in the small non-germinating 'rods,' and archicarps in the swollen extremities of the branches, but without clearly making out the participation of the two kinds of organs in the formation of the compound sporophores. Van Tieghem[2] studied the species of Coprinus about the same time, and found spermatia in the small rod-like gonidial cells, and also discovered archicarps, which were fertilised by conjugation with the spermatia and then developed in very characteristic manner into compound sporophores; he even obtained a hybrid form by crossing the two species, but he very soon withdrew all these statements[3] and has adhered from that time to the views expressed above. It was unfortunate that the declaration of this change of opinion coincided to the day with the first publication of Brefeld's researches, and we must allow him the merit of finally clearing up the matter, because, as regards Van Tieghem, it would always have been a question which of the two diametrically opposite results arrived at within a space of ten months was after all the right one.

Section XCIII. We have no complete observations on the course of the development in the many other Hymenomycetes and Gastromycetes. But all that is known of them favours the view that *their history is in general the same as that of the species with which we are perfectly acquainted in the essential points above described*, nor is anything known which could be a sufficient reason for objecting to this view. Besides the agreement in form and structure in the mature state, there is the unvarying fact that the commencements of the compound sporophores wherever discovered are formed, without the interposition of intermediate members, from hyphal bundles of the mycelium which have exactly the same origin as those which remain purely vegetative. R. Hartig's careful observations on the wood-destroying Polyporeae, Thelephoreae, and Hydneae should be especially mentioned in this place in addition to the cases noticed in the foregoing descriptions.

Secondly, the little that is known of the germination of the spores and of the first products of germination accords with the observations to which attention has been drawn above. This is not much; the spores of many Hymenomycetes when sown put out simple or branched germ-tubes; non-germinating 'rods' are formed, according to Eidam's observation[4], on Agaricus coprophilus, Bull. on the first commencement of mycelia grown in a nutrient fluid, and they appear in strings that are crowded together in tufts and twisted into curls and entangled with one another. In the great majority of the Gastromycetes the first products of germination have not been hitherto observed or the supposed observations are very doubtful[5]; germination may in their case depend on special and hitherto unknown conditions, or the circumstances may perhaps be the same as in Sphaerobolus.

No doubt would be raised as to the agreement of all the Basidiomycetes in the

---

[1] Phys. Med. Ges. z. Erlangen, 14 Dec. 1874.—Pringsheim's Jahrb. X, p. 179.
[2] Comptes rendus de l'Acad. d. Sc. Paris, 80 (1875), p. 373.
[3] Ibid., 81, p. 879 (15 Nov. 1875).
[4] Bot. Ztg. 1875, p. 649.
[5] See Hoffmann in Bot. Ztg. 1859, p. 217.

rhythm of their development, in accordance with the remarks in the preceding paragraphs, if the statement that gonidia occur on the mycelia and compound sporophores of the Hymenomycetes proved to be true. The statement has been made only in a few isolated cases. The supposed gonidia mentioned by Oersted in **Agaricus variabilis**, P. are strictly of this kind, also those of **A. racemosus**, P., **A. vulgaris**, P., **Fistulina hepatica** and **Polyporus Ptychogaster**, Ludwig. My own older remarks on Nyctalis perhaps do not strictly or perhaps do not at all belong to this connection, but they may nevertheless be mentioned here.

In the case of all these statements it is distinctly to be observed, that no one of them puts the real nature of these gonidia or spores as they may also be called, and their connection with the particular species of Hymenomycetes which forms basidia, beyond doubt, for in no case is it clear whether they reproduce the hymenomycetous form or something else, or, as may possibly happen, nothing at all; usually the first beginnings of germination in the presumed gonidia have not been observed. The possibility is nowhere excluded that they belong to parasites of the particular Hymenomycetes.

The slender stipe of **Agaricus racemosus**, Pers. which springs from a sclerotium terminates in fully developed specimens in a pileus, which according to old existing descriptions and figures has the typical structure of the Agaricineae. The stipe is beset throughout its whole length with short hair-like spreading branches, which were compared by Fries [1] and Berkeley [2] to the sporophores of the form-genus **Stilbum**, because, like them, they abjoint at their extremities numerous spores (gonidia) arranged in rows and forming together a small gelatinous head. In other specimens the branching is more irregular and the primary stipe also ends in a head of gonidia. The gonidia according to Tulasne [3] are ellipsoid or elongated in form and produce long germ-tubes when sown in water.

On the extremities of short mycelial strands of **Agaricus vulgaris** Fr. Hoffmann [4] saw here and there small cylindrical cells serially abjointed, and called them spermatia.

Short non-septate erect simple sporophores rise according to Oersted [5] from the mycelium of **Agaricus variabilis**, P. and abjoint ellipsoid spores simultaneously in a small head at their extremities, after the manner of Corda's form-genus **Cephalosporium**.

De Seynes observed a formation of 'gonidia' often in large quantities on the pileus of **Fistulina hepatica**; these were abjointed singly or in tufts close beside one another or in rows, and were of ellipsoid form, usually about 8 μ in length, and furnished with very thick brownish-red membranes. They were borne usually on relatively slender copiously branched hyphae, which appeared in many cases to spring as branches from the thicker hyphae of the substance of the pileus, and were found chiefly on the upper side of the pileus, sometimes on its surface, sometimes at a depth of a centimetre in its tissue, spreading through it oftentimes in enormous quantities. They were found in every one of the many specimens examined by De Seynes from Europe, America, and Asia. De Seynes observed only very feeble and almost doubtful attempts at germination in old material. My own early notes on the subject, which I have had no opportunity of completing, confirm De Seynes's statements to a great extent, but add, that the hyphae which produce the gonidia may also occur on the under surface of the pileus and between the tubuli of the hymenium, and that they were not found

---

[1] Epicris, p. 90.   [2] Crypt. Bot. p. 365.
[3] Fung. Carpol. I, p. 110.   [4] Bot. Ztg. 1856, p. 158.
[5] Oversigten d. Verhandl. d. k. Dän. Ges. d. Wissensch. Jan. 1865.

on all specimens. We can arrive at no certainty in the matter without a complete account of their development and especially of their germination.

Corda[1] has described a Fungus growing on old pine-wood under the name of **Ptychogaster albus**, a round body of the size of a hazel-nut or even much larger, which has the appearance of a Lycoperdon and is white when young, but when the spores are ripe is of the colour of a brown clay. Its structure, however, is not that of a Lycoperdon. It consists at first of a loose soft weft of hyphae, which radiate for the most part from a more dense basal portion, their free sterile extremities terminating in the periphery. The hyphae have many transverse septa and clamp-connections at the septa. The spores are formed in the interior on the hooked or spirally twisted ends of the branches of the hyphae ; these branches are converted into mucilage when the spores are developed and disappear, so that the brown spore-dust lies among the decaying remains of the hyphae. According to Tulasne[2] the spores are formed in large numbers as round lateral protuberances on the filaments and in no fixed order ; according to Cornu[3] the twisted hyphae break up each into a single row of spores. Tulasne compares this Fungus to **Pilacre Petersii**, Berk. et Br. and **Onygena faginea**, Fr. and suspects it to be a gonidial form of an Ascomycete, mentioning especially **Poronia**. E. Fries[4] on the contrary considers Ptychogaster to be a monstrous product of **Polyporus borealis**, and Cornu thinks this probable on account of the clamp-connections which are so common in the Hymenomycetes. F. Ludwig[5] has in fact recently found specimens of Ptychogaster which had a hymenium of Polyporus on the under side, the side towards the substratum. The elements of the hymenium appeared to spring directly from those of the Ptychogaster, and no similar form of Polyporus was found in the vicinity. No more is stated, and careful artificial cultivation only

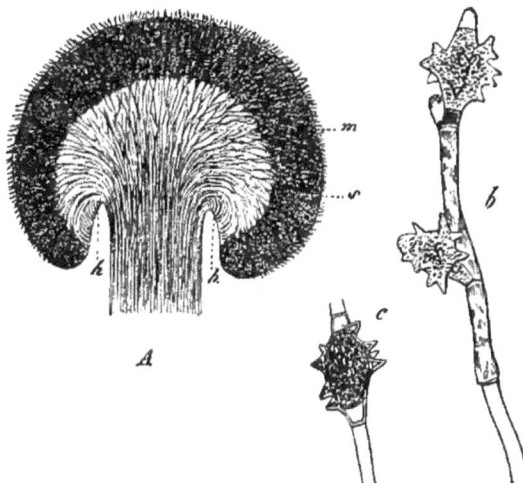

FIG. 162. *Nyctalis asterophora*, Fr. *A* a young specimen, in vertical section, and in transmitted light ; *m* hymenophorum, *s* layer of chlamydospores, *h* primordium of hymenium. *b* hypha with half-ripe chlamydospores. *c* a ripe chlamydospore. *A* slightly magnified, *b* and *c* magn. 390 times.

can show whether the two forms really belong to the same species, which Ludwig names **Polyporus Ptychogaster**, or whether it is a case of cohabitation of two species, or parasitism.

The ' conidia ' reported by Richon[6] in **Hydnum Erinaceus** and **Corticium dubium** require more thorough investigation in all points.

I have elsewhere[7] represented the occurrence of two kinds of spores in the sporophores of Fries' **Nyctalis asterophora** as belonging to the cases which we are considering here. The sporophores of this Fungus are developed in the same way as in the gymnocarpous Agaricineae. The loose air-containing weft of delicate

[1] Icon. II, p. 23.

[2] Ann. d. sc. nat. sér. 5, IV, p. 290, and XV, p. 228, t. 12.

[3] Bull. Soc. bot. de France, XXIII, p. 362.

[4] Summa Veg. Scand. p. 564.

[5] Zeitschr. f. d. gesammten Naturwiss. 53 (1880), p. 424, tt. 13, 14.

[6] Bull. Soc. bot. de France, 1881, p. 180.

[7] Bot. Ztg. 1859, p. 385.

radiating hyphae, which forms the whole of the upper side of the pileus, produces at an early stage and from all parts numerous stellate spores or *macrogonidia* of a yellowish brown colour, which I named chlamydospores ; when fully developed it is a yellowish brown layer, which may be 1 mm. in thickness and ultimately decays (Fig. 162). Lamellae are formed on the under surface of large pilei, and on them basidia which produce four spores, and are usually few in number. The tissue of the under side of the pileus, which bears the lamellae when fully developed, is distinctly different from that of the layer of chlamydospores both in the shape and size of its cells. In young specimens, on the other hand, the Fungus consists entirely of similar hyphae, and there is an uninterrupted connection between the hyphae of the under side of the pileus and those of the chlamydospore-layer, the latter appearing as branches from the former. In some cases no basidia or lamellae are formed. Tulasne states that a third kind of spore is found in the layers of macrogonidia, namely, small colourless cylindrical cells or microgonidia which are abjointed in long rows.

A second species, Nyctalis parasitica, Fr., which like N. asterophora grows on the larger Agarici, especially Russula adusta, Fr., forms narrowly ellipsoid smooth macrogonidia in the interior of the whole of the tissue of the thick swollen lamellae, the other parts of the pileus remaining free from them. Tulasne not unfrequently found typical four-spored basidia isolated in the same lamellae as the chlamydogonidia, but I never saw any in my specimens. The view that the gonidia are organs of Nyctalis is founded with regard to both species on the fact, that the hyphae from which they are produced are evidently branches from the rest of the tissue, as seems to have been ascertained beyond doubt in younger specimens.

Tulasne[1] has declared against this view. Like Corda, Bonorden, and other writers before him [2] he holds that the chlamydospores (with the microgonidia of N. asterophora) are organs of two Fungi which are parasitic on Agaricus parasiticus, a distinct species which itself lives upon Russula and other forms, and produces more or less degeneration in them, and he places these parasites in Hypomyces, a genus of the Sphaeriaceae which lives on other Fungi, under the names of H. asterophorus and H. Baryanus. The grounds of his judgment are the resemblance between these organs and those of the same name in other species of Hypomyces which are known to be parasitic on the Agaricineae, the fact, which I can vouch for, that the chlamydospores sometimes arise isolated on the mycelium of Nyctalis which grows in or on Russula, and lastly the occurrence of indubitable perithecia in company with the gonidia on Nyctalis asterophora. He has not detected any defect in the investigation of the anatomy and life-history of the plant, on which I founded my opinion.

Tulasne's view is so very probable, that I have taken much pains to find out where my mistake is, but repeated examinations have hitherto always produced the same result. If we simply adhere therefore to the facts before us, we must abide by my first view, especially since Agaricus parasiticus has never been found, as far as I know, without a chlamydospore-apparatus, except perhaps where it is replaced by Nyctalis microphylla of Corda[3]. The structure also of the sporophore of N. parasitica is so very different from that of N. asterophora, that the conjecture that the two forms of Nyctalis are the same Agaricus parasiticus which has been differently altered by different parasites, seems to me much more rash than my own opinion as expressed above. Moreover an experiment of Krombholz by artificial cultivation of the plant seems to afford support to my explanation. Krombholz[4] sowed the stellate spores of N. asterophora on a young Russula adusta, and as he states with all due precautions,

---

[1] Ann. d. sc. nat. sér. 4, XIII, p. 5, and Sel. Fung. Carpol. III, 54, 59.
[2] See Bot. Ztg. 1859, p. 385.
[3] Icon. IV, Fig. 134.
[4] Essbare Schwämme, Heft I, p. 3.

and observed in a few days the first beginnings of a mycelium make their appearance exactly on the spots which had been sown, and on the ninth day the first commencements of the compound sporophores, which reached their normal development and maturity between that and the twentieth day. But even here the possibility of parasitism is not absolutely excluded.

A notice of Sautermeister [1] to the effect that **Exidia recisa** may also have ascocarps in old and shrivelled compound sporophores has never been confirmed, and must have been founded on the settlement of a species of Ascomycete in old plants belonging to the Tremellineae.

SECTION XCIV. If we cast a glance backward at sections LXXIV to XCIII, we shall see that we must assume a direct affinity or phylogenetic connection throughout the whole assemblage of the Basidiomycetes. The course of development is the same in its main features wherever it has been ascertained. The organs which have been designated by the same name in the foregoing account, especially the basidia and basidiospores, must from the data before us be regarded as *strictly homologous*. The Hymenomycetes on the one side and the Gastromycetes on the other are evidently two closely connected series. They are in general very different from one another, and the difference lies chiefly in their compound sporophores, but they sometimes also approach each other very nearly. Gautieria, and we may say also some forms of Secotium, are evident connecting links between the groups of the Hymenogastreae and Polyporeae. Gautieria which has all the characters of the Hymenogastreae, but has its chambers open and covered with no peridium, may be compared to a curled Merulius; the question naturally arises, whether the interior chambers have been formed by differentiation or in some way directly corresponding to this comparison.

If we could attribute a decisive value to the habit of the plants, we should dwell upon the great resemblance between the stalked Hymenogastreae, like Secotium erythrocephalum (Fig. 144) and a veiled Boletus, or still more perhaps that of Polyplocium [2] to the same species, though Polyplocium is too little known in its earlier states. But among the Polyporeae there is a remarkable form Polyporus volvatus, Pk., the Polyporus obvallatus of Berk. and Cook [3], which considered by itself must be placed with or close to the Hymenogastreae. Its sporophore, which lives in the bark of trees, is a hollow spherical body flattened at the poles and about the size of a hazel-nut, with a thick closed wall of leathery texture; its interior surface is covered with the hymenium of a Polyporus on the part next the substratum and is sterile on the opposite side.

That all the groups of the Gastromycetes converge towards the Hymenogastreae directly or indirectly through the Lycoperdaceae, and that they may therefore be deduced phylogenetically from the Hymenogastreae, is a necessary conclusion from the account which has here been given of them. Corda and Tulasne long since drew attention to this connection, and to the affinity between the Hymenogastreae and the Hymenomycetes which has been brought into prominence in the preceding pages.

---

[1] Bot. Ztg. 1876, p. 819.
[2] See Berkeley in Hooker's Journ. II. 201, and Corda, Icon. VI.
[3] Ellis, North American Fungi, No. 307.

It would be possible to assume another point of connection between the Hymenomycetes and the Gastromycetes, if we regard the mode of development of the compound sporophore from one side only. Then Amanita among the Hymenomycetes would approach nearest to the Gastromycetes, because the first development of the parts in one and the other is the result of differentiation *in the interior of the primordial coil of hyphae.* Brefeld, in connection with some former suggestions of my own of this kind, has recently given decisive weight to the above consideration. But then Amanita is closely connected in all other respects with the series of the Agaricineae ; the agreement between their propagative and especially their hymenial apparatus and that of the Gastromycetes is the very smallest possible. We might disregard this fact, and venture a jump across the intervening space, if no better means of connecting these groups could be found. But since we have such a mode before us, and at the same time there is no ground for assuming the existence of two points of union, the jump need not be made. The facts lead to the other conclusion, that the development of the compound sporophores by internal differentiation, called above the *angiocarpous,* makes its appearance at two widely separated points within the group of the Basidiomycetes, namely, in the series of the Agaricineae and in that of the Polyporeae. From the Polyporeae it leads further on to the formation of the Gastromycetes.

After the special descriptions which have already been given of the complex Lycoperdaceae and especially of the Phalloideae with their strange forms and differentiations of tissue, nothing further is required to show that they are the most highly developed of the Basidiomycetes. It is apparent also that all the series of the Basidiomycetes converge towards the more simple Thelephoreae and Tremellineae and meet in them. These forms may therefore be a connecting link between the Basidiomycetes and the rest of the Fungi, and since no such link is known elsewhere, there is a strong antecedent probability that this exact point of connection is to be found there. Comparison shows an evident close affinity between the Tremellineae and the tremelloid Uredineae (see page 283). In fact there is no essential difference between the two groups either in the course of development or in the structure of the sporophores. It is unnecessary to repeat here what has been already said respecting the former. If special stress is to be laid on the formation of gonidia and sprouts in the germination of the Tremellineae, this phenomenon is undoubtedly repeated in the abjunction, so common in the Uredineae, of ' secondary' and even tertiary sporidia on primary sporidia which do not at once find a substratum favourable to nutrition when they first germinate. These sporidial sprouts are produced, as far as is known, only in small numbers, but this constitutes only a quantitative difference and has an evident relation, as may be remarked in passing, to the strictly parasitic mode of life of the Uredineae. The conformation of the compound sporophores in the members of both divisions is to a certain extent the same. The organs termed basidia in Auricularia (page 306) do not differ in any essential point from the teleutospores of Chrysomyxa and Coleosporium ; the forms too of the basidiospores or sporidia may be almost said to be the same. Uredineae like Leptochrysomyxa Abietis are simply Tremellineae, and would be placed with them in the natural system if their allies which form aecidia were not known. The same may be said without exaggeration of the Leptopuccinieae, for the teleutospores of the latter, which must form one term of the comparison, are not more unlike the basidia of the Tremellineae than many of these are unlike one another. The gelatinous character also of the membranes in many Uredineae might also be mentioned, if such an argument were not too weak for serious consideration. The statements made in former paragraphs show that the teleutospores, or in some cases perhaps only the promycelia developed from them, in the tremelloid Uredineae, are not only very like the basidia of the Tremellineae, but *strictly homologous* with them. There is the same relation between the sporidia and the basidiospores.

Further there can be no doubt, according to the facts stated above on page 283, that the teleutospores of the tremelloid Uredineae are strictly homologous phylogenetically with the teleutospores of the forms which produce aecidia, and that the two sections together form the sharply defined group of the Uredineae. The rhythm of the development in the species which form aecidia shows that this group belongs to the series of the Ascomycetes (see page 132). The Basidiomycetes are thus connected with the series of the Ascomycetes through the Tremellineae and the tremelloid Uredineae.

The teleutospores of the Uredineae, if referred to the general course of development in the Ascomycetes, come under the definition of gonidia, and the basidiospores of the Basidiomycetes, which are phylogenetically homologous with them, necessarily come under the same idea. If those members of the development are said to be homologous which occupy corresponding places in the course of the ontogenetic development, the homology between the series of the Ascomycetes and the tremelloid Uredineae and Hymenomycetes is *interrupted and not restored* in the sense explained on page 123.

And now if in deciding with regard to the phenomena of relationship which arise out of the similarity of the development we hold to the view that the forms in question are all phylogenetically connected, the question arises, whether in passing from the tremelloid Uredineae and Basidiomycetes to the Uredineae with aecidia we should suppose that the latter proceeded from the tremelloid Uredineae or Basidiomycetes and therefore acquired the sporocarp as a new member of the development which was wanting in the earlier series; or whether the converse is not the more probable course of the phylogenetic development, and the sporocarp known as the aecidium has been ejected from the ontogenetic development of the tremelloid Uredineae and Basidiomycetes. One of the two things must obviously have taken place. In the first case the phylogenetic formation of the Uredineae bearing aecidia would be an act of *progressive development*, for a new and highly differentiated member would be added to those previously existing in the development: in the other case the development would be *retrogressive*, since this section of it would have disappeared. It was remarked above on page 285 and on a former occasion [1] that the absence of every kind of intermediate form is opposed in this case to the theory of a progressive development; there would be a jump from the tremelloid Uredineae to the perfect species with aecidia, and we must put up with it if it cannot be avoided, but our experience of known cases of progressive development does not add to its plausibility. On the other hand, there

---

[1] Bot. Ztg. 1879, p. 825.

are analogous cases known in the Ascomycetes which illustrate the opposite theory of a retrogressive development.

An attempt was made in former sections to show that the series of the Ascomycetes connects with the Algae which form oospheres through the Peronosporeae and Zygomycetes and their allies, and may be derived from them in progressive succession; also that the Uredineae are united as a member to this series. This view may at present be said to be the one which accords best and most naturally with the facts as known to us. But if this is the accepted view, we must necessarily adopt that one of the alternative assumptions, which makes the tremelloid Uredineae and Basidiomycetes degenerate descendants of those members of the series of the Ascomycetes, to which observation directly points; otherwise they must have a double phylogenetic origin, which can hardly be admitted in so comparatively narrow a group as the Uredineae. The series of the Basidiomycetes when once parted off advanced to a high and peculiar development, as is shown by the Amanitae, Phalloideae, Sphaerobolus, and other forms.

The views which have been developed in the preceding paragraphs explain the intimation given above on page 331 respecting the terminology of the reproductive organs in the Basidiomycetes. The word *gonidia has a different meaning here* from what it has in the series of Ascomycetes, where it was used with reference or in opposition to the *carpospores* (page 129); in the Basidiomycetes, where there are no carpospores, the term gonidia is applied to certain spores other than the *basidiospores*, and these are themselves phylogenetically homologous with certain gonidia of the Uredineae. The question whether the gonidia of the Tremellineae can be considered to be homologous with the uredogonidia of the Uredineae may fairly be termed a needless one; certainly there is little real objection to be made to this view. Still the question necessarily connected with it, whether the 'rods' observed hitherto without germination in some Hymenomycetes (page 332) are homologous with the gonidia of the Tremellineae, or are spermatia derived through the descent from the Ascomycetes, had better be left for the present without an answer. Further observations will perhaps determine these matters.

The views here expressed respecting the genealogy of the Fungi are of course only an attempt to bring the separate facts at present known into the unity of a single scheme; every displacement in the basis of facts may necessitate an alteration in the scheme.

The near connection which has been pointed out between the Hymenomycetes and the Uredineae may well be regarded as an established fact that is not likely to be set aside. There are certain alleged reasons why the former should be considered to be derived phylogenetically from the latter, and not *vice versa*. These are also good reasons in the present state of our knowledge, but the progress of inquiry may affect them, especially as they are derived, not only directly from the subject-matter in question, but partly also from remoter sources.

Starting from the same facts we may even now arrive at other results than these, as appears from Brefeld's views[1], though not without the help of some very bold hypotheses. What is to be said concerning these views will be found in a former treatise of my own[2] to which the reader is referred.

---

[1] Schimmelpilze, III and IV.

[2] Beitr. IV, p. 131.

The balance of argument has been already shown in the text to be in favour of considering the Hymenomycetes as united in a single series with the Tremellineae and these with the series of Ascomycetes through the Uredineae. The details of the connection, the question whether it is effected by several species and which those species are, have yet to be settled.

Certain observations gave rise some time since [1] to the conjecture that there were close relations between the Basidiomycetes and the Ascomycetes, the Basidiomycetes being the gonidial forms of perfect existing Ascomycetes. The conjecture originated in the observation of indubitable asci in pustule-like swellings in old lamellae of Agaricus melleus, and in phenomena observed in Nyctalis and mentioned on page 335. The finding of perithecia, which according to Tulasne are connected with the star-shaped spores of N. asterophora, was rather in favour of this view than against it. I regarded these Nyctalis-spores and the asci of A. melleus as organs of the particular species of the Hymenomycetes, because they seemed to proceed directly from hyphae of the pileus. At present I agree with Tulasne, as was intimated at page 336, in thinking that it is most probably a case of parasitism, and I spoke of the ascogenous hyphae of A. melleus at page 266 by their parasitic name of Endomyces. But since the parasitic nature of Endomyces and the defect in the former observations have not yet been shown, this brief further notice of the subject may be permitted in this place. In case the earlier statements should be confirmed, and it should be shown that Tulasne's perithecia belong to Nyctalis asterophora, the views here advanced would be subject to some alteration, since it would be necessary to assume that the Basidiomycetes are phylogenetically connected with the Ascomycetes, not at one point only, but at several. But in the present state of our knowledge it is difficult to admit any more points of connection in the ontogenetic cycle of the species of Ascomycetes with the greater number of Basidiomycetes.

Literature (see also p. 116).

a. Hymenomycetes.

The works of Persoon, E. Fries, Corda (Icones fungorum), Tulasne (Carpologia I), are the chief works to be consulted. Only the more important original treatises are cited below. Some works have been noticed already in the text.

F. NEES V. ESENBECK, Plantar. mycetoidarum, &c. Evolutio (Entw. d. Agaricus volvaceus) in N. Act. Acad. Leop. Carol. XVI, pars I.

JOS. SCHMITZ, Ueber d. Bildung neuer Theile bei d. Hymenomyceten ;—Id., Ueber d. Längen-Ausdehnung bei d. Pileaten (Linnaea, XVI, 1842) ;—Id., Ueber Entw. Bau u. Wachsthum von Thelephora sericea u. hirsuta (Linnaea, XII, 1843, p. 417).

BONORDEN, Allgem. Mykologie, pp. 156–196 et passim ;—Id., Beobachtungen ü. d. Bau d. Agaricinen (Bot. Ztg. 1858, p. 201).

H. HOFFMANN, Pollinarien u. Spermatien von Agaricus (Bot. Ztg. 1856, p. 137) ;—Id., Beitr. z. Entwicklungsgesch. u. Anatomie d. Agaricinen (Bot. Ztg. 1860, p. 389) ;— Id., Icones analyticae fungorum, I–IV, Giessen, 1861–1865.

DE BARY, Zur Kenntn. einiger Agaricinen (Bot. Ztg. 1859, p. 385).

J. DE SEYNES, Organisation des Champignons supérieurs (Ann. d. sc. nat. sér. 5, I, p. 269) ;—Id., Recherches sur l. végétaux inférieurs, I. Des Fistulines, Paris, 1874.

TULASNE, Obs. sur l'organisation des Tremellinées (Ann. d. sc. nat. sér. 3, XIX, p. 194) ;—Id., Nouvelles notes sur les fungi Tremellini et leurs alliés (Ann. d. sc. nat. sér. 5, XV). See also sér. 5, IV.

M. WORONIN, Exobasidium Vaccinii (Ber. d. naturf. Ges. Freiburg, 1867).

O. BREFELD, Bot. Unters. ü. Schimmelpilze, III ;—Id., Bot. Unters. ü. Hefepilze, V, Leipzig, 1883.

---

[1] Bot. Ztg. 1859, p. 404.

VAN TIEGHEM, Sur le développement du fruit, &c. des Basidiomycètes et des Asco-
    mycètes (Bull. Soc. bot. de France, 1876, p. 99, and Bot Ztg. 1876, p. 161).

R. HARTIG, Wichtige Krankh. d. Waldbäume, Berlin, 1874 ;—Id., Die Zersetzungs-
    erscheinungen d. Holzes d. Nadelholzbäume u. d. Eiche, Berlin, 1878.

b. Gastromycetes.

The chief sources of our knowledge of the morphology and anatomy of the Gastro-
    mycetes are the two following works :—

CORDA, Icones fung. II, V, VI.

TULASNE, Fungi hypogaei, Paris, 1851.

In Corda will also be found many reproductions especially of short and interesting
descriptions of exotic forms by Berkeley in Hooker's Journal.

After these—

MICHELI, Nova Plant. Genera, 1729 (Phallus, Clathrus).

BERKELEY in Annals and Mag. of Nat. Hist. 1839, and Ann. d. sc. nat. sér. 2,
    XII, 1842, p. 160.

VITTADINI, Monographia Lycoperdineorum (Mem. Acad. Torino, V, 1842).

J. SCHMITZ, Ueber Cyathus (Linnaea, XVI, 1842).

TULASNE, De la fructification des Scleroderma comparée à celle des Lycoperdon et
    des Bovista (Ann. d. sc. nat. sér. 2, XVII) ;—Id., Sur les genres Polysaccum et
    Geaster (Ann. d. sc. nat. sér. 2, XVIII, 1842) ;—Id., Recherches sur l'organ. d.
    Nidulariées (Ann. d. sc. nat. sér. 3, I, 1844) ;—Id., Description d'une nouvelle
    espèce de Secotium (Ann. d. sc. nat. sér. 3, IV, 1845) ;—Id. in Exploration
    sc. d'Algérie, p. 434, Tab. 23 (Development of Clathrus).   On the same subject
    see Berkeley in Hooker's Journ. of Bot. IV, p. 68.

V. SCHLECHTENDAL u. MÜLLER, Mitremyces Junghuhnii (Bot. Ztg. 1844, p. 401).

BONORDEN, Mycologische Beobachtungen in Bot. Ztg. 1851, p. 18 (Phallus, Sphaero-
    bolus) ;—Id., Die Gattungen Lycoperdon u. Bovista (Bot. Ztg. 1857, p. 593).

ROSSMANN, Beitr. z. Entw. d. Phallus impudicus (Bot. Ztg. 1853, p. 185).

JUL. SACHS, Morphologie d. Crucibulum vulgare, Tul. (Bot. Ztg. 1855).

V. SCHLECHTENDAL, Phalloideen (Linnaea 31, 1862), with copious notices of the
    literature of the subject.

HOFFMANN, Icon. analyt. Fungor. II, p. 33 (Hymenogaster).

DE BARY, Beitr. z. Morph. u. Phys. d. Pilze, I, 1864 (Phallus).

R. HESSE, Mikroskop. Unterscheidungsmerkmale d. Lycoperdaceengenera in Pring-
    sheim's Jahrb. X, p. 384 ;—Id., Keimung d. Sporen v. Cyathus striatus in Pring-
    sheim's Jahrb. X, p. 199.

N. SOROKIN, Développement du Scleroderma verrucosum (Ann. d. sc. nat. sér. 6, III).

E. EIDAM, Keimung u. Entwickl. d. Nidularieen in Cohn's Beitr. z. Biol. II.

O. BREFELD, Bot. Unters. ü. Schimmelpilze, III, p. 174.

A. PITRA, Ueber Sphaerobolus (Bot. Ztg. 1870, p. 681).

SCHRÖTER, Entw. v. Tulostoma in Cohn's Beitr. II, p. 65.

ED. FISCHER, Zur Entwicklungsgesch. d. Gastromyceten in Bot. Ztg. 1884 (Sphaerobolus,
    Mitremyces).

DIVISION III. MODE OF LIFE OF THE FUNGI.

CHAPTER VI. Phenomena of Germination.

1. CAPACITY OF GERMINATION AND POWER OF RESISTANCE IN SPORES.

SECTION XCV. The greater number of known spores in the Fungi, the word spore being taken in the meaning assigned to it on page 128, are capable of germination from *the moment that they are ripe*. In a smaller number of known forms this is not the case; the spores do not germinate till after they have passed through *a period of rest* which follows upon their maturity.

Examples of the first kind, that is, of spores which **germinate immediately after maturity,** are all ascospores, most spores of the Hymenomycetes, the majority of the forms known as gonidia, the oospores of some Saprolegnieae, as Achlya spinosa, A. apiculata, and Aplanes (section XL).

Some spores are capable of germinating even before maturity, that is, before they have reached the condition which according to empirical rules indicates maturity (see page 59). The gonidia of the Saprolegnieae, for example, may omit their regular swarm-cell period in cultivated specimens and germinate beneath the cover-glass[1]; the ascospores of Sordaria fimiseda in Woronin's[2] experiments showed themselves capable of germination long before the completion of their definitive membrane and their discharge from the ascus, and similar cases are not unfrequent among the Ascomycetes.

The period of time during which spores of this category retain the power of germination, where the conditions are unfavourable to germination and the spores are not subjected to any serious injury, varies in different species and individuals and according to the form of the spore. It is short in the comparatively watery and turgescent gonidia of the Peronosporeae, and in the uredospores, aecidiospores, and sporidia of the Uredineae, lasting where the spores are not quite dried up for a few weeks, seldom for a few months, and coming to an end, as far as we know at present, with the summer in which the spores were matured. Gonidia, for example, of Cystopus candidus which were not perfectly air-dry continued capable of germination 6–8 weeks, those of Phytophthora infestans for 3 weeks, perfectly air-dry gonidia of the latter species (caught on glass-plates) lost their power of development in 24 hours. Similar results were obtained in the case of the spores of the Uredineae mentioned above. The ripe swarm-spores of the Saprolegnieae and Peronosporeae remain capable

---

[1] Thuret in Ann. sc. nat. sér. 3, XIV, t. 22.
[2] Cited on page 262.

of germination for a short time only after completing their brief period of swarming and die if the conditions are unfavourable to germination. But the diplanetic spores in these groups (see pages 108, 143) are to some extent exceptions to this rule, since they may continue alive and unaltered for days and weeks, perhaps longer, after entering upon the transitory period of rest, if the conditions required for their further development, especially a sufficient supply of oxygen, are excluded. Further development begins if they are placed in fresh water containing oxygen, and first of all the second stage of swarming.

But many spores of this first category, belonging to Fungi which do not live in water, retain their vitality and power of germination for a long time, if they are protected and kept in an air-dry state.

It has been proved that the spores of many common Fungi will remain alive for a period varying 1–2 years; for example the ascospores of Penicillium glaucum have germinated when two years old, the gonidia after $1\frac{1}{2}$–$1\frac{3}{4}$ years; the spores of Coprinus stercorarius[1] when more than 1 year old, the gonidia of Aspergillus niger when of the same age, of Sordaria curvula after 28 months, of Botrytis Bassii and of allied forms of Isaria after 1–2 years, of Mucor stolonifer after 1 year. The gonidia of Phycomyces nitens, a species nearly related to Mucor Stolonifer, is a good example of specific and individual variation. Van Tieghem[2] found that they hardly retained their vitality for 3 months, and I once observed the germination of gonidia which were 10 months old, while others had lost their vitality after the space of 1 month.

The resting-spores of some Ustilagineae are very long-lived according to Hoffmann and v. Liebenberg[3]; the latter observer found the spores of Tilletia Caries from a herbarium specimen capable of germination after $8\frac{1}{2}$ years, those of Ustilago Carbo after $7\frac{1}{2}$ years, of U. Tulasnei and Urocystis occulta after $6\frac{1}{2}$, of Ustilago Kolaczeckii, U. Crameri and U. destruens after $5\frac{1}{2}$ years, of U. Rabenhorstiana after $3\frac{2}{3}$ years. In some of the forms here mentioned, as U. Carbo, the maximum of possible duration appears to be given; in others as Tilletia Caries the vitality may continue beyond the time specified, which is given for the oldest material at command, and this showed such active power of germination that it may be assumed that still older spores would have retained the power.

Gonidia of Aspergillus flavus[4] proved in one case to be capable of germination after having been kept in a dry state for 6 years, those of A. fumigatus[5] for as many as 10 years.

Well-known examples of spores which **require to go through a period of rest** before they are capable of germination are found in the oospores of the Peronosporeae and most Saprolegnieae, the zygospores of the Mucorini, the organs described above as resting-spores in the Chytridieae, Protomyces, and other forms, the teleutospores of most Uredineae—with the exception however of Leptopuccinia,

---

[1] Brefeld, Schimmelpilze, II, 76; III, 15.

[2] Van Tieghem et Le Monnier in Ann. d. sc. nat. sér. 5, XVII, 288.

[3] Hoffmann, in Pringsh. Jahrb. II, 267.—v. Liebenberg, Ueber d. Dauer d. Keimkraft d. Sporen einiger Brandpilze (Oesterr. Landw. Wochenblatt, 1879, Nr. 43, 44).

[4] Brefeld, Schimmelpilze, IV, p. 66.

[5] Eidam, in Cohn's Beitr. z. Biolog. III, 347.

Coleosporium, Leptochrysomyxa, Chrysomyxa, Cronartium, Hemileia, and others, in which the teleutospores belong to the first category,—the megalogonidia or resting gonidia of Hypomyces and its allied forms, some spores of the Ustilagineae and some others.

The period of rest in many of these cases is observed to last for a fixed time, and to coincide with certain periods of vegetation and seasons of the year. The teleutospores, for instance, of the Uredineae, the most frequent representative of which is Puccinia graminis, are dormant during the winter. The teleutospores of Puccinia ripen at the end of summer and during the autumn and germinate under favourable conditions in the following spring, and it is difficult to procure their germination before that time even under cultivation. This appears also to be the history of the oospores of the Peronosporeae which do not live in water, and of the resting-spores of Protomyces macrosporus and some species of Synchytrium, S. Anemones for instance and S. aureum, &c. (see page 167), which all ripen in the summer and do not germinate till the next spring, and in their case also it is found difficult to shorten the resting period in plants under cultivation.

It is different, according to Woronin [1], with the resting-spores of Tuburcinia Trientalis among the Ustilagineae and of Sorosporium Saponariae, which ripen in the course of the summer and are not capable of germination before the end of September of the same year.

This coincidence with a fixed period of the year is at least not a general rule in the zygospores of the Zygomycetes, in the oospores of many Saprolegnieae [2] (the Pythieae), and in the resting-spores of Synchytrium Taraxaci, though it may occur in some cases (Sporodinia, Synchytrium). In the absence of this coincidence we can only speak of a necessary period of rest which lasts some weeks or months and varies in different species and individuals.

We have no exact observations on the maximum of the resting time possible under favourable circumstances in the case of species which are not confined to fixed seasons of the year. It may extend to a year and probably may exceed a year in the spores of Pythium proliferum kept under water. Our knowledge of the Zygomycetes and Saprolegnieae would lead us to suspect that, when the conditions required for germination are excluded, the power of germination ceases at an earlier period in them than in the more hardy forms, which retain their vitality more than a year without being obliged to go through a period of rest.

The forms in which the resting time coincides with fixed periods of the year appear as a rule not to retain their powers beyond the favourable time for germination which follows their maturity, or not to retain them long beyond that time. Teleutospores of Puccinia graminis which have lasted during the winter germinate with great readiness in the spring which succeeds their period of ripeness, more slowly and more infrequently during the following summer months, and I was unable to procure their germination after August or in the spring of the second year. It is the same with other allied species. Woronin failed to make the spores of Tuburcinia Trientalis germinate when they had remained without germinating from

---

[1] Beitr. V. See also before on page 180.
[2] See De Bary, Beitr. IV.

the time of their maturity till the late autumn; and it would appear, from the observations which we possess, that the regularly hibernating oospores of the Peronosporeae and the winter-spores of the Chytridieae, &c. behave in a similar manner, though no exact researches have been made into the possible maximum duration of their resting time.

The power possessed by the spores of Fungi of **withstanding the effects of injuries caused by agencies from without** varies very much in individual cases. Disregarding the effect of poisonous bodies, we have to consider under this head the withdrawal of water, extreme temperatures, and mechanical attacks.

As regards *mechanical injury*, Van Tieghem[1] found that some ripe spores—gonidia of Phycomyces, Pilobolus oedipus, Mortierella reticulata, and zygospores of Sporodinia and Mucor fusiger—when they are wounded or cut into not too small pieces, have the power of healing the surfaces of the wound by forming a new protoplasmic pellicle and a new membrane, or, if the lesion is slight and partial, they can divide into daughter-cells, and each of these cells and each of the cicatrised fragments continues or becomes capable of germination if the circumstances are favourable.

Some spores are very sensitive to the *withdrawal of their water* which is effected by drying them in moderately warm air (20–22° C). This is the case with swarm-spores developing beneath the water as well as the gonidia abscised in the Peronosporeae mentioned above, and with the short-lived spores of the Uredineae. The example of Phytophthora infestans shows that these spores lose their powers of germination quickly in the air-dry state even after a period of 24 hours. It would appear also that the oospores of the Saprolegnieae which ripen under water cannot bear being dried in the air at an ordinary temperature. It is estimated that at least 95 per cent. of the ripe ejected spores of Sclerotinia ciborioides lose their power of germination if they are kept dry in the air on glass plates for 12 days in a temperature of about 20° C.

Occasional observations seem to show that the spores of many Lichen-fungi are among those which will not bear desiccation, but the point requires more thorough investigation.

It is evident that the spores which we have been describing form a very small part of the whole number; the great majority of the spores of Fungi retain their power of germination for a long time in the air-dry state. Some display great power of resisting high temperatures and must therefore be able to bear the artificial extraction of a large proportion of the water which they contain.

Resistance to extreme temperatures is shown more especially in relation to a high degree of cold. The spores which were recently spoken of as regularly hybernating in our temperate climates will endure a cold of $-15°$ to $-25°$ C. without having their power of germination in the least affected. This experience justifies us in assuming the same or a similar power in the majority of the other long-lived spores mentioned above. This was confirmed by Hoffmann in the case of Ustilago Carbo and U. destruens, Trichothecium roseum, Fusarium heterosporum, Penicillium glaucum, Botrytis cinerea, and by Schindler[2] in that of Tilletia Caries.

---

[1] Ann. d. sc. nat. sér. 6, IV, p. 315.
[2] Wollny's Forschungen a. d. Agriculturphysik, III, p. 288.

The behaviour of Saccharomyces Cerevisiae, though not strictly belonging to this connection, may be adduced to show that Fungus-spores are capable of bearing very low degrees of temperature, for some only of its cells, according to Schumacher[1], were killed by a temperature of $-113.75°$ C., while others lived and retained their powers of growth.

On the other hand, many spores, especially those that are long-lived, can endure extremely high temperatures without losing their vitality. Some observations show that the length of the time to which they are exposed to the heat is an important element in the case. The dry spores of a certain number of Fungi are not killed by a heat which considerably exceeds $100°$ C. According to Nägeli it may be said that a temperature of $130°$ C. is necessary to secure their destruction. But the death-point of the spores of Fungi is often much lower than this in water or watery vapour, and it has not been shown that any can under these circumstances survive a temperature of $100°$ C.

According to H. Hoffmann the dry spores of Ustilago Carbo and U. destruens support unharmed a temperature of from $104°$ to $128°$ C.; the spores of U. Carbo are killed in a chamber filled with vapour at a temperature between $58°.5°$ and 62^{v} C., those of U. destruens by a temperature of from $74°$ to $78°$ C. in the space of an hour, and by one of from $70°$ to $73°$ C. in 2 hours. Tilletia Caries, on the other hand, will bear according to Schindler a temperature of $95°$ C. if dry, but not more than that. According to Payen[2] the spores of a mould found in bread, Oidium aurantiacum, can endure a temperature of $120°$ C.; at $140°$ C. they were discoloured and killed.

Pasteur's researches showed that the spores of Penicillium glaucum remain unaltered in dry air at a temperature of $108°$ C. Many, but not all, lose their power of germination at $119°$ to $120°$ C.; all lose it very quickly if the temperature is raised to between $127°$ and $132°$ C. The limit is similar in 'Ascophora elegans' (a species of Mucor). Similar results were obtained from experiments on some spores of uncertain origin mixed with dust, among which Botrytis cinerea, P. (gonidia of Peziza Fuckeliana) or a closely allied form may be certainly distinguished; these bore a temperature of $120°$ C. Spores suspended in a fluid which was heated up to $100°$ C. were always killed in Pasteur's experiments.

Against these statements must be placed the results of Tarnowsky's researches communicated by Sachs[3], according to which spores of Penicillium glaucum and Mucor stolonifer when heated in air to a temperature of from $70°$ to $80°$ C. during 1–2 hours very seldom germinated, and were killed by a temperature of from $82°$ C. to $84°$ C.; they lost all vitality when warmed in a fluid up to from $51°$ to $55°$ C. We may just mention some older statements of J. Schmitz[4] on this subject, which certainly stand much in need of investigation.

The foregoing statements show that the spores of the Fungi are similar in character to the organs with analogous functions in other plants. They afford

---

[1] Sitzgsber. d. Wiener Acad. 70, 1 Abth. Juni, 1874, where other literature is cited. See also Pfeffer, Physiol. II, 438.

[2] Compt. rend. XXVII, p. 4.

[3] Lehrb. Auflage 4, p. 699.

[4] Verhandl. d. naturh. Ver. d. Rheinlande, II (1845).

special examples of the general rules regarding capability of germination, resistance, &c. which hold good in those organs.

The theoretical explanation of the phenomena is therefore the same as in analogous cases outside the Fungi and presents the same difficulties. This is not the place to discuss this question, and the reader is therefore referred to physiological works, and especially to the chapter which deals with it in Pfeffer's Physiology.

It is hardly necessary to call attention once more to the consideration, that the particular phenomena and characters change or may change from one case to another, and that consequently the phenomena to be expected in one Fungus-spore cannot be concluded with certainty from those of another, though it may be an allied form; each must be investigated for itself. This applies as much to spores of the same name in different species of Fungi as to differently named spores in the same species. More-over the results of enquiries into persistence and resistance show individual differences in similar spores. For we not only see germination more difficult and slower under otherwise similar conditions near the limits of persistence and resistance—the gonidia of Aspergillus flavus, for example, germinate at once when fresh, but not till the conditions for germination have been in operation for fourteen days when they have been kept in a dry state for six years,—but some spores die sooner when approaching those limits than other similar ones. These differences may be due to internal causes which may perhaps be summed up in the words dissimilar maturation; or possibly, after similar maturation, some spores are better protected than others against slowly operating and modifying causes acting from without, and this it is extremely difficult or impossible to determine with certainty[1]. But we must certainly not disregard these differences when dealing practically with the question, and it is possible that they have been the occasion of some of the discrepancies in the foregoing statements.

There is another point which is also sufficiently obvious, namely, that the characters described above imply adaptations, sometimes of a rougher, sometimes of a more delicate kind, to definite modes of life. Two instances of this may be given. The spores of ordinary Moulds,—of Penicillium, Eurotium, Aspergillus, and the Mucorini, and the gonidia of many Ascomycetes,—which are capable of germination as soon as they are ripe and are also of a hardy nature, continue when they are once matured to be for a long time always ready for further development if they meet with the necessary conditions, and this, it may be said, may happen at any spot and at any time in the case of these Fungi, because they have a very wide field of choice as regards the nature of the substratum (see section XCVII). The teleutospores of Puccinia graminis, which were described above as passing normally through the winter in a dormant state, find the most favourable conditions in the ensuing spring, when their power of germination is at its best, not only for germination but also for the further development of the short-lived sporidia, which are the products of their germination; the conditions are favourable to the sporidia, because the danger of desiccation is comparatively small in the moist cool spring-time, and there is young foliage on the Berberis-plants into which the germ-tubes from the sporidia must penetrate to secure their further development (see section CX). In autumn, when the teleutospores are

---

[1] See on this point for example v. Liebenberg, as cited on page 344.

mature, this condition for their development is almost entirely wanting, and the height of summer which follows upon their regular time of germination is equally unfavourable; the phenomena connected with their powers of germination are closely accommodated to these circumstances. The teleutospores of this Fungus are at the same time the only means, or if we take some doubtful and exceptional cases into consideration almost the only means, by which the plant lives through the winter in these latitudes, for all the other organs die as a rule at the beginning of winter. Exactly similar conditions prevail *mutatis mutandis* in the case of the spores which normally live through the winter, of other Uredineae, of Protomyces, Synchytrium, and the Peronosporeae, of Cystopus Portulacae in a remarkably striking manner in Germany, of Tuburcinia Trientalis and some others.

These cases of exact adaptation to the function of hibernation are connected with a long series of other cases within the affinity of the forms mentioned in which the above characters in the spores are subject to very manifold modifications to meet new adaptations in the life-conditions of the plants. Thus, for instance, in the alliance of Puccinia graminis we come to the Leptopuccinieae with a mycelium which lives through the winter, and spores which are usually short-lived and germinate as soon as they are ripe, and still further to the Chrysomyxae with a mycelium which also lives through the winter, and a copious supply of spores which are all short-lived and can germinate only in the summer in which they are formed (see sections LXXXII, CX). Similar cases occur everywhere in other groups.

2. EXTERNAL CONDITIONS OF GERMINATION.

Section XCVI. The external conditions necessary for the commencement of germination in the spore are in general the same as are required for the germs and seeds of other plants ; *a certain temperature of the environment, a supply of oxygen and water*, and *in some cases also of nutrient substances.*

The cardinal points in the **germination-temperature** of spores have been exactly ascertained for a few Fungi only. Wiesner[1] found the minimum to be from 1.5° to 2° C. in the gonidia of Penicillium glaucum, the maximum 40° to 43° C., and the optimum about 22° C.; the cardinal points are probably similarly situated in very many Fungi in our latitudes. Thus Ustilago Carbo germinates according to Hoffmann at a temperature of from +0.5 to 1° C., Botrytis cinerea at +1.6° C., Ustilago destruens requires a temperature above +6° C., but continues to germinate at 38.75° C.; I observed Cystopus candidus develope zoospores and their germ-tubes as well at +5° as at 25° C.

More exact observations will bring to light many variations in this respect in species and individuals such as are found in other regions of the vegetable kingdom. Aspergillus fumigatus, Fresen., which was carefully examined by Lichtheim[2], may be mentioned as a striking instance of this kind among the Fungi; in this plant it was approximatively determined that the minimum did not fall much below +15° C.

---

[1] Sitzgsber. d. Wiener Acad. 67, I (1873).
[2] Berliner klinische Wochenschrift, 1882, Nr. 9.

Fungi which are in the habit of germinating in the digestive canal of warm-blooded animals, the Mucorini, Piloboli, Ascoboli, and indeed also the Sordarieae and Coprini, have an optimum corresponding, as might be expected, with the temperature of the body of the animal. Brefeld[1] even gives the minimum as from 35° to 40° C. in the case of one species of this group, which he does not determine more nearly. He also gives from 35° to 40° C. as the minimum in the case of Crucibulum vulgare, which is supposed to germinate regularly in the intestines of animals, while Eidam[2] has observed its germination at a temperature between 20° and 25° C.

The **supply of oxygen**, which must be regarded as necessary for general reasons, has never been specially investigated in the germination of the Fungi. Its necessity is shown when spores are sown in drops of liquid beneath a covering glass. The spores germinate better the nearer they lie to the edge of the covering glass where there is a free supply of air; they often do not germinate at all in the middle.

The necessity of a **supply of water** is evident in all cases and is intelligible. It is plain from the swelling of the germinating spore and the formation of successively growing vacuoles that the process of germination begins with absorption of water. It depends on the species or the form of the spore whether pure water is necessary, or whether the water must contain certain nutrient substances or whether germination takes place both in water and in nutrient solutions.

The first category includes a number of strictly parasitic forms, the Peronosporeae especially and Uredineae, and also the Erysipheae, Polystigma rubrum and species of Rhytisma. Many of the Lichen-fungi may also belong to it, but this is a point which requires investigation. The spores especially of the two first groups germinate best when they are not in but on drops of water or in an atmosphere saturated with vapour, which no doubt supplies them with the requisite amount of water in the shape of slight precipitations. Nutrient solutions may even hinder germination. Observation shows that the normal processes in germination are effected entirely at the expense and by the transference of the substances in the spore (see sections LXXIX—LXXXII).

The gonidia of Mucor stolonifer, Chaetocladium and most of the Mucorini and the ascospores of Sclerotinia Fuckeliana are examples of the third category which represents the other extreme. These Fungi emit at best only weak rudimentary germ-tubes in pure water, and it is not certain that these are not due to small admixtures of nutrient substances which entered the water with the spores. The ascospores of the Sordarieae and the spores of the Coprini also belong to this list. Their germ-tubes have been obtained hitherto only in nutrient solutions, and we have no exact account of their behaviour in pure water, because it has but little practical interest.

The spores of most Fungi belong to the second and intermediate category, with an inclination to one or other extreme according to the species. Examples are seen in the gonidia of Penicillium, Sclerotinia Fuckeliana, and other common forms of Moulds, and in the spores of the Ustilagineae and Tremellineae, with respect to which the reader should compare the statements in the sections of Chap. V which refer to them.

---

[1] Schimmelpilze, IV, p. 20.
[2] In Cohn's Beitr. z. Biol. II.

In the cases in which a supply of nutrient substance is *necessary* for germination, the commencement of the process of germination implies a previous commencement of a process of nutrition, which is only another way of expressing the facts. Accordingly germination is attended by the permanent and constant increase in the amount of protoplasm which was described on page 113. The germinating spore behaves in the same way as a growing vegetative cell. The same thing evidently takes place in the intermediate cases connected with this extreme. It has yet to be enquired how far water only may be taken up from the solution in other cases, for instance at the commencement of germination.

The nutrient substances suited for germination are in most cases of the same kind as those adapted to the vegetation of each species. Yet there appear to be exceptions to this rule. The spores of Ascobolus furfuraceus [1] which vegetates in the faeces of herbivorous animals have only at present been made to germinate on the mucous membrane of the stomach of living animals. They did not germinate in water or in nutrient solutions or in juice obtained by artificial means from the stomach of a pig, neither in the ordinary temperature of the room nor in that of the animal body. It remains to enquire how far the germination depended on special material qualities of the surface of the mucus membrane. I observed similar phenomena in the case of Onygena corvina which vegetates normally on the feathers of birds of prey. No result was obtained when spores were sown on microscopic slides with the variations mentioned above; but there was a fine formation of compound sporophores of Onygena on the hairs cast up by a white owl which had been kept a long time in captivity and which had received the spores of the Fungus with a mouse which it had eaten; the Fungus developed on the hair from the mouse on which the spores had been strewed and from no other. The development of Onygena came to an end indeed on most of the hairs thus obtained, and a rich growth of Gymnoascus, not intentionally sown, took its place, which perhaps drove out the Onygena. Further investigation of this case is necessarily deferred.

The formation of the germ-tubes of Protomyces macrosporus has hitherto been observed only on the epidermis of Aegopodium, not in nutrient solutions. It remains to be seen whether this is a similar case to those which we have been considering, or one where a particular nutrient substance is required for the process of germination.

It appears from the facts which have now been enumerated, that the necessity for the presence of nutrient matter in germination, as well as for certain temperatures and for the changes not exactly ascertained which must take place during the state of rest, are connected with analogous specific differences or adaptations in the plants themselves. We can scarcely suppose that the necessity for a nutrient substance depends simply on the *quantity* of reserve material present in the spore. This is shown by the recorded observations on the ascospores of Sclerotinia Fuckeliana as compared with the very similar spores of the nearly related Sclerotinia Sclerotiorum. The latter germinate readily in pure water. On the other hand, it is obvious that the spores which only need a supply of water for normal germination must be provided also with the requisite quantity of reserve material.

There are still many spores of Fungi, entire groups of them, which have never been observed to germinate. Putting the spermatia mentioned in the morphological portion

---

[1] Janczewski in Bot. Ztg., 1871, p. 257.

of this work (page 240) altogether out of the question, we may quote here, as examples only, the basidiospores of most Gastromycetes (Phalloideae, Lycoperdaceae, Hymenogastreae), and the ascospores of Tuber, Elaphomyces, and allied genera,—organs which in contrast with the more or less doubtful spermatia are quite certainly homologous with allied forms which germinate readily. To these may be added the zygospores of Mucor stolonifer, Brefeld's Mortierella Rostafinskii. The germination of Agaricus campestris also has never been certainly observed[1]. The number of cases of this kind has gradually diminished, the more it was recognised that the conditions of germination or adaptations might vary from one species to another and the more these conditions were ascertained in particular cases. This must be considered in judging of unsuccessful experiments in germination, and may lead us to expect success from further investigations in cases where it has hitherto been wanting. On the other hand, phenomena such as those pointed out on page 332 in connection with Sphaerobolus must not be disregarded.

The sporocarps of Erysiphe which were closely studied by Wolff[2] show that in germination as elsewhere adaptations may occur which deviate considerably from those ordinary cases of adaptation on which our views are apt to be founded. These arrive at maturity, that is, separate from the nutrient substratum with the termination of the period of vegetation. At this time the asci are formed inside them and in most species the spores also are formed in the asci, but in some (E. graminis, E. Galeopsidis) no spores are yet formed. In this condition the sporocarp enters upon its winter's rest. When this is over and the sporocarp is placed in water, the spores are formed and ripen in E. graminis and E. Galeopsidis, in other species, as E. communis, they ripen only; then the spores are at once discharged from the sporocarp which bursts by the swelling of its contents, and the spores germinate. In this case therefore the mature sporocarp hibernates, provided, as minute investigation shows, with reserve material for the subsequent formation of the spores, and the spores, even those which are apparently and which were formerly supposed to be ripe before hibernation, do not really mature and acquire the capacity for germination till after the winter rest is over. In contrast to this proceeding most other homologous sporocarps form mature spores with the power of germination either without passing through a resting stage or before they enter it, or else they become dormant before the formation of the asci when they are in form like a sclerotium or xyloma. It is not yet ascertained whether there are not species even of Erysiphe which dispense with a period of rest. Tulasne's[3] account of E. guttata seems to point in this direction.

CHAPTER VII. Phenomena of Vegetation.

1. GENERAL CONDITIONS AND PHENOMENA.

Section XCVII. Fungi resemble all other plants in the main features of their organisation and material composition; it may be assumed therefore beforehand, and the assumption is confirmed by experience, that the most general conditions of vegetation are the same in both classes, which may therefore in general terms be said to be dependent in the same manner on light, heat, gravitation, and the chemical

---

[1] See Nylander, Flora, 1863, p. 307.
[2] As cited on p. 294. See above, pages 76 and 202.
[3] Carpol. I.

nature of their environment, while each individual according to its particular organisation has a special reaction on the influence exercised by these agencies[1].

In judging of the phenomena which present themselves to our notice, it is at least as necessary to consider the influence of temperature on the vegetation and growth of Fungi as the temperatures of germination noticed above. In this relation also plants are subject to fixed rules. Every vegetative (and fructificative) process has certain limits of temperature and a fixed optimum in each species. In Wiesner's series of experiments cited above on page 349, all other conditions being the same, the optimum in the growth of the mycelium of Penicillium glaucum was about 26° C., that in the formation of gonidia, like that of germination, was about 22° C. Many Fungi which are natives of our temperate zone probably exhibit the same relations to temperature as Penicillium. That there are differences, however, between one species and another is shown at once by the circumstance, that some Moulds make their appearance spontaneously in closed places, *ceteris paribus* in the hottest time of the year, and it may almost be said at no other time. I observed this for instance years ago in the case of Aspergillus clavatus, Desm. According to Siebenmann's[2] statements, which, it is true, require further testing, Eurotium repens flourishes in a temperature of from 10° to 15° C., and disappears at 25° C.; the same is said to be the case with E. Aspergillus glaucus; A. albus and A. ochraceus do well at from 15° to 20° C., but suffer if the temperature rises above 25° C. In A. niger, on the contrary, Raulin[3] found that the optimum in the formation of mycelium and gonidia was *ceteris paribus* 34° C.; in A. fumigatus Lichtheim[4] places it at from 37° to 40° C. These statements contain limits which may at all events be turned to account. All but the two last are imperfect, because they give no exact information as to the nature of the substratum and any other forms which may have been growing with those observed.

Transgression of the limits of the temperatures of vegetation leads at first in the Fungi, as in all plants, to rigidity whether arising from heat or cold, without destroying life. There are of course individual and specific differences in the power of resisting unfavourable influences, but it may be assumed that the higher limit of endurance in most Fungi in a state of vegetation, as in other plants, is about 50° C., though it is sometimes higher than this in a stage of rest when they contain little water, as is the case with spores; on the other hand daily experience tells us, that many growing Fungi can stand a hard frost. The excessive power of resistance displayed by Saccharomyces has been already noticed at page 347.

The time also which is required for extreme temperatures to take effect is of course to be considered in all cases.

SECTION XCVIII. Chemical analysis and an examination of the organisation of Fungi teach us that they have the same need of food as other plants, and like them

---

[1] On the observed effects of light on the growth of Fungi and on the phenomena of etiolation, geotropism, heliotropism, hydrotropism, and thermotropism see Pfeffer's Physiologie and the references there to other works. Also, Wortmann, in Bot. Ztg. 1881, p. 368 and 1883, p. 462, and Van Tieghem in Bull. Soc. Bot. de France, Febr. 11, 1876, and in Ann. d. sc nat. sér. 6, IV, p. 364, also in his Traité de Botanique, pp. 116, 301, and Molisch, in Bot. Ztg. 1883, p. 607.

[2] Die Fadenpilze Aspergillus, &c. u. ihre Beziehungen zur Otomycosis, Wiesbaden, 1883, p. 24.

[3] Ann. d. sc. nat. sér. 5, XI, p. 208. [4] As cited on page 349.

they can only take up their food in a fluid or gaseous condition. The general difference between the process of nutrition in the Fungi and in plants which contain chlorophyll and similar substances consists in their inability from want of chlorophyll to decompose carbon dioxide. They must obtain their carbon by taking up organic carbon-compounds previously formed in some other bodies, as is the case with all organisms or parts of organisms which do not contain chlorophyll.

According to experiments with Moulds and Saccharomycetes in artificial nutrient solutions, the nitrogen is taken up and disposed of to the benefit of the plant in the form of inorganic compounds, provided the carbon is obtained as some organic compound such as sugar ; Moulds like Penicillium, Mucor racemosus, and Aspergillus niger (Pasteur, Fitz, Raulin) take up nitrogen from ammonia-compounds as well as from nitrates, while ammonia is a good and nitric-acid a very indifferent source of nitrogen for the Yeast-saccharomycetes (Mayer, Nägeli)[1]. Moreover the need of carbon as well as that of nitrogen may be supplied in the same Fungi by means of organic compounds which they are able to take up ; some of these supply more nutrition than others, and some supply none at all. According to Nägeli almost all compounds of carbon afford nourishment with the addition of oxygen, provided they are soluble in water and not too poisonous. Urea, formic acid, oxalic acid and oxamide (Nägeli) besides CO_2 and cyanogen are exceptions to this rule. A large number of compounds may serve as sources of nitrogen, if they are in a soluble state or can be made soluble by the Fungus. Free nitrogen and cyanogen cannot by themselves supply nourishment. Some compounds containing nitrogen may serve at the same time as sources of nitrogen and of carbon, others, as oxamide and urea, only as sources of nitrogen. According to Nägeli, Penicillium grows best in a solution of proteid (peptone) and sugar ; then in the following solutions arranged in descending order according to their nutritive capabilities : 1. leucine and sugar ; 2. ammonium tartrate or sal-ammoniac and sugar ; 3. proteid (peptone) ; 4. leucine ; 5. ammonium tartrate, ammonium succinate, asparagin ; 6. ammonium acetate.

As regards the constituents of the ash the requirements of the Fungi are essentially the same as those of other plants, but with this limitation according to Nägeli[2], that Fungi are comparatively less particular in their selection.

The amount of available food-material in the substratum is not the only point of importance ; its chemical nature also has to be considered, as was intimated above in the account of the conditions required for germination. Dutrochet[3] discovered some time since that the development of Moulds was affected by the acid or alkaline reaction of the fluids in which they grew, and more recent investigations, dating from the year 1860, have shown the existence of important specific differences in this respect in the Fungi. The common Moulds flourish in nutrient solutions which are more or less acid ; they do not grow so well or refuse to grow in neutral or slightly alkaline fluids. The Schizomycetes behave as a rule in the reverse way (vid. infra, Chap. XI). A trace of acid is sufficient to retard the develop-

---

[1] A review of the subject and its literature will be found in Pfeffer, Physiol. I, 242. See also Nägeli, Ernährung d. niederen Pilze (Untersuch. &c. 1882, 1, and Sitzgsber. d. Münchener Acad. Juli, 1879), and Raulin in Ann. d. sc. nat. sér. 5, XI (1869), 220.

[2] Untersuchungen, 1882.   [3] Ann. d. sc. nat. sér. 2, I, p. 30.

ment of some Basidiomycetes [1]. Shades of difference varying from one species to another are found within these average limits.

Respect must also be had to certain physical conditions in the food-material as well as to its chemical qualities. This was shown by the fact that the plants thrive differently according to the difference in concentration of the same good nutrient solutions [2]. Dependence on other things, such as cohesion and conditions of imbibition, may also require to be taken into account in many cases, especially among parasitic Fungi.

It follows of course from universal physiological laws that a process of respiration accompanies that of nutrition in the vegetating Fungus-cell, as free exhalation of oxygen or as intramolecular respiration.

Since Fungi take up food-material from the substratum, and cause fermentations or more or less perfect combustion of the substratum by their processes of respiration, they must necessarily produce chemical changes in the organic bodies in which they live; they also give rise in numerous cases to unorganised ferments with specific modes of operation. Species of Saccharomyces, Penicillium, and Aspergillus niger—not however Mucorini which excite alcoholic fermentation in a solution of grape-sugar [3]—produce invertin which splits cane-sugar into dextrose and laevulose. The mycelial hyphae of many Fungi and the germ-tubes of many parasites on other plants grow in thick even lignified or cuticularised cellulose-membranes and in starch-granules, and make passages through the parts which may cause, as Hartig has so well shown [4], wide-spread destruction in the woody tissue. The lignin first of all in the walls of the tracheides in pine-wood, then the cellulose, and finally the middle lamella is dissolved according to Hartig by Trametes radiciperda and T. Pini. Similar effects are produced by other wood-destroying Hymenomycetes. The hyphae of Cordyceps spread widely in the thick chitinous investment of the larvae of insects. These facts distinctly prove the secretion of solvents, and we can scarcely conceive of these in any other form than that of ferments.

We have become acquainted during the last twenty or thirty years with a very large and varied series of phenomena connected with Fungi and their substrata and with a corresponding number of specific adaptations between them. It will be necessary now to take a somewhat closer view of these adaptations and therefore of the habits of the Fungi as they have been observed and their effects on the bodies which they inhabit. With respect to special chemical questions the reader is referred to treatises on the chemistry of fermentation, and to pathological works for some questions which arise on the subject of the aetiology of diseases.

---

[1] Brefeld, Schimmelpilze, IV, p. 7.

[2] See especially Raulin, as cited on last page.

[3] See Pfeffer, Phys. I, 282. Béchamp, in Comptes rendus, 36 (1833), p. 44. Gayon in Comptes rendus, 86, p. 52.

[4] Hartig, Die Zersetzungserscheinungen d. Holzes, Berlin, 1878, and Lehrb. d. Baumkrankheiten, p. 78. Among the earlier literature may be cited: Unger in Bot. Ztg. 1847. Wiesner, in Sitzgsber. d. Wiener Acad. Bd. 49. Schacht, in Monatsber. d. Berl. Acad. 1854, and Lehrb. d. Anat. I, 160, and in Pringsheim's Jahrb. III, 442, &c.

2. NUTRITIVE ADAPTATION.

SECTION XCIX. Fungi have long been divided into two main sections founded on their nutritive adaptation Those which constitute the first category feed on living organisms whether plants or animals, and are termed *parasites*. Their relationship with their hosts is that of a common life, a *symbiosis*. The others inhabit decaying bodies and feed on dead organic substances, and have been named therefore since 1866 *saprophytes*. The two kinds of adaptation are sharply distinguished from one another till we come to cases in which it may be doubtful whether a body should be said to be alive or dead, but with these we are not further concerned ; both are distributed unequally amongst the different species, and are developed in varying degrees in different individuals with many intermediate gradations. The extreme cases and the first to be distinguished are species of purely and strictly saprophytic and species of strictly parasitic mode of life. Others lie between these two extremes ; there may be firstly species which both can and do normally go through the whole course of their development as saprophytes, but which have also the power of going through their course of development wholly or in part as parasites ; these may be called with Van Tieghem *facultative parasites*. Secondly, species which, as far as we know, do as a rule go through the whole course of their development as parasites, but at the same time are able, at least in certain stages, to vegetate as saprophytes : these may be termed *facultative saprophytes*. As far as we fully and certainly know, the parasitic mode of life is always indispensable to the complete development of this second kind. Keeping in view the latter circumstance we obtain the following grouping according to the life-condition : 1. **Pure saprophytes**. 2. **Facultative parasites**. 3. **Obligate parasites**, that is, species to which a parasitic life is indispensable for the attainment of their full development. The latter are again divided into (a) **strictly obligate parasites**, that is, species which, as far as we know, live only as parasites, and (b) **facultative saprophytes**. Most species in category 2 may be able to attain their full development in both modes of life. It is at present uncertain whether there are transitional forms between them and 3 (b), parasitism being the rule in some species, but the full development being possible in exceptional cases in a saprophytic mode of life ; the many gradations between them of other kinds make it not improbable.

In presence of these gradations the foregoing division can only be a frame, such as is necessary for obtaining a clear survey of the phenomena, and it should be distinctly observed that it has in view the ascertained adaptations and arrangements which hold good in the natural, and, as we are accustomed to say, the spontaneous course of things. There may be possibilities of existence in a species which transcend these adaptations. Artificial conditions may in some cases be established which may result, for example, in the development of a spontaneously and strictly parasitic Fungus in a way not parasitic, just as it is possible to rear normal bean or maize-plants by water-culture. We can conceive such a possibility in every case, even in much more difficult cases than any which occur in this work, such for instance as those of trichinae and tape-worms. It must also be allowed that plants cultivated under artificial conditions of such a kind may be of the greatest help to the understanding of the phenomena. How far a state of things produced for the purpose of experiment does actually occur among the natural conditions is a distinct question requiring a special enquiry for its determination, and the observed fact is as little altered by any experimental result in the present case, as in the instance referred to above of land-plants which thrive when grown in water.

3. SAPROPHYTES.

Section C. By far the larger part of the Fungi, according to our present knowledge, are **saprophytes**; our fifth chapter has already shown that this is the case.

We at present have only an imperfect acquaintance with life-conditions and adaptations in these species, yet there must be a great variety of them in all the many cases described in the books, in which a particular species of Fungus is always actually found on a distinct, and, as it may be termed, a specific substratum. Indications of this fact, to which attention has been already called on page 351, are afforded by the Fungi which grow on the dejecta of warm-blooded animals, dung, feathers, &c.

The spores of these species, as was first shown by Coemans[1] in the case of Pilobolus, easily find their way from their regular places of origin to the food of the animal, are provided with the necessary conditions for germination in the intestinal canal, and complete the development which they have commenced there on or in the voided dejecta. Experiments in cultivation such as may be carried out without difficulty show that other methods beside this one are not excluded, at least in the case of many species which live on dung (Mucor, Pilobolus, Sordaria, and some species of Coprinus).

The observations of Pasteur[2] and E. Hansen[3] have brought to light some peculiar arrangements in the life of the Saccharomycetes which excite alcoholic fermentation when growing spontaneously, though these arrangements are not yet quite intelligible. According to the careful investigations of the latter writer Saccharomyces apiculatus appears in the open air on garden fruits containing sugar as soon as they are ripe, finding nourishment and growing on them, especially if the outer rind is broken. It very rarely or never occurs on the fruit before it is ripe, and if it has shown itself on early-ripening fruits, such as currants, gooseberries, or cherries, it is wanting on others which ripen later, plums and grapes for instance, as long as they are still green. In the interval between two periods of ripening of the fruit, even in winter, it is found capable of development in the soil beneath the plants whose ripe fruit it attacks, and very rarely in any other place. Distinct spores are not known in this species, only the vegetative cells produced by sprouting. The actual life-history of the plant is therefore very simple; it is easy to conceive how it finds its way with or from the ripe fruit by the aid of wind or rain to the ground and is carried back with dust from the ground to the fruits, and its living through the winter in the ground is not at all surprising; but it has still to be explained why it is so rarely or never found on green fruit or in some other place. Pasteur[4] had shown before that cells of

---

[1] Monogr. d. Pilobolus. See before on page 158.
[2] Études sur la bière, especially at p. 155.
[3] Meddelelser fra Carlsberg Laboratoriet, I, Résumé français, p. 159.
[4] Études sur la bière.

Saccharomyces which excite fermentation, but the forms or species of which he does not determine, are found in abundance at harvest time on grapes and on their stalks, while they are rarely or never found at a later time on grapes which have remained through the winter and on young grapes in summer; this means that those cells which may happen to have survived have at least become incapable of development.

Few saprophytic Fungi are known to be specific ferment-organisms, if we judge of them by their effect on the substratum. Several species of Saccharomyces are the Yeast-fungi of alcoholic fermentation, and near them come species of Mucor which produce a similar kind of fermentation. The power of producing fermentation is a specific peculiarity, as has already been pointed out on page 271, and not confined to any particular growth-form, as that of the Sprouting Fungi for example. It is wanting among the Saccharomycetes in the flowers of wine, Saccharomyces Mycoderma or mesentericus, and perhaps in some others; it varies in the forms which excite fermentation according to the species, all other conditions being the same. Of the Mucorini, Mucor racemosus, M. circinelloides, and M. spinosus cause a tolerably active fermentation in the sugar, while the activity of M. Mucedo is small, and that of M. stolonifer is scarcely greater[1]. Van Tieghem[2] showed that the mycelium of Penicillium and of Aspergillus niger when growing in solution of tannin breaks up the tannin into gallic acid and glycose.

The ferment-secretions have already been noticed on page 355. It is almost certain that further investigations will show the existence of fermenting power in other saprophytic Fungi.

It is known that the final result of the process of vegetation in most of the saprophytes which have been examined is a combustion of the organic substratum. Penicillium also and Aspergillus niger cause combustion of the tannin when they vegetate on the surface of the solution with an unlimited supply of oxygen.

4. PARASITES.

SECTION CI. We have little exact knowledge of the chemico-physiological processes in the life of the **parasitic Fungi**, because the symbiotic relation puts great complications and difficulties in the way of their precise investigation.

We encounter on the other hand in these Fungi a very long and varied series of phenomena of one-sided or reciprocal adaptation between the parasite and the living organism on which it feeds, and some of these phenomena are of a very obvious character. In contemplating them we have to set out from the following general considerations.

The plant or animal on which a parasite lives is termed its *host* or feeder. Every parasite species lives on certain host-species, and the limits within which it can choose its host are different in different species. Some parasites have never been observed on more than a single host, Peronospora Radii for example on Pyrethrum inodorum, Uromyces tuberculatus on Euphorbia exigua; so Cystopus Portulacae, Rhytisma Andromedae, Triphragmium Ulmariae and T. echinatum, and many other species that

---

[1] Brefeld, Ueber Gährung, as cited on page 188.—Gayon in Comptes rendus, 86 (1878), p. 52, and in Ann. Chim. et. Phys. XIV (1878), p. 258.

[2] Ann. d. sc. nat. sér. 5, VIII (1867), p. 210.

are parasitic on plants; of parasites on animals Laboulbenia Baeri is found only on house-flies. Very many kinds thrive on a larger or smaller circle of nearly allied species which serve them as hosts; among these are many Uredineae, Ustilagineae, and Peronosporeae, Epichloe typhina which lives on the Gramineae and Claviceps purpurea; Cordyceps militaris grows on insects of various orders, especially Lepidoptera; C. cinerea, as far as is known, only on species of Carabus, other kinds only on wasps, and so on. Some kinds make specific exceptions within the immediate circle of affinity of their host, or they occasionally travel beyond that circle; I succeeded for instance in transferring Puccinia suaveolens, which usually lives on Cirsium arvense and Centaurea Cyanus, to Taraxacum but not to Tragopogon; Phytophthora infestans, which is usually confined to the Solanaceae, is found excep-- tionally on the Scrophularineae (Anthocercis viscosa, Schizanthus Grahami), Perono- spora parasitica of the Cruciferae on Reseda luteola.

This exceptional power of accommodation forms the passage to the third category, that is, to parasites which attack plants and animals of very different cycles of affinity either without any distinction whatever or with a preference for certain species. Examples of parasites of this kind living on plants are species of Erysiphe, as E. guttata which lives on the leaves of Corylus, Carpinus, Fagus, Betula, Fraxinus, and Crataegus, Phytophthora omnivora which attacks Fagus, Sempervivum, the Oenothereae and other plants, but not Solanum tuberosum [1], and Sclerotinia Sclerotiorum, which can penetrate as a parasite into the most diverse juicy parts of plants. Of cases of parasites on warm-blooded animals may be mentioned Lichtheim's Mucor rhizopodi- formis, one of the pathogenous Moulds which will not develope on the dog, but grows vigorously in the rabbit; Aspergillus fumigatus attacks both of these animals; no others have been tried. For further examples the reader is referred to descriptive and pathological treatises.

These facts and gradations would lead us to expect that there must also be differences in the aggressive behaviour of a parasite to the different varieties and in- dividuals of a host; or, to express the matter in the converse way, in the *predisposition* of the individuals for the attacks of the parasite. In this direction also there are all possible gradations. On the one hand there are parasites which, as far as we know, show no preferences of the kind, for instance all the strictly parasitic species of the genus Peronospora and of the group of the Uredineae in which this point has been examined. The other extreme is represented by the Saprolegnieae for example, which attack fishes, and by the Sclerotinieae and Pythieae, which as facultative parasites attack Phanerogams. These will be discussed at greater length in a suc- ceeding page (see p. 380). The physiological reason for these predispositions cannot in most cases be exactly stated; but it may be said in general terms to lie in the material composition of the host, and therefore to be indirectly dependent on the nature of its food. In the case of the Pythieae, for example, it is easy to see that the host displays degrees of susceptibility or power of resistance in presence of the parasite proportioned to the amount of water which it contains [2]. It must on the whole be

---

[1] Bot. Ztg. 1881, p. 595.

[2] On the disposition of plants see Soraner, Landw. Versuchsstationen, XXV (1880), p. 327, and the discussion in Bot. Ztg. 1882, pp. 711 and 818. The questions of disposition and immunity in the case of diseases of animals caused by parasites are fully discussed in medical literature.

conceded that a predisposition for the attack of a parasite may in some cases be a *sickly* one, especially if there are deviations at the same time from the condition which we are accustomed from experience to consider the healthy condition in the particular species. But it is also evident, on the other hand, that the predisposition to the attacks of parasites does not always show a sickly condition of the plant, not for instance when there is no parasite present. The real state of things must be investigated and determined case by case.

It was shown by many examples in the morphological portion of this work, that the parasite is either an *endophyte* and lives inside the organs or even the cells of the host, or is to a great extent an *epiphyte* on its outer surface. A purely epiphytic mode of life, in which the parasite rests on or is attached to the outer surface of the host, is comparatively rare if we disregard the case of the Lichen-fungi to be described in section CXV; the Laboulbenieae (page 263) and Melanospora parasitica [1] and *perhaps* also those Chytridieae which are said only to rest on their host may be mentioned as examples of this kind. Other epiphytes, as Erysiphe, Piptocephalis, and Syncephalis, enter the interior of the host at least by the haustoria which they send into it. Chaetocladium adhering to its host and with its tubes in open communication with those of the Mucor which serves as its host does not strictly come into either of the two categories (see on page 20). Either designation may be applied to the Fungi which spread in the deric tissues of the higher animals.

After these preliminary remarks we may proceed to consider the phenomena of adaptation above indicated under three general heads: 1. The *attack of the parasite on its host*, that is, the first beginning of the occupation. 2. The *course taken after occupation* by the further growth of the parasite. 3. The *reactions of the host after its occupation* and the *results of the reciprocal action of the two symbionts*. It is owing to the nature of the subject-matter, that though the three questions are kept theoretically distinct from one another, the answers to them must necessarily travel out of the domain of one question into that of another, and especially from the second to the third.

SECTION CII. **The parasitic Fungus attacks its host** by means of its spores, or of the germ-tubes emitted by the spores, or of the hyphae developed from the germ-tubes.

The first case, in which the first attack is made by the spore before germination, is confined to a comparatively small number of epiphytic species, which will be noticed again in another connection at the conclusion of the paragraph, and to certain facultatively parasitic and facultatively endophytic species of Moulds, namely, the pathogenous Aspergilli and Mucor-forms (M. rhizopodiformis and M. corymbifer) which have been studied by Lichtheim [2]. These Fungi are developed in the internal organs of warm-blooded animals, when their spores find their way into the blood-passages and are carried by the blood to suitable spots; wounded places therefore, though of very small extent, are always in the natural course of things the parts where the endophytically developed Fungus first makes its attack. These forms are actually known as true endophytes only from artificial injection usually of a large number of

---

[1] O. Kihlman, as cited on page 262.
[2] As cited on page 349, and in Zeitschr. f. klin. Medicin, VII, Heft 2.

spores; they appear in nature rather as epiphytic growths on the walls of cavities in the bodies of animals which are easily accessible from without, such as the passages of the ear and the bronchi.

In most cases the spore of the parasite begins the emission of a germ-tube independently of the host, either after simple absorption of water or by appropriation at the same time of food-material produced outside the host. If the tubes or the hyphae which proceed from them then come into contact with the host, they fasten upon it in the way peculiar to each species. The most common case of the kind is when the spore finds its way by some mode of dissemination or other to the surface of the body of the plant or animal, and puts out germ-tubes which penetrate into the body. Parasites which, like Ancylistes Closterii and Polyphagus Euglaenae, attack unicellular organisms living in societies, send out mycelial branches from the individual first attacked, and these can fasten upon fresh individuals and by degrees on entire aggregates of the host-cells. Some facultative parasites of higher organisms, Sclerotinia for example, Agaricus melleus and others of R. Hartig's tree-destroying species, behave in the same manner, since any of the hyphae or mycelial strands are able to make their way into new individual hosts.

The act of penetration is accomplished in two ways; the germ-tube or branch of the mycelium either grows into the interior of the host *through a natural opening* in it, or it pierces *through the firm membranes* of the surface of the body of the host. The one or the other mode is adopted according to the species and the kind of spore; it is seldom that both occur promiscuously.

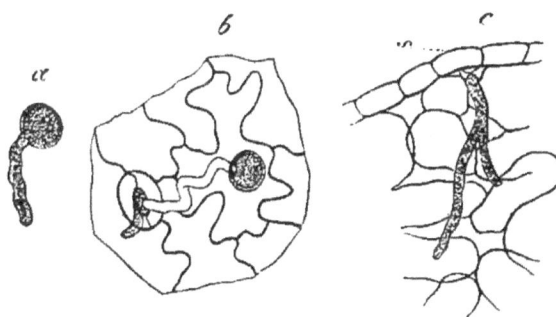

FIG. 163. *Uromyces appendiculatus. a* uredospores germinating in water. *b* uredospores which have germinated on the epidermis of *Faba vulgaris*, the germ-tube penetrating into a stoma. *c* germ-tube which has passed through the stoma *s* into the parenchyma of a leaf of *Faba* and there ramified: *c* is a portion of a transverse section through a leaf of *Faba*, the cell-wall of the spore and the piece of the germ-tube which is outside the leaf not being shown. Magn. 195 times

Many examples of the first kind are supplied by those endophytic parasites on plants, in which the germ-tubes *enter the host by the stomata only*. All the uredospores and aecidiospores of the Uredineae for instance germinate on the moist epidermis of phanerogamous plants. The germ-tube grows in a curve on the surface of the epidermis, and when its tip reaches a stoma it descends into it, usually after it has first become vesicularly swollen outside the stoma, and then passes on into the air-space which lies beneath it. Here it increases rapidly in size and receives the entire protoplasm of the germ-tube, while the rest of the germ-tube outside the spore-membrane dies away. The extremity of the tube which has thus penetrated into the host can now put forth branches which develope into mycelial hyphae (Fig. 163). These germ-tubes enter the stomata of any phanerogam, but only develope further in the species which is the proper host of the particular parasite; they wither away in all other species in the subepidermal air-space. The short germ-tubes from the sporidia of Leptopuccinia Dianthi, DC. proceed in a similar manner. If a sporidium of this plant germinates in the neighbourhood of a stoma of the host, its germ-tube grows towards it, enters it and developes into a mycelium.   If germination, which occurs readily everywhere in a damp atmosphere, takes place on some other substance, the tubes grow irregularly in every direction and perish after a short increase in length. The entry through the stomata has been observed also in species of Entyloma[1] and Kuhn's Polydesmus exitiosus[2].   Further instances will be found in pathological literature.

Among endophytic parasites on animals I mention here the germ-tubes of the aerial gonidia of Cordyceps militaris ('Isaria farinosa'), which I only saw enter the stigmata of caterpillars on which they had developed from the germinating spores[3]; but this observation requires to be revised.

The second case in which the **germ-tubes or hyphae pierce through the firm membranes** of the uninjured host is probably the more common.   It is of course the form which occurs in all endophytes on unicellular organisms.   Examples of it are seen in the case of parasites on higher plants in the germ-tubes from the sporidia of the Uredineae, excepting always Leptopuccinia Dianthi just mentioned, and in those of most of the Peronosporeae and Ustilagineae[4]; Polystigma rubrum[5] together with many other Pyrenomycetes and Discomycetes, Claviceps also and the facultatively parasitic Sclerotinieae (see section CVIII) may be added to the list.   It is to be particularly observed that the germ-tubes of these parasites on higher plants *never* penetrate into the host by a stoma.   Even if the spore lies on or near a stoma, the germ-tube either pierces through a guard-cell, or crosses the cleft as it grows and pierces the wall of an adjacent cell.

The germ-tubes of most of the insect-killing Cordyceps, Botrytis Bassii, and the Entomophthoreae belong to this class; their tubes pierce through the chitinous skin of the body of the host, and may begin to ramify in the substance of the thick chitinous skin of the larger caterpillars.

Some parasites on plants show both modes of proceeding, for the same germ-tubes may penetrate through the stomata and through the membrane of epidermal cells; this is the case in Peronospora parasitica, Phytophthora infestans, and Exobasidium Vaccinii[6]; species also of the mode of life of Sclerotinieae can enter the host by the stomata.

Finally, there are a certain number of parasites whose germ-tubes and hyphae penetrate into woody plants, not through uninjured surfaces, but **where some wound has been received,** and from thence make their way into open spaces, such as injured vessels (Nectria cinnabarina), or pierce through the cell-membranes.   This is the case with most of the tree-destroying Hymenomycetes studied by Hartig, with Peziza Willkommii and the species of Nectria which are parasites on trees.   See section CVIII.

From this series of phenomena which constitutes the general rule deviations occur in two directions, but the deviations are connected with the rule by intermediate forms.

One of these deviations is found chiefly in endophytes which vegetate intracellu-

---

[1] Bot. Ztg. 1874, pp. 93, 103.
[2] Krankheiten d. Culturgewächse, p. 152.
[3] Bot. Ztg. 1869, p. 590.
[4] Wolff, as cited on page 185.—Kühn, in Sitzgbr. d. Naturf. Ges. Halle, 24 Jan. 1876.
[5] Fisch, in Bot. Ztg. 1882, p. 851.
[6] Woronin, as cited on page 341.

larly, and consists in extreme cases in this, that germination does not take place independently of the host, but only when the spore capable of germination has reached the surface of the proper host. When it has done this it at once puts out a germ-tube at the point of contact which penetrates directly through the membrane, otherwise it perishes without germinating. This is the history of many Chytridieae, Synchytrium especially, which are entirely or partially intracellular in their vegetation, of Completoria [1] also, and, as it appears, of Protomyces macrosporus. This mode of penetration is also the normal one in some Chytridieae and Pythieae, though they are able also to put out small short-lived germ-tubes without contact with the surface of the host. A quite peculiar mode of proceeding, but approaching the above, has been observed in the swarm-spores of Cystopus and Peronospora nivea (Umbelliferarum); these spores put out germ-tubes in water which soon die away; in drops of water on the surface of their host they come to rest usually on or close to the stomata of the latter and send their germ-tubes into them and then proceed with their further development.

The other deviation from the general rule is observed in epiphytic parasites on plants, which continue their chief growth outside the host during the whole of their life, but send haustoria into its cells. Here the spores form germ-tubes independently of the host, but where the tubes are in contact with a cell of the host, they send out peculiarly shaped branches, which pierce through the wall of the cell and develope into haustoria. In the Mucorini, which are more or less facultatively epiphytic (Piptocephalis, Syncephalis, &c.), a copious formation of mycelium and gonidia may take place independently of the host if sufficient food is supplied to the plant. The germ-tubes of the Erysipheae [2], after a short increase in length, send a haustorium at once into an epidermal cell of the host and develope on the food thus supplied to them from it into mycelial hyphae, which successively form new haustoria similar to the first. If the young germ-tube does not encounter the epidermis of a suitable host it dies after a slight elongation.

When the germ-tube penetrates through a membrane, which usually happens after it has grown for a short time in some other direction, its extremity bends round towards the wall which is to be pierced, presses upon it and then grows transversely or obliquely through it. In doing this it may maintain nearly the same breadth in the perforated membrane as it had outside it, or it may be considerably narrowed and contracted. But in certain cases, as for example in the sporidia of the Uredineae, the portion of the tube which passes through the outer wall of the epidermal cell is a very slender process, usually appearing only like a simple line even when highly magnified; as soon as this process has entered the cavity of the cell its tip swells at first into a roundish and then into an elongated tube-like vesicle, and the entire protoplasmic content of the spore streams into it; the spore itself and the portion of the germ-tube which is outside the epidermis of the host are seen to be filled only with a watery fluid and soon disappear. The filiform process also which passes through the cell-wall then becomes indistinguishable, and the opening which it produced in the wall appears to become closed up again; in a short time after the

---

[1] Leitgeb as cited on page 160.
[2] De Bary, Beitr., and Wolff, Beitr., as cited on pages 261, 262. See also Fig. 6.

perforation of the wall every trace of the proceeding has disappeared with the exception of a small projection which attaches the tube within the cell to the place of entrance. The tube now grows and ramifies inside the epidermal cell, and ultimately pierces through the inner wall of the cell and developes a mycelium in the tissue beneath it (Fig. 164).

The majority of the intracellular Chytridieae, especially the Synchytrieae, show the same extremely slender perforating process, the same transference of the protoplasm of the spore, and the same ultimate disappearance of the wall of the empty spore and the perforating process.

In some species the penetration begins with an indentation in the membrane, which must be accompanied with a corresponding local extension of the surface ; the indentation forms a sheath of a certain depth round the tube, and is subsequently pierced at the apex, showing sometimes characteristic structural peculiarities. This is the process in the case of Leitgeb's Completoria (see page 166), Peronospora Radii and some other species.

The above phenomena of penetration on the part of germ-tubes and haustoria take place only in the membranes of the host which happen to be suitable to the parasite. The germ-tubes when placed on other species usually perish without penetrating into the cells. I have only once observed an exception to this

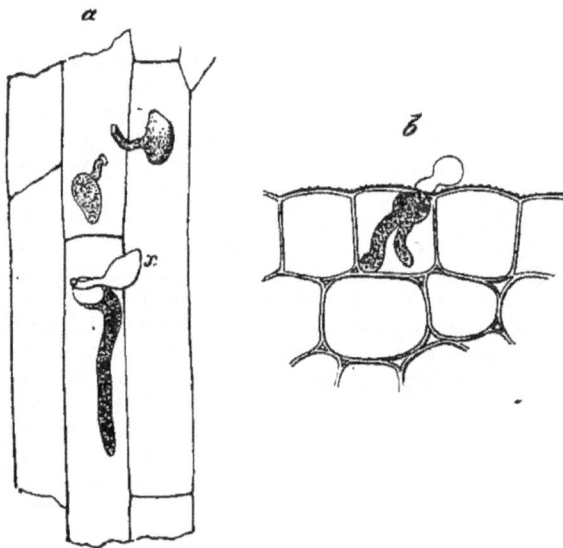

FIG. 164. *a Uromyces appendiculatus.* Sporidia germinating on the epidermis of the stem of *Faba vulgaris,* Mch. ; the germ-tube of one sporidium *x* has penetrated into a cell of the epidermis and grown considerably. *b Phytophthora infestans* ; zoospore germinating and germ-tube penetrating into an epidermal cell (cut through transversely) of the stem of a potato. The preparation made seventeen hours after the dissemination of the spores. Magn. 390 times.

rule ; in this case the germ-tubes of Peronospora pygmaea, Ung. which lives on species of Anemone penetrated into the epidermal cells of Ficaria ranunculoides, but died away there at once. The thickness or other structural characters of the membranes of the host, which vary at different ages and in different individuals, are in most cases of little moment, though young and delicate membranes are more easily and more rapidly pierced than those which are strongly thickened. In certain cases, however, perforation is possible only in certain states of development of the membranes of the host, and these states have some relation to the age. The Synchytrieae for example only penetrate into the epidermal cells of young leaves of their host which are not fully unfolded; the sporidial germ-tubes of Endophyllum Euphorbiae only into the epidermal cells of the young foliage of Euphorbia amygdaloides which are formed in the same summer with themselves, not into the leaves of the previous year which have gone through the winter ; and many Ustilagineae only into parts of young germinating host-plants.

In some endophytes, Phytophthora omnivora [1], Tuburcinia Trientalis [2], Protomyces macrosporus [3], the entrance of the germ-tube of the parasite into the cells of the host is more narrowly localised within the limits assigned above. While most perforating endophytes make their way into the interior of the cells of the host at any spot on their outer surface, the three Fungi above named make their entrance at the outer edge of the side wall which divides two epidermal cells, and then grow on in the middle lamella of this wall, splitting it in two and so pressing transversely or obliquely through the epidermis; ultimately they produce both an intracellular and an intercellular mycelium. This is at least the prevailing mode of penetration in these species; perforation of the outer wall and lumen of an epidermal cell occurs exceptionally in Phytophthora.

In some of the purely epiphytic Fungi which do not penetrate into the host, some for example of the Chytridieae [4] and the Laboulbenieae, the spores when conveyed by some method of dissemination adhere simply to the surface of the host, which is large in comparison to the parasite. The Lichen-fungi which live on small and usually unicellular Algae put out germ-tubes which embrace the cells of the host, as will be described in section CXV, when they encounter them in their elongation. It has never been observed that the direction of this growth is influenced by the host before contact.

Kihlman [5] has recently observed a very remarkable arrangement for fastening on the host in the case of Melanospora parasitica, which is epiphytic on species of Isaria. The almost cylindrical brown-walled spore, which is 5–6 μ in length, germinates by the emission of a germ-tube at each extremity, the tubes, whether grown in water or in nutrient solutions, being scarcely longer than the transverse diameter of the spore. If the spore lies against or on a hypha of Isaria, which is most frequently the case in a state of nature, the germ-tube becomes firmly attached to the hypha of the host and then developes into a mycelium. If the germ-tube comes into contact with an older hypha of Melanospora, the membrane which separates them is dissolved and they coalesce with one another. But if a germinating spore lies at some distance from a growing hypha of Isaria, and it is not difficult to procure this in plants grown on a microscopic slide, the direction of its growth in length is deflected towards the spore till it comes in contact with the germ-tube, which then unites with it and begins to develope. The greatest distance at which the germinating spore can influence the direction of growth of the hypha is from four to five times the length of the spore.

The physiological analysis and explanation of all these phenomena of aggression, adhesion, and penetration through openings and membranes has yet to be undertaken. We can here only notice briefly some of the chief points to be considered.

The facts which have been stated above with regard to the perforation show, on the one hand, distinct effects produced by the germ of the parasite on the host. The

---

[1] R. Hartig, Arbeiten d. forstbot. Instit. München, I.
[2] Woronin, as cited on page 185.
[3] Wolff in Bot. Ztg. 1874.
[4] See above, p. 171.
[5] As cited on page 262.

germ-tube causes a solution of continuity in the membrane which is being perforated in the line of the perforation. Where the perforation is a hole which is permanently fitted by the tube, as for instance in the chitinous skin of insects and in many plant-cell-membranes, it is natural to suppose that the hole is caused by partial dissolution of the membrane, and that this dissolution is the result of a fermentation which proceeds from the Fungus (see page 355). The case is somewhat different where, as in the perforating germs of Uredineae and Synchytrieae, the holes soon close up again. Here it may at least be asked, whether the perforating process from the tube or spore *attached* to the epidermis of the host does not split the membrane by purely mechanical means, much in the same way as a sharp needle divides the plate of caoutchouc which is pierced by it, and whether the small split does not close up again purely in consequence of the elasticity of the membrane, as is the case in the plate of caoutchouc after the needle is withdrawn, when the turgescence in the perforating process sinks to nothing in consequence of the growth of the germ-tube. The deflection of Isaria by Melanospora can only be explained by assuming that some substance is secreted by the germinating spore which exercises a specific attraction on the growing hyphae of Isaria.

On the other hand, the effects produced by the host on the parasite which is germinating or about to germinate are more varied than those of the parasite on its host, and a greater number of physiological questions are connected with them. We do not know the cause why germ-tubes penetrate through stomata, why some spores attach themselves to stomata, others to the surfaces of membranes, why the hyphae of the Lichen-fungi clasp the cells of the Algae in their embrace, and so on. We may make an attempt to explain the fact that certain parasitic germ-tubes perforate the epidermis of one species of Phanerogams and not of others by assuming the secretion of specific ferments and specific differences in the structure, firmness or cuticularisation of the membranes of the host; but it is scarcely possible to explain from the data before us why a germ-tube bends its extremity towards the membrane of the proper host and not towards every membrane or moist surface, or even turns only towards the outer edges of the lateral walls of the epidermal cells, as in the cases mentioned above. Are specific physical irritations brought into play in these cases, or chemical stimulations, which may be supposed to operate through unknown secretions from the surface of the host, with certain specific reactions on the part of the parasite? Questions of this kind present themselves here at every turn, as they do in some similar phenomena which occur in the saprophytes, and offer very promising subjects for experimental enquiry [1].

SECTION CIII. The **parasite pursues its further development** as soon as it has attached itself to the host, while **the living host reacts on the plant** which occupies it permanently and sometimes continues to spread through it; this reaction varies extremely in different cases, being everywhere determined by the specific qualities of the symbionts. The chief phenomena attending these processes will be briefly

---

[1] Pfeffer's work on chemical stimuli, which appeared some time after the above words were written, has shown more distinctly the way to the answering of these questions, and has made some advance upon it. See Ber. d. Deutschen Botan. Ges. 1883, and Unters. d. Bot. Instit. zu Tübingen, I, Heft 3 (1884).

described in the following paragraphs and will be illustrated by examples. Some of them have been already noticed in passing in the sections of Chapter V, where they will be readily found with the help of the index. Further details must be sought in the different monographs and in pathological treatises.

Among the phenomena which are of quite general occurrence it may be mentioned in connection with the **growth of the parasite**, that in extreme cases it either continues to be confined to the spot where it first attacked its host and to its immediate neighbourhood, or spreads far beyond that spot; in the latter case it may grow through or over the existing parts of the host for considerable distances, or *pari passu* with the growth of the host, as is specially seen in many Lichen-fungi. In smaller hosts consisting of one or few cells, with the exception of the Lichen-fungi which will be described at length in the sequel, the difference between these cases is of course small; in larger plants on the contrary it is very striking. The Laboulbenieae, for instance, which are parasites on insects are narrowly confined to the part which they first attack ; the species of Cordyceps which belong to the Entomophthoreae grow through the entire body of the insect. Many corresponding examples might be mentioned from parasites on plants, and it need scarcely be added that there is no want of intermediate forms between the two extremes.

Parasites which spread through the whole of the host, or over large portions of it, may either show the same behaviour and the same development on every or almost every part of the body, or they may have certain phases of their development confined to certain parts, and this latter rule may be invariable or be very generally observed. Parasites on insects, species of Cordyceps for example, spread almost through the entire body of the creature ; C. militaris puts forth its stromata at any part without distinction of the surface of the caterpillar which it attacks, often at many places at the same time ; C. sphecocephala only on the under surface of the thorax between the first or between the two first pairs of feet of the West Indian wasp (Polistes Americanus) which is its host [1]. The same rule will be exemplified below in the case of very many parasites on plants, and has been already noticed to some extent in Chapter V.

According to their **effect on the host and the reactions of the host** on this effect, two chief classes of parasitic Fungi may be distinguished, namely, a *destructive* and a *transforming* or *deforming* class; the two extremes are united by a large number of intermediate forms.

When a parasite of the **destructive class** attacks and occupies its host the parts attacked by it become sickly, die, and are decomposed in a longer or shorter time without previously showing any signs of abnormal growth. It depends on the particular species whether these phenomena in large plants are confined to the parts directly attacked by the parasite or whether the whole body of the host becomes sickly and dies. All facultative parasites may be placed in this class, as will be shown in detail below; of obligate parasites on plant-forms the species of Phytophthora almost without exception, many Uredineae, such for example as the species of Puccinia which live on the Gramineae, or at least those portions of their life-cycle which inhabit the grass, and with some exceptions the Ustilagineae (species of Tilletia, Ustilago

---

[1] See Tulasne, Carpol. III.

Carbo, Claviceps, and many others) belong to the same category, together with all those that live on animals, unless we choose to reckon the phenomena of inflammation, suppuration, and formation of tumours caused by the presence of the Fungus in warm-blooded animals as cases of abnormal growth, a point which may for the present remain undecided.

The occupation by the **deforming parasite** is followed immediately by anomalous processes of growth in the host or in the parts of the host, the word anomalous being here understood to mean every condition different from that which is found in the plants or the parts not attacked by the Fungus. Countless examples of this class of parasites are to be found among those which live on plants. The phenomenon necessarily presupposes a power of growth in the parts to be deformed, and in the higher plants therefore it usually implies that they were attacked in the young state when their growth is still incomplete.

The extremes of deformation, which pass, it is true, readily into one another, consist on the one hand in an abnormal increase of growth and abnormal enlargement of parts of the tissue, which are in other respects normal and normally disposed, and hence in the swelling of the individual cells, as in the case of epidermal cells which are occupied by Synchytrium and the adjoining cells, or else in a monstrous enlargement and inflation of entire members and aggregates of members in the higher plants, such as the swelling of the flower-stalks and the enlargement, often to an enormous size, of the flowers of the Cruciferae when attacked by Cystopus. These may be said to be cases of *hypertrophy*. On the other hand the parts may be deformed with very slight or with no hypertrophy worth mentioning; such are well-known deformations of the shoots of herbaceous species of Euphorbia by Uromyces Pisi, U. scutellatus, and Endophyllium Euphorbiae, and the 'witches' brooms' on the branches of the fir and cherry-tree when attacked by Peridermium elatinum or Exoascus. In the fir (Abies pectinata), for example, these branches grow vertically upwards, like small trees, from the horizontal limbs, with branches spreading in every direction, and leaves which spread in the same manner and fall off year by year, while the entire excrescence continues to grow for years [1]. In the flowers of Knautia arvensis, when occupied by Peronospora violacea, the stamens often, though not always, acquire the characters of normal petals of a beautiful violet colour, and the blooms are filled by them. These phenomena of deformation by Fungi may be termed *mycelogenetic metamorphosis*. The processes in the formation of Lichens, to which we shall recur in a later page, have a considerable resemblance to them.

Lastly, the **new formation** of members, such as are not seen in any form on the plant when free from the Fungus, are caused on parts of some of the higher plants by the presence of the Fungus. The most striking instances of the kind are the delicate round bodies of the size of a cherry which Exobasidium Vaccinii produces on the leaves of the alpine rose, and the excrescences on the stem of Laurus canariensis, L. caused by Exobasidium Lauri and described at length by Geyler [2],—club-shaped for-mations with blunt edges, of the length of a finger, and branched like an antler, which Schacht even mistook for aerial roots. But the strangest example of the kind is found in the Saprolegnieae when attacked by Rozella, which will be described below.

---

[1] Bot. Ztg. 1867, p. 257.

[2] Bot. Ztg. 1874, p. 321.

Excrescences of the kind just described and local hypertrophies caused by Fungi have been fitly compared with *galls* and have sometimes received that name.

It is obvious that all these mycetogenous deformations and new formations and the phenomena also of simple destruction are in direct causal connection with the process of feeding the Fungus. In the latter case we see directly that the Fungus grows at the expense of the parts which are destroyed, the substance of its own body constantly increasing. In the case of tumours and hypertrophies there is often at first a striking over-production of building material, as starch, and this is afterwards used for the completion of the development of the Fungus. In connection with this it often happens that the parts deformed by the Fungus are also killed prematurely; they die and are decomposed sooner than the same parts when free from the Fungus and not deformed. But every conceivable gradation is found in this respect in different species and sometimes in different individuals between the parasitism which quickly destroys its victim and that in which parasite and host mutually and permanently further and support one another,—the relation which is most conspicuous in the formation of Lichens and which Van Beneden [1] has termed *mutualism.*

The phenomena here touched upon have not been submitted in any case to a strict physiological analysis; but the general nature of the enquiry is so obvious that it is unnecessary to discuss it here.

In the following summary of the chief phenomena and combinations which actually occur we must keep the Fungi which live on animals distinct from those which inhabit plants.

Fungi which are Parasitic on Animals.

Section CIV. The Fungi which attack the bodies of living animals furnish a series of instructive examples of the phenomenon of **facultative parasitism** (see page 356).

A number of species of **Eurotium** and **Aspergillus** (Sterigmatocystis), which all occur chiefly as saprophytes and in that mode of life reach their full development, in some cases even forming sporocarps, are able to migrate to the bodies of warm-blooded animals and live at their expense, producing an abundance of typical gonidia, but not, as far as we know, arriving at the formation of sporocarps. Their vegetation causes or promotes a diseased state of the parts, known to physicians as *mycosis.* Aspergillus flavus, A. niger and A. fumigatus, Eurotium repens and Aspergillus glaucus are characteristic promoters of the disease of the human ear which bears the name of *otomycosis aspergillina* [2]. The Fungi find a nidus in the diseased (serous) or excessive normal secretions of the skin, and their rapid growth causes inflammation and excoriation of the parts. But in these cases, as Siebenmann urges, they do not penetrate through the epidermis and are not developed in the healthy ear, so that they virtually retain their saprophytic character, however decidedly they must be considered to be promoters of disease.

---

[1] Animal Parasites and Messmates (Internat. Scientific Ser. xix). See also De Bary, Die Erscheinnng d. Symbiose, Strasburg, 1879.

[2] Siebenmann, Die Fadenpilze Aspergillus, &c. u. ihre Beziehungen z. d. Otomycosis aspergillina, Wiesbaden, 1883; many special treatises on the subject are enumerated in this publication.

Aspergilli of this kind, one of which has been certainly determined as Aspergillus fumigatus, have been found since 1815, and especially since Virchow's more recent and more exact account of them, spontaneously developed in the human lungs and in the air-passages of birds. They find their way there no doubt, as they reach the ear, with the dust which mingles with the atmospheric air, and meet with similar conditions and adopt a similar mode of development. Gaffky and others, Lichtheim especially, obtained characteristic phenomena of development, in this case phenomena of disease, when the gonidia of Aspergillus fumigatus and A. flavescens, Eidam, two species distinguished by the high optimum of their vegetative temperature, over 37° C., were introduced by injection into the blood of animals, such as rabbits and dogs. On the other hand, Eurotium Aspergillus glaucus, E. repens, Aspergillus niger, and Penicillium glaucum, the latter of which was once unjustly suspected by Grawitz, were proved by similar experiments to be incapable of development in animal bodies, and therefore harmless.

Lichtheim found that the facultative parasitism of the two species of **Mucor** noticed above on page 359 was perfectly analogous with that of the two species of Aspergillus just mentioned, with the limitation only which was also noticed before, that Mucor rhizopodiformis has no endophytic development in the dog.

The spores of these Fungi introduced by injection into the blood-passages are carried through them into all parts of the body. It would appear that they do not germinate in the blood-current itself, but in certain organs of the animal into which they are conveyed by the blood. The living organs show different degrees of liability to the attack of the Fungus, especially when the spores are injected in smaller quantities, and Lichtheim arranges them in the following descending series for the Mucoreae: kidneys, Peyer's patches, mesenteric glands, spleen, marrow, liver. Similar results appeared with Aspergillus fumigatus, but with less regularity and with a characteristic localisation of the Fungus in the membranous labyrinth. The living brain remained free in all cases from the development of the Fungus. But when the organs are dead the germination and development of the Fungi takes place in all in the same degree. The spores develope mycelia in living bodies, but it is only in exceptional cases that a fresh formation of spores takes place. The development of the Fungi is attended by characteristic local derangements, and these produce disturbance of the general health, for a fuller account of which the reader is referred to medical works [1] and especially to Lichtheim.

Spontaneous Aspergillus-mycosis and Mucor-mycosis in internal organs removed from direct access of air is to say the least a doubtful occurrence.

Many Fungi living in insects are **obligate parasites.** Everything of importance that is known of the development of the epiphytic **Laboulbenieae** which belong to this category has been stated at page 263. Their dissemination by means of the spores conveyed from one insect which has been attacked by the Fungus to another, especially during the act of conjugation, has been clearly described

---

[1] See especially Virchow, Archiv, IX (1856), p. 557,—Fresenius, Beitr. p. 84,—Lichtheim, in Berliner klinische Wochenschrift, 1882, Nr. 9 and in Zeitschr. f. klin. Med. VII, Hft. 2.—Gaffky, Mittheil. aus d. k. Reichsgesundheitsamt, I, 526. These papers contain further notices of the literature of the subject.

by Peyritsch[1] in the case of the house-fly. The health of the insect attacked by these epiphytes seems to be very little disturbed.

The development of the **Entomophthoreae** which attack insects has also been given above (page 158). We may add here that the body of the insect is occupied in essentially the same way as by the species of Cordyceps which will be described below. The Entomophthoreae, like the Laboulbenieae, are, as far as is known, strictly obligate parasites, and go through the whole course of their development, with the exception only of a brief stage of germination, in and on their host while it is either still alive or recently killed by their vegetation.

The life-history of the species of **Cordyceps** which attack insects is more complicated. Cordyceps militaris, as examined in caterpillars, may be taken

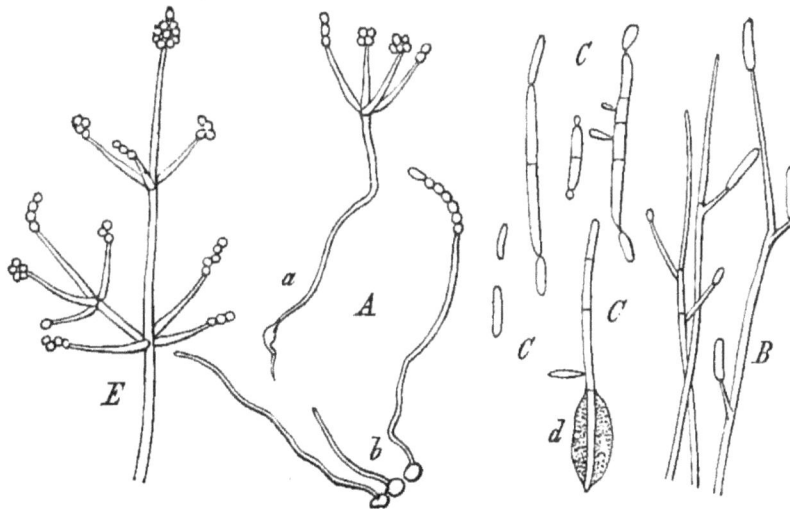

FIG. 165. *Cordyceps militaris*, Fr. *A* secondary spores from the asci germinating in water on a microscopic slide. *a* a single spore with one of its germ-tubes erect and branched, its extremity and the branches having formed chains of gonidia. *b* three secondary spores germinating; the germ-tube of one of them has risen into the air and formed a chain of gonidia on its apex. *B* extremities of hyphae which have penetrated through the chitinous skin of a caterpillar, have reached its inner surface and are abjointing cylindrical gonidia. *C* cylindrical gonidia with sprouts, from the blood of a caterpillar attacked by the Fungus; one extremity of *d* is fixed in a blood-cell. *E* extremity of a filiform gonidiophore which has grown out of the skin of a caterpillar of Sphinx Euphorbiae killed by the Fungus and converted into a sclerotium. Magn. about 400 times.

as an example[2]. The ascospores formed in the orange-coloured club-shaped stromata are ejected as narrowly filiform or rod-shaped bodies divided by transverse walls before they leave the ascus into a row of many shortly cylindrical secondary spores, which are at least 160 in number. When placed in any fluid, they usually separate from one another, swell slightly, become rounded in shape and then put out germ-tubes (Fig. 165, *b*); sometimes, but not always, the spores become partially united together again by means of short connecting tubes before they germinate. Germination takes place on the surface of the skin of a caterpillar if it is only slightly moist. The germ-tubes penetrate at once, and at any part of the surface, into the chitinous skin of the insect. Here they enlarge into somewhat stouter fungal hyphae, which ramify and in the simplest case make their way by a sinuous course into the deeper layers of the skin, at length reaching the inner surface and insinuating themselves between the bundles of muscles and lobes of fatty substance of the creature.

---

[1] As cited on page 273.

[2] Bot. Ztg. 1867, p. 1, and 1869, p. 590.

Here their further growth in length ceases; but now begins, sometimes even within the soft inner layers of the skin itself, the successive abjunction apparently in small quantities of longish cylindrical gonidia, known from their shape as cylinder-gonidia, partly on the extremities of the primary branches, partly on short lateral branchlets (Fig 165 *B*). From the place of their formation these pass at once into the blood which fills the cavity of the body, where they elongate to twice or several times their original size, and divide repeatedly by transverse walls, and then begin to develope like Sprouting fungi, i. e. they produce repeated orders of similiar cells by terminal and lateral sprouting (Fig. 165 *C*). These cells are disseminated through the blood by the movements of the insect and fill it by degrees in a dense mass. They also penetrate into the blood-cells or are embraced by them in the course of the amoeboid movement of the latter (Fig. 165 *C, d*). They grow at the expense of the blood, which diminishes in quantity to such a degree that the insect at length loses its normal turgidity, becomes soft and relaxed and in this state dies. As soon as death has taken place all the sprout-cells begin to develope rapidly at the expense of the substance of the dead body into copiously branching hyphae, which not only fill the entire cavity of the body which till now contained the blood with a dense weft and expand it to its former size in the turgescent state, but grow in a dense mass through all parts of the body, except the intestinal canal which remains empty, and to a great extent absorbs them. A body is thus formed in 1–2 days' time which retains the shape of the living insect, but consists of a close weft of fungal hyphae with some small remains of the body of the insect. This Fungus-body with the form of an animal has the biological peculiarities of a sclerotium. It can give rise directly to fresh stromata, and can do this in a few weeks after its formation if it lies in a moist state; if it is dried, it passes into a resting-state the maximum duration of which is not exactly determined, but it may certainly continue for some months without prejudice to the power of further development. Such is the course of development of Cordyceps in its simplest form.

But deviations from this course and complications of it occur not unfrequently, of which the following are the most important. If its ascospores are sown in water or in nutrient solutions without a living host, they germinate and the germ-tubes develope hyphae which branch with more or less copiousness according to the amount of nourishment supplied; in water only small plants are produced with few or no branches (Fig. 165 *A, a, b*). Some of the branches spread as a mycelium in the nutrient solution, and have the power, like the hyphae on the inner surface of the caterpillar's skin, of abjointing cylinder-gonidia. It is true that this has not been observed in the species in question, but it may be safely assumed since it has been observed in Botrytis Bassiana, which agrees with Cordyceps in all these biological relationships. Other branches of the germ-plants rise erect from the fluid into the air and branch, forming whorls of ramifications on the extremities of which they serially and successively abjoint gonidia (see p. 66). The first gonidia on the young germ-plants are cylindrical like those in the body of the insect (Fig. 165 *A, b*), only usually shorter. All the succeeding ones, even the second in a row which began with a cylindrical gonidium, are spherical in form; they may therefore be called round or aerial gonidia. The mycelium also which is developed in the dead body of the caterpillar very often produces gonidia of this kind only and no cylindrical ones.

The gonidiophores of these plants on most of the caterpillar-bodies examined, which also bore stromata, were observed to be small hyphal branches with whorls of branchlets like the germ-plants just described (Fig. 165 *E*), and these together formed a delicate down on the surface. But on other insects they grow into a dense mould-like covering some millimetres in height and white with a dust of countless gonidia, or else like the Coremium-form of Penicillium they form club-shaped Fungus-bodies 1-2 centimetres in height and covered all over or in the upper part, which is borne on an orange-yellow stalk, with a felt of branchlets which abjoint gonidia. The last-named bodies are known as form-species under the name of Isaria farinosa. Both the Isaria-form and mould-covering are found commonly on a sclerotioid insect-body by themselves, i. e. without the stromata. I once succeeded in obtaining two poorly developed stromata with some large Isarieae from a caterpillar of Spinx Euphorbiae infected by ascospores, which had changed to the chrysalis state after infection.

Gonidiophores with the round aerial gonidia are also obtained if cylinder-gonidia from the still living insect or portions of the mycelium from the insect converted into a sclerotium are cultivated in the air on a suitable substratum; the amount of luxuriance with which they are developed varies with the supply of food. They are formed too under the same circumstances on small plants, which proceed from germ-tubes produced at once in a fluid from the aerial gonidia themselves. These germ-tubes do not ultimately penetrate into the skin of the insect, at least not in the experiments made with the caterpillar of Sphinx Euphorbiae. When the insects are sprinkled with the spores the germ-tubes are seen to enter the tracheae through the stigmata, and then to bore through the wall of the tracheae and so reach the cavity of the body, where the cylinder-gonidia are then abjointed and disseminated, and sprout; at length the insect dies and becomes a sclerotium, exactly in the way described above in the case of the direct products of the ascospores. Stromata have never been known to be formed on insects killed by infection with aerial gonidia, only a fresh crop of aerial gonidia especially of the Isaria-form.

The Fungus therefore which we are describing can only arrive at its full development, that is, can only form perithecia, as an obligate parasite, and to this mode of life it is closely adapted. When it ceases to be a parasite and the dead insect becomes a sclerotium, we see a saprophytic stage of the existence of the plant follow upon the strictly parasitic stage. When the conditions for a parasitic mode of life are withdrawn, facultative saprophytism takes its place (see page 356), and a new form of adaptation may make its appearance with the formation of aerial gonidia accompanied also with the normal formation of perithecia. The entrance upon the parasitic mode of life is however comparatively easy, because the formation of all kinds of spores takes place abundantly in nature, especially in wooded places frequented by the insect-hosts. All that is known of the mode of life of other insect-killing Fungi allied to Cordyceps militaris agrees entirely with the account here given of that species.

The foregoing account of the life-history of Cordyceps is founded chiefly on the facts observed on infecting the caterpillars of Sphinx Euphorbiae with the Fungus, and on investigations some of which have not hitherto been published. I make this latter remark, because the statements agree so exactly with the account given by myself[1] and supplementing that of Vittadini[2] of Botrytis Bassii, the Fungus of the 'muscardine' of the silk-worm caterpillar which was formerly only known to produce gonidia, that it might be thought that I had simply transferred the observations on the one species to the other. The truth is that the two agree perfectly with one another.

Slight deviations from the course described in one species of insect occur in the case of other species both in the reaction of the insect on the effects produced by the parasite as well as in the development of the latter, as I have shown in another place in connection with Botrytis Bassiana. A case of such deviation may be mentioned here as occurring in Cordyceps. The caterpillars of Sphinx Euphorbiae when infected with the ascospores were killed in from fifteen to twenty days, and the spots in the skin infected by the Fungus showed, as in Botrytis Bassii, nothing beyond a brown discoloration varying with the individual and spreading all round from the intruded hyphae. The process was somewhat different when the caterpillars of Gastropacha Rubi were infected. First of all it was much slower; of seventeen specimens infected the first died in about thirty days, the last not till after the lapse of seventy days, and death was preceded by slowly increasing weakness. In the second place the skin showed signs of disease in the spots where the spores had been sown after the Fungus had penetrated into the live insect, but long before its death; it swelled up and became of a darker colour and hard, and was covered with a delicate white mould composed partly of the regular verticillately branched gonidiophores of the species, partly of accumulations of small roundish colourless cells and shortly cylindrical pieces of hyphae divided by a few transverse walls. These hyphae not unfrequently abjointed normal gonidia on narrowly conical lateral branchlets. It could not be determined whether these structures were produced by the breaking up of mycelial hyphae or by the development of normal gonidia. They showed themselves to belong to Cordyceps by the fact that they all gave rise to mycelia with characteristic verticillately branched gonidiophores when cultivated on microscopic slides.

The above statements respecting the entrance of the Isaria farinosa-form into the host are reproduced from my paper in the Botanische Zeitung for 1869, which should be consulted for further details. In this paper and in that of 1867 I gave expression to some doubts with respect to the view maintained by Tulasne[3], that Isaria farinosa belonged to the cycle of development of Cordyceps militaris; these doubts were suggested partly by the failure of attempts to obtain stromata and Isaria-forms in turn from one another in specimens under cultivation, partly by differences, which it is true were only quantitative, in the ramification of the branches producing gonidia. The latter objection might easily be dismissed, and as has already been remarked in the text I do not now think it necessary to maintain the first. A caterpillar of Sphinx Euphorbiae, which had become a sclerotium as usual after infection by ascospores, when laid on moist sand produced first of all two small stromata provided with normal perithecia. These died before the asci were fully formed, and then Isaria was produced in abundance. Portions of mycelium cultivated on microscopic slides had already afforded Isaria. In this case therefore either Isaria was ultimately produced from the ascospores, or the insect had been infected with them and unintentionally also with Isaria, which in the end stifled and supplanted the form with perithecia. I have no reason for assuming such accidental mingling of the two forms, and have framed my opinion accordingly; at the same time the possibility of the mixture is not excluded and it was necessary to call attention to that fact.

For some further remarks on Cordyceps, Botrytis Bassii and some other forms see above, page 253.

---

[1] Bot. Ztg. 1867.

[2] Della natura del Calcino o mal del segno (Giorn. Instit. Lombard. III (1852), p. 142, c. t. 2)

[3] Carpol. III.

SECTION CV. Our knowledge of the Fungi which are parasitic on animals, other than those contained in the groups which have now been considered, is so small from the botanical point of view, that, with all due acknowledgment of the medical interest of these plants and medical research, we can only touch upon them briefly in this place. In doing this we shall refer especially to the medical works on the subject and to important special treatises, from which the reader will obtain further directions if he wishes to examine the somewhat profuse literature in greater detail[1]. We cannot of course enter here into purely medical questions.

The most important of the plants in question are the **parasitic Saprolegnieae**, the **Fungi which cause diseases of the skin in warm-blooded animals**, men included, the **Fungus of thrush or aphthae**, and **Actinomyces**. Some kinds comprised in this class are quite doubtful.

Parasitic Saprolegnieae. Numerous cases are recorded in which living fish, such as gold-fish, and other creatures living in water, as salamanders and frogs, were attacked by Saprolegnieae, grew sick and died[2]. Destructive epidemics among salmon have recently been reported, especially in the English and Scottish rivers, and these epidemics are characterised by the development of Saprolegnieae[3]. We learn from Huxley's investigations that the Fungus settles on portions of the skin of an apparently healthy fish where there are no scales and sends mycelial or rhizoid-branches through the epidermis into the inner layers of the skin, causing at first local and then general disturbance of the system. Similar statements are made in other cases. The examination of the Fungus has only shown that it is some form of Saprolegnia. The formation of oospores, on which the determination of the species depends, was either not observed or imperfectly described. Disregarding Huxley's results for the moment, we may gather from the statements before us that the Fungi in question are ordinary Saprolegnieae, which must have migrated to the living animal as facultative parasites, since they usually vegetate as saprophytes (see page 141). If this is so, there must have been some peculiarity in the fishes before they were attacked by the Fungus, which is not found in the same fishes in the natural state; there must be some special reason for their being attacked by the Saprolegnia, perhaps a disease of some other kind which we must not enquire further into here; for the ordinary species of Saprolegnia are so abundant in our streams and lakes, that if they could attack the fish indiscriminately as facultative parasites, not one could possibly be free from the Fungus. Direct experiments also have shown me that healthy gold-fish may continue lively and free from the Fungus for months in water, in which Saprolegnieae kept purposely in large quantities were forming an abundance

---

[1] The material collected from time to time will be found in the following publications :—

Ch. Robin, Hist. nat. d. végétaux parasites qui croissent sur l'homme et les animaux vivants, Paris, 1858.

Knchenmeister, Die aff n. in d. Körper d. Menschen vork. Parasiten, II, Leipzig, 1855.

Steudener in Volkmann's Samml. klinischer Vorträge, Nr. 38, Leipzig 1872.

Banmgarten, Pathogene Mikroorganismen, I. (Deutsche Medic. Zeitung, 1884, 1).

[2] Hoffmann in Bot. Ztg. 1868, p. 345, and the older works on the Saprolegnieae noticed above on page 145.

[3] Huxley in Nature, Vol. XXV (1881–1882), p. 437. See also the English reports in Just's Jahresber. V, 96, 456, IX, 253.

of spores. It would of course alter the case if there were distinct parasitic species of Saprolegnia different from the common ones, but we know of no such species at present.

The remarks here made apply on the whole to the epidemic among salmon investigated by Huxley, but some points require further explanation. The Fungi in this case appear on the outer surface of the skin in the form of ordinary Saprolegnieae; they could be transferred by tapping to dead flies and be made to develope further on them, but their gonidia are described as being always without motion. This at once raises the question whether they really belong to Saprolegnia, and at any rate it is quite uncertain whether we are dealing with a case of facultative parasitism in species which are usually saprophytic or with one or several peculiar and specifically parasitic forms.

Section CVI. The following are the best-known species of **Fungi of diseases of the skin. Achorion Schoenleinii**, Remak, the Fungus of favus, **Trichophyton tonsurans**, Malmsten, the Fungus of ringworm or tinea (herpes) tonsurans which is identical according to Köbner with that of sycosis or mentagra parasitica (**Microsporon Audouini** and **M. Mentagrophytes**, Rob.; **Microsporon furfur**, Rob., the Fungus of pityriasis versicolor[1]). These Fungi are parasitic on the skin of different mammals and birds. They grow luxuriantly in and beneath the epidermis, in the hair-follicles and hairs. Their appearance on the human skin is characterised by the forms of disease enumerated above. Trichophyton tonsurans has also been observed on horned cattle, horses, dogs, and rabbits, Achorion on the domestic mouse, the rabbit and the head of domestic fowls; Microsporon furfur was seen by Köbner on rabbits after inoculation. They may all be conveyed from one individual to another, from men to other animals and *vice versa* by sowing their spores, and as these develope, the characteristic disease in each case makes its appearance. Transference by inoculation can be successfully performed on sound individuals, but certain forms of predisposition in the patient, which we cannot discuss here, appear to favour or to hinder the development of the Fungus.

Of these Fungi as they appear in and on the portions of the skin attacked by them we know only the septate mycelial hyphae, the branches of which divide transversely into rows or chains of spores capable of germination and resembling those of Oidium lactis or the chain-gemmae of Mucor (see pages 67 and 155). When

---

[1] Remak, Diagnost. u. Pathogen. Unters. Berlin (1845), p. 193.

Köbner, Ueber Sycosis, &c. in Virchow's Arch., XXII (1861), p. 372;—Id., Klinische u. experimentelle Mittheil. ans d. Dermatologie u. Syphilidologie, Erlangen, 1864.

Strube, Exanthemata phyto-parasitica eodemne fungo efficiantur (Diss. inaugur. Berolini, 1863).

J. Lowe, On the identity of Achorion Schönleinii and other veg. parasites with Aspergillus glaucus (Ann. mag. nat. history, ser. z, XX (1857), p. 152).

W. Tilbury Fox, Skin diseases of parasitic origin, London, 1863.

Kleinhans, Die parasitären Hautaffectionen, Erlangen, 1864.

P. J. Pick, Unters. ü. d. pflanzlichen Hautparasiten (Verhandl. d. Zoolog-Bot. Ges. in Wien, XV, 1865).

J. Peyritsch, Beitr. z. Kenntn. d. Favus (Medicin. Jahrb. Bd. XVII, Wien (1869), Heft II, p. 61).

P. Grawitz in Virchow's Archiv, 70, p. 546.

Ed. Lang, Vers. einer Benrth. d. Schnppenflechte (Vierteljahrschrift f. Dermatologie u. Syphilis, 1878, p. 333);—Id., Vorläuf Mittheil. ü. psoriasis (Ber. d. naturw. Med. Vereins z. Innsbruck, VIII, 1878).

fresh spores also are sown in nutrient solutions the germ tubes which are at once emitted develope only mycelia producing spores in the manner just described. In this respect, according to Grawitz, Achorion, Trichophyton, and Microsporon are exactly alike, but they differ from one another in size. This difference is attributed by Grawitz to differences in the food, which can no doubt give rise to great diversity of size in the same species, and he therefore considers the three forms as belonging to one and the same species; and he further identifies this species with Oidium lactis, partly on the ground of the resemblance of the three forms when grown in a nutrient solution to Oidium, and partly because inoculation with pure Oidium will produce diseases of the skin which resemble a mild herpes.

The view that these four forms belong to one another cannot on our part be summarily rejected, but at the same time it requires further proof. In any case the comparison with the Mucor which forms gemmae shows that those forms do resemble imperfectly developed states of other known species of Fungi with typical gonidia and carpospores. Hence the question arises whether organs of this description are to be found also in skin-parasites. On the answer that may be given to this question will depend the determination of the special qualities of these Fungi as parasites. At present the question is still unanswered, though many attempts have been made to solve it in past times by means of artificial cultivation; but Saccharomyces, Penicillium, Eurotium and all sorts of Moulds made their appearance in the impure material used for these experiments, and then the skin-parasite was introduced in one way and another without reasonable ground into the cycle of forms of these species, as Peyritsch long since clearly showed.

It has been proved by experiment that **Saccharomyces albicans**, Reess (Oidium albicans, Robin) causes a formation of pustules and scab, known as thrush or aphthae, on the mucous membrane of the mouth, throat, and oesophagus especially in young individuals. Grawitz and Reess[1] have recently shown that the plant is a form of Sprouting Fungus with long cells resembling Saccharomyces Mycoderma; its ascospores have never been observed; it does well as a saprophyte, but excites weak alcoholic fermentation in saccharine solutions and is therefore a facultative parasite. It has yet to be determined whether it is identical with S. Mycoderma (the flowers of wine) or with some similar form.

SECTION CVII. The name of **Actinomyces Bovis** has been given by Harz to a remarkable growth discovered by Bollinger and Israel which occurs in peculiar swellings on the jaw-bone, especially in cattle, and is in causal connection with them, but is also found inside certain parts of the body in pigs and men[2]. In the swelling,

[1] Grawitz in Virchow's Arch. 70, p. 566, and 73, p. 147.—Reess, Ueber d. Soorpilz (Sitzgsber. d. Phys. Med. Ges. zu Erlangen, 9 Juli 1877 and 14 Jan. 1878). The literature of the subject is given by Kehrer, Der Soorpilz, Heidelbg. 1883.

[2] Bollinger, Ueber eine neue Pilzkrankheit beim Rinde (Centralbl. f. med. Wiss. 1877, Nr. 27).
J. Israel, Nene Beob. v. Mycosen d. Menschen in Virchow's Arch. 74 (1878), and 78.
O. Harz, Actinomyces Bovis. (Deutsche Zeitschr. f. Thiermedicin, 1. Supplementheft (1878), p. 125). See also in the same publication, p. 45.
E. Ponfick, Die Actinomycose d. Menschen, Berlin 1882.
Johne, Die Actinomycose (Deutsche Zeitschr. f. Thiermedicin, VII, (1882), p. 141, tt. 8–10).
Pusch, Ueber Lungenactinomycose (Arch. f. wiss. u. pract. Thierheilkunde, IX (1883), p. 447). In this paper the different works on the subject are most fully enumerated.

which we must not describe at greater length in this place, Actinomyces forms yellow bodies like sand-grains about 1 mm. in diameter. The larger of these bodies, which are visible to the naked eye, always consist of a number of single growths of Actinomyces united into a mass by the soft swollen tissue.

Each Actinomyces may be best described as a round or less commonly elongated hollow body sometimes pressed flat with a relatively thick wall and narrow cavity. The wall looks like a dense hymenomycetous or discomycetous hymenium with very slender elements, being composed of filaments which are copiously branched and have their branches at right angles to the surface and therefore radially disposed when the form of the body is round, and are crowded close together and difficult to separate from one another. Many of these crowded branches are club-shaped at their outer extremity, and in this point again therefore may be compared with asci or narrow hymenomycetous basidia; some are constricted and torulose. They mostly end at a uniform height in the smooth outer surface of the body, though single ones sometimes extend a long way beyond the rest according to the figures which are given (see Ponfick, t. vi).

The inner cavity of the body, which is surrounded by this kind of wall and as was said is comparatively narrow, is filled with a dense tangled mass of slender much-branched filaments, the branches of which are continuous with those of the wall. Roundish or elongated grains of about the thickness of the filaments and not unlike small spores are found between the filaments, at least in some specimens.

The filaments appear to be filled with a homogeneous protoplasm, in which single granules or perhaps vacuoles are rarely to be distinguished; it is uncertain and a disputed point whether they are septate. They attain at most a breadth of $2-3\,\mu$ in the broadest parts of the club-like swellings which I could find, in other parts scarcely a third of that measurement.

The Actinomyces is sometimes incrusted with lime.

The structure of Actinomyces certainly favours the view that it is of the nature of a Fungus, but it has no closer resemblance than this to well-known Fungi. It is not possible therefore to assign it a place in the system, or to form any clear idea of the history of its growth and development from the analogy of other Fungi. Experiments in its cultivation outside the body of the animal have yielded no results of importance to our knowledge of its development. All that can be said about it is founded entirely on the state of things observed in the creature attacked with actinomycosis either living or dead. From the experiments of Ponfick and especially of Johne it would seem possible that Actinomyces grows, because when fresh matter was introduced by inoculation beneath the skin or in hollow places in the bodies of horned cattle, the specific swellings containing Actinomyces were produced in them even in parts of the body at a distance from the places where the inoculation was performed. That the latter were a new growth and not the individuals introduced by the inoculation is certainly not proved, but is not to be disputed; at any rate there can be no doubt that actinomycosis is produced by inoculating with Actinomyces. We can frame conceptions of our own as to how growth is eventually brought about, but none of them rest on a secure foundation.

The same writers found often in fresh material the club-shaped elements of the wall-layer separated from one another, and a large number of club-shaped sprouts shooting from them, especially from their basal portion, so that the whole was palmately lobed in form; these clubs, or also, as is stated, portions of them delimited by transverse divisions, may then be regarded as spores ('gonidia, conidia'), the sprouting or branching of which may give rise to a new plant. On the other hand we must not forget the round bodies which are sometimes found in the inner cavity and which may be spores, nor the occurrence of small plants which consist almost entirely of the slender filaments and the origin of which from the sprouting of the clubs is not fully explained. But all this still leaves us in entire ignorance of the real history of growth.

The failure of the attempts which have been made to cultivate Actinomyces outside the body of an animal, supposing always that it is really a plant, leads naturally to the assumption that it is an obligate parasite. But even this may be doubtful. From the experience of the pathologists who relied chiefly on the local occurrence of the swellings, it is probable that the surface of the mouth and throat and in some cases small wounds on them are the parts where the presumed parasite makes its entry and attacks the animal, and that it is conveyed there with the food. Johne frequently found in the pockets of the tonsils of pigs, even when the animals were quite free from actinomycosis, small bits of plants rough with spikes, such as bits of the awns of grain and the like, and Fungus-bodies resembling growths of Actinomyces attached to them in considerable quantities. Further investigation is required to explain the true meaning of all these observations.

A peculiar disease known as the madura disease, which is endemic in some districts of India and causes dangerous swellings and degeneration in the feet and hands, has been ascribed to a parasitic Fungus, named by Berkeley **Chionyphe Carteri**[1]. More thorough investigation has shown that it is at least doubtful whether there is any causal connection between the disease and the growth of a Fungus. Fungus-elements are found, but not invariably according to more recent accounts, in the swellings, and there is no ground whatever for supposing it to belong to the form obtained by cultivation on rice-pap which bears the name of Chionyphe. It is hard to say what this Chionyphe itself is.

The often described occurrence of Fungi in eggs is a special case of saprophytic vegetation, and is not therefore one for consideration here.

PARASITES ON PLANTS.

a. **Facultative parasites.**

SECTION CVIII. Parasites on plants display much greater variety in their adaptations to their peculiar mode of life than those which live on animals.

A quite gradual passage from saprophytes to parasites is effected especially by the facultative parasitism of certain saprophytic Moulds which cause rottenness in orchard fruits; the softening of pears, it should be observed, is not due to the action of a Fungus. These phenomena were investigated by Davaine[2] in 1866, and

---

[1] H. J. Carter, in Ann. mag. nat. Hist. IX (1862), p. 442, and in Journ. Linn. Soc. VIII, 1865. M. J. Berkeley, in Journ. Linn. Soc. VIII, 1865.
H. V. Carter, Mycetoma or the fungus disease of India, London 1874.
Hirsch in Virchow's and Hirsch's Med. Jahresber. X, 1 (1875), p. 437, XI, 1 (1876), p. 382.
Lewis and Cunningham, The Fungus Disease of India, Calcutta 1875.
[2] Recherches sur la pourriture des fruits (Comptes rend. 83, pp. 277, 344).

afterwards by Brefeld[1]. Species of Mucor (M. stolonifer and M. racemosus), Penicillium glaucum, Trichothecium roseum, and other species are able to make their way into sound juicy fruits, and vegetate and cause rottenness in them. These do not rot without the Fungi. If spores are sown on the uninjured surface of thick-skinned fruits like the apple and pear, where there is sufficient moisture for germination, the germ-tubes are unable to penetrate into them or do so with difficulty; but they enter with ease if the spores are sprinkled on wounded places where the skin is broken. Mycelia which have already acquired some strength are better able to force their way through the unbroken skin. The softer the fruits have become from other causes the more easily are they penetrated by the Fungus; fruits therefore like strawberries and raspberries with thin skins are very liable to be attacked. Davaine found that the vegetative organs of succulent plants, such as Sempervivum, Mesembryanthemum, and Stapelia show the same phenomena as thick-skinned fruits. Observation of the fruits shows that the Fungi develope more easily the nearer the vital powers of the parts attacked are to their lower limit, and at this point the conditions of saprophytic vegetation make their appearance.

The parasitic phase of vegetative life is seen in its more characteristic form in many other facultative parasites on plants, and with many shades of difference in different species. The **Sclerotinieae, Pythieae, Nectrieae,** and Hartig's **tree-destroying Hymenomycetes** may be taken as examples for closer consideration. Many other Fungi, species of Pleospora and Cladosporium for example and allied forms will have to be added to this if the group is more thoroughly investigated.

Of the Sclerotinieae, **Sclerotinia Sclerotiorum** (see pages 30, 52, 218) may go through the whole course of its development as a saprophyte and finds opportunity for this in the natural state on dead plants. But it can also attack certain living and healthy plants and parts of plants as a parasite and destroy them. But according to our present experience it always requires to go through a previous stage of existence as a saprohyte in order to be capable of parasitism. The allied **S. ciborioides** which preys on clover behaves in a similar manner.

Sclerotinia Fuckeliana inclines more in the direction of saprophytism; both the mycelia which produce gonidia and those which form sclerotia are found chiefly in dead parts of plants, especially on decaying leaves, &c. At the same time it is one of the chief agents in the production of decay in juicy fruits, and more thorough investigation will confirm the experience drawn from every conservatory, that the mycelium when it has once reached a certain degree of strength becomes parasitic on living plants and kills them.

I have previously[2] given a very imperfect account of the circumstances connected with the vegetation of **Peziza Sclerotiorum,** and at that time I also misunderstood to some extent the facts which were stated about it. More recent observations have given me a clearer understanding of the matter and it will be well to describe them here in greater detail. The ripe ascospores germinate in pure water, emitting short tubes which soon cease from further development; in a suitable nutrient solution, in must of grapes for instance and on ripe juicy berries of any kind, they develope into a vigorous mycelium which forms sclerotia, and the same result is obtained if they are sown on dead vegetable

---

[1] Sitzgber. d. Naturf. Freunde zu Berlin, Dec. 21, 1875. See Bot. Ztg. 1876, p. 281.
[2] See the first Edition of this work, p. 215.

substances, and even according to Brefeld on bread. If the spores for example are sown on a piece of carrot which has been killed by hot water, a vigorous fungal growth is obtained; but on the moist surface of living portions of the same plant only short germ-tubes are produced, as in simple water, and these do not penetrate into the living tissue, even where the surface has been injured; the parts which have been sown remain for weeks free from the Fungus. If on the contrary the infection of the sound part is due to germ-tubes which have developed to a small amount only in a nutrient solution—how much cannot be exactly stated, but it is sufficient if the germ tubes are scarcely visible to the naked eye,—they penetrate at once into the living tissue and kill it, and form mycelium and sclerotia; pieces of older mycelium behave in the same way. The results are obtained with all parts of the plant, according as they are alive or dead and are inoculated with spores or with germ-tubes which have reached a certain stage of development. I never saw a germ-tube make its way into living tissue without having been previously nourished as a saprophyte; some statements to the contrary will be noticed in the sequel.

But Sclerotinia Sclerotiorum is also found as a parasite on living cultivated plants, not to mention the injury which it does to turnips in store. I observed it destroy the beans (varieties of Phaseolus vulgaris) in a garden in the neighbourhood of Bregenz two years successively, and a similar occurrence has been reported recently by Prillieux from Algeria[1]. It is also very fond of attacking Zinnia elegans and the Petunias. If we try to infect sound specimens of these favourite species, even quite young seedlings, with spores which are germinating in pure water, we always get the same negative result as in the carrot; the plants remain uninjured. If an extremely small amount of some nutrient solution is supplied to the germ-tubes emitted by the spores they at once become strong enough to penetrate into the plants at any place and then to develope into a mycelium which will spread through them and destroy them and form sclerotia, unless the amount of food which it obtains is insufficient, as in the case of seedlings. The same results were obtained with older vigorous mycelia. In the case of plants growing naturally and rooted in the soil we can see how the Fungus as a rule makes its way into the stem from the surface of the ground, and leaving the roots untouched ascends in the tissue of the aerial parts, especially in the masses of parenchyma. In this way the whole plant is killed and dried up and becomes of a pale straw colour. During this process it is not necessary for the Fungus to appear on the surface; in fact it often remains quite inside and then forms its sclerotia in the shape of cylindrical or prismatic bodies inside the dead pith especially in the neighbourhood of the nodes, or, as in Phaseolus, in the fruits also between the ovules; in Zinnia it often fills the receptacle with a sclerotium which like it is conical in shape. In a very moist environment however the mycelium may come out to the surface of the plant which it has attacked in smaller or larger quantity in white flakes and tufts, and can also form its sclerotia there; it may also pass over to the foliage of neighbouring plants with which it comes into contact, and destroy them, proceeding from above downwards. This may be observed in a very striking manner where beans stand close together in a plot.

All these phenomena may easily be reproduced by artificial cultivation in pots. It is only necessary to place some mycelium, grown from spores and made capable of infection in the way described above, at the base of the plant to be infected, and keep the whole sufficiently moist. Experiments have shown that commencements of mycelia capable of infecting other plants may be obtained from spores on a small bit of dead vegetable substance, a piece for instance of a dead leaf. The mycelia therefore may be formed on every bit of moist ground covered with vegetation to which the spores find their way. Sporocarps formed spontaneously from sclerotia of the previous year were to be found in the bean-garden just spoken of, and these supplied the spores.

As a saprophyte the Fungus developes on all dead parts of plants employed in

---

[1] Comptes rend. 99 (1882), p. 1368.

its cultivation, though they may supply unequal amounts of nutrient material.   On the other hand it by no means attacks every kind of Phanerogam as a parasite.   I was unable to find, after repeated and careful search, any trace of the Fungus on the plants of a moist meadow close to the beans above mentioned.   Among plants under cultivation one variety of Phaseolus vulgaris in the same garden was very slightly infected with the Fungus in spite of the immediate proximity of the others.   Experiments in inoculation also showed that Phaseolus multiflorus was scarcely ever attacked ; in other specimens the Fungus developed only scantily, but in single young seedlings kept very moist its growth was vigorous.   Living plants of Brassica (B. rapa, B. Napus, and B. oleracea) were never attacked, either as young seedlings or as plants ready to blossom.   Further details must be reserved for another place.

It should be added, that the hyphae of the Fungus when once they have become capable of infecting make their way into the superficial living cells by piercing their walls, and grow indiscriminately in and between and through the cells of the living tissue and soon kill them.   The power of infecting is shown by the power of penetrating the membranes, which are evidently dissolved at the points of penetration.   Hence it is very probable that this power depends on the presence of a substance which can dissolve a membrane, a ferment in fact, and that this substance is not formed and discharged in sufficient quantity till the germ-tube from the spore is properly nourished and developed.

The foregoing statements with regard to the power possessed by Sclerotinia Sclerotiorum of infecting are in opposition to some which have been published in other works and especially in Frank's Pflanzenpathologie, p. 530 ff. ; according to these accounts the young germ-tube has the infecting power without previous preparation, and plants of Brassica were at once attacked by the Fungus.   The observations may be correct in both cases and the difference may arise from the fact, that several species resembling one another in appearance, but differing in the mode of vegetation, are confounded together under the name of Sclerotinia Sclerotiorum.   The discussion of this point would occupy us too long and must be reserved for another occasion.

Among other species of Pythium [1] which have been carefully examined Hesse's **P. de Baryanum** is a parasite on living and healthy plants, and attacks them both with its germ-tubes which are formed in water and with the branches of the mycelium. when it has acquired strength.   It attains its full development equally well on dead vegetable substances and on the dead bodies of animals, and is therefore equally a saprophyte and a parasite.   In the latter character it penetrates into the cells of a great variety of Dicotyledons and Monocotyledons and into the prothallia of Ferns, but leaves plants of Spirogyra and Vaucheria untouched.   Some Phanerogams are also said to be secure from its assaults, but this statement requires confirmation.   It attacks with especial readiness and frequency the young and watery seedlings of Phanerogams such as the Cruciferae and Amarantus, taking possession of them and destroying them rapidly and completely.   Full-grown land-plants are usually less readily attacked and the injury done to them is more local; but they too may be rapidly destroyed by the Fungus if they are placed in water.

Other and nearly allied species of Pythium are, as far as is known, partly pure saprophytes, partly facultative parasites within narrow limits.   P. intermedium and P. megalacanthum grow as saprophytes on dead parts of plants.   They leave living Phanerogams, even young seedlings which are so liable to the attacks of P. de Baryanum, always and absolutely untouched ; but P. intermedium readily attacks the

---

[1] Bot. Ztg. 1881, p. 531.

prothallia of Ferns and quickly kills them, while P. megalacanthum was only rarely induced to penetrate into them, and its development in them was slow.

Nectria cinnabarina[1] is one of the most common saprophytes on dead twigs. Their bright red cushions burst forth in abundance from the rind of branches which have been killed by frost in the previous winter. Their germ-tubes do not attack the surface of living branches, or living rind or bast tissue when laid bare. But if they find their way to a wounded surface where the *wood* is exposed, they penetrate into it, and the mycelium grows rapidly upwards in the vessels, causing decomposition of the wood-substance; this is followed by the death of larger or smaller portions of a branch or stem, and the further development and the formation of perithecia by the Fungus then takes place in the dead rind. This at least is the course of proceeding in Acer, Tilia, and Aesculus. Other woods gave an uncertain or negative result.

Nectria Cucurbitula, Fr.[2] developes a mycelium producing gonidia when raised from spores in a nutrient solution. It is uncertain whether this species can arrive at the formation of perithecia in this saprophytic mode of life. But its germ-tubes penetrate through wounds in the *living rind* of pines into the living tissue, and spread rapidly through it during successive periods of vegetation, covering a distance of 10 cm. in the longitudinal direction in one season; they at last kill the tissues which they have attacked and then form perithecia which come out upon the surface. This highly pernicious Fungus usually makes its attack where wounds have been caused by hail-stones or by fractures under a weight of snow, and especially where the bark has been eaten away by Grapholitha pactolana.

Nectria ditissima, Tul., the Fungus which causes canker in deciduous trees especially in the apple-tree[3], agrees closely with N. Cucurbitula as regards the arrangements which have been described. The Fungus penetrates at some injured spot into the living rind and spreads slowly in it and also in the adjoining wood; the parts attacked are dried up and destroyed so far as they consist of juicy tissue, while cushion-like formations appear all round them advancing centrifugally in successive periods of vegetation; in consequence of these and of the successive partial drying up of the parts, malformations are produced in the shape of swollen places with a depressed dead centre, which themselves also in course of time partially die away. Gonidia and perithecia emerge from the periderm at the margin of these swollen places where the tissue is still succulent.

The wood-destroying **Hymenomycetes** occupy a prominent position among facultative parasites. **Agaricus melleus** may be taken first as a typical example of this group. We know from Hartig's researches supplemented by Brefeld that the spores of this plant germinate on dead vegetable substances and produce the mycelium which is characterised by its strands or rhizomorphous form. This mycelium may develope spontaneously as a saprophyte and bear sporophores in and on dead wood,

---

[1] H. Mayr, Ueber d. Parasitismus v. Nectria cinnabarina (Unters. a. d. Forstbot. Instit. z. München, III).

[2] R. Hartig, Unters. a. d. Forsthot. Instit. z. München, I, p. 88.

[3] R. Hartig, l. c. I, 109.—R. Göthe in Thiel's Landw. Jahrb. IX (1880), p. 837.

trunks of trees, wooden conduit-pipes, &c.   But the strands also make their way from the soil through the uninjured living rind into the roots of healthy living trees, especially our Coniferae; there they destroy the inner rind and then grow at its expense into the subcortical expansions described above, from which hyphae push on further through the medullary rays into the wood.   While the mycelium is spreading at these spots and mounting in the stem, it kills the living tissue and ultimately the whole tree.   The propagation of the Fungus from tree to tree by means of the mycelial strands spreading far and wide in the soil has been already described.   The symptoms of disease which precede death in fir-trees are known as '*resin-flux*'[1] ('Harzsticken, Harzüberfülle') and the phenomena attending the decomposition of the wood, which advances with the spread of the Fungus, should be learnt from Hartig's excellent descriptions[2].

Trametes radiciperda of Hartig (Polyporus annosus, Fr)., which attacks the wood of fir-trees from the roots and kills it, comes nearest to Agaricus melleus in its mode of life and operation, but it has not the rhizomorphous strands.   The filamentous mycelium penetrates from without into the uninjured rind of the roots, whether it has developed directly from the spores or, as is most conducive to the spreading of the Fungus, has grown out of an infected root in moist soil and has encountered a sound root in contact with the first.   It has been observed to put out germ-tubes capable of infecting in less than twenty-four hours in a moist environment.   Whether it can arrive at its full development when growing as a pure saprophyte on dead wood is still uncertain, but according to some observations of Hartig it is probable; at any rate it would appear to be adapted rather for strict than for facultative parasitism.

R. Hartig has also made us acquainted with a number of other Hymenomycetes which produce decompositions of the wood of living trees in forms varying with each species, and thus kill the plants; Trametes Pini, Polyporus fulvus, P. vaporarius, P. mollis, and P. borealis in fir-trees; Hydnum diversidens, Thelephora Perdix, Polyporus sulphureus, P. igniarius, P. dryadeus, and Stereum hirsutum in the oak.   It is probable that there are many other wood-destroying Fungi which approach the above in the qualities just indicated.

All these Fungi attack the wood from places exposed by wounds and do not penetrate into it through the uninjured rind, with the exception of Polyporus mollis, in which this point has not yet been cleared up.

It is probable therefore that they are fed and made capable of infecting chiefly by the products of the decomposition of the superficial wound-layers which have been laid bare and killed, in the same way therefore as the Sclerotinieae are facultative parasites; but no satisfactory experiments have been made to ascertain this point. On the other hand it is tolerably certain that Stereum hirsutum often attains to its perfect development as a saprophyte in dead wood.   We have no certain knowledge on this point in the case of the other species which have been named.

---

[1] [There appears to be in use no English equivalent of the expression Harzsticke and Harzüberfülle; the word resin-flux is therefore introduced as indicating a prominent symptom of the disease, although it is not an exact rendering of the German terms.]

[2] See above, page 23.

b. Obligate Parasites.

SECTION CIX. Most of the groups described in Chap. V contain a large number of Fungi which are obligate parasites, as will be seen by reference to the accounts there given; but within the class itself we meet with every gradation of adaptation, from the strictly parasitic mode of life to arrangements in which we may almost as well speak of facultative parasites, as of obligate parasites which are at the same time facultative saprophytes. **Phytophthora omnivora**[1] in the group of the Peronosporeae well exemplifies the transition from facultative parasitism. This Fungus is a destructive endophyte in many kinds of living Phanerogams, as Fagus and especially in young seedlings, in Sempervivum, the Oenothereae, and others. Other species of Phanerogams it leaves uninjured, as Solanum tuberosum particularly and Lycopersicum. Its development is rapid in proportion to the amount of water contained in the host, even when that amount exceeds the limits of a normal and sound state of the plant and becomes pathological, as when the land-plants just named are immersed in water. The development of the Fungus culminates in the formation of oospores, usually in large numbers, after the host has been killed by its assailant. The Fungus can also grow on dead organic, even animal, bodies in water and form an abundance of gonidia, but without arriving, as far as is known, at the formation of oospores. On this latter account it is better to place it among facultative saprophytes. The same may be said of its nearest relative Phytophthora infestans the Fungus of the potato-disease, with the limitation that the adaptation to a parasitic mode of life is more marked.

For the same reasons the members of the Mucorini, **Piptocephalis, Syncephalis,** and **Chaetocladium,** to which the expression facultative parasites was first applied, may be termed facultative saprophytes. They may be grown as saprophytes from spores in a nutrient solution and produce an abundance of gonidia, but according to present observations they only attain to their full development in the formation of zygospores when they live as parasites on other Mucorini in the manner described in a former page.

The species also of the **Ustilagineae** (see page 179), in which the young plants can vegetate by sprouting or in the manner of the Hyphomycetes in solutions made from organic bodies, must be termed facultative saprophytes. But the saprophytic faculty is much less important to them than the parasitic mode of life, for it is as parasites only that they are able to produce the resting spores and sporophores which are peculiarly characteristic of their development. This is so, even if the round cells of intercalary origin obtained by Brefeld[2] in large quantities from the mycelium of Tilletia Caries, when it is grown as a saprophyte in nutrient solutions, really have the characters of resting-spores; this can only be proved by observing their germination, and in default of this observation it remains unproved; moreover the characteristic sculpture on the surface of the spore-membrane of T. caries 'could not be clearly seen' on the products of cultivation. In the majority of the species which have been examined the saprophytic development does not go beyond the

---

[1] Bot. Ztg. 1881, p. 585.
[2] Hefepilze, p. 159.

production of gonidia; and it may be assumed, though it has not been yet distinctly proved, that they may form germs capable of infecting, and that by means of these germs they attack the host-plant and thus return to a parasitic life, somewhat in the way described at page 373 in the case of Cordyceps.

The facts stated in pages 265, 339 show that the behaviour of Exoascus and Exobasidium is quite similar to that of the Ustilagineae.

The example of Cordyceps leads to the mention of Claviceps, and Epichloe typhina also approaches near to the latter genus in the points in question. The gonidia (and possibly also the ascospores) of these Fungi may develope small mycelia producing fresh gonidia in the manner described on page 227. That this facultative saprophytism in its various degrees is a frequently recurring phenomenon in parasitic Ascomycetes is to be expected, though more stringent proof of it is in most cases still to be desired.

Of all the Ascomycetes the **Lichen-fungi**, according to our present knowledge, must be mentioned first as examples of strictly obligate parasites, after them the Erysipheae and Polystigma (section LXIII). It has still to be ascertained what are the exact conditions in this respect in the large number of parasites in the groups of the Hysterineae and Phacidieae. The Peronosporeae also contain excellent examples of the class which we are considering, for most species of Peronospora and all of Cystopus are strictly parasitic, the first stages only of germination being completed outside the host. This is the case also with Protomyces and many Chytridieae, some of which even commence germination on the surface of the host (see Chapter V). Lastly, only the strictest parasitism is known in the group of the Uredineae so rich in forms; they germinate if sufficiently supplied with water, and their further development takes place only on the proper host.

Section CX. The Fungi which are parasitic on plants naturally exhibit within the limits of the chief phenomena of parasitic vegetation and its effects, which were pointed out on page 359, a variety of special adaptations in respect of their choice of a host and their spreading in, upon, or along with it. The reaction of the host itself which varies in each case corresponds again to the spread of the Fungus. We call attention to the following facts which are of general interest in relation to these points, again referring the reader to the former sections of this work and to the special literature of the subject.

As regards the **choice of the species to serve as a host**, the rules stated on page 359 are of the first importance. Most parasites living on plants require a single proper host for the completion of their whole course of development, though they may enjoy a larger or smaller room for choice between different species more or less nearly allied to one another. Of all the hosts that are possible for a species of Fungus some may be more favourable to their development than others. Cystopus cubicus for example flourishes and forms abundance of gonidia on the leaves of species of Tragopogon, Podospermum, and Scorzonera, but forms oospores almost exclusively on Scorzonera, especially on S. hispanica; oospores are extremely rare on Tragopogon in my experience. So it is with the Uredineae and species of Erysiphe. The best known Erysiphe is the Fungus of our grape-vine[1] which in Europe only forms gonidia

---

[1] See De Bary u. Woronin, Beitr. III.

(Oidium Tuckeri, Brk.); its sporocarps are *perhaps* the objects described as Uncinula spiralis, which grow in North America on a native vine. This dissimilar promotion by hosts of different species of the otherwise similar course of development of the Fungus makes no difference in the phenomena in question. Parasites which go through their whole course of development on a single host of a particular species are termed *autoecious* or *autoxenous*. All the parasites described in preceding chapters are examples of this class, with the exception of one group to be named presently. It is not perhaps altogether superfluous to remark, that the larger part, or at any rate very many species, of the Uredineae, in which the alternation of generations is most copiously differentiated, are autoecious.

Thus the entire development of Uromyces Phaseolorum is completed on species of Phaseolus, that of U. appendiculatus on the Vicieae, of Puccinia Tragopogonis on Tragopogon, of P. Pimpinellae on Myrrhis or Chaerophyllum, of P. Falcariae on Falcaria Rivini, of P. Violarum on species of Viola, and so on.

The contrary is the case with a number of Uredineae which form aecidia. These are obliged to change their host with the separate sections of their alternating generations in order to complete the course of their development, like the Cestodes and other parasitic worms. They are accordingly termed *heteroecious*, or still better *metoecious* or *metoxenous* as changing their place of habitation or host [1].

I was myself the first to establish this metoecism in the case of Puccinia graminis, in which the phenomenon or at least its consequences were recognised more than a hundred years ago by agriculturists, who rightly maintained against the botanists that grain grown in the neighbourhood of shrubs of Berberis were liable to be attacked by rust, that is by Puccinia graminis. This parasite exhibits the pleomorphism and alternation of generations of the Uredineae which form aecidia in its greatest variety of form (see page 279). Its teleutospores pass the winter on old stems of wild and cultivated grasses, especially Triticum repens, while the germ-tubes from the sporidia which are developed in the ensuing spring penetrate into the epidermal cells of Berberis vulgaris, more rarely into those of a Mahonia, and never into a grass. They develope rapidly in Berberis into a mycelium which produces aecidia but never forms uredospores or teleutospores, and if the germ-tubes of the aecidiospores find their way into the stomata of suitable Gramineae they develope there and there only into a mycelium which produces uredospores and teleutospores. The germ-tubes of the uredospores in their turn develope only on the Gramineae, and in the manner which is common to all uredospores.

Later investigations have shown that an analogous change of hosts takes place in many other species.

The aecidia of Puccinia Rubigo vera and P. coronata which also live on Gramineae or Cyperaceae in the other sections of their development are confined, the former to the Boragineae, the latter to species of Rhamnus; those of P. Moliniae to Orchis; those of P. Caricis to Urtica and of P. (Caricis) limosae to Lysimachia thyrsiflora; Uromyces Dactylidis forms its aecidia on the leaves of common species of Ranunculus [2], and its

---

[1] See on the terminology Bot. Ztg. 1867, p. 264; on other points see the literature of the Uredineae enumerated above on page 286.

[2] See also Cornu, Comptes rend. 1882, 94, p. 1731.

uredospores and teleutospores on Gramineae; Uromyces Pisi forms uredospores and teleutospores on Vicieae and its aecidia on Euphorbia Cyparissias, the well-known Euphorbia-aecidium. Next to these species all the Gymnosporangieae are examples of the phenomenon in question, as Oersted was the first to show from gardeners' traditions; their teleutospore-layers inhabit species of Juniperus, and migrate to Pyrus and other Pomaceae for the formation of aecidia, which were once known by the name of Roestelia. The aecidia of several species which live on the Ericaceae are formed on leaves of the Abietineae which are entering their first year, those of Melampsora Göppertiana, as Hartig showed, on Abies pectinata; those of Chrysomyxa Rhododendri, the Fungus of the alpine rose, and of C. Ledi on the spruce, Abies excelsa. The Coleosporium of species of Senecio migrates according to Wolff to the leaves of Pinus sylvestris, and there produces its aecidia which were known by the old name of Peridermium Pini. The rest of the known cases of the kind will be found collected together in Winter's Pilzflora.

The same work also enumerates the forms bearing teleutospores and those bearing aecidia, of which it is known that the germs which are capable of infecting do not proceed to further development on the hosts on which the forms themselves grew. The aecidia-forms never propagate themselves on their hosts; teleutospore-forms are produced only in certain circumstances through the medium of the uredospores which accompany the teleutospores. From the analogy of the cases in which metoecism is certainly known these forms must be separate sections in the development of metoecious species. Their complete form-cycle has yet to be ascertained. To this group belong on the one hand most of the species of Melampsora and Coleosporium, the Cronartieae and also Hemileia vastatrix, the Fungus of the coffee-plant, on the other the aecidia of fir-cones, the aecidium known as Peridermium elatinum which causes the 'witches' brooms' in Abies pectinata (see page 368), the aecidium of species of Clematis, and many others.

Metoecism, that is, enforced change of the living host, is not known outside the group of the Uredineae; its supposed occurrence in other species has not yet been confirmed. There is another phenomenon which must of course be kept quite distinct from it, and which may be termed *lipoxeny* or deserting the host in opposition to a change of the host. Many Fungi which inhabit plants spend a certain period of their life in strictly parasitical fashion on the host, and then separate from it and complete the other sections of their development independently without a living host, and entirely at the expense of the reserve of food which they have appropriated from the host. The separated thallus may be compared as regards the economy of the metabolism to a ripe spore which is able to germinate at the expense of the reserve of food which it contains. This phenomenon is most striking in Claviceps which continues strictly parasitic up to the ripening of its sclerotia, and produces the stromata from them after they have fallen from the plant under favourable conditions of temperature and supply of water in the next period of vegetation (see on page 227). A similar course of events may take place in Peziza Durieuana which forms its sclerotia on stems of Carex, and Peziza Curreyana which forms them on stems of Juncus and Scirpus, and produces ascocarps from them in the next period of vegetation. The Sclerotinieae may also be mentioned in this connection; and the long series of Ascomycetes which inhabit leaves, such as Polystigma, Rhytisma, Phyllachora,

Phacidium &c., must be added to the list,—species which commence their vegetation on the living leaf, and complete their development by forming their sporocarps on it when fallen and decayed in the next period of vegetation, and here too at the expense of the reserve of food obtained from the living host. How far the building-material required for the last section of the life of the plant is partly taken from the dead decaying leaf has never been thoroughly investigated and can scarcely be determined with entire certainty. But this supply of material can be at most only a small addition, since on the one hand the Fungus-body, separated from the surrounding substance of the fallen leaf, also reaches its normal and complete development, and on the other hand it can be seen directly that the reserve material, stored up in the Fungus-body is used up in the course of this development. It is possible that in this case also intermediate forms and shades of difference may occur, in which a final stage of enforced saprophytism succeeds the parasitic vegetation, in the way described above on page 373 in the case of Cordyceps.

SECTION CXI. The purely local circumstances connected with the **spread of parasites on plants** from the place of attack need no further discussion in the case of hosts formed of one or few cells, because the parasite must necessarily remain narrowly localised in and on so small a body; and the remark equally applies to those which like the Synchytrieae live in single cells of larger plants.

Parasites which form mycelia in more highly organised and especially in phanerogamous plants behave very differently according to the species or to the segment of their development; in the one extreme they are *confined to the immediate neighbourhood of the point of attack*, in the other they *spread widely or unlimitedly from that point over or through the host.*

From among the many species which have been noticed from time to time in former sections of this book and in addition to them, we may name here as examples of the **first category** the parasites which form narrowly circumscribed spots on the leaves of Phanerogams, such as many Uredineae, and among them Puccinia graminis, P. Rubigo vera, Uromyces Phaseolorum, Peronospora viticola and P. nivea (Umbelliferarum), Protomyces macrosporus, Entyloma Calendulae, species of Polystigma and Rhytisma. Each distinct spot inhabited by the Fungus is the result of the growth of one or occasionally of several spores. Fresh spots are added one after another on a surface in proportion as new spores from any quarter, for instance from those first established on the leaf, germinate there and assail it. The species of Claviceps are confined to the flowers of Gramineae and Cyperaceae during the whole of the parasitic portion of their life, and destroy the young ovary in the manner described above. The germ-tube of one spore at least is required to infect each ovary, and it attacks the ovary directly.

Among the Fungi of the **second category** which spread far from the point of attack are the often mentioned Sclerotinieae, Pythieae and Phytophthora, which assail the host at any point and grow through it to an unlimited extent in every direction, provided the external conditions are favourable. Cystopus candidus is instructive as an instance of strictly obligate parasitism. Its germ-tubes find their way into every stoma in Lepidium sativum and Capsella on which germinating spores fall (page 363). But those germs only develop further which have penetrated into the cotyledons, and the mycelium may spread from them through the entire host as it developes and form a multitude of gonidia on it.   These gonidia cannot attack plants when once they have lost their cotyledons.   Hence when a number of these plants are growing together, some are often found thickly covered with Cystopus and surrounded by others of the same age which are quite untouched by the Fungus.

Endophyllum Sempervivi makes its way in spring into any leaf of the host, spreads through all parts of the plant and produces its aecidia in the succeeding spring in the younger of the leaves which have lived through the winter, and may then persist for years in the same state of diffusion in a rosette of leaves, forming fresh aecidia every spring.   The spot where Endophyll m Euphorbiae enters its host has been mentioned on page 364 ; the mycelium spreads from there through the entire plant and produces aecidia in the young leaves of the next year, producing deformities in their petioles.   Melampsora Göppertiana penetrates in summer into the shoots of Vaccinium Vitis-Idaea, and its mycelium spreads through the parenchyma but does not deform it.   After the succeeding spring it enters every year into the new terminal and lateral branches from the infected shoot, causing peculiar deformities in them, and forms its teleutospores not in the epidermis of the leaves but in that of the stem[1].

Kühn and R. Wolff have shown that the germs from the sporidia of many Ustilagineae which attack grasses penetrate into the host when it is young and germinating, sometimes into the first sheathing leaf, sometimes into the lowest node of the young stem and even into the base of the young roots.   The mycelium then grows with the growing stem and its lateral shoots, and at length makes its way into the organs in which the Fungus prefers to produce its sporophores, and consumes them and forms its spores.   The mycelium which has grown with the growing internodes is not entirely destroyed when the elongation is completed ; intercellular branches are preserved in the nodes, and these send new branches into the axillary buds which may be produced at the nodes, to repeat the process of growth already described.

We are acquainted with a number of endophytic parasites, which so far resemble those last described in their behaviour, that their mycelium spreads through large portions of the host and then produces its sporophores in or at certain spots, and this may happen once only, or is repeated every year from the perennial mycelium in new branches and leaves, &c.   Though the act of entering the host has not been observed in these cases, there can be no doubt that the circumstances are quite or nearly the same as those which are certainly known either in Endophyllum, or in the Ustilagineae which have just been described.   Among these forms are most of the remaining Ustilagineae, the Aecidium of Euphorbia Cyparissias, the Peridermium elatinum of the 'witches' brooms' of Abies pectinata, the Exoasci which also form 'witches' brooms,' Peronospora Radii of Pyrethrum inodorum and Peronospora violacea, Brk. of Knautia arvensis which produce spores only in the flowers of the host, Epichloe typhina which lives on the Gramineae and many others.

Species in which the segments of their development differ from one another and stand to one another in the relation of alternate generations spread in the host in the same or in a different manner in the different segments.   The first is the case for example in the Uredineae which have been already adduced as examples of narrow

---

[1] Hartig, Lehrbuch d. Baumkrankheiten, p. 56.

localisation. Other species partly belonging to the same genera show the opposite behaviour. The mycelium of Puccinia Tragopogonis which bears aecidia for instance, spreads all through the host and produces its aecidia over the entire surface of the leaves, while that which grows from the aecidiospores and forms teleutospores is confined to small spots on the leaf. The same is the case also with Uromyces Pisi, the converse with Melampsora Göppertiana.

The varying extent to which Tuburcinia Trientalis spreads (see page 180) according to the kind of spore from which it grows is a nearly allied phenomenon.

But mycelia grown from similar spores appear in some cases to be narrowly localised or to spread widely according to the species of the host. Cystopus candidus for example often appears to be limited to narrow spots in the leaves of species of Brassica, into which it must have penetrated through the stomata of the full-grown leaves, while it behaves in exactly the opposite way in Capsella and Lepidium. Its mode of entrance in Brassica would be in accordance with experience, for I once observed that the germ-tubes of this Fungus were able to develope a mycelium in the full-grown leaves of Heliophila crithmifolia. It is not improbable that some Uredineae are affected in the same way by a difference in their hosts, but the point requires further examination.

In many Fungi the mycelia which spread through the host have everywhere the same characters, or at least they have no distinct and special characters at distinct places in the host; they may form their spores at any spots where the external conditions, as supply of air or peculiar mechanical relations, permit, and actually do so form them.

On the other hand there are many species which assume different characters in different organs of the host, showing structural differences in the hyphae themselves, and especially capacity or incapacity for forming sporocarps and spores. It is obvious that conditions of nutrition must in general be the cause of this phenomenon, but we have at present no precise physiological account of the matter. We can only therefore speak provisionally in such cases of favourite places for the formation of one and another kind of spore. Some remarkable examples have been already briefly given in Chapter V, but we may add a few more in this place.

Cystopus Bliti forms its gonidia only on the leaves of Amarantus Bliti, and its oospores only in the stems; Cystopus candidus forms gonidia in great abundance on all the aerial organs of its hosts, but I never found its oospores on the leaves; some species of Peronospora behave in the same way; P. Arenariae, Brk., for example, produces gonidia on all parts of the leaves of Möhringia trinervia, oospores almost exclusively in the parts of the flowers. Very many Ustilagineae form their spores only in or on the parts of the flowers of the host, some in the anthers, others in the ovary or ovule, others again (Sorosporium Saponariae, Ustilago Tragopogonis) on the whole surface of the corolla and of the parts inclosed by it. It has already been remarked that Peronospora Radii and P. violacea are also confined to the flowers of the host for the formation both of oospores and gonidia. In other Ustilagineae, as Urocystis occulta, certain parts of the leaves are the places where the spores are formed, in Ustilago hypodites chiefly the surface of the stem where it is covered by the leafsheath. Epichloe typhina always forms its stromata on the outer surface of the sheath of a foliage-leaf which stands underneath an inflorescence, generally the next but one or two beneath it; the leaf or leaves above it and the inflorescence itself wither away in consequence.

In this and in almost all the cases of the Peronosporeae and Ustilagineae which have been mentioned the mycelium spreads through the entire stem of the plant but cannot be seen in it, and causes no change in it which is visible from without. A superficial examination only would lead us to suppose that only the parts chosen for the formation of spores are occupied by the Fungus, and that the rest of the plant which looks quite healthy is free from it. But in most cases of the kind all or nearly all the similar parts of a plant where spores are formed by preference are seen even in their young state to be uniformly occupied by the spore-forming organs of the Fungus. This enables us to conclude in doubtful cases that the Fungus is one which is widely spread and not confined to a limited spot, because in the latter case it scarcely ever happens, for obvious reasons, that all similar young parts are attacked in this uniform manner.

Section CXII. Independently of these coarser phenomena of localisation the **histological distribution,** as we may briefly term the behaviour of the Fungus to the cells and tissues of the host, is also liable to variations in endophytic parasites on plants and in the endophytic haustoria of parasites which are otherwise epiphytic. In considering these variations we must of course also keep in view the reactive influence of the host, which is necessarily connected with them.

In the higher plants with their various kinds of tissue the wood-destroying Fungi described at page 383 spread in the tracheal and sclerenchymatous tissues and destroy them. But most parasites exclusively or almost exclusively *attack living cells containing protoplasm,* the parenchyma therefore, soft bast and epidermis, where the tissues are differentiated, and even the wood-destroying Fungi just named do the same. Accordingly a number of histological localisations result from conditions such as have been already described, and further details must be sought in special treatises on the subject. But as a topic of general interest we must briefly notice the differences in the behaviour of endophytic parasites to the living cell.

In unicellular hosts an endophyte is necessarily *intracellular,* that is, vegetates inside the cell. It may also be intracellular in pluricellular tissues, but it may also spread in the *intercellular spaces,* or it may exhibit both modes of distribution. When its growth is intercellular, it stands of course in the relation of an epiphyte to the individual cell, and of that we need say no more here. The entry of the parasite into the cell of the plant or into parts of it takes place in very dissimilar forms and with very dissimilar consequences.

To begin with the extreme which lies nearest to the superficial view of the subject, there are many parasites whose hyphae penetrate through the membrane and into the protoplasmic body of the cell, and kill and destroy the latter at once; examples of such Fungi are seen in all the facultative parasites described above, Sclerotinia and Pythium and their near allies the Phytophthoreae and others; Ancylistes is an instance of this in unicellular plants (page 139).

Other parasites also penetrate the membrane of the cells and pass through them and through the protoplasmic body. But the latter does not at first succumb to the assault of the parasite, but retains the qualities which it has in the living cell, and is fenced off from the parasite by a pellicle which forms a close sheath round the intruder. Assimilation, metabolism, and even growth of the cell continue often for weeks and months, the latter in some cases longer even than when there is no parasite present. The cell succumbs comparatively slowly to the influence of the parasite. This phenomenon, which again shows many shades of difference in different cases, is everywhere common, as in the Peronosporeae and Uredineae, in the haustoria of the Erysipheae, in epidermal cells attacked by Synchytrium, in the Saprolegnieae when attacked by Olpidiopsis (see on page 166), and some others. Leitgeb in his work on Completoria [1] calls attention to this subject, which has often been observed and mentioned. He finds also in the continued activity of the protoplasmic body when thus attacked the explanation of the fact, which has likewise been often described, that mycelial hyphae, which grow transversely through a cell, are enclosed in a sheath of cellulose which is continued without interruption into the membrane of the cell attached, or springs from it. This sheath is often developed in a remarkable manner on the endophytic mycelia of the Ustilagineae [2].

The piercing of the membrane after the Fungus has made its first attack on the host often begins with an indentation in it, as was described above in page 364 in the case of the penetration of many germ-tubes. In some cases, as in the club-shaped haustoria of Peronospora densa [3], the indentation is all that is effected; further investigation of these matters which have hitherto been little regarded may bring to light other instances of the same kind.

Lastly there are endophytes, whose hyphae do not pierce through the membranes, but grow in them in the direction of their surface and there ramify. Some species of Exoascus, according to Sadebeck's observations noticed above on page 265, spread in this way all their life long in the outer wall of the epidermis beneath the cuticle of the parts attacked, which they do not burst through till they form their asci. Other species, like Exoascus Pruni, behave in a similar way, only their mycelium grows through the inner tissue into the subcuticular wall, taking its way as is usually said 'between the cells;' but it would be more correct and more in accordance with known facts to say in the cell-membranes. The mycelium of Rhytisma Andromedae, which forms sporocarps, spreads beneath the cuticle in the outer wall of the epidermis in the same way as that of Exoascus; the mycelia which produce spermogonia of Puccinia Anemones, of Phragmidium and of some other Uredineae, and, according to Cornu [4], the mycelium of Cladosporium dendriticum in the fruits of the Pomaceae, pursue a similar course. These Fungi burst through the cuticle and come to the surface when they form their spores or spermogonia. The path of the Fungus must really be intercellular to reach this position in the Uredineae; how it arrives there in other cases is not known. In the Fungus which lives in the vine and which I named Sphaceloma ampelinum, the germ-tubes of the gonidia penetrate into the outer wall of the epidermis and spread in it beneath the cuticle, which is only burst by the small tufts of hyphae which grow up vertically to the surface and abjoint fresh gonidia.

---

[1] Cited on page 160.
[2] See R. Wolff, and Fischer v. Waldheim as cited on page 185.
[3] Ann. d. sc. nat. sér. 4. XX, p. 29.
[4] Comptes rend. 93 (1881), p. 1162.

Later there follows also a penetration of the Fungus into and between the dead and dying cells of the deeper-lying tissue-layers [1].

In connection with this subject it will be well to recall the fact, that the hyphae of certain Fungi which are parasitic on other Fungi unite with those of the host, and the two protoplasmic bodies enter into continuity and coalesce by disappearance of the cell-membrane in such a manner that there is no longer a clear distinction between host and parasite. The fact has been observed in Chaetocladium and has been described above; it probably occurs in other cases also, for example in Artotrogus [2]. The obscure and disputed phenomena observed in Nyctalis and some other like cases may perhaps find their explanation in a similar relation between the two plants.

SECTION CXIII. **The effect which endophytic and epiphytic plant-parasites exercise on their host**, when they do not at once put an end to its life, is seen in a richly graduated series of phenomena, from the slow exhaustion and wasting away of the host to the characteristic transformations which in extreme cases even promote the health and growth of the host, and which take place when certain parasites attack plants that are sufficiently young and capable of growth. Examples of the first kind are to be found on all sides and may be studied in pathological and descriptive treatises. The first indication of a hypertrophic transformation may be said to be given by the fact to which Cornu [3] has called attention, that spots on leaves and fruits occupied by such Fungi as Erysiphe guttata (Aceris), Cladosporium dendriticum and the Uredineae have a more abundant and more persistent supply of chlorophyll than their neighbours which are free from the Fungus. Next to this comes the normal accumulation of products of assimilation in cells attacked by a Fungus which has been already noticed in a former page, and lastly the mycetogenetic growth of the cells themselves and of the tissues which they compose. For the hypertrophies and transformations proceeding from the latter cause, so far as they occur in Phanerogams, it is sufficient to refer to the examples adduced on page 368 and in earlier chapters. We may however add to them in this place the description of the peculiar transformations produced in the Saprolegnieae by certain Chytridieae, and especially the one which results in the formation of the plant-form described as a Lichen.

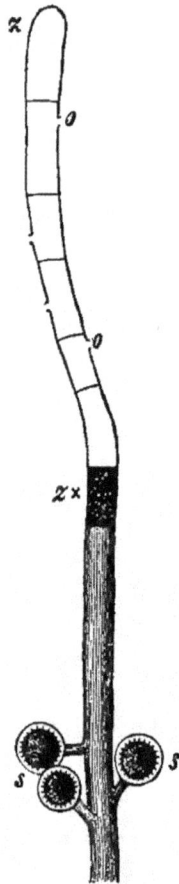

FIG. 166. Species of *Saprolegnia*. Extremity of a tube attacked and deformed by *Rozella septigena*. The upper part is divided by transverse walls into seven sporangia of *Rozella*, six of which *z* are empty, some showing the spot *o* in the lateral wall where the spores escaped; the lowest *z×* is still full of spores. Further down the tube has three outgrowths *s* resembling oogonia, which contain each of them a resting-spore of *Rozella*. A diagrammatic representation after Cornu (Ann. d. Sc. nat. S.r. 5, XV, pl. 6). Magn. about 100 times.

---

[1] See Bot. Ztg. 1874, and R. Göthe, Ueber d. schwarzen Brenner u. d. Grind d. Reben. Leipzig 1878.
[2] Bot. Ztg. 1881, p. 575.   [3] Comptes rend. 93 (1881).

The Chytridieae of the genera Rozella and Woronina, which were discovered by Cornu and more thoroughly examined by A. Fischer[1], are parasites on the stout tubes of the Saprolegnieae which are destined to form spores (see Figs. 68 and 69). Rozella septigena, Cornu lives on species of Saprolegnia. Its swarm-spores are ellipsoidal in shape and 6–8 μ in length, and bore their way into the growing tubes of Saprolegnia. When a spore has effected an entrance in this way, it remains visible for a short time in the protoplasm of the tube and then can no longer be distinguished; it seems to lose its individuality, to distribute itself and be lost in the protoplasm of the tube. The tube at once swells out as a large quantity of fresh protoplasm streams into it, and it divides by transverse walls formed for the most part in basipetal succession into swollen cylindrical or barrel-shaped segments, which are from two to several times longer than broad (Fig. 166). The transverse walls are formed in the same way as in the delimitation of sporangia in Saprolegnia. Then the protoplasm in each segment becomes invested with another special membrane lying in close contact with the first and thus the cell becomes the sporangium of a Chytridium (see section XLVI). About 60–80 hours elapse between the entry of the spore and the emptying of the sporangium. The process is the same if more than 1 spore has entered the tube of the Saprolegnia; when a single spore entered 2–14 sporangia were observed to be formed, 21 sporangia when 4 spores entered the tube. Rozella appears to form resting-spores when many of its swarm-cells have made their way into the Saprolegnia. In this case also the swarm-cells entirely disappear in the protoplasm of the host, which then puts out stalked spherical branches resembling the oogonia of Saprolegnia, and a resting-spore is formed from the protoplasm of each branch (Fig. 166, *s*). The phenomena observed in Woronina are quite similar to those in Rozella apart from certain differences in the form of the parts.

If no later investigations bring to light facts at variance with those here given, and this according to Fischer's statements is not probable but cannot however be considered to be impossible, the above case is an instance of a parasite surrendering its individuality, so far as that can be recognised morphologically, after its entrance into the host; it becomes changed into a part of its host, and the two can no longer be distinguished from one another by our present means of investigation. The parasite also communicates new properties to the host which is some two hundred times its size, and through them the host developes into its own parasite. This transmutation of the host by the parasitic spore which has coalesced with it, though peculiar, is yet analogous with the fertilising effect of the spermatozoid on the oosphere of organisms which in the sexual sense are highly differentiated from one another.

Fungi which form Lichens.

SECTION CXIV. A large number of species of parasitic Fungi employ Algae as their hosts, either unicellular Algae which live isolated, or pluricellular Algae, or Algae which unite together in gelatinous colonies. The germ-tube of the Fungus

---

[1] Ueber d. Parasiten d. Saprolegnieen. See also citations on page 171.

grows round the cells of the Algae and developes into a thallus. The Alga follows this growth, inclosed in a definite form between the hyphae of the Fungus, and thus forms an integral part of the thallus. As the Alga grows, the chlorophyll in its protoplasm is constantly engaged in assimilating carbon dioxide and forming organic carbon compounds for the use of the Fungus. On the other hand the rhizoid-branches of the Fungus (see page 45) spread in and on the substratum and contribute the needful mineral food. The two processes of vegetation mutually support and supplement each other. The Alga can live alone, as a plant containing chlorophyll, though it may be open to question whether it can do so in all cases; but its vegetation is not hindered by the Fungus, it can often be shown to be permanently promoted by it. The Fungus as a strictly obligate parasite is dependent for its vegetative growth on the Alga, without which it cannot attain to its full development, and in most cases can scarcely get beyond the first stages of germination.

The power of assimilating carbon dioxide which the Alga possesses in virtue of its chlorophyll makes it unnecessary that organic carbon compounds should be supplied as food to the thallus formed of itself and the Fungus in combination. Very many of the forms in this group do in fact live on places like bare stone or the ground, which do not supply these compounds. The question whether other species, which live on the bark of trees, humus-soils and similar substances, do partly vegetate as saprophytes has not been thoroughly examined; but Frank's observations on species living upon the bark of trees, which will be further noticed below, are in favour of the existence of a saprophytism of the kind.

In one point the Fungus is the superior in the common household; it alone produces spores, the Alga with a few exceptions remaining barren as long as it is combined with the Fungus[1].

The bodies thus composed of Fungus and Alga combined are known by the name of **Lichens**; it is equally well known that this name indicates an important phenomenon in nature, and is applied to a very varied series of forms, large enough to fill the pages of many descriptive works and local floras.

It follows necessarily from the above that two component parts, *a Fungus and an Alga, sometimes even several Algae, can be distinguished in every Lichen.*

The general morphological characters of the Fungi which form Lichens have been already stated in previous sections of this book. In the vegetative part they consist of hyphae connected together in the various ways described in section I, and in most cases united into a Fungus-body. According to the course of development indicated by the sporiferous structure the great majority of Lichen-fungi belong to the Ascomycetes, both **Discomycetes** and **Pyrenomycetes**, a few recently become known to us to the **Hymenomycetes**.

The number of algal forms which enter into combination with Fungi is according to our present knowledge not inconsiderable. There is one corresponding Alga to each species of Lichen, in other words, each species of Lichen-fungus is confined to a certain species of Alga. It is rare that two are found together in a Lichen, and then one of them always predominates over the other. On the other

---

[1] **Synalissa symphorea** Nyl.   See Bornet in Ann. d. sc. nat. sér. 5, XVII, p. 50.

hand the same species of Alga may serve as host to different species of Fungi, and serve accordingly as a component part in very different forms of thallus.

The following list contains the genera and groups of Algae which are known to form Lichens; the reader is at the same time referred to Schwendener's and Bornet's special works and to some others also which will be named below, and to the general works on the Algae[1]. Figures 167–169 represent some of these Algae as they appear in the thallus of the Lichen and in most cases in connection with the hyphae of the Fungus by which they have been attacked.

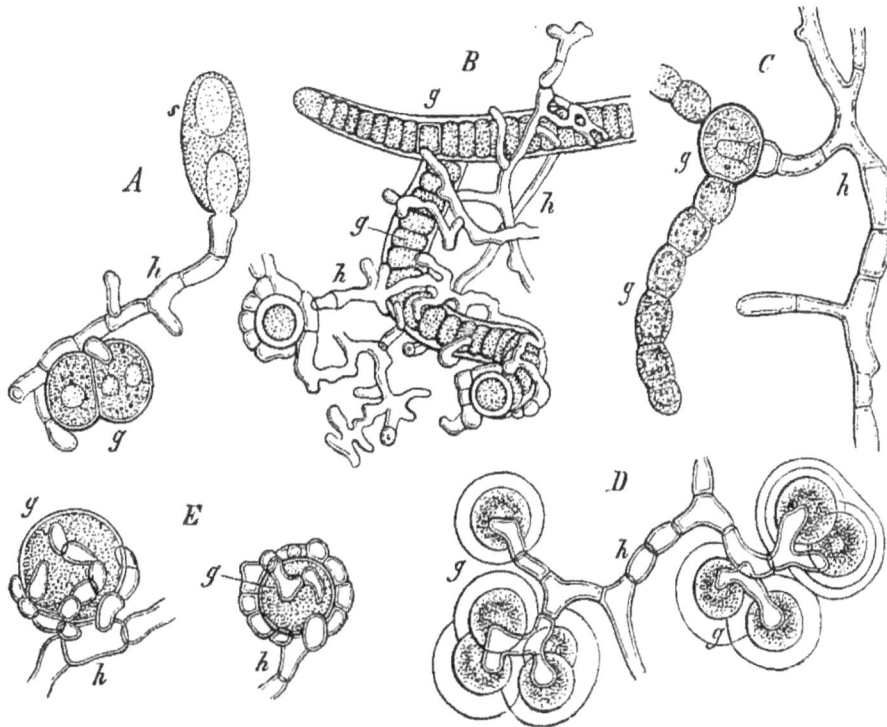

FIG. 167. Lichen-forming Algae. The Alga is in all cases indicated by the letter *g*, the assailing hyphae by *h*. *A Protococcus viridis*, Ag. (*Cystococcus*, Näg.) attacked by the germ-tube from a spore of *Physcia parietina*. *B Scytonema* from the thallus of *Stereocaulon ramulosum*. *C Nostoc* from the thallus of *Physma chalazanum*, Mass. *D Gloeocapsa* from the thallus of *Synalissa Symphorea*, Myl. *E Protococcus* sp. (*Cystococcus*) from the thallus of *Cladonia furcata*, P. *ACDE* magn. 950. *B* 650 times. After Bornet.

1. Chlorophyll-green Algae.

(*a*) With short usually round cells living free or loosely united, but not forming filaments or cell-surfaces, the group of 'Palmellaceae': **Protococcus**, Kg., **Pleuro-coccus**, Menegh., **Cystococcus**, Näg. (Figs. 167 *A*, *E*, 168), **Dactylococcus**, Näg., **Stichococcus**, Näg.

(*b*) With cells firmly connected together into filaments and cell-surfaces, the group of 'Confervae': possibly species of **Ulothrix** and especially the **Chroolepus**-forms (**Trentepohlia**, Mart., Bornet) (Fig. 169) distinguished by the orange colour of the contents (haematochrome) in addition to the chlorophyll. To these may be added the genus **Phyllactidium** said by Bornet to be the Alga of Opegrapha

---

[1] See especially Nägeli, Gattungen einzelliger Algen; also Kirchner, Algenflora von Schlesien.

filicina, and Cunningham's[1] **Mycoidea parasitica** which with some other allied Algae helps to form the species of Strigula common on evergreen leaves in the tropics.

2. **Algae which are blue-green, violet and other colours, owing to the presence of phycochrome**, and are often united together into large bodies by means of their gelatinous membranes.

(*a*) Nostocaceae with their cells forming filaments: **Calothrix**, Ag. (Schizo-siphon, Kg.), **Scytonema**, Ag. (Fig. 167, *B*), **Lyngbya**, Ag., **Nostoc**, Vauch. (Fig. 167, *C*), **Stigonema**, Ag. (Sirosiphon, Kg.).

(*b*) Chroococcaceae with their cells not forming filaments: **Gloeocapsa** (Fig. 167, *D*), **Chroococcus**, **Aphanocapsa**, Näg.

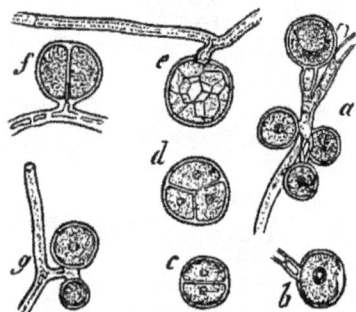

FIG. 168. *Cystococcus.* *a—e* from the thallus of *Imbricaria tiliacea.* *g* from the thallus of *Sphaerophoron coralloides.* *f* from the thallus of *Usnea barbata.* *c*, *d* isolated algal cells, the rest with hyphae attached to them. *e—f* cells dividing. *a—e* and *g* magn. 390. *f* 700 times. *f* after Schwendener.

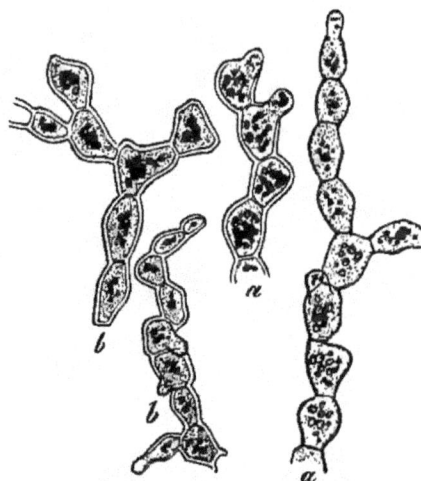

FIG. 169. *Trentepohlia (Chroolepus umbrinum, Kg.).* *a* from the thallus of *Lecanactis illecebrosa*, Duf. *b* from the thallus of *Graphis scripta*. Magn. 390 times.

SECTION CXV.. **Origin of the Lichen-thallus.** If the ripe ascospore of the Lichen-fungus is placed on a moist substratum, it in most cases readily puts out germ-tubes; these may in some cases form numerous branches (see page 113), but they always perish after a certain time if they do not encounter suitable Algae, even when the germination has taken place on a substance favourable to the nutrition of the Lichen.

Supposing the substratum to be favourable and the right Alga to be within reach (see Fig. 170), the germ-tube puts out branches on one side, which seize on the Alga. If the Alga consists of single isolated cells, as for instance in the Pleurococcus-lichens observed by Stahl, smaller branchlets are formed at the points of contact with an algal cell, which closely embrace it and inclose it in fresh ramifications. If the Alga is a compact pluricellular body, as in the Nostoc-lichens examined by Reess, in which the cells united together in rows are surrounded by broad gelatinous membranes and the protoplasmic bodies therefore are imbedded in a large quantity of an intermediate gelatinous substance, the hyphal branches penetrate into the jelly and put out branches in it which grow luxuriantly through the algal body and unite with the protoplasmic bodies of the cells. At the same time in all cases known to

---

[1] Trans. Linn. Soc. London, ser. 2, I, p. 301.

us the germ-tube from the spore puts out other branches, which penetrate into the nutrient substratum and evidently obtain the necessary mineral matter from it. If there is no substratum of this kind, as when the plants are grown on glass plates, the above processes do not go beyond the very first stages.

The two observers just named carried the cultivation of their plants, with the precautions suggested naturally by the circumstances stated above, from germination to the production of the fully grown thallus, which in some cases also produced sporocarps. The same results were obtained in a series of other cases by the investigations of Bornet and Treub at least as regards the earliest stages, and these investigations and a comparison of fully developed thallus-forms show distinctly that the first stages of the development have a close resemblance to one another in all or almost all Lichens, though accompanied with certain specific modifications which were naturally to be expected throughout. Even the peculiarities observed by Frank in the development of the thallus of certain Graphideae, as Arthonia vulgaris and Graphis scripta, which grow on the bark of trees, are really only specific modifications. In this case the hyphae of the thallus, which can only have come from the germ-tubes of the spores, though the observations on this point are still imperfect, make their way into the outer layers of the periderm in smooth stems of oaks and ashes and there grow as saprophytes independently, that is without Algae, into a thallus formed of an abundance of slender hyphae which spread through the cells of the periderm. Then its proper

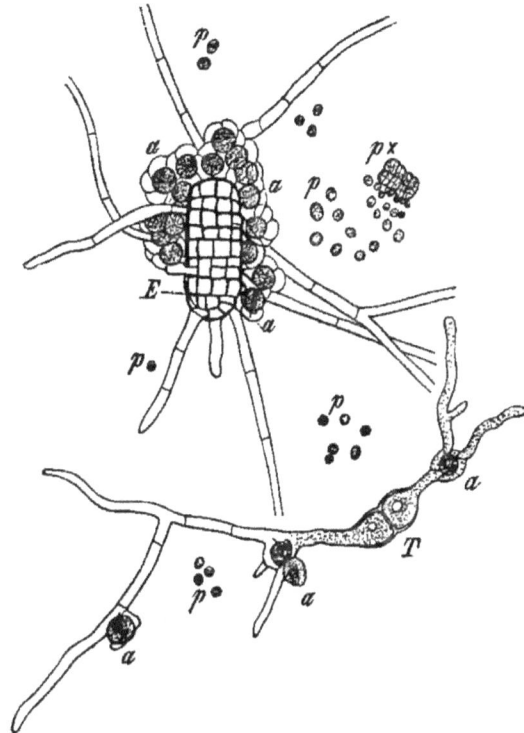

FIG. 170. *E* a compound pluricellular ascospore of *Endocarpon pusillum*, cultivated on a microscopic slide and putting out numerous germ-tubes. *T* a bicellular compound spore of *Thelidium minutulum* also germinating on a microscopic slide. *p* cells of *Pleurococcus* ejected with the spores from the hymenium of *Endocarpon* and vegetating on the slide; *px* a group of several cells of *Pleurococcus* formed at the spot by the growth of single cells. *a* cells of *Pleurococcus* attacked and partly grown round by branches of the germ-tubes, and consequently considerably larger than those which are vegetating in freedom. After Stahl. Magn. 320 times.

Alga, Chroolepus umbrinum (Fig. 169), finds its way from without through the cell-walls of the peridermis into the previously formed hyphal thallus and is seized by it. The cells of the Chroolepus are in rows forming filaments with apical growth, and it is by means of this growth that they penetrate into the thallus, in the same way as mycelial hyphae pierce through membranes. The Alga is a frequent inhabitant of the bark of trees and makes its way into the periderm for its own purposes. Its penetration into the thallus of the Fungus can scarcely be supposed to be caused by the Fungus, but is merely an adaptation which favours the formation of a Lichen.

Lecanora pallida commences its growth according to Frank's observations with the development of a hyphal thallus in the periderm like the Graphideae just mentioned, while the Alga which belongs to it, a form of Cystococcus (?) with free round cells, comes 'casually' into contact with the hyphal thallus, its cells being caught in the superficial crevices of the periderm in which it is spreading and there seized upon by the Fungus. This is the origin no doubt of many Lichens which occur in the same situation.

The nicest adaptation between Alga and Fungus, and one which goes so far as to exclude 'accident,' is displayed in a small number of species of Lichens which Stahl especially has closely examined: Stigmatomma cataleptum, Endocarpon pusillum, Polyblastia rugulosa, etc. Here the cells of the Alga departing from their usual behaviour make their way regularly into the hymenium of the sporocarps of the Fungus, and increase to such an extent between the paraphyses, that they lie in great quantities round the asci as they ripen. Whenever any spores are discharged a certain number of algal cells adhere to them and are ejected with them from the sporocarp, and are ready at once to be seized upon by the germ-tubes.

The morphological characters of the Alga when it is assailed by the Fungus experience a more or less profound change from its condition in the free state. Size, structure and arrangement of cells are affected in a manner which varies in the same species of Alga according to the species of Fungus to which it serves as host. In this point also specific differences which are noticed in descriptive works on the Lichens make themselves apparent in different cases. We will only mention here as an example of these differences the smooth round gelatinous colonies of Nostoc lichenoides and similar Algae, which are changed by the intrusive Fungus into lobed thallus-bodies with progressive growth in the direction of the margin, but usually without the structure and arrangement of the algal cells which they contain being affected. With these may be compared the hymenial Algae of Stahl just mentioned. These in Endocarpon pusillum are a form of Pleurococcus. Within the hymenium the cells of the Alga are very small, roundish in form and a pale green, and they multiply by successive bipartitions, the new cells separating from one another. When seized by the hyphae of Endocarpon their diameter increases to six times the original size, their parietal chlorophyll-corpuscles assume a bright green colour, and the bipartitions are continued with irregular arrangement of the successive partition-walls and a corresponding change in the grouping of the cells themselves. The Alga retains these characters as the thallus developes. If the same Pleurococcus-cells when ejected from the hymenium of Endocarpon pusillum are seized upon by the hyphae of Thelidium minutulum, they assume the same green colour, but they do not increase so much in size, and within the surrounding hyphal weft the divisions take place by partition-walls disposed alternately in three directions and cutting one another at right angles. In accordance with these divisions the cells are arranged in rectangular packets with blunt angles and divided into squares, just as they are arranged when vegetating free and alone (Fig. 170, 175).

Lastly it is of importance to call attention to a fact noticed by Frank in the Lichens mentioned above as studied by him, that a part of the algal cells in the Lichen may to a great extent or entirely lose their chlorophyll without apparently dying in consequence of the loss; but this point requires closer investigation.

If the Alga is released from its union with the Fungus by natural or artificial means, it may, as we learn from the researches of Famintzin and Baranetzki, Schwendener, Bornet, Woronin, and Stahl, continue to vegetate, recovering its typical morphological characters as an Alga, if the necessary conditions are present. See for example Fig. 170.

Where the Alga is much altered in the Lichen, careful examination of liberated cells is often necessary for the certain determination of the species. Such an examination has yet to be made in the case of many of the Lichen-algae, especially the Palmellaceae, and consequently these forms cannot at present be named with certainty and exactness.

SECTION CXVI. **Form and structure of the Lichen-thallus.** The seizure of the Alga by the Fungus is followed by the development of the combined body into its ultimate form, its growth into the perfect Lichen-thallus (the blastema of Wallroth), on which organs of reproduction of the Fungus make their appearance when the highest development has been reached. These organs in the majority of Lichens are *sporocarps* (*apothecia* and *perithecia*) together with *spermogonia* (see page 211); in a few, which will be briefly noticed again at the close of this account, they are *basidia-bearing hymenia*; in many they are small *brood-buds* which separate spontaneously from the thallus and are known as *soredia*. As a general rule the Alga takes part only in the construction of the thallus and soredia and not in that of the sporocarps and spermogonia or the basidia-bearing hymenium. The 'thallodic' margin of the apothecia which is characteristic of many genera and contains Algae (see Figs. 86, 87, 89) belongs, as the name rightly expresses, to the thallus and not to the apothecium. It has often been observed indeed at the first beginning of the formation of the apothecia (see page 213), that algal cells are included in the primordial hyphal weft; but they die away in most species without increasing in size or number in the sporocarp itself.

Nylander[1] was the first who called attention to a remarkable exception to this rule in the case of the pyrenocarpous species just mentioned, in which the Algae, —Pleurococcus and Stichococcus in the best ascertained cases,—which are taken into the primordial sporocarp, grow with it and multiply their cells to such an extent that they gather thickly round the asci. This is the origin of the hymenia containing Algae and the hymenial algal cells of Stigmatomma cataleptum, Sphaeromphale, and especially Endocarpon pusillum and Polyblastia rugulosa which have been especially investigated by Stahl, as was mentioned on page 400[2]. The algal cells are distinguished by their considerably smaller size from their sister-cells in the thallus to which they belong.

The fully developed thallus as regards its outward configuration appears in three chief forms, which are not however sharply distinguished from one another: the *fruticose* (thallus fruticulosus, filamentosus, thamnodes, Fig. 86 A), which rises with a narrow base from its substratum, and is either simple or more usually branched like a shrub; the *foliaceous* (thallus foliaceus, frondosus, placodes, Fig 86 L), in form a flat, leaf-like, usually lobed and crisped body which spreads over the surface of the substratum, but is only attached to it at one or at several scattered points and can

---

[1] Synopsis meth. Lichenum, p. 47.
[2] See Beitr. z. Entwicklungsg. d. Flechten, where the literature is cited.

therefore be separated from it without much injury; and the *crustaceous* (thallus crustaceus, lepodes), a flat crust on or in the substratum and adhering firmly to it at least by its whole under surface, so that it cannot be separated from it without injury. The genera Cladonia and Stereocaulon are peculiar, having shrub-like formations (*podetia*) rising from scaly or granular foliaceous bodies (the thallus or protothallus of the Lichenologists). These forms all agree in two points of structure; they are composed, as has been said, of hyphae and of algal cells inserted between them, and from the first commencement of their formation hyphal branches from the base or surface turned towards the substratum enter the subtratum as *rhizoid-hyphae*, which sometimes, especially in foliaceous species, are united into rhizoid-strands. These organs by which the plant attaches itself to the substratum and takes up food-material are termed by Lichenologists *rhizines*, and they often penetrate, especially in species living on rock or soil, as much as a centimetre deep into the substratum and spread through it in dense ramifications. The rhizoids of thespecies which live in the rind of trees seem never to penetrate far into it, at all events they do not reach the living tissue of the rind. In many species, which, like the Graphideae examined by Frank and Lecanora pallida, form thin crusts in the rind of trees, we find in place of distinct rhizoid-branches the system of hyphae described above, which spreads in superficial periderm-layers without going further inwards. These hyphae may form a persistent thallus which grows all its life entirely in the periderm along with the Algae which it has attacked, as in the above-mentioned and similar Graphideae and in Pyrenula nitida, and it may be covered by one or more layers of periderm-cells, in which

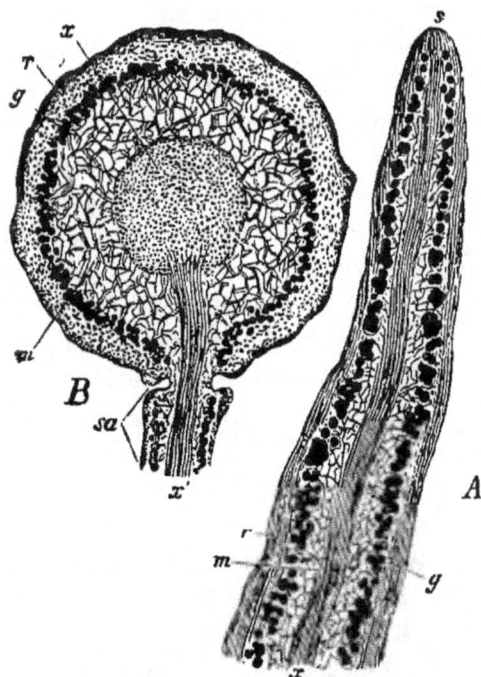

FIG. 171. *Usnea barbata. A* optical longitudinal section of the extremity of a thin branch of the thallus which has become transparent in solution of potash (see Fig. 86 *A*). *B* transverse section through a stronger branch with the point of origin of an adventitious branch *sa*; *r* cortical layer, *m* medullary layer, *x* stout axile strand, *g* the algal zone (*Cystococcus*), *s* apex of the branch. After Sachs. Magn. nearly 100 times.

case it is termed by Lichenologists *hypophloeodic*; or it makes its way out to the outer surface of the periderm in consequence of the subsequent growth in thickness of the parts which contain the Algae, and forms there an *epiphloeodic* crust.

In other points of structure differences appear, which have no simple relation to these growth-forms, but may be repeated in each of them. Lichenology since Wallroth's time has accordingly distinguished between the *heteromerous* and the *homoiomerous* thallus. The former is peculiar to the 'true Lichens' (Lichenes of Fries, Lichenaceae of Nylander, Gnesiolichenes of Massalongo), the latter to the Phyco-lichenes of Massalongo (Collemaceae of Nylander, Byssaceae of Fries). The hetero-merous thallus occurs in the large majority of species, and displays in fact a structure the main features of which can be clearly defined. The homoiomerous on the contrary can at present only be described as not heteromerous, and under this head certain types may be distinguished by positive characters, the most suitable of which are drawn from the Algae which form part of each species, as will be explained below.

1. **The growth of the Fungus-body in the heteromerous thallus** (Fig. 171) is progressive in the direction of the apex in the fruticose Lichens, in the direction of the margin in the others (see section XII). The apex or margin which takes the lead in growth is formed of the terminal portions of the hyphal system, and these are the form-elements which determine the course of the development. The Alga follows in its growth the apex or margin which precedes it, moving after it, but always continuing at a certain distance behind it. This growth by gradual advance may be followed in successive transverse zones by intercalary surface-growth and growth in

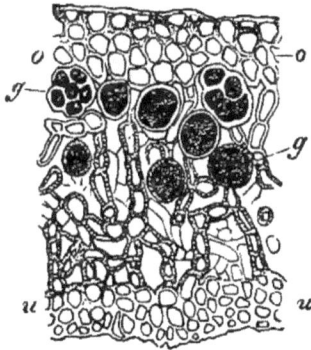

FIG. 172. *Physcia parietina,* Kbr. Section through the young thallus; *o* upper rind-layer, *u* under rind-layer, *g* algal zone (*Cystococcus*). After Schwendener. Magn. 500 times.

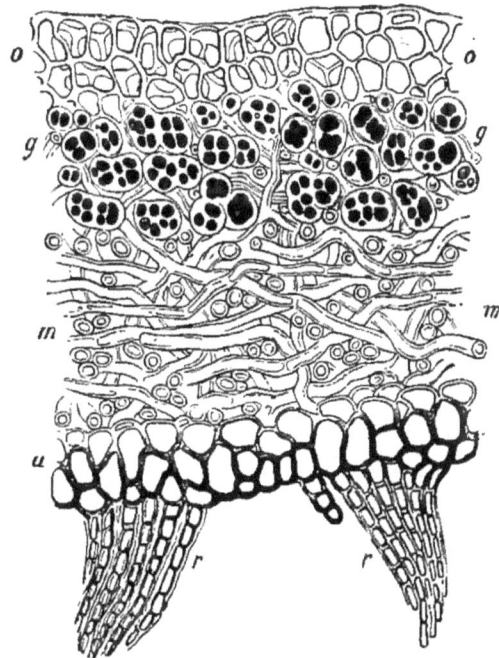

FIG. 173. *Sticta fuliginosa.* Transverse section through the foliaceous thallus; the letters as in Fig. 172, also *m* medullary layer, *r* rhizoid strands. The algal zone here consists of a species of *Chroococcus* with thick colourless gelatinous membrane and bluish green protoplasm which is black in the figure. After Sachs. Magn. 500 times.

thickness, caused by the formation and insertion of new hyphal branches and algal cells and also by the increase in volume of those previously formed.

The ramifications also in the foliaceous and fructicose thallus proceed from the hyphal system. They are partly bifurcations, and partly arise 'adventitiously' at points not in most cases very exactly and morphologically defined.

The structure also of the mature heteromerous thallus (Figs. 171-173) is in its ground-plan that of the Fungus-body of the Lichen. We may as a rule distinguish in it a comparatively thin (its average thickness is about 10 μ) usually transparent dense peripheral *rind-layer* (stratum corticale) clothing the free surface and a weft beneath the rind which is usually loose and everywhere furnished with interstices containing air, the *medulla* or *medullary layer* (stratum medullare). Both layers belong to the hyphal system, their constituents are ramifications of the same hyphae. The cells of the Alga are in almost all cases inserted where the medulla and the rind meet. Together they form a green zone of varying size projecting into the medullary tissue to a different depth at different points, everywhere traversed by single hyphae of the medulla running to the rind, and in some places showing larger interruptions. This is the *algal layer* or *algal zone* which is commonly termed the third tissue-layer. But single algal cells or groups of cells are often found scattered through the medulla, as in Solorina and Placodium, or the whole mass of algal cells is distributed with tolerable uniformity through the medulla, as in Bryopogon.

Sundry modifications of this type and small deviations from it occur in different species and groups of species. The following are the most important ones :—

Some **fruticose** thalli, especially if cylindrical in form, are covered all round with a uniform rind, as in Usnea (Fig. 171), Bryopogon, Roccella, Sphaerophoron and others. In many fruticose forms with a flat thallus (species of Evernia and Cetraria) and in most **foliaceous** forms the rind of the upper surface which is towards the light is different from that of the lower surface (Figs. 172, 173); in Anaptychia, Peltigera, Solorina and most of the foliaceous portions of the thallus in the Cladonieae the upper side only has the rind up to the margin, the lower surface has no rind.

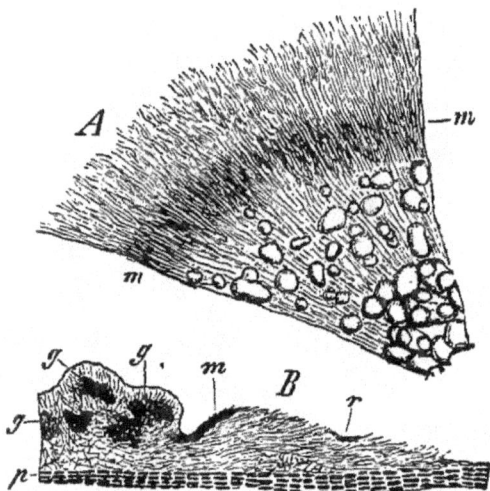

FIG. 174. *Lecidella enteroleuca*, Kbr. on the rind of *Tilia*. *A* surface-view of the margin of a young growing thallus. *B* radial vertical longitudinal section through the same. *m* black marginal bands, *g* groups of Algae, *p* periderm. *A* slight'y magnified. Circumference of *B* magn. 45 times, but the whole completed from greater enlargement.

When there is this difference in the two surfaces, the algal zone apart from the scattered cells only appears on the side towards the light. Even in the almost cylindrical thallus of Sphaerophoron with a uniform rind all round it, it is often more strongly developed on the side towards the light than on the lower one. The podetia of Cladonia, which according to Krabbe[1] are to be regarded as strongly developed parts (envelope-apparatus) of apothecia but which always have the structure and growth of a fruticose thallus, are always without a rind when young, and in some species, as C. rangiferina, never have a rind, their surface being formed of a loose hyphal weft containing Algae. Most species are at a later period of their life partially covered with small wart-like rind-scales, and some are completely covered with a rind, as C. furcata.

Some of the **crustaceous** forms, for example the genera Psora, Psoroma, Placodium, Endopyrenium &c. and Thalloidima candidum are only distinguished from the foliaceous by having their under surface covered all over with rhizoid-hairs which penetrate into the substratum. These hairs either spring as branches from the elements of a distinct rind, as in species of Placodium and Endopyrenium, or where the rind is wanting on the under surface they are direct continuations of the medullary hyphae.

Other crustaceous Lichens, those namely with **areolate surfaces** or which have a **granular appearance**, have essentially the same structure in the fully developed areolae as the species mentioned above which are without a rind on the under surface. Some of

[1] See on page 222.

them which have a very thick thallus, as Haematomma ventosum, Mass. and Lecanora Villarsii, are distinguished by a very thick medullary layer, sometimes forming according to Tulasne $\frac{28}{38}$ of the whole thickness of the thallus. The granulation or areolation is produced by the course of growth which has already been described. This is the case at least in a number of species examined by Frank and myself which live in bark, Lecidella enteroleuca, Kbr. (Lecidea parasema, Nyl.), Ochrolechia pallescens, Lecanora pallida and Pertusaria, and it is probable that a large number of other species of similar habit agree with them in this respect.

In these Lichens (Fig. 174) the margin of the thallus consists simply of several layers of hyphae, which radiate in the direction of the surface (*A*) and run parallel to one another in a vertical and radial longitudinal section (*B*). These together form a broad marginal zone often of different thickness in different radial strips, and with the extremities of single free hyphae projecting beyond its circumference. The peripheral portion of this zone, as Frank has pointed out, always grows through outer layers of the periderm, and first emerges from them at some distance from the periphery where it is thicker. The margin advances by marginal progressive growth and the surface of the thallus therefore enlarges. The algal cells which grow and multiply move forwards in the same direction behind the margin, and *within* the layers of hyphae. They lie collected together into groups, and as the algal cells constantly increase in number the hyphae are continually forming fresh branches in and around each group, and becoming more and more densely interwoven till their originally radial course is at length obliterated ; this happens especially on the side towards the substratum. Thus a coil of closely interlacing hyphae inclosing the algal cells is formed at the points indicated, the extremities of the filaments nearest the surface running chiefly at right angles to the surface and forming a rind which is usually very thin (Fig. 174 *B*).

The spots in which these formations take place rise above the surface of the marginal zone in the form of arched projections, small warts, which increase rapidly in number and size with the increase of the distance from the margin till they touch one another and become confluent. The mature thallus appears variously uneven, warted, granular or areolated according to the shape of the projections, their elevation and the extent to which they become confluent. In some forms with a very thick thallus, as Haematomma ventosum, Mass. and Lecanora Villarsii, Ach., the areolation of the surface of the thallus may be caused according to Schwendener by the rind becoming fissured in consequence of excessive dilatation of the medulla ; this may of course happen independently of the first formation of the areolae as here explained.

Some crustaceous Lichens, as Lecidea (Rhizocarpon, Kbr.) geographica, L. confervoides and Urceolaria cinerea according to Tulasne's figures, show an edging round their distinctly areolated thallus, formed of radiating branched confervoid strands of hyphae firmly united together and adhering closely to the substratum. The areolae of the thallus are formed in centrifugal succession on this marginal expansion, at first as small scales or warts which gradually increase in size till they are in contact with each other on every side. We need not assume that there are other differences between this mode of development and that first described, besides the one difference that in the present case the marginal zone is divided into the separate strands, while in the other it is coherent. The structure of the areolae shows no special peculiarities. The marginal strands of the last-mentioned forms as well as the marginal zone of those which grow in the manner of Lecidella enteroleuca, Kbr., are often under the name of *hypothallus* confounded with rhizoids or root-hairs. The term *protothallus* as it indicates a commencing state is more suitable but is at the same time unnecessary.

As regards the **minuter structure of Lichens**, the hyphae of most of the heteromerous species are comparatively slender. The **hyphae of the medulla** are usually slenderer than those of the rind, and their cells may be very long, as in Usnea, sometimes reaching according to Schwendener a length of 200 μ ; but the hyphae may also be short-celled, as in Fig. 172, and their structure may even be pseudo-parenchymatous as in Endopyrenium and Catopyrenium. Details on these points will be found in descriptive works and especially in Schwendener's earlier writings. It has been already observed that the medulla is in most cases furnished with broad interstices containing air ; in some fruticose forms, as Cladonia and Thamnolia, there is a broad axile cavity occupying the largest part of the total volume of the thallus. In Usnea a dense axile hyphal strand passes through the middle of the thallus and is surrounded by a loose weft of hyphal branches running to the rind (Fig. 171) ; in Evernia vulpina and E. flavicans the loose medulla is traversed by several similar longitudinal strands which approach and separate alternately in an irregular manner.

The **hyphal branches which compose the rind** are united together without interstices in all cases except in some species of Roccella. They are either unmistakeable hyphal filaments, with the lumina of their segments at least evidently elongate-cylindrical, though they may not be as long as the cells of the medulla ; or they are composed of short isodiametric rounded prismatic cells, so that the rind often has a very regular and delicate pseudo-parenchymatous structure, as in Parmelia, Physcia (Fig. 172), Endocarpon, Sticta (Fig. 173) and Peltigera. It is only in Bryopogon and Anaptychia ciliaris that the course of the much elongated hyphae of the rind is almost strictly longitudinal even in the mature thallus. In all other cases the hyphae lie in every direction and form an irregularly woven weft or pseudo-parenchyma ; or they are nearly exactly vertical to the surface, as in the pseudo-parenchymatous cell-rows of Endocarpon and Peltigera and in a very beautiful form in tufted branches of Roccella, in which also the lateral connection is somewhat loose and the extremities are free. The single branched hyphae may be clearly distinguished in thin sections especially of Roccella fuciformis, and a slight pressure will isolate them.

The thickness of the membranes, the breadth of the lumina in the medulla and rind, and the mutual relations of the two layers vary extremely in the different genera and species ; the details will be found in Schwendener. The cell-walls are often unusually thick especially in rind-hyphae with long segments (Usnea, Bryopogon, Sphaerophoron &c.), and from their close connection with one another they seem to form a homogeneous mass, in which the lumina appear as narrow canals. But this homogeneous mass is seen in thin sections, especially when treated with dilute solution of potash or ammonia, to consist of separate obscurely stratified membranes. The structure of these rind-layers has much resemblance to that of some sclerotia (see page 31).

The differences between the rind of the upper and under side lie in the thickness of the rind-layer, the size and arrangement of the cells, the colour and similar points (see for example Figs. 172, 173).

The surface of the rind is sometimes covered with a fine *felt of hairs* which are small projecting hyphal branches, as on the upper side of Anaptychia ciliaris, Peltigera malacca, P. canina also in the young state, and on the lower side of Sticta and Nephroma.

Among the **wart-like prominences** which occur on the upper surface of some foliaceous and fruticose species, those of Peltigera aphthosa and the smaller ones of Usnea are thickenings or exuberant growths in the rind ; the larger ones in Usnea and the warts and scales of Evernia furfuracea in the matured state are a kind of projections of the thallus, consisting of algal cells inclosed by an air-containing tissue connected with the medulla and invested by the rind. The blackish branched enlargements on the upper side of Sticta fuliginosa and Umbilicaria pustulata are formed of a dense hyphal weft containing Algae surrounded by a single layer of pseudo-parenchymatous rind with a brown membrane. A similar structure has been observed in the wart-like and rod-like excrescences, which often form a thick covering on the surface of many crustaceous Lichens and cause the coralline appearance of the forms known by the name of Isidium ; they require further investigation.

Peculiar **interruptions of the rind** occur on the under side of the thallus of Sticta. They are either largish flat spots with a somewhat indistinct outline, as in Sticta pulmonacea, or small circumscribed pits, the bottom of which is formed by the exposed medulla and the margin surrounded by a raised portion of the rind. The pits are formed according to Schwendener by the cortical layer being first thrust outwards in the form of a wart by exuberant growth of the medulla and then ceasing to grow in the protuberant part, while the surface of the thallus continues to enlarge all round it. In some species, as Sticta macrophylla, the interruption in the rind is preceded by the formation of a hollow space in the enlargement in the medulla. The pits are termed *cyphellae*; the older view, that they are small *gemmae-cups*, receptacles of soredia, is unfounded, or is true only in the case of certain species, as Sticta aurata (Schwendener).

There is little to add to what has been already said of the crustaceous Lichens. Most of them are distinguished by the comparative thickness of their rind.

The margin of many species, especially those which live in the bark of trees, is often marked by a **black line** which has been noticed by many writers ; Lecidea enteroleuca, Kbr. among the species specially mentioned above is an instance of this phenomenon, which is evidently caused by the dark colouring of the cell-membranes of the few *uppermost* layers of hyphae ; this colouring appears at a certain period in the life of the plant in the marginal zone where there are still no Algae. Where the thallus is engaged in active growth, the black line, if it is present, runs within the young colourless margin. The black colour is absent from the prominences which contain algal cells, doubtless because the few cell-layers which are so coloured are forced apart by the dilatation of the prominence and soon thrust aside. When the thalli of many individuals of the species in question come in contact with one another, the black lines often seem to form persistent boundaries between the different individuals like the lines on a geographical map. The reason of this may be, that the marginal growth of the thallus ceases where it encounters others, and the colour then appears in the marginal zone and there remains (see Fig. 174).

The thallus of some foliaceous Lichens, Umbilicaria pustulata for instance and of many crustaceous forms, is capable, as we learn from Schwendener, of a **growth in thickness which lasts a long time**, often for years, and with this is connected a decay and death of the rind which advances from without inwards. The dead layer often remains lying on the living one as an almost structureless transparent mass ; in other cases it is quickly destroyed by atmospheric agencies and becomes indistinguishable. The loss is replaced by regeneration from the inner surface, so that the rind always remains of about the same thickness. The medullary hyphae whose course is in the algal zone ramify and intertwine and form a tissue without interstices, which becomes like the rind and lies close against it on the inside. The outer algal cells are inclosed in this secondary rind-weft, but they gradually shrivel up and die while new ones are formed inside by division. If the process has been going on for some time, the whole of the rind is studded with dying or dead algal cells. They may be shown by the cellulose-colouring of their membranes with iodine, especially if first treated with potash.

The heteromerous thallus exhibits some remarkable peculiarities in respect of the **material constitution of the hyphae**. The membranes, as far as they have been examined, show the characters of the lichenin mentioned on page 10. They are transparent, almost always colourless in the medulla and in the inner portions of the rind, brittle when dry, swelling considerably and becoming soft and flexible in cold water. Some of the cells of the rind at least contain air when the plant is in the dry state, and the whole layer is thus rendered opaque. Absorption of water expels the air and increases the transparency of the membranes, so that the green algal zone is seen through it ; hence the change of colour in the surface caused by moistening the plant. The interstices in the medulla continue to contain air even when moistened, and the tissue therefore remains opaque. The membranes of most hyphae are not coloured by iodine ; they dissolve on addition of sulphuric acid either without assuming any colour, as in Usnea barbata, or after first turning brown, as in Anaptychia ciliaris according to Speerschneider, or after acquiring a violet hue which is sometimes very intense. The membranes of the rind of Cetraria islandica, except those of the superficial coloured cells, turn a beautiful blue with solution of iodine alone, as Schleiden[1] has observed, owing to the presence of Dragendorff's Lichen-starch in the cells (see page 10). The membranes of the medulla of Sphaerophoron coralloides show the same colour with iodine; Schwendener found that the rind of Cornicularia tristis and of some other species not precisely determined became blue with iodine only here and there.

The membranes of the rind, especially the superficial layers, in many Lichens are of different hues owing to the presence of a **colouring matter uniformly diffused** through the substance of the rind, as in Cetraria islandica, Bryopogon jubatus and many others; the dark colour of the under rind of Evernia furfuracea comes from the colouring of the outer lamellae of the membranes; the narrow lumina are seen in section surrounded by thick colourless membranes, between which are dark brown bands forming boundary lines; the matters imparting this uniform colouring have not yet been thoroughly investigated, at least not microchemically.

A second series of colourings is caused by **granular imbeddings** or **incrustations.** The tissue of very many Lichens is seen to contain small round or elongated granules of some **organic substance** lying partly on the free surface and on the medullary hyphae, partly imbedded in the membranes which are united together without interstices. Granules of this kind, which are either colourless or only of a faint yellow colour even in the mass, occur for instance in the upper rind of Placodium cartilagineum and other species, in Imbricaria caperata, Dill., I. incurva, P. and other species, on the medullary hyphae of Peltigera, Solorina saccata and some species of Sticta (Schwendener), on the rind of Roccella, especially R. fuciformis and of Thamnolia and in the inner rind of Sphaerophoron coralloides; distinctly or intensely yellow granules are found outside the rind of Evernia vulpina and Physcia parietina, or imbedded in the peripheral region of the rind of Cetraria straminea (Schw.) and Usnea barbata and in the whole of the membranes of the rind of Psoroma gypsaceum, or scattered in groups through the rind of Bryopogon divergens (Ach.) and B. sarmentosus and especially of B. ochroleucus where they form an uninterrupted layer near the surface, and in the medullary hyphae of Sticta aurata and S. crocata and some other species; red incrustations are found on the medullary hyphae of Solorina crocea. The presence of these bodies in the substance of the thallus and on it causes either a lively coloration, or, as in Thamnolia and Roccella, a dull yellow aspect and opaqueness in the parts in which they are found. They are all readily soluble in alkali either without change or with loss of colour, or at least deliquesce and become turbid under the influence of these agents, as in Physcia parietina and Solorina crocea, in the latter also without change of colour. It was not improbable therefore that the granules consisted of *lichinic acid*, and Fr. Schwarz has recently proved that this is the case in a number of species.

Another series of infiltrations and imbeddings is composed of **inorganic matter.** First and foremost is the rust-colour not unfrequently assumed by individuals ('formae oxydatae') of many crustaceous Lichens which are typically of another colour; it has often been stated and has now been proved by Gümbel, that this colour is due to the infiltration of a *salt of iron*, perhaps of a vegetable acid. Still more remarkable is the occurrence of *calcium oxalate* in many Lichens especially of the crustaceous division, either in the form of octahedral crystals, or of irregular crystalline bodies, or of small granules. In accordance with the rule which prevails in the rest of the Fungi it is never found inside the cells, but either on the surface of the rind, or on the medullary hyphae or in the interstices between them, or imbedded in the shape of minute granules in the membranes of the compact tissue of the rind. The latter is the case for instance in Psoroma lentigerum, in which the entire rind is wholly opaque and white in reflected light, owing to the close array of granules of the salt. Besides the above species, in which the medullary hyphae are also incrusted with granules, they occur also in Ochro-

[1] *Grundzüge*, II.

lechia tartarea, Mass., in Urceolaria scruposa which has an enormous number of crystals occasionally of large size in the interstices of the medulla, and in Thalloidima candidum where granules are found both on and in the upper side of the rind ; the thallus of the Pertusarieae, especially Pertusaria fallax, contains large, irregular, interstitial crystalline masses ; granular incrustations are found in the medulla of Chlorangium Jussuffii, and smaller isolated crystals are scattered about in the interior of the thallus of Megalospora sanguinea and M. affinis, Mass., Ochrolechia pallescens, Mass. and other species.

Yet it would be incorrect to say that calcium oxalate occurs in all Lichens, or always in the octahedral form. It does not even occur in all crustaceous forms ; I have as yet looked in vain for it in Lecanora pallida and in Lecidella enteroleuca, Kbr. Neither Schwendener nor myself have found it in any foliaceous species except Placodium and Endocarpon monstrosum mentioned by Schwendener, and we both noted its absence in most fruticose forms. In the young branches only of Roccella fuciformis I have seen tolerably large crystals which were not however chemically examined, and groups of small rods and granules composed of the salt in question in the rind and medulla of Thamnolia vermicularis.

This is not the place for a full enumeration and description of the substances organic and inorganic which analysis has detected in Lichens ; the facts have been collected and the literature noticed by Rochleder[1] and by Husemann and Hilger[2]. It is remarkable that the ash-constituent of Lichens is often very large, but on this point also information must be sought in special treatises[3].

The **Algae of the heteromerous thallus** in the majority of species are **chlorophyll-green** 'Palmellaceae' ; **Cystococcus** is the Alga in most of the Parmelieae ; **Pleurococcus** has been determined by Stahl in Endocarpon pusillum and Thelidium minutulum, and **Stichococcus** in Polyblastia rugulosa ; Bornet has determined the Alga in Solorina saccata and possibly also in S. crocea, Nephroma arcticum and Psoroma sphinctrinum, Nyl. to be **Dactylococcus**, Näg., and in Sticta glomulifera, Ach. he conjectures it to be an **Ulothrix**. It will be seen that a considerable variety of species are known to occur in Lichens, and it is desirable that the Algae should be isolated and more exactly determined in other forms which have only been hastily examined. Roccella has a species of **Chroolepus** for its Alga. The **Nostocaceae** and **Chroococcaceae** which owe their blue-green colour to the presence of phycochrome are peculiar to many heteromerous Lichens ; species of **Nostoc** according to Bornet's more or less certain determinations occur in Nephromium, Nyl., Peltigera, Stictina, Nyl, and in the species of Sticta which have blue-green Algae, the remaining species being formed with Algae from the group of the **Palmellaceae** (Fig. 173) ; **Scytonema** is the Alga of Coccocarpia molybdaea, Pers., and **blue-green Algae** of undetermined species are the Algae of the Lichen-genera Psoroma and Verrucaria.

The thallus of some heteromerous Lichens is peculiar in including a second Alga in addition to the species which is its principal and constant constituent. The chief Algae in Solorina saccata and S. crocea, some species of Stereocaulon, Sticta glomulifera and some others[4] are **Palmellaceae** ; but **Nostoc** occurs with them in the Solorinae

---

[1] See Gmelin's Handb. d. Chemie, VIII.

[2] Die Pflanzenstoffe.

[3] Thomson in Ann. d. Chem. u. Pharm. 53, p. 254.

  Knop in Erdm. Journ. f. pract. Chem. 38, p. 46 ; 40, p. 386 ; and in Ann. d. Chem. u. 49, p. 108.

  Gümbel, Ueber Lecanora ventosa (Denkschr. d. Wiener Acad. Math. Naturw. Cl., XI).

  Lindsay, Popular Hist. of Brit. Lich. p. 51.

  Uloth in Flora 1861, p. 568.

  Th. Fries, Genera heterolichenum, pp. 8–12.

[4] Forssel has recently published an elaborate account of these occurrences. See Studier öfver Cephalodierna, in Abhandl. d. Schwed. Acad., Anhang. 8, Nr. 3, Stockh. 1883, and in Flora, 1884.

and in Sticta, and **Scytonema** or **Stigonema** in Stereocaulon. In many forms, for example in Sticta and Stereocaulon, the Nostocaceae are localised in peculiarly shaped branched or convex outgrowths from the thallus, which have received the name of *cephalodia.*

The Alga is firmly attached to the adjoining hyphae; where the structure is a dense pseudo-parenchyma as in Endocarpon it is squeezed in between the hyphae; where it is filamentous and lacunose the hyphae send out branches, which attach themselves firmly to the algal cells; it not unfrequently happens that the hypha comes into contact with the new cells formed by division of the algal cells, and in that case it sends out small branches from the point of contact which embrace the Alga closely in the way

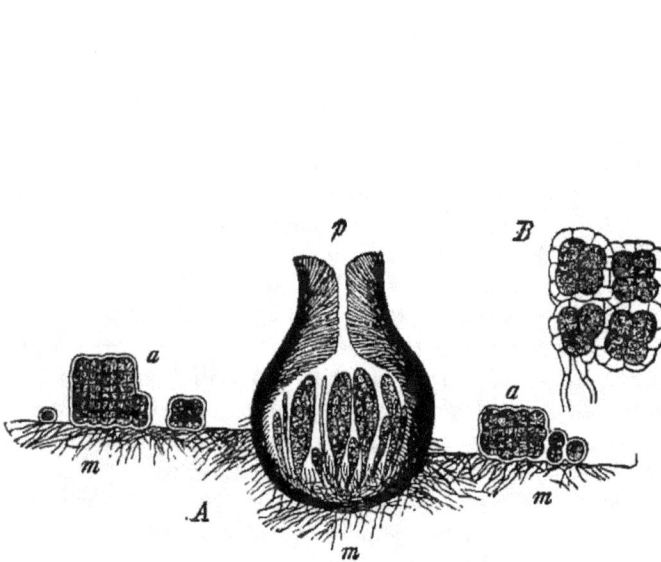

FIG. 175. *Thelidium minutulum.* *A* perithecium on the thallus; *a* groups of Algae, *m* the part of the thallus without Algae spreading in the substratum, *p* the perithecium cut through the middle, represented diagrammatically and slightly magnified. *B* a group of Algae with hyphae growing round them, magn. 480 times. After Stahl.

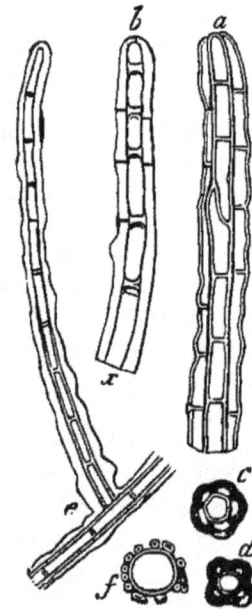

FIG. 176. *v—d Cystocoleus ebeneus,* Thw. *a* extremity of a branch seen from without. *b* a similar one in optical longitudinal section; *x* the Algae; *a* and *b* from preparations made transparent by Schulze's solution. *c, d* transverse sections.

e, f Coenogonium Linkii, Ehrb.; *e* slender branch of the thallus with a lateral branch in optical longitudinal section. *f* transverse section through a stronger branch. After Schwendener. *a, b, c, d, e* magn. 390, *f* 300 times.

described above in the case of germ-tubes (Fig. 167 *E*). The Stigonema which is peculiar to the cephalodia of Stereocaulon furcatum is seized by the hyphae in the same way as in Ephebe which will be noticed further on.

2.   There are a considerable number of Lichens, which are reckoned in descriptive works among the heteromerous species and agree with them on the whole in the manner of their growth, but which depart from the scheme of the heteromerous thallus in **having no differentiation of a rind,** and having the Algae scattered among the hyphae. Of this kind are the marine fruticose Lichinae which have Calothrix scopulorum, Ag. for their Alga[1], the frondose Pannarieae with Algae from the groups of Nostoc and Scytonema, a large number of Graphideae of different genera,

---

[1] See Kny in Sitzgsber. d. Naturf. Freunde 1874, and Bornet in Ann. d. Sc. Nat. sér. 5, XVII.

Graphis (Opegrapha) varia, O. plocina, Kbr., O. saxatilis, Schaer. and many other species, together with Lecanactis illecebrosa, Kbr., species of Arthonia, and Pyrenula nitida, in which the crustaceous thallus with progressive marginal growth consists of an air-containing weft of slender hyphae, which grow round filaments of Chroolepus as their Alga without any differentiation into medulla and rind (see above on p. 399). For further details in these cases which require more exact investigation the reader is referred to special treatises.

3. Another special form of structure must now be mentioned, the **granular green thallus of certain Lichens living on the ground** which Stahl has examined carefully in Thelidium (Fig. 175). In this case the thallus is formed of a loosely woven mass of hyphae which vegetate beneath the surface of the ground. Branches of these hyphae rise to the surface and attach themselves as in germination to the proper algal cells there present; the particular Alga of Thelidium is Pleurococcus. The hyphal branches grow completely round the cells forming a thin layer, and so follow the growth of the Alga that this relation between the two elements is constantly maintained, the hyphae pushing in between the products of the division of the algal cells as they are formed one after another, and twining round them as at the first. In this way groups of Algae are formed lying on the hyphal thallus and surrounded and held fast by the hyphae. No further differentiation takes place in them. It is possible that the process is the same in Biatora vernalis and similar forms, which should be examined on this point.

4. The thallus in **Chiodecton nigrocinctum,** Mont., **Byssocaulon niveum,** Mont. and **Coenogonium Linkii** (Fig. 176 *e, f*) consists of a branched filamentous Alga, belonging according to Bornet to the genus Trentepohlia and probably the same as Kützing's Chroolepus flavum, and surrounded by a dense weft of fungal hyphae; **Coenogonium confervoides,** Nyl. consists of a larger species nearly approaching according to Bornet to Chroolepus villosum and wrapped in a similar fungal weft. In these forms of thallus the Alga either does not alter its growth at all, or very little. In Coenogonium the Lichen has the conferva-form and the Alga advances by longitudinal growth, so that the tips of its branches are often found free from the hyphae. A similar structure is seen in the European species **Racodium rupestre,** Fr. (Cystocoleus of Thwaites) which is known only in the sterile state, with the difference however that here the hyphae which acquire thick and brown membranes surround the apex also of the branches of the Chroolepus in a single layer which has no interstices (Fig. 176, *a–d*). We learn from Bornet that Opegrapha filicina, Mont. behaves in the same way as Coenogonium, only that there the Alga is one of the species of Phyllactidium which form scutiform cell-layers on the leaves of tropical Ferns.

5. The **Collemaceae** or **gelatinous Lichens** are a group in which the Algae are Nostocaceae or Chroococcaceae with gelatinous membranes or membrane-sheaths; the Fungus makes its way into the gelatinous membrane, and, with the exception of the rhizoids and the sporocarps which issue from it, grows entirely inside it and ramifies in it. Fungus and Alga are differently related to one another in the common growth in the different species, but the Fungus never dominates the Alga so decidedly as in the heteromerous forms. The form and structure of the Alga is accordingly different in different cases, but the structure always varies less than the form.

Ephebella Hegetschweileri[1] in the fresh condition has quite the look of a Scytonema (Fig 167, *B*, *g*); but if the plant is heated in solution of potash the gelatinous sheath of the Scytonema-filaments is seen to be traversed by a compact weft of very delicate hyphae running chiefly in a longitudinal direction, out of which apothecia are sometimes, but rarely, developed.

The thallus of **Ephebe** (Fig. 177, 178), **Spilonema, Gonionema,** Nyl. and **Lichenosphaeria,** Bornet has the structure and the branched shrubby form of

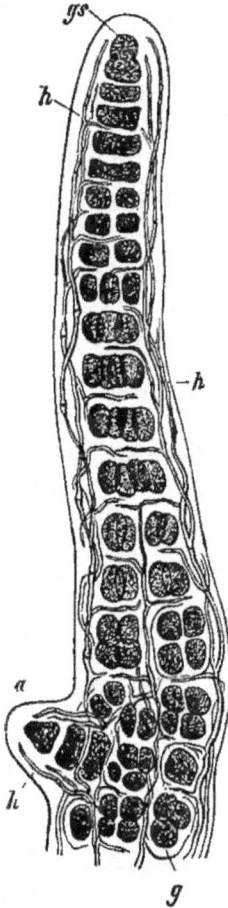

FIG. 177. *Ephebe pubescens,* Fr. A branched filiform thallus of *Stigonema* with the hyphae of the Fungus growing through its gelatinous membranes. Extremity of a branch of the thallus with a young lateral branch *a*; *h* hyphae, *g* cells of the Alga, *gs* the apex of the thallus After Sachs. Magn. 500 times.

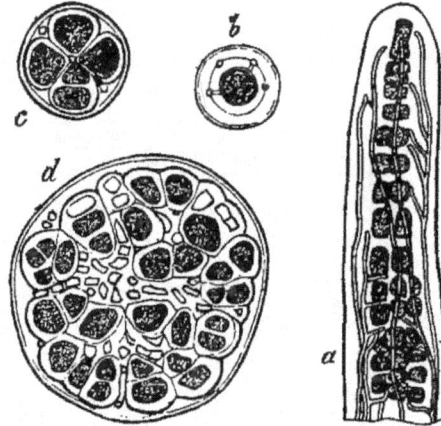

FIG. 178. *Ephebe pubescens,* Fr. A branched filiform thallus of *Stigonema* with hyphae of the Fungus growing through its gelatinous membranes. *a* tip of the thallus after being boiled in solution of potash. *b, c* transverse sections through the uppermost portion of a branch. *d* transverse section through the lower and older portion of a branch. The cells of the Alga are indicated by the dotted shading. Magn. 390 times.

Stigonema, Ag. (Sirosiphon, Kg.) with the addition of the hyphae which pass in the longitudinal direction through the outer wall of the Stigonema-filament and grow in length behind its advancing apex; but their branches are seen in older transverse sections of the Alga to intrude in numbers between its cells. Single branches of the Alga which is attacked by the Fungus may continue quite free from it.

The genus **Collema** and its nearest allies Synechoblastus, Leptogium, Mallotium, Obryzum, Plectopsora (Arnoldia), Lempholemma (Physma) and some others are

---

[1] Itzigsohn in Hedwigia, I, 123.

Nostoc-colonies attacked by the Lichen-forming Ascomycete. The hyphae of the Fungus intrude between the cell-rows of the Alga, which are strung together like the beads of a rosary interrupted by heterocysts and imbedded in a firm jelly; they branch in a manner which varies according to the species, and determine the general growth of the compound thallus.

In most forms numerous peripheral branches of the hyphae run vertically into the surface of the gelatinous colony and end blindly in it. Leptogium, Obryzum and Mallotium (Fig. 179) are exceptions to this, in which the extremities of the peripheral branches pass into a *rind* or outer membrane, a simple or in places a double layer of polyhedric tubular cells without interstices with pellucid contents and colourless or brown walls, which covers the whole thallus. The membrane of the cells is often stronger on the upper side of the thallus and thickened more on the outer than on the inner side, recalling the epidermis of the higher plants. Numerous multicellular hairs spring from all parts of the cells of the lower rind in Mallotium, sometimes short and isolated, sometimes longer and united in bundles; they serve to fix the thallus to the substratum.

No direct and intimate connection can be perceived in most species between the branches of the hyphae and the single algal cells. But the species of Plectopsora and

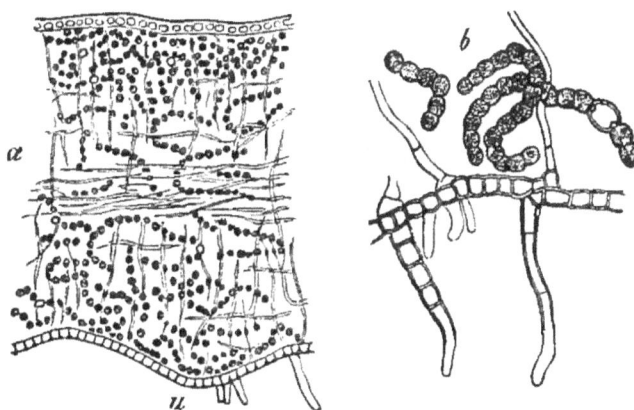

FIG. 179. *Mallotium Hildebrandii*, Garov. *a* radial longitudinal section through the thallus; *u* the under side. *b* portion of a very thin section through the under side showing the rind, hairs, hyphae and Nostoc-filaments. *a* magn. 190, *b* 390 times.

Physma are distinguished by the presence of haustoria, short hyphal branches which grow in the direction of any Nostoc-cells and either lay their conical extremity close upon them (Arnoldia minutula, Born.), or penetrate into them (species of Physma). The cells thus attacked become at first larger and fuller of protoplasm than those which are untouched; but according to Bornet they die prematurely (see Fig. 167 *C*).

Of these Nostoc-Lichens which have no rind the Gloeocapsa-lichens or Omphalarieae (Omphalaria, Synalissa, Thyrea, Paulia, Fée, Peccania, Mass., Enchylium and Phylliscum, Nyl.) are really distinguished only by the Algae, which belong to the genus Gloeocapsa and its allies. Their cells are round and multiply by division in three directions, and after division they separate from one another inside the broad gelatinous membranes, which are stratified as a result of their mode of formation and hold the cells together in one mass.

Of the branches of the hyphae some only thrust themselves into the gelatinous membranes, some attach themselves by their blunt extremity directly to the individual cells, so that most of the cells are borne on them as on a slender stalk. Each division of the algal cell is followed by a branching of the stalk, which thus supplies the new stalk that is wanted (Fig. 167 *D*, 180). Hence the cymose grouping of the stalked algal cells, which is a very striking feature in some species, and which

Thwaites was the first to describe[1]. Neither changes of structure nor premature death have been observed in the algal cells on account of this attachment to the hyphae. But the effect of the Fungus on the Algae is shown in a very remarkable manner by the formation of a thallus of a fixed shape, which in some species is of comparatively large size and has a progressive marginal growth, while the gelatinous colonies of Gloeocapsa which are not attacked by the Fungus have a very indefinite shape and are only loosely connected together.

6. We conclude this account of the Lichens with a special consideration of **those which are formed with Hymenomycetous Fungi**, a mode of proceeding which is not indeed consistent with the arrangement hitherto adopted but which will be found to conduce to perspicuity. The few species of the kind that are known belong to the tropical zone and are distributed among the genera **Cora**, Fr., **Rhipidonema**, Mattirolo, **Dictyonema**, Mont., and **Laudatea**, Johow. They grow on dead parts of plants and are attached to them by a copious growth of rhizoids or mycelium. Cora Pavonia,

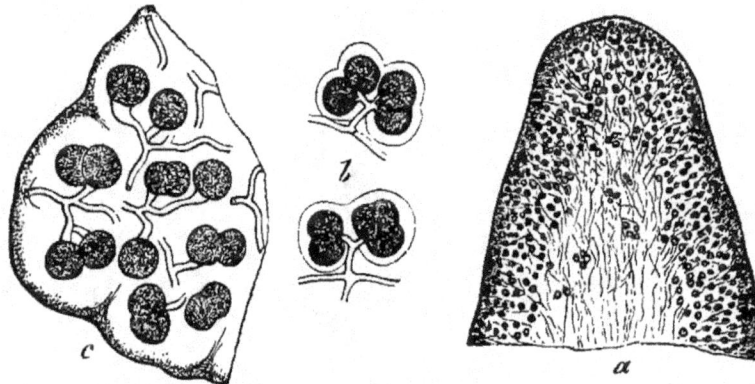

FIG. 180. *a, b Thyrea pulvinata*, Mass. *a* vertical longitudinal section through the margin of the thallus, *b* groups of Algae. *c Synalissa sp.* (*Plectopsora botryosa*, Jack, Leiner and Stitzenberger, Krypt. Bad. Nr. 301), portion of a thin transverse section through a small lobe of the thallus. The surface of the gelatinous substance which is violet-red in nature is shaded in the figure. Circumference of *a* magn. 90 times, the other parts diagrammatically represented, *b* magn. 390, *c* 720 times.

Fr. developes its thallus in the form of a flat semicircular fan resembling Stereum or Thelephora (see p. 53) with marginal progressive growth. The thallus consists of a loose air-containing weft of stout hyphae without evident differentiation into medullary tissue and rind; its middle layer contains an abundance of groups of Chroococcus-cells closely surrounded and embraced by branches of the hyphae.

The thallus of Dictyonema and Rhipidonema, of similar construction to that of Cora, is constructed out of filaments of Scytonema, which are attacked by the hyphae of the Fungus, as in Ephebe and Ephebella, but by greater numbers of them, and are thus taken up into a thallus-tissue formed of hyphae only. Johow's Laudatea spreads over the substratum like a crustaceous Lichen, being attached to it by a close mycelial weft, and winds its hyphae round Scytonema-filaments on the free surface in the manner of Ephebe.

Hymenomycetous sporiferous layers (see page 300) with numerous paraphyses and comparatively few four-spored basidia are formed on the under side of Cora and Dictyonema and on the free surface of Laudatea ; in Cora in the form of broadly

---

[1] Ann. and Mag. of Nat. Hist. ser. 2, vol. III (1849).

conical bodies like small flat apothecia of Peziza with the apex of the cone towards the thallus and produced by suitable branching of a tuft of hyphae; these bodies when they are present in large numbers unite together at the margins and form patches of some size; in Dictyonema and Laudatea they are smooth expanded layers usually resembling those of the Thelephoreae. The sporiferous structure of Rhipidonema is not clearly understood. Further details will be found in Mattirolo and Johow.

Section CXVII. The **thallus of very many Lichens forms small brood-buds** which separate spontaneously from the thallus and under favourable conditions may develope into a Lichen-thallus similar to the parent structure. These have been known since Acharius' time by the name of *soredia.* They appear to be wanting in some species, as in Lecidea geographica and Endocarpon pusillum; in others they are very abundant, as in many species of Usnea, Bryopogon, Ramalina, Evernia, Imbricaria, Parmelia, Pertusaria and others.

In their most general characters the soredia are small portions of the thallus which have a distinct form in each case and consist of one or more algal cells and the hyphae which have grown round them. As the soredium developes into a new thallus the two constituents behave in the manner which has been described above in the special types of thallus. Their structure at the moment of separation is in many cases that of the mother-thallus in a rudimentary form, as in Collema where they are roundish branchlets, prolifications of the upper surface or margin of

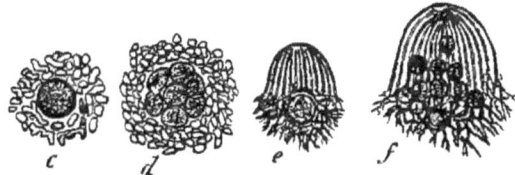

FIG. 181. *Usnea barbata.* c. an isolated mature soredium, with an algal cell (*Cystococcus*) in the envelope of hyphae. *d* another with several algal cells in optical longitudinal section. *e, f* two soredia in the act of germinating; the hyphal envelope has grown out below into rhizoid-branches, and above shows already the structure of the apex of the thallus (see Fig. 171). After Schwendener. Magn. more than 500 times.

the thallus appearing like granules to the naked eye, in the Graphideae also which contain Chroolepus, in Roccella and others. Detailed investigations in many species have not yet been made.

The soredia have been carefully studied by Tulasne and especially by Schwendener in the typical heteromerous forms which contain Palmellaceae, especially Cystococcus, and there their development and structure usually exhibit the following peculiar features (Fig. 181).

They are formed at spots in the algal zone beneath the rind which are distributed over the surface or margin of the thallus and vary according to the species. Branches of the hyphae twine round an adjacent algal cell or group of cells which have arisen by division and form a soredium; if a group of cells is thus attacked the hyphae also insinuate themselves between the cells and twine round each of them. The roundish body thus formed consists entirely of one or more algal cells and an envelope of hyphae which differs in compactness and colour in different species, and in some, as in Bryopogon, is not perfectly closed; how it is separated from the thallus has not been clearly ascertained.

By unlimited repetition of this process at one spot the soredia accumulate there beneath the rind, which swells up and at length bursts, the soredia emerging from the fissure as a powdery mass loosely held together by the hyphae of the thallus, which are drawn out with it. These powdery and easily dissipated accumulations of the separate bodies which have now been described are termed by Schwendener *soredia-heaps* or *sori*, and it will be well to retain his terminology, though the word soredium was originally applied to the whole heap.

A soredium can give rise to an unlimited number of new soredia in the manner just described after a sorus has emerged from the rind, and even after the soredia have separated from one another and been dispersed as dust. This uniform multiplication leads to the formation of layers and accumulations of soredia which are not unfrequent in shaded spots, as in the case of Physcia parietina. Where the conditions are favourable to vegetation a thallus is formed and fully developed from them (see Fig, 181, *e, f*). Such a thallus may even be frequently produced in Usnea according to Schwendener on the parent-thallus, forming *soredial branches* which remain firmly united to it.

Excessive development of roundish soredia-heaps with a remote resemblance to apothecia on a crustaceous or foliaceous thallus, on which apothecia are not at the same time produced, gives rise to the forms on which Acharius has founded his pseudo-genus **Variolaria.**

Pseudo-lichens. The above account necessarily excludes from the class of Lichens those forms which are ranked with them in our books because they have been collected by Lichenologists, but which do not possess the one mark of the true Lichen, namely, the presence of Algae in their thallus. Among these are the **Celidieae, Abrothallus** and others, which are parasitic on the thallus of Lichens, the genus **Myriangium** mentioned above on page 107, the **Arthopyrenieae** examined by Frank [1], Körber's **Naetrocymbe,** and **Atichia** [2]. All these forms except the last are simple Ascomycetes; Atichia, according to Millardet's able investigations, is a Fungus of uncertain position which forms clustered spore-groups. Those species also will of course be excluded which have been placed in the same genus with others having the true Lichen-thallus, and are really nearly related to them but have not the mode of life of a Lichen. Frank found that Arthonia vulgaris has the Lichen-thallus (see on page 399), but Arthonia epipasta, Kbr. has only hyphae in its thallus. Both species live on the bark of trees; the former is a Lichen-fungus with that habitat, the latter simply a saprophytic Fungus. There is nothing strange in this, for there are natural genera in other parts of the system, in which some of the species are usually parasitic, while others are adapted to a saprophytic mode of life, for example Sclerotinia. Norman's **Morioleae** (Moriola, Spheconisca) also appear to me, from the descriptions which I have read of them [3], to be to say the least doubtful Lichens; they deserve further investigation. If they really have Algae in their thallus, they are simply Lichens which have also the power of inclosing parts of Liverworts, pollen-grains and probably other strange bodies in their thallus.

Historical notice of the Lichens. It is well known that the views of botanists respecting the thallus of the Lichens up to the year 1868 were not those which have been given in the preceding pages. It was supposed to be a body whose constituent parts all proceeded from one another, and ultimately from the germinating spore; the algal cells especially, so far as any clear conception had been formed about them, were regarded as

---

[1] Biol. Verhältn. einiger Krustenflechten in Cohn's Beitr. z. Biol. II.
[2] See Millardet, as cited on page 262.
[3] See Botaniska Notiser, 1876, No. 6 a. Also Just's Jahresber. 1875, p. 105. I have not been able to consult the paper Allelotismus in K. norske Vid. Selsk. Skrifto. VII, 1872.

products of development of the hyphae. That conversely the hyphae also might proceed from the Algae, was a view which was at any rate often expressed, but which could not be maintained by those who looked fully into the facts on which it was supposed to rest. Wallroth expressed this view most decidedly, though he could not in his day draw from it the histological conclusions which have been stated above, when he named the algal cells *gonidia, brood-cells*, and expressly understood by this name asexual organs of reproduction produced from the thallus and capable of developing under favourable conditions into a new and perfect Lichen-thallus[1]. This view of Wallroth was in fact derived from the correct observation of the origin of the thallus from soredia, and from confounding small soredia of heteromerous species with the Algae of their thallus, and even with other free green algal cells, a mistake quite conceivable in his day. But after this confusion and the non-reproductive character of Wallroth's gonidia had long been recognised, the expression was still retained in an altered sense for the Algae of the Lichen-thallus, and with it the terms *gonidial layer* or *gonimic layer* (stratum gonimon), *hymenial gonidia*, and others of the same kind. The term *chromidia* proposed by Stitzenberger[2] met with little favour. I have avoided the word gonidia in the foregoing account, because it is a convenient expression in its original acceptation for the reproductive organs of the Fungi, and has been so used with all possible consistency in my former chapters, and to apply it in a different sense to the thallus of the Lichen-fungi would necessarily have caused confusion. At the same time it is unnecessary to introduce a new term, because the old word *Alga* declares the real nature of the objects briefly and distinctly, and satisfies all the requirements of the terminology.

After the discovery of the fact that the Algae of the thallus are not reproductive organs but assist in the vegetation of the plant by means of their chlorophyll, a long time elapsed before clear insight was obtained into their true character. They continued to be regarded as parts of a simple organism, the Lichen, and this view was confirmed by the observations first of Bayrhoffer and after him especially of Schwendener, which showed that they are often so placed at the extremities of branches of the hyphae in heteromerous species containing Cystococcus that they may very well be, supposed to be the swollen terminal cells of the hyphal branches forming chlorophyll. I myself adopted this view in the case of the heteromerous Lichens in my first edition in 1865. In the case of other species, especially the gelatinous Lichens containing Chroococcaceae and Nostocaceae, there appeared to me to be objections to it which could not be removed, and I was led to propose the following alternative: either the Lichens in question are the perfectly developed and fruitful states of plants, the imperfectly developed forms of which have hitherto been placed with the Algae under the names of Nostocaceae and Chroococcaceae, or the Nostocaceae and Chroococcaceae are typical Algae which assume the form of the Collemeae, Ephebeae, &c., when they are invaded by certain parasitic Ascomycetes which spread their mycelium in the growing thallus and often attach themselves to the algal cells containing phycochrome (Plectopsora, Omphalarieae). Then Schwendener not only adopted the latter alternative at the close of his work on the Lichen-thallus in 1868, but extended it to all Lichens and thus founded the view which has been set forth in the preceding paragraphs.

The arguments which he advanced in support of this doctrine on the subject of the Lichens are drawn from the *purely anatomical* data given above. They may be briefly stated in the following manner: All Lichen-gonidia, as they have been hitherto termed, are so like certain Algae that every one of them can be directly placed under some well-known genus or even species of Algae; the 'gonidia' have never been distinctly proved to have been formed from branches of the hyphae, but all observations tend to show that the Alga has been attacked by a Lichen-fungus. Schwendener did not go beyond this point even in his later and more elaborate work on the subject. Bornet's investi-

---

[1] Wallroth, Naturgeschichte d. Flechten, I (1825), especially at p. 46.
[2] Flora, 1860, p. 216.

gations which followed quickly on those of Schwendener were so extremely clear that they were calculated to raise the probability of the view and the conviction of its truth to the highest point, especially as Famintzin, Baranetzki and Woronin had previously isolated Cystococcus and Nostocaceae from Parmelieae, Cladonieae and Peltigera and cultivated them further as Algae with independent power of vegetation,—a mode of proceeding which was carried out by Bornet in some other species. Cystococcus was reproduced in abundance in these experiments from swarm-spores.

Still Schwendener's view continued to encounter a strong opposition. It is not so easy to shake off the yoke of the traditions in which men have been brought up. The same early instruction which led me to a guarded expression of my hypothesis of the Collemaceae and made Schwendener at first regard it as rash, had so fixed the convictions of many who had devoted themselves specially to lichenology, that they rejected sometimes with indignation a view which 'pitilessly robbed their favourite Lichens of their independent existence and turned them by the stroke of a magician's wand into a spider-like tyrant Fungus and a captured and enslaved Alga.' These words of Crombie briefly indicate the point of view of the 'lichenologists,' and a paper by the same writer, translated by von Krempelhuber in Flora, 1875, gives a condensed account of the line of argument pursued by the opposition, and to this paper the reader is referred.

The arguments advanced scarcely touched the heart of the question. Appeal was made to earlier statements respecting the origination of hyphae from 'gonidia' and gonidia from hyphae, which could not however stand the test of critical examination. Körber[1] himself, the most acute in the search for effective arguments, admits the decisive facts. He says that the hypha produced by the germination of the spore must meet with the 'gonidia' specifically belonging to it, that is, coming from the particular Lichen-species, if it is to give rise to a normal Lichen. But this hypha and whatever else there is in the Lichen-thallus except the 'gonidia' do not belong to a Fungus, but to the Lichen, and the 'gonidia' which are specifically necessary are no Algae, but free independent Lichen-gonidia which have become 'asynthetic.' It is therefore a simple case of change of name. That this is not the case as regards the Fungus-portion of the Lichen need not be restated here. The view that the lower Algae which vegetate where Lichens are found are Lichen-gonidia escaped from the thallus originated with Wallroth and was often expressed after him[2]. It was excusable in the year 1825; but our knowledge of the lower Algae had reached a point in 1874 which no longer permitted this summary mode of dealing with them. With the needful limitations they must be admitted as Algae in the sense which accords with our present views. The principal limitation is, that the gonidium is an Alga which enters into the Lichen-thallus and may in some cases leave it again, and this position has never been shaken.

The orthodox lichenologists had always one objection in reserve. They had the right to require and could not help requiring that it should be shown how a Lichen-thallus proceeds from the germ-tube of the spore and from the supposed Alga. Without complete proof that this takes place Schwendener's view was only a hypothesis, and the more surprising the hypothesis was the more reasonable was it to say, that perhaps there were things still unknown to us at the bottom of the whole matter, the discovery of which might settle the question in another and unexpected manner.

Schwendener's observation of the entrance of the hyphae of an unknown Fungus into the jelly of algal colonies, which appear as constituents of Collemaceae, could not be accepted as decisive, because it is not only possible but certain that this may happen with several Fungi which are not Lichen-Fungi. Bornet's observations and his and Treub's experiments in cultivation made some advance towards the decision of the question, but their results were still imperfect. It was not till it was understood that the

---

[1] Zur Abwehr d. Schwendener–Bornet'schen Flechtentheorie, Berlin, 1874.

[2] See E. Fries, Lichenogr. Europ. XX.—Kützing in Linnaea, 1833;—Id., Phycol. generalis, p. 167.
—Hicks in Quarterly Journ. Micr. Sc. VIII, p. 239, and new series, I, p. 157.

young plant grown from the spore must meet at the same time with the Alga and with suitable nutriment of other kinds that the experiments became decisive. Reess produced Collema from its component parts with the thallus fully formed, but time was not allowed for the development of the sporocarp. Then Stahl by judicious selection of the objects to be experimented on effected the synthesis of three species of Lichens; he gave a complete demonstration of the entrance of the Alga into the composite Lichen, the changes which it undergoes there and its possible ultimate liberation, and he succeeded especially in proving that the same Pleurococcus may appear as a constituent of more than one species of Lichen.

These results satisfied all demands and the Lichen-question was settled once for all against the old traditional view.

The discovery by Mattirolo that the thallus of Lichens may also be composed of Hymenomycetous Fungi was a welcome confirmation and illustration of the new view.

There have in fact never been any real difficulties in the way of accepting Schwendener's theory; the only obstacle to be encountered was the tradition which had been cherished for some centuries. It must be allowed that it is perhaps difficult to realise, how when Lichens are growing socially in large numbers each one of them is formed by synthesis of spore and free Alga. But in most cases it is not necessary to suppose that this is the case; the primary synthesis may take place only seldom; the large number of soredia produced will account for the most copious propagation of the species which can be imagined. On the other hand Stahl's observations on the hymenial Algae show that there are adaptations which almost ensure the synthesis every time that spores germinate. Considering the great variety of forms in the Lichens it is possible that many special adaptations may yet be discovered in the course of time. Frank's work cited above supplies evidence in support of this view. It is moreover quite conceivable that there are species of Algae which have become so adapted to lichenism, that they can no longer attain their full development outside the Lichen-combination, perhaps not even vegetate independently any longer[1]; examples of the kind are not at present known, but the possibility of their occurrence should be borne in mind.

Our historical sketch would not be complete without a record of the fact that since the year 1878[2] Dr. Arthur Minks has attempted to rescue the old tradition by the discovery of a new organ termed a microgonidium which takes part in the formation of a Lichen-thallus of quite different structure from anything described above. More than this short notice cannot be expected from the author of a serious botanical work.

LITERATURE OF THE LICHEN-THALLUS.

'Geschichte und Litteratur der Lichenologie von den ältesten Zeiten bis zum Schlusse des Jahres 1870,' by A. von Krempelhuber, a very remarkable work, gives in three volumes containing altogether more than 1650 large octavo pages an account of the subject indicated by the title-page, and a very careful notice of all the works connected with it. The Regensburg 'Flora' is a rich repertory of Lichenology since 1855. The reader is referred once for all to these two publications for all literary and historical details.

It is unnecessary to say anything more in this place about the various publications on the sporocarps of Lichens and matters connected with them; a sufficient account has already been given of them in Chapters III and V. It remains only to name the authorities from which our account of the Lichens has been derived, so far as they have

---

[1] See Frank, in Cohn's Beitr. z. Biol. d. Pfl. II.

[2] Flora, 1878, p. 209. Also A. Minks, das Microgonidium, ein Beitr. z. Kenntn. d. wahren Natur d. Flechten, 8°. Basel, 1879. See also Just's Jahresber. 1876 and succeeding years.

not been already mentioned in the text.   Further information with regard to the literature of the subject will be found in the works here cited.

G. F. W. MEYER, Entw., Metamorphose u. Fortpfl. d. Flechten, Göttingen, 1825.

E. FRIES, Lichenographia Europaea reformata, Lundae, 1831.   Introduction.

WALLROTH, Naturgeschichte d. Flechten, Frankfurt, 1825–1827.

KÖRBER, De gonidiis Lichenum (Diss. inaug. Berol. 1839) ;—Id., Ueber d. individuelle
    Fortpflanzung d. Flechten (Flora, 1841, Nos. 1, 2).

BAYRHOFFER, Einiges ü. d. Lichenen u. deren Befruchtung, Bern, 1851.

L. R. TULASNE, Mém. p. servir à l'hist. organogr. et physiol. d. Lichens (Ann. d. sc. nat.
    sér. 3, XVII), with 16 plates.

SPEERSCHNEIDER, Anatomie u. Entw. d. Hagenia ciliaris (Bot. Ztg. 1853, p. 705, 1854,
    p. 593) ;—Id., Anat. u. Entw. d. Usnea barbata dasypoga (Bot. Ztg. 1854, p. 193) ;—
    Id., Anat. u. Entw. d. Parmelia Acetabulum (Bot. Ztg. 1854, p. 481) ;—Id., Anat.
    u. Entw. d. Ramalina calicaris (Bot. Ztg. 1855, p. 345) ;—Id., Anat. u. Entw. d.
    Peltigera scutata (Bot. Ztg. 1857, p. 521).

NYLANDER, Syn. meth. Lichenum, I, Paris, 1858–60.

TH. M. FRIES, Genera heterolichenum recognita, Upsala, 1861.

S. SCHWENDENER, Ueber d. Bau u. d. Wachsthum d. Flechtenthallus (Vierteljahs-
    schrift d. naturf. Ges. Zürich, 1860) ;—Id., Unters. ü. d. Flechtenthallus in Nägeli,
    Beitr. z. wiss. Bot. Heft. 2, 3 u. 4, Leipzig, 1860–68;—Id., Ueber Ephebe (Flora,
    1863) ;—Id., Die Algentypen d. Flechtengonidien, Basel, 1869 ;—Id., Erörterungen
    z. Gonidienfrage (Flora, 1872);—Id., Die Flechten als Parasiten d. Algen (Verh. d.
    Baseler naturf. Ges.), 1873.   See also Bot. Ztg. 1868, p. 289, and 1870, p. 59.

S. BORNET, Recherches sur les gonidies d. Lichens (Ann. d. sc. nat. sér. 5, XVII,
    tt. 6–16, and XIX, No. 5).

BARANETZKI, Beitr. z. Kenntn. &c. d. Flechtengonidien (Mélanges Biol. Acad. Peters-
    bourg, VI, Dec. 1867, and Pringsheim's Jahrb. VII, 1868).

FAMINTZIN U. BARANETZKI, Zur Entwicklungsgeschichte d. Gonidien u. Zoosporenbil-
    dung d. Flechten &c. (Bot. Ztg. 1867, p. 189, and Mém. Acad. Petersbourg, sér.
    7, XI ; also Bot. Ztg. 1868).

M. WORONIN, Sur les gonidies du Parmelia pulverulenta (Ann. d. sc. nat. sér. 5, XVI,
    317).

M. REESS, Ueber d. Entstehung d. Flechte Collema glaucescens (Monatsber. d. Berlin.
    Acad. Oct. 1871) ;—Id., Ueber d. Natur. d. Flechten (Samml. wiss. Vorträge von
    Virchow u. v. Holtzendorff, Heft 320, 1879).

A. BORZI, Intorno agli officii d. Gonidii d. Licheni (Nuov. Giorn. Bot. Ital. VII, 1875).

M. TREUB, Lichenencultur (Bot. Ztg. 1873) ;—Id., Onderzoekingen over de Natuur der
    Lichenen (Diss. Leiden, 1873), with an accurate account of the history of the subject
    in addition to the author's investigations.

A. B. FRANK, Biol. Verhältn. einiger Krustenflechten in Cohn's Beitr. z. Biol. d.
    Pfl. II.

E. STAHL, Beitr. z. Entwickelungsgesch. d. Flechten, II, Leipzig, 1877.

O. MATTIROLO, Contribuzione allo studio del genere Cora (N. Giorn. Bot. Ital. XIII,
    1881).

F. JOHOW, Ueber westind. Hymenolichenen (Sitzgsber. d. Berlin. Acad. 21 Feb. 1884).

SECOND PART.

MYCETOZOA.

CHAP. VIII. MORPHOLOGY AND COURSE OF DEVELOPMENT.

THE name Mycetozoa is here applied to a group of Fungus-like organisms amounting at the present time to nearly three hundred species, the larger number of which are contained in the division **Myxomycetes** or Slime-Fungi (the Myxogasteres of Fries) together with the smaller one distinguished by Van Tieghem under the name of **Acrasieae.**

The resemblance of the Mycetozoa to the Fungi is due partly to their mode of life and nutrition, partly to the close agreement in structure and biological characters between their organs of reproduction and the spores of Fungi. A spore-terminology corresponding to that of the Fungi will therefore be applied to the present group. With this word of preliminary direction we now proceed to a more particular examination of the group.

MYXOMYCETES.

SECTION CXVIII. The **ripe spore** of the Myxomycetes is round and ellipsoidal with the structure of a simple Fungus-spore (see Figs. 182, 183, 193). An episporium colourless or coloured, smooth or marked by characteristic surface-sculpture and varying in thickness in each species, incloses a dense and homogeneous turbid protoplasm, which contains one, or sometimes in abnormally large specimens, two nuclei in the shape of round transparent bodies with a small central less transparent nucleolus. Other bodies of definite shape are sometimes but rarely inclosed in the protoplasm; these have not been very closely examined, but are usually spoken of as oil-drops or lumps of mucilage.

The spores are capable of germination as soon as they are ripe in most of the species in which this point has been examined; the moment of maturity will be described more exactly in section 120. It is only in the Cribrarieae and Tubulinae that all attempts to procure the germination of the spores have hitherto been without success. Germination takes place under conditions which will be more particularly related in a subsequent page; in most species when the spores are placed in water.

The germinating spore (Fig. 182) swells first of all by absorption of water, and one or two small vacuoles, which disappear and reappear alternately, are seen near the upper surface of the protoplasm in which rotating movements are often observed; at length, and usually 12–24 hours after the scattering of the spores, the membrane bursts and the protoplasm oozes or creeps slowly out of the opening. The protoplasmic body then either at once, as is the rule, or after a transitory period of rest, during which it assumes a spherical form, commences amoeboid movements, undulating changes of outline and protrusions and withdrawals of pointed processes, and in this way becomes elongated into a body which moves about in the water like a swarm-spore and is known by the name of *swarm-cell* (Fig. 182 *d–f*).

The swarm-cell has the same structure as the protoplasmic body before it emerged from the spore, only that the granules in the protoplasm are collected together in the larger part of the cell, which is the hinder part in the movement, while the anterior part is free from granules and also contains the nucleus. From one to three vacuoles lie in the posterior part, one of which at least is known as the contractile vesicle, because in about a minute's time it grows smaller and disappears, and then reappears and enlarges till it is one third or one half the breadth of the protoplasmic body. The granules or lumps of mucilage in the spore either continue in the swarm-cell, or they are dissolved before the protoplasm leaves the spore, or they are extruded and left behind within the membrane. The swarm-cell has no firm membrane, but careful observation shows that it is surrounded by a tolerably broad, pellucid and indistinctly defined envelope of the consistence of mucilage.

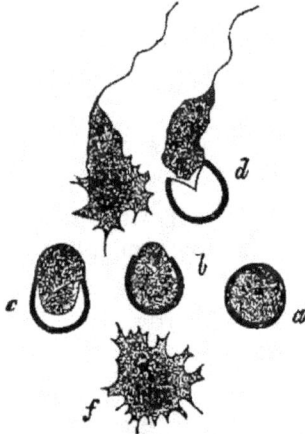

FIG. 182. *Trichia varia.* Spores in water. *a* before germination. *b—d* escape of the swarm-cell from the ruptured spore-membrane. *e* an older swarm-cell with cilia. *f* an amoeboid swarm-cell without cilia. The fine wart-like dots on the spore-membrane such as occur in Fig. 193 *b* are omitted in the drawing. Magn. 390 times.

The movements of the swarm-cells are of two kinds: a *hopping* and an amoeboid *creeping* movement.

In the hopping movement the cell floats freely about the water with its anterior extremity usually turned upwards. This extremity is finely pointed, the point being drawn out into a long cilium or flagellum with an undulating and swinging movement; in exceptional cases only there are two cilia. The posterior extremity is usually broad and rounded off, and the presence of a cilium there is quite abnormal. The body thus constituted rotates round its longitudinal axis in the circumference of a cone the apex of which is formed by the posterior extremity. The cilium swings with an undulating motion from side to side, making the swarm-cell move in a similar manner and advance in one direction; sometimes there is no rotation. The body at the same time exhibits constantly varying undulatory movements of its surface, with bending and contraction and recurrent expansion of the parts.

In the creeping movement the swarm-cell lies on the firm substratum, and either advances in one direction with a vermicular movement and with the cilium stretched out in front; or it assumes a roundish form and thrusts out processes, *pseudopodia*, in every direction and then draws them in again. The two kinds of movement, the hopping and the creeping, often pass into one another, and may frequently be observed alternating with one another in the same individual with apparently alternating retraction and protrusion of the cilium. Swarm-cells with purely amoeboid motion have been unnecessarily distinguished by the name of *myxamoebae.*

The swarm-cells multiply by bipartition, which, to judge by the vast numbers sometimes obtained from a sowing of the spores, may be repeated through several generations. The movement becomes more sluggish before division, the swarm-cell contracts into a spherical form and the cilium and the vacuoles disappear. This is followed by the appearance of an annular constriction in the middle which speedily becomes deeper and in a few minutes divides the body into two spherical halves which at once resume the characters of motile swarm-cells. The nucleus becomes indistinct during the division but does not entirely disappear, and it may be presumed from analogy that it also is divided. Exceptions to the rule here laid down have been observed in Chondrioderma difforme and Didymium praecox, in which the protoplasm was divided inside the spore-membrane and issued from it in the shape of two swarm-cells, about as often as it escaped from it in the form described above. Famintzin and Woronin too found

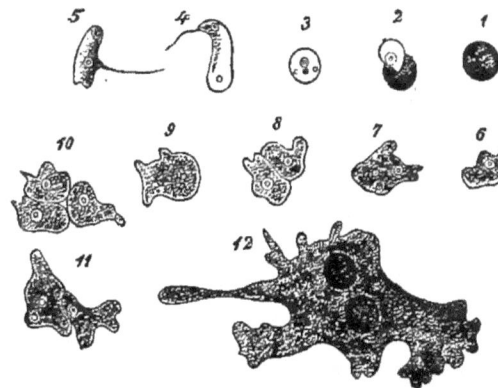

FIG. 183. *Chondrioderma difforme.* 1 ripe spore. 2 the same germinating. 3—5 swarm-cells. 6, 7 the same in the amoeboid state. 8 two amoeboid swarm-cells in close contact. 9 the same when coalesced to form the beginning of a plasmodium. 10 three swarm-cells in contact with one another. 11 two of them after coalescence, the third still free. 12 young plasmodium, which has taken up two spores into its substance. After Cienkowski from Sachs' Lehrbuch. Magn. 350 times.

that the protoplasmic body which came out of the spore in the Ceratieae divided by successive bipartitions into eight portions, which separated from one another as swarm-cells provided with cilia.

SECTION CXIX. The further development of the swarm-cells consists in their uniting together to form the large motile protoplasmic bodies, which Cienkowski named **plasmodia**. The course of events in this process was directly and completely observed under the microscope by Cienkowski in Didymium leucopus, Fr., Chondrioderma difforme (Didymium Libertianum) and Perichaena liceoides, Rost. (Licea pannorum, Cienk.). A number of less perfect observations of Lycogala, Fuligo and Stemonitis which have been communicated to me, and the resemblance to one another of all fully formed plasmodia, justify the assumption that the course of development is essentially alike in all Myxomycetes.

The phenomena directly observed in the development of plasmodia are of the following kind (Fig. 183). In a few days after the spores are sown the divisions become less frequent, most of the swarm-cells have the creeping form in which they are without cilia, and many have increased in size and contain single large strongly refringent granules. Then they approach close to one another in groups composed of two or more and again separate, till at length two or three are seen to come into close contact with one another and become fused together into a single body, the young plasmodium. This body exercises an attraction in a way not yet explained upon other swarm-cells of the same species, and they attach themselves one by one to its surface and coalesce with it. The newly formed plasmodium is distinguished by its greater size from the swarm cells without cilia, while it exhibits essentially the same movements and changes of shape.

The plasmodia when once formed by the coalescence of the swarm-cells begin to increase in size, and as they grow assume the form usually of branched strands, their dimensions far exceeding those of the swarm-cells (Fig. 183, 12, and Fig. 185). The latter is especially the case in some of the Physareae (Calcareae of Rostafinski); the stouter branches may in these species even exceed the thickness of a strong bristle, and the plasmodium in the form of a copiously branched reticulated or frill-like expansion covers surfaces varying in extent from an inch to a foot. This is the case in Fuligo varians (Aethalium septicum of authors), Leocarpus vernicosus, Didymium Serpula and D. praecox and Diachea elegans. Other species of Physareae as Chondrioderma difforme and Didymium leucopus, Fr. have usually much smaller plasmodia, just large enough to be clearly seen with the naked eye or invisible to it, and this is the case with all the other forms which have been hitherto examined not belonging to the Physareae.

The plasmodia of the Physareae if sufficiently supplied with water spread over the surface of the substratum, usually decaying parts of plants, in the form of veins and net-works of veins which have been well known since Micheli's time and especially from Fries' excellent descriptions, and from their mesentery-like appearance have been termed **Mesentericae.** As they spread also readily on the microscopic slide, their structure and conformation have been specially investigated (Figs. 184, 185). They are chiefly composed of a soft protoplasm of the consistence of cream, which may be readily spread out into a shapeless smear and is usually colourless but sometimes yellow, as in Fuligo, Leocarpus vernicosus, and Didymium Serpula, or a reddish yellow as in Physarum psittacinum. The microscope enables us to distinguish in them a colourless homogeneous slightly turbid fundamental hyaloplasm which usually appears distinct as a bounding layer of varying breadth, and a granular protoplasm which forms the chief mass of the plasmodium. Vacuoles are formed not unfrequently in the granular plasm and sometimes also in the hyaloplasm, some of which alternately disappear and reappear while others are more stable. The granules are more or less numerous in different species and individuals, but are always present in sufficient quantities to make the great mass of the plasmodium highly turbid, and in the stouter branches of some species (Fuligo) quite opaque. The composition of the granules is in some cases not precisely determinable; in others they consist of calcium carbonate. These granules of lime form the larger part of the granular mass in the Physareae, where they are spherical with dark contours, glistening, and of tolerably uniform size. Where the yellow colouring matter occurs it generally accompanies the granules of lime, partly appearing as a thin coating on the single granules, partly forming somewhat large roundish bodies with a delicate outline in which one or more lime granules are imbedded; it is soluble in alcohol at least in Fuligo and Didymium Serpula.

Nuclei were not at first observed in the plasmodia. Cienkowski even stated expressly that the nuclei present in the swarm-cells disappear when they coalesce. But Schmitz[1] and Strasburger[2] have recently established the presence of numerous nuclei in the plasmodium, and it may be presumed that they are the persistent nuclei of the swarm-cells and products of their division.

Besides the proper constituents of the plasmodium strange bodies of very various kinds are often found inside it, such as spores of Fungi and Myxomycetes (see Fig. 183, 12), parts of plants, &c. These objects are taken up from without into the interior of the growing and moving plasmodium, one may say engulfed by it, in a way which will be noticed again below, and they may be provisionally termed the *solid ingesta*.

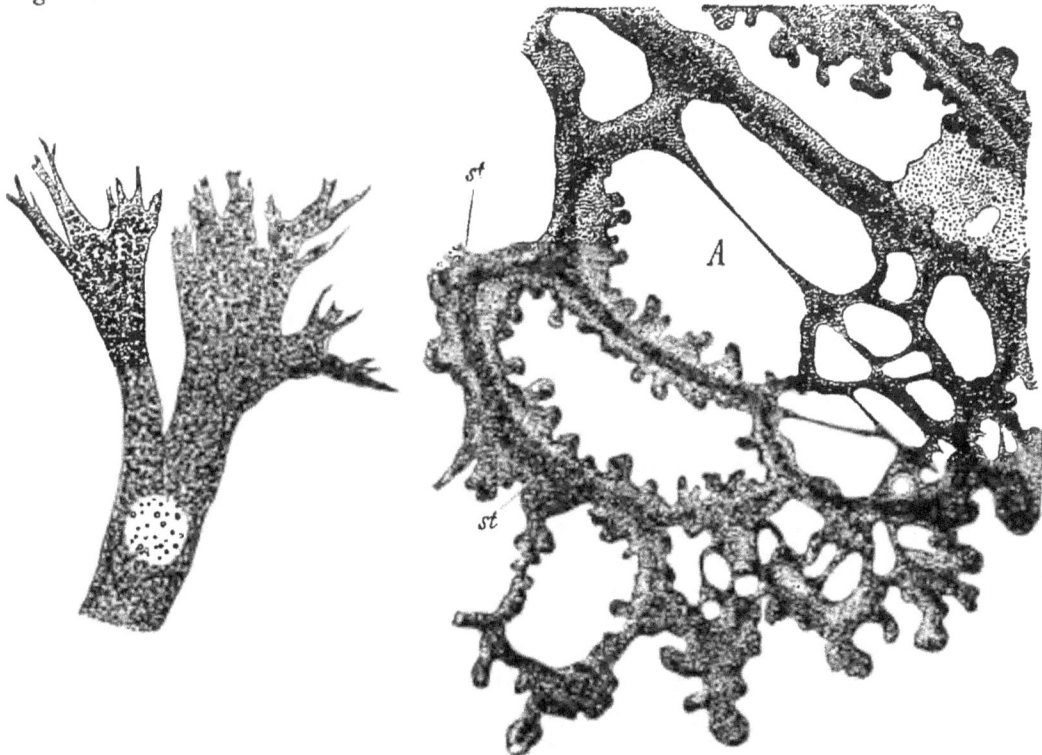

FIG. 184. *Chondrioderma difforme.* Extremities of branches of a plasmodium. Magn. 390 times.

FIG. 185. *Didymium leucopus.* Portion of the margin of a small reticulated plasmodium. After Cienkowski from Sachs' Lehrbuch. Magn. 100 times.

Reinke has recently published an elaborate investigation of the various substances which enter into the composition of the plasmodium of Fuligo[3], to which the reader is here referred.

The amoeboid movements of the swarm-cells are continued in the plasmodia. They may be seen in larger specimens by continuous observation even with the naked

---

[1] Sitzgsber. d. Niederrhein. Ges. 4 Aug. 1879.

[2] Zellbildung u. Zelltheilung, 3 Aufl. p. 79.

[3] Studien ü. d. Protoplasma von J. Reinke u. H. Rodewald in Untersuch. aus d. bot. Laborat. d. Univers. Göttingen, II, Berlin, 1881.

eye. The microscope reveals a constant change of outline in all the branches, sometimes in the form of a slight undulating movement, sometimes of an unceasing
protrusion and withdrawal of small pointed tentacle-like processes or pseudopodia.
Some of these pseudopodia or single flat projections of the main branches swell into
a knob at the free extremity and presently grow into larger branches, while in another
part branches diminish in size and gradually sink back into the main stem. Here
two branches grow out towards each other till they touch one another, and then
coalesce and anastomose; there a branch becomes constricted at some point and
divides into two. By these processes a plasmodium may separate into several parts,
and several plasmodia may be united into one; but according to Cienkowski's and
my own observations union never takes place between plasmodia of different species.
Branches of every degree and every size participate in the movements, and the smaller
the branch the more active is its movement. The alternation in the movements
takes place at all points alike in the plasmodium, but the protrusion of branches
predominates on one side, the retraction on the opposite. Hence there is often an
active advancing movement, a locomotion of the plasmodium in the direction of the
greater protrusion, and the anterior portion of the whole body which leads the way
in the advance assumes the form of a system of branches with swollen extremities,
which spread out like a fan and are connected together into a reticulated structure
by numerous and constantly changing anastomoses; in other words it becomes a
flat surface pierced so as to form a sieve or net and traversed by the stronger
branches like swollen veins, its margin being thickened and uneven (Fig. 185).

The inner substance of the plasmodium is also subject to a variety of active
movements and displacements which are seen to be directly connected with the
amoeboid movements, but sometimes appear to be independent of them. First there
are the previously mentioned swellings and sinkings of the marginal hyaloplasm, the
locality of which is constantly changing. Secondly stream-like displacements of the
inner granular plasm with change of speed and direction. The varying breadth of
the marginal hyaloplasm shows that there is a constantly varying pressure of the
granular mass in the direction of the periphery. The movement in the interior takes
place in the form of streams which occupy the whole breadth of a branch, or run in
narrow threads through the surrounding substance which is apparently motionless.
The movements are chiefly directed towards the swelling and advancing extremities
of the branches, into which the granular mass streams in, the alternating backward
flow being weaker and less copious. The reverse takes place in branches which are
being withdrawn. But movement and streaming may be constantly alternating with
rest in the interior of the plasmodium without this prevailing and directly perceptible
connection with the amoeboid change of shape.

Further details with respect to these phenomena must be sought in monographs
and in the physiological treatises which have appeared since my first work on this
subject. See also below in section CXXVII.

The surface of the plasmodia of the Physareae which I have examined is
covered with a soft shiny *envelope*, which is not distinctly defined on the outside but
is quite distinguishable from the marginal layer. It forms a border round the thicker
branches which is often more than 0,01 mm. in thickness, and is in itself colourless
and pellucid but is often covered with small particles of soil which adhere to it.

It consists of a sticky substance which swells in water and contracts in alcohol and is scarcely coloured by iodine, and must therefore be different from protoplasm. It passively follows the movements of the plasmodium. Portions of it often remain adhering in the form of thin films of mucilage to spots from which a plasmodium has moved away. This envelope is often very thin round the rapidly swelling extremities of the branches, and cannot be distinguished at all round slender pseudopodia, having been either pierced through by them or extenuated by their advance till it is no longer perceptible.

The plasmodia of the Stemoniteae, Trichiaceae, Ceratieae and Lycogala have in the main the same structure and power of movement as those of the Physareae, but they never have the granules of calcium carbonate and therefore usually appear to be much more finely granular than in the Physareae. The dark-blue or violet-brown plasmodia of the Cribrariae and Dictydium contain large brown granules of some organic substance, but have as yet been very insufficiently examined. The plasmodia of Lycogala which live in rotten wood are surrounded by a thick colourless membrane; I observed a similar membrane some time ago in Arcyria punicea. It is not yet ascertained how this membrane behaves in the movements. I was unable to see it in former years in specimens of Lycogala grown in water. I found the plasmodia of Stemonites fusca, when it issues from the substratum to form its sporangia, surrounded by a stout envelope, the inner and thicker layer of which is coloured a dark blue by iodine while an outer thin layer remains colourless. All the plasmodia last mentioned are inconspicuous bodies, the stouter branches of which in Arcyria punicea are not more than 16 μ in thickness, in Lycogala not more than 24 μ. They live for the most part in the interior of rotten parts of plants, especially rotten wood, and are not visible to the naked eye till they come to the surface to form sporangia.

SECTION CXX. **Transitory resting-states.** Those stages of the development in Myxomycetes which have the power of motion are able to pass into resting-states and to return again under favourable conditions to the state of movement. Three resting-forms are at present known: *microcysts, thick-walled cysts* (Cienkowski) and *sclerotia*.

It appears from cultures of Chondrioderma difforme that these transitory resting-states are not necessary members of the course of development. Their formation would seem to take place only when the development of the swarm-cells into plasmodia or of the plasmodia into sporangia is interrupted by insufficiency of food, by slow desiccation, or by slow cooling to below a certain minimum. But there are a number of observations which also point to other at present unknown causes. The state of movement is restored when the bodies after desiccation are again placed in water of the proper temperature.

The term **microcyst** was given by Cienkowski to the resting-state of the swarm-cells. Under the conditions above mentioned these cells assume the form of spheres which are smaller than the spores and are surrounded by a very delicate colourless membrane, as in Perichaena liceoides according to Cienkowski, or are without a membrane but provided with a very firm marginal layer. In other respects their structure remains the same as that of motile swarm-cells, only that the vacuoles in many cases disappear and the protoplasm becomes more dense. The swarm-cells of Didymium praecox and D. difforme encysted in this way and perfectly dry retain their vitality for more than two months; how soon life becomes extinct in them has not been ascertained. If placed again in water, they return to the motile swarming state, and the more quickly the shorter the period of desiccation. Those of Perichaena liceoides cast off their outer membrane under these conditions.

The **thick-walled cysts** and **sclerotia** are resting-states of plasmodia. The former were examined by myself in isolated cases in Fuligo, and Cienkowski followed them through the whole course of their development in Perichaena liceoides. The cysts were formed by young plasmodia in both species. According to Cienkowski's observations the plasmodium divides by the rending of its branches into pieces of very unequal size, which draw in their processes and assume the shape of smooth spheres. Then a membrane of considerable thickness forms on the surface and becomes rough and wrinkled and assumes a dark-brown colour. Within this membrane the protoplasm contracts still more and forms on its surface a second coat with a double-contour. If placed in water after drying for several weeks the round bodies remain first of all unchanged for some weeks, and then the protoplasm begins to display slow undulating movements and at length swells up, makes a hole in the surrounding coats, and slowly emerges from them with all the characters of a plasmodium.

The sclerotia are the resting-states of full-grown plasmodia. They have been observed in Didymium leucopus and D. difforme, D. Serpula, Fuligo, Physarum sinuosum, Perichaena liceoides and in a number of Physareae which have not been more precisely determined, and perhaps also by Corda in Stemonitis[1]. Some of them are the forms on which Persoon based his fungal genus Phlebomorpha.

When their formation begins the plasmodium draws in its slenderer processes and assumes the shape of a sieve-like plate, or in Fuligo of a small tuber a few millimetres in diameter and with irregular prominences on its surface; the granules become distributed uniformly through the fundamental substance, the solid ingesta are extruded, the movement gradually ceases, and the whole body breaks up into an innumerable quantity of roundish or polyhedric cells with an average diameter of from 25 to 40 μ. The body becomes at the same time of a wax-like consistence and dries into a brittle horny mass, resembling the sclerotia of many Fungi.

Each cell chiefly consists of a firm protoplasm which incloses vacuoles varying in size and number, and a pigment and granules distributed in the same manner as in the motile plasmodia, and shows a sharply limited marginal layer. Nuclei are there no doubt, though they have not as yet been observed. In the strongly developed sclerotia of some species (Fuligo, Didymium Serpula) the protoplasm is surrounded by a distinct colourless membrane, which in both the species mentioned shows the reaction of cellulose with iodine and sulphuric acid or with Schulze's solution. The membranes are firmly attached to one another, either immediately or, as in Fuligo, by a homogenous intermediate substance which softens in water. In small weakly developed specimens of the above species and in all sclerotia that have as yet been examined in other species, as for instance in Didymium difforme, no distinct membranes can be seen round the protoplasm.

---

[1] Icon. Fung. II, Fig. 87 b.

The outer surface of the sclerotia is usually covered by a layer of the same homogeneous substance with a capacity for swelling which is found between the cells in Aethalium. Upon it there are also in many cases (Fuligo, Didymium) scales or grains or crystals of calcium carbonate which must have been excreted during the formation of the sclerotia.

If a mature and dry sclerotium is placed in water it at once swells up, and its cells coalesce once more into a motile plasmodium often in from six to fifteen hours, in older specimens after a longer interval which may last some days. Where membranes of cellulose are present they are first dissolved. The process begins at the surface and advances towards the centre.

If single cells of a sclerotium are watched, contractile vacuoles are seen to form in them a few hours after they are moistened, and protrusion of motile branches and pseudopodia and the creeping forward movement all begin as in plasmodia. Where moving cells meet and touch they coalesce; if moving cells encounter cells that are still at rest, they absorb them. In this way a large plasmodium is gradually formed containing many sclerotium-cells which it has engulphed. These phenomena, which were first observed by Cienkowski in Didymium difforme, explain the formation of the plasmodium from the compact sclerotium. In plasmodia recently formed from sclerotia in which the cells have not separated from one another we always see a number of unaltered or evidently dead sclerotium-cells carried along by the stream of granules; these become presently less frequent and at length entirely disappear; they must therefore be either dissolved or they coalesce with the substance of the plasmodium.

Sclerotia are known to retain their vitality in a dry state for 6–8 months. Fuligo and Didymium Serpula are known from several direct observations to persist during the cold and dry season of the year in the condition of a sclerotium, and pass again into the motile state with damper and warmer weather. Vitality did not last more than 7–8 months in most of the observed cases, though sclerotia of Didymium Serpula lived more than a year (others only 7 months), and Léveillé[1] quotes an observation to the effect that a sclerotium of a Myxomycete had been known to return to the motile condition after having been kept for 20 years.

SECTION CXXI. **Development of sporophores and sporangia.** The development of the plasmodia closes with the formation of spores within receptacles, *sporangia*, or on the outside of *sporophores*. The latter are confined to the Ceratieae, the former are common to the rest of the Myxomycetes. We may therefore with Rostafinski distinguish the Ceratieae as *exosporous*, all other Myxomycetes being *endosporous*.

The sporangia of the **endosporous forms** are vesicles, which are usually about 1 mm. in height but may considerably exceed that size, and rise with or without a stalk above the substratum or lie upon it in the form of round or flat tubes. Their structure when they are fully formed will be more minutely described in section CXXII. Their development from the plasmodium is divided into the successive sections of *forming, development of wall, separation of the spore-plasm,* and lastly

---

[1] Ann. d. sc. nat. sér. 2, XX, p. 216.

formation of the spores and of the capillitium which in many genera accompanies the spores. These processes also have been most thoroughly and satisfactorily examined in the lime-containing Physareae, Rostafinski's Calcareae, and will be described from them in this place.

In the simpler cases of *forming of sporangia*, which we may examine first, either an entire plasmodium spread out on its substratum becomes transformed into a sporangium, or it divides into a number, often a large number, of pieces, each of which suffers transformation. This transformation is the result of the same kind of amoeboid movement of the protoplasm as that which causes the change of shape and place in the vegetating plasmodium, with the difference only that in the forming of sporangia the masses of protoplasm by drawing in their branches become constantly broader and rounded off and at length assume stable forms. The sporangia which lie flat on their substrata are in conformation simply portions of a plasmodium which have been thus contracted and thickened. Erect sporangia on a narrow or stalk-like base begin as node-like swellings on a branch of the plasmodium, and gradually rise to their ultimate form as the surrounding protoplasm flows into them and assumes an upward direction.

The soft envelope which surrounded the retracted branches of the plasmodium remains sticking to the substratum and there dries up; but there appears in its place round the young sporangium a firm membrane evidently produced by its further growth, and this membrane constitutes the wall of the sporangium which often grows to a considerable thickness. In the stalked forms the sporangium-wall begins to strengthen and get firm at the base of the stalk, and the process advances upwards. The zones of the membranes successively growing stronger serve as a support for the upward moving protoplasm.

During the forming of the sporangia the solid ingesta which are present in the plasmodium begin to be expelled from it. When a sporangium has assumed its definitive form the *spore-plasm* is separated off inside it. The granules of calcium carbonate, the pigment and perhaps other substances also, which at present are not precisely determined, are eliminated from the true protoplasm. In Physarum and its nearest allies they move to the wall and become imbedded in it or attached to it; or they collect into lumps of various shapes which are arranged differently inside the sporangium in different species, and by the formation of a membrane round them are soon converted into vesicles containing calcium carbonate or pigment (Fig. 191). In Didymium and the forms nearest to it the calcium carbonate granules in the plasmodium are dissolved and in this state are expelled from the sporangium, and while the granules disappear inside, the outer surface becomes covered with crystals of the salt. In Didymium Serpula the only well-known species of this group with coloured plasmodia, the yellow colouring matter is gathered up at the same time inside the sporangium into round lumps, which are then inclosed in pigment-vesicles by the formation of membranes round them. The protoplasm which remains after these substances have been removed from it is for the most part a colourless body with small granules uniformly distributed through it. By staining numerous nuclei are made visible in it. This plasm has been termed spore-plasm, because much the larger part of it is used for forming spores; as soon as it has got rid of the foreign substances in it its nuclei rapidly increase in number, and at length the whole mass divides simultaneously into polyhedral portions with rounded corners, each of which encloses a nucleus, and becoming invested with a firm membrane ripens into a spore. The spores when first formed are somewhat larger, never smaller, than when they are mature.

A comparatively small portion of the spore-plasm is expended in forming the *capillitium*, a filamentous structure to be described more fully in section CXXII, which spreads through the cavity of the sporangium and is connected both with the wall of the sporangium and with the vesicles containing pigment and calcium carbonate. The capillitium always appears before the spores, and all the parts in the earliest state in which they have been examined are seen to be formed and arranged exactly as when they are mature, only they are at first extremely delicate and gradually acquire their subsequent firmness. These points will be noticed again below.

Fuligo, the ' flower of tan,' agrees with Physarum in all other points, but shows a variation in the phenomena described above in the forming of the sporangia ; here a number of plasmodia collect together from every side and become fused together into a narrow dense reticulum with the shape of a cushion, which may come to be twelve inches in breadth and one in thickness or may remain quite small. The strands of this reticulum anastomose in every direction with one another. Then all the spore-plasma passes after its separation into the inner portion of the reticulum which swells proportionately, and then acquires the structure of the sporangia of Physarum ; the vesicles with calcium and colouring matter remain behind in the peripheral layers, where they collapse and dry up forming a calcareous crust or rind from one to several millimetres in thickness.

Of the rest of the endosporous Myxomycetes, those, that is to say, which do not belong to the Physareae, it may be confidently affirmed that the development of the sporangia from the beginning of forming is thoroughly like that which has just been described. The development of the spores is in all cases the same. Strasburger succeeded in following the multiplication of the nuclei, which precedes the formation of the spores, in Trichia fallax through all stages of the division, and ascertained their morphological agreement with the divisions of the nucleus in many other vegetable and animal cells. The case is the same also with the formation of the capillitium, with some limitations naturally arising from generic differences. The elimination of contained substances which leads to the formation of the spore-plasm is less copious or there is none at all to be seen, as is the case in Stemonitis in all states which have at present been examined, because the substances secreted in the one case are wanting from the first in the other. The very first stages in the forming of sporangia are in most instances imperfectly known because of the difficulty of procuring plasmodia in the vegetating state. What is known of it in most genera with simple sporangia, as Trichia, Arcyria, Dictydium, &c., corresponds with the foregoing description of the Physareae. On the other hand the large and often thick-walled spore-receptacles of Lycogala, Reticularia, Lindbladia and others are formed by such an accumulation and intermingling of numerous plasmodia as has been described in the case of Fuligo. Rostafinski applies the collective name *aethalium* to all bodies which originate in these combinations of plasmodia. Little more is known of the processes of differentiation in their development than can be concluded from the mature state.

The conformation of the sporangia in Stemonitis runs differently in one respect from that of all other known forms. The slender threads of the plasmodium, which lives in rotten wood, unite at first into large cylindrical or ellipsoid bodies of homogeneous protoplasm, which rest their broad surface on the substratum. Then a hollow cylindrical firm central column is separated off in the protoplasmic body, and rises vertically from a membranous base resting on the substratum, advancing by acropetal growth (Fig. 186 *a, b*). The mass of protoplasm, the longitudinal axis of which is traversed by the column, stretches at first at the same time in the same direction; but it afterwards loses hold of the substratum at its base and clinging to the central column moves on it a certain distance upwards till at

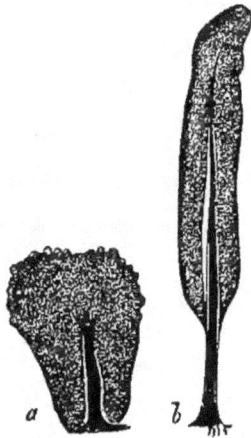

FIG. 186. *Stemonitis ferruginea. a* a commencing sporangium with the first beginning of the central column. *b* sporangium which has reached its mature form, capillitium and spores not yet formed: Both figures represent specimens in optical longitudinal section hardened in alcohol and then rendered transparent in glycerine. *a* magn.12 times, *b* 15 times.

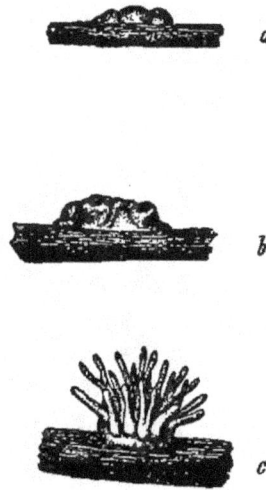

FIG. 187. *Ceratium hydnoïdes.* Forming of sporophores on plasmodia which have come to the surface of a piece of wood. Successive stages of development according to the letters *a—c; c* the mature state. After Famintzio and Woronin, about 3 times the natural size.

length it becomes stationary and developes into a sporangium in the usual manner (Fig. 186 *b*). The sporangium is supported on the lower portion of the central column which is laid bare by the upward movement of the protoplasm, as on a stalk. Other genera allied to Stemonitis behave in a similar manner. Further details must be sought in the different monographs.

The plasmodia of the **ectosporous Ceratium hydnoïdes** come forth to the surface from the interior of the rotten wood which is their habitation to form their spores (Fig. 187). Here they appear at first to the naked eye as white cushion-shaped bodies (Fig. 187 *a*); examination with the microscope shows that these cushions are formed of countless microscopically slender plasmodium-branches, united together in every direction into a net-work of narrow meshes, such as is shown in Fig. 188 *a*. The meshes are filled with a hyaline homogeneous gelatinous substance of watery consistence, which forms a thin coating on the surface of the net-work. To form the sporangia cylindrical often dichotomous outgrowths, resembling the spikes of a Hydnum and growing to be a few millimetres in length, rise erect from the surface of the cushion (Fig. 187 *b, c*). The whole body of protoplasm moves into these outgrowths, leaving only a thin flat layer to connect it with the substratum. During these movements the branches of the plasmodia all change their previous distribution through the whole of the thickness of the body and creep to the periphery, where they form a net-work stretching only in the direction of the surface and formed of threads which become successively broader and meshes growing narrower and narrower (Fig. 188 a). The whole is covered on the outside by a thin layer of the hyaline protoplasm which alone forms the whole of the inner portion of the body. When these processes are completed, the net-work of protoplasm breaks up simultaneously into numerous polyhedral portions of nearly uniform size (Fig. 188 b). They contain each a nucleus, become flattened from without inwards, and remain grouped in a simple layer which follows the surface like an epithelium. Then each of these proto-plasmatic bodies begins at once to grow convex towards the outside, and to lengthen out at right angles to the surface of the whole body into the shape of a sphere borne on a slender conical stalk (Fig. 188 b). A delicate membrane is formed at the same time within which all the protoplasm passes through the stalk into the spherical expansion at the extremity. The latter then becomes invested all round with a some-what thicker membrane with an ellipsoid outline, and thus becomes a mature spore which is readily detached from the empty hyaline stalk. The entire gelatinous sporophore undergoes no further changes, but in most cases soon dissolves and disappears.

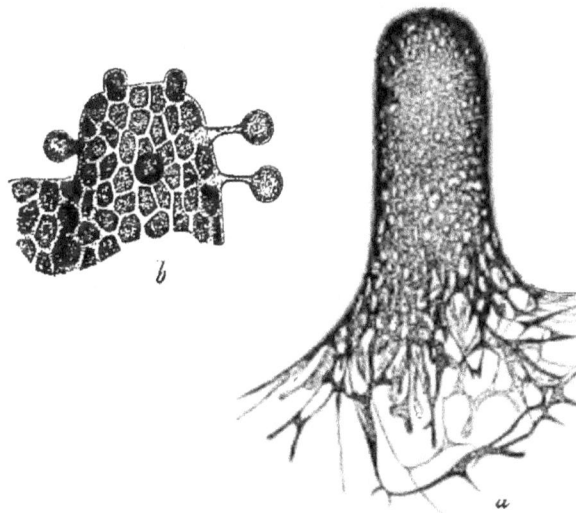

FIG. 188. a *Ceratium hydnoides* Piece of a sporophore in the act of forming : the branches of the plasmodium in the upper part are already beginning to be arranged into a dense peripheral net-work. b *Ceratium porioides*. Piece of the margin of a sporophore, in the beginning of spore-formation ; to the right two spores, which subsequently become slightly ellipsoid, on their stalks. After Famintzin and Woronin. a magn. about 68 times, b 120 times.

The other known ecto-sporous Myxomycete, Ceratium porioides, is distinguished from the species just described only by the yellow colour and by the shape of its sporophore, which resembles in form the hymenium of a Polyporus (see page 288).

The development of the sporophores and receptacles as just described runs its course rapidly if the conditions are favourable. According to a series of observations on spontaneously developed Physareae, species of Trichia, Stemonitis and others, the complete development from the commencement of the formation to maturity requires an interval on the average of about 12 hours ; the development is quicker or slower in particular species, or according to the temperature and moisture of the environment.

The sporophores of Ceratium are perfectly formed according to Famintzin and Woronin in the course of a summer night. The entire development of the species may also be accomplished very rapidly. Cienkowski obtained in four days fully formed plasmodia of Chondrioderma difforme when grown on microscopic slides, and these formed sporangia on the fifth day.

Delay may of course occur where the conditions are unfavourable. That the several species behave very differently in these respects is shown by the fact, that many of them, Trichia rubiformis, T. clavata and T. varia for example, are observed to form their sporangia almost entirely during a short portion of the yearly period of vegetation. The biological relations of most of the species require further examination.

SECTION CXXII. The **structure of the mature sporophores** is n all cases essentially the same as in the Ceratieae. The **ripe sporangia** in the majority of the endosporous genera, which show a great amount of variation in different species, must be described from a few of the typical forms, which have been known for some time. For special peculiarities the reader is referred to monographs and especially to that of Rostafinski.

We must first distinguish between the *simple sporangium*, which proceeds from *one plasmodium* or from a part of one plasmodium, and the *aethalium*, as Rostafinski understands that term, which is formed from large *combinations of plasmodia*.

1. It has been already said that the **mature sporangia** in most Myxomycetes are round or elongated, stalked or sessile vesicles one to a few millimeters high; less frequently, as in Didymium serpula, Trichia serpula and Licea flexuosa, P., they are cylindrical or flattened tubes forming a network and lying on the substratum.

FIG. 189. *Didymium squamulosum,* A.S. (*D. leucopus,* Fr.) A ripe sporangium divided longitudinally near the middle with the spores removed. Magn. about 25 times.

The wall of the sporangium is formed of a membrane which in constitution resembles the cellulose-membranes of plants. It is either a structureless hyaline and sometimes, as in Diachea and some species of Physarum, an extremely delicate membrane, or it is thick and firm and evidently stratified, as in Leocarpus vernicosus, Craterium, Trichia varia and others; in the Physareae included in the old genus Diderma it is even double, that is, it is differentiated into two layers which may be easily separated from one another and which do often separate of themselves. Projecting thickenings of different dimensions in the shape of warts and ridges occur in some cases, on the whole of the surface for example of the thick olive-brown outer layer of Licea flexuosa, and on the inner surface of the base of the sporangium of Arcyria incarnata, A. punicea and A. nutans. The whole of the inner surface of the membrane in Cribraria and Dictydium shows projecting thickenings in the form of flattened ridges connected together into a delicate net-work. The membrane is colourless or coloured in various shades of violet, brown, red and yellow according to the genus and species; it is usually continued at the point of attachment of the sporangium into an irregular membranous expansion formed of the dried envelopes of the plasmodium, and securing the sporangium to the substratum.

The stalks, except in the Stemoniteae, are tubes usually with a thick wall, which is wrinkled and folded lengthwise and is continued above into the wall of the sporangium. Its lumen is either in open communication with that of the sporangium, as in Trichia and Arcyria, or is separated from it by a transverse wall; if the septum is convex upwards it is termed the *columella* or central column (Fig. 189). The cavity of the stalk varies in breadth in different species and contains nothing but air, as in Physarum hyalinum, or is filled in the manner which will be described below.

The structure of the membrane of the sporangium in most of the species which contain no calcium, Licea, Perichaena, Cribraria, Arcyria, Trichia, &c., is usually such as has now been described. In some of these it also contains coloured granules of an organic substance, the nature and origin of which have yet to be ascertained; these granules are imbedded in the stronger ridges of thickening matter in Cribraria and Dictydium; in Perichaena liceoides they lie singly or in groups on the outer surface. The olive-brown outer layer of Licea flexuosa shows an irregularly granular structure throughout the entire thickness.

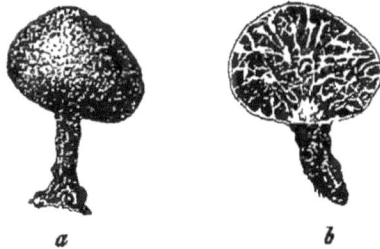

FIG. 190. *Physarum leucophaeum*, Fr. *a* sporangium seen from without. *b* sporangium divided in half and the frame-work of the capillitium exposed by removal of the spores. Magn. 25 times.

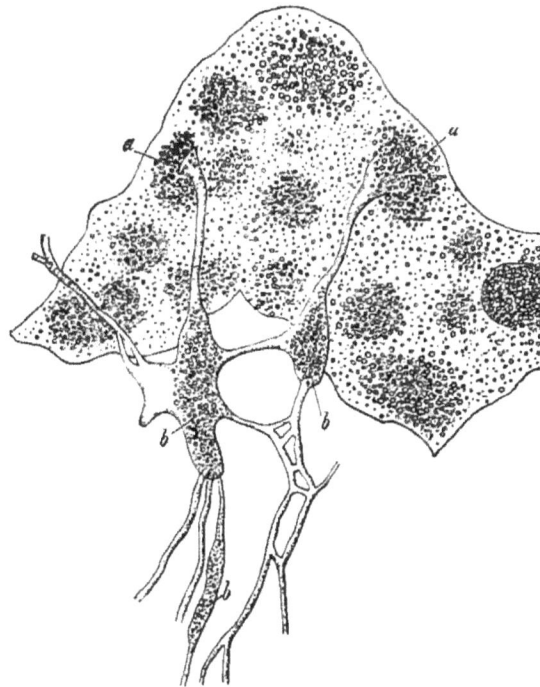

FIG. 191. *Physarum leucophaeum*, Fr. Piece of the wall of a sporangium with tubes of the capillitium attached and spread out in water; *a* points of attachment of the tubes of the capillitium; *b* calcium carbonate-vesicles; to the right oo the margin a calcium carbonate-vesicle on the wall. The rest of the wall is covered with calcium carbonate-granules singly or in groups. Magn. 390 times.

On the other hand the wall of the sporangium in most of the Physareae and their allies (Calcareae of Rostafinski) is incrusted, wholly or in part according to the genus and species, with the calcium carbonate eliminated in the separation of the spore-plasm. In a number of genera, of which Physarum may be taken as a representative (Figs. 190, 191), the calcium carbonate appears in the form of small round granules which either lie isolated and more or less deeply sunk in the wall, or form dense irregular accumulations on its inner surface. In many species the granules of calcium carbonate, especially those that are collected together in heaps, are surrounded by the extruded colouring matter mentioned above, which is yellow in Physarum aureum, P., Ph. sulphureum, A.S. and other species, or more rarely reddish-yellow as in Ph.

psittacinum, Ditm. The heaps of granules appear in this case to the naked eye as small coloured patches or warts on the dry sporangium; where there is no pigment they are white.

Didymium (Fig. 189) is distinguished by a crystalline covering of calcium carbonate like hoar-frost formed of stellate glands and small single crystals on the outer surface of the sporangia. The Diderma-forms mentioned above, which partly approach Physarum and partly Didymium, have the wall of the sporangium differentiated into a delicate inner layer which is free from or contains very little calcium carbonate, and an outer layer, a brittle incrustation of lime, consisting of round or crystalline fragments of calcium carbonate closely crowded and held together by a small quantity of organic matter, which when the calcium carbonate is dissolved remain behind as a delicate membrane.

Calcium carbonate eliminated from the plasmodium in a granular or crystalline form is imbedded in unusually large quantities in the basal wall of sessile sporangia and in the wall of the stalk in stipitate sporangia in many of the Calcareae. In the latter case a considerable amount of the salt often occurs inside the stalk and columella, where it is not unfrequently associated with irregular lumps of some organic substance and to a great extent fills up the cavity as in Didymium leucopus, Fr. (Fig. 189) and Diachea.

The cavity of the sporangium is either filled exclusively with the numerous spores, as in Licea and Cribraria, or, as in most genera, tubes or threads of different forms occur between the spores and constitute the *capillitium*. The capillitium of Physarum and its nearest allies (Figs. 190, 191) consists of somewhat thin-walled non-septate tubes which spread their branches in every direction and combine them into a network. Many branches run from the periphery of the net-work to the wall, and are firmly attached to it by their usually funnel-shaped extremities. The tubes are swollen and inflated at the nodes of the net-work, forming the calcium carbonate-vesicles mentioned above and sometimes also containing a pigment. All the Calcareae have a capillitium which is everywhere firmly attached by the extremities of its branches to the wall of the sporangium in the manner here described. In Didymium (Fig. 189), with which genus Spumaria and Diachea agree in this respect, the capillitium consists of very slender threads, from 1·3 to 2 μ in breadth in D. nigripes, Fr. and D. leucopus, Fr., as much as 2·4 μ in D. farinaceum, which are cylindrical or slightly flattened, solid or with an indication of a cavity in the form of a single line in the longitudinal axis, and usually of a dirty violet-brown colour in the broader parts. The threads are usually quite free from calcium carbonate; in a few cases, as in D. physaroides, they inclose single angular granules or crystals of calcium carbonate. They run in Didymium from below upwards most commonly straight or sinuous in a radial direction from the insertion in the stalk to the upper and lateral wall, their anastomoses usually forming an acute angle. D. Serpula is remarkable for the numerous round pigment-vesicles which adhere to the threads of the capillitium, and like the spores have a violet-brown membrane, but are from four to six times their size, being sometimes 50 μ in diameter.

The capillitium of the Trichiae and Arcyriae consists of tubular threads which have no deposits of calcium carbonate, and are either not attached at all to the wall of the sporangium or only at a few definite points. In Arcyria (Fig. 192) it is a non-septate tube separating into countless branches, which form a net-work by their anastomoses. The thick homogeneous wall has the same colour as the sporangium-membrane, and its outer surface is usually furnished with projections which take the form either of small spikes or warts, or of annular or semi-annular transverse ridges according to the species. In Arcyria punicea and A. cinerea the capillitium is anchored by the blind ends of branches of the net-work which are grown to the lower part of the wall of the sporangium. In most species (A. incarnata, A. nutans) it is nowhere in connection with the wall, but is fastened loosely by a few branches from the tubes, which descend into the stalk and are squeezed in between the cells which fill it up and which will be described presently. So long as the capillitium is inclosed in the sporangium its branches are all bent in every direction and folded up, and the meshes with their four, five or more sides are narrow and irregular. When the sporangium

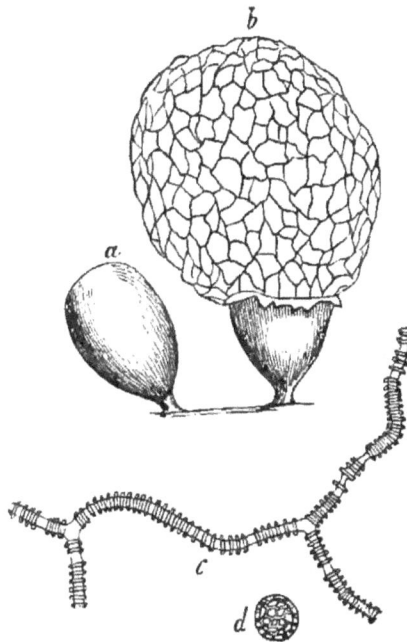

FIG. 192. *a, b Arcyria incarnata*, P. in outline. *a* a ripe sporangium closed. *b* the same open with the expanded net-work of the capillitium. *c, d Arcyria Serpula*, Wigd. (*A. anomala*, De Bary.) *c* portion of a capillitium. *d* a spore. *a* and *b* magn. 20, *c* and *d* 390 times.

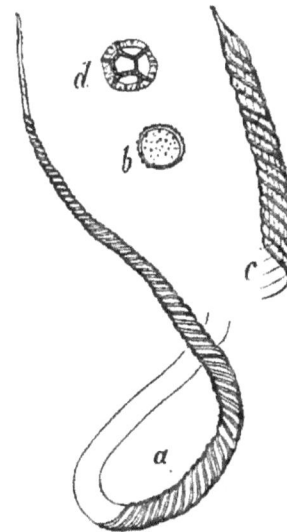

FIG. 193. *a, b Trichia fallax*, Fr. *a* half of a capillitium-tube. *b* superficial view of a spore. *c, d Trichia chrysosperma*, DC, De By. *c* extremity of a capillitium-tube. *d* spore. Magn. 390 times.

opens on reaching maturity, the branches straighten themselves in most species (A. cinerea is an exception), the meshes become broader and the circumference of the net-work many times larger (Fig. 192 *a, b*), and the structure never returns in any degree to its original form.

The capillitium-tubes of Trichia and Hemiarcyria which have frequently been described are united in the latter genus (H. rubiformis, H. clavata, &c.) into a net-work with branches which at the same time have free extremities. In Trichia (Fig. 193) they are quite free, and either simple or furnished with single short branches, their extremities being usually pointed, and in some species, as T. fallax, long and attenuated. The length of the free tubes varies usually between ·3 mm. and 7 mm., the average thickness being from 5 to 7 μ; longer and much shorter tubes occur here and there. The transverse section is usually circular. Their contents appear to be pellucid, but treatment with potash discloses an axile thread of an opaque substance, which turns yellow with iodine and is a remainder from the contents of the young tubes. The membrane of the tubes is thick but not evidently stratified, with a different degree of flexibility in different species, and coloured with different shades of yellow, red and reddish brown. It shows in all cases ridge-like projections or thickenings on the outer surface which run spirally round the tubes and often appear as folds in the membranes, since the lumen of the tubes is broader along their course and constricted in the intervals between them. The direction of the spiral as it ascends is with few exceptions, one only of which has come under my notice in Trichia varia, the same as that of the hand of a watch. The number of ridges varies with the species, being 2 in T. varia and 3–5 in T. fallax and T. chrysosperma. Variations of number in the same tube are partly due to bifurcation of the ridges, partly to the fact that some of them do not reach the end of the tube. In some species, as Hemiarcyria rubiformis, the back of the ridge is beset with spike-like processes. Trichia chrysosperma has a number of smaller slender ridges running between the spiral ridges and parallel to the longitudinal axis of the tube, and connecting each pair of spiral ridges in a scalariform manner (Fig. 193 c). The tubes of the capillitium lie mixed up together in great numbers and folded many times in the sporangium. If they are dried or their water is extracted by alcohol, they stretch themselves out but never become quite straight; if moisture is restored they acquire still stronger curvatures, and the same phenomena are repeated each time that the moisture is changed. This hygroscopic motility and the spiral ridges recall the elaters of the Hepaticae, and the tubes of Trichia have therefore received the name of *elaters*.

Certain species of Trichia and Arcyria, some of which have been already named, have the cavity of the stalk of the sporangium filled with vesicles or cells, which resemble spores but are larger and incapable of germination. They may very well be receptacles of excreta like the calcium carbonate-vesicles of the Calcareae.

The sporangia of Stemonitis, Comatricha, Enerthenema, and their allies are also distinguished in the mature state by certain peculiarities. They are borne on the stalk described above as being of the thickness of a hair or bristle and narrowing gradually upwards; this stalk passes through the base of the sporangium and occupies its longitudinal axis as a central column (columella), and either reaches the apex where it expands in Enerthenema into a membranous disk passing into the wall, or comes to an end below the apex and splits as it were into threads of the capillitium. Stalk and columella are hollow tubes, the cavity of which contains air and lumps of some organic substance. The wall is thick and marked with longitudinal wrinkles, and is coloured a dark brown throughout or has an outer layer colourless. The base of the stalk expands into an irregular membranous disk which rests on the substratum (Fig. 186). The primary branches of the dark-brown capillitium spring with a broad base from the whole outer surface of the columella, or as in Enerthenema only from the disk-like expansion at its extremity. These branches ramify repeatedly in every direction, and the ramifications become united into a net-work which has everywhere a large number of meshes. A large number of slender branches run from the meshes of the circumference only, and become attached by their free extremities to the wall of the sporangium. The structure of the stronger branches of the capillitium is like that of the columella, but their lumen is not in communication with the lumen of the columella; the slenderer branches resemble those of Didymium and Diachea. The wall of the sporangium is a simple and usually very delicate membrane, and like all other parts is free from deposits of calcium carbonate.

The capillitia which have now been described, though of many and apparently very different kinds, must nevertheless from the data before us be all regarded as peculiar membranous or parietal formations secreted or excreted from the protoplasm. Their material composition is not accurately known and will be briefly noticed again in the sequel, but it would appear to be not essentially different from that of the thicker portions of the outer wall of the sporangium to which they are sometimes attached. In the Calcareae (Physarum, Didymium, &c.) their substance or wall passes continuously into the wall of the sporangium of which they appear simply as processes, and thus closely resemble the branching bands of cellulose which spring from the outer wall of a cell of Canlerpa and stretch into the interior of the cell. That they are at the same time quite or partially hollow and take up excreted matter into their cavity, especially into the calcium carbonate-vesicles before described, is in favour of this conception. The wall-processes accordingly serve partly as supports partly as receptacles of the excreta. All that is known of their development (see page 431) is likewise in harmony with this view, which finds important support in Strasburger's recent observation that the development of the capillitium is essentially the same in Trichia as in the Physareae, though the tubes in the latter lie free and with a blind termination in the spore-plasm or between the spores, and are so like some vegetable cells that they were long taken for cells. According to Strasburger the tubes of the capillitium of Trichia fallax are formed by secretion of a membrane from the round spaces in the protoplasm which are of the same shape as the tubes, and are filled with a fluid (?) and have no nucleus. This would not prevent their also serving as receptacles of any excreta. What is true of Trichia may certainly be assumed of Arcyria and the species allied to it. The exceptional features in the Stemoniteae require therefore no further discussion.

2. The structure of the **aethalium in the mature state** has been sufficiently explained in the description of the history of the development of the most common form Fuligo varians, the Aethalium septicum of authors, which will be found on page 431. The body, which has the shape of a cake or cushion, is covered with a brittle rind some millimeters in thickness, which is at first of a golden yellow colour but afterwards becomes pale or cinnamon-coloured, and is continued all round the margin into a membranous expansion lying on the substratum. This rind consists of portions of the combined plasmodia which have collapsed and retain only the calcium carbonate and pigment; within it is a dark-gray mass finely speckled with yellow and readily crumbling to powder, and formed of a fine net-work of tubes, which may be nearly 1 mm. in thickness and which have exactly the same structure as the capillitium of the sporangium of Physarum.

Other aethalia are connected by their structure with Didymium, Diachea, Licea, &c., as Fuligo is with Physarum; but some like Reticularia umbrina require further examination.

The round sessile aethalia of Lycogala epidendron, Fr., which are as large as a pea or hazel-nut, have a very peculiar structure. They resemble small sporophores of the Lycoperdaceae. Their surface is covered by a paper-like membrane or rind (peridium), the outer side of which is irregularly warted, while numerous tangled threads, a capillitium, stretch from its inner surface into the cavity of the body which is filled with spores. The rind is composed of two portions having a layer of finely granular mucilage between them and separating readily from each other. The inner portion in the surface-view is perfectly homogeneous or finely punctated; seen in section it is an evidently stratified membrane about 8 μ in thickness and of a bright brown colour. The outer and much thicker portion on the other hand is formed chiefly of a weft of cylindrical tubular branched threads disposed in several irregular layers; the thickness of the threads is usually 20–33 μ, and their walls are stratified and thick, sometimes 10 μ in thickness, the outer lamellae consisting of a homogeneous jelly while the innermost layer is of firmer consistence and provided with slit-like pits or reticulate thickenings. Numerous branches of these rind-threads bend inwards, and piercing through the inner rind appear as threads of the capillitium in the central cavity. Here they have only the innermost lamellae of their membrane which is pitted or thickened in a reticulate or sometimes annular manner, the outer lamellae coming to an end in the inner rind. The thickenings project outwardly in the form of ridges which have the appearance of wrinkles and vary in height and breadth, being often very flat. The threads of the capillitium, which are often compressed and riband-like, branch and anastomose copiously. Finally the warts on the outer surface of the rind are thick-walled closed vesicles filled with a densely granular substance. These vesicles are undoubtedly remains of plasmodia filled with excreted substances, the whole body having been composed when young of their dense and uniform reticulum. The threads of the outer rind appear to be the thickened and subsequently emptied membranes of other peripheral plasmodial strands; the development of the inner part is imperfectly known.

The space not occupied by the capillitium is entirely filled with *spores* in all the sporangia of the Myxomycetes. All parts are kept moist with water till they are mature; then the water evaporates, and the wall of the sporangium dries up and opens in various ways to release the spores. The mode of dehiscence is generally very irregular; the wall as it dries becomes brittle and breaks up into small pieces at the least touch or quite of its own accord. This is the case in almost all the Physareae, and in Fuligo, Spumaria, Stemonitis and others. In the Cribrarieae the portions of the membrane which have not been thickened fall to pieces, the thickened portions remaining as a delicate lattice work. The rind in Lycogala and Reticularia tears irregularly and perhaps spontaneously at the apex. In Chondrioderma floriforme the outer layer of the wall of the sporangium splits from the apex into stellately diverging lobes. In Trichia, Hemiarcyria and Arcyria the dehiscence and the extrusion of the spores is assisted by the expansion of the capillitium caused by desiccation, and in the first genus by the hygroscopic movements also. The wall either opens spontaneously by an annular fissure in the lowest part of the sporangium, as in Arcyria punicea and A. cinerea, or in the upper part as in Hemiarcyria rubiformis, or by irregular fissures (Fig. 192 *a, b*) either spontaneously or when subjected to a slight amount of violence. The various monographs must be consulted for further details.

The **ripe spores** vary in size in the different species, their diameter being about 5 μ in Lycogala epidendron and 15 μ in Trichia chrysosperma. In many species single abnormally large spores often occur amongst the typical ones. They are always roundish in form when saturated with water, but as they dry they often become concave or boat-shaped, like the spores of many Fungi (see page 106). They are provided with an episporium, a thick unstratified membrane which in a few cases, as in Trichia fallax and some species of Didymium, is two-layered, and is provided in many species with a thinner spot which is perforated in germination and may therefore be termed a broad germ-pore. The outer surface of the membrane is smooth or furnished with projections of definite shape (Figs. 192, 193) according to the species and genus, and is usually of a deep colour, being violet or violet-brown for instance in all Calcareae and Stemoniteae, and yellow and red in the Trichiaceae. The protoplasm enclosed by the episporium has been already described.

As regards the **material composition of the membrane in sporangia, capillitium and spores**, we know that it behaves towards reagents in a *similar* manner to cuticularised plant-cell-membranes and to spore-membranes in the Fungi, but analyses of it are still wanting. The blue or violet colour of cellulose with iodine does not usually appear; but Wigand and myself found exceptions to this rule in Trichia furcata, Wigd., T. pyriformis and T. varia, in which the innermost layers of the walls of young sporangia become a dirty blue with iodine and sulphuric acid; the membranes also of the spores and of the cells which fill the stalk of Arcyria cinerea, A. punicea and A. nutans and those of the spores of Lycogala epidendron all turn a beautiful blue with the same reagents. Further details will be found in the special treatises which will be named below.

In the foregoing account of the Myxomycetes the nomenclature which Rostafinski introduced in his monograph has been substituted for the older nomenclature, or unequivocal synonyms only of Rostafinski's names have been retained. The most frequently recurring names are:

    Calcareae = Physareae of my earlier works.
    Fuligo varians = Aethalium septicum of former authors.
    Perichaena liceoides = Licea pannorum, Cienk.
    Chrondrioderma difforme = Didymium Libertianum, Fres., the ' Physarum album ' of Cienkowski.
    Licea flexuosa = Licea serpula, Fr., &c.

Further details respecting the nomenclature and the structure of the spore-receptacles should be sought in Rostafinski's writings: see also Just's Jahresbericht, for 1873.

ACRASIEAE.

Section CXXIII. The Acrasieae, which live on the excrements of animals and on decaying parts of plants, commence their development like the Myxomycetes with the escape of a swarm-cell from a spore. The swarm-cell always remains in the form which has amoeboid creeping movements, never assuming that which has cilia and the hopping movement. After multiplying greatly by successive bipartitions the swarm-cells unite again, many hundreds in number oftentimes if the development has been vigorous, in order to form spores. But they do not for this purpose coalesce into plasmodia. The swarm-cells are piled up one on another without coalescing, remaining distinct and artificially separable from one another though closely crowded together, and forming bodies of definite shape which rise perpendicularly above the surface of the substratum, and in which every one or the majority of the swarm-cells assumes the structure of a spore of a Myxomycete, having usually a delicate membrane and being of the average size of from 5 to 10 μ. These spore-heaps resemble small sporangia of a Myxomycete, but have no distinct wall; the spores are only held together and surrounded by a slight structureless enveloping substance.

In the **Guttulinae**, which have been carefully examined by Cienkowski and Fayod, the development, apart from the possible resting-states which will be described below, is limited to the above phenomena. The ripe spore-heaps are formed only of spores. They lie on the substratum as round or elongated bodies of the size of a pin's head and with the appearance of small white, yellow or red drops.

In the other genera of the group **Dictyostelium** and **Acrasis** a division of labour appears among the heaped up swarm-cells. When the body which is formed by them begins to rise above the substratum, those which are in its central line form firm outer membranes at the expense of their protoplasm, and are gradually changed into chambers formed of cellulose and filled with a transparent substance. They remain in close union with one another without interstices, and being built up on or against one another in a single row or in several rows in large Dictyostelia, they form a stalk which rests on the substratum and traverses the middle of the body somewhat in the manner of the stalk in the sporangium of Stemonitis (Fig. 186 *a*). This stalk grows for a time acropetally by addition of new elements. The aggregate of swarm-cells surrounding the stalk lengthens in proportion as the stalk grows and becomes correspondingly narrower, and ultimately separates from the substratum as the stalk ceases to grow; it then creeps up the stalk leaving it bare and proceeds to form its spores at the apex. This process also is to some extent illustrated by Fig. 186 which represents Stemonitis. In Dictyostelium the spore-masses borne on the stalk resemble in the main those of the Guttulineae. According to Van Tieghem the spores are arranged in Acrasis in rows one above another like beads in a rosary. For further details the reader is referred to Van Tieghem's and Brefeld's publications.

In Guttulina protea, Fayod found that the swarm-spores grown in a fluid and remaining isolated may undergo the change into spores of the form and structure which they have normally in the state of aggregation. When the conditions are unfavourable for the development these swarm-cells form a thick outer membrane in a complicated manner, and pass into a state which corresponds to the encystment of the swarm-cells of the Myxomycetes (page 427) and may therefore be termed a transitory resting-state. Van Tieghem observed a different mode of encysting in Acrasis and Dictyostelium when the conditions were unfavourable; a swarm-cell put out a number of processes or arms one after another, which separate from the parent-cell, round themselves off and become invested with a membrane.

In Brefeld's first work on Dictyostelium the development was to some extent incorrectly described, the dense aggregations of swarm-cells which develope into the stalked spore-masses being supposed to be plasmodia, that is, products of the coalescence of swarm-cells, and the rest of the phenomena being interpreted in accordance with that supposition. Van Tieghem's correction of this mistake in 1880 was cautious but at the same time clear and complete, and Brefeld has recently (1884) published a full confirmation of this correction.

Affinities of the Mycetozoa.

Section CXXIV. In investigating the affinities and homologies of the Mycetozoa which have now been described, we must distinguish between the **Myxomycetes which**

form plasmodia by coalescence of swarm-cells and the **Acrasieae which do not form plasmodia.** The two groups are evidently closely related to one another, the only important difference between them being the coalescence of swarm-cells in the one group and their firm aggregation only in the other. It is easy to conceive that the one form of development has proceeded directly from the other, either the Myxomycete form from that of the evidently more simply Acrasieae, or in the converse order.

The group of the Mycetozoa differs distinctly from the Fungi which have been the subject of the first part of this book in all such characteristics as do not belong to all organisms alike, and the descriptions already given of both kinds of plants render any further explanation of the point unnecessary ; their connection also with other known plants is still more remote. The difference would not be less decided, if the Mycetozoa were without their remarkable amoeboid movements, for such movements are observed in other vegetable cells which have not a firm membrane. The characteristic mark of separation lies in the formation of plasmodia or aggregation of swarm-cells.

It is obvious moreover according to our present knowledge that the Mycetozoa are the superior terminal member or the two terminal members of a series of forms or developments which commence elsewhere. The most highly differentiated sections of the group, the Calcareae, Trichiae, Lycogala and others, give evidence of no close affinity with any more highly differentiated group ; in other words, like the Gastromycetes with which they were classed by earlier botanists, they do not connect with any group above them. Hence in enquiring after their affinities we must be content with searching for a possible connection with an inferior group, and for the simpler forms from which they have proceeded.

When we seek for such a connection among the forms with which we are acquainted, we find it impossible to establish any strict homologies, and we are limited to the observation of resemblances in form, structure and mode of life. Such a course of unprejudiced comparison leads us by a very short step to the naked ' Amoebae ' of the zoologists, especially in Bütschli's sense[1], as the starting point,— organisms with bodies having the amoeboid movements of the swarm-cells of the Myxomycetes, which multiply, as far as we at present know, by successive bipartitions without forming plasmodia, and which may pass singly and without aggregation or coalescence into states of rest not essentially different in their characteristics from those of the spores of the Myxomycetes. Guttulina is really a naked Amoeba of this kind, and is distinguished from other known forms only by the aggregation of its spores. Guttulina protea, mentioned above as forming solitary spores, differs in this respect from the Amoebae ; it may be classed as well with the naked Amoebae as with the Acrasieae, and forms therefore a perfect connecting link with the Amoebae. Thus on the one hand the more highly differentiated Acrasieae join on at once to Guttulina, and on the other a short step further brings us to the type of Myxomycetes which form plasmodia, in which coalescence of swarm-cells into a plasmodium and redivision of the plasmodium into spores take the place of their aggregation. Forms like Guttulina may have developed phylogenetically in two divergent directions, on the one hand into the more

---

[1] Bronn's Thierreich, see below, page 454.

highly differentiated Acrasieae, on the other into forms which produce plasmodia. Plasmodiophora, which will be further noticed below, is perhaps one of the simplest representatives of the latter kind, though this must remain uncertain for reasons which will be stated. In the group of the Myxomycetes the type becomes highly differentiated.

From these naked Amoebae with which the Mycetozoa are connected in the ascending line, the zoologists with reason derive the copiously and highly developed section of the shell-forming 'Rhizopods,' as they are understood by Fr. E. Schulze and Bütschli, though the course of their ontogenetic development is still imperfectly known. And since there are sufficient grounds for placing the Rhizopods outside the vegetable and in the animal kingdom, and this is undoubtedly the true position for the Amoebae which are their simpler and earlier forms, the Mycetozoa which may be directly derived from the same stem are at least brought very near to the domain of zoology. It has been already pointed out that the Mycetozoa show only a slight agreement, either in the general course of their development or in the characteristic features of its separate stages, with organisms which are of undoubted vegetable origin, whether they be Fungi or plants other than Fungi; the agreement, with the exception of the few cases in which cellulose makes its appearance, is confined to phenomena which are common to all organised bodies. It is exactly in the species, which like Lycogala and Fuligo are most like the Fungi, that the agreement is of the smallest possible amount, being confined to purely external marks such as those between birds and winged insects. On these various grounds, which have been worked out at different times with greater or less clearness according to the state of our knowledge, I have since the year 1858 placed the Myxomycetes under the name of Mycetozoa outside the limits of the vegetable kingdom, and I still consider this to be their true position.

We may further enquire whether closer ties of relationship do not appear at some point or other between the group of the Mycetozoa at their lower limits and members of the vegetable kingdom. In the search for these and judging by known facts, we find that the only forms which we can take into consideration are the Chytridieae which have no mycelium, as Synchitrium, Olpidiopsis, Rozella and Woronina (see sections L–LII). It has been already more or less distinctly stated that these forms are nearly related to the Mycetozoa. They agree with the Myxomycetes, first in the peculiar circumstance that the entire vegetative body is finally transformed into one many-spored sporangium, secondly in the fact that their spores and the vegetative body itself in the young state have the power of amoeboid movement for a longer or shorter time. But these are phenomena which are common to them and many other Thallophytes, with which no one ever has supposed or ever will suppose them to have any near affinity, Botrydium for example or Porphyra; it is plain also that they have been appealed to from a wish to find some group of undoubtedly vegetable forms in which the Myxomycetes could be included. Of the characteristic phenomena of development in the Mycetozoa, the Chytridieae mentioned above show *neither the aggregation of the Acrasieae, nor the formation of plasmodia by coalescence of swarm-cells.* If the term plasmodium has in their case been used to describe bodies originating in the growth of a single spore, this arose either from an erroneous idea (Cornu), or it is a misuse of the word, for though Chytridium in its young state often shows the amoeboid movement of a plasmodium, it has not the character which is involved in the origin of a plasmodium; this is the case too with the spore of Porphyra which also has the power of amoeboid motion.

There is therefore no real ground for assuming a direct relationship with these Chytridieae, whether they do or do not form a natural group with the other species which produce mycelia, a question which, as was explained in Chap. V, must for the present remain undetermined.

It is a totally different question whether it is possible to suppose a common origin for these particular Chytridieae and the Mycetozoa, and consequently a more remote and indirect relationship. The comparison of the facts known to us shows it to be probable, as Bütschli points out, that the starting point of the naked Amoebae of the zoologists is to be sought in the group of very simple organisms known as the Flagellatae, and a study of the swarm-spores of the Mycetozoa leads to the same view, for in the stage of their existence when they are furnished with cilia they have all the characters of the simpler Flagellatae. But not only the Chytridieae which produce no mycelium but all the group show such close affinity to the Flagellatae, that they might if necessary be phylogenetically derived from them. But this is true also of the entire assemblage of the simple Algae, with which it was sought to connect the Fungi in the sections of Chapter V. We may as well place the Volvocineae with the Flagellatae as with the Chlorophyceae, if we prefer that arrangement, and no one will doubt the close affinity which exists between them and the rest of the undoubted Chlorophyceae.

If then we distinctly separate the Mycetozoa from the Fungi, and are prepared even to draw the boundary line which divides the two organic kingdoms between these groups, we do not thereby deny that members of the two divisions may approach very near to the group of the Flagellatae, towards which all the evidence shows that the two kingdoms converge, and thereby approach also very near to one another.

The purpose of the foregoing remarks has been to do for the Mycetozoa what was previously done for the Fungi, namely to establish their affinities on the foundation of the facts of which we are at present in possession, or speaking more boldly to give the true account of them. Such an attempt whenever made must be made with the material then at hand. If the foundation of facts changes with the progress of our researches a fresh attempt must be made.

The views of botanists as to the position of the Mycetozoa in the system, have already varied much in the course of time. The older view just noticed above, which placed the Myxomycetes with the Gastromycetes on the strength of a mere resemblance between the mature sporangia in the two groups, has now only a historical interest. Further remarks on this point will be found in my monograph of 1864.

The ideas with regard to the place of the Mycetozoa in the natural system which were expressed by Famintzin and Woronin in their beautiful work on Ceratium do not at the present day call for discussion. Cornu in 1872 endeavoured to connect them with the Chytridieae, chiefly by assuming the formation of plasmodia in the genera of Chytridieae which do not produce a mycelium; but we have already shown that this assumption is without foundation. The opinion represented in Brefeld's work on Dictyostelium (page 20), that this organism might connect the group of the Myxomycetes with the Fungi through the Mucorini, is refuted by a comparison of the course of development in the two groups. The more recent utterances of the same writer on the point under consideration presuppose on the one hand their connection with the Fungi as something ascertained, and on the other hand in dealing with the affinities of the Fungi so leaves the firm ground of definite facts for the regions of speculation, that they cannot be admitted into a discussion which keeps close to facts. The view lately promulgated by J. Klein in his treatise on Vampyrella is essentially the same as that which would derive the Myxomycetes and the Chytridieae from a common stem lying outside the series of Fungi. It sees the form, from which Myxomycetes, Chytridieae and Rhizopods all descended, in the Vampyrellae which belong either to the Rhizopod-type (or Heliozoa-type) or to that of the Myxomycetes, but it can hardly be said to have any foundation in facts; it should have gone somewhat further back to the Flagellatae, as has been suggested above.

DOUBTFUL MYCETOZOA.

SECTION CXXV. I here exclude from the ranks of the true Mycetozoa a few forms or groups of forms, some of which have been occasionally mentioned in the preceding sections. These forms so far as they are known have many points of resemblance with the Mycetozoa, but either our knowledge of them is imperfect, or else they depart so far in certain points from the typical Myxomycetes and Acrasieae, that it is better to have their position in the system for the present undetermined. At the same time we may properly give a brief enumeration and description of them in this place.

Bursulla crystallina, Sorokin, is, according to the author's account of it, a very small Myxomycete growing on horses' dung, with an ovoid stalked sporangium 0.03 mm. in height, and forming eight spores by simultaneous division. The spores before they become invested with a firm membrane escape from the swollen apex of the sporangium in the form of swarm-cells without cilia but capable of amoeboid movement, and subsequently coalesce in indefinite numbers and form plasmodia, which in their turn become fashioned each into a single sporangium or into a group of several sporangia according to their size. Sorokin saw no nucleus in the swarm-cells at the ordinary vegetative temperature; on the other hand a nucleus was observed when the sporangia were exposed to a very low temperature (as low as $-27°$ C). The development was in other respects the same; we may conclude therefore that the nuclei of the swarm-cells had been overlooked in the first-mentioned case. We may venture therefore a step further and assume that when a swarm-cell with a nucleus encounters one which is supposed to be without a nucleus, the two coalesce into a cell which forms a membrane and passes into a resting state as a kind of oospore; and that after hibernation the protoplasm issues forth from the membrane and becomes fashioned into an ordinary sporangium. This may in fact be simply a case of the encystment of small plasmodia. Apart from the peculiar features which require examination we may really have a small Myxomycete before us in Bursulla.

The course of development in Haeckel's pelagic **Protomyxa aurantiaca** entirely resembles that of a Myxomycete. 'Protoplasmic body, a plasmodium of an orange-red colour, (always?) formed by coalescence of several swarm-spores, 0.5–1 mm. in diameter; with very many thick arborescently branched pseudopodia which form a net-work by frequent anastomoses. Resting state, a spherical lepocytode with a diameter of 0.15 mm., and a thick structureless envelope (cyst). Swarm-spores pear-shaped, conical at the smaller end and running out into a very strong flagellum; movement that of the swarm-cells of the Myxomycetes. The spores when they come to rest creep about in the manner of the Amoebae.' Such is Haeckel's diagnosis. This organism differs from the Myxomycetes chiefly in the absence of firm spore-membrane, and in the circumstance that neither cell-nucleus nor division of the swarm-cells has been observed. **Myxastrum radians**, Haeckel, also a marine form and distinguished by the presence of silica in the spore-membranes, appears to be nearly allied to Protomyxa.

Cienkowski's **Vampyrellae** are organisms with amoeboid movement which live on Algae and Diatoms. Some like Vampyrella Spirogyrae and V. pendula suck the protoplasm and chylorophyll-corpuscles from out of the living cells of species of Spirogyra or Oedogonium, when they have pierced their walls, while V. vorax embraces the entire cells of Diatoms, Desmids and similar forms with their pseudopodia, and absorbs them into its own substance. In both cases the reception of a certain quantity of food is followed by a period of rest, a smoothing of the surface of the body and the excretion of a delicate firm membrane. In this state of rest the bodies which have been absorbed are digested, that is are dissolved till there remain only comparatively minute portions of the protoplasm which have assumed a brown colour, and in the case of Vampyrella vorax of the membranes. Next follows the excretion of the undigested substance from the living protoplasm, the division of the latter into usually 2–4 swarm-cells and their escape from the membrane; the two processes go on simultaneously, the division being effected while the spores are escaping at 2–4 separate points. Then according to J. Klein from 2–4 swarm-cells, seldom more, at once coalesce again and form a plasmodium, which repeats the process just described of absorption of food and subsequent formation of swarm-cells. In addition to this course of development resting cysts may also be formed, in which case the body which has come to rest inside the membrane excretes undigested remains of the food, and then without forming swarm-cells excretes a new membrane. The subsequent fate of these cysts is still unknown. Other transitory states of rest, as in the small cysts of the Myxomycetes, may occur within the periodic course of development described above, and no coalescence may take place, the cells passing singly through the swarming state as above but not forming plasmodia.

Cienkowski's **Nucleariae** appear to be just like the Vampyrellae in the course of their development and in their manner of life. They are distinguished from them by the presence of nuclei, which are said to be wanting in the Vampyrellae. Coalescence into plasmodia has not been observed in them, but it is not excluded by the ascertained facts.

Cienkowski's **Monas Amyli** has motile swarm-cells provided with two cilia, and a number of these cells surrounding a starch-grain may coalesce into small plasmodia. The plasmodium forms a membrane, and after its substance has increased in size at the expense of the starch-grain it produces a large number of new swarm-cells by simultaneous division. It is said also that a single swarm-cell may spread itself round a grain of starch without uniting with others to form a plasmodium, and thus become the starting-point of the development which was described above.

An exactly similar course of development has been observed in Klein's **Monadopsis** and Cienkowski's **Pseudospora** and **Colpodella**, except that the latter two genera, as far as is at present known, do not form plasmodia, but each swarm-cell after absorbing food becomes the mother-cell of a new generation of swarm-cells.

Plasmodiophora Brassicae is parasitic on the roots of cruciferous plants, especially species of cabbage, and causes large swellings on them. An amoeboid swarm-cell with cilia escapes from the round thin-walled spore in water, and penetrates without first undergoing division into the epidermis of the young root and from thence into the parenchymatous tissue. Then the cells of the host become greatly enlarged and large bodies with amoeboid movements make their appearance in them; but it could not be certainly determined whether these bodies were due to the growth of one swarm-cell, or to the coalescence of several swarm-cells as in the Myxomycetes, or possibly to a modification of the protoplasm similar to that which occurs in Rozella (see page 395). Ultimately the entire protoplasm contained in a cell of the parenchyma becomes motionless and divides simultaneously into a very large number of spores of the character stated above, and in this case also without previously forming a special membrane.

Finally Zopf[1] appears to include all sorts of lower organisms with amoeboid movements together with some of the forms last described under the name of 'lower slime-Fungi.' This use of the term does certainly not correspond with the meaning hitherto assigned to it, and to avoid any misunderstanding I say very distinctly that this application of the term and therefore also the discussion of any other forms than those which have now been mentioned cannot be admitted in this place.

CHAPTER IX. MODE OF LIFE OF THE MYCETOZOA.

SECTION CXXVI. **Germination.** The spores of the Mycetozoa, in which the germination has been observed, are able to germinate from the moment that they are ripe. Some retain the power of germination for a long time if protected from injury; many Calcareae, for example, Physarum, Didymium, Chondrioderma, Perichaena liceoides, retain their vitality 2–3 years, some, as Physarum macrocarpum according to Hoffmann, eve for 4 years. In Trichia varia and T. rubiformis vitality lasted according to express observations only 7 months; in other species of Trichia and allied forms it appears to be extinguished at a still earlier period.

The requisite conditions for germination in most known forms are the usual spring and summer temperature of our temperate climates and a sufficient supply of water. The majority germinate readily when placed in pure water, well developed fresh material often in a few hours. Nutrient substances dissolved in the water do not hinder germination; this at least was found to be the case in Fuligo and Chondrioderma. The Ceratieae and such Acrasieae as have been examined do not germinate in pure water, but only in a suitable nutrient solution. The like necessity and the use of unsuitable solutions may account for the want of success which has attended the attempts hitherto made to procure the germination of their spores of the Cribrarieae and Tubulina.

---

[1] See Biolog. Centralblatt, Bd. III, Nr. 22.

The requisites for germination are the same in sclerotia and cysts as in spores, as was stated above on page 428, where also will be found all that is known of the external causes which lead to the formation of these states.

SECTION CXXVII. Some account has necessarily been already given in section CXIX of the **phenomena attending the life of the plasmodia.** For many general questions which here come under consideration, the reader is referred, in accordance with the purpose of this book, to works on general physiology, and especially to Pfeffer's Physiology, vol. II, chap. 8 and Stahl's latest publication on the subject, and the account here given must be confined to a short review of their mode of life. This has been investigated chiefly in the plasmodia of the Physareae, Fuligo especially, which are readily procured. What is known of other forms appears to agree with the accounts given of the Physareae, but requires more exact investigation.

Movement of plasmodia. The **internal causes** of the changes of shape, of the protrusion and withdrawal of processes and the interior streaming of granules, which are attendant on the organisation of the protoplasmic body and are to a great extent unknown to us, cannot of course be discussed in this place.

The most important **external causes** of the movements of the plasmodium and of the changes in its form are, the *illumination*, the *distribution and movement of the water* in the substratum, the *chemical nature of the environment*, and the *conditions of temperature*. It is uncertain to what extent purely mechanical influences are also operative. Rosanoff's former assumption of geotropic movements has proved to be without foundation. We may therefore, in accordance more or less with the general terminology of the movements of growth, speak of phenomena of *heliotropism*, *hydrotropism* and *rheotropism*, *trophotropism* and *thermotropism*.

Heliotropism. A vegetating plasmodium stretches out its branches and reticulations uniformly in every direction on uniformly moistened surfaces, such as paper steeped in water and kept in a dark or equably but faintly illuminated chamber. If the intensity of the illumination is increased, the power of movement, according to Baranetzki, is generally diminished, and if the amount of illumination is different in different portions of the expanded surface, the branches are drawn in from the bright side and others are put forth towards the darker side; the plasmodium also moves towards the darker side. The direction of these movements is independent of the direction of the beams of light which fall on the plasmodium, being determined only by the intensity of the illumination.

Hydrotropism. If while all other conditions are uniformly favourable the water is unequally distributed in the substratum, the vegetating plasmodia, if not on the point of forming spores, withdraw from the drier spots when the dryness has reached a certain degree and wander towards the moister.

Rheotropism. If a stream of water is made to flow slowly through a moistened porous substratum, such as filtering paper or strips of linen cloth, the plasmodia which are vegetating on the moist surface wander in the direction of the stream, without regard to the particular direction in space in which it moves.

Trophotropism. Vegetating plasmodia spread out on surfaces which yield little or no nutriment move towards bodies which contain nutrient substances as soon as they are offered to them, here too without regard to the direction in space in which the movement has to be made. If the plasmodium of Fuligo which usually lives in tan is spread out on the moist vertical wall of a glass, it remains in this position, other things being the same, as long as the surface of the glass is covered with a film of pure water. If an infusion of tan is added to the water, in such a way that the plasmodium is touched by it at one spot only, it begins to move rapidly towards this spot and gradually puts out numerous branches which dip into the infusion. A small piece of tan placed close to a plasmodium under similar conditions is quickly seized by a number of freshly protruded branches. The similarity in the effect of the fluid containing the infusion of tan and the solid piece of tan shows that it can only be due to chemical constituents of the tan; what these are has not been precisely ascertained.

If a plasmodium comes into contact on one side with other bodies dissolved in water, the opposite effect is produced, namely repulsion of the plasmodia. Even a solution containing $\frac{1}{2}$ or $\frac{1}{4}$ per cent. of grape sugar produced this effect at first in Stahl's experiments, but the plasmodium by degrees became accustomed to it and behaved to it as to the infusion of tan. A sudden change in the concentration of the saccharine solution, either by increasing it to a certain amount (2 per cent.) or diminishing it, gives rise to similar phenomena to those first described. Stahl observed the same repulsions in experiments with saline solutions.

If oxygen is excluded on one side, a movement takes place, as might have been expected beforehand, towards the side where oxygen is admitted.

Thermotropism. If the substratum, other conditions being the same, is unequally warmed on different sides, the plasmodium moves, at least within the limits of temperature observed in Stahl's experiments ($+7°$ to $30°$ C.), towards the side which is most highly warmed.

Most of the phenomena observed in spontaneously vegetating plasmodia, especially their creeping hither and thither and in and out, according to the time of the year and the state of the weather, on the substratum of vegetable remains, such as leaves, tan and the like, may be explained very simply from the experimental results here recorded.

A further fact established by Stahl must be added here in explanation of another and very remarkable phenomenon, namely that in the plasmodia of Fuligo and some species of Physarum, in which the point could be examined, the reaction against locally unequal distribution of water in the environment changes with the age. The plasmodia are *positively hydrotropic*, that is wander from the dry to the moister spots, other things being equal, during the vegetative stage, but become *negatively hydrotropic* near the moment of formation of sporangia, that is they move from the moister to the drier spots.

This movement also takes place without regard to the mere direction in space, and so may be upwards or downwards &c., and it explains the general fact that almost all plasmodia, as soon as they are ripe for forming sporangia, move to comparatively very dry spots on the surface of the moist substratum, often travelling a considerable distance, before being transformed into sporangia; it explains also, according to Stahl's observations, the elevation of the commencing sporangium in a direction at right angles to the comparatively moist substratum.

Further investigation is necessary to determine whether other causes also may not assist in certain cases to give rise to the latter phenomenon ; the question also, whether the peculiar characters of plasmodia under discussion may not change at a certain period of their development in relation to other things, as well as to hydrotropism, has still to be examined, especially with reference to a statement of Hofmeister[1] that certain plasmodia moved towards the side of strongest illumination.

To the movements which have now been described must be added one more which requires a brief consideration. It was stated above on page 425 that small solid bodies are engulphed in the substance of the plasmodia, at least in the Calcareae or Physareae. This is effected by definite movements; the surface of the plasmodium rises cushion-like round the bodies which are in contact with it, and the margins of the raised part gradually run together over them and cover them.

This phenomenon occurs in the plasmodia, as soon as they have been formed by coalescence of swarm-cells, but not in the swarm-cells themselves, if we put aside certain isolated observations on Dictyostelium which have yet to be confirmed. It is not confined to any particular spot of the plasmodium, and may continue till sporangia begin to be formed; then the foreign bodies which have been absorbed and are still present are all ejected, some of them even at an earlier period. All this shows, that the solid bodies are not simply squeezed into the soft and passive substance of the plasmodium; but that there is a reaction of the plasmodium in response to the stimulus which it experiences from contact with them.

Substances of various kinds are taken in this way into the substance of the plasmodium : fragments of dead vegetable cells, spores of Fungi and of the Myxomycetes themselves, sclerotium-cells of Myxomycetes, grains of starch, small portions of colouring matters if brought near the plasmodium.

All these substances it should be observed consist of organic compounds, and it is highly probable that some of them at least supply food to the plasmodium which has engulphed them. It is not certainly ascertained whether entirely indifferent inorganic substances are absorbed by it. The question therefore remains unanswered, whether the movements of engulphing are caused by the purely mechanical stimulus of contact, or by certain chemical qualities of the substance to be engulphed. In the latter case the phenomenon would rank immediately with the movement in the direction of nutrient bodies described above, and both would be special cases of a more general law of reaction in response to chemical irritants. An old observation of my own supports the view that the reaction not only is or may be dependent on a definite chemical quality in the body which causes the stimulus, but that plasmodia of different kinds react unequally on the same stimulation. A number of pieces of carmine were absorbed by Didymium Serpula, scarcely any by Chondrioderma difforme.

SECTION CXXVIII. The **process of nutrition** takes place only in the amoeboid states of the Mycetozoa, in the swarm-cells therefore and the plasmodia. All the better known Myxomycetes in their actual primary adaptation are saprophytes ;

---

[1] Pflanzenzelle, p. 20.

they live on dead organic and especially vegetable substances, of course with the necessary ash-constituents, and are found therefore chiefly in accumulations of dead parts of plants—leaves, tan, rotten wood, and the like. What definite chemical substance does actually and usually serve or is fitted to serve as nutrient material to the Myxomycetes is a question which has not yet been thoroughly examined.

The facts recorded above show that the food is taken in during the swarm-cell condition only in the fluid state or state of solution, and this is also the case, at least in most instances, with the plasmodia. That it is so appears, to say the least, extremely probable by the behaviour of the plasmodia of Fuligo to the extract of tan, as shown by Stahl's experiments quoted above. This agrees with the observation that plasmodia of Chondrioderma difforme may be obtained from spores in watery infusions of vegetable substances though no solid bodies are supplied to them, and lastly with the fact that solid ingesta have never been found in the plasmodia of certain species, as Lycogala, though it must be allowed that these have not been very thoroughly examined.

On the other hand we see solid bodies taken up by the plasmodia which have been more particularly described above, among others by Chondrioderma, and some of them at least again thrown out. The body taken into the plasmodium is often more or less perfectly dissolved. It has already been stated that the sclerotium-cells of a species engulphed by its own plasmodium gradually disappear and pass into the substance of the plasmodium, but it is uncertain whether this is a case of actual dissolution of the body introduced, or of a coalescence with the body which absorbs it, like that of the swarm-cells or the branches of the plasmodium. In the plasmodium of Didymium Serpula which were fed with carmine, the fragments of carmine were to some extent at least dissolved; they were repeatedly carried along in the stream of granules, and in twenty-four hours' time were enclosed each in a vacuole filled with a clear red solution. This continued for several days. On the other hand the Chondrioderma mentioned above showed no signs of dissolving the few fragments of carmine which it received into its substance. In experiments instituted by Dr. Wortmann a number of starch-grains were taken in by plasmodia of Fuligo, and they showed deep corrosions in the course of from two to three days. This shows the presence of a ferment capable of dissolving starch and confirms Kühne's previous determination. A ferment which acts upon cellulose must be present, at least during the passage of the sclerotia of Fuligo into the motile condition, because the cellulose membranes are rapidly dissolved during that time. Krukenberg has ascertained the presence of a peptonising ferment[1].

These facts all point to the conclusion that the solid ingesta are to some extent at least appropriated as food and digested, the undigested remainder being then cast out. We have no exact physiological investigations of these questions and of others which are connected with them. So far as plasmodia devour and digest living bodies, the name of saprophytes can only be applied to them with some modification of its ordinary mean.

Some of the forms which were classed above as doubtful Myxomycetes, Bursulla for example, are saprophytes in their mode of life. The account given above of

---

[1] Unters. d. physiol. Instit. z. Heidelberg, II, p. 273. See also Reinke cited on p. 52.

Plasmodiophora shows that it is, in the terminology employed in the Fungi, an endophytic parasite which greatly deforms its host. Vampyrella and the forms with a similar mode of life must no longer be termed parasites; they devour other organisms wholly or partially, absorbing the objects which they admit into their substance by the same or nearly the same movements as those by which plasmodia take in their solid ingesta.

Literature :—

E. FRIES, Systema mycologicum, III, 1829.

A. DE BARY, Die Mycetozoen (Zeitschrift f. wiss. Zoologie, Bd. X, 1859, and 2nd ed. Leipzig, 1864). For a variety of details and full accounts of the literature of the subjects the student is referred to the second edition of my work and to the following publications :

L. CIENKOWSKI, Zur Entwicklungsgeschichte der Myxomyceten, and Das Plasmodium in Pringsheim's Jahrb. f. wiss. Bot. III, 325 and 400.

J. T. ROSTAFINSKI, Versuch eines Systems der Mycetozoen (Diss. Strassburg, 1873) ;— Id., Sluzowce (Mycetozoa), eine Monographie, Paris, 1875 (in Polish), with full lists of works on the subject down to 1875.

J. ALEXANDROWITSCH, Ueber Myxomyceten (in Russian), Warsaw, 1872.

STRASBURGER, Zur Entwickelungsgeschichte der Trichia fallax (Bot. Zeit. 1884, p. 305).

A. FAMINTZIN u. M. WORONIN, Ueber Ceratium hydnoides u. C. porioides (Mém. Acad. Pétersbourg, XX, No. 3, 1873).

O. BREFELD, Dictyostelium mucoroides (Abh. d. Senckenb. Ges. VII, 1869) ;—Id., Untersuchungen aus der Gesammtgebiete der Mycologie, I, Leipzig, 1884.

L. CIENKOWSKI, Ueber einige protoplasmatische Organismen (Guttulina). See Just's Jahresber. for 1873, p. 61.

VAN TIEGHEM, Sur quelques Myxomycètes à plasmode agrégé (Bull. Soc. bot. France, 27 (1880), p. 317.

V. FAYOD in Bot. Ztg. 1883. Guttulina protea.

L. CIENKOWSKI, Beiträge zur Kenntniss der Monaden in M. Schultze's Archiv f. Mikrosk. Anatomie, I, p. 203, tt. XII–XIV. See also Regel in Bot. Ztg. 1856, p. 665.

M. WORONIN, Plasmodiophora Brassicae in Pringsheim's Jahrb. XI, p. 548, tt. 29-34.

E. HAECKEL, Monographie der Moneren (Jenaische Zeitschr. IV, p. 64).

F. E. SCHULZE, Rhizopodenstudien (Arch. f. Mikrosk. Anatomie, XI and XIII, p. 9).

J. KLEIN, Vampyrella (Bot. Ztg. 1882 and Bot. Centralbl. XI (1882), Nr. 5-7).

N. SOROKIN, Bursulla crystallina (Ann. d. sc. nat. sér. 6, II, p. 40, t. 8).

S. ROSANOFF, De l'influence de l'attraction terrestre sur la direction des Plasmodia des Myxomycètes (Mém. Soc. de Cherbourg, XIV, p. 149).

J. BARANETZKI, Influence de la lumière sur les Plasmodia des Myxomycètes (Mém. Soc. de Cherbourg, XIX, p. 321).

E. STRASBURGER, Wirkung des Lichtes und der Wärme auf Schwärmsporen. Jena, 1878, p. 69.

E. STAHL, Zur Biologie der Myxomyceten (Bot. Ztg. 1884).

The zoological material, which lies on the confines of the domain of the Mycetozoa and encroaches upon it, together with a list of works on the subject will be found in the copious treatise of H. G. Bronn entitled Klassen und Ordnungen des Thierreichs, I, Protozoa, edited by O. Bütschli, Leipzig and Heidelberg, 1880. The student is specially referred to this work.

Third Part.

BACTERIA OR SCHIZOMYCETES.

Chapter X. Morphology of the Bacteria.

Section CXXIX. The forms which we have now to consider are termed by Nägeli[1] **Schizomycetes** or **Fission-Fungi**; they are known also by an older name, **Bacteria**, which was restored by Cohn in 1872 as the designation of the entire group. I prefer the latter appellation for a group which not only includes Fungi in Nägeli's sense, namely Thallophytes which have no chlorophyll, but has among its most characteristic members forms which contain chlorophyll and cannot therefore with any propriety be termed Fungi. I avoid the use of the term Bacterium as a generic name. To denote the species which constitute the genus Bacterium of authors, I use partly the generic name Bacillus which will be more precisely defined in the sequel, and partly the name Arthrobacterium, the latter being applied to all species in which endogenous formation of spores, to be described presently, has not yet been observed. It must not be supposed that we have in this way effected a final reform of the nomenclature; we gain only a short expression for the present state of our still very imperfect knowledge.

The Bacteria consist of minute cells often less than $1\,\mu$ in breadth, and are either isodiametric or roundish in shape or else cylindrical and rod-like; they multiply if supplied with a sufficient amount of food by successive bipartitions, each cell dividing into two similar daughter-cells through an unlimited number of generations. The successive divisions are in most cases all in the same direction, and hence all the cells which have proceeded from one initial cell are arranged in a single simple filiform row if they remain united to one another. All the members of the row are alike capable of growth and division. It less frequently happens that, without any other change in the behaviour of the cells, the successive divisions take place in alternately varying directions, so that the arrangement of the generations which continue connected together is from the first other than that of a simple row.

---

[1] Verhandl. d. Deutschen Naturforscher-Versammlung zu Bonn. See Bot. Ztg. 1857, p. 760.

Little is known of the more intimate construction of the cells of the Bacteria, owing to their minute size. All that can with any confidence be affirmed of the greater part of them is founded less on direct observation than on the analogy of the larger cells of other organisms, with which they agree in their chief characteristics so far as these can be recognised, and with which they are also connected by intermediate forms.

The **protoplasm of the cell** in most forms and even in the larger ones appears, when in a state of active vegetation, to be a homogeneous and faintly refringent mass filling the cell-cavity. Distinct little granules (microsomata), the constitution of which is still undetermined, may be distinguished occasionally in the larger forms in this condition. They appear in greater abundance as the vegetative activity diminishes, and the protoplasm may then be still more frequently seen to form a parietal layer inclosing a pellucid central cavity. Highly refringent (crystalline) granules of sulphur, which owe their origin to the decomposition of the sulphates by the plant, are often imbedded in considerable quantity in the protoplasm of the species of Beggiatoa which grow in springs containing sulphur, as was first pointed out by Cramer and Lothar Meyer.

The protoplasm of some species which appear to belong to this group forms **chlorophyll**, and seems to be coloured green throughout by this substance. Van Tieghem found two forms of this kind living in water, which he distinguishes as Bacterium (Arthrobacterium) viride and Bacillus virens[1]; W. Engelmann a third marked by the very pale tint of its chlorophyll, which he names Bacterium (Arthrobacterium) chlorinum[2].

Most species are distinguished by the absence of chlorophyll and analogous colouring-matters. In this respect they agree with the Fungi, and it is owing to this fact and its physiological consequences that they have received the name of Fungi. In some species, Zopf's Beggiatoa roseo-persicina for example and its subordinate forms, the protoplasm is uniformly tinged with a red colouring-matter, which Lankester has carefully examined and named bacteriopurpurin[3]. It is not yet certainly ascertained whether the colouring matters, which often give an intensely red, blue, yellow, or other tint to the gelatinous accumulations of some small forms, such as Micrococcus prodigiosus, are contained in the membranes only or are attached also to the protoplasm.

Some species which contain no chlorophyll form a substance in their protoplasm, which from its behaviour with reagents and the physiological relationships observed in certain cases, must be considered to be more or less like starch, or more correctly granulose. The cells of Prazmowski's Bacillus (Clostridium) butyricus (Amylobacter Clostridium, Trécul) and Spirillum amyliferum, van Tieghem[4], become more highly refringent in the stages which precede the formation of spores (section CXXX), and their protoplasm then assumes a blue or violet colour with solution of iodine, either

[1] Bull. Soc. bot. de France, 27 (1880), p. 174. The figure there given by Van Tieghem from Perty is interesting, but it must remain a question whether it belongs to this group.

[2] Bot. Ztg. 1882, p. 323.

[3] Quart. Journ. of Micr. Sc., New Series, XIII (1873), p. 408.

[4] See Prazmowski, Unters. ü. d. Entwickgsg. u. Fermentwirk. einiger Bacterienarten, Leipzig, 1880, and Van Tieghem in Bull. Soc. Bot. de France, XXVI (1879), p. 65.

throughout or with the exception of certain transverse zones which do not turn blue ; and in both cases the substance which has become blue spreads uniformly through the protoplasm, without forming bodies in it of definite shape. This phenomenon has been observed when the nutrient substratum contains starch, and when it is entirely free from starch. The amyloid substance disappears with the formation of spores. This amyloid reaction with iodine occurs in Hansen's vinegar-forming Arthrobacterium (Bacterium) Pastorianum, and occasionally in Leptothrix buccalis [1], but without proof of any connexion with spore-formation; it is not found in the majority of the forms which have been examined, nor is there any report of the occurrence of amyloid bodies in the species which contain chlorophyll.

Nuclei have not as yet been observed in Bacteria.

The protoplasm of the Bacteria is surrounded in all cases, so far as we can determine, by a **membrane**. In the case of cells or cell-rows which vegetate actively in a fluid as isolated bodies and do not become cemented together into large masses the cell-wall appears on the lateral faces of the protoplasmic body as a thin bounding surface; on the boundary lines of cylindrical cells closely united in rows it is a septum, which is in many cases only to be distinguished by the use of desiccating and colouring reagents, and is so entirely invisible in the living specimen that a cell-row composed of several cells looks like a homogeneous unsegmented cylinder.

This delicate membrane immediately clothing the protoplasmic body must, in some forms at least, and especially in some species of Spirillum, be highly extensible and at the same time elastic. The straight cylindrical body in these species is often seen to bend strongly backwards, and then to recover its former direction. According to the views which prevail at the present day it is the protoplasm only that can be supposed to be the active cause in this phenomenon, and the investing membrane must possess the qualities just mentioned to be able to follow its movements.

Few or more probably no vegetating cells of Bacteria are clothed with this delicate membrane alone at the highest stage of their development. This is only the innermost lamella of a membrane which increases in thickness, and in doing so swells and becomes gelatinous in its outer portions. Such gelatinous outer layers or investments are found wherever care is taken to observe them, and direct examination shows that they are either connected in the manner indicated with the delicate inner membrane or are formed from it.

The particular character of the gelatinous envelope varies in the different species within wide limits. In the freely moving rod-like cells of the typical Bacteria it is invisible, but it can be recognised in the flakes of slimy matter formed by larger accumulations of these forms. In other cases it is of greater thickness and firmer consistence, and may either form distinct gelatinous sheaths round isolated cells and aggregates of cells, or unite and cement the cells together into larger gelatinous masses.

The chemical composition of these gelatinous membranes would appear to be very different in different species. Löw [2] found that the membranes of the mother of

---

[1] See Zopf, Spaltpilze.

[2] Nägeli, Ueber d. chem. Zusammensetzung d. Hefe (Sitzgsber. d. Münchener Acad., Mai, 1871); —Id., Theorie d. Gährung, p. 111. .

vinegar, and Scheibler and Durin [1] that those of Leuconostoc mesenterioides, were chiefly composed of the carbohydrate which comes nearest to cellulose; but it appears probable from the researches of Nencki and Schaffer [2] that in the gelatinous masses (zoogloeae) of purtrefactive Bacteria it consists chiefly of the albuminoid compound which is the principal constituent of the protoplasm of the cell, and to which these writers have given the name of *mycoprotein*, in combination with infinitesimal quantities of cellulose-like substance. I speak of this as probable only, because it is always a little doubtful how far the substances discovered by macrochemical examination have belonged to the one or the other portion of these minute bodies.

The membranes are in very many cases colourless; but in some instances, as has been already said, it is supposed that the intense blue, red, and other hues assumed by some bacteria-masses, and due to colouring-matters resembling anilin dyes, do really belong to the gelatinous membranes, provided they are not excretory products which have found their way into the substratum [3]. The sheaths round the filament of Cladothrix and Crenothrix are often rust-coloured or dark brown from the presence of ferrous hydrate disseminated through their substance.

Many forms of Bacteria have the **free movement of swarm-cells** in fluids. Their rapid forward motion is accompanied with rotation round their longitudinal axis, and in many cases with apparent curvature of their bodies. But many observations under the most favourable circumstances have failed to detect in these forms anything like a distinct organ of locomotion. There are however other swarming forms in which extremely delicate filiform processes described as cilia or 'flagella' have been observed since Cohn's, or even perhaps since Ehrenberg's time; these processes appear at one or both extremities, one usually but sometimes two or even three together, proceeding from the same point. It would appear to be uncertain whether these formations, like the cilia of other vegetable swarm-cells, are parts and processes of the protoplasm and project through the membrane, or whether they belong to, and are appendages of the membrane itself. The grounds which Van Tieghem [4] alleges for the latter view, namely that no direct connection can be traced between these processes and the protoplasm of the cell, while they behave towards colouring reagents in the same way as the membrane and not as the protoplasm, are against their being true cilia. It is to say the least questionable whether they function as organs of locomotion, considering the irregularity of their occurrence in the forms which are endowed with motion; and it would also be well to enquire whether the so-called flagella or cilia may not vary in character according to the species, and belong in some instances, as for example in Bacillus subtilis, to the membrane, in others, as in the larger arthrosporous species, to the protoplasm.

According to the shapes in which they appear in the vegetative states a series of principal forms are distinguished :—

---

[1] See Van Tieghem in Ann. d. sc. nat. sér. 6, VII, p. 180.

[2] Journ. f. pract. Chemie, neue Folge, 20 (1879), p. 443.

[3] See Schröter, Ueber einige durch Bacterien gebildete Pigmente, and Cohn's Beiträge z. Biol., Heft 2, p. 109; also Nägeli, Untersuch. ii. niedere Pilze, p. 20.

[4] Bull. Soc. Bot. de France, XXVI (1879), p. 37.

a. **Regard being had solely to the shape of the isolated cells or to their simplest genetic union,—**

1. **Cocci** : isolated cells which are isodiametric or at least very slightly elongated in one direction. These are distinguished when necessary according to their dimensions into *micrococci, macrococci* and *monad-forms.*

2. **Rod-like forms** : cells elongated in one direction and cylindrical, rarely fusiform, isolated or in a short chain. These again are distinguished into *short rods* (*Bacteria*), *long rods* (*Bacilli*), *fusiform rods* (*Clostridia*) and some others.

3. **Spiral forms**: spirally twisted rods, some with *narrrow coils* (*Spirillum*, *Spirochaete*), some with *distant and very steep coils* (*Vibriones*).

It follows necessarily from what has been already stated that no sharp line of distinction as regards their shape can be drawn between short rods, for instance, and cocci, or between a slightly twisted Vibrio and a Bacillus which departs to a trifling extent only from a mathematically straight line; nor can they at present be always clearly distinguished by their structure.

This is especially the case with respect to the rod-like forms, since a rod may be a single cell of the corresponding shape, or a number of cells firmly connected together and closely related to one another genetically. In the latter case while the cell is dividing repeatedly, the partition-wall may be of so delicate a structure that the compound body, if not carefully examined, may be taken for a simple homogeneous body. Hence when these organisims are simply spoken of as rods we must understand that the writer is alluding to their outward appearance only, unless the structure also is exactly described.

To these three kinds must be added a fourth, namely the **swollen bladder-like forms.** Individuals of this kind are found in company with the other three and are evidently produced from them; they are distinguished by having their cell swollen to several times the size of the other form, with a knobbed and irregular outline. These inflated forms have been observed in artificial cultivations where the nutrient substances are in insufficient quantity or are exhausted; Zopf and Cienkowski found them in Cladothrix and Crenothrix, Buchner and Prazmowski in forms of Bacillus, and Neelsen in Bacterium cyanogenum. They are therefore considered to be diseased states of the other forms, and have been termed by Nägeli and Buchner *involution-forms.* Hansen, on the other hand, has shown that they occur very frequently, indeed almost invariably, with the Bacteria of mother of vinegar; we do not know whether they have not some further meaning in this and perhaps also in most other cases.

b. **According to the mode of the connection between the individual cells,** each of the above form-groups may be,—

1. **free,** that is not firmly joined together, though occurring in the society of great numbers of like individuals.

2. **in the form of filaments,** that is joined together and forming long filiform rows. These filaments are unbranched in most Schizomycetes and then the form is known as *Leptothrix* or *Mycothrix*; they are branched in a few cases only (*Cladothrix*). In the latter form one extremity of a cell bends outward from the row in which it occurs, and continues its growth and divisions in a divergent direction. The

terminology adopted in the case of the Scytonemeae is also applied to this form of branching in the Schizomycetes, which has therefore been designated by the really incorrect or at least unnecessary name of *false branching (pseudo-ramification).*

In some of the larger forms of this group, *Crenothrix* for example, *Cladothrix,* and species of *Beggiatoa,* the filaments attach themselves by one extremity to fixed bodies, while the other extremity stretches free into the surrounding fluid; here therefore there is a distinction between base and apex, and this is accompanied by certain corresponding phenomena of growth, such as the direction of the branches and some others.

The formation of filaments occurs in those Schizomycetes in which growth and divisions advance only or chiefly in one, and that the longitudinal direction. If these take place alternately in two or three directions while the genetic connection is maintained, then

3. Groups of cells are produced forming **flat surfaces or masses.** The dice-shaped pockets of Sarcina ventriculi are the best-known examples of this kind. Figs. 170 *p* x and 175 *a* will give an idea of their appearance.

4. The isolated and connected forms of each of the kinds described above may again be united by coherent mucilage into larger **gelatinous masses,** which are known by the older and more general name of *Palmella,* or by the more recent term of *Zoogloea.* These masses form gelatinous layers or pellicles according to the species or culture-form, and cover the surface of the solid or fluid substratum; or if suspended in a fluid they form lumpy bodies of very various shapes and are often lobed and branched. The gelatinous cell-membranes in these masses are either fused together into a homogeneous structure, or show a stratification which varies in the different isolated cells or aggregates of cells. In the larger and more firmly united masses the cells, whether isolated or connected together, have not the power of locomotion which many of them, as we have seen, possess in the free state.

All these varieties of shape and connection are merely *growth-forms* like those designated Filamentous Fungi, Sprouting Fungi, Compound Fungus-body, &c. (see section I). But the Bacteria were at first distinguished into species and genera according to these forms of growth, and on the too hasty assumption that the forms produced from them were always like the parents, and since the year 1872 these distinctions have been precisely defined by Cohn. But it is obvious from what has now been stated that we are dealing in the present case with *form-species* and *form-genera* only, using these words in the sense assigned to them on page 120; the names Micrococcus, Bacillus, Spirillum, Spirochaete, Vibrio, Leptothrix, Zoogloea and others, applied above to the Schizomycetes, were in this sense used originally as names of genera and not as designations of forms of growth. The relations of these form-genera to the natural genera, that is to the genera founded on the entire course of development, will be considered presently.

SECTION CXXX. The forms comprised under the name of Bacteria or Schizomycetes may be distributed, in accordance with the **course of their development** and with the facts as at present known to us, into two groups, and such is to some extent the course adopted by Van Tieghem in his new text-book. The first group will contain the species which have their spores formed endogenously, the *Endo-*

sporous Bacteria; the second those which have no endogenous spore-formation, the *Arthrosporous Bacteria.* It has yet to be seen whether this distinction will be permanently maintained. It is evident from the gaps in our present knowledge that many forms are met with whose behaviour in this respect has not yet been ascertained. The distinction therefore is not a convenient one for a purely practical classification of the Bacteria.

a. ENDOSPOROUS BACTERIA.

The forms included under this term are chiefly known in the growth-form of single rods consisting of one or a few cells, or of rods joined together and forming long filaments; they may also be collected together into larger gelatinous masses or membranes. In some forms the rods are spirally twisted, and these I name here Spirillum of Van Tieghem. Others do not show these curvatures, but are either straight or very slightly bent; all these I include under the term Bacillus and place under that genus all the endosporous forms which have been hitherto known either as Bacillus or as Clostridium, Bacteridium, Vibrio, or by some other name. All non-endosporous forms bearing these names on account of their growth-form are of course excluded from the group.

The Bacteria in question are distinguished by the peculiar mode of spore-formation. At the commencement of the process the protoplasm of each cell which has hitherto been homogeneous becomes somewhat darker and in some cases visibly granular, and in the forms enumerated on page 455, which however are the smaller number, it gives the amyloid reaction. Then a darker and comparatively very small body makes its appearance in the interior of each cell and soon increases rapidly in volume, acquiring a distinct outline some time before it reaches its ultimate size and becoming strongly refringent. It has now the aspect of a glistening bluish sharply defined dark granule, and continues to grow till it has reached its definite size and shape, which it does in the space of a few hours. As it enlarges, the surrounding protoplasm or the amyloid substance disappears, and the body when fully developed is surrounded only by a pellucid substance inside the very delicate membrane of the mother-cell; this body may be termed a *spore* or *resting spore.*

In the great majority of observed cases one spore only is formed each time in a cell. The supposed exceptional case [1] of two spores being formed in a single cell is rare, and is moreover said to occur in forms which as a rule produce only one spore; it is possible that the partition-wall between two sporogenous cells may have been overlooked. The sporogenous cell is according to the species either not different from the vegetating cells of the same species or form, or is distinguished by being somewhat thicker and of a different shape; the change of shape is often caused by the appearance of a fusiform or club-shaped enlargement at one extremity, in which the spore is formed. In this case the mature spore is usually much shorter than the mother-cell, and is seen as a glistening body in the enlarged portion of the parent-cell; the apparently empty part of the cell, and in some cases also sterile

---

[1] Prazmowski, as cited on page 455.—E. Kern, Ueber ein Milchferment &c. (Bot. Ztg. 1882, p. 264, and Bull. Soc. Hist. Nat. Moscou, 1882).

cylindrical sister-cells, is attached to the spore as a longer or shorter appendage. Such structures with one swollen sporiferous extremity are the 'capitate Bacteria' of older writers. In other species there is less difference in size between the spore and the mother-cell, though the latter is never quite filled by the spore. In Spirillum amyliferum and Bacillus (Clostridium) butyricus, which show the granulose-reaction before the formation of the spore, the spot where the comparatively small spore begins and completes its formation is, according to Van Tieghem, a terminal portion of the mother-cell in which there is no granulose.

The motile forms may continue their movement during the development of the spore; they become stationary as the spore matures, and finally in all cases the membrane of the mother-cell dissolves sooner or later and the spore is set at liberty.

In most of the species which have been examined the formation of spores coincides with the moment when the substratum has expended its nutrient material or from other causes, such as an accumulation of products of fermentation, has become unfitted to support the vegetation of the species. At the same time the phenomenon does not always depend on the quality of the substratum. Prazmowski has shown that Bacillus butyricus may be in active process of vegetative multiplication while some of its cells are forming and maturing their spores.

The formation of spores when once begun extends usually to the greater number of the isolated cells and aggregates of cells in a pure culture; but a certain number of the cells remain sterile, and no definite rule determining their distribution has yet been discovered. The parts which remain sterile are seen to break up and disappear if fresh nutriment is not supplied in time; if it is supplied they may continue to vegetate. Plants grown in quantity in a pure medium and left to themselves often produce enormous masses of ripe spores.

The ripe spore varies from round to elongate-ellipsoidal or cylindrical according to the species. It has the appearance, as has been already said, of a highly refringent usually colourless body (reddish in Bacillus erythrosporus, Cohn) with a dark and sharply defined outline; in some cases it looks like an oil-drop, but reagents show that the resemblance to the latter is only superficial. It consists of a highly refringent mass of protoplasm, which with our present means of investigation is perfectly homogeneous. This protoplasmic body, as is shown in germination, is closely surrounded by a thin but firm and often apparently brittle membrane; outside the cell-wall may often be seen a pale slightly refringent envelope with lightly marked contour and of apparently gelatinous consistence, the material composition of which cannot be exactly ascertained, but which forms a delicate covering to the spore, and sometimes also appears to be prolonged at each extremity of the spore into a small tail-like appendage. Pasteur[1] was the first who described these appearances but he did not distinctly recognise their significance.

The bodies in question are proved by their germination to be spores. They are in a condition to germinate, as soon as they have reached the development described above at the expense of the mother-cell; and they retain the power of

---

[1] Études sur la maladie des vers à soie, I, 228.

germination for a considerable time, showing a remarkable degree of resistance to the effects of desiccation, extreme temperatures and the like (see section CXXXIV).

Germination commences as soon as the spore is subjected to the conditions required for the nutrition and vegetation of the species, that is, as soon as it is placed in a suitable nutrient solution at a proper temperature. It is completed in a few hours when the conditions are favourable, and consists chiefly in the development of the spore into a cell which assumes all the characters of the parent-cell as regards conformation and vegetation. The spore at first enlarges in size, loses its high refringent power and becomes pale and turbid, like a bacterium-cell when in an active state of vegetation; it then elongates and assumes the shape characteristic of the species and at once begins to divide like the vegetative cell, and locomotion may commence at the same time. When the elongation has reached a certain small amount, which is moreover different in different individuals, a membrane dividing usually into two regular valves of equal size is seen in most cases to separate gradually from the growing cell, being evidently raised off from it by the hyaline gelatinous outer layer of the membrane of the growing cell. The valves are usually thin and pale-coloured; but in Bacillus subtilis they have nearly the same amount of refringent power as the ripe spore, so that it is probable that the latter owes its characteristic appearance to the membrane which is thrown off in germination. The pieces of the detached membrane gradually disappear in the surrounding fluid. In spores which have elongated in the direction of the longitudinal axis of the mother-cell the membrane splits in the same or in the transverse direction. The direction varies with the species; the membrane for example of Bacillus butyricus according to Prazmowski, and of other species, parts longitudinally, that of Bacillus subtilis transversely.

The membrane is not thrown off in the above manner in all cases in germination, but is sometimes seen to swell up and finally disappear. I observed this repeatedly in Bacillus Megaterium and Buchner [1] saw it in the Bacillus of anthrax.

The direction of growth in length of the vegetative cell first developed from the spore in relation to the longitudinal axis of the spore or its mother-cell is the same in all observed cases as that of the spore, whether the spore-membrane bursts longitudinally or transversely, or swells up and disappears. This is the case also with Bacillus subtilis, as will be described hereafter at greater length, where according to Brefeld and Prazmowski the first cell usually issues transversely at right angles to the longitudinal axis of the spore from the spore-membrane which has burst on one side.

The above is the course of development observed especially by Brefeld, Van Tieghem [2] and Prazmowski in many of the species which contain no chlorophyll. It occurs also in Van Tieghem's species containing chlorophyll which have been mentioned above. In these the chlorophyll disappears during the formation of the spores and reappears in germination. Whether the bacterium of blue milk is one of this kind is still uncertain after Neelsen's account [3] of it and requires further investigation.

---

[1] Nägeli, p. 272.   [2] See Van Tieghem in Bull. Soc. Bot. 26 (1879), p. 141.
[3] Cohn's Beitr. III, Heft 2.

All the above phenomena are in themselves sufficiently simple, and their course is essentially the same in all the species; but it is nevertheless desirable that we should study a few examples more closely, and see in what light the parts in question present themselves and the form which the specific differences assume.

Our first example shall be the large species long known in our laboratories by the name of **Bacillus Megaterium**. This exceedingly instructive form (see Fig. 194) was first observed in boiled cabbage-leaves used for the cultivation of Myxomycetes and species of moulds, and was afterwards studied in pure cultures in water or gelatine mixed with 7–10 per cent. of grape sugar and a small quantity of meat-extract and also in a pure 2–3 per cent. solution of meat-extract. The gelatine is liquefied by the Bacillus. Most of the cultures to be described below were carried out in the summer-temperature of an ordinary room, that is, not much above or below 20° C.

This species forms small rods 2–5 μ in thickness and cylindrical in shape with the ends rounded off. The rods, which do best when obtained from spores, grow rapidly in a fresh nutrient solution where they have no competitors to disturb them, and become usually 4–6 times longer than they are broad; then they separate by transverse division into two halves or into two unequal parts, which again grow to rods of the size above mentioned (Fig. 194 a, b). A single rod floating in the solution usually appears in these circumstances even under high magnifying power to be unsegmented, and to be filled with a slightly refringent protoplasm in which only a few separate granules can be distinguished. But the application of desiccating and colouring reagents, alcohol and tincture of iodine for example, shows that the rods even in this

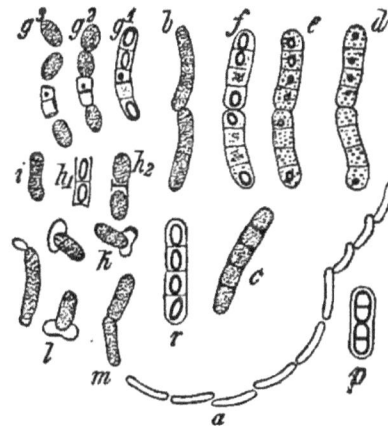

FIG. 194. *Bacillus Megaterium. a* outline of a motile chain of rods in active vegetation. *b* a pair of motile rods in active vegetation. *p* a quadricellular rod in the state of *b* after treatment with alcoholic solution of iodine. *c* a five-celled rod in the first preparation for forming spores. *d—f* successive stages of a pair of rods while forming spores. *d* about two o'clock in the afternoon. *e* about one hour later, *f* an hour later than *e*. The spores in formation in *f* are ripe towards evening; no others were formed, the one which apparently began to be formed in the third cell from the top in *d* and *e* has disappeared; the cells in *f* which did not contain spores perished by about nine o'clock in the evening. *r* a quadricellular rod with ripe spores. *g¹* a five-celled rod with three ripe spores placed in a nutrient solution after several days' desiccation, half-an-hour after noon. *g²* the same specimen at about half-past one. *g³* the same about four o'clock. *h₁* two spores with the walls of the mother-cells dried and then placed in a nutrient solution about forty-five minutes after eleven. *h₂* the same about half-past twelve. *i, k, l* later stages of germination as explained in the text. *m* a rod formed from a spore placed eight hours before in a nutrient fluid and in the act of splitting transversely. *a* magn. 250, the rest of the figures 600 times.

state consist of short members which become twice as long as broad or a little longer, and then divide by the formation of a transverse septum into two members (*p*). The transverse septa are extremely delicate when young, but when the water is withdrawn by a reagent they and the lateral wall stand out clearly distinguished from the shrunken protoplasm. Older transverse septa swell to a greater thickness in the living individual, and acquire at the same time a soft gelatinous consistence; this causes the transverse splitting of the longer rods mentioned above, the pieces remaining loosely coherent or separating entirely from one another, according to the relation between the cohesion of the jelly and the forces from without which promote a separation.

The rods are usually not quite straight but slightly curved. The curvature often appeared to me to change from one direction to another in the same individual, but it was impossible to feel sure of this point in presence of other movements to be described presently. When two sister-rods begin to separate transversely from one another, the curvature usually becomes somewhat more pronounced at the extremities where the division takes place, and the ends of the rods become slightly oblique to one another and overlap each other a little, or one thrusts itself laterally past the other, like the short commencement of a so-called false branch in Scytonema and similar genera of the Nostocaceae (Fig. 194 *b, m*). In these conformations the rods may either separate from one another in rapid succession, or remain united but always loosely united by their extremities into chains of varying numbers of members, seldom more than ten (Fig. 134 *a*). It follows from what has been already said, that a chain of this description is never quite straight; its general outline, when it is supposed to be at rest, is different in individual cases; it is irregularly undulated, and kneed where the members join one another, often forming more or less acute angles, and never approaches the form of a regular spiral. But in the stage of development of which we are now speaking the chain of rods is never in a state of rest; the rods are in constant though comparatively slow movement. There is first a movement of rotation, each rod rotating about its longitudinal axis. Secondly there is a swinging movement; one or both extremities of the rod describes a nearly circular path in such a manner, that in the first case the outer surface of the rod moves in the path of the side of a cone the apex of which is formed by the other extremity of the rod; in the second case it moves in the path of the sides of two cones the apices of which meet in the middle of the rod after the fashion of an hour-glass. Thirdly, there is a movement of progression, the rod moving through the fluid in varying and still undetermined directions. If the rods are united together into a chain, all the cells move together through the fluid in the same direction, but each separate rod shows the tendency to the rotation and swinging mentioned above. These movements are disturbed or prevented in proportion to the resistance encountered from the cohesion of the gelatinous substance which binds the cells together, and from the mass and movement of the cohering cells. Hence chains formed of several members move forward in irregularly varying directions and with not less irregularly varying undulations, and with varying degrees of articulate movement at the points of junction. Vibratile cilia or flagella have never been observed in this species.

If a pure culture is obtained on a microscopic slide from the smallest possible number of spores, the Bacillus will be found in 24–28 hours to have multiplied to such a degree, that the nutrient fluid, which at first was perfectly clear to the naked eye, has become turbid and milky; the organisms themselves are diffused with tolerable uniformity through the fluid, or some of them are settling to the bottom. Most but not all of the little rods now separate from one another, and each consists of about 4–6 isodiametric cells. The transverse walls between the cells, though always delicate, become gradually more defined. The protoplasm fills the cell-cavity less perfectly, and appears to line the wall of the cell and to inclose a clearer central space; it is also dotted with a number of granules, some of which are highly refringent (Fig. 194 *c, d*). In this stage the cells have ceased to increase in size and to divide, and are preparing to *form spores*. The commencement of the formation of a spore in a cell is indicated by the apppearance of a small roundish highly refringent body in the protoplasm, usually close to the surface of one extremity of the cell. The nature of this body cannot be clearly made out, but the first impression which it conveys is the same as if one of the highly refringent bodies just mentioned as disseminated through the protoplasm had increased somewhat in size. The body then grows perceptibly larger while the protoplasm round it diminishes (Fig. 194 *d*, *e*, *f*). In the course of a few hours it grows into a longish cylindrical object, which is shown by its subsequent behaviour to be a *spore*. The spore is slightly shorter than the cell which produces it, but its breadth is only from a third to one half of it. It has a sharply defined outline, appears to be perfectly homogeneous, is very highly refringent, and has a bluish tint. This is the appearance which it presents long before it attains to its full size. As it grows, the protoplasm which surrounds it becomes continually clearer and more transparent; when full-grown it lies straight or oblique in a pellucid substance within the membrane of the mother-cell, which is persistent for a time, but which if kept in the fluid at length disappears entirely and leaves the spore at liberty. Only one spore is ever formed in each cell. It is remarkable how often the formation of spores begins in the terminal cells of a rod, though this is not always the case, and proceeds rapidly from them to the adjacent cells. It takes place not unfrequently in all the cells of a rod, and always in the majority of them in successful cultures; but single cells often prove an exception to this rule, and spores sometimes begin to form in a cell but are never fully developed. In all these cases the cells which do not produce spores die off unless a timely supply of fresh nutriment excites some of them to renewed vegetation.

The movement of the rods becomes slower perhaps when spores begin to be formed, but does not at once cease, and this is highly inconvenient for those who are watching the process of spore-formation. It is only when all the cells of a rod have formed spores, or those which produce no spores are dead, that the rods lose the power of motion.

The *germination* of the spores (Fig. 194 *g–m*) has been observed in material kept dry for at least 24 hours after maturation. If material in this state is placed in a fresh nutrient solution, the cells in which no spores have formed cease to grow, and drop off one by one. The membranes of some of the sporiferous cells may have burst before desiccation, otherwise they still remain round the ripe spore; in the latter case they are now seen to swell up gradually and dissolve, so that the spores are released from the previous combination. At the same time the dark outline and the high refringent power of the spore itself disappear, and it assumes the uniformly pale aspect of a rod when actively vegetating, having its outline, especially at the extremities, at most only a little more distinctly defined than in the rest of its body. In this condition it increases slowly in volume during several hours (8–12 in the observed cultures) till it has become as broad as a normal rod without essentially deviating from its original shape. Then not unfrequently a thin membrane divides transversely or obliquely into two portions and is all at once separated from its surface, and the delicately circumscribed cell slips out, having the breadth, shape and structure of a short rod, and then grows rapidly in the direction of its previous longer axis and commences the active vegetation described above.

[4]   H h

The removal of the transversely-opened membrane is the best and only certain demonstration of the fact that the spore is actually invested with a special membrane. This membrane is easy of observation, especially in cultures in very flat drops of the fluid. It sometimes divides quite across into two halves, which are attached at first like caps to the extremities of the elongating rod and are afterwards torn from them. I often failed to observe the final removal of the membrane in larger quantities of the fluid. There each cap, distinguished by its more sharply defined outline, was seen to be attached to the extremities of the growing spore and gradually ceased to be distinguishable; it would seem therefore that in these cases the membrane, after splitting transversely, is not removed from the spore in connected pieces, but swells up or is dissolved and so disappears. The membrane often clings to the rod-cell which has slipped out of it in such a manner that the longer diameters of the two bodies cross one another, and hence the appearance is produced as if the growth in length in germination is in a direction at right angles to the longer diameter of the spore or parent-rod; but this is not really the case.

The species which we are considering occurs also in another form, namely in long curved chains, in which a segmentation of the smooth cylindrical rods is only obscurely distinguishable, while the isodiametric cells on the other hand are distinctly defined, are slightly swollen and protuberant, and are in many cases themselves curved at the points of greatest curvature. Torulose chains of this kind or, to use the terminology noticed above and founded on growth, *cocci* grouped in rows, are usually developed in great abundance, being loosely intertwined into thick convoluted masses and sometimes breaking up here and there into separate cells. They are motionless or show a very slight movement. There was little or no formation of spores. I almost invariably discovered this form when the cultures were rendered impure by other small Bacterium-forms, but the Bacillus formed the majority. It may be a question how far this fact points to the cause of the appearance of the torulose chains, but it is certain that they become transformed into the even rods when the culture is pure.

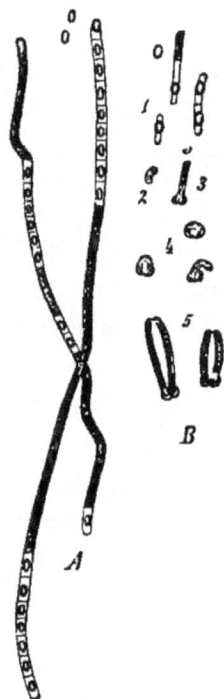

FIG. 195. *A Bacillus Anthracis.* Two filaments grown on a microscopic slide in a solution of meat-extract, partly in an advanced state of spore-formation; at the top two ripe spores which have escaped. The spores are drawn somewhat too narrow; they are nearly as broad as the transverse diameter of the mother-cell. *B Bacillus subtilis.* 1. fragments of a filament with ripe spores. 2. spore beginning to germinate; the outer wall ruptured transversely. 3. young rod projecting from the wall of the spore in the usual transverse position. 4. Germ-rods curved in a horse-shoe shape and with the extremities connected; one of them with one extremity subsequently released. 5. Germ-tubes with the two extremities remaining connected and already greatly increased in size. Magn. 600 times.

The cells of the **Bacillus of anthrax, B. Anthracis,** Cohn (Fig. 195 *A*), approach 1 μ in thickness in vigorous specimens grown in the blood of animals attacked by it, and grow to about 3–4 times that length. They are connected together in the blood into straight rods of different lengths; in cultures on some suitable dead substance they form long pluricellular filaments; these may be much twisted or bent into acute angles, and are at the same time often collected together in great numbers into bundles or sheaves and coiled round one another; such formations appear as flaky precipitates in larger quantities of fluid. The mode of spore-formation in the filaments is the same as in the preceding species. The ripe spores are as broad as the mother-cells but much shorter, and roundish to ellipsoidal in outline; they are usually set at liberty soon after reaching maturity by the disorganisation of the membrane of the mother-cell. They behave in germination like the spores of the preceding species, with the difference only that they have not been observed to part with an empty membrane; I have at most seen a form of cap on the extremities of the spore, Fig. 194 *i*, and sometimes it seemed as if a small delicate cap was at length removed from them.

In some cultures, as for example in solutions of peptone, I have seen the filaments of this Bacillus break up over a wide area into round cells, cocci, which collect together into groups forming lumps or clusters. These with some doubtful exceptions were found to be dead. When removed into a good fresh nutrient solution they showed no signs of growth but gradually perished, while a few initials gave rise to a new and luxuriant growth of filaments; but it remained uncertain whether these initials came from coccus-cells which had retained their vitality, or from spores present in the solution.

According to the statements of the pathologists[1], which however require morphological proof, the Bacillus of anthrax introduced into an animal appears to vegetate there during its first period of growth, and during the first stages of the disease as shown by the fever-temperature, in the form of small round isolated cells or cocci; after some time it developes into the elongated rods which fill the blood-passages in the last stage of the disease. I did not find this statement confirmed by my experiments with guinea-pigs; in 20 hours after the introduction of spores into the skin I found an abundant growth of rods in the blood-vessels near the point of inoculation.

This Bacillus shows no power of independent movement in any stage of its development apart from the slight oscillatory motions of doubtful origin which have been frequently observed during the first stages of germination.

Bacillus subtilis (Fig. 195 *B*), known as the hay-bacillus from being commonly obtained from an infusion of hay, behaves much in the same manner as B. Anthracis when forming its spores; but its filaments are as a rule narrower, and when grown in largish quantities of fluid they are closely compacted together into a soft gelatinous pellicle covering the surface of the fluid while its own upper surface is dry. If the plant grows vigorously in a comparatively narrow vessel, the surface of the pellicle becomes wrinkled and folded in consequence of the opposition offered to its surface-growth by the wall of the vessel.

The formation of spores is exactly the same as in B. Anthracis; the ripe spore with the zone of the mother-cell which surrounds it is often somewhat broader than the cell was originally.

It has been shown by Brefeld after a skilful examination of this species, that the outer wall of the spore, which is comparatively thick and continues to be highly

---

[1] See Roloff in Archiv f. Thierheilkunde, IX, Heft 6 (1883), p. 459.

refringent, splits across the middle in germination after its first distension into two portions, which however remain firmly united at one side. The protoplasm invested with a delicate membrane elongates in the direction of the longer axis of the spore and of the longer axis of the mother-cell which coincides with it, and at the same time usually bends through about 90°, and one extremity is thus thrust out of the opening in the outer wall of the spore while the other extremity remains in the spore. The protoplasmic body then developes into a rod. It appears from these phenomena as if the direction of longitudinal growth in this case in germination is at right angles to that of the parent-filament, but this is not really so. The wall of the spore which has opened on one side is evidently very elastic; it manifests so considerable a resistance to the elongating rod as it bends, that the rod with both its extremities fixed in the spore becomes curved before the one extremity is set free. The resistance sometimes goes so far that both extremities remain in the spore, and in that case the elongating rod assumes the form of a horse-shoe, the limbs of which may be of considerable length. In other respects its growth is entirely the same as the ordinary growth which begins with the formation of a bend in the protoplasm, and if the rod subsequently divides into partial rods, two separate rods answering to the two limbs of the horse-shoe may often be seen to project side by side from the ruptured wall of the spore (Fig. 195 *B*, 4, 5).

The cells produced in germination grow and divide by the formation of transverse walls; the products of this growth do not however remain united in a filament, but separate one after another into rods consisting of a few or in many cases of a single cylindrical cell, which may be 4–5 times longer than broad. These rods, which multiply very rapidly and abundantly in fluids containing a good supply of food, display active swarming movements all the time of the kind described above. I was unable to satisfy myself with regard to the presence of cilia or flagella during the swarming stage even in Bacillus subtilis. The rods become disseminated through the fluid as they swarm and render it turbid. The last stage of vegetation is indicated by the entrance into a state of rest, in which the cells increasing greatly in size remain united together in filaments and the filaments form pellicles, till at last formation of spores begins afresh.

b. **ARTHROSPOROUS BACTERIA.**

Section CXXXI. In the course of the development of the species in this group single members may simply separate from their connection with others, and under suitable conditions become the initial members of new combinations; they may therefore claim to be called spores. In other respects there is no general characteristic distinction between them and the purely vegetative members.

In connection with the fact that the species in this group are less like one another than the Endosporous Bacteria, and that some species have a greater variety of growth-forms, the mode of formation of the cells which may be termed spores varies greatly in the different species.

It is to the forms of this group that I give the name Arthrobacterium proposed for them in my introduction on page 454; they resemble Bacillus in their vegetative form, and are usually known as species of Bacterium, but require a more exact.

morphological investigation. The other genera are distinguished by the difference in their conformation.

Leuconostoc mesenterioides, the 'frog-spawn' of sugar-factories, consists in the vegetative state of coiled rosary-like chains of small round cells inclosed in firm sheaths of mucilage, and accumulated in great numbers into large compact gelatinous masses ('zoogloeae'). Many of the cells perish at the close of the vegetative season when the nutrient substratum is exhausted. Single cells of the rows, disposed at irregular intervals, become somewhat longer than the rest and have a thicker wall, and their contents are apparently denser and more highly refringent. They are at length set at liberty by the dissolution of the mucilage, and they develope in a fresh nutrient solution into new rosary-like chains resembling the parents.

Arthrobacterium (Bacterium) Zopfii is the name given by Kurth to a species which was originally found in the intestinal canal of fowls, and which vegetates at first in the form of small rods when placed on a fresh nutrient substance. If the substratum is solid, gelatine for instance with meat-extract, the rods remain connected together and form long filaments, often with intercalary spiral coils and knots; in a fluid substratum short resting filaments are formed at high temperatures only, and at 20° C. the rods separate and swarm. When the nutrient substratum is exhausted the filaments and rods break up into isodiametric members, cocci, which do not themselves divide again, resist unfavourable external agencies, and develope into rods or filaments if supplied with fresh nutriment; they may therefore be termed spores.

A similar course of development is described by Zopf in the case of **Arthrobacterium (Bacterium) merismopoedioides**, a species which vegetates in muddy water; it is complicated however by the circumstance that the isodiametric cocci which are the product of the disruption of the rods do not pass into a state of rest, but after swarming for a time lie without motion on the surface of the water; then they divide first in one and then in two directions which cross one another alternately at right angles, and arrange themselves in flat tabular gelatinous expansions in a manner corresponding to the divisions. If the cocci are now placed in a fresh portion of muddy water they will again develope, after swarming for a brief time, into rods or filaments.

Crenothrix, Cladothrix and **Beggiatoa** are closely comparable in their development with the preceding species, but they appear to exhibit a greater variety of growth-forms. The agreement consists in the copious and uniform multiplication of the coccus-form, which is alternately aggregated into zoogloeae or is free and even motile; the cocci are the product of filaments and rods and may develope into them again. The greater complication of the cycle of forms is due to the assumption by portions of the filaments and rods of the form of motile spirilla, and to their multiplying abundantly in this form. The return from the spiral to the straight forms has, as far as I know, never been observed. The following brief description from Zopf may assist in the understanding of these species.

Crenothrix Kühniana (Fig. 196) is a Schizomycete which occurs frequently in water containing some amount of organic substances, and sometimes in quantities which are dangerous to health. Its cocci *a* are spherical in shape and 1–6 μ in size. They multiply by successive bipartitions and in so doing are united into zoogloeae (*f, g*), which grow from microscopic minuteness to more than a centimetre in size and become aggregated in the water into slimy masses a foot in depth. The gelatinous matter is at first colourless, but may assume a colour varying from brick-red to dark brown by admixture of hydrated oxide of iron. When grown in bog-water the cocci develope into rods or filaments (*h*) of unequal thickness, which at a certain age become invested with a continuous thin but firm gelatinous sheath with the same admixture of iron as is found in the jelly of the zoogloea-forms. The single rod-like cells within their sheaths pass by repeated transverse bi-partitions into the form of nearly isodiametric members, which then round themselves off. The members in the thicker filaments often assume a flat disk-like shape, and then divide into 2–4 small cells by walls parallel to the longitudinal axis of the filament. Both these cells and the rounded members of the slenderer filaments ultimately escape in the form of cocci from the sheath, being set free partly by the swelling of the sheath along its whole length, partly by its rupture at the apex (*r*). In the latter case some of the cocci slip of themselves out of the opening in the sheath, while others are passively thrust out of it by the growth in length of the other parts which remain in the sheath. The cocci may, though they rarely do, become motile, and pass again out of this state into the resting zoogloea-form; they also develope once more into rods and filaments in the manner

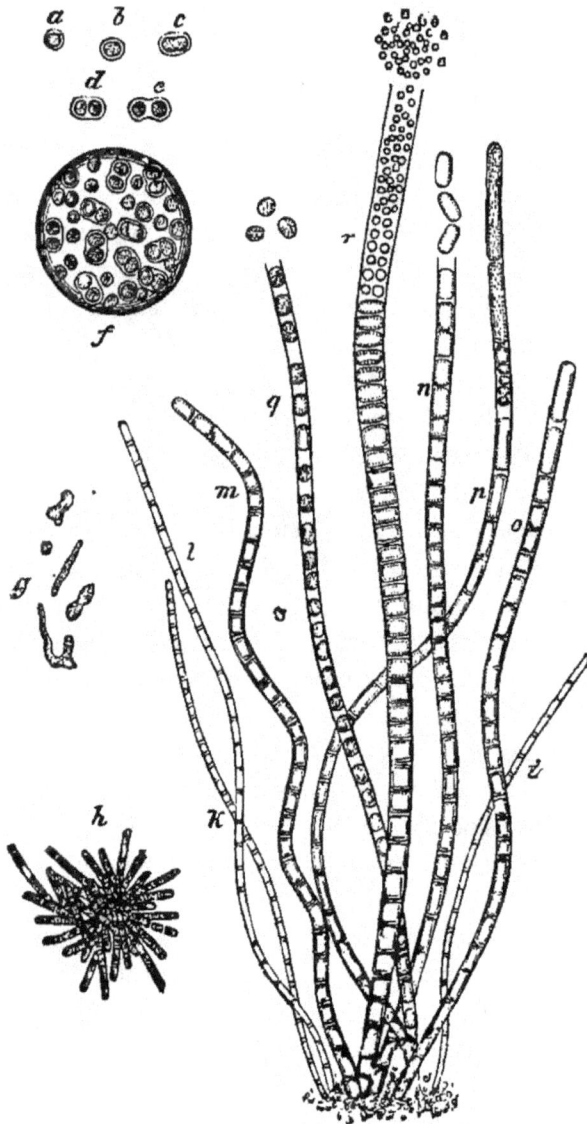

FIG. 196. *Crenothrix Kühniana.* *a—e* cocci or spores. *c—e* cocci dividing. *f* cocci collected into a group and connected together by a gelatinous substance ('zoogloea'), the contour dark. *h* a group of cocci developing into filaments. *i—r* filaments of various forms and stoutness attached below to a substratum; *m—r* show the formation of the common sheath round the single members; *q* and *n* separating above into members; *r* with the upper members becoming successively broader and comparatively shorter, the uppermost members having separated by longitudinal divisions into round spores ('cocci'), which have escaped at the upper end from the sheath. *g* cocci-zoogloeae. After Zopf. *g* natural size, the rest magn. 600 times.

which has already been described. In addition to these forms curved spirilla-like forms are also found, which may also break up into pieces, but without passing, as far as has been at present observed, into the motile state.

Beggiatoa alba (Figs. 197,198) forms filaments which in an intact state are attached in an erect position to fixed objects in dirty water, in water discharged from factories, and in hot sulphur springs. The filaments vary in thickness from 1 to more than 5 μ, and consist of a single row of cells, the protoplasm of which contains granules of sulphur in quantities differing in each cell (see page 455); when the sulphur is very abundant it may be difficult to perceive the boundaries between the cells. The filaments have no separate common sheath and readily divide transversely into pieces. Their cells pass successively from the lengthened rod-form into the isodiametric form, and these in the case of the thicker filaments into flat discs which finally divide by longitudinal walls into four quadrants (Fig. 127, 6–8). The disc-like cells as well as the isodiametric members of the slender filaments separate after a time from one another (Fig. 197, 9) and round themselves off, and then become active swarm-cells (Fig. 197, 10); at length they come to rest and attach themselves to fixed objects. They multiply abundantly by bi-partition and form irregu-larly-shaped zoogloea-heaps.

FIG. 197. *Beggiatoa alba.* 1 group of attached filaments. 2—5 portions of filaments of different stoutness, 5 in the act of breaking up into fragments. The small dark circles in the interior are granules of sulphur; in the parts of the filaments where the granules are abundant the transverse segmentation is indistinct, in others it is more clearly seen. 6—8 fragments rich in sulphur showing the transverse septation clearly after treatment with methyl-violet solution; in 8 the longitudinal division also is shown in the separate members (formation of cocci or spores). 9 filaments which have broken up into spores. 10 spores in states of movement. The dark circles are in all cases granules of sulphur. After Zopf. 1 magn. 540, the rest of the figures 900 times.

They may also develope into rods and these again into the filaments above described after the rods have in many cases themselves passed through the swarming-state. In this species also spirally twisted filaments are found as well as the straight ones which we have been hitherto considering,

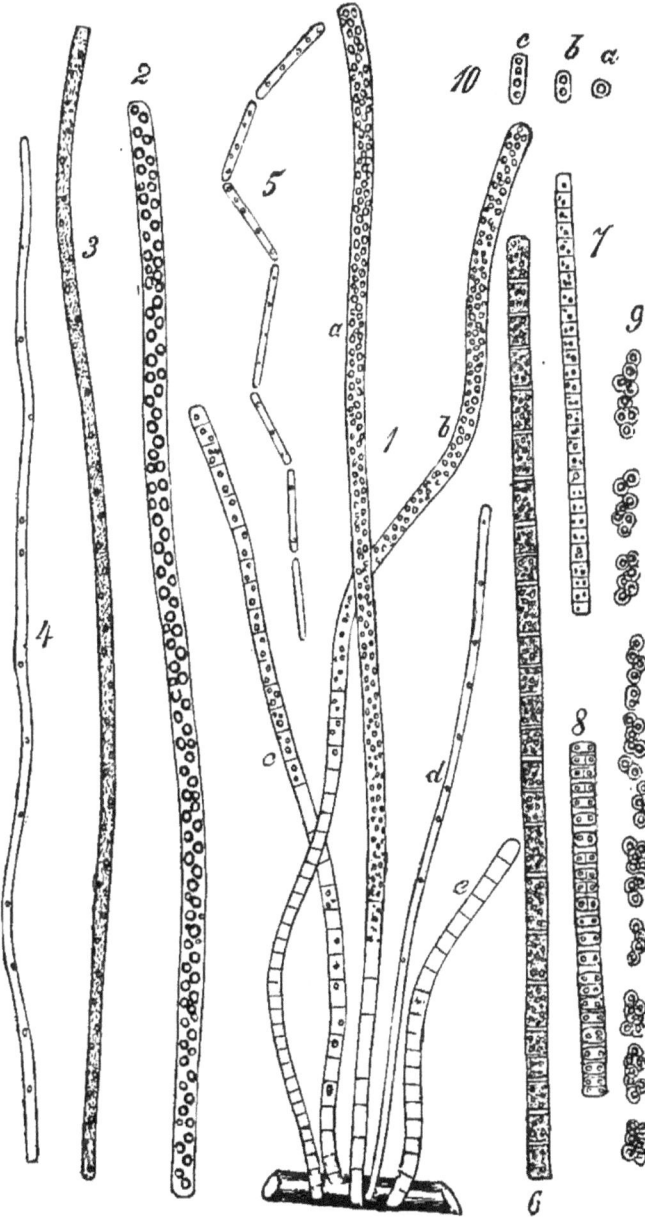

and these break up into fragments containing from two to several coils and exhibiting active movement; they were formerly known by the name of Ophidomonas, and are said to have a long oscillating cilium at each extremity (Fig. 198 *E*).

The same states have been observed in **Beggiatoa roseo-persicina** as in B. alba; the net-like zoogloea-form, once known as **Clathrocystis**, is a peculiar and remarkable feature in this species.

Cladothrix and **Leptothrix buccalis** of tooth-caries resemble each other in their development. Further details will be found in Zopf's descriptions. The Fungi of mother of vinegar, **Arthrobacterium aceti** and **A. Pastorianum** (Hansen), must also be placed in the arthrosporous group. They are distinguished it is true, as Hansen has observed, by the occurrence of many large vesicular cells between the small cocci or rod-cells of a chain, and the almost constant appearance of these cells at once suggests that they are connected with some process of spore-formation. But the observations afford no distinct support to this view, and the phenomenon must for the present be classed with those of involution which were mentioned on a former page. The Micrococcus also of Pasteur's fowl-cholera may also belong to this group [1].

SECTION CXXXII. The foregoing review of the Bacteria will supply us with some safe means of determining the **question of the specific value of observed forms,** a question which is at present the subject of much discussion and which must not therefore be ignored in this place.

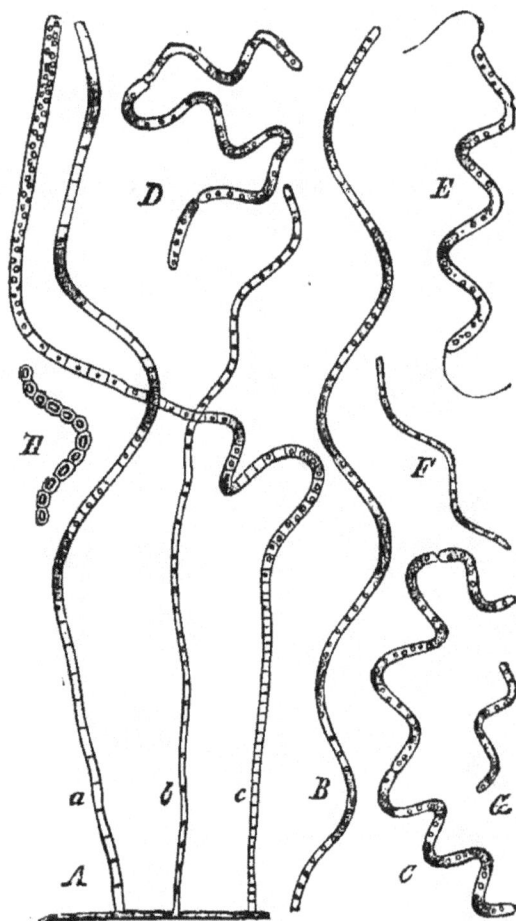

FIG. 198. *Beggiatoa alba.* Curved and spiral forms. *A* group of attached filaments. *B—H* portions of spiral filaments ; *C, D, F—H* in the act of dividing into smaller fragments and without movement; *H* with the separate cells distinctly shown. *E* swarming fragment ('Spirillum') with a cilium at each end. The sulphur-granules here as in Fig. 197. After Zopf. Magn. 540 times.

There are two views on this subject which appear at least to be diametrically opposed to one another. One of these is, as it seems to me, incorrectly ascribed to Cohn, and maintains that every Bacterium which occurs in the same growth-form and produces the same effects of decomposition, though this latter point does not strictly fall within our limits, represents a species in the sense in which the word

---

[1] Pasteur in Comptes rendus, 90 (1880), pp. 239, 952, 1030, and 92 (1881), p. 430.

is used in natural history. The fact is that Cohn in his publication of the year 1872, which laid the foundation for the morphological treatment of the group, distinguished a certain number of genera, Micrococcus, Bacterium, Bacillus, Vibrio, Spirillum, &c., by a series of marks, and especially by the shape of the individual cells and their simplest forms of connection, and gave the name of species to the several forms which recur regularly within each of these genera, and have a characteristic shape, decomposing effect and other qualities. It appears therefore that what Cohn distinguishes is that which we have named above form-genera and form-species.

The other view goes so far in the opposite direction as to deny the existence of distinct species of Bacteria, and to regard their forms as modifications of one species, or, as it may be expressed in other terms, it supposes that they are modifications which may be transformed into one another by breeding. Earlier allusions to this view are to be found, but it was distinctly opposed to Cohn's classification by Lankester [1] and Lister in 1873 [2], and Billroth in 1874 included all forms of Schizomycetes with which he was acquainted in one collective species Coccobacteria septica. It subsequently received support from the views which Nägeli expressed in 1877 in the words, 'I have in the last ten years examined some thousands of Schizomycetes and I could not maintain, except in the case of Sarcina, that there is any necessity for distinguishing them into so many as two specific forms [3];' he adds however, *that he is far from asserting that all the forms do belong to a single species, and that it would be rash to express a decided opinion in a matter in which morphological observation and physiological examination are both so defective.* He gave utterance to similar sentiments in 1882 [4]. He accepts in fact the principle which led Cohn to establish his form-genera and form-species and the species which he founded on physiological characters, namely the necessity for a provisional arrangement, whilst expressly declining to say whether the forms as distinguished by him do actually correspond to real natural history species.

Nägeli's words quoted above in full contain a pregnant criticism of the whole point in dispute as far as it has at present been explained. Neither side rests on the only sure foundation, an exact observation of the continuity or non-continuity of the development of the supposed forms or species, and this is especially apparent in Billroth's work. Without this observation the question cannot be decided; it is more necessary in this case because the forms in question are small and very like one another, and are often mixed up together and liable therefore, unless very carefully observed, to be mistaken one for another. Lankester made some approach to an exact observation of continuity in one case only, in which the characteristic tints of his Bacterium rubescens (Beggiatoa roseo-persicina) showed the connection between the forms with more than usual distinctness. We have before us at present some

---

[1] [Professor Ray Lankester in a letter published in Nature, vol. xxxiii. p. 414 (March 4, 1886), pointing out the significance of his observations upon Bacterium rubescens published in 1873 in relation to the pleomorphism of Bacteria and criticising the statement in the text, says, ' I cannot think that he [De Bary] gives a correct statement of my relation to the conclusion which he finally adopts. The view which I put forward in 1873 is *precisely* that which Professor De Bary now espouses.' For further particulars the reader is referred to Professor Ray Lankester's letter.]

[2] Both in the Q. J. Micr. Sc., new series, XIII.　　　　　　[3] Die niederen Pilze &c., p. 20.

[4] Unters. ü. niedere Pilze, p. 130.

exact investigations into the morphology and life-history of these plants. It results from these investigations that the forms above described as cocci, rods, filaments, are growth-forms, like a tree or shrub, a Filamentous Fungus, Sprouting Fungus or Fungus-body. It has been shown by Cienkowski, Neelsen, Hansen and Zopf, that there are species which can assume the different growth-forms one after another, sometimes with astonishing rapidity. R. Koch, Brefeld, Prazmowski, and Van Tieghem, have made us acquainted with other species with greater uniformity of growth, and Buchner has shown that external causes give rise to variations in the same growth-form in one or more of these species. These investigations all confirm the view advanced by Cohn, that there are species of Bacteria corresponding to the species of higher organisms, but the distinctions between them are not those of Cohn's form-species. There are comparatively *uniform* species, though they may be capable of some variation, such as Bacillus subtilis, B. Anthracis and B. Megaterium; on the other hand there are *pleomorphous* species, especially among the arthrosporous Bacteria, which may appear first in one and then in other quite different growth-forms. It may be assumed in cases of variation and great differences of form, that external causes operate to determine the form, and that the growth-form may be the result of an adaptation to varying external agencies, as happens in cases like that of Mucor (see page 154), though the operation of these external form-determining causes has not yet been demonstrated in all cases. In connection with this point it may be considered probable, that the vegetative process of the different growth-forms of a species may cause different results of decomposition in the substratum, and that the decomposing effects of the same form may vary with the substratum. It is with due attention to these considerations that the determination of the species of the Bacteria must now proceed, we may almost say, begin; it is obvious that this must rest essentially on the morphology of the organisms, while at the same time the physiological relationships must not be disregarded in the course of the investigation.

SECTION CXXXIII. As regards the **place of the Schizomycetes in the natural system**, it is apparent from the foregoing statements that the course of their development does not point to any close affinity with the Fungi. To say that they are offshoots of the Fungi is to 'contradict all trustworthy observations'[1] so flatly, that the view need not be seriously discussed in this place. They are termed Fungi only in the sense of their being Thallophytes which contain no chlorophyll, and with reference only to the vegetative process implied by the absence of chlorophyll, while the course of their development and their classification are entirely disregarded. The species of Bacillus and Spirillum therefore which have been mentioned above as containing chlorophyll are at all events no Fungi.

The forms included above under the name of **Arthrosporous Schizomycetes** show an unmistakeably close affinity with the chlorophyllaceous and phycochromaceous Algae, which form the group of Nostocaceae in the wider use of the term as including the Nostocaceae and Chroococcaceae. This has been generally allowed since Cohn drew attention to the point in 1853, and it has been recently and very fully worked out by Zopf. According to our present knowledge the Arthrosporous Bacteria are

---

[1] See Cohn, Beitr. II, p. 188.

simply Nostocaceae or Schizophyta which contain no chlorophyll; at the same time the position of the entire group in the general system remains still undetermined[1].

Of the forms which are here termed **Endosporous Bacteria** it can only be said at present that they come nearest of all known forms to the Arthrosporous Bacteria, and apparently are very nearly allied to them. It may be repeated in this place that the sharp separation between the two groups rests on the knowledge of them which we at present possess, and may disappear with its extension. Till then the separation must be maintained, and it must remain an open question whether the *resemblance* between the two groups really implies close *affinity*, or whether the endosporous forms may not stand in still nearer relationship to other members of the system. If we look around us under the guidance of ascertained facts for such affinities, we are led once more, as Bütschli has also pointed out[2], to the Flagellatae. The arthrosporous form such as Beggiatoa, with their generations alternating between a state of rest and one of swarming movement by aid of cilia, show an unmistakeable resemblance to the simpler forms of that varied group. The mode of forming their spores which is characteristic of the Endosporous Bacteria finds its *analogue*, as far as we can venture to speak in the present state of our knowledge, only in the formation of the spores or cysts, to use the customary phrase, in the simple Flagellatae[3] known as Spumella vulgaris, Cienk., and Chromulina. In these species the spore is formed, as in the Bacteria in question, inside and from a portion of the protoplasm of a cell, and this mode of formation occurs nowhere else among the lower Thallophytes. We may at least suspect that a homology is indicated by this in itself only analogous phenomenon, but the facts which have been observed lend no support to the assumption. In dealing with this question we must keep well in mind that we are not equally well acquainted with all portions of the range of organisms included under the name Flagellatae, especially as regards the course of development of the species, and that their affinities must be to some extent obscure and uncertain. For this reason we will not add anything further to the remarks which have now been made.

But if we assume for a moment a connection between the Bacteria and the Flagellatae, it is evident that as a consequence the following series of forms converge to the Flagellatae: *firstly*, the series of Bacteria and Nostocaceae; *secondly*, that of the Mycetozoa (see page 445); *thirdly*, that of the chlorophyllaceous Algae, with which are connected in ascending line the main series of the vegetable kingdom and of the Fungi as one or more lateral branches[4], and perhaps also side by side with the chlorophyllaceous Algae some smaller groups which are now placed with the Chytrideae; *fourthly* and lastly the Rhizopoda and the Protozoa with the Animal Kingdom which connects with these in an ascending line. On the above assumption the position of the series of Bacteria in the whole system would be the definite one of a group of organisms linking with the Flagellatae as a common point of commencement and departure, and coordinating with the series of the Algae

---

[1] Bot. Ztg. 1881, 1.   [2] Bütschli (as cited on page 453), p. 808.
[3] Cienkowski in Schulze's Archiv f. mikr. Anat. VII, p. 434.—Bütschli, (as cited on page 453), pp. 797, 816, t. 45.
[4] Bot. Ztg. 1881.

or Mycetozoa. This would not affect the relations between the Endosporous and Arthrosporous Bacteria, or between the latter and the Nostocaceae; nor can there be a doubt that the Nostocaceae which contain chlorophyll and phycochrome are in any case further removed from the Flagellatae than the allied species of Beggiatoa and other arthrosporous forms, and that they therefore occupy the other extremity of the whole series, which has received the name of Schizophytes and is the more remote from the supposed point of departure.

With respect to the coordination of the Endosporous Bacteria and the rest of the Schizophytes, we can only repeat what has been already said, that the final determination of the point must be deferred till we are in possession of more perfect knowledge of single forms.

CHAPTER XI.—MODE OF LIFE OF THE BACTERIA.

SECTION CXXXIV. **Capacity of germination and power of resistance in the spores.** All spores of Bacteria in which the point has been investigated are capable of germination from the moment that they are mature, provided that the conditions are favourable. If prevented from germinating they show wonderful power of resisting the external agencies which are usually pernicious or fatal to living organisms, and individual spores show this power in different degrees in different species.

These points have not yet been sufficiently investigated in the Arthrosporous Bacteria. Kurth[1] found that the spores ('cocci') of his Bacterium Zopfii are killed in from 17–26 days, when dried in a moderately high temperature (37° C.) and then kept in an air-dry state at the ordinary summer temperature, while the vegetating rods of the same species died in 7 days when subjected to the same treatment. In a heated fluid their death-point is about 56° C. Similar small powers of resistance to desiccation and high temperatures would probably be found in most of the forms, such as Beggiatoa and Crenothrix, which are adapted to vegetate in water.

On the other hand the spores of many endosporous forms are instances of the highest powers of resistance. Those of Bacillus subtilis retain their vitality for years when kept in an air-dry condition, and those of B. Anthracis will remain alive, according to Pasteur[2], in absolute alcohol and after being exposed for 21 days to the influence of pure oxygen compressed by a pressure of ten atmospheres. We have no precise observations extending over larger periods of time, but Brefeld found the power of germination unimpaired after the lapse of three years when the spores were kept in an air-dry condition, after the lapse of one year when they had lain in water; and from these facts as well as on account of other characteristics to be noticed presently, we may safely assume that their powers of resistance are at least equal to those of the most resistent spores of Fungi (page 344). The spores of

---

[1] Bot. Ztg. 1883, 409.
[2] Charbon et septicémie (Comptes rendus, 85 (1877), p. 99).

the Bacillus in question also possess wonderful powers of withstanding extremely high temperatures. Even in a fluid they are proof against a temperature higher than that of boiling water. Brefeld[1] found that they all germinated after being boiled for a quarter of an hour in a nutrient solution, the greater part of them after half an hour of the same treatment, a smaller number after the space of one hour, none when the boiling had continued for three hours. When spores were heated in nutrient solutions up to 105° C. they were killed in 15 minutes, up to 107° C. in 10, and up to 110° C. in 5 minutes.

Fitz found in 1882 that the spores of his Bacillus butylicus (? B. butyricus of Prazmowski) endured a temperature of 100° C. for a period varying from 3 to 20 minutes, according to the special quality of the spores and the medium employed. Temperatures below 100° C. were sufficient to kill the spores when they were exposed to them for a longer continuance. In solution of glycerine death ensued in 2–6 hours at a temperature of 95° C., in 6–11 hours at 90° C., in 7–11 hours at 80° C.; vitality was not destroyed after exposure for 12 hours to a temperature of 70° C. The power of resistance was less in a solution of grape-sugar; the spores were killed in 6 hours at a temperature of 90° C.

Buchner[2] found that some of the spores of the Bacillus of anthrax were killed by being boiled in water for two and three hours, and that after four hours' time they were all dead. Those of Bacillus Megaterium retain their power of germination after being boiled in water for a few minutes. Of spores of less certain derivation occurring in ordinary waters Pasteur states that they can even withstand a temperature of 130°.

These facts, and statements of a similar kind occurring in publications on Bacteria, make it probable that the death-point for the spores of the Endosporous Bacteria is generally very high, though it varies with the nature of the medium. But here as elsewhere results obtained in one case must not be at once assumed to be certainly true in others, since Brefeld in the work from which we have quoted has shown that the spores of a form of Bacillus, not B. subtilis, will not live in boiling water.

That the spores of Bacteria are able to bear extremely low temperatures has not been proved by direct experiment, but may be concluded from the behaviour of the vegetating cells which will be considered below.

The account given in section XCVI of the **external conditions of germination** is also generally true of the Bacteria, and to this the reader is referred.

The minimum and optimum **temperature required for germination** especially in the Endosporous Bacteria appear to be usually high, other conditions being equally favourable; at least most of the experiments show that germination does not take place or is very slow in the temperature of an ordinary room, and becomes active only when the temperature is raised. In the case of Bacillus subtilis the minimum is certainly below the temperature of an ordinary room, for germination proceeds in it, though slowly. According to Prazmowski from 30°–35° C. is near the optimum; I myself have seen spores some days old germinate in the most vigorous manner in a temperature of 40° C. three hours after they were sown. Bacillus

---

[1] Schimmelpilze, IV.
[2] See Nägeli, Unters. ü. niedere Pilze (1882), p. 220.

Anthracis will not germinate, as far as is known, in the temperature of a room oscillating about 20°; the minimum is said to be from 35°–37° C., and the optimum can scarcely lie much higher. On the other hand dried spores of Bacillus Megaterium several days old germinated without exception in a summer temperature of 20°–25° C. in 8–10 hours after they were sown. The cardinal point is not yet completely established, but the foregoing data suffice to justify the general statements in the introduction to the chapter, and also to show that specific differences exist in the Bacteria as in other groups, and must be investigated in each case.

All spores of Bacteria which have been examined resemble those of the Fungi described in page 351 in requiring for their germination a supply of proper nutrient substances in addition to the water which they absorb; they germinate therefore only in nutrient solutions or on a substratum containing water and a nutrient substance. Observation has shown hitherto that the food required to induce germination is qualitatively the same as that which is necessary for vegetative development; at least germination takes place when this food is present, and we do not yet know whether it can take place under other conditions.

Section CXXXV. **The general conditions and phenomena of vegetation** in Bacteria are, as might be expected, analogous with or like those in other plants, especially the Fungi (see section XCVII).

Although comparatively few cases have been carefully examined to determine the **relation of temperature** to vegetation in Bacteria, it would seem that the range of temperature is great and the optimum usually high.

Brefeld[1] determined the activity of the vegetative process at different temperatures by observing in specimens that were well supplied with nutriment how long a time elapsed before a division took place in a rod. He found that with a parity of conditions a rod formed a division every half-hour when the temperature of the air was 30° C., every three-quarters of an hour at 25° C., every hour and a-half at 18.75° C., and every 4 or 5 hours at 12.5° C., while at a temperature of 6.2° C. the vegetative process was extremely slow. The formation of spores required about 12 hours at a temperature of 30° C., an entire day at 22.5° C., two days at 18.75° C., and several days when the temperature was not above 12.2° C.; below 6° C. no spores were formed. Vegetation continues active, according to Cohn[2] and Prazmowski, at a temperature of 40°–50° C., and is accompanied with energetic movement of the rods.

'Bacterium Termo' grows and vegetates according to Eidam[3] between the temperatures of 5.5° C. and 40° C.; its optimum is 30°–35° C. Koch[4] states of the Bacillus of anthrax, that in gelatine cultures its growth and spore-formation are finest and most vigorous at a temperature of 20°–25° C. Between 30° C. and 40° C. its growth and the formation of new spores usually come to an end in 24 hours; up to 25° C. the time required for this increases till it reaches 35–40 hours. Below 25° C. the decrease in the temperature is very marked in the negative sense; at 23° C. 48–50 hours are required for forming the spores, and at 21° C. 72 hours. At 18° C. the first spores appear after about 5 days, at 16° C. after 7 days, and the

---

[1] Schimmelpilze, p. 46.   [2] Beitr. z. Biol. II, 271.   [3] Cohn's Beitr. z. Biol. I, 3, p. 208.
[4] Mittheil. aus d. k. Gesundheitsamte, I, p. 64.

spore-formation becomes constantly more sparing. Growth and formation of spores cease below a temperature of 15° C. Fitz by comparing the forms in a mixture which had completed its fermentation in the same time at different temperatures found that the optimum of his Bacillus butylicus in solution of glycerine was 40° C., that of another species unnamed and grown by itself 37°–40° C. The maximum in both species was 45°–45.5° C.

The upper limit of temperature at which vegetating Bacteria can continue to exist appears to be little higher than that observed in the case of most other plants. Cohn found that it approached near to the maximum of vegetation in Bacillus subtilis, being 50–55° C. Fitz found it in the second of his two species just mentioned at about 56° C. According to Eidam, 'Bacterium Termo' was killed when the liquid in which it was vegetating was heated during 14 hours up to 45°, and during 3 hours to 50°. From Buchner's experiments mentioned above with the Bacillus of anthrax it appears that the dried rods are killed at the same degree of temperature as the spores. Further details will be found in special works on the Bacteria[1].

The vegetative forms of the Bacteria are able to bear the lowest temperatures to which they can well be exposed. Frisch[2] found putrefactive Bacteria and species of Bacillus, among them B. Anthracis, still retaining the power of development after being frozen in liquid at a temperature of −111° C.

It is in accordance with the analogy of other organisms that the temperatures at which Bacteria vegetate and lose their vitality should vary with the character of the substratum. Nägeli says[3] that by making changes in the nutrient solution the death of the Fission-fungi may be ensured within a certain time at any temperature between 30° and 110°, but he does not distinguish here between germination of spores and vegetative states.

The Bacteria differ much from one another in respect of the necessity for a **supply of oxygen.** At one end of the series the vegetation is promoted in the highest degree, other conditions being the same, by the largest possible supply of atmospheric air containing free oxygen; this is the case with Bacillus subtilis, and Arthrobacterium aceti; at the opposite end, as in B. butyricus, it is promoted by the exclusion of free oxygen. Accordingly Pasteur distinguishes between **aerobiotic** and **anaerobiotic** vegetation or forms[4].

Cases have been observed by Engelmann[5] lying between these two extremes, in which a less pressure of oxygen is required than that which is afforded by the composition of the atmosphere. According to Nägeli[6] the aerobiotic forms will also vegetate when deprived of a supply of free oxygen.

As regards the proper **nutriment of Bacteria** it is to be presumed that those which have the green colour of chlorophyll, if they really contain chlorophyll

---

[1] See also Pfeffer, Physiol. II.

[2] Sitzgsber. d. Wiener Acad., Mai, 1877.

[3] Die niederen Pilze &c. (1877), p. 30, and see also p. 200.

[4] See also on this point Nencki in Journ. f. pract. Chemie, nene Folge, XIX, XX; also Genning in the same publication.

[5] Bot. Ztg. 1882, 321.

[6] Die niederen Pilze (1877), p. 28.

assimilate carbon dioxide; and this is confirmed by Engelmann's observation [1] that Bacterium chlorinum gives off a small quantity of oxygen in the sunlight.

Nägeli's researches show that the Bacteria which contain no chlorophyll generally require the same kind of food as the Moulds which have been examined, and that the organic compounds employed as food exhibit the same scale of feeding power in the one case as in the other (see section XCVIII); but of inorganic nitrogenous compounds nitric acid serves as food for the Bacteria, though in a less degree than ammonia.

Bodies which are contained in solution in the substratum but are not nutrient substances may greatly influence the vegetation of the Bacteria. It was observed above (p. 354) that an acid reaction in the nutrient solution is as a rule unfavourable to the development of the Bacteria. Some species however can put up with this state of things to a certain point if they are by themselves in the solution. But their vegetation may be entirely stopped by it, and so Moulds or sprouting Fungi present with them in the fluid may thrive on the acid and overpower the Bacteria. Other bodies, which are products of the decomposition of the substratum caused by the Bacteria and pass into the solution, may also impede the vegetation as soon as they acquire a certain degree of concentration in the solution. According to Fitz's observation in 1882, the vegetation of Bacillus butylicus in glycerine solution under otherwise optimum conditions was stopped by 2.7–3.3 per cent. of ethyl alcohol, by 0.9–1.05 of butyl alcohol and by 0.05–0.1 of butyric acid, &c.

It should be observed here that bodies which serve as food, and in the case of the aerobiotic forms the oxygen also, act as stimulants on the Bacteria, awakening or accelerating their powers of movement and determining its direction. Engelmann [2] has shown that sensitive aerobiotic Bacteria are brought to rest by cutting off the supply of oxygen, and are at once set in motion again by renewing the supply, and the movement is directed towards the source of the oxygen, for example a cell containing chlorophyll which is reached by the sun's rays. An infinitesimal portion of an oxygen-compound, according to Engelmann's calculation the trillionth part of a milligram, is sufficient to set very sensitive aerobionts in motion, and they are therefore the most delicate of reagents for the evolution of oxygen. Such sensitive forms, when the oxygen is supplied to them from one direction only, move as close as possible to the source of the oxygen, such as a cell containing chlorophyll or the edge of the cover-glass when a specimen is grown in drops of a fluid on a microscopic slide. Other forms under the same conditions only approach within a certain distance of the source of the oxygen, and this distance increases as the oxygen diminishes. From this it is concluded that these forms can do with a pressure of oxygen less than that of the atmosphere. Anaerobionts behave in this respect in the reverse way to that of the sensitive aerobionts. The movement of motile forms in fluids is similarly quickened by a supply of proper soluble nutrient material, and it is directed, if the supply comes from one side, towards the diffusion-stream of the nutrient solution which flows into the liquid substratum from the source of the supply. Hence the Bacteria gather in dense swarms round solid bodies containing nutrient substances when placed in the fluid in which they happen to be present [3].

---

[1] Bot. Ztg. 1882.   [2] Bot. Ztg. 1881, p. 441; 1882, pp. 663, 419.
[3] Further details will be found in Pfeffer in Unters. d. Bot. Inst. z. Tübingen, I, Heft 3.

SECTION CXXXVI. The Bacteria, apart from the forms which contain chlorophyll and which have not yet been carefully studied, are distinguished according to their actual vegetative adaptation into *saprophytes* and *parasites*, in the sense in which the words were employed in section XCIX.

The adaptation of the **saprophytic forms** presents the same general points of view as that of the saprophytic Fungi. Many Bacteria are, like these Fungi, to some extent organisms which produce *oxidations*, combustions of the substratum. The Micrococcus of vinegar or mother of vinegar (Arthrobacterium aceti, Mycoderma aceti) oxidises ethyl alcohol in atmospheric air and converts it into acetic acid; but it may also convert it by combustion into carbonic acid and water[1]. Bacillus subtilis and as it would appear B. Megaterium also cause similar combustion of organic compounds and produce carbonic acid and water. Many others excite characteristic *fermentations*, lactic acid fermentation, butyric acid fermentation and the viscous fermentation of sugar, &c., they also act as inciters of *putrefactive processes*. For the details of these phenomena, which are the subject of so much discussion at the present day, the reader is referred to the special literature of the Bacteria and of the chemistry of fermentation, to the excellent researches of A. Fitz especially, and those of Nägeli and Duclaux, and to Pfeffer's Physiologie, I, chap. 8.

Many Bacteria on the other hand are **parasitic** in and on living organisms. In the determination and description of their mode of life the same points of view must be taken, and the same divisions and nomenclature applied, as those which were explained at length in connection with parasitic Fungi in speaking of their relations to their host and the effects they produce in it, for the same or quite analogous phenomena occur in both cases. In the succeeding remarks therefore there is a tacit reference throughout to sections CI–CXIII.

All parasitic Bacteria live as endophytes in the cavities of the body or in the substance of the tissue of the host. Their structure and growth determine the mode in which they attack the host; they find their way either as spores or in the vegetative form into normal cavities of the body accessible from without or into wounds, and in both places they continue a process of vegetation; they may also be passively conveyed from wounded surfaces in the bodies of animals into the blood and lymphatic passages, or else they penetrate into the cells and tissue from any surface to which they have been conveyed. The Bacillus of anthrax for instance penetrates into the mucous layer of the intestinal canal[2], when it has been carried to it in the animal's food. The effects of fermentation will have something to do with the perforations thus produced, and the direction of the movement will depend on the co-operation of the chemical and physical qualities of the substratum and possibly of the spontaneous motion of the parasite.

All known parasitic Bacteria are simply and for the most part vigorously *destructive* in their effect upon their host, if we do not in their case also reckon inflammatory processes (the formation perhaps of tubercles) among phenomena of diseased growth and new formation.

Bacteria **parasitic on plants** have scarcely ever been observed, a fact to which

---

[1] Pasteur in Comptes rend. 54, p. 265, and 55, p. 28.—Nägeli, Theorie d. Gährung, p. 111.

[2] See Koch, Mittheil. d. Reichsgesundheitsamts, I, p. 61.

R. Hartig has already drawn attention.   One reason for this may be that the parts of plants have usually an acid reaction.   At the same time Wakker [1] has recently described a disease in the hyacinth known in Holland as the yellow sickness, the characteristic symptom of which is the presence of yellow slimy masses of Bacteria in the vessels.   In the resting (autumnal) bulb the masses of Bacteria are confined to the vascular bundles of the bulb-scales; at flowering time they are found also in the leaves, and not in the vessels only but in the parenchyma also, where they fill the intercellular spaces, destroy the cells, and ultimately emerge through the ruptured epidermis and appear on the outside.   The case demands a thorough investigation.

The Bacteria on the other hand which are **parasitic in living animals** are, according at least to prevailing views, comparatively numerous, and the most prominent feature in them is their *facultative parasitism.*   Of this we have instances which have been investigated with some care and may be considered as well established, and it will be well to give here a special account of one of the most important of the number, namely **Bacillus Anthracis.**

The structure and development of this species have been already portrayed in Fig. 195.   It attacks the Mammalia, especially rodents and ruminants, with the exception of some species and individuals; mice, guinea-pigs, rabbits, sheep and cattle are unequally liable to be infected by it in the descending order.   It will also attack human beings.   It is communicated with difficulty to dogs, more readily to cats.   Observers are not agreed as to the degree of liability of birds, frogs and fishes to be infected with it, and we cannot further discuss their statements in this place. We know from Rayer, Pollender and Davaine that it causes the disease known as *anthrax* in the animals first mentioned.   My own experiments on which this account is partly based were chiefly made on guinea-pigs, and on material obtained from them.

When the Bacillus has gained admission into the blood of an animal capable of the infection, it grows and multiplies in the rod-form described above to such a degree that the entire mass of the blood is permeated by these organisms.   The animal sickens as the Bacillus multiplies and the result is usually fatal.   The Bacillus may find its way into the blood directly by intentional introduction of rods or spores or from accidental wounds; a prick of a needle charged with rods or spores, so slight as not to draw blood, is sufficient to give the infection to a sensitive animal.   But it may also reach the blood from the intestinal canal, into which it is conveyed in the natural way only, that is through the mouth with the food.   Rods introduced in this way have no further effects, if the digestive passages are without a wound; they probably perish in the acid contents of the stomach.   But if spores are introduced the animal takes the infection.   The spores pass unharmed through the acid stomach and germinate in the alkaline contents of the intestinal canal, and the rods which are the product of germination are found in the mucous membrane of the canal, having forced their way probably through the lymph-follicles and Peyer's patches, as Koch supposes.   From hence the way is open through the capillaries into the blood-passages.

According to Koch's investigations the infection comes much more frequently

---

[1] Bot. Centralblatt, 14, p. 315.

from the intestinal canal in cases of anthrax arising from natural causes, that is, not artificially produced. But to understand the life-history of the Bacillus and with it the aetiology of the disease, we must first enquire how the spores find their way into the intestinal canal. It cannot be directly from a diseased or lately deceased animal, because the Bacillus forms its spores neither in the living creature nor inside the unopened carcase [1]. But it is evident from what has been said on page 466, that the Bacillus may not only germinate and vegetate luxuriantly outside the body of the animal, but that the formation of spores takes place there almost exclusively, or at least, as is proved by every culture-experiment, in the greatest abundance, if there is a proper supply of oxygen and a temperature of 20°–25° C. A sufficient further supply of nutrient substances must also be presupposed, and experiment has shown that these are found in abundance in every variety of dead organic bodies, and not only in substances of animal origin, such as the solid and fluid parts of animals themselves that have died of anthrax, or the bloody excreta of those that are ill of the disease, but also in vegetable bodies in which the reactions are not too acid, such as potatos, beetroot or crushed seeds, &c. It is evident from this, that the Bacillus is not only able to live as a *saprophyte*, but that it must adopt that mode of life in order to arrive at a section of its existence of the greatest importance morphologically and biologically, namely that in which it forms its spores. It appears further that it can and does readily find the necessary conditions for its vegetation as a saprophyte on the surface of a moist pasture-ground, when it has once found its way there, and can maintain itself there from year to year by means of its spores and of the rods which dry up or are frozen in unfavourable vegetative periods; it is not necessary that the place should be visited by animals. We may readily conceive, how graminivorous animals liable to infection may take in the spores of the Bacillus with their food in such places and become infected, for the Bacillus *has the capacity of parasitism*. In the case of cattle that feed in herds, if one falls sick others quickly take the infection and the disease becomes epidemic, because the number of Bacilli on the ground is increased by the addition of those in the bloody excreta of the sick animals, the pasture being thereby rendered more dangerous for the herds, and because stinging flies and the like may directly inoculate one animal with the Bacilli contained in the blood of another.

It is obvious that under these conditions an animal is in greater danger of infection if it has wounded surfaces whether of the skin or of the mucous membrane of the mouth and digestive canal.

Our experience with the domestic animals has taught us that anthrax is endemic in certain localities, and breaks out there spontaneously, at first attacking single animals, apparently without direct infection from others but usually starting from the intestine, and afterwards spreading to other individuals. It is not easy to explain why separate districts should thus be the favoured home of anthrax, and why an organism which seems to be so capable of dissemination should not be found everywhere and be everywhere alike capable of producing disease. The reason may be, as Koch supposes, that the dangerous localities are wet and liable to be flooded, and that the Bacillus grows more abundantly on wet ground than on dry, and is also

---

[1] See Koch, Mittheil. d. Reichsgesundheitsamts, I, pp. 60, 147.

raised above the surface of the soil by the floods and spread over the plants which are subsequently eaten by the cattle. It is at the same time possible that the Bacillus may also be conveyed to the localities by the bodies of animals which have died of anthrax. This would at all events be more likely to happen in a district which has once been much infected with the fever than in any other. Mice are very susceptible to the disease, and the dead bodies of these creatures and of other small rodents would be especially calculated to propagate the infection. But Pasteur's sensational hypothesis, that the Bacillus is introduced into the soil by the burial of the bodies of infected animals, and that its spores are then conveyed by earth-worms from beneath to the surface, is not necessary for the explanation of the phenomena in these or any other localities; it is moreover open to the objection that the formation of spores never or scarcely ever takes place, as is urged by Koch, in the unopened body of an animal in the temperature of the deeper layers of soils and with the small amount of oxygen in the air which they contain.

The Bacillus here described is shown by its life-history to be a strictly facultative parasite, which only reaches the highest stage of its development in the non-parasitic state, and not only can but actually often does go through the entire course of its development in this state during many generations and even many years. It has been shown that it has a virulent effect at least on the animals above mentioned. Whether it can vegetate in other species of animals without doing them harm is not yet ascertained. But its virulent effects on those animals which suffer from its presence may be diminished by certain methods of breeding, and indeed be weakened till they become quite innocuous. Pasteur first discovered through his experiments on fowl-cholera, and Koch[1] confirmed his results, that this takes place when the Bacillus is grown in a neutralised nutrient solution, such as a meat-broth, with a plentiful supply of oxygen and at a high temperature. The attenuation of the effects may be carried so far that a mouse, the most susceptible animal with which experiments have been made, can suffer inoculation without being made ill; the attenuation is induced rapidly when the temperature is raised to nearly 43° C. and may be completed in 6 days; at a temperature of 42° C. the cultivation may require to be continued during 30 days, and the process is still slower in the temperature of an ordinary room. The Bacillus vegetates under these conditions and multiplies without alteration of its morphological characters, but *it does not produce spores.* Cultures kept at a temperature of 42°–43° C. perish in about a month's time, but fresh cultures can be obtained from them from 1–2 days before that time has expired. The Bacillus may recover its virulent properties after a certain degree of attenuation, if it finds its way into an animal which is susceptible to the infection and kills it. There is a degree of attenuation in which it is innocuous to full-grown guinea-pigs, but not to very young ones; if the latter are inoculated with the attenuated matter, the Bacillus returns to its state of greater virulence. The data before us do not show whether a return to virulence is possible from the highest degree of attenuation, nor have we any distinct experiments to prove whether the attenuated Bacillus developes at all in the animal which is inoculated with it but continues healthy, or in what manner it developes. It has been assumed that it does develope there, but there are no precise facts on

---

[1] See his essay, Ueber d. Milzbrandimpfung, Cassel, 1883, p. 17.

which to ground the assumption; but enough is known for the determination of the practical question of protective inoculation by means of the attenuated Bacillus. We cannot however enter further into this point here, but must refer the reader to medical works on the subject.

Be this as it may, it is easy to conceive, in the case of a facultative parasite which is able to adapt itself to nutrient solutions of different concentration and qualitative composition at a temperature of 15°–20° C. and to the blood of a mammal at one of 37°–40° C., that changes in the adaptation and food may be followed by changes in the deleterious effects, which may be supposed to be due to the production of some kind of ferment. An analogous though quantitatively different case is that of Sclerotinia Sclerotiorum described on page 380, in which the capacity for a parasitic life depends on the food supplied to the plant in its young state. We may also compare the Mucorini (page 358), which vary their form and the decomposition which they produce with the medium in which they live, and the Bacterium described by Wortmann[1], which gives rise to a ferment capable of dissolving starch if it is supplied with nothing but starch-grains for its food, but ceases to produce the ferment if fed with carbohydrates in solution or with ammonium acetate.

In the foregoing description it has been tacitly assumed that Bacillus Anthracis is a distinct species, and the present state of our knowledge requires the assumption. The Bacillus of anthrax has a resemblance to other species, and among them to Bacillus subtilis, which is not a facultative parasite or at least may under certain conditions be a harmless parasite; it varies also in the breadth and length of its cells and in other respects; but it always remains within the limits of the specific characters, the most important of which are given on page 466, and which distinguish it from other species and especially from B. subtilis also described above.

Buchner has maintained, in opposition to this view, that the Bacillus of anthrax and the hay-bacillus may be made to pass one into the other by breeding, and that they are therefore only states of one and the same species. He has not supplied us with the strictly morphological proof necessary to establish this opinion, since he has not taken into consideration the behaviour of the spores of his altered forms in germination, at least in his published communications, and yet this is one of the characteristic marks of distinction. He adopted also the macroscopic mode of cultivation, in which it is not possible to ensure an uninterrupted control of the continuity of the development, or of the accidental mingling of different species. His transformation of the virulent Bacillus of anthrax into the supposed innocuous hay-bacillus was effected in cultures at a high temperature and with a more than ordinary supply of oxygen; the temperature was 36°C., the apparatus employed ensured constant shaking in air, and the solution contained 0.5 per cent. of meat-extract. The transformation was not obtained at a temperature of 25°C. and when the apparatus was not kept in motion. It was evidently Pasteur's innocuous state which was produced in this case, but that is by no means Bacillus subtilis. The results of the reverse transformation appear, according to Buchner's own account, extremely doubtful. Now that the facultative parasitism and the possible change of virulence in Bacillus Anthracis have been demonstrated, and since

[1] Zeitschrift f. physiol. Chemie, VI, p. 287.

it has been further established that species can be distinguished in the Bacteria as in other organisms, the whole controversy has lost the importance which it was once supposed to have.

In Buchner's experiments on the change of the hay-bacillus into the Bacillus of anthrax, Bacilli from an infusion of hay were bred with certain precautions in fresh blood. The macroscopic character of the masses of Bacilli was changed, and intermediate forms were obtained between the hay-bacillus and the Bacillus of anthrax, and strange to say *no reversion* was obtained in the intermediate forms when solution of meat-extract or infusion of hay was substituted as the nutrient fluid. Mice and rabbits were inoculated with the altered matter, and some of the creatures experimented on sickened and died of anthrax, but *much the larger number did not take the infection.* Koch will not allow that the observed disease was true anthrax, and maintains that it may have been a disease which is common in mice and cannot always be at once distinguished from anthrax—a disease known as *malignant oedema*, and produced by a Bacillus morphologically very like the Bacillus of anthrax, which must have found its way into the culture with the hay-bacillus. If we allow that the disease in the cases actually ascertained was anthrax and disregard Koch's doubt on the point, the things to be remembered are chiefly these. By far the largest part of the original material used for the experiment may have been ascertained to consist of Bacillus subtilis ; but we have no proof that other Bacilli, lost at first in the overwhelming mass of B. subtilis and practically not distinguishable from them, were not contained along with B. subtilis in the hay-infusion. It would indeed be wonderful if one species, B. subtilis, were on all occasions *the only form* obtained without admixture of any others from a material like hay treated according to a definite procedure, especially as the apparently exceptional security of the procedure, the boiling the hay, offers no certain guarantee, because the spores of other Bacteria as well as Bacillus subtilis are capable of withstanding that temperature. Bacillus subtilis then may be present in the hay-infusion in much the larger numbers, and the forms mixed with it may be comparatively few. But we have to ask whether this numerical relation may not be altered or even reversed in other nutrient fluids, for instance in blood, whether single spores of Bacillus Anthracis may not have been present in the original material and only after change of cultivation have been in a condition to produce a small quantity of infectious material among the individuals of the other species, and thus occasional cases of infection may have occurred among the instances of failure ; to these questions the data before us afford no certain answer. Allusion has been already made to Koch's views, and we will not draw attention here to other points of difficulty. The reader is referred for further particulars to the original publications.

The Bacillus of anthrax has been discussed at some length because it is at present the best-known example of the Bacteria which inhabit the bodies of animals and incite disease in them. Modern pathology resting on older observations and experiments, among which the researches into anthrax itself occupy a prominent position, and supported by Nägeli's theoretical considerations, endeavours to refer all infectious diseases in animals, excepting the few that are caused by Fungi (p. 376) and some that were not formerly supposed to be infectious, to the invasion of Bacteria as their proximate cause. These organisms have been sought for sometimes with great, even with excessive zeal, and some have been found. The parasitic qualities, in virtue of which many of these organisms incite disease, have been sufficiently proved in a number of cases, for example in septicaemia, erysipelas, recurrent fever in warm-blooded animals, Pasteur's fowl-cholera, the flacherie of the silk-worm, though our botanical knowledge of the plants themselves is still very defective. Some forms are still the subjects of lively discussion on the part of experimental pathologists. Bacteria occurring in a living or dead body are not necessarily the inciters of disease, and the attempt to decide the point experimentally often encounters great difficulties. To expatiate further in the domain of pathology would carry us beyond the limits to which we are confined. But we shall perhaps contribute something of value to the above-mentioned discussion as well as to the determination of the recognised cases if we append a few short general remarks to the foregoing account of anthrax. For further details the reader is referred to medical works and to the compilations of Marpmann and Zopf, which are not however as complete as might be wished.

So far as can be judged from the accounts before us, all the Bacteria which are suspected of being or are proved to be parasites with the power of inciting disease, with one exception which will be noticed again below, are capable of vegetating and being bred in dead organic matter; some form their spores chiefly or exclusively in this saprophytic stage of their development. The Bacteria of the latter category are therefore facultative parasites like the Bacillus of anthrax, and the rest perhaps are so too; if not, they are at any rate facultative saprophytes. Hence they can both vegetate, like the Bacillus of anthrax, outside the living animal; the localities must be ascertained in each separate case, and it follows that the danger of infection is different in this case and in that of obligate parasitism.

Further it will depend on the species, race and individual among the Bacteria in question in their quality of parasites what hosts they will choose, as is the case with the Bacillus of anthrax; or conversely, some species of animals or some individuals will be more liable to be attacked by a given species of Bacterium, while others will be secure from it. It is at the same time conceivable that this condition may vary in individuals, an individual not before susceptible may for instance become susceptible; this may be due to external causes whether otherwise injurious or apparently indifferent. We have sufficient proof that such changes do actually occur.

As the disposition of the host may vary, so also a change may take place in the qualities, and specially in the virulence of the parasite, as we see in the case of the attenuated Bacillus of anthrax. This may perhaps be assumed to be the general rule in the forms which approach near to that species and of which we are here speaking. The change may be in the direction of loss of virulence, or on the contrary of its recovery. It may therefore also happen, that experiments in artificial infection with the same forms of Bacterium may, cæteris paribus, yield different results, some of a positive, others of a negative character. It is possible that the great difficulty or impossibility of obtaining animals that are exactly alike for experiments *may at least help* to heighten the apparent contradictions.

The investigation of the Bacillus of anthrax has further shown that the changes just mentioned may be accomplished in a form which would be considered by a naturalist as distinctly *specific*, and which at the same time maintains its specific characters within the limits of variation to which it is subject. Such changes therefore afford no ground for doubting the existence of distinct parasitic species. From all other trustworthy sources of knowledge we obtain the same testimony, that real species can and must be distinguished in the Bacteria exactly as in other groups of plants and animals, and that the parasitic forms which incite disease do not differ in this respect from the rest. Nägeli's words[1], 'if my view is correct, the same species

[1] Niedere Pilze (1877), p. 64.

assumes in the course of generations various morphologically and physiologically unlike forms one after another, and these forms, in the course of years and decades of years, may produce souring in milk, butyric acid in sauerkraut, ropiness in wine, putrefaction in albumen and decomposition in urine, may turn articles of food containing starch red, and give rise to diphtheria, to typhus, to recurrent fever, to cholera, to intermittent fever'—these words or the view which they embody would not have been published even in 1877, if their author had studied the forms in question, and especially the parasitic forms, with greater attention.   At the present day, when our knowledge of the facts is still more advanced, such a position can no longer be maintained.   It is specially in the domain of the parasitic Bacteria that investigation has established more and more distinct species, and shown that every disease incited by a parasite which has been thoroughly examined may be traced to a definite form of Bacterium, of the specific character of which there can be as little doubt as of that of a large Fungus or of a worm.   The assertion that there are distinct species of parasitic Bacteria, and that every special disease caused by Bacteria is the work of a distinct species, is not merely a convenient form of statement, as Nägeli thinks, but it is the only one which is in unison with the *facts* as at present known.

If a species like the Bacillus of anthrax also vegetates as a saprophyte, it is obvious that it may set up different processes of decomposition in the dead substratum from those which constitute disease in a living body.   Further, if diseases supposed to be due to the presence of Bacteria 'have a limited duration in the history of the human race, change, arise and disappear,' this is no objection to the observed facts but merely a reason for making special efforts to explain them ; and accepting the fact, equally well observed, that men as well as Bacteria may change some of their qualities in the course of time and yet retain their specific characters unaltered, we may suppose that the attempts at explanation will possibly in course of time be successful.

It is uncertain whether there are obligate as well as facultative parasites among the Bacteria, either species that are strictly obligate or some that have also a narrowly limited power of saprophytic vegetation.   The forms which may possibly excite diseases that are strictly contagious, smallpox for example, should be tested on this point.   Mention must be made in this connection of **Spirochaete Obermeyeri,** one of the most characteristic parasites and an undoubted inciter of disease, which appears invariably in the blood of those who are suffering from recurrent fever.   It has been successfully transferred from men to apes, but to no other species of mammal on which the experiment has been tried.   It has been attempted to cultivate it outside the body of a living animal, but as yet without success [1].

It is very doubtful whether the minute organism, **Nosema Bombycis**, Nägeli, **Panhistophyton**, Lebert, which accompanies and according to Pasteur's experiments causes the destructive disease in the caterpillar of the silkworm known as pébrine or gattine, belongs to this group.   It appears in the form of small ellipsoid or somewhat elongated peculiarly refractive bodies *resembling* Bacteria, which may penetrate through all parts of the caterpillar and butterfly.   We learn from Pasteur that it may find its

---

[1] v. Heydenreich, Unters. ii. d. Parasiten d. Rückfalltyphus, Berlin, 1877.—Lachmann in Deutschen Arch. f. klin. Medicin, 27, p. 52 b.

way into the membrane of the intestinal canal if supplied to a healthy caterpillar with its food, appearing there first singly and then multiplying rapidly and spreading through other organs. Its development and even its mode of multiplication, which is said to be by bipartition, is not yet clearly ascertained, and we can only affirm with Pasteur that it is a highly dangerous parasite, and expect more distinct conclusions from further investigations[1].

The above case must not be confounded with the forms of disease included under the name of flacherie. These are due, according to Pasteur[2], to the disturbances in the digestive process caused by decomposition or fermentation of the food in the intestinal canal, through the presence of an endosporous rod-shaped Bacterium and a chain-forming Micrococcus, the M. Bombycis of Cohn[3]. Doubtless this is a case of facultative parasitism, though further investigation is desirable.

LITERATURE OF THE BACTERIA.

The literature of the Bacteria has increased to an enormous size in the last ten or fifteen years. I have taken some pains to make myself acquainted with it, but I cannot affirm that my efforts have been entirely successful. It is at present quite impossible, especially in the medical part of the subject, for scientific criticism to keep pace with the eager study of the Bacteria, while on the other hand it is not the object of this work to supply a mere index.

For these reasons I have avoided first of all touching further on the medical side of the question than was required to complete the account of the morphology and biology of the Bacteria; and in the second place I abstain from attempting a complete enumeration of the literature of these organisms. Copious notices of it are to be found in the following works :—

A. MAGNIN, Les Bactéries, Paris, 1878.

W. Zopf, Die Spaltpilze, 2nd ed., Breslau, 1884, in Schenk's Encyclopädie.

G. MARPMANN, Die Spaltpilze, Halle, 1884.

DUCLAUX, Chimie biologique (Vol. IX of the Encyclop. Chim. of Frémy, Paris, 1883).

The lists of works in the last three books are far from being complete, but by consulting them and the works which will be cited presently every student will find his way to whatever part of the subject is of immediate interest to him. The reader therefore is referred to these publications as the most important, and after them to the medical Journals, Annual Reports and recent Text-books, and finally to Just's Botanischer Jahresbericht; and in the subjoined list I confine myself to noticing the chief sources of information, which with my own researches have served as the foundation for the account of the morphology and biology of the Bacteria given in the text. A few works already quoted in the notes to the text and referring to special points are not mentioned again below.

1. General literature of the Bacteria.

L. PASTEUR, Examen de la doctrine des gén. spontanées (Ann. Chim. sér. 3, 64, and Ann. d. sc. nat., Zoologie, sér. 4, XVI, extracted in Flora, 1862, p. 355);—Id., Études sur le vin, Paris, 1866;—Id., Maladies des vers à soie, Paris, 1870;—Id., Études sur la bière, Paris, 1876.

---

[1] Pasteur, Études sur la maladie des vers à soie, Paris (1870), I, p. 207. The earlier literature of the subject will be found there. See also Frey u. Lebert in Vierteljahrsschrift naturf. Ges, Zürich, 1856.—De Quatrefages, Mém. de l'Acad. des Sciences, XXX, 1860.—Leydig in Du Bois-Reymond's u. Reichert's Arch., 1863, p. 186.—Hoffmann, Mycol. Ber. (Bot. Ztg. 1864, p. 30).

[2] Études sur la maladie des vers à soie.

[3] Beitr. z. Biol. I, 3, p. 165.

Also communications of Pasteur, his pupils and opponents since 1858 in the Comptes rendus of the Paris Academy.   Among these the remarkable and finished treatise, Sur la choléra des poules (Cptes. rend. 90 (1880), pp. 239, 952, 1030) should be especially noticed.   See also below in the second list.

F. COHN, Unters. ü. d. Entwicklungsgesch. d. mikroskop. Algen u. Pilze (Nov. Act. Acad. Leop. 1854, 24, p. 1;—Id., Unters. ü. Bacterien (Beitr. z. Biol. d. Pflanzen, I, Heft 2, p. 127, Heft 3, pp. 141, 208, II, p. 249.)—Koch, Schröter, Eidam, in the same publication, Vols. I, II, Wernich, Miflet, Mendelsohn, Neelsen, in the same publication, Vol. III).

L. CIENKOWSKI, Zur Morphol. d. Bacterien (Mém. Acad. St. Pétersbourg, XXV, No. 2, 1877).

E. WARMING, Obs. sur quelques Bactéries qui se rencontrent sur les côtes du Danemark (Videnskab. Meddelelser fra Nat. Forening, Kjobenhavn, 1875-6).

R. KOCH, Zur Aetiologie d. Wundinfectionskrankheiten, Leipzig, 1878.

C. v. NÄGELI, Die niederen Pilze in ihren Beziehungen zu d. Infectionskrankheiten, München, 1877 ;— Id., Unters. ü. niedere Pilze a. d. pflanzenphysiol. Instit. z. München, 1882.

P. VAN TIEGHEM in Bull. de la Soc. bot. de France, 26 (1879), pp. 37, 144, and 27 (1880), pp. 148, 174 ;— Id. in Ann. d. sc. nat. sér. 6, VII (Leuconostoc) ;—Id., Traité de Botanique (1883), p. 1108.

E. C. HANSEN, Meddelelser fra Carlsberg Laboratoriet, I, Kopenhagen, 1878-82.

BREFELD, Bot. Unters. ü. Schimmelpilze, IV.

A. PRAZMOWSKI, Unters. ü. d. Entwicklungsgesch. u. Fermentwirkung einiger Bacterien-Arten, Leipzig, 1880, and in Bot. Ztg. 1879, p. 409.

A. FITZ, Ueber Spaltpilzgährungen in Ber. d. Deutschen Chem. Ges. I. Jahrg. 9 (1876), p. 1348.—II. J. 10 (1877), p. 276.—III. J. 11 (1878), p. 42.—IV. Ibid. p. 1890.—V. J. 12 (1879), p. 474.—VI. J. 13 (1880), p. 1309.—VII. J. 15 (1882), p, 867.—VIII. J. 16 (1883), p. 844.—IX. J. 17 (1884), p. 1188.

W. ZOPF, Unters. ü. Crenothrix polyspora, d. Urheber d. Berliner Wassercalamität, Berlin, 1879 ;—Id., Zur Morphol. d. Spaltpflanzen, Leipzig, 1882.

KURTH, Bacterium Zopfii (Bot. Zeitung, 1883).

MITTHEILUNGEN d. kais. Gesundheitsamts, I (1881), II (1884).

2. Anthrax.

O. BOLLINGER in Ziemssen's Handb. d. speciellen Pathologie u. Therapie, III (1874), and Pasteur, Comptes rend. 1877, 84, may be consulted for the earlier literature.

PASTEUR, Maladie charbonneuse (Cptes. rend. 84 (1877), p. 900) ;—Id., Charbon et septicémie (Cptes. rend. 85 (1877), p. 99) ; see also Cptes. rend. 87 (1878), p. 47, and Bull. de l'Acad. de Médecine, 1878, pp. 253, 497, 737 ;—Id., Chamberland et Roux, Cptes. rend. 92 (1881), pp. 209, 429, 266, &c.

R. KOCH, Die Aetiologie d. Milzbrandkrankheit (Cohn's Beitr. z. Biol. II, 277) ;—Id., Mittheilungen a. d. k. Reichsgesundheitsamt, I.

H. BUCHNER in Nägeli's Unters. ü. niedere Pilze, 1882 (see previous list).

OEMLER, Experimentelle Beitr. z. Milzbrandfrage (Arch. f. Thierheilkunde, II-VI).

ARCHANGELSKI, Beitr. z. Lehre v. Milzbrandcontagium (Centralblatt f. d. medicin. Wissensch. 1883, p. 257).

ROLOFF, Ueber Milzbrandimpfung u. Entw. d. Milzbrand-Bacterien (Archiv. f. Thierheilkunde, IX (1883), p. 459).

EXPLANATION OF TERMS.

Abjection (Abschleuderung) of spores. Throwing off with force of spores from a sporophore.

Abjoint. To joint off or delimit by septa.

Abjunction (Abgliederung). Delimitation by septa of portions of a growing hypha as spores.

Abscise. To cut off or detach by solution of a zone of connection.

Abscision (Abschnürung) of spores. Detachment of spores from a sporophore by disappearance through disorganisation or otherwise of a connecting zone.

Accessory gonidia. In Mucorini : gonidial formations found in some species in addition to the typical ones of the group.

Actinomycosis. A disease in animals and man characterised by the development of tumours in the jaw-bone, vertebrae, lymphatic glands and other places within which sulphur-yellow bodies like sand-grains occur, each consisting of an aggregate of an organism, Actinomyces, which is supposed to be a Fungus.

Acrogenous. (*a*) Producing at the summit. (*b*) Produced at the summit.

Acrogonidium. Gonidium formed at the summit of a gonidiophore.

Acropetal. In the direction of the summit. Comp. **basipetal.**

Acroscopic. Looking towards the summit, i. e. on the side towards the summit. Comp. **basiscopic.**

Acrospore. Spore formed at the summit of a sporophore.

Adventitious. Produced out of normal and regular order.

Aecidiospore. Spore formed in an aecidium.

Aecidium. In Uredineae : sporocarp consisting of a cup-shaped envelope (peridium) and a hymenium occupying the bottom of the cup from the basidia of which spores (aecidiospores) are serially and successively abjointed.

Aerobiotic. Organisms which require for their vegetation a supply of free oxygen are aeriobiotic. Comp. **anaerobiotic.**

Aethalium. In Myxomycetes : compound sporiferous body formed from a large combination of plasmodia.

Algal layer. In heteromerous Lichens : green band at the line of junction of the rind and medulla of the thallus in which the cells of the Alga of the Lichen are aggregated. Same as **algal zone, gonidial layer, gonimic layer, stratum gonimon.**

Algal zone. Same as **algal layer.**

Alveolate. Pitted so as to resemble honeycomb.

Amoeboid. Like an amoeba, i. e. a small portion of protoplasm exhibiting creeping movement by putting out and drawing in pseudopodia.

Amylogenesis. Formation of starch.

Amylum-grain. Starch grain.

Analogous. Having the same function. Comp. **homologous.**

Androgynous. Having male and female sexual organs developed on the same branch of the thallus. Comp. **diclinous.**

Androspore. Male spore, i. e. spore which on germination produces a body bearing a male sexual organ.

Anaerobiotic. Organisms which can vegetate without a supply of free oxygen are anaerobiotic. Comp. **aerobiotic.**

Angiocarpous. Having a hymenium developed by internal differentiation within the sporophore and from the first covered by a special envelope. Comp. **gymnocarpous.**

Annulus. In Hymenomycetes : portion of ruptured marginal veil or of tissue of the stipe forming a collar or frill or sheath upon the stipe after the expansion of the pileus. Frequently used to designate the special form distinguished as **annulus inferus.** Same as **ring.**

Annulus inferus. In Hymenomycetes: collar attached to the stipe below the apex formed by rupture of marginal veil round the margin of the pileus. See **annulus.**

Annulus mobilis. In Hymenomycetes: portion of ruptured marginal veil remaining as a moveable annular sheath upon the stipe after expansion of the pileus. See **annulus.**

Annulus superus. In Hymenomycetes: same as **armilla.**

Anther. In Hymenomycetes: old term for **cystidium.**

Antheridium. (*a*) Male sexual organ. (*b*) In Hymenomycetes: old term for cystidium.

Anthrax. Disease in animals and man excited by Bacillus Anthracis.

Aphthae. Same as **thrush.**

Apogamy. Loss of sexual function without suppression of the normal product of the sexual act.

Apothecium. Same as **discocarp.**

Appendicula. In Erysipheae: branching hair-like process at the summit of sporocarp.

Archegonium. Female sexual organ with narrow upper portion (neck) pierced by a canal usually enclosing one or more cells (neck-canal-cells) and leading to a basal dilated portion (venter) containing one oosphere (ovum) and a smaller cell at the entrance of the neck-canal (ventral canal-cell). After fertilisation the embryo is developed within the venter.

Archicarp. Beginning of a fructification, i. e. cell or group of cells fertilised by a sexual act. Same as **ascogonium, carpogonium.**

Areolate. Marked out into small areas or spaces.

Armilla. In Hymenomycetes: plaited frill suspended from apex of stipe formed by a layer of tissue separated from the surface of the stipe except at apex, and forming at first a covering membrane of the hymenium, from which it is detached on expansion of the pileus. Same as **annulus superus, frill.**

Arthrosporous. In Schizomycetes: species which have no endogenous spore-formation are arthrosporous.

Asciferous. Bearing asci.

Ascocarp. In Ascomycetes: sporocarp producing asci and ascospores ; its three kinds are apothecium or discocarp, perithecium or pyrenocarp, and cleistocarp.

Ascogenous. Producing asci.

Ascogonium. In Ascomycetes: same as **archicarp.**

Ascophore. Sporophore bearing an ascus. See **sporophore.**

Ascospore. Spore formed in an ascus. Same as **thecaspore.**

Ascus. In Ascomycetes: large cell, usually the swollen extremity of a hyphal branch, in the ascocarp within which spores (typically 8) are developed. Same as **theca.**

Ascus-apparatus. In Ascomycetes: portion of the sporocarp consisting of the asci together with the ascogenous cells.

Ascus suffultorius. Corda's term for **basidium.**

Autoecious. A parasite which goes through the whole course of its development on a single host of a particular species is autoecious. Same as **autoxenous.** Comp. **metoecious, lipoxenous.**

Autoxenous. Same as **autoecious.**

Axile. In the axis of any structure.

Azygospore. In Mucorini: apogamously formed spore resembling a zygospore.

Basidiogenetic. Produced upon a basidium.

Basidiophore. Sporophore bearing a basidium. See **sporophore.**

Basidiospore. Spore acrogenously abjointed upon a basidium.

Basidium. Mother-cell from which spores are acrogenously abjointed. Same as **ascus suffultorius, sterigma.**

Basipetal. In the direction of the base. Comp. **acropetal.**

Basiscopic. Looking towards the base, i. e. on the side towards the base. Comp. **acroscopic.**

Bion. An individual morphologically and physiologically independent.

Blastema. Wallroth's term for the lichen-thallus.

Brood-bud. (*a*) In Lichens: same as **soredium.** (*b*) In Archegoniatae: same as **bulbil.**

Brood-cell. Propagative cell, naked or with a membrane, produced asexually, separating from the parent and capable of developing directly into a new bion. Same as **gonidium, conidium.** It passes without demarcation into the **brood-gemma** and **bulbil.**

Brood-gemma. Pluricellular propagative body without differentiation, produced asexually, separating from the parent and capable of developing directly into a new bion. Same as **gemma.** It passes without demarcation into the **brood-cell** on the one side, and into the **bulbil** on the other.

Bulbil. (*a*) In some Fungi doubtfully considered Ascomycetes: small pluricellular bodies incapable of germination. (*b*) In Archegoniatae: deciduous leaf-bud capable of developing directly into a new bion. Same as **brood-bud**.

Bulbus. In Hymenomycetes: swollen base of the stipe of the sporophore.

Canker. Disease in deciduous-leaved trees caused by Nectria ditissima, Tul., and characterised by malformation of the rind, exhibiting a swollen cushion-like margin and a depressed dead centre.

Cap. In Hymenomycetes: same as **pileus**.

Capillitium. Sterile thread-like tubes or fibres, often branched or combined in a net, interpenetrating the mass of spores within a ripe sporogenous body.

Capitate. Having the form of a head.

Carpogonium. Same as **archicarp**.

Carpophore. Stalk of a sporocarp.

Carpospore. Spore formed in a sporocarp.

Cellular spore. Same as **sporidesm**.

Cementation (Verklebung) of hyphae. Union of membranes by a narrow slip of cementing substance, so that hyphae are inseparably grown together. Same as **concrescence**.

Cephalodium. Peculiarly shaped branched or convex outgrowth of a lichen-thallus in which algal cells are localised.

Chain-gemma. In Mucoreae: gemma having the form of a septate confervoid filament, the segments of which are capable of sprouting. Same as **sprout-gemma**.

Chlamydospore. Spore with a very thick spore-membrane.

Chromidium. Term proposed by Stitzenberger for an algal cell in a lichen-thallus. See **gonidium**.

Clamp-cell. See **clamp-connection**.

Clamp-connection (Schnallen-verbindung). Small semicircular hollow protuberance attached laterally along its whole length (or leaving an eye-hole) to the walls of two adjoining cells of a septate hypha and stretching over the septum between them, either communicating with one or both cells of the hypha or completely delimited from both and then forming a **clamp-cell (Schnallen-zelle)**.

Cleistocarp. Ascocarp in which the asci and ascospores are formed inside a completely closed envelope from which the ascospores escape by its final rupture.

Coalescence (Verschmelzung) of hyphae. Complete fusion of the membranes of two originally separate hyphae or hyphal branches.

Cochleariform. Spoonshaped.

Collenchyma. Form of thick-walled parenchyma in which the middle of the lateral walls of the prismatic cells are thin but the angles strongly thickened so as to round off the cavity of the cell.

Columella. Sterile axile body within a sporangium.

Compound Fungus-body (zusammengesetzter Pilzkörper). Growth-form in which the thallus is constituted by the cohering of the ramifications of separate hyphae. Comp. **Filamentous Fungus, Sprouting Fungus.**

Compound spore. Same as **sporidesm**.

Compound sporophore (Fruchtkörper). Sporophore formed by the cohesion of the ramifications of separate hyphal branches. Comp. **simple sporophore**.

Concatenate. Linked together in a chain.

Conceptacle. General expression for a superficial cavity opening outwards within which gonidia are produced.

Concrescence. Same as **cementation**.

Conidiophore. Same as **gonidiophore**.

Conidium. Same as **brood-cell**.

Conjugation. Union of two gametes to form a zygote.

Conjugation-cell. Same as **gamete**.

Cortex. Same as **rind**.

Cortina. In Hymenomycetes: marginal veil ruptured at its connection with the stipe and hanging from the margin of the pileus as a shreddy membrane. Same as **curtain, velum** in narrower sense of Persoon.

Cross-septation (Querzergliederung). Division of the terminal portion of a hypha or hyphal branch by transverse septa into a number of spore-cells.

Crustaceous thallus (thallus crustaceus). In Lichens: a thallus is crustaceous when it forms a flat crust on or in the substratum, adhering firmly to this by its whole under surface, so that it cannot be separated without injury. Same as **thallus lepodes**.

Crystalloid. Crystal of proteid.

Cup. In Ascomycetes: same as **discocarp**.

Curtain. Same as **cortina**.

Cutis. Same as **pellicula**.

Cyphella. In Lichens: circumscribed pit in the rind on the under surface of the thallus.

Cystidium. In Hymenomycetes: large unicellular; often inflated, structure projecting beyond the basidia and paraphyses of the hymenium. See **anther, antheridium, pollinarium.**

Dichotomy. Forking in pairs, i.e. cessation of previous increase in length at an apex with continuation equally in two diverging directions. Comp. **monopodium**.

Diclinous. Having male or female sexual organs developed on different branches of a thallus. Comp. **androgynous**.

Dioecious. Having male and female organs on different individuals. Comp. **monoecious**.

Discocarp. In Ascomycetes: ascocarp in which the hymenium lies exposed whilst the asci are maturing. Same as **apothecium, cup.**

Discus. Hymenium of a discocarp. Same as **lamina proligera, lamina sporigera.**

Dissepiment. Same as **trama.**

Dorsiventral. Horizontally extended so as to have a dorsal and a ventral surface.

Ectosporous. Having exogenously formed spores. See **exosporous.** Comp. **endosporous.**

Ejaculation of spores. Same as ejection of spores.

Ejection (Ausschleuderung) of spores. Throwing out with force of endogenously formed spores from a sporangium. Same as **ejaculation of spores.**

Elater. In Myxomycetes: a free capillitium thread.

Encarpium. Trattinick's term for **sporophore.**

Endogenous. Produced inside another body. Comp. **exogenous.**

Endogonidium. Gonidium formed within a receptacle (gonidangium).

Endophyte. Plant growing inside another plant and parasitic upon it or not parasitic. Comp. **epiphyte.**

Endosporium. Innermost coat of a spore. Comp. **exosporium, episporium.**

Endosporous. Having endogenously formed spores. Comp. **exosporous, ectosporous.**

Entozoic. Living inside an animal.

Envelope-apparatus. In Ascomycetes: all the parts of the sporocarp except the ascus-apparatus which consists of asci and the ascogenous cells.

Epinasty. That state of a growing dorsiventral organ in which the dorsal surface grows more actively than the ventral surface. Comp. **hyponasty.**

Epiphloeodic. Of Lichens: living upon the surface of the periderm of a plant. Comp. **hypophloeodic.**

Epiphragm. In Nidularieae: delicate membrane closing the cup-like sporophore.

Epiphyte. Plant growing upon the outside of another plant and either not parasitic upon it or parasitic. Comp. **endophyte.**

Epiplasm. Same as **glycogen-mass.**

Episporium. Outer (second) coat of spore. See **exosporium.** Comp. **endosporium.**

Ergotised. Attacked by ergot.

Excipulum. Outer envelope of a discocarp developed as part of the envelope-apparatus.

Exogenous. Produced on the outside of another body. Comp. **endogenous.**

Exosporium. (*a*) Same as **episporium.** (*b*) In Peronosporeae: thick coat developed from periplasm around the oospore.

Exosporous. Having exogenously formed spores. Comp. **endosporous.**

Extracellular. Outside of a cell. Comp. **intracellular.**

Extramatrical. Outside of a matrix or nidus. Comp. **intramatrical.**

Facultative. Occasional, incidental. Comp. **obligate.**

Facultative parasite. An organism which can and normally does go through the whole course of its development as a saprophyte, but which may also go through its development wholly or in part as a parasite. Comp. **obligate parasite, facultative saprophyte.**

Facultative saprophyte. An organism which normally goes through the whole course of its development as a parasite, but which can at certain stages vegetate as a saprophyte. Comp. **obligate parasite, facultative parasite.**

Favus. Disease of the skin caused by Achorion Schönleinii, Remak.

Felted tissue. Same as **tela contexta.**

Fertilisation-tube. In Peronosporeae: tube put out by the antheridium which pierces the oogonium and is the channel through which gonoplasm passes from the antheridium to the oosphere.

Fibrillose. Having a finely lined appearance as if composed of fine fibres.

Fibrillose mycelium. Same as **fibrous mycelium.**

Fibrous mycelium. Mycelium in which the hyphae form by their union elongated branching strands (mycelial strands). Same as **fibrillose mycelium.** Comp. **filamentous mycelium, membranous mycelium.**

Filamentous Fungus (Fadenpilz). Growth-form in which the thallus is constituted by a branched hypha alone, i.e. without union with other hyphae. Comp. **Compound Fungus-body, Sprouting Fungus.**

Filamentous mycelium. Mycelium of free hyphae which are at most loosely interwoven with one another, but without forming bodies of definite shape and outline. Same as **floccose mycelium.** Comp. **fibrous mycelium, membranous mycelium.**

Filamentous sporophore. Same as **simple sporophore.**

Filamentous thallus (thallus filamentosus). Same as **fruticose thallus.**

Flabelliform. Spread out like a fan.

Flacherie. Disease of the silkworm due to

fermentation of food in intestinal canal caused by Micrococcus Bombycis, Cohn.

Flagellum. (*a*) Solitary long swinging process of the protoplasm of a swarmspore. (*b*) Long whip-like process on the cells of some Schizomycetes.

Floccose mycelium. Same as **filamentous mycelium.**

Foliaceous thallus (thallus foliaceus). In Lichens: a flat, leaf-like, usually lobed and crisped thallus which spreads over the surface of the substratum, but is only attached at one or several scattered points and can be separated therefore from it without much injury. Same as **frondose thallus (thallus frondosus), thallus placodes.**

Form-genus. A genus constituted by similar form-species.

Form-species. Species constituted by a single stage of the life-cycle of a pleomorphous species and supposed of itself to be the complete representative of a species.

Formae oxydatae. In Lichens: crustaceous forms which have acquired a rust-colour owing to infiltration of a salt of iron.

Frill. Same as **armilla.**

Frondose thallus (thallus frondosus). Same as **foliaceous thallus.**

Fructification. Unicellular or pluricellular body developed as a result of the sexual act from an archicarp alone or from adjacent hyphae as well. In the unicellular form it is a **zygospore** or **oospore**; in the pluricellular form a **sporocarp.**

Fruticose thallus (thallus fruticulosus). In Lichens: a thallus attached by one point only and by a narrow base to the substratum from which it grows upwards as a simple or more usually branched shrub-like body. Same as **filamentous thallus, thallus thamnodes.**

Fuliginosus. Sooty.

Funiculus. In Nidularieae: cord of hyphae attaching peridiolum to the inner surface of the wall of the peridium.

Gamete. Sexual protoplasmic body, naked or invested with a membrane, motile (zoogamete or planogamete) or non-motile, which on conjugation with another gamete of like or unlike outward form gives rise to a body termed zygote. Same as **conjugation-cell.**

Gattine. Same as **pébrine.**

Gelatinous felt (Gallertfilz). Same as **gelatinous tissue.**

Gelatinous tissue (Gallertgewebe). Tissue which is slimy owing to the cell membranes being soft and mucilaginous. Same as **gelatinous felt.**

Gemma. Same as **brood-gemma.**

Germ-cell. First product of commencing germination of a spore.

Germ-tube (Keimschlauch). Tubular process put out by a spore in tube-germination at one or more points of its surface which by continued progressive apical growth developes into a hypha forming either a promycelium or a mycelium.

Germ-pore. Pit on the surface of a spore-membrane through which a germ-tube makes exit.

Gill. Same as **lamella.**

Gleba. Chambered sporogenous tissue within a sporophore.

Glycogen-mass. Protoplasm permeated with glycogen, especially in asci. Same as **epiplasm.** Sometimes shortly termed **glycogen.**

Gonidial layer. (*a*) Aggregation of simple gonidiophores to form a cushion-like layer or crust. (*b*) In heteromerous Lichens: same as **algal layer.**

Gonidiophore. Sporophore bearing a gonidium. Same as **conidiophore.** See **sporophore.**

Gonidium. (*a*) Same as **brood-cell.** (*b*) In Lichens: algal cell of thallus. Same as **chromidium.**

Gonimic layer. Same as **algal layer.**

Gonoplasm. In Peronosporeae: portion of protoplasm of antheridium which passes through the fertilisation-tube and coalesces with the oosphere. Comp. **periplasm.**

Green-rot. Disease in wood characterised by the tissues becoming a verdigris green. Peziza aeruginosa, Pers., is commonly associated with this condition, but its connection with the prominent feature of the disease is still uncertain.

Growth-form. A vegetative structure marked by some easily recognised feature of growth characterising individuals or stages in the life-cycles of types which have no necessary genetic affinity. Thus Sprouting Fungus, Filamentous Fungus, &c. are growth-forms.

Gymnocarpous. Having the hymenium exposed when the spores are maturing. Comp. **angiocarpous.**

Gynandrosporous. In Oedogonieae: dioecious forms in which the female plant produces androspores are gynandrosporous.

Haustorium. Special branch of a filamentous mycelium serving as an organ of attachment and suction.

Heliotropism. Phenomena induced in a growing organ by the influence of illumination.

Herpes tonsurans. Same as **ring-worm.**

Heteroecious. Same as **metoecious.**

Heteromerous. In Lichens: a thallus with stratified tissue owing to algal cells forming an algal layer and dividing the hyphal tissue into an outer (rind) and an inner (medullary) stratum is termed heteromerous. Comp. **homoiomerous.**

Heterosporous. Having asexually produced spores of more than one kind. Comp. **homosporous.**

Homoiomerous. In Lichens: a thallus with algal cells and hyphae distributed uniformly and in about equal proportion is termed homoiomerous. Comp. **heteromerous.**

Homologous. Having the same position and development.

Homosporous. Having asexually produced spores of only one kind. Same as **isosporous.** Comp. **heterosporous.**

Hydrotropism. Phenomena induced in a growing organ by the influence of moisture.

Hymenial Alga. In Lichens: algal cell in a sporocarp. Same as **hymenial gonidium.**

Hymenial gonidium. Same as **hymenial Alga.**

Hymenial layer. Same as **hymenium.**

Hymenium. Aggregation of spore-mother-cells, with or without sterile cells, in a continuous stratum or layer upon a sporophore. Same as **sporogenous layer, hymenial layer.** See also **discus, lamina proligera, lamina sporigera.**

Hymenophorum. Portion of a sporophore which bears a hymenium.

Hypha. The element of a thallus in most Fungi; a cylindric thread-like branched body consisting of a membrane enclosing protoplasm, developing by apical growth and usually becoming transversely septate as it developes.

Hyponasty. That state of a growing dorsiventral organ in which the ventral surface grows more actively than the dorsal surface. Comp. **epinasty.**

Hypophloeodic. Of Lichens: living in the periderm of a plant. Comp. **epiphloeodic.**

Hypothallus. In crustaceous Lichens: marginal outgrowth of hyphae, often strand-like, from the thallus. Same as **protothallus.**

Hypothecium. Layer of hyphal tissue immediately beneath a hymenium. Same as **subhymenial layer.**

Inception (Anlegung). First beginning.

Inner peridium. See **peridium internum.**

Intracellular. Inside a cell. Comp. **extracellular.**

Interweaving (Verflechtung) of hyphae. Union by intertwining without firm cohesion with one another.

Intralamellar tissue. In Hymenomycetes: same as **trama.**

Intramatrical. Inside a matrix or nidus. Comp. **extramatrical.**

Involucrum. Persoon's term for **velum.**

Involution-form. Swollen bladderlike form of Schizomycete, supposed to be a diseased condition of the form with which it is found associated.

Involution-period. Same as **resting period.**

Involution-stage. Same as **resting stage.**

Isogamy. Conjugation of two gametes of similar form. Comp. **oogamy.**

Isosporous. Same as **homosporous.**

Kernel. In Pyrenomycetes: old term for the whole of the softer part of the pyrenocarp within the firm outer wall. Also termed **nucleus.**

Lamella. In Hymenomycetes: vertical radial plate on the under surface of the pileus upon which the hymenium is extended. Same as **gill.**

Lamina proligera. Same as **discus.**

Lamina sporigera. Same as **discus.**

Lipoxenous. A parasite that leaves its host and completes its development independently, and at the expense of reserve of food appropriated from the host is lipoxenous. Comp. **metoecious.**

Lipoxeny. Desertion of a host.

Lysigenetic. Formed by disorganisation or dissolving of cells.

Macrogonidium. Large gonidium compared with others produced by the same species. Same as **megalogonidium.** Comp. **microgonidium.**

Madura. Disease in man characterised by swelling and degeneration in feet and hands and supposed to be due to Chionyphe Carteri, Berkl., but the causal connection is not definitely made out.

Malignant oedema. Disease in animals like anthrax and due to a Bacillus in form resembling Bacillus Anthracis.

Marginal veil. In Hymenomycetes: a membrane stretching from the margin of the pileus to the surface of the stipe in the young sporophore and covering over the hymenium. Same as **velum partiale.** See **velum.**

Medulla. Central tissue within the rind of a Fungus-body. In Lichens: same as **stratum medullare.**

Megalogonidium. Same as **macrogonidium.**

Membranous layer. Same as membranous mycelium.

Membranous mycelium. Mycelium in which the hyphae form by interweaving a membranous layer. Same as membranous layer, mycelial layer. Comp. filamentous mycelium, membranous mycelium.

Mentagra parasitica. Same as sycosis.

Merispore. Segment of a sporidesm.

Meristem. Actively dividing cell-tissue.

Meristematic. Consisting of meristem.

Meristogenetic. Produced by a meristem.

Metabolism. The chemical processes inseparably associated with the vital activity of protoplasm.

Metoecious. Forms which pass through separate sections of their complete development upon different hosts are metoecious. Same as metoxenous, heteroecious. Comp. autoecious, lipoxenous.

Metoxenous. Same as metoecious.

Microcyst. In Myxomycetes: a resting state of the swarmcells.

Microgonidium. Small gonidium compared with others produced by the same species. Comp. macrogonidium.

Micropylar. Belonging to the micropyle.

Microsoma. Small granule embedded in the hyaline plasm of protoplasm and constituting an essential portion of its substance.

Monoecious. Having male and female organs on the same individual. Comp. dioecious.

Monopodium. An axis of growth which continues to grow at the apex in the direction of previous growth, while lateral structures of like kind are produced beneath it in acropetal succession. Comp. dichotomy, sympodium.

Multilocular spore. Same as sporidesm.

Muscardine. Disease of the silkworm caused by Botrytis Bassii.

Mutualism. Symbiosis in which two organisms living together mutually and permanently help and support one another.

Mycelial layer. Same as membranous mycelium.

Mycelial strand. See fibrous mycelium.

Mycelium. Vegetative portion of thallus of Fungi composed of one or more hyphae.

Mycetogenetic. Produced by Fungi.

Mycetogenetic metamorphosis. Deformation of parts due to Fungi.

Mycosis. A disease of animal tissues due to the vegetative activity of species of Eurotium.

Myxamoebae. In Mycetozoa: swarmcells with purely amoeboid creeping motion.

Neck. In Pyrenomycetes: conical or cylindrical prolonged apex of pyrenocarp through which runs the canal leading to the ostiole. Same as tubulus.

Nucleus. See kernel.

Obligate. Necessary, essential. Comp. facultative.

Obligate parasite. An organism to which a parasitic life is indispensable for the attainment of its full development. Comp. facultative saprophyte, facultative parasite.

Ontogeny. Development of an individual.

Oogamy. Conjugation of two gametes of dissimilar form. Comp. isogamy.

Oogonium. Female sexual organ usually a more or less spherical sac, without the differentiation into neck and venter of archegonium, and containing one or more oospheres (ova). The oospore does not divide to form a proembryo within the cavity of the oogonium on the parent plant.

Oosphere (egg, ovum). Naked nucleated spherical or ovoid mass of protoplasm which, after its nucleus has coalesced with the sperm nucleus, developes the oospore.

Oospore. Immediate product of fertilisation in oosphere.

Ostiole (ostiolum). In Pyrenomycetes: aperture in pyrenocarp through which discharge of spores takes place. Same as pore.

Outer peridium. See peridium externum.

Ovule. In Phanerogams: macrosporangium.

Panicle. Twice or more branched structure in which the base of each branching is elongated.

Paniculate. Having the form of a panicle.

Paraphyses-envelope. In Uredineae: same as peridium.

Paraphysis. Sterile capilliform hyphal branch accompanying spore-mother-cells in a hymenium. Applied by Phoebus especially to a cyetidium.

Parasite. Organism living on or in and at the expense of another living organism (host). Comp. saprophyte.

Parthenogenesis. Form of apogamy in which the oosphere (ovum) itself developes into the normal product of fertilisation without a preceding sexual act.

Pathogenous. Producing disease.

Pébrine. Disease of the silkworm caused by Nosema Bombycis, Näg., a bacterioid organism. Same as gattine.

[4]   K k

Pellicula. The separable rind-layers of some compound sporophores. Same as **cutis.**

Penicellate. Having the form of a pencil of hairs.

Periderm. The cork-cambium and its products.

Peridiolum. In Nidularieae: chamber of the gleba, forming a nest of spores, free or attached by a funicle within the peridium of the sporophore.

Peridium. General term for the outer enveloping coat of a sporophore upon which the spores develope in a closed cavity. In Uredineae it envelopes the aecidium and is also termed **pseudoperidium, paraphyses-envelope.** In Gastromycetes termed also **uterus,** and may be differentiated into **peridium externum (outer peridium),** the outermost layer which opens in various ways and separates from the **peridium internum (inner peridium)** a layer directly enclosing the gleba.

Peridium externum. See **peridium.**

Peridium internum. See **peridium.**

Periphysis. In Pyrenomycetes: sterile capilliform hyphal branch projecting from the wall of the pyrenocarp where there is no hymenium into its cavity.

Periplasm. In Peronosporeae: protoplasm in the oogonium and the antheridium which does not share in the conjugation. Comp. **gonoplasm.**

Perithecium. Same as **pyrenocarp.**

Phototactic. Taking up a definite position with reference to the direction of incident rays of light.

Phylogeny. Development of a species or larger group.

Pileus. In Hymenomycetes: primarily, the conical or dome-shaped upper portion of the compound sporophore bearing a hymenium on its under side; now extended to all compound sporophores in which the hymenium looks to the ground. Same as **cap.**

Pityriasis versicolor. Disease of the skin caused by Microsporon furfur, Rob.

Plasmodium. In Mycetozoa: body of naked plurinucleated protoplasm exhibiting amoeboid motion.

Plasmatoparous. In Peronosporeae: forms are plasmatoparous when in germination the whole protoplasm of a gonidium issues as a spherical mass which at once becomes invested with a membrane and then puts out a germ tube.

Plastid. Small variously shaped portion of protoplasm of a cell differentiated as a centre of chemical activity.

Pleomorphism or **Pleomorphy.** The occurrence of more than one independent form in the life cycle of a species.

Pleuroblastic. In Peronosporeae: forms producing vesicular lateral ontgrowths serving as haustoria are pleuroblastic.

Pluricellular. Composed of two or more cells.

Plurisporous. Having two or more spores.

Podetium. In Cladonieae: stalk-like or shrubby branched outgrowth of the thallus bearing exposed hymenia.

Pollinarium. Same as **cystidium.**

Pore. (*a*) In Pyrenomycetes: same as **ostiole.** (*b*) In Hymenomycetes: same as **tubulus.**

Primary lamella of spore. Outermost layer of the coats of a spore representing the original delicate wall of the primordial spore.

Primordium. First beginning of any structure.

Procarp. An archicarp with a special receptive apparatus, the trichogyne.

Promycelium. Short and short-lived product of tube-germination of a spore which abjoints acrogenously a small number of spores (sporidia) unlike the mother-spore and then dies off.

Prosporangium. In Chytridieae: vesicular cell the protoplasm of which passes into an outgrowth of itself, the sporangium, and becomes divided into swarmspores.

Prothallium. A thalloid oophyte or its homologue.

Protothallus. Same as **hypothallus.**

Pseudoparenchyma. Symphyogenetic cellular tissue.

Pseudoperidium. See **peridium.**

Pseudopodium. In Mycetozoa: a protrusion of the protoplasm of an amoeboid body which may be drawn in or into which the whole mass may move.

Puffing (stäuben). Sudden discharging of a cloud of spores.

Pullulation. Same as **sprouting.**

Pulvinate. Having the form of a cushion.

Pycnidiophore. Compound sporophore bearing pycnidia.

Pycnidium. In Ascomycetes: a variously shaped cavity resembling a pyrenocarp formed on the free surface of a thallus and containing gonidia which are termed pycnogonidia. See **receptaculum.**

Pycnogonidium. Gonidium produced in a pycnidium. Same as **pycnospore, stylospore.**

Pycnospore. Same as **pycnogonidium.**

Pyrenocarp. Cup-shaped ascocarp with the margin incurved so as to form a narrow-mouthed cavity. Same as **perithecium.**

Receptacle (receptaculum). Term of varying signification, most usually implying a hollowed-out body containing other bodies. Has the following special applications in this book :—(*a*) Leveillé's term for sporophore. (*b*) Same as **stroma.** (*c*) In Ascomycetes : stalk of a discocarp. (*d*) In Ascomycetes: same as **pycnidium.** (*e*) In Phalloideae : inner portion of sporophore supporting the gleba. (*f*) In Lichens: cup of the thallus containing soredia.

Rejuvenescence. Transformation of whole of protoplasm of a previously existing cell into a cell of a different character.

Resin-flux (**Harzsticke, Harzüberfülle**). Disease in conifer characterised by copious flow of resin with ultimate death of the tree, due to attack of Agaricus melleus.

Resting period. Period during which a dormant or quiescent state is exhibited. Same as **involution-period.**

Resting-stage. Stage of dormancy or quiescence. Same as **involution-stage.**

Resting state. Quiescent or dormant condition.

Rheotropism. Phenomena induced in a growing organ by the influence of a current of water.

Rhizine. Same as **rhizoid.**

Rhizoid. Delicate filiform or hair-like organ of attachment. Same as **rhizine.**

Rhizomorphous. Having delicate branching form like rootlets.

Rind. (*a*) The outer layer or layers of a Fungus-body. Same as **cortex.** In Lichens: same as **stratum corticale.** (*b*) The outer layers of the bark in a tree with secondary thickening and sometimes all the tissue outside the active phloem.

Ring. Same as **annulus.**

Ringworm. Disease of the skin due to Trichophyton tonsurans, Malmsten. Same as **tinea tonsurans, herpes tonsurans.**

Rudimentary (**rudimentär**). An organ or member is rudimentary which remains stationary at a stage of development in which it is in every respect immature.

Saprophyte. Plant living on dead organic substance. Comp. **parasite.**

Schizogenetic. Formed by separation of tissue elements owing to splitting of the common wall of cells.

Sclerosed. Exhibiting sclerosis.

Sclerosis. Induration of a tissue or a cell-wall either by thickening of the membranes or by their lignification, i. e. formation of lignin in them.

Sclerotioid. Resembling a sclerotium.

Sclerotium. Pluricellular tuber-like reservoir of reserve material forming on a primary filamentous mycelium from which it becomes detached when its develop-

ment is complete, usually remains dormant for a time, and ultimately produces shoots which develope into sporophores at the expense of the reserve material. In Mycetozoa the sclerotium is formed out of a plasmodium and after its period of rest developes a plasmodium again.

Secondary mycelium. Rhizoid attachments developed from the base of a sporophore which are somewhat like the normal mycelium of the species.

Semen multiplex. Tulasne's term for **sporidesm.**

Septate spore. Same as **sporidesm.**

Simple sporophore (**Fruchthyphe, Fruchtfaden**). Sporophore consisting of a single hypha or branch of a hypha Same as **filamentous sporophore** Comp. compound sporophore.

Soredial branch. Branch produced by the development of a soredium into a new thallus while still on the mother-thallus.

Soredium. In Lichens: single algal cell or group of algal cells wrapt in hyphal tissue, which, when set free from the thallus, is able at once to grow into a new thallus. Same as **brood-bud.**

Soredium-heap. Same as **sorus.**

Sorus. Heap or aggregation. (*a*) In Synchitricae: heap of sporangia developed from a swarmcell. (*b*) In Lichens : heap of soredia forming a powdery mass on the surface of thallus.

Spermatiophore. Structure bearing a spermatium.

Spermatium. Male non-motile gamete-cell which conjugates with a trichogyne. The male sexual function of all spermatia is not yet demonstrated.

Spermatozoid. Male motile gamete.

Spermogonium. Receptacle in which spermatia are abjointed.

Spora cellulosa. Same as **sporidesm.**

Spora composita. Same as **sporidesm.**

Spora multilocularis. Same as **sporidesm.**

Sporangiolum. In Mucorini: small sporangium produced in some genera in addition to the large sporangium.

Sporangiophore. Sporophore bearing a sporangium. See **sporophore.**

Sporangium. Sac producing spores endogenously.

Spore. Single cell which becomes free and is capable of developing directly into a new bion.

Spore-group. Same as **sporidesm.**

Spore-plasm. Protoplasm of a sporangium devoted to the formation of spores.

Sporidesm. Pluricellular body becoming free like a spore and in which each cell is an independent spore with power of germination. Same as **spore group,** compound spore, spora composita, septate spore, semen multiplex, multilocular spore (spora multilocularis), cellular spore (spora cellulosa), pluricellular spore.

Sporidium. Spore abjointed on a promycelium.

Sporiferous. Bearing spores.

Sporoblast. Körber's term for a merispore.

Sporocarp (sporocarpium). Pluricellular fructification, i. e. pluricellular body developed as the result of the sexual act from an archicarp alone or from adjacent hyphae as well, unlike the body which produced the archicarp, and essentially serving to the formation of spores. See **fructification.**

Sporogenous. Producing spores.

Sporogenous layer. Same as **hymenium.**

Sporophore (Fruchtträger). Branch or portion of thallus which bears spores or spore-mother-cells. Same as **receptaculum** of Leveillé, **encarpium** of Trattinick. Its forms are distinguished as **gonidiophore, sporangiophore, basidiophore, ascophore,** &c.

Sporophyte. In Archegoniatae: the segment in the life-cycle which produces spores.

Sporula. Old term for what is designated above as **spore.**

Sprout-cell. Cell produced by sprouting.

Sprout-chain. Chain of cells produced by sprouting.

Sprout-gemma. In Mucoreae: same as **chain-gemma.**

Sprout-germination (Sprosskeimung). Germination of a spore in which a small process (germ-cell), with a narrow base, protrudes at one or more points on the surface of the spore, then assumes an elongated cylindrical form, and finally is abjointed as a sprout-cell. Comp. **tubegermination.**

Sprouting. Formation of an excrescence with a narrow base, and of the same character as the parent, at one or more points on a cell, which after enlargement is delimited by a transverse wall either before or after reaching its proper size. Same as **pullulation.**

Sprouting-fungus (Sprosspilz). Growthform in which the thallus consists of a sprouting-cell or chain of sprouts. Comp. **Filamentous Fungus, Compound Fungus-body.**

Sterigma. (*a*) Same as **basidium.** (*b*) Stalk-like branch of a basidium bearing a spore. (*c*) Cell from which a spermatium is abjointed.

Sterile basidium. In Hymenomycetes: body in the hymenium like a basidium but non-sporiferous and possibly a paraphysis.

Stipe (stipes). General term for the stalk of a sporophore.

Strand. See **mycelial strand.**

Stratum corticale. See **rind.**

Stratum gonimon. Same as **algal layer.**

Stratum medullare. See **medulla.**

Stroma. Compound Fungus-body having the form of a cushion, crust, foliaceous expansion, or erect unbranched or branched shrub-like body. Same as **receptaculum.**

Stylospore. Same as **pycnogonidium.**

Suberification. Same as **suberisation.**

Suberisation. Transformation of a cell-wall into suberin, i. e. conversion into cork. Same as **suberification.**

Subhymenial layer. Same as **hypothecium.**

Subulate. Awl-shaped.

Suspensor. In Mucorini ; club-shaped or conical portion of hypha adjoining a gamete-cell after its delimitation. Same as **zygosporophore.**

Swarmcell. Motile naked protoplasmic body.

Sycosis. Disease of the skin due to the attack of Microsporon Audouini, Rob., and Microsporon mentagrophytes, Rob. Same as **mentagra parasitica.**

Symbion. Organism which lives in a state of symbiosis.

Symbiosis. Living together of dissimilar organisms.

Symphyogenetic. Formed by union of previously separate elements.

Sympodial. Of the nature of a sympodium.

Sympodium. An axis made up of the bases of a number of successive axes arising as branches in succession one from the other. Comp. **monopodium.**

Tela contexta (Filzgewebe). Weft of distinct hyphae. Same as **felted-tissue.**

Teleutogonidium. Same as **teleutospore.**

Teleutospore. In Uredineae: spore formed by abjunction on, but not separating from, a sterigma, producing in germination, which takes place after a resting period, a promycelium.

Tetraspore. In Rhodophyceae: one of the spores formed by division of a mother cell into four parts.

Thallodic. Belonging to the thallus.

Thallus. A vegetative body without differentiation into stem and leaf. In Fungi is the whole body of the plant not serving directly as an organ of reproduction.

Thallus lepodes. Same as **crustaceous thallus.**

Thallus placodes. Same as **foliaceous thallus.**

Thallus thamnodes. Same as **fruticose thallus.**

Theca. Same as ascus.

Thecaspore. Same as ascospore.

Thermotropism. Phenomena induced in a growing organ by the influence of conditions of temperature.

Thrush. Disease of the mucous membrane of mouth, throat and oesophagus in children characterised by formation of pustules due to Saccharomyces albicans, Reess. Same as aphthae.

Tinea tonsurans. Same as ringworm.

Torulose. Swollen at intervals.

Trama. In Basidiomycetes: middle tissue in the projections or septa of the sporophore which bear hymenium. Same as dissepiment, intralamellar tissue.

Tremelloid. Resembling Tremella.

Trichogyne. Thread-like receptive portion developed as part of an archicarp.

Trophoplast. Same as plastid.

Trophotropism. Phenomena induced in a growing organ by the influence of the chemical nature of its environment.

Tube-germination (Schlauchkeimung). Germination of a spore in which the first product is a germ-tube. Comp. sprout-germination.

Tubulus. (a) In Pyrenomycetes: Same as neck. (b) In Hymenomycetes: tube lined with hymenium on the surface of a pileus. Same as pore.

Unicellular. Formed of one cell.

Uredo. Hymenium producing uredospores only. Termed also uredo-layer.

Uredogonidium. Same as uredospore.

Uredospore. In Uredineae: spore formed by acrogenous abjunction on a sterigma from which it separates when mature and on germination produces a mycelium bearing uredospores or uredospores and teleutospores.

Uterus. In Gastromycetes: same as peridium.

Veil. Same as velum.

Veines aërifères. Same as venae internae.

Veines aquifères. Same as venae externae.

Velum. In Hymenomycetes: special envelope within which the growth of the whole or a portion of the sporophore takes place. Same as veil, involucrum of Persoon. By Persoon applied to what is defined above as cortina. See marginal veil, velum universale.

Velum partiale. Same as marginal veil. See velum.

Velum universale. In Hymenomycetes sac enclosing the whole of a sporophore as it grows and ultimately ruptured at the apex by the unfolding pileus. Same as volva. See velum.

Venae externae. In Tuberaceae: white veins seen on section of the sporophore produced by dense tissue containing air and filling the asciferous chambers. Same as veines aërifères. Comp. venae internae.

Venae internae. In Tuberaceae: dark-coloured veins seen on section of the sporophore indicating the walls of asciferous chambers, which are composed of tissue containing no air. Same as venae lymphaticae, veines aquifères. Comp. venae externae.

Venae lymphaticae. Same as venae internae.

Volva. Same as velum universale.

Witches' broom. Disease on the silver-fir, birch, cherry, and other trees characterised by the development of a tangle of shoots in a tuft and due to the attack of Peridermium elatinum or of Exoascus.

Woronin's hypha. In Ascomycetes: a coiled hypha found in some forms at the place where the sporocarp subsequently developes and probably homologous with an archicarp.

Xyloma. Sclerotioid body of varying shape which does not send out branches developing into sporophores but produces sporogenous structures in its interior.

Yeast-fungus. Species of Saccharomyces. Sometimes used as equivalent to the growth form distinguished as Sprouting Fungus, but this misuse leads to confusion.

Zoogloea. In Schizomycetes: colony imbedded in a gelatinous substance.

Zoospore. Motile spore.

Zygospore. Immediate product of conjugation of two similar gametes.

Zygosporophore. In Mucorini: same as suspensor.

INDEX.

Names, as 'Acrasieae,' 'Abrothallus,' 'Achlya Braunii,' with no further addition, refer to the course of development of the Orders, Genera, &c., as described in the second division of the first part and in the second and third parts; those which are followed by some additional word, as 'Achlya apiculata, capacity of germination,' refer to the remaining sections of the book. An asterisk after the number of a page indicates a Figure.

Abjection of spores, 68, 72.
Abrothallus, 416.
Abscision of spores, 68, 69.
Absidia, 147, 150, 152.
— capillata, 150.
—, discharge of spores, 83.
— septata, 150.
Acarospora, number of spores, 79.
Achlya apiculata, capacity of germination, 343.
— Braunii, 142, 144.
—, discharge of spores, 82.
—, formation of swarm-spores, 143.
—, 'pleomorphism,' 127.
— polyandra, 142, 144.
— —, germination, 142*.
— prolifera, 144, 145.
— —, germ-plant, 141*.
— racemosa, 144.
— —, fertilisation, 142*.
—, simple sporophores, 46.
— spinosa, 143.
— —, capacity of germination, 343.
—, swarm-spores, 107.
Achlyogeton, 140.
—, discharge of spores, 83.
—, swarm-spores, 107.
Achorion Schoenlinii, parasitism, 376.
Acolium ocellatum, discharge of spores, 98.
Acrasieae (Acrasiea), 421, 441, 443.
Acrasis, 442.
Acrocordia gemmata, 246.
— tersa, 246.
Acrogonidia, 249.
Acroscyphus, discharge of spores, 96.
Acrospores, 129.
Acrostalagmus cinnabarinus, abjunction of spores, 71.
—, formation of gonidia, 65.
—, germination, 111.

Acrostalagmus cinnabarinus, mycelial strands, 22.
—, structure of spores, 103.
Actinomyces Bovis, development and parasitism, 377.
Actinomycosis, 377.
Adaptations, different, or metamorphosis, 256, 259.
Aecidia, 274.
—, abscision of spores, 68.
—, formation of spores, 66.
—, structure of spores, 100.
Aecidiospores, capacity of germination, 343, 344.
Aecidium Sedi, 282.
Aethalieae (Aethaliei), 431, 434, 439.
Aethalium, 429, 431, 434, 439.
— septicum, 424, 439, 441, See also Fuligo varians.
Affinities of Bacteria, 474.
— of Fungi, 337, 340.
— of Mycetozoa, 442.
Agaricineae (Agaricini), 288, 289, 296, 300, 303, 338.
—, clamp-connections, 2, 19.
—, mycelial strands, 22.
—, structure of compound sporophores, 57.
Agaricus, 301.
—, abjection of spores, 68, 72.
— aeruginosus, mycelial strands, 22, 23.
— androsaceus, mycelial strands, 22.
— arvalis, sclerotia, 42.
— balaninus, 304.
— campestris, 291.
— —, cellulose, 8.
— —, development of sporophore, 289*, 291.
— —, excretion of calcium, 11*.
— —, mycelial strands, 22, 23.
— —, cementation of hyphae, 4.
— cirrhatus, 297.

Agaricus cirrhatus, sclerotia, 32, 39, 42.
— cyathiformis, development of sporophore, 56.
— deliciosus, 300.
—, development of sporophore, 55*.
— dryophilus, 297.
— —, development of sporophore, 55*.
— —, mycelial strands, 22.
— fumosus, 304.
— fusipes, 297.
— —, sclerotia, 42.
—, gelatinous membranes, 9, 10, 13.
— grossus, germination of sclerotia, 40, 42.
— laccatus, 304.
— melleus, 302, 329, 341.
— —, commencement of sporophore, 49.
— —, development of the veil, 290*, 291.
— —, gelatinous membranes, 9.
— —, membranes, 12.
— —, mycelial strands, 22, 23, 24*.
— —, parasitism, 361, 384.
— metatus, cellulose-membrane, 8.
— —, mycelial strands, 22.
—, mycelial strands, 22, 23.
— olearius, 301.
— platyphyllus, mycelial strands, 22, 23.
— Pluteus, 304.
— praecox, 300.
— —, mycelial strands, 22, 31.
— racemosus, 334.
— —, branching of compound sporophore, 51.
— —, germination of sclerotia, 40
— —, sclerotia, 42.

Agaricus Rotula, mycelial strands, 22, 23.
—, sclerotia, 32.
— stercorarius, germination of sclerotia, 40.
—, structure of sporophore, 57.
— tuber regium, sclerotia, 42.
— tuberosus, 297.
— —, germination of sclerotia, 40.
— —, sclerotia, 32, 42.
— variabilis, 333, 334.
— velutipes, 297.
— viscidus, 304.
— volvaceus, sclerotia, 42.
— vulgaris, 334.
— —, development of sporophore, 54.
— —, structure of lamellae, 301 *.
— —, structure of sporophore, 57.
Aglaospora, 243.
— profusa, germination, 114.
— —, number of spores, 79.
—, number of spores, 79.
Alcoholic fermentation, 270.
Algae, course of development, 121.
— of the Lichen-thallus, 397, 398.
— of the heteromerous Lichen-thallus, 409.
Algal layer of the Lichen-thallus, 404.
Algal zone of the Lichen-thallus, 404.
Alternaria, 229, 252.
—, formation of spores, 67*.
—, spores, 68.
Alternation of bions, 124, 125.
— of generations, 123, 124.
Amanita, 292, 296, 297, 298, 338, 340.
—, growth of compound sporophore, 50.
—, hyphal weft, 3.
— muscaria, 292, 293, 298, 300.
— —, cellulose, 8.
— —, coloration of membranes, 10.
— —, gelatinous membranes, 9, 13.
— phalloides, cellulose membrane, 8.
— rubescens, development of sporophore, 293*.
— vaginata, 295.
Amoebae, 443.
Amoebidium parasiticum, 170.
Amylobacter Clostridium, 455.

Amylocarpus, 105.
Amylum, 7.
— in Bacteria, 455.
Anaptychia ciliaris, apothecium, 188*, 189*.
— —, chemical properties, 407.
— —, spores, 98.
— —, structure of thallus, 404, 406.
Ancyclisteae (Ancylistei), 132, 139, 170.
Ancyclistes, 139.
— Closterii, 139.
— —, parasitism, 361, 392.
Anixia truncigena, discharge of spores, 97.
Annual layers in pilei of Polyporeae, 57.
— in hymenia, 307.
Annulus, 290.
— inferus, 291.
— mobilis, 292.
— superus, 295.
Antheridial branch, 198, 202, 239.
Antheridium, 202, 305.
Anthers, 305.
Anthina, 29.
— flammea, cellulose-membrane, 8.
— —, mycelium, 29.
— pallida, cellulose-membrane, 8.
— —, mycelium, 29.
— purpurea, cellulose-membrane, 8.
— —, mycelium, 29.
Aphanocapsa in Lichen-thallus, 398.
Aphanomyces, 143.
—, discharge of spores, 83.
—, formation of spores, 74, 75.
— scaber, 142.
—, swarm-spores, 108.
Aphthae (thrush), 377.
Aplanes, 144.
— Braunii, 142.
—, capacity of germination, 343.
Apogamy, 123.
Apothecia, 187, 239, 401.
Appendages of spores, 102, &c.
Archicarp, archicarpium, 49, 121, 198, 199, 201, 238, 239.
Arcyria, 431, 434, 436, 440.
— anomala, capillitium, spore, 437*.
— cinerea, 437, 441.
— incarnata, 434, 437.
— —, capillitium, spore, 437*.
— nutans, 434, 437, 441.
— punicea, 427, 434, 437, 441.

Arcyria Serpula, capillitium, spore, 437*.
Areolation of spore-membrane, 100.
Armilla (frill, annulus superus), 295, 300.
Arnoldia, structure of thallus, 412.
— minutula, 413.
Arthonia epipasta, 416.
—, structure of thallus, 411.
— vulgaris, 416.
— —, origination of thallus, 399.
Arthopyrenia, 416.
Arthrobacterium, 454, 468.
— aceti, 472, 479, 481.
— merismopoedioides, 469.
— Pastorianum, 456, 472.
— Zopfii, 469.
Arthrobotrys, 252.
—, formation of spores, 64.
—, gonidia, 98.
— oligospora, simple sporophores, 47*.
—, spores, 68.
Arthrosterigmata, 240.
Artotrogus, 135, 232.
—, parasitism, 394.
Asci, 45, 60, 76, 191, 192.
Ascobolus, 190, 198, 199, 206, 232, 235, 239.
— furfuraceus, 224.
— —, compound sporophores, 186*, 207*.
— —, conditions of germination, 350, 351.
— —, ejection of spores, 85, 92*.
— —, structure of spores, 102.
—, conditions of germination, 350.
—, development of spores, 78.
—, ejection of spores, 85, 86.
—, germination, 111.
—, immersus, structure of spores, 102.
—, number of spores, 79.
—, puffing, 90.
— pulcherrimus, ejection of spores, 86.
— sexdecimsporus, 79.
—, structure of spores, 105.
Ascochyta, 252.
Ascodesmis, 186, 201, 221.
Ascogenous cells, 186.
Ascogenous hyphae, 186, 209, 214.
Ascogonium, 198, 213.
Ascomycetes, 120, 185, 285.
— as Lichen-fungi, 396.
—, capacity of germination, 348.
—, compound spores, 98.

Ascomycetes, course of development, 223.
—, development of sporophore, 197.
—, doubtful species, 132, 263.
—, envelope - apparatus of sporocarp, 186.
—, functionless organs, 256.
—, germination, 115.
—, hairs, 59.
—, imperfectly known species, 238.
—, inception of sporophore, 49.
—, lipoxeny (desertion of host), 388.
—, structure of sporophore in endophytic, 57.
Ascomycetous series, 120.
Ascophora elegans, resistance of spores, 347.
Ascospores, 60, 129, 232.
—, capacity of germination, 343.
—, development, 76.
Ascotricha, 211.
Ascus, 60.
— suffultorius, 61.
Ascus-apparatus, 186.
Aseroe, 312, 325.
— rubra, compound sporophores, 326*.
Aserophallus, 312, 325, 326*.
Aspergillus, 252, 256.
— albus, conditions of vegetation, 353.
—, capacity of germination, 348.
— clavatus, 253.
— —, conditions of vegetation, 353.
— —, mycelial membranes, 21.
— flavescens, parasitism, 370.
— flavus, 256.
— —, capacity of germination, 344, 348.
— —, parasitism, 369.
— fumigatus, capacity of germination, 344.
— —, conditions of germination, 349.
— —, parasitism, 359, 370.
— glaucus, parasitism, 397.
— niger, 206, 257.
— —, conditions of vegetation, 353.
— —, effect on the substratum, 358.
— —, mycelial membranes, 21.
— —, parasitism, 370.
— —, secretion of ferment, 355.

Aspergillus ochraceus, conditions of vegetation, 353.
—, parasitism, 360.
— purpureus, 206.
—, sclerotioid perithecia, 42.
—, simple sporophores, 46.
Assimilation of carbon dioxide by Bacteria, 479.
Athelia, 22.
Atichia, 416.
Atractium, 252.
Auricularia, 338.
— Auricula Judae, 306.
— —, basidia, 305*.
— —, formation of spores, 62*.
— mesenterica, structure of sporophore, 58.
— sambucina, 306.
Autoecious, 387.
Autoecism, 387.
Autoxenous, 387.
Azygites, 150.
Azygospores, 150, 159.

Bacidia, 223.
Bacillus, 454, 458, 460, 468, 473.
— Anthracis, 466*, 476, 477, 478, 482, 486.
— butylicus, 477, 480.
— butyricus, 455, 461, 462, 479.
— erythrosporus, 461.
— Megaterium, 463*.
— subtilis, 457, 462, 466*, 467, 476, 477, 479, 486.
— virens, 455.
Bacteria, 454.
—, aerobiotic, 479.
—, affinities, 474.
—, anaerobiotic, 479.
—, arthrosporous, 460, 468.
—, assimilation of carbon dioxide, 479.
—, capacity of germination, 476.
—, cell-forms, 458.
—, course of development, 459.
—, endosporous, 459, 460, 475.
—, formation of spores, 460.
—, mode of life, 476.
—, parasitic, 481, 487.
—, resistance of spores, 476.
—, saprophytic, 481, 487.
Bacteridium, 460.
Bacteriopurpurin, 455.
Bacterium, 454, 458, 468, 469, 473.
— butyricum, 477.
— chlorinum, 455.
— cyanogenum, 458.
— merismopoedioides, 469.
— Pastorianum, 456.
— Termo, 478.

Bacterium viride, 455.
— Zopfii, 469, 476.
Bactrospora, number of spores, 79.
Baeomyces, 200.
— roseus, 222.
Balsamia, 195.
Basidia, 45, 61, 286, 302, 305.
—, definitive, 306.
—, primary (initial), 306.
—, secondary, 306.
—, sterile, 302.
Basidiomycetes, 120, 132, 286.
—, angiocarpous, 338.
—, clamp-connections, 2, 19.
—, course of development and relationships, 328.
—, formation of spores, 61, 63, 67, 68.
—, germ-pores, 100.
— gymnocarpous, 289.
Basidiospores, 339, 340.
—, capacity of germination, 343.
Basidium, 60, 61, 306.
Battarea, 311.
—, capillitium, 8.
— Steveni, 317.
— —, fibres of capillitium, 317*.
— —, thickenings of membrane, 8.
Beer-yeast, 267.
Beggiatoa, 455, 459, 469, 475.
— alba, 471*, 472*.
— roseo-persicina, 455, 472, 473.
Bion, 123.
Blastema, 401.
Blastenia, 223.
— ferruginea, 223.
Boletus, 288.
—, assumption of a blue colour in the air, 15.
—, coloration of membranes, 10.
— edulis, 301.
— elegans, 297.
—, gelatinous membranes, 9, 10.
— luteus, 297.
Botryosporium, formation of spores, 63.
—, thallus, 1.
Botrytis, 252.
— Bassii, 252.
— —, capacity of germination, 344.
— —, formation of spores, 65, 65*.
— —, parasitism, 362.
— cinerea, 224, 238, 252, 254.
— —, cell-nucleus, 6.

Botrytis cinerea, conditions of germination, 349.
— —, development from a sclerotium, 38, 41.
— —, gonidiophores, 48.
— —, membrane, 12.
— —, organs of attachment, 21.
— —, resistance of spores, 347.
— —, simple sporophores, 48.
— —, thallus, 1.
— erythropus, development from sclerotia, 41.
Bovista, 11, 309, 310, 315.
—, formation of spores, 64.
— plumbea, capillitium, 12, 314*, 315.
Brachycladium, 252.
Branching in Bacteria, 458.
— of the thallus, 1.
Bristles on compound sporophores, 59.
Brood-buds, 124, 401, 415.
Brood-cells, 124, 129, 154.
— of Lichens, 417.
Bryopogon divergens, incrustations, 408.
— jubatus, 246.
— —, colouring matter, 408.
—, distribution of Algae in the thallus, 404.
— ochroleucus, incrustations, 408.
— sarmentosus, incrustations, 408.
—, soredia, 415.
—, structure of thallus, 404, 406.
Bulbils, 124, 263.
Bulbus, 292.
Bulgaria, ejection of spores, 89.
—, gelatinous membranes, 9, 13.
— inquinans, ejection of spores, 91.
— —, germination, 115.
— sarcoides, 242.
— —, ejection of spores, 86.
Bursulla crystallina, 446.
Byssaceae (Byssacei), 402.
Byssocaulon niveum, structure of thallus, 411.
Byssus, 29.

Caeoma, 282.
— Euonymi, 282.
— Mercurialis, 282.
— pinitorquum, structure of spores, 100.
Calathiscus, 312.
Calcareae (Calcariei), 424, 430, 435, 436, 441, 448, 451.

Calcium carbonate in Mycetozoa, 435, 436.
— — -vesicles, 430.
— oxalate, 11*, 408.
Callopisma, 223.
Calocera, 288, 305.
—, branching of compound sporophore, 51.
—, formation of spores, 63.
—, gelatinous membranes, 12.
—, structure of sporophore, 58.
Calosphaeria biformis, 239.
—, number of spores, 79.
— princeps, 241.
— verrucosa, number of spores, 79.
Calothrix in Lichen-thallus, 398.
Calycieae (Calyciei), discharge of spores, 96, 98.
Canker in deciduous trees, 383.
Cantharellus, 288, 297.
— infundibuliformis, development of sporophore, 56.
Cap-fungi, 287.
Capacity of germination, 343.
— in Bacteria, 476.
— in Mycetozoa, 448.
Capillitium, 194, 310, 314.
— in Mycetozoa, 431, 436.
Capitate Bacteria, 461.
Carbon dioxide, assimilation by Bacteria, 479.
Caries in teeth, 472.
Carpogonia, 212.
Carpospores, 129, 232, 340.
Catopyrenium, structure of thallus, 406.
Cauloglossum, 319.
— transversarium, 310.
Celidieae (Celidiei), 416.
Cell-forms in Bacteria, 458.
Cell-membrane, 8.
—, gelatinous, 9.
— in Bacteria, 456.
—, lignified, 9.
—, mucilaginous, 9.
—, sclerosed, 9.
—, suberised, 9.
Cell-nucleus, 6.
— in Bacteria, 456.
— in Mycetozoa, 425.
Cellulose in Bacteria, 457.
— in Mycetozoa, 428, 441, 442.
Cellulose-membrane, 8.
Cellulose-reaction, 13.
Cementation of hyphae, 4.
Cenangium Frangulae, 243.
— fuliginosum, spores, 99.
Cephalodia, 410.
Cephalosporium, 334.
Cephalotheca tabulata, discharge of spores, 97.

Cephalothecium, 252.
Ceratieae (Ceratiei), 423, 427, 434, 448.
Ceratium hydnoides, development of sporophores, 432*, 433*.
— porioides, development of sporophores, 433, 433*.
Ceratonema, 29.
Cetraria islandica, chemical properties, 408.
— —, colouring matter, 408.
— —, Lichen-starch, 10.
— —, structure of thallus, 404, &c.
— straminea, incrustations, 408.
Chaetocladieae (Chaetocladiei), 148, 151.
Chaetocladium, 117, 147, 149, 150, 151, 152, 153, 155.
—, conditions of germination, 350.
— Fresenii, 150.
— Jonesii, 150.
— —, haustoria, 20.
—, parasitism, 360, 385.
Chaetomieae (Chaetomiei), 260.
Chaetomium, 192, 211, 243, 260.
—, discharge of spores, 97.
— fimeti, 193.
— —, discharge of spores, 97.
—, germination, 111.
—, hairs, 59.
—, resin-excretion, 10.
Chaetostylum, 152.
Chain-gemmae, 155.
Chalara, 267.
— Mycoderma, 267.
Chalara-form, 250.
Change of host, 387.
Cheilaria, 252.
Chiodecton nigrocinctum, structure of thallus, 411.
Chionyphe Carteri, 379.
Chlamydospores, 154, 249, 336.
Chlorangium Jussuffii, calcium oxalate, 409.
Chlorophyll, 6.
— in Bacteria, 455, 462, 474.
Chlorosplenium aeruginosum, colouring matter, 14.
Choanephora, 150, 153, 154.
Cholera in fowls, 472, 486.
Chondrioderma, 448.
— difforme, 423, 427, 433, 441, 452.
— floriforme, 440.
—, germination and formation of plasmodia, 423*, 425*.
Chromidia, 417.
Chroococcaceae, 474.

Chroococcaceae in Lichen-thallus, 398.
Chroolepus in Lichen-thallus, 397.
— umbrinum, 398 *.
Chrysochytrium, 167.
Chrysomyxa, 277, 282, 338.
— Abietis, 284.
—, abscision of spores, 69.
—, capacity of germination, 344, 349.
—, formation of spores, 66.
— Ledi, 284.
— —, parasitism, 388.
— Rhododendri, 284.
— —, abscision of spores, 71 *.
— —, germination, teleuto-spore-layer, 284 *.
— —, parasitism, 388.
— —, spore-chain, 279 *.
— —, structure of spores, 101 *.
—, structure of spores, 101 *.
Chytridieae (Chytridiei), 132, 160, 445, 475.
—, capacity of germination, 344.
—, discharge of spores, 82.
—, doubtful species, 170.
—, parasitism, 360, 363, 364, 365, 386, 395.
—, resting state of spores, 344.
—, swarm-spores, 107.
—, thallus, 5.
Chytridium Brassicae, 162.
— macrosporum, 161.
— Mastigotrichis, 161.
— Olla, 161, 162.
— —, propagation, 165 *.
— roseum, 161.
— vorax, 161.
— —, swarm-spores, 107.
Cicinnobolus, 247, 252, 253.
— Cesatii, pycnidia, 247 *.
Cilia, 107, 457.
Circinella, 152.
Cladochytrieae (Cladochytriei), 165, 169.
Cladochytrium, 170, 184.
— Iridis, propagation, 166 *.
— —, swarm-spores, 108 *.
—, formation of spores, 61.
—, germination, 109.
— Menyanthidis, 166.
Cladonia, 200, 222.
—, cellulose, 13.
— decorticata, 222.
— furcata, Algae of the thallus, 397 *, 404.
— Novae Angliae, spermatia, 211 *.
— Papillaria, 221, 222.
—, podetia, 402.
— pyxidata, 222.

Cladiona rangiferina, 222.
— —, structure of thallus, 404.
—, structure of thallus, 404, 406.
Cladosporium, 252.
— dendriticum, parasitism, 393, 394.
—, formation of gonidia, 67.
— herbarum, 229.
—, spores, 68.
Cladothrix, 457, 458, 459, 469, 472.
Clamp-cells, 3.
Clamp-connections, 2.
Clathrocystis, 472.
Clathrus, 322.
— cancellatus, 312, 324.
— —, compound sporo-phores, 324.
— hirudinosus, 312.
—, mycelial strands, 23.
Clavaria, branching of com-pound sporophore, 51.
—, compound sporophores, 48.
— juncea, cellulose-mem-brane, 8.
— minor, sclerotium, 42.
Clavarieae (Clavariei), 288, 303.
—, structure of compound sporophore, 58.
Claviceps, 186, 191, 192, 200, 235, 239, 244, 246, 248, 260.
—, cellulose, 13.
—, fatty matter, 7.
—, formation of gonidia, 65.
—, gonidiophores, 36.
—, inception of sporophore, 49.
—, lipoxeny, 388.
— microcephala (sclero-tium), 41.
— nigricans (sclerotium), 41.
—, parasitism, 362.
— purpurea, 220.
— —, development of sporo-phore, 227 *, 228 *.
— —, germination of sclero-tia, 38, 38 *.
— —, parasitism, 359.
— —, sclerotia, 36 *, 41.
— pusilla, sclerotia, 41.
—, resting state, 37.
—, sclerotia, 30, 31, 33, 35, 36 *, 39.
—, secondary mycelium, 45.
—, spores, 99.
—, structure of ascus, 95.
Cleistocarpous Ascomycetes, 186, 193.
Clitocybe, 297.
Clostridium, 458, 460.
— butyricum, 455, 461.

Club-shaped Hymenomy-cetes, 287.
Coalescence of hyphae, 2.
Cocci, 458.
Coccobacteria septica, 473.
Coccocarpia molybdaea, Algae of thallus, 409.
Coemansia, 156.
Coenogonium confervoides, structure of thallus, 411.
— Linkii, structure of thal-lus, 410 *, 411.
Coleosporium, 281, 282, 338.
—, abscision of spores, 69.
— Campanularum, 282.
—, capacity of germination, 345.
— formation of spores, 66.
— Senecionis, parasitism, 388.
Collema, 198, 212, 224, 231, 237, 239, 240, 241.
— cheileum, number of spores, 79.
— microphyllum, fertilisa-tion, 212 *.
— —, development of sporo-phore, 214 *.
—, soredia, 415.
—, structure of thallus, 412.
Collemaceae (Collemacei), 211, 214, 231, 234, 258, 402.
—, structure of thallus, 44.
Collenchyma, 316, 328.
Collybia, 297.
Colouring matters, 14.
— of Bacteria, 455, 457.
— of fatty substances, 7.
— of Lichens, 408.
— of Mycetozoa, 424, 434.
Colpodella, 447.
Columella, 152, 173, 435, 438.
Colus, 312.
— hirudinosus, 325.
— —, receptacle, 326 *.
Comatricha, 438.
Commencement of fructifi-cation (archicarp), 121.
Companion-hyphae, 215.
Completoria, abjection of spores, 73.
— complens, 160.
—, parasitism, 363, 364, 393.
Compound sporophore, 46, 48, 186, 288.
—, development, 49.
— in Ustilagineae, 178, 181, 184.
—, structure, 57.
Concrescence of hyphae, 4.
Conditions of vegetation, general, 352.
Confervae in Lichen-thallus, 397.
Conidia, 129, 141.

Conidiobolus utriculosus, 160.

Conidium, 131.

Conjugation in Ascomycetes, 198.

— in Ustilagineae, 178, 181, 184.

— in Zygomycetes, 145.

Conjugation-cells, 145, 148.

Conoplea, 252.

Conversion of membranes into mucilage, 9.

— in abscision of spores, 69.

— in Bacteria, 456.

Coprinus, 291, 295, 296, 303, 304, 305, 306, 329, 330, 331.

—, abjection of spores, 73.

—, abjunction of spores, 64.

—, cementation of hyphae, 4.

—, clamp-connections, 2.

— comatus, 306.

—, compound sporophores, 49.

—, conditions of germination, 350.

—, duration of growth, 51.

— ephemeroides, 291, 295, 297, 332.

— ephemerus, 297, 303.

—, germination, 111.

—, germ-pore, 101.

—, growth of compound sporophore, 50.

— lagopus, 291, 295, 296, 332.

— —, gonidia, 332*.

— micaceus, 295, 304, 306.

— —, development of sporophore, 292*.

— —, hymenium, 303*.

—, mode of life, 357.

— niveus, sclerotia, 42.

—, secondary mycelium, 45.

— stercorarius, 295, 329, 332.

— —, abjection of spores, 73.

—, capacity of germination, 344.

— —, germination of sclerotia, 38, 39, 42.

— —, sclerotia, 30, 32, 34, 35, 38, 39, 42.

Coprolepeae, structure of spores, 102.

Cora Pavonia, structure of thallus, 414.

Corallofungus, 29.

Cordyceps, 186, 191, 229.

— capitata, 221.

— cinerea, parasitism, 359.

—, ejection of spores, 96.

—, formation of gonidia, 66.

—, growth of compound sporophores, 50.

— militaris, 94, 221, 253, 255, 367.

Cordyceps militaris, germination of gonidia, 371*.

— —, parasitism, 359, 362, 367, 371.

— ophioglossoides, 221.

—, parasitism, 362, 367, 371.

—, secretion of ferment, 355.

— sphecocephala, 367.

—, structure of ascus, 95.

Coremium, 48.

Cortex, 193.

Cortical layer of compound sporophore, 58.

Corticium, 287, 304.

— amorphum, 302.

— —, chemical behaviour of spore-membranes, 105.

— —, formation of spores, 64*.

— —, maturity of spores, 68.

— —, structure of spores, 100, 101.

— calcareum, secretion of calcium, 11.

— calceum, formation of spores, 64.

— dubium, 335.

— quercinum, 307.

— —, periodical growth, 57.

Cortina (curtain), 291.

Coryneum, 252.

—, spores, 68.

Course of development in Fungi, 118.

— in Bacteria, 459.

Craterium, 434.

Crenothrix, 457, 458, 459, 476.

— Kühniana, 469, 470*.

Crepidotus, 333.

Cribraria, 434, 435, 436.

Cribrarieae (Cribrariei), 427, 448.

—, experiments in germination, 421.

Cronartium, 280.

—, capacity of germination, 345.

—, parasitism, 388.

Crucibulum, 311, 320, 321, 330.

— vulgare, 319.

— —, conditions of germination, 350.

— —, development of sporophore, 319*, 320*.

— —, membranes, 12.

Crustaceous Lichens, 402.

—, growth in thickness, 407.

Cryptospora, number of spores, 79.

— suffusa, 239.

Crystalloids, 7.

Ctenomyces, 206.

Cucurbitaria, 221, 245.

— elongata, 247, 248.

— Laburni, 246, 248, 259.

— —, development of spores, 78.

— —, ejection of spores, 95.

— —, germination, 114.

— macrospora, 246.

— —, germination, 114.

— —, layer of gonidia, 246*.

Cutis, 58.

Cyathus, 311, 320, 321, 329.

—, clamp-connections, 2.

— striatus, 321.

Cyclomyces, 288.

Cylinder-gonidia, 371.

Cylindrosporium, 252.

Cyphella, 287.

Cyphellae, 407.

Cystidia, 303.

Cystococcus in Lichen-thallus, 397, 398*.

— viridis, 397*.

Cystocoleus, 411.

— ebeneus, structure of thallus, 410*.

Cystopus, 138, 233.

—, abscision of gonidia, 66*, 69*.

— Bliti, parasitism, 391.

— candidus, 135, 138.

— —, capacity of germination, 343.

— —, conditions of germination, 349.

— —, fertilisation, 136*.

— —, germination, 136*, 138*.

— —, gonidia, 138*.

— —, haustoria, 20*.

— —, parasitism, 389, 391.

— cubicus, abscision of gonidia, 69.

—, development, 134.

—, endogenous formation of spores, 73.

—, layer of gonidia, 48.

—, parasitism, 363, 386.

— Portulacae, abscision of gonidia, 66*, 69*.

— —, capacity of germination, 349.

— —, parasitism, 358.

— —, ripeness of spores, 68.

—, structure of spores, 106, 107.

—, swarm-spores, 107.

Cysts, 427.

—, thick-walled, 427.

Cytispora, 252.

Cytisporeae (Cytisporei), 251.

Cyttaria, gelatinous membranes, 9, 13.

Cyttarieae (Cyttariei), 186.

Dacryomitra, 288, 305.
Dacryomyces, 288, 305, 306, 329, 331.
— deliquescens, 331.
—, formation of spores, 63, 64.
—, gelatinous membranes, 13.
Dactylium, 249.
—, formation of spores, 65*.
— macrosporum, coloration of membrane, 8.
—, pits, 13.
Dactylococcus in Lichen-thallus, 397.
Daedalea, 288.
—, hyphal weft, 3.
—, membranes, 12.
—, mycelial membranes, 22.
— quercina, cellulose, 8, 13.
—, suberisation of membranes, 9.
Delastria, 195.
Dematieae (Dematiei), membrane, 12.
—, spores, 68.
Dematium, 29.
—, germination, 114.
— herbarum, formation of gonidia, 67.
— pullulans, sprouting, 271*.
Dendryphium, 252.
Dermatea amoena, 243.
— carpinea, 243.
— Coryli, 243.
— dissepta, 243.
Dermatocarpon, 192.
Desertion of host (lipoxeny), 388.
Diachea, 434, 436.
— elegans, 424.
Diatrype, 186, 192, 218, 240.
—, number of spores, 79.
— quercina, spermogonia, 241*.
— —, number of spores, 79.
— verrucaeformis, number of spores, 79.
Dictydium, 427, 431, 435.
Dictyonema, structure of thallus, 414.
Dictyostelium, 442, 451.
Dictyuchus, 143.
— clavatus, 144.
— —, formation of spores, 75.
—, formation of spores, 74.
— monosporus, formation of spores, 75.
—, swarm-spores, 108.
Diderma, 434, 436.
Didymium, 430, 436, 441, 448.
— difforme, 427, 428, 429.
— farinaceum, 436.
— leucopus, 423, 424, 428, 436.

Didymium leucopus, plasmodium, 425*.
— —, sporangium, 434*.
— Libertianum, 423, 441.
— nigripes, 436.
— physaroides, 436.
— praecox, 423, 424, 427.
— Serpula, 424, 425, 428, 429, 430, 434, 436, 452.
— squamulosum, sporangium, 434*.
Dimargaris crystalligena, 156.
Diplodia, 247.
—, germination, 111.
—, secretion of resin, 10.
Discocarp, 187.
Discomycetes, 186, 187, 189.
—, as Lichen-fungi, 396.
—, ejection of spores, 85, 86, 89.
—, glycogen, 6.
—, puffing of spores, 90.
Discus, 187.
Diseases in silk-worms, 486, 489.
Dispira cornuta, 156.
Dissepiment, 301.
Doassansia, 173.
Dothidea, 191.
— Melanops, 246.
— Ribesia, germination, 114.
— Sambuci, number of spores, 79.
— Zollingeri, 245.

Effect of parasite on host, 367.
Ejection of spores, 84.
Elaphomyces, 193.
—, asci, 77.
—, conditions of germination, 352.
—, discharge of spores, 81, 97.
— granulatus, 193.
— —, development of spores, 80.
—, mycelial strands, 22, 23.
—, number of spores, 79.
Empusa, 158.
—, abjection of spores, 73.
— Grylli, 159.
— macrospora, 158.
— Muscae, 158, 159.
Encarpium, 17.
Enchylium, structure of thallus, 413.
Endocarpon, 192.
— miniatum, 222.
— monstrosum, calcium oxalate, 409.
— pusillum, 224, 409.
— —, Algae of thallus, 409.
— —, formation of thallus, 400.

Endocarpon pusillum, germination, 339*.
— —, hymenial Algae, 400, 401.
— —, number of spores, 79.
— —, structure of thallus, 406.
Endomyces, 341.
— decipiens, 266.
Endophyllum, 278, 281, 285.
— Euphorbiae, 281.
— —, parasitism, 364, 368, 390.
— Sempervivi, 277, 281.
— —, germination, 279*.
— —, parasitism, 390.
Endophytes, 360.
—, behaviour to the living cell, 393.
Endopyrenium, structure of thallus, 404.
Endosporium, 100.
Enerthenema, 438.
Entomophthora, 158.
—, abjection of spores, 73.
— curvispora, 158, 159.
—, formation of spores, 62.
— ovispora, 158, 159.
—, 'pleomorphism,' 126.
— radicans, 158, 159.
Entomophthoreae (Entomophthorei), 132, 158, 184.
—, parasitism, 362, 367, 371.
Entyloma, 172, 173, 174, 178, 179, 180, 181.
— Calendulae, germination, 175*.
— —, parasitism, 389.
—, formation of spores, 61.
— Magnusii, 179.
—, parasitism, 362.
— Ranunculi, 179, 180.
— serotinum, 180.
— Ungerianum, germination, 175*.
Envelope-apparatus in Ascomycetes, 186.
Envelope of plasmodia of Mycetozoa, 426.
Ephebe pubescens, 242.
— —, structure of thallus, 412*.
— structure of thallus, 412.
Epichloe, 186, 191, 192, 200, 221, 244, 246, 260.
—, abscision of gonidia, 71.
—, structure of sporophore, 57.
— typhina, formation of gonidia, 65.
— —, parasitism, 359, 386, 390, 391.
— —, structure of ascus, 95.

Epinasty of pileus in Agari-cineae, 50, 289.
Epiphragm, 320.
Epiphytes, 360.
Epiplasm, 77.
Episporium, 100.
Eremascus, 187, 232, 233, 234, 236, 237.
— albus, 197, 224.
—, development of ascus, 198*.
Ergot, 35.
Erysipelas, 486.
Erysiphe, 202, 232, 236, 238, 239, 245, 352.
— Aceris, parasitism, 394.
—, asci, 76.
— communis, conditions of germination, 352.
—, discharge of spores, 86.
— Galeopsidis, 203.
— —, conditions of germina-tion, 352.
— graminis, 203.
— —, conditions of germina-tion, 352.
— —, haustoria, 19.
— guttata, number of spores, 79.
— —, parasitism, 359, 394.
—, hairs, 59.
—, number of spores, 79.
—, parasitism, 359, 363.
— spiralis, 225.
— Tuckeri, haustoria, 19*.
Erysipheae (Erysiphei), 193, 198, 199, 201, 225, 235, 244.
—, conditions of germination, 350.
—, discharge of spores, 85.
—, formation of gonidia, 66.
—, haustoria, 19*.
—, mycelium, 19*.
—, parasitism, 386, 393.
Euchytridieae (Euchytridei), 164.
Eurotium, 193, 198, 204, 226, 232, 234, 236, 237, 239, 245.
—, abscision of gonidia, 70.
—, asci, 76.
— Aspergillus glaucus, con-ditions of vegetation, 353.
— —, formation of gonidia, 70*.
— —, parasitism, 369.
—, capacity of germination, 348.
—, compound sporophores, 49.
—, discharge of spores, 81, 97, 98.
—, formation of spores, 62.
—, parasitism, 369.

Eurotium repens, conditions of vegetation, 353.
— —, development of sporo-phore, 203*.
— —, parasitism, 369.
—, resinous excretion, 10.
—, simple sporophores, 46.
—, structure of spores, 100.
—, taking up of colouring matter, 14.
Eusynchytrium, 167.
Eutypa, 191, 218.
—, discharge of spores, 97.
Evernia flavicans, 406.
— furfuracea, colouring mat-ter, 408.
— —, structure of thallus, 406.
—, soredia, 415.
—, structure of thallus, 404.
— vulpina, incrustations, 408.
— —, structure of thallus, 408.
Excipulum, 188.
Exidia, 287, 305.
—, gelatinous membranes, 13.
— recisa, 337.
— spiculosa, basidia, 305*.
— —, formation of spores, 62*.
Exoascus, 265, 266.
— alnitorquus, 265, 266.
— aureus, 265.
— bullatus, 266.
— deformans, 265.
—, ejection of spores, 89.
— epiphyllus, 266.
—, germination, 114, 115.
—, number of spores, 79.
—, parasitism, 368, 386, 390, 393.
— Populi, 266.
— Pruni, 265, 266.
— —, development of spores, 79.
— —, ejection of spores, 86, 92.
— —, inception of sporo-phore, 49.
— —, parasitism, 393.
—, structure of spores, 100.
— Ulmi, 266.
Exoascus-group, 269.
Exobasidium, 271, 287, 329, 331, 368.
— Lauri, parasitism, 368.
—, parasitism, 386.
— Vaccinii, parasitism, 362, 368.
Exosporium, 100, 135, 252.
—, spores, 68.
— Tiliae, germination, 114.

Fatty matters, 7.
Favolus, 288.

Favus, 376.
Feeder (host), 358.
Felted tissue, 5.
Ferment, secretion, 355, 452.
Fermentation caused by Bac-teria, 481.
— — by Fungi, 384.
Ferns, course of develop-ment, 120.
Fertilisation-tube, 134.
Fibrillaria, 29.
Filamentous Fungi, 1.
Filamentous sporophore, 46.
Fistulina, 302.
— hepatica, 300, 302, 334.
Flacherie (in silk-worms), 486, 489.
Flagella, 107, 422, 457.
Flagellatae, 445, 475.
Flower of tan, 431.
Flowering plants, course of development, 121.
Flowers of wine, 267, 358.
Foliaceous Lichens, 401.
Food, a condition of ger-mination, 351.
Food-material of Fungi, 353.
Form-genera, 119, 459, 473.
Form-species, 119, 459, 473.
Fowl-cholera, 472, 486.
Frill (in Hymenomycetes). See Armilla.
Frog-spawn (Bacterium), 469.
Fructification, 120.
Fruticose Lichens, 401.
Fuligo, 423, 424, 425, 428, 429, 431, 440, 448, 449, 450, 452.
— varians, 424, 439, 441.
Fumago, 245, 247, 249, 251, 253, 254, 271.
—, formation of gonidia, 66.
— salicina, 244.
Fungi, affinities, 337, 340.
—, course of development, 118.
—, genealogy, 337, 340.
Fungus-body, 2.
—, compound, 2.
—, sclerotioid, 190.
—, slimy mucilaginous, 9.
Fungus-cellulose, 8, 13.
Funiculus, 321.
Fusarium heterosporum, re-sistance of spores, 346.
Fusiform rods in Bacteria, 458.
Fusisporium, 249, 252.
— Solani, 245.
—, spores, 68.

Galls, 369.
Gametes, 120, 148, 150.

Gastromycetes, 286, 308, 337.

Gastromycetes, compound sporophores, 49.

— conditions of germination, 352.

—, development of sporophore, 50.

—, formation of spores, 63*.

—, gelatinous membranes, 9.

—, membranes, 11, 12.

—, structure of sporophore, 58.

Gattine, 488.

Gautieria, 308, 337.

Geaster, 308, 309, 311, 315, 316, 317.

—, capillitium, 12.

— coliformis, 314.

— —, capillitium - threads, 314*.

— fimbriatus, 314, 316, 317.

— fornicatus, 314, 316, 317.

— hygrometricus, 308, 309, 313, 315, 316, 317.

— —, compound sporophores, 316*.

— —, formation of spores, 63*.

— —, gelatinous membranes, 9, 12.

— mammosus, 314, 317.

— rufescens, 317.

— tunicatus, 309.

Gelatinisation of membranes, 9.

— in abscision of spores, 69.

— in Bacteria, 456.

Geminella Delastrina, 174.

Gemmae, 60, 61, 154, 155, 230, 328, 330, 331.

Gemmae-cups (in Lichens), 407.

Genabea, 195, 196.

Genea, 196.

—, mycelial strands, 22.

—, structure of spores, 100.

Genealogy of Fungi, 337, 340.

Geoglossum, 189.

—, development of spores, 78.

—, ejection of spores, 86.

—, hirsutum, development of spores, 78.

Germ-pore in Mycetozoa, 441.

Germ-pores, 100, 111.

Germ-tube, 1, 110.

Germination, 109, 130.

— in Bacteria, 462, 465, 467.

— in Mycetozoa, 448.

—, conditions of, 349.

— — in Bacteria, 462, 477.

— — in Mycetozoa, 448.

— phenomena of, 343.

Germination of spores, 109.

Gleba, 193, 309.

Gloeocapsa in Lichen-thallus, 397*, 398.

Gloeosporium, 252.

Glycogen, 6, 77.

Gnesiolichenes, 402.

Gomphidius, 297.

Gonatobotrys, 252.

—, formation of spores, 63.

—, gonidia, 98.

—, simple sporophores, 47.

Gonidia, 45, 60, 129, 131, 179, 180, 239, 244, 331, 334, 338, 340.

—, accessory, 146.

—, capacity of germination, 343.

— in Lichens, 417.

Gonidial layer (gonimic layer), 417.

Gonidiophore, 45, 245.

Gonionema, structure of thallus, 412.

Gonoplasm, 134.

Granulose in Bacteria, 455, 461.

Graphideae (Graphidei), origin of thallus, 399.

—, soredia, 415.

—, structure of thallus, 410.

— subcortical thallus, 402.

Graphiola, 173.

Graphis scripta, origin of thallus, 399.

— —, Algae of thallus, 398*.

— —, structure of thallus, 410.

Graphium, 29.

Green rot in wood, 14.

Growth of parasites, 367.

— of Fungus-bodies, 3, 50.

Guepinia, 287, 305.

— contorta, structure of sporophore, 58.

—, gelatinous membranes, 13.

— helvelloides, 287.

Guttulina, 443.

— protea, 442, 443.

Guttulinae, 442.

Gymnoascus, 198, 199, 206, 224, 232, 235.

Gymnomycetes, 251, 252.

Gymnosporangium, 279.

—, parasitism, 388.

— Sabinae, 276.

Gyrocephalus, 287.

Gyrophora, 189.

— cylindrica, spermogonium, 240.

Haematomma ventosum, structure of thallus, 405.

Hairs of compound sporophores, 59.

Haplocystis mirabilis, 170.

Haplomycetes, 252.

—, thallus, 1.

Haplotrichum, 252.

—, formation of spores, 62, 63.

—, simple sporophores, 47.

Haustoria, 18, 19*.

Hay-bacillus, 467.

Helicosporangium, 263.

Helicostylum, 152.

Heliotropism in plasmodia, 449.

Helminthosporium, 252.

—, spores, 68.

Helotium, ejection of spores, 86.

Helvella, 189.

— crispa, ejection of spores, 86.

— —, puffing of spores, 92.

— —, structure of sporophore, 58.

— elastica, development of spores, 77.

— —, structure of sporophore, 58.

— —, structure of spores, 106.

— esculenta, development of spores, 77.

— —, germination, 113*.

— —, structure of sporophore, 58.

— —, structure of spores, 106.

—, structure of sporophore, 57.

Hemiarcyria, 437, 440.

— clavata, 437.

— rubiformis, 437, 438, 440.

Hemileia, 274.

—, capacity of germination, 345.

— vastatrix, 282.

— —, parasitism, 388.

Hemipuccinia, 279.

Hendersonia, 252.

Himantia, 29.

Homologies of stages of development, 119.

— restored, 123.

— interrupted, 123.

Host (vegetable and animal), 358.

—, reaction on parasite, 366.

Hyacinth, yellow sickness, 482.

Hydneae, 288, 303, 333.

Hydnobolites, 195.

Hydnocystis, 196.

Hydnotria, 196.

Hydnum auriscalpium, hairs, 59.

Hydnum cirrhatum, 301.
— diversidens, 307.
— —, parasitism, 384.
— Erinaceus, 335.
— —, gelatinous membranes, 9.
— gelatinosum, 301.
— zonatum, 301.
Hydrotropism in plasmodia, 449, 450.
Hymenial Algae, 400, 401.
Hymenial gonidia, 417.
Hymenium, 49, 191, 300.
Hymenochaete, 303.
Hymenogaster, 308.
— —, clamp-connections, 2.
— —, formation of spores, 63.
— Klotzschii, 309, 313.
— —, formation of spores, 63.
— —, germ-pores, 101.
Hymenogastreae (Hymenogastrei), 308, 309, 310, 313, 337.
—, conditions of germination, 352.
—, formation of spores, 63.
—, mycelial strands, 22, 23.
Hymenomycetes, 286, 287, 337.
—, abjection of spores, 73.
— as forming Lichens, 396, 414.
—, capacity of germination, 343.
—, compound sporophores, 48.
—, ferment-secretions, 355.
—, formation of spores, 63.
—, gelatinous membranes, 9, 12.
—, glycogen, 6.
—, growth of compound sporophores, 51, 55.
— —, periodic, 51.
—, gymnocarpous, 289, 296.
—, membranes, 12.
—, mycelial membranes, 21, 22.
—, parasitism, 384.
—, sclerotia, 32.
—, structure of spores, 106.
—, veiled, 289, 296.
Hymenophorum, 300.
Hypertrophy, 368.
Hypha (proper name), 29.
Hypha, 1.
—, Woronin's, 199, 218.
Hyphae, ascogenous, 186, 208, 214.
—, cementation, 4.
—, coalescence, 2.
—, concrescence, 4.
Hyphal weft, 3.

Hyphasma, 29.
Hypholoma, 197.
Hyphomycetes, 251.
—, abscision of spores, 71.
—, absence of calcium, 11.
—, thallus, 1.
Hypochnus, 287.
— centrifugus, basidia, 301*.
— —, gonidial layer, 48.
— —, sclerotia, 32, 40, 42.
—, clamp-connections, 2.
—, mycelial membranes, 22.
— purpureus, 306.
Hypocopra, 198, 210.
— finicola, 210, 224, 261.
—, structure of spores, 102.
Hypocrea citrina, number of spores, 79.
—, formation of gonidia, 65.
— gelatinosa, number of spores, 79.
— lenta, number of spores, 79.
— rufa, 253.
—, number of spores, 79.
Hypodermii, 184.
Hypomyces, 245, 249, 254.
— armeniacus, sclerotia, 41.
— asterophorus, 336.
— Baryanus, 336.
—, capacity of germination, 345.
— chrysospermus, 249.
—, detachment of spores, 68.
—, formation of gonidia, 65.
— rosellus, 249.
— Solani, 246, 249.
Hyponasty of pilei of Agaricineae, 56, 289.
Hypothallus, 405.
Hypothecium, 188.
Hypoxylon, 186, 218, 244, 248.
— concentricum, discharge of spores, 104.
Hysterangium, 308, 310.
— clathroides, gelatinous membranes, 12.
— gelatinous membranes, 9.
Hysterineae (Hysterinei), 190.
—, parasitism, 386.
—, structure of spores, 102.
Hysterium macrosporum, gelatinous membranes, 9.
— nervisequum, structure of spores, 102.

Ileodictyon, 312.
Imbricaria caperata, incrustations, 408.
— incurva, incrustations, 408.
— saxatilis, 246.
— sinuosa, 246.

Imbricaria soredia, 415.
— tiliacea, Algae of thallus, 398*.
Incipient mycelium, 178.
Incrustations of Lichens, 408.
Inner membrane of spores, 100.
Interstitial substance in sporangium of Mucorini, 75.
Intralamellar tissue, 301.
Invertin, 355.
Involucrum, 289.
Involution-forms, 458.
Irpex, 288.
Isaria, 48, 273.
— brachiata, branching, 51.
—, capacity of germination, 344.
— farinosa, 255.
— —, parasitism, 362.
— strigosa, 253.
Isidium, 406.

Kickxella, 156.
Kneiffia, 287.

Laboulbenia Baeri, parasitism, 359.
— flagellata, 263*.
— Nebriae, 264.
— vulgaris, 264.
Laboulbenieae (Laboulbeniei), 263.
—, parasitism, 360, 365, 367, 370.
—, thallus, 5.
Lactarius, 298, 299, 301, 304.
— chrysorrhoeus, 298, 301.
— deliciosus, 298, 303, 304.
— mitissimus, 298, 302.
— pallidus, 298.
—, pseudo-parenchyma, 3.
— subdulcis, 301.
— —, structure of sporophore, 299*.
Lagenidium, 139.
Lamellae, 288.
Lamia culicis, 159.
Lamina proligera, 187.
Lamina sporigera, 187.
Lateral branches in Saprolegnieae, 141.
Laticiferous tubes, 299.
Laudatea, structure of thallus, 414.
Layer (stroma, receptaculum), 48.
Lecanactis illecebrosa, Algae of thallus, 398*.
— —, structure of thallus, 411.
Lecanora, 223.

Lecanora pallida, development of spores, 78.
— —, thallus, 400, 402, 405.
— subfusca, apothecium, 190*.
— Villarsii, structure of thallus, 405.
Lecidea, 223.
— confervoides, structure of thallus, 405.
— enteroleuca, structure of thallus, 405.
— formosa, 222.
— geographica, structure of thallus, 405.
— parasema, structure of thallus, 405.
— sabuletorum, 245.
Lecidella enteroleuca, development of spores, 78.
— —, growth of thallus, 404*.
Lempholemma, structure of thallus, 412.
Lenzites, 288, 301.
— betulina, membranes, 12.
—, growth of compound sporophores, 57.
Leocarpus vernicosus, 424, 434.
Leotia, 189.
— lubrica, development of spores, 78.
Lepiota procera, 297.
Leptochrysomyxa, 283.
— Abietis, 338.
—, capacity of germination, 345.
Leptogium, 198.
—, structure of thallus, 412, 413.
Leptomitus brachynema, 144.
— lacteus, 144.
—, formation of spores, 74.
Leptopuccinia, 283, 344.
— annularis, 283.
—, capacity of germination, 344, 348.
— Circaeae, 283.
— Dianthi, 284.
— —, parasitism, 361, 362.
— Malvacearum, 284.
— Veronicae, 283.
Leptopuccinieae (Leptopucciniei), 283, 338.
Leptosphaeria Doliolum, 247.
Leptothrix, 458, 459.
— buccalis, 456, 472.
Leptothyrium, 252.
Lesions in spores, effect on capacity of germination, 346.
Leucochytrium, 167.

Leuconostoc mesenterioides, 457, 469.
Libertella, 252.
Licea, 435, 436.
— flexuosa, 434, 435, 441.
— pannorum, 423, 441.
— Serpula, 441.
Lichen-acids, 10, 408.
Lichen-fungi, 188, 224, 242.
—, conditions of germination, 350.
—, ejection of spores, 93.
—, gelatinous membranes, 9.
—, mode of life, 395.
—, parasitism, 360, 366, 367, 386.
—, resistance of spores, 346.
—, swelling of membrane, 9.
Lichen-sporocarps, 222.
Lichen-starch, 10, 408.
Lichen-thallus, 59.
—, chemical constitution, 407.
—, crustaceous, 401.
—, epiphloeodic, 402.
—, foliaceous, 401.
—, fruticose, 401.
—, growth, 401.
—, heteromerous, 402.
—, homoiomerous, 402.
—, hypophloeodic, 402.
—, mode of origination, 398.
—, subcortical, 402.
—, structure, 401.
Lichens, 396.
—, cell-membranes, 8.
—, colouring matter, 408.
—, ejection of spores, 87.
—, historical remarks, 416.
—, hyphal weft, 3.
—, number of spores, 79.
—, pseudo-parenchyma, 3.
—, swelling of membrane, 9.
—, thallus, 1.
—, true, 402.
Lichenaceae (Lichenacei), 402.
Lichenin, 10, 407.
Lichenosphaeria, structure of thallus, 412.
Lichina, discharge of spores, 96.
—, structure of thallus, 410.
Lichinic acid, 408.
Lignification of membranes, 9.
Lindbladia, 431.
Lipoxeny (desertion of host), 388.
Long rods (Bacilli), 458.
Loss of capacity in sporophores of Ascomycetes, 254.
Lycogala, 423, 427, 431, 440, 452.
— epidendron, 440, 441.

Lycoperdaceae (Lycoperdacei), 308, 310, 311, 313, 337.
—, experiments in germination, 352.
—, mycelial strands, 22, 23.
—, pseudo-parenchyma, 3.
Lycoperdon, 11, 308, 309, 310, 314, 315.
—, abjunction of spores, 64.
— Bovista, 314.
— giganteum, 314.
— perlatum, 316.
—, pits, 13.
— pyriforme, formation of spores, 63*.
Lyngbya in Lichen-thallus, 398.
Lysurus, 325.

Macrococci, 458.
Macrogonidia, 225.
Macrosporium, 252.
— Sarcinula, 229.
Madura-disease, 379.
Main series of Fungi, 120.
Malignant oedema, 486.
Mallotium Hildebrandtii, structure of thallus, 413*.
—, structure of thallus, 412, 413.
Marginal veil (in Hymenomycetes), 290.
Martensella, 156.
Massaria, 192.
— Platani, 258.
—, structure of spores, 102.
Mechanism for abjection of cells, 72, 85.
Medulla (Medullary layer) of Lichen-thallus, 403.
— of compound sporophores, 58.
Megalogonidia, 225.
Megalospora affinis, calcium oxalate, 409.
— —, germination, 112*.
—, number of spores, 79.
— sanguinea, calcium oxalate, 409.
Melampsora, 281.
— Göppertiana, parasitism, 388, 390, 391.
— populina, 280, 282.
— salicina, 282.
Melanconis, germination, 114.
—, structure of spores, 102.
Melanconium, 252.
—, spores, 68.
Melanogaster, 309, 310.
—, gelatinous membranes, 9, 13.
Melanospora, 191, 192.

Melanospora parasitica, 191, 198, 199, 210, 211, 226, 235, 251, 365.
— —, discharge of spores, 97.
— —, parasitism, 360, 365, 366.
— —, structure of spores, 106.
Melanotaenium, 176.
Melogramma Bulliardii, germination, 114.
Membranes of vegetative cells, 11.
Mentagra parasitica, 376.
Merispores, 98.
Merulius, 288.
Mesentericae, 424.
Metamorphosis, 256, 259.
—, mycetogenetic, 368.
Metoecious, 387.
Metoecism, 388.
Metoxenous, 387.
Micrococcus, 459, 473.
— Bombycis, 489.
— prodigiosus, 455.
— —, taking up of colouring matter, 14.
Microcysts, 427.
Microgonidia, 225.
Microgonidium, 419.
Micropuccinia, 285.
Microsomata, 455.
Microsporon Audouini, parasitism, 376.
— furfur, parasitism, 376.
— Mentagrophytes, parasitism, 376.
Milk, blue, 462.
Mitremyces, 312, 326.
—, gelatinous membranes, 9, 13.
Mode of life of Bacteria, 476.
— of Mycetozoa, 448.
— of Fungi, 343.
Monad-forms, 458.
Monadopsis, 447.
Monas Amyli, 447.
Monoblepharis, 132, 140, 141.
— sphaerica, fertilisation, 140*.
—, swarm-spores, 107, 109.
Montagnites, 297.
Morchella, 189.
—, ejection of spores, 89.
— esculenta, development of spores, 77.
— —, ejection of spores, 85.
—, structure of sporophore, 37.
Moriola, 416.
Morioleae (Moriolei), 416.
Mortierella, 146, 149, 150, 151, 154, 155.

Mortierella, discharge of spores, 83.
—, haustoria, 20.
— nigrescens, 150.
— reticulata, resistance of spores, 346.
— Rostafinskii, 150.
— —, experiments in germination, 352.
Mosses, course of development, 120.
Mother of vinegar, 456, 458, 481.
— -fungus, 472.
Moulds, fatty matters, 7.
Mucor, 149, 152, 155.
— circinelloides, ferment-production, 358.
— corymbifer, parasitism, 360.
—, discharge of spores, 83.
—, ferment-production, 358.
—, formation of spores, 74, 75.
— fusiger, 150, 152.
— —, cellulose-membrane, 8.
— —, resistance of spores, 346.
—, mode of life, 357.
— Mucedo, 150, 154.
— —, cellulose-membrane, 8.
— —, ferment - production, 358.
— —, germination, 113.
—, parasitism, 380.
— plasmaticus, formation of spores, 75.
—, pleomorphism as alleged, 126.
— racemosus, 150.
— —, conditions of vegetation, 354.
— —, ferment - production, 358.
— —, parasitism, 380.
—, resistance of spores, 347.
— rhizopodiformis, parasitism, 359, 360, 370.
—, simple sporophores, 46.
— spinosus, ferment - production, 358.
— stolonifer, 147, 150, 152.
— —, conditions of germination, 350, 352.
— —, conjugation, 148*.
— —, ferment - production, 358.
— —, germination, 113.
— —, parasitism, 380.
— —, resistance of spores, 347.
— —, secondary mycelium, 46.
—, taking up of colouring matter, 14.

Mucor tenuis, 150.
Mucor-yeast, 156.
Mucoreae (Mucorei), 148, 151.
Mucorin, 7.
Mucorineae (Mucorini), 132, 145, 169, 232, 233, 236, 271.
—, calcium oxalate, 11.
—, capacity of germination, 348.
—, cell-nucleus, 6.
—, conditions of germination, 350.
—, crystalloids, 7.
—, doubtful, 156.
—, ferment-secretion, 358.
—, formation of spores, 75.
—, gemmae, 60, 61.
—, germination, 113.
—, glycogen, 6.
—, parasitism, 363, 385.
—, resting-state of spores, 344.
—, simple sporophores, 46.
—, thallus, 1.
Muscardine, 374.
Mushrooms, cell-membranes, 8.
—, thallus, 1.
Mutualism, 369.
Mycelial membranes, 18, 21.
Mycelial strands, 18, 22.
Mycelium, 18.
—, filamentous, 18.
—, fibrillose, fibrous, 18.
—, sclerotioid, 42.
—, secondary, 45.
Mycena, 297.
—, cementation of hyphae, 4.
—, gelatinous membranes, 9, 13.
Mycenastrum, 315.
— corium, 315.
— —, capillitium - threads, 314*.
Mycetozoa, 411.
—, affinities, 442.
—, doubtful, 446.
Mycoderma, 22.
— aceti, 481.
Mycoderma-form, 250.
Mycogone, 245, 252.
Mycoidea parasitica in Lichen-thallus, 398.
Mycoprotein, 457.
Mycosis, 369, 370.
Mycothrix, 458.
Myelin, 300.
Mylitta, 42.
Myriangium, 193, 416.
— Durieui, 197.
Myriocephalum botryosporum, structure of spores, 103.

Mystrosporium, 229, 252.
Myxamoebae, 423.
Myxastrum radians, 447.
Myxocyclus, 252.
— confluens, structure of spores, 102.
Myxogasteres, 421.
Myxomycetes, 421, 475.
—, affinities, 442.
Myzocytium globosum, 139.

Naemaspora, 252.
Naetrocymbe, 416.
Neck in perithecia, 191.
Nectria, 200, 215, 221, 239, 245.
— cinnabarina, 244, 246.
— —, parasitism, 362, 383.
— cucurbitula, parasitism, 383.
— development of spores, 78.
—, discharge of spores, 97.
— ditissima, 228.
— —, parasitism, 383.
—, germination, 114, 115.
— inaurata, 114, 115.
— Lamyi, germination, 115.
— Solani, 246.
— —, abscision of gonidia, 71.
Nephroma arcticum, Algae of thallus, 409.
—, structure of thallus, 406.
Nephromium, Algae of thallus, 409.
New formation of members caused by Fungi, 368.
Nidularia, 311, 320.
Nidularieae (Nidulariei), 308, 312, 319, 332.
—, gelatinous membranes, 13.
—, mycelial strands, 22, 23.
Nosema Bombycis, 488.
Nostoc in Lichen-thallus, 397*, 398.
Nostocaceae, 475.
— in Lichen-thallus, 398.
Nucleariae, 447.
Nuclei in spores, 106.
Nuclein, 6.
Nucleus in perithecia, 193.
Nummularia, 218.
—, discharge of spores, 97.
Nutritive adaptation, 356.
Nyctalis, 297, 334, 341.
— asterophora, 335, 341.
— —, compound sporophores, 335*.
— mycrophylla, 336.
— parasitica, 336.
— —, development of sporophore, 55.

Obelidium, 164.
Obryzum, structure of thallus, 412.
Ochrolechia pallescens, calcium oxalate, 409.
— —, germination, 112.
— —, structure of thallus, 405.
— tartarea, calcium oxalate, 408.
Octaviania, 308.
— asterosperma, compound sporophores, 308*.
— —, hymenium, 309*.
— carnea, 309, 313.
— —, formation of spores, 63*.
—, structure of spores, 100.
Oedema, malignant, 486.
Oidium, 238, 252.
— albicans, parasitism, 377.
— aurantiacum, resistance of spores, 347.
— erysiphoides, 252.
— fructigenum, 252.
— lactis, 252, 253, 377.
— —, cell-nucleus, 6.
— —, formation of spores, 67.
— leucoconium, 252.
— Tuckeri, 225.
— —, parasitism, 387.
Oil-drops in spores, 106.
Olpidieae (Olpidiei), 166, 167, 168, 169, 170.
Olpidiopsis, 444.
— fusiformis, 166.
—, parasitism, 393.
— Saprolegniae, 161, 166.
Omphalaria, structure of thallus, 413.
Omphalarieae (Omphalariei), structure of thallus, 413.
Omphalia, 297.
Onygena, 193.
— corvina, 196.
— —, conditions of germination, 351.
—, discharge of spores, 97.
— equina, 197.
— faginea, 335.
Oogonia, 120, 133.
Oosphere, 120, 133.
Oospores, 129, 232.
—, capacity of germination, 343.
—, resting state, 344, 345.
Opegrapha filicina, 397.
— —, structure of thallus, 411.
— plocina, structure of thallus, 411.
— saxatilis, structure of thallus, 411.

Opegrapha varia, 246.
— —, structure of thallus, 411.
— vulgata, 246.
Ophidomonas, 472.
Organs, functionless in Ascomycetes, 256.
— of attachment, 45.
— of propagation, 124.
Ostiole of perithecium, 191.
Otomycosis aspergillina, 369.
Outer envelope of ascocarp, 188.
Oxygen as a condition of germination, 349.
Ozonium, 29.

Palmellaceae in Lichenthallus, 397.
Palmellae-forms in Bacteria, 459.
Panhistophyton, 488.
Pannaria, 223.
—, structure of thallus, 410.
Panus stypticus, structure of sporophore, 58.
— —, gelatinous membrane, 13.
Papulospora, 2.
Paraphyses, 48, 76, 187, 192, 286, 302.
— envelope, 275.
Parasites, 356, 358, 481.
—, autoecious, 387.
—, autoxenous, 387.
—, facultative, 356, 369.
—, inhabiting animals, 369.
—, inhabiting plants, 379.
—, metaxenous, 387.
—, metoecious, 387.
—, obligate, 356.
— which change their host, 387.
Parmelia, Algae of thallus, 409.
— pulverulenta, 222.
—, soredia, 415.
— stellaris, 222.
—, structure of thallus, 406.
Paulia, discharge of spoes, 97.
—, structure of thallus, 413.
Paxillus, 297.
Pébrine, 488.
Peccania, structure of thallus, 413.
Pellicula (cutis), 58.
Peltigera, 240.
—, Algae of thallus, 409.
— aphthosa, structure of thallus, 406.
— canina, structure of thallus, 406.
—, incrustations, 408.

Peltigera malacea, structure
    of thallus, 406.
—, structure of thallus, 404,
    406.
Penicillium, 193, 198, 199,
    226, 232, 234, 239, 245,
    251, 254.
—, abscision of gonidia, 66,
    70*.
— aureum, 206.
—, capacity of germination,
    348.
—, conditions of germination,
    350.
—, conditions of vegetation,
    353, 354.
—, discharge of spores, 81,
    97.
—, effect on substratum,
    358.
—, fatty matters, 7.
—, formation of spores, 62.
—, germination, 111.
— glaucum, 204, 226.
— —, abscision of gonidia,
    70*.
— —, capacity of germina-
    tion, 344.
— —, cell-nucleus, 6.
— —, conditions of germina-
    tion, 349.
— —, mycelial membranes,
    21.
— —, parasitism, 370, 380.
— —, resistance of spores,
    347.
—, gonidiophores, 48.
—, sclerotioid compound
    sporophores, 43.
—, secretion of ferment,
    355.
—, simple sporophores, 46.
—, thallus, 1.
Perichaena, 435.
— liceoides, 423, 427, 428,
    435, 441, 448.
Periconia, 252.
—, formation of spores, 66.
Peridermium elatinum, 277,
    282.
— —, parasitism, 368, 388,
    390.
— Pini, parasitism, 388.
— —, structure of spores,
    100.
Peridiola, 311, 320.
Peridium, 48, 193, 275,
    308.
—, inner, 311.
—, outer, 311.
Periphyses, 192.
Periplasm, 133.
Perithecia, 76, 187, 190, 239,
    401.

Peronospora, 138, 232.
—, abscision of spores, 72.
— Alsinearum, fertilisation,
    133*.
— arborescens, fertilisation,
    133*.
— Arenariae, parasitism, 391.
— calotheca, haustoria, 20*.
— densa, germination, 112.
— —, haustoria, 20.
— —, parasitism, 393.
—, development, 134.
—, discharge of spores, 82.
—, membrane, 12.
— nivea, haustoria, 20.
— —, parasitism, 363, 393.
— parasitica, haustoria, 20.
— —, parasitism, 359, 362.
—, parasitism, 358, 386, 391.
— pygmaea, germination,
    112.
— —, haustoria, 20.
— —, parasitism, 364.
— Radii, parasitism, 358,
    364, 390, 391.
—, simple sporophores, 46,
    47.
—, thallus, 1.
— Umbelliferarum, parasi-
    tism, 363, 389.
— Valerianellae, 135.
— violacea, parasitism, 368,
    390, 391.
— viticola, parasitism, 393.
Peronosporeae (Peronospo-
    rei), 132, 183, 232, 233,
    236.
—, absence of calcium, 11.
—, capacity of germination,
    343, 344, 349.
—, cell-nucleus, 6.
—, cellulose-membrane, 8.
—, conditions of germina-
    tion, 350.
— discharge of spores, 82.
—, formation of spores, 74,
    75.
—, germination, 113.
—, haustoria, 20*.
—, mycelium, 20*.
—, parasitism, 359, 362, 386,
    393.
—, pleuroblastic, 20.
—, resistance of spores, 345.
—, resting state of spores,
    344.
—, swarm-spores, 109.
—, thallus, 1.
Pertusaria communis, ger-
    mination, 114*.
— de Baryana, germination,
    112*.
— fallax, calcium oxalate,
    409.

Pertusaria lejoplaca, develop-
    ment of spores, 78.
— —, germination, 114*.
—, number of spores, 79.
—, soredia, 415.
—, structure of thallus, 405.
Peziza (see also Sclerotinia),
    189.
— abietina, 86.
— —, ejection of spores, 86.
— —, structure of spores,
    106.
— Acetabulum, develop-
    ment of spores, 106.
— —, ejection of spores, 85.
— —, puffing of spores, 89,
    92.
— —, structure of spores, 77.
— aeruginosa, colouring mat-
    ter, 8.
— Aglaospora, 243.
— arduennensis, 243.
— aurantia, fatty matters,
    7.
— —, structure of spores,
    100.
— baccarum, sclerotium, 30,
    41.
— benesuada, doubtful sper-
    matia, 243*.
— bolaris, 243.
— —, germination, 115.
— calycina, development of
    spores, 78.
— Candolleana, sclerotium,
    30, 31, 32.
— ciborioides, sclerotium,
    30, 41.
— confluens, 208.
— —, development of spores,
    76, 77*.
— —, ejection of spores, 86.
— convexula, ejection of
    spores, 85, 86.
— —, structure of spores,
    102.
— cupularis, ejection of
    spores, 85, 86.
— Curreyana, desertion of
    host, 388.
— —, sclerotia, 33, 37, 41.
— Cylichnium, 243.
— —, germination, 115.
—, development of asco-
    spores, 76.
— Duriaei, sclerotia, 37.
— Durieuana, 243.
— —, desertion of host, 388.
— —, sclerotia, 41.
—, ejection of spores, 89.
— Fuckeliana, 245, 254.
— —, abscision of spores,
    63, 72.
— —, cell-nucleus, 6.

Peziza Fuckeliana, development of spores, 62, 78.
— —, doubtful spermatia, 243.
— —, gonidiophores, 38.
— —, resistance of gonidia, 347.
— —, sclerotia, 31*, 37, 41.
— fulgens, colouring matters, 8.
— —, mycelial strands, 22.
— fusarioides, 242.
— granulata, 215.
— —, ejection of spores, 86.
—, hairs, 59.
— hemisphaerica, structure of spores, 58.
— —, structure of sporophore, 106.
—, inception of sporophore, 49.
— melaena, development of spores, 77.
— —, ejection of spores, 85, 86.
— —, number of spores, 79.
— —, structure of spores, 102, 106.
— melanoloma, 215.
— nivea, development of sporophore, 53.
— pitya, development of spores, 76.
— —, ejection of spores, 85.
— Rapulum, mycelial strands, 22.
— ripensis, sclerotium, 41.
— Sclerotiorum, 243.
— —, clamp-connections, 19.
— —, compound sporophores, branching, 51.
— —, development of spores, 78.
— —, development of sporophore, 53*.
— —, ejection of spores, 87*.
— —, puffing of spores, 89.
— —, sclerotia, 30, 31*, 35, 37, 41.
— —, structure of spores, 106.
— —, structure of sporophore, 58.
— scutellata, 215.
—, structure of spores, 99.
—, structure of sporophore, 58.
— Tuba, sclerotia, 41.
— tuberosa, 243.
— —, development of spores, 78.
— —, germination, 115.
— —, sclerotia, 30.
— —, structure of spores, 106.

Peziza vesiculosa, ejection of spores, 86.
— —, structure of spores, 106.
Phacidiaceae (Phacidiacei), 190.
—, parasitism, 386.
Phacidium, development of spores, 83.
—, lipoxeny, 388.
— Pinastri, development of spores, 83.
Phalloideae (Phalloidei), 308, 309, 312, 322, 338, 340.
—, compound sporophores, 49.
— —, growth, 51.
—, conditions of germination, 352.
—, formation of spores, 63.
—, gelatinous membranes, 9, 13.
—, mycelial strands, 22, 23.
—, pseudo-parenchyma, 3.
—, secretions of calcium, 11.
Phallus, 312.
— caninus, calcium oxalate, 11*.
— —, development of sporophore, 322, 323*.
— —, formation of spores, 63*.
— —, mycelial strands, 23.
— impudicus, development of sporophore, 322, 323*.
— —, duration of growth, 51.
— —, mycelial strands, 23.
Phellorinia, 327.
Phelonites strobilina, 282.
—, structure of spores, 100.
Phlyctidieae (Phlyctidiei), 164.
Pholiota, 297.
Phoma, 248, 252.
Phragmidium, 276, 277.
—, detaching of spores, 68.
—, parasitism, 393.
—, spores, 68, 98.
Phragmotrichum, spores, 68.
Phycolichenes, 402.
Phycomyces, 146, 147, 150, 152, 155.
—, discharge of spores, 83.
— microsporus, 150.
— nitens, 150.
— —, capacity of germination, 344.
— —, membrane, 8.
— —, mycelium, 146*.
— —, organs of propagation, 146*.
—, resistance of spores, 346.
Phycomycetes, 120, 132.
—, formation of spores, 74.

Phycomycetes, swarm-spores, 107.
Phyllachora, lipoxeny, 388.
—, stroma, 43.
— Ulmi, 216.
— —, ejection of spores, 94.
Phyllactidium in Lichen-thallus, 397.
Phylliscum, structure of thallus, 413.
Phyllosticta, 252.
Phyllosticteae (Phyllostictei), 251.
Physareae (Physarei), 424, 426, 428, 430, 433, 434, 439, 440, 441, 449, 451.
Physarum, 430, 434, 436, 448.
— album, 441.
— aureum, 435.
— hyalinum, 435.
— leucophaeum, capillitium, sporangium, 435*.
— macrocarpum, 448.
— psittacinum, 424, 435.
— sinuosum, 428.
— sulphureum, 435.
Physica parietina, Algae of thallus, 397*.
— —, incrustations, 408.
— —, soredia, 416.
— —, structure of thallus, 403*, 406.
Physma, 108, 214, 231, 240, 259.
— chalazanum, Algae of thallus, 397*.
—, structure of thallus, 412, 413.
Physoderma Butomi, 166.
— Heleocharidis, 166.
— maculare, 166.
— pulposum, 164.
— vagans, 166.
Phytophthora, 137.
—, development, 134.
— discharge of swarm-spores, 82*.
—, formation of swarm-spores, 75.
—, germination, 109.
— infestans, abscision of spores, 72.
— —, capacity of germination, 343.
— —, discharge of swarm-spores, 82*.
— —, germination of gonidia, 137*.
— —, germination of swarm-spores, 364*.
— —, gonidia, 137*.
— —, gonidiophores, 47*.
— —, haustoria, 20.
— —, parasitism, 359, 362, 385.

Phytophthora infestans, resistance of spores, 346.
— —, simple sporophores, 47*.
— —, swarm-spores, 108*.
— —, taking up of colouring matter, 14.
— omnivora, 135.
— —, parasitism, 359, 365, 385.
—, parasitism, 367, 389, 392.
—, simple sporophores, 47.
Pietra fungaja, 42.
Pilacre Petersii, 335.
Pilaira, 150, 152.
—, discharge of spores, 83.
Pileus, 287.
— in Hymenomycetes, 48.
Pilobolus, 152, 155.
— anomalus, 150.
— —, detaching of sporangia, 83.
—, conditions of germination, 350.
— coloured fatty matters, 7.
— crystallinus, 150.
— —, abjection of sporangia, 72, 83.
—, formation of spores, 74, 75.
—, membrane, 12.
—, mode of life, 357.
— oedipus, 163.
— —, abjection of sporangia, 72*, 83.
— —, resistance of spores, 346.
—, structure of spores, 106.
Piptocephalideae (Piptocephalidei), 149, 151, 153.
Piptocephalis, 147, 149, 153, 155.
—, formation of spores, 67.
— Freseniana, 150, 153*.
— —, conjugation, 149*.
— —, haustoria, 20.
—, parasitism, 360, 363, 385.
Pistillaria hederaecola, sclerotia, 42.
— micans, sclerotia, 42.
Pits, 13.
— in spore-membrane, 100.
Pityriasis versicolor, 376.
Placodium, 223.
—, calcium oxalate, 409.
— cartilagineum, incrustations, 409.
—, distribution of Algae in thallus, 404.
—, structure of thallus, 404.
Plasmodia, 423.
—, movement, 449.
—, nutrition, 451.
—, phenomena of life, 449.

Plasmodiophora, 444.
— Brassicae, 448.
Plastids, 7.
Plectopsora, structure of thallus, 412, 413.
— botryosa, structure of thallus, 414*.
Pleomorphism, 126, 238.
Pleospora, 186, 191, 200, 235, 239, 245, 248, 271.
— Alternariae, 230, 255.
— —, development of pycnidia, 247*.
— Clavariarum, 244.
—, ejection of spores, 95.
—, formation of gonidia, 66.
— herbarum, 220, 229, 230, 247, 255.
— —, compound spores, 98.
— —, development of spores, 79.
— —, ejection of spores, 95*.
— —, germination, 114.
— polytricha, 245, 247.
— sarcinulae, 230.
—, sclerotioid compound sporophores, 43.
Pleurococcus in Lichenthallus, 397.
Pleurostoma, 191.
Pleurotus, 297.
Podaxon, 317, 318.
— carcinomatis, capillitium-threads, 318*.
— pistillaris, 318*.
Podetia (podetium), 222, 402.
Podosphaera, 198, 201, 233, 235, 236, 237.
— Castagnei, development of spores, 79.
— —, development of sporophores, 201*, 226*.
— —, haustoria, 19*.
— pannosa, 226*.
Pollinaria, 305.
Polyactis, 252.
Polyblastia, 223.
— rugulosa, Algae of thallus, 409.
— —, hymenial Algae, 400, 401.
— —, structure of thallus, 400.
Polydesmus, 229.
— exitiosus, parasitism, 362.
—, spores, 68.
Polyphagus, 169.
— Euglenae, 162*.
— —, parasitism, 361.
— parasiticus, 164.
—, swarm-spores, 107.
Polyplocium, 337.

Polyporeae (Polyporei), 288, 333, 337.
—, annual layers in compound sporophores, 57.
—, duration of growth in compound sporophores, 51.
Polyporus, 288, 301, 303, 307.
— abietinus, mycelial membranes, 22.
— annosus, parasitism, 384.
— —, structure of sporophore, 57.
— borealis, 335.
— —, parasitism, 384.
—, cellulose, 13.
—, clamp-connections, 2.
— destructor, abjection of spores, 73.
— dryadeus, parasitism, 384.
— fomentarius, cellulose, 8.
— —, membranes, 12.
— —, periodic growth of compound sporophores, 57.
— —, structure of sporophore, 58.
— fulvus, 307.
— —, parasitism, 384.
— —, structure of sporophore, 58.
—, growth of compound sporophore, 50, 57.
— hirsutus, hairs, 59.
— hispidus, hairs, 59.
— igniarius, 303, 307.
— —, cellulose, 8.
— —, parasitism, 384.
— —, periodic growth of compound sporophore, 57.
— lucidus, structure of sporophore, 59*.
—, membranes, 12.
— mollis, parasitism, 384.
— obvallatus, 337.
— officinalis, cellulose, 8.
— —, membranes, 12.
— —, secretion of resin, 10.
— ptychogaster, 334, 335.
— Ribis, periodic growth, 57.
— sulphureus, parasitism, 384.
— tuberaster, mycelium, 42.
— umbellatus, 304.
— vaporarius, parasitism, 384.
— versicolor, membranes, 12.
— volvatus, 337.
— zonatus, membranes, 12.
— —, periodic growth, 57.
Polysaccum, 309, 326, 327.
—, formation of spores, 63.

Polystigma, 191, 200, 235, 236, 239, 240, 241, 258, 259, 285.
— fulvum, 199, 215.
—, gelatinous membranes, 9.
—, lipoxeny, 388, 389.
—, parasitism, 386.
— rubrum, 199, 215, 224, 226, 240.
—, conditions of germination, 350.
— —, parasitism, 362.
— stellare, mycelial strands, 22, 23.
—, stroma, 43.
—, structure of sporophore, 57.
Pore, 191.
Pores in Polyporeæ (Polyporei), 288.
Poronia, 244.
Precursors of ascocarps in Ascomycetes, 244.
Predisposition for parasites, 359.
Primordium of the mycelium, 110, 178.
— of sporophore, 199, 215, 218.
Procarpium (procarp), 120.
Promycelium, 111, 177.
Prosporangium, 163.
Protagon-mixtures, 300.
Prothallium, 121.
Protococcus in Lichenthallus, 397*.
Protomyces, 132, 171.
—, capacity of germination, 349.
—, germination, 109.
—, ejection of spores, 85.
—, formation of spores, 61.
— macrosporus, 171.
— —, cellulose-membrane, 8.
— —, conditions of germination, 351.
— —, development of spores, 172.
— —, ejection of spores, 89, 91.
— —, parasitism, 363, 365, 386, 389.
— —, resting spores, 172*.
— —, structure of spores, 106.
— Menyanthidis, 166.
— pachydermus, 172.
—, resting state of spores, 345.
Protomyxa aurantiaca, 446.
Protoplasm, 6.
— in Bacteria, 455.
— in Mycetozoa, 422, 423, 425, 426.
Protothallus, 405.

Protozoa, 475.
Psalliota, 291, 297.
Pseudo-lichens, 416.
Pseudo-parenchyma, 3, 5, 405.
Pseudo-peridia, 275.
Pseudopodia, 423, 426.
Pseudospora, 447.
Psora, structure of thallus, 404.
Psoroma, Algae of thallus, 409.
— gypsaceum, incrustation, 408.
— lentigerum, secretion of calcium, 11.
— —, calcium oxalate, 408.
— sphinctrinum, Algae of thallus, 404.
—, structure of thallus, 404.
Pterula, 29.
Ptychogaster albus, 335.
Puccinia, 279, 282.
— Aegopodii, 285.
— Alliorum, 277.
— Anemones, 277.
— —, parasitism, 393.
— Asari, 285.
— Berberidis, 279, 283, 284.
— Caricis, parasitism, 387.
— coronata, parasitism, 387.
— —, structure of spores, 100.
—, detaching of spores, 68.
—, Falcariae, 279.
— —, parasitism, 387.
—, formation of spores, 62*.
— fusca, 277, 281.
— Galiorum, 277.
— graminis, development of aecidia, 275*, 276*.
— —, capacity of germination, 345, 348.
— —, development of spores, 62*.
— —, germination, 110*, 280*.
— —, germ-pores, 100.
— —, parasitism, 387, 389.
— —, rest of spores, 345.
— —, spermogonia, 276*.
— —, structure of spores, 101.
— Malvacearum, 284.
— Moliniae, parasitism, 387.
— Pimpinellae, parasitism, 387.
— Pruni, 285.
— reticulata, structure of spores, 100.
— Rubigo vera, 283.
— —, germination, 280*.
— —, parasitism, 387, 389.
—, spores, 68, 98.
—, structure of spores, 100.

Puccinia suaveolens, 277.
— —, parasitism, 359.
— Tragopogonis, parasitism, 387, 391.
— Violarum, parasitism, 387.
Putrefactive processes incited by Bacteria, 481.
Pycnidia, 49, 225, 239, 246.
—, large-pored, 248.
—, small-pored, 248.
Pycnis sclerotivora, 247.
Pycnochytrium, 167.
Pycnogonidia, 225, 239, 246.
Pycnospores, 225, 239, 246.
Pyrenocarp, 187.
Pyrenomycetes, 186, 187, 200.
— as Lichen-fungi, 396.
—, development of spores, 78.
—, ejection of spores, 91, 97.
—, hyphal weft, 4.
—, number of spores, 79.
—, stromata, 50.
—, structure of ascus, 96.
Pyrenula, 192, 223.
— minuta, 246.
— nitida, structure of thallus, 411.
— —, thallus, 402.
— olivacea, 246.
Pyronema, 190, 198, 224, 231, 234, 237, 239.
— confluens, 208.
— —, development of sporophore, 209*.
Pythium, 234.
— de Baryanum, 135, 137.
— —, parasitism, 382.
—, development, 135, 137.
—, discharge of spores, 82.
— endophytum, 139.
— gracile, fertilisation, 133*.
— intermedium, 137.
— —, parasitism, 382.
— megalacanthum, parasitism, 382.
—, mode of life, 132.
—, parasitism, 359, 363, 382, 389, 392.
— proliferum, 135.
— —, rest of spores, 345.
—, rest of spores, 345.
—, swarm-spores, 107.
— vexans, 135, 232.

Quaternaria, 218.
—, discharge of spores, 97.
Queletia, 318.

Racodium cellare, 22.
— rupestre, structure of thallus, 411.

Ramalina, soredia, 415.
Reaction of host on parasite, 366.
Receptaculum, 17, 48, 188, 312.
Recurrent fever, 486, 488.
Reduction in the course of development, 125.
Resin in Boletus, 15.
Resin-excretions, 10.
Resin-flux, 384.
Resistance in spores, 343.
— in Bacteria, 476.
Resting gonidia, 144.
— mycelia, 230, 245.
— sporangia, 144.
— spores, 345.
— — in Bacteria, 460.
— states, transitory in My-cetozoa, 427.
Reticularia, 431, 440.
— umbrina, 439.
Rheotropism in plasmodia, 449.
Rhipidium, 144.
Rhipidonema, structure of thallus, 414.
Rhizidieae (Rhizidiei), 162, 169, 170.
Rhizidium, 164, 165.
Rhizines, 402.
Rhizocarpon, structure of thallus, 405.
Rhizoids, 45, 59, 402.
Rhizomorpha fragilis, 28.
—, parasitism, 383, 384.
— subcorticalis, 28.
— subterranea, 28.
Rhizophydium, 164.
Rhizopoda, 444, 475.
Rhizopogon, 308.
—, formation of spores, 63.
Rhizopus, 152.
— nigricans, 152, 154.
— —, conjugation, 148*.
—, discharge of spores, 83.
Rhytisma, 186, 240.
— acerinum, ejection of spores, 87, 92.
— Andromedae, 224.
— —, germination, 111*.
— —, parasitism, 358, 393.
— —, structure of spores, 102.
—, asci, 76.
—, conditions of germina-tion, 350.
—, lipoxeny, 388.
—, parasitism, 389.
—, stroma, 43.
— —, structure, 58.
—, structure of sporophore, 57.
Rind of compound sporo-phore, 58.

Rind (rind-layer) of Lichen-thallus, 403.
Roccella, Algae of thallus, 409.
— fusiformis, incrustations, 408.
— —, calcium oxalate, 409.
— —, structure of thallus, 406.
— Montagnei, 246.
—, soredia, 415.
—, structure of thallus, 404, 406.
Rod-like forms in Bacteria, 458.
—, gonidia, 331.
Roesleria hypogaea, dis-charge of spores, 96.
Roestalia, 388.
Root-hairs, 45, 54, 59.
Rosellinia Aquila, structure of ascus, 96.
— quercina, 215.
— —, mycelium, 42.
Rotting in orchard fruit, 379.
Rozella, 169, 444.
—, parasitism, 368, 395.
— septigena, parasitism, 395.
Russula, 297, 304.
— adusta, 298.
—, calcium oxalate, 11.
— foetens, var. lactiflua, 300.
— integra, 298.
— —, structure of sporo-phore, 58.
— olivacea, 298.
—, pseudo-parenchyma, 3.
Rutstroemia (see Sclero-tinia), 41.
—, ejection of spores, 86.
Ryparobius, 208.

Saccharomyces, 132, 263, 267, 270.
— albicans, 267.
— —, parasitism, 377.
— apiculatus, 271, 272.
— —, mode of life, 357.
—, cell-nucleus, 6.
— Cerevisiae, 267.
— —, resistance to effects of heat, 347.
— —, sprouting, 4*, 267*.
— —, swelling of membrane, 10.
—, conditions of vegetation, 353.
— ellipsoideus, 267.
— —, formation of spores, 268*.
—, germination, 114.
— glutinis, 272.
— mesentericus, 358.

Saccharomyces, Mycoder-ma, 267, 268, 269, 358, 377.
— Pastorianus, 267.
—, pleomorphism, alleged, 270.
—, secretion of ferments, 355.
—, thallus, 5.
Saccobolus, ejection of spores, 92.
—, structure of spores, 102.
Sagedia aenea, 246.
— callopisma, 246.
— carpinea, 246.
— netrospora, 246.
— Thuretii, 246.
— Zizyphi, 246.
Salts of iron in Lichen-thal-lus, 408.
Sap-cavities (vacuoles), 6.
Saprolegnia, 143.
—, discharge of spores, 82.
—, formation of spores, 74.
— hypogyna, 141, 142.
— monoica, 141.
—, parasitism, 359, 393.
—, simple sporophores, 46.
—, sporangia, 46.
—, swarm-spores, 107, 108.
Saprolegnieae (Saprolegniei), 141, 232.
—, capacity of germination, 343.
—, cell-nucleus, 6.
—, cellulose-membrane, 8.
—, discharge of spores, 82.
—, parasitism, 375, 395.
—, resistance of spores, 345.
—, rest of spores, 344.
—, simple sporophores, 46.
Saprophytes, 356, 357, 480.
—, facultative, 356.
—, obligate, 356.
Sarcina, 473.
— ventriculi, 459.
Sarcinula, 229.
Sarcogyna, number of spores, 79.
Scales on compound sporo-phores, 59.
Schizomycetes, 454.
Schizonella, 176.
Schizophyllum, 302.
Schizophytes, 474, 476.
Schizosiphon in Lichen-thallus, 398.
Sclerangium, 315.
Scleroderma, 309, 311, 313, 315.
—, formation of spores, 63.
—, mycelial strands, 23.
Sclerosis of membranes, 9.
Sclerotia, 18, 30.

Sclerotia, development, 34.
—, further development, 37.
—, gelatinous membranes, 10.
—, hyphal weft, 3.
— in Mycetozoa, 427, 428.
—, membranes, 12.
—, pseudo-parenchyma, 3.
—, resting state, 37.
Sclerotinia, 41, 200, 260. (See also Peziza.)
— ciborioides, organs of attachment, 21.
— —, parasitism, 380.
— —, resistance of spores, 346.
—, development of spores, 78.
— Fuckeliana, 219, 224, 238, 251.
— —, compound sporophores, 38*.
— —, conditions of germination, 350, 351.
— —, germination, 113.
— —, gonidiophores, 48.
— —, organs of attachment, 21.
— —, parasitism, 380.
— —, sclerotia, 31*, 34, 37, 38*, 39.
—, gelatinous membranes, 9.
—, lipoxeny, 388.
—, membranes, 12.
—, organs of attachment resembling haustoria, 21.
—, parasitism, 359, 360, 380, 389, 392.
—, sclerotia, 30, 34, 39.
— Sclerotiorum, 200, 218, 224.
— —, compound sporophores, development, 53*, 219*.
— —, conditions of germination, 351.
— —, ejection of spores, 87*.
— —, germination of sclerotia, 38, 39.
— —, organs of attachment, 21.
— —, parasitism, 359, 380.
— —, sclerotia, 30, 31*, 34, 37, 39.
— tuberosa, 243, 260.
— —, organs of attachment, 21.
Sclerotium, 40.
— areolatum, 41.
— Clavus, 41.
— Cocos, 42.
— compactum, 41.
— complanatum, 42.
— cornutum, 42.

Sclerotium crustuliforme, 42.
— Cyparissiae, 41.
— durum, 251.
— echinatum, 41, 238.
— fulvum, 41.
— lacunosum, 42.
— laetum, 42.
— muscorum, 32.
— mycetospora, 42.
— pubescens, 42.
— Pustula, 41.
— roseum, 41.
— scutellatum, 42.
— Semen, 41.
— stercorarium, 32, 42.
— stipitatum, 42.
— sulcatum, 41.
— truncorum, 42.
— vaporarium, 42.
— varium, 41.
— vulgatum, 42.
Scytonema in Lichen-thallus, 397*, 398.
Sebacina incrustans, 306.
Secotieae (Secotiei), 310, 313.
Secotium, 310, 313, 337.
— erythrocephalum, 337.
— —, compound sporophore, 310*.
Semen, 130.
— multiplex, 98.
Sepedonium, 245, 252.
Septicaemia, 486.
Septoria, 252.
Sexual organs, 49, 233.
—, assumed in Basidiomycetes, 332.
Sexuality, 120.
— in Ascomycetes, 237.
Short rods (Bacteria), 458.
Silk-worms, diseases in, 486, 489.
Simblum, 326.
Simple sporophores, 46.
— —, germination, 114*.
— —, structure of thallus, 404.
Sirosiphon in Lichen-thallus, 398.
Solorina crocea, Algae of thallus, 409.
— saccata, Algae of thallus, 409.
— —, incrustation, 408.
Sordaria, 191, 198, 199, 210, 235, 243, 260.
— Brefeldii, attachment of spores to ascus, 88.
—, conditions of germination, 350.
— coprophila, 243.
— —, structure of spores, 103.
— curvula, 243, 261.

Sordaria curvula, capacity of germination, 344.
— decipiens, 243.
—, ejection of spores, 86, 88, 91*.
— fimicola, 210, 261.
— fimiseda, 192.
— —, capacity of germination, 343.
— —, development of spores, 78.
— —, ejection of spores, 85, 92.
— —, germ-pores, 101.
— —, number of spores, 79.
— —, structure of spores, 103, 104*.
—, germination, 111.
— minuta, 243, 261.
— —, ejection of spores, 91*.
—, mode of life, 357.
—, number of spores, 79.
— pleiospora, number of spores, 79.
—, structure of spores, 103, 106.
Soredia, 244, 401, 415.
Soredia-heaps, 416.
Sorosporium, 175, 176.
— Saponariae, 172, 179.
— —, parasitism, 391.
— —, rest of spores, 345.
Sorus, 167, 416.
Spathulea, 189.
Spermatia, 199, 211*, 239, 240, 257, 276*.
—, supposed in Basidiomycetes, 333.
—, doubtful, 242.
Spermogonia, 199, 211*, 239, 240, 241, 257, 276*, 401.
Sphacelia, 227, 252.
Sphacelotheca, 173.
— Hydropiperis, compound sporophores, 174*.
Sphaerella Plantaginis, 220.
Sphaeria discretia, fatty matters, 7.
— eutypa, fatty matters, 7.
— inquinans, ejection of spores, 95.
— Lemaneae, 214.
— —, ejection of spores, 85, 94.
— obducens, development of spores, 79.
— —, ejection of spores, 95.
— oblitescens, 245.
— praecox, germination, 115.
— —, structure of spores, 104.
— Scirpi, development of spores, 79.

Sphaeria Scirpi, ejection of spores, 85, 93*.
— —, structure of spores, 99*, 102, 103.
— Stigma, fatty matters, 7.
Sphaeriaceae (Sphaeriacei), ejection of spores, 85.
—, formation of gonidia, 66.
—, structure of spores, 102.
Sphaeriae compositae, 186.
Sphaerobolus, 326, 329, 330, 331, 332, 340.
—, coloured fatty matters, 7.
—, mycelial strands, 22.
—, stellatus, 328.
Sphaeromphale, 192.
Sphaerophoron, 193.
— coralloides, development of spores, 78.
— —, discharge of spores, 96*.
— —, gonidia, 398*.
— —, incrustations, 408.
—, discharge of spores, 96*, 97.
—, structure of thallus, 404, 406.
Sphaeropsideae (Sphaeropsidei), 251, 252.
Sphaeropsis, 252.
Spheconisca, 416.
Sphere-yeast, 156.
Sphyridium, 200.
— fungiforme, 221.
— placophyllum, 221.
Spicaria, 245.
Spilonema, 242.
—, structure of thallus, 412.
Spilosphaeria, 252.
Spinellus, 152.
Spiral forms in Bacteria, 458.
Spirillum, 456, 458, 459, 473.
— amyliferum, 455, 461.
Spirochaete, 458, 459.
— Obermeyeri, 488.
Spora, 128.
Sporae cellulosae, 98.
— compositae, 98.
— multiloculares, 98.
Sporangia (sporangium), 45, 73, 130, 319.
— in Mycetozoa, 429, 434.
Sporangiola, 152.
Spore, 130.
Spore-development (see also Spore-formation), 60.
— —, acrogenous, 61.
— by free cell-formation, 61.
— —, cell-division, 61.
— intercalary, 61.
Spore-formation, branched concatenate, 66.

Spore-formation, by cross-septation, 67.
—, endogenous, 73.
— in Bacteria, 460, 465.
— in Mycetozoa, 429.
—, simple concatenate, 66.
—, simultaneous, 63.
—, successive, 63.
—, sympodial, 65.
Spore-germination, 109.
Spore-groups, 99.
Spore-heads, 63.
Spore-initial cells, 98.
Spore-membrane, chemical properties, 104.
Spore-membranes, 83.
Spore-mother-cells, 48, 60, 98.
Spore-primordia, 98.
Spore-receptacles, 48.
Spores, 45, 59, 121, 124, 128.
—, abjection, 68, 72.
—, abjunction, 61.
—, abscision, 68.
—, appendages of, 102.
—, capacity of germination, 343.
—, compound, 98.
—, conditions of germination, 349.
—, delimitation, 61, 68.
—, detaching, 68.
—, discharge, 82.
—, fatty matters, 7.
—, germination, 109.
— of Bacteria, 461.
— of Mycetozoa, 421, 440, 441.
—, pluricellular, 98.
—, resistance, 346.
—, resting state, 345.
—, ripeness, 68.
—, septate, 68, 98.
—, structure, 99.
Sporidesmieae (Sporidesmii), 68.
Sporidesmium, 68, 229.
—, spores, 68.
Sporidesms, 99.
Sporidia, 111, 130, 177.
—, capacity of germination, 343.
—, whorl of cylindrical or subulate (Kranzkörper), 177.
Sporoblasts, 99.
Sporocadus, 252.
Sporocarpium (sporocarp), 121, 185, 239.
Sporodinia, 147, 148, 149, 150, 151, 152.
— grandis, 147, 150, 154.
—, resistance of spores, 346.

Sporodinia, resting time of spores, 345.
Sporogenous layer (hymenium), 49.
Sporophore, 45.
— in Mycetozoa, 429.
Sporophyte, 121.
Sporormia fimctaria, spores, 99.
Sporula, 130.
Spread of parasite in host, 389.
Sprout-gemmae, 155.
Sprout-germination, 110.
Sprouting Fungi, 4*, 5, 60.
—, form, 4*, 5, 267.
—, germination, 114.
— —, in Ustilagineae, 177.
—, growth, 155.
—, mucilage, 9.
Spumaria, 436, 440.
Spumella vulgaris, 475.
Steganosporium, 252.
Stemoniteae (Stemonitei), 427, 434, 439, 441.
Stemonitis, 423, 431, 432, 433, 438, 440.
— ferruginea, formation of sporangia, 432*.
— fusca, 427.
Stephensia, 196.
Stereocaulon, Algae of thallus, 410.
—, podetia, 402.
— ramulosum, Algae of thallus, 397*.
Stereum (see also Thelephora), 303, 304.
— hirsutum, development of sporophore, 53*.
— —, fatty matters, 7.
— —, parasitism, 384.
— —, periodic growth, 57.
— —, structure of sporophore, 57, 58.
— rubiginosum, 303.
— tabacinum, 303.
Sterigma, 61.
Sterigmata in spermatia, 240.
Sterigmatocystis, 206, 256.
—, parasitism, 369.
Stichococcus in Lichenthallus, 397.
Sticta, Algae of thallus, 409.
— aurata, incrustation, 408.
— fuliginosa, structure of thallus, 403*.
— glomulifera, Algae of thallus, 409.
—, incrustation of thallus, 408.
— pulmonacea, thallus, 187*.
— —, structure of thallus, 406.

Sticta, structure of thallus, 406.

Stictina, Algae of thallus, 409.

Stictosphaeria, 191, 218, 241.

—, discharge of spores, 97.

— Hoffmanni, 258.

Stigmatea, 244.

Stigmatomma cataleptum, origination of thallus, 400.

— —, hymenial Algae, 400, 401.

Stigmatomyces Baeri, 263.

— —, development, 263*.

— Muscae, development, 263*.

Stigonema in Lichen-thallus, 398.

Stilbospora, 252.

—, spores, 68.

Stilbum, 29, 252, 334.

Stipes in sporophore of Hymenomycetes, 287.

Stratum corticale, 403.

— gonimon, 417.

— medullare, 403.

Striation of spore-membrane, 100.

Stromata, 48, 186.

Stylospores, 154, 225, 239, 246, 279.

Stysanus, 252.

Suberisation of membranes, 9.

Subhymenial layer or tissue, 301.

Sulphur in Bacteria, 455, 471.

Supply of food, a condition of germination, 351.

Suspensor in zygospores, 148.

Swarm-cells, 107.

— in Mycetozoa, 422.

Swarm-spores, 60, 107, 129.

—, capacity of germination, 343.

—, discharge, 82.

—, resistance, 346.

Symbiosis, 356.

Synalissa, structure of thallus, 413, 414*.

— symphorea, Algae of thallus, 397*.

Syncephalis, 146, 153, 154.

— curvata, 150.

—, formation of spores, 67, 117.

— furcata, 153.

—, haustoria, 20.

— nodosa, 149, 150.

— parasitica, 360, 363, 385.

—, secondary mycelium, 45.

Synchytrium, 162, 167, 168, 169, 444.

Synchytrium, rest of spores, 345.

— aureum, 160.

— —, rest of spores, 345.

—, capacity of germination, 349.

— Oenotherae, 168.

—, rest of spores, 345.

— Stellariae, 168, 169.

— —, propagation, 168*.

— Succisae, 169.

— Taraxaci, 168.

— —, rest of spores, 345.

Synechoblastus, 198.

—, structure of thallus, 412.

Systematic arrangement of Fungi, 132.

Syzygites, 151.

— ampelinus, 151.

Taphrina, 265.

—, germination, 115.

Tarichium, 159.

Teeth, caries, 472.

Tela contexta, 5.

Teleutogonidia, 281.

Teleutospores, 279, 282, 339.

—, resting state, 345.

Terfezia, 195.

Terminology of Fungi, 128, 130.

Tetrachytrium triceps, 170.

Thalloidima candidum, calcium oxalate, 409.

— —, structure of thallus, 404.

Thallus, 1.

—, branching, 1.

—, crustaceus, 402.

—, differentiation, 17.

—, fertile (stroma), 186.

—, filamentosus, 401.

—, foliaceus, 401.

—, frondosus, 401.

—, fruticulosus, 401.

—, lepodes, 402.

—, placodes, 401.

—, thamnodes, 401.

Thamnidium, 152.

—, discharge of spores, 82, 83.

— elegans, 150.

Thamnolia, incrustation, 408.

—, structure of thallus, 406.

— vermicularis, calcium oxalate, 409.

Thamnomyces, 186.

—, branching, 51.

Thecae (asci), 76.

Thecaphora, 176.

— hyalina, 177.

— Lathyri, 177, 179.

Thecaspores, 129.

Thelephora crocea, mycelial membranes, 22.

Thelephora hirsuta, membranes, 12.

— —, mycelial membranes, 22.

—, membranes, 12.

— mesenterica, gelatinous membranes, 13.

—, mycelial membranes, 22.

— Perdix, 307.

— —, parasitism, 384.

— setigera, mycelial membranes, 22.

— suaveolens, mycelial membranes, 22.

Thelephoreae (Thelephorei), 288, 333, 338.

—, mycelial membranes, 22.

Thelidium minutulum, 224.

— —, Algae of thallus, 409.

— —, germination, 399*.

— —, origin of thallus, 400.

— —, perithecia, 190*.

— —, structure of thallus, 410*.

Thermotropism of plasmodia, 450.

Thrush (aphthae), 377.

Thyrea pulvinata, structure of thallus, 414*.

—, structure of thallus, 413.

Tilletia, 172, 174, 177, 178, 180, 181.

— Caries, 179.

— —, capacity of germination, 344.

— —, germination, 177*.

— —, parasitism, 385.

— —, resistance of spores, 346.

—, parasitism, 367.

Tinder-fungus, 3.

Tinea (herpes), 376.

Tissue, hymenial, of Léveillé, 302.

—, intralamellar (trama), 301.

—, subhymenial, 301.

Tolyposporium Junci, 177.

Torula, 252.

Trama, 301, 309.

Trametes Pini, 301, 303, 304, 307.

— —, duration of growth, 51.

— —, membranes, 12.

— —, mycelial membranes, 22.

— —, parasitism, 304.

— —, periodical growth, 57.

— radiciperda, secretion of ferment, 355.

— —, parasitism, 384.

Transmutation of host, 395.

Tremella, 305.

— Cerasi, 306, 331.

Tremella, foliacea, 306.
—, formation of spores, 62.
—, gelatinous membranes, 13.
— mesenterica, 331.
— violacea, 306.
Tremellineae (Tremellini), 271, 287, 288, 298, 301, 303, 305, 329, 331, 338.
—, conditions of germination, 350.
—, fatty matters, 7.
—, gelatinous membranes, 9.
—, gemmae, 60, 61.
Tremellodon, 288, 305.
— gelatinosus, hair-formations, 59.
Trentepohlia in Lichen-thallus, 397, 398*.
Trichia, 431, 433, 435, 437, 439, 440.
— chrysosperma, 438, 440.
— —, capillitium, spores, 437*.
— clavata, 434.
— fallax, 431, 437, 438, 439, 441.
— —, capillitium, spores, 437*.
— furcata, 441.
— pyriformis, 441.
— rubiformis, 448.
— Serpula, 434.
— varia, 434, 438, 441, 448.
— —, germination, swarm-cells, 422*.
Trichiaceae (Trichiacei), 427, 441.
Trichiae, 436.
Trichogyne, 198, 209, 213, 215, 234, 237.
Trichophyton tonsurans, parasitism, 376.
Trichothecium roseum, parasitism, 380.
— —, resistance of spores, 346.
—, spores, 68.
Triphragmium, 281, 282.
— echinatum, parasitism, 358.
— —, structure of spores, 100.
—, spores, 98.
— Ulmariae, parasitism, 358.
Trophoplasts, 7.
Trophotropism in plasmodia, 449.
Truffles, 195.
—, glycogen, 6.
Tube-germination, 110.
Tuber, 195.
— aestivum, 195.
— —, development of spores, 80.
— —, structure of spores, 100.

Tuber, asci, 76.
— brumale, development of spores, 80*.
—, conditions of germination, 352.
— dryophilum, 195.
— excavatum, 195.
— melanosporum, 195.
— —, development of spores, 80.
— —, structure of spores, 100.
— mesentericum, 195, 196.
—, number of spores, 79.
— rapaeodorum, 195, 196.
— rufum, compound sporophore, 196*.
Tuberaceae (Tuberacei), 193, 195.
—, clamp-connections, 19.
—, discharge of spores, 97.
—, hyphal weft, 3.
Tubercularia, 252.
— vulgaris, 244.
Tubuli in sporophore of Polyporeae, 288.
Tubulina (Tubulinae), 448.
—, experiments in germination, 421.
Tubulus of perithecia of Pyrenomycetes, 191.
Tuburcinia, 175, 176, 177.
— Trientalis, 178, 180, 181.
— —, capacity of germination, 345, 349.
— —, parasitism, 365, 391.
— —, rest of spores, 345.
Tulostoma, 287, 326.
—, capillitium, 12.
— mammosum, basidia, 310*.
— —, compound sporophore, 327*.
— pedunculosum, sclerotia, 42.
Tympanis, 245.
— conspersa, spermogonia, 240*.
— —, number of spores, 79.
— saligna, number of spores, 79.
Typhula, 329.
— caespitosa, sclerotium, 41.
—, clamp-connections, 2.
— erythropus, 41.
— Euphorbiae, 33, 41.
— graminum, 33, 41.
—, growth of compound sporophore, 52.
— gyrans, inception of sporophore, 49.
— —, gelatinous membranes, 9.
— —, membranes, 12.
— —, sclerotium, 30, 33*, 34, 35, 37, 39, 42.

Typhula gyrans, lactea, sclerotium, 41.
— phacorrhiza, sclerotium, 33*, 37, 42.
— Todei, sclerotium, 41.
— variabilis, development of sporophore, 52.
— —, sclerotium, 30, 33, 34, 37, 38, 42.

Ulothrix in Lichen-thallus, 397.
Umbilicaria pustulata, growth in thickness of thallus, 407.
—, number of spores, 79.
Uncinula spiralis, 225.
Urceolaria cinerea, structure of thallus, 405.
— scruposa, calcium oxalate, 409.
Uredineae (Uredinei), 120, 132, 274, 287.
—, aecidia-forming, 274.
—, capacity of germination, 343, 349.
—, coloured fatty matters, 7.
—, conditions of germination, 350.
—, germination, 113.
—, germ-pores, 101.
—, gonidiophores, 50.
—, haustoria, 20.
—, parasitism, 359, 361, 363, 366, 367, 386, 387, 389, 390, 393.
—, spores, abscision, 71*.
— —, resistance, 346.
— —, rest, 344.
— —, structure, 100, 106.
—, structure of sporophore, 57.
—, tremelloid, 274, 283, 339.
Uredo, 279.
— Symphyti, 282.
Uredo-layer, 279.
Uredogonidia, 281.
Uredospores, 279.
—, capacity of germination, 343.
—, formation, 62.
—, germination, 111.
—, structure, 101*.
Urocystis, 172, 175, 177.
— occulta, 180.
— —, capacity of germination, 344.
— —, parasitism, 391.
—, structure of spores, 104.
— Violae, 178, 181.
Uromyces, 281, 282.
— appendiculatus, germination, 361*.
— —, germination of sporidia, 364*.

Uromyces appendiculatus, parasitism, 387.
— Belienis, 279.
— Cestri, 279.
— Dactylidis, parasitism, 387.
— Phaseolorum, 281.
— — parasitism, 387, 389.
— Pisi, parasitism, 368, 388, 391.
— Scrophulariae, 279.
— scutellatus, parasitism, 368.
—, spores, 98.
— —, detaching, 68.
— —, formation, 62.
— tuberculatus, parasitism, 358.
— Viciae Fabae, 277.
Usnea barbata, Algae of thallus, 398*.
— —, chemical properties, 407.
— —, incrustation, 408.
— —, soredia, 415.
— —, structure of thallus, 402*.
— —, thallus, 187*.
—, structure of thallus, 404, 405, 406.
Ustilagineae (Ustilaginei), 120, 132, 172, 176, 179.
—, capacity of germination, 344.
—, conditions of germination, 350.
—, haustoria, 19.
—, parasitism, 359, 362, 364, 367, 385, 390, 391, 393.
—, spores, rest, 345.
— —, structure, 100.
Ustilago, 174.
— antherarum, 179.
— Carbo, 177, 179, 180, 181.
— —, capacity of germination, 344.
— —, conditions of germination, 349.
— —, germination, 178*.
— —, parasitism, 367.
— —, resistance of spores, 346.
— Cardui, 177.
— Crameri, capacity of germination, 344.
— destruens, 177, 179, 180.
— —, capacity of germination, 344.
— —, conditions of germination, 349.
— —, resistance of spores, 346. .
— flosculorum, 177.
— hypodytes, 172, 175.

Ustilago flosculorum, parasitism, 391.
— Ischaemi, 175.
— —, formation of spores, 68.
— Kolaczeckii, capacity of germination, 344.
— Kühniana, 177, 179.
— longissima, 173, 179, 183.
— —, germination, 178*.
— Maidis, 179, 181.
— olivacea, 173.
— Rabenhorstiana, capacity of germination, 344.
— receptaculorum, structure of spores, 102.
— Tragopogonis, 172, 182.
— —, development of spores, 175*.
— —, germination, 178*.
— —, parasitism, 391.
— Tulasnei, capacity of germination, 344.
— utriculosa, 177.
— Vaillantii, 177, 179.
Ustulina, 186, 218, 244, 248, 260.
Uterus, 308.

Vacuoles (sap-cavities), 6, 424.
Valsa ambiens, 258.
— —, number of spores, 79.
— nivea, spermogonia, 240*.
— —, number of spores, 79.
— salicina, number of spores, 79.
—, structure of spores, 102.
Valseae (Valsei), 191, 192.
—, discharge of spores, 97.
Vampyrella, 447, 453.
— pendula, 447.
— Spirogyrae, 447.
— vorax, 447.
Variolaria, 416.
Vegetation, general conditions, 352.
Veines aërifères, 195.
— aquifères, 195.
Velum, 289, 291.
— partiale, 290.
— universale (volva), 290.
Venae externae, 195.
— internae, 195.
— lymphaticae, 195.
Vermicularia, 252.
— minor, fatty matters, 7.
— —, sclerotium, 41.
Verpa, 189.
Verrucaria, 192, 222.
—, Algae of thallus, 409.
— carpinea, 246.
— Gibelliana, 246.

Verrucarieae (Verrucariei), 242.
Verticillium, 245, 249, 252.
Vibrio, 458, 459, 473.
Vibriones, 458.
Volva, 290.
Volvaria, 292, 295, 297.

Warts on compound sporophores, 59.
Water-content, 6.
— -supply, as a condition of germination, 375.
Whorl of sporidia (Kranzkörper), 177.
Wine-yeast, 269.
Witches' brooms, 266, 368, 390.
Withdrawal of water, influence on germination, 346.
Woronina, 161, 169, 444.
—, parasitism, 395.
Woronin's hypha, 199, 218.

Xylaria, 186, 192, 199, 236, 244, 260.
—, branching, 51.
— bulbosa, sclerotium, 41.
—, compound sporophores, development, 56.
— —, growth, 50, 51.
— pedunculata, structure of spores, 102.
— —, chemical behaviour of spore-membrane, 105.
— polymorpha, development of sporophore, 216*.
— —, development of spores, 78.
Xylarieae (Xylariei), 186, 236, 248, 260.
—, discharge of spores, 97.
Xyloma, 43, 190.
Xylostroma, 22.

Yeast-fungi, 4*, 5, 267, 270, 358.
Yeast-fungus of alcoholic fermentation, 267.
Yeast-mucilage, 10.

Zeora, 223.
Zoogloea, 457, 459.
Zoospores, 107.
Zygochytrium, 146, 151, 169.
— aurantiacum, 156.
Zygomycetes, 132, 145.
—, rest of spores, 345.
—, thallus, 1.
Zygospores, 129, 145.
—, formation, 147.
—, resting state, 344, 345.

THE END.

Reprint Publishing

For People Who Go For Originals.

This book is a facsimile reprint of the original edition. The term refers to the facsimile with an original in size and design exactly matching simulation as photographic or scanned reproduction.

Facsimile editions offer us the chance to join in the library of historical, cultural and scientific history of mankind, and to rediscover.

The books of the facsimile edition may have marks, notations and other marginalia and pages with errors contained in the original volume. These traces of the past refers to the historical journey that has covered the book.

ISBN 978-3-95940-093-0

Made in Germany

www.reprintpublishing.com

www.ingramcontent.com/pod-product-compliance
Lightning Source LLC
Chambersburg PA
CBHW082118210326
41599CB00031B/5807